Jochen Brüning, Matthias Staudacher (Eds.)
Space – Time – Matter

Also of Interest

Invariant Differential Operators.
Volume 1: Noncompact Semisimple Lie Algebras and Groups
Vladimir K. Dobrev, 2016
ISBN 978-3-11-043542-9, e-ISBN (PDF) 978-3-11-042764-6
e-ISBN (EPUB) 978-3-11-042780-6

Invariant Differential Operators.
Volume 2: Quantum Groups
Vladimir K. Dobrev, 2017
ISBN 978-3-11-043543-6, e-ISBN (PDF) 978-3-11-042770-7,
e-ISBN (EPUB) 978-3-11-042778-3

Minkowski Space. The Spacetime of Special Relativity
Joachim Schröter, 2017
ISBN 978-3-11-048457-1, e-ISBN (PDF) 978-3-11-048573-8,
e-ISBN (EPUB) 978-3-11-048461-8

Quantum Invariants of Knots and 3-Manifolds
Vladimir Turaev, 2016
ISBN 978-3-11-044266-3, e-ISBN (PDF) 978-3-11-043522-1,
e-ISBN (EPUB) 978-3-11-043456-9

Introduction to Topology
Min Yan, 2016
ISBN 978-3-11-037815-3, e-ISBN (PDF) 978-3-11-037815-3,
e-ISBN (EPUB) 978-3-11-037815-3

Space – Time – Matter

Analytic and Geometric Structures

Edited by
Jochen Brüning and Matthias Staudacher

DE GRUYTER

Editors
Prof. Dr. Jochen Brüning
Humboldt-Universität zu Berlin
Institute of Mathematics
Unter den Linden 6
10099 Berlin
Germany
bruening@mathematik.hu-berlin.de

Prof. Dr. Matthias Staudacher
IRIS Adlershof
Humboldt-Universität zu Berlin
Zum Großen Windkanal 6
12489 Berlin
Germany
matthias@mathematik.hu-berlin.de

ISBN 978-3-11-045135-1
e-ISBN (PDF) 978-3-11-045215-0
e-ISBN (EPUB) 978-3-11-045153-5

Library of Congress Control Number: 2018936388

Bibliographic information published by the Deutsche Nationalbibliothek
The Deutsche Nationalbibliothek lists this publication in the Deutsche Nationalbibliografie; detailed bibliographic data are available on the Internet at http://dnb.dnb.de.

Typesetting: Integra Software Services Pvt. Ltd.
Printing and binding: CPI books GmbH, Leck
Cover image: Claudia Bachmann, Dipl.-Designerin, Berlin
♾ Printed on acid-free paper
Printed in Germany

www.degruyter.com

Contents

Jochen Brüning and Matthias Staudacher

Introduction

1 Space Time Matter

Under this title, Hermann Weyl published a famous book in the spring of 1918 [22]. Together with his other book, "*Gruppentheorie und Quantenmechanik* (*The Theory of Groups and Quantum Mechanics*)" [23] from 1928, Weyl created a balanced correspondence between mathematics and physics that was quite unique at that time, by alternating chapters of mathematical and physical developments converging to a synthesis in the last chapter. He was well aware of the rareness of his work[1] but he was guided by the desire to understand the intricate intertwining of philosophical, mathematical, and physical thought. He came quite close to such a synthesis but in the end, the mathematician Weyl perhaps prevailed over the philosopher and the physicist.

After the demise of Hilbert's formalism from the hands of Gödel, the axiomatic approach nevertheless persisted, as demonstrated by the formidable work of Bourbaki, starting in 1934. Proceeding strictly from the general to the concrete, the various volumes presented entire fields of mathematics but apparently without viable connections. This was due to the presentation: The flawless precision of the axioms and the proofs made it very difficult to recognize the original problems and the intuitive ideas that had brought the subject to life in the first place. Even worse, it became principally impossible to distinguish logically between important and unimportant questions, as long as they were pertinent.

Fortunately, new fields were born – like complex function theory in several variables, algebraic geometry, and algebraic and geometric topology – that were driven by wild thinking[2] and only slowly grew into "decent" mathematical fields, to be eventually enshrined in the Bourbaki palace.[3] However, there were no obvious connections to physics in most of the new theories and so they met with little interest from the physics community at first.

On the physics side, the necessity to account for the birth and death of particles and the theory of special relativity in quantum physics led to quantum field theories that did not admit a sound mathematical basis, and hence did not attract large numbers of mathematicians. Nevertheless, there were big achievements, such as

1 "Ich kann es nun einmal nicht lassen, in diesem Drama von Mathematik und Physik – die sich im Dunkeln befruchten, aber von Angesicht zu Angesicht so gerne einander verkennen und verleugnen – die Rolle des (wie ich genugsam erfuhr, oft unerwünschten) Boten zu spielen." [23, p. V/VI]

2 Solomon Lefschetz, one of the fathers of algebraic geometry but originally an electrical engineer, was said to have never stated a wrong theorem and never given a correct proof.

3 The Bourbaki group still exists and keeps working but the speed of development has become too high for a definite treatment. Therefore not all of the subjects mentioned are covered yet.

DOI 10.1515/9783110452150-201

Yang–Mills theory in the 1950s, explaining eventually the weak and strong interaction, or the standard model of particle physics in the 1960s (which would have delighted Weyl). At the same time, the field of global analysis matured on the mathematics side, culminating in the index theorem of Atiyah and Singer [3], one of the most important theorems proved in the second half of the twentieth century. In a single identity, it combines topology, differential geometry, and analysis; this phenomenal result implied a wealth of new connections, reviving a feeling of the unity of mathematics. The new techniques developed for this proof were greatly expanded and allowed to attack new problems. On the other hand, Yang–Mills theory now inspired work in low-dimensional differential topology that ultimately led, for example, to completely unexpected facts on four-dimensional manifolds. Conversely, index calculations were also used in physics, notably in computing anomalies, but these connections appeared accidentally, not systematically. This fact confirmed Weyl's statement [22, p. 6] once more that "there is in the development of mathematics and physics an unmistakable mysterious parallelism" which is not easily recognized. Nevertheless, Freeman Dyson stated in [12]: "As a working physicist, I am acutely aware of the fact that the marriage between mathematics and physics, which was so enormously fruitful in past centuries, has recently ended in divorce."

With the appearance and rapid evolution of string theory, a certain reapproachment of mathematics and physics became visible. On the one hand, physicists reanimated the old and almost forgotten theory of open Riemann surfaces, asking new and difficult mathematical questions. But they also invented new mathematical subjects, such as mirror symmetry, and predicted mathematical results that had not even been conjectured so far. Mathematicians were interested in these results as such and tried hard to supply valid proofs, especially thrilled by the amazing work of Edward Witten. Witten applied ideas from quantum field theory and supersymmetry to well-known mathematical concepts and stated powerful mathematical results in many areas, like three-dimensional knot theory or Morse theory. But since he made frequent use of the non rigorous Feynman integral, other methods of proof had to be invented. The importance of the new fact of "physical mathematics"[4] for mathematics proper was gradually acknowledged by the community, culminating in the award of a Fields Medal to Edward Witten in 1990: the "Nobel Prize in Mathematics" was given to a physicist for the first time ever.[5]

Another astonishing aspect of this development was the enormous increase in complexity of notions and technology on both the physics and the mathematics side, while many fundamental results could be deduced much more easily with the new tools (as shown, e.g., by Witten's derivation of the positive mass conjecture [24] or the Atiyah–Singer index theorem [25]). This exciting development posed at the same

4 See the brilliant essay by Greg Moore: "Physical Mathematics and the Future," available at https://www.physics.rutgers.edu/~gmoore.

5 See the appraisal of Witten's work by Atiyah, in the Proceedings of the ICM 1990, vol. I, p. 31.

time a massive challenge for education, partially compensated by the fact that the new material was highly attractive for gifted newcomers who are looking for a challenge.

2 The scientific concept of CRC 647

The German reunification in 1990 opened the way for collaboration between mathematicians and physicists from East and West Germany, with Berlin a natural place to start geographically, but also in view of the great Berlin tradition in mathematical physics as shown by the work of Helmholtz, Kirchhoff, Planck, and Einstein. To relate to these heroic times, the CRC 288 "Differential Geometry and Quantum Physics" was started in 1992 as a joint application of scientists from both parts of Germany, oriented towards new horizons but including also modern developments of traditional topics of mathematical physics. This CRC worked successfully for the maximum time of 12 years, and established connections that lasted beyond its existence.

Based on the experience of CRC 288, in 2003 a small group of scientists began to explore the framework of a new collaborative project. The main points of discussion were, naturally, the scientific program and the modes of collaboration, both viewed in relation to the possibly interested scientists in the Berlin area. Integrable systems formed one of the main objects of study in CRC 288, from many different perspectives. This topic integrates Analysis, Geometry, and Theoretical Physics in a natural way, and these scientific fields were well represented in the Berlin area. Hence such a combination was welcomed, not in the least since some new players had arrived in Berlin, notably the mathematicians Klaus Altmann (Algebraic Geometry, FU), Christian Bär (Differential Geometry, UP), Klaus Ecker (Geometric Analysis, FU), Gerhard Huisken (Geometric Analysis, AEI), Klaus Mohnke (Symplectic Geometry, HU), and Dorothee Schüth (Spectral Geometry, HU), while from the former members of CRC 288, five mathematicians (and principal investigators) were inclined to participate in the new project: Helga Baum (Differential Geometry, HU), Alexander Bobenko (Discrete Geometry, TU), Jochen Brüning (Geometric Analysis and Spectral Theory, HU), Thomas Friedrich (Differential Geometry, HU), and Ulrich Pinkall (Differential Geometry, TU). The physics part was mainly advocated by Dieter Lüst and Albrecht Klemm (both HU), who naturally favored String Theory as an essential point of the future research program.

This proposal was met with sympathy by the mathematicians, in view of the surprising mathematical propositions mentioned before that had been formulated by string theorists and still caused hard work on the mathematical side in verifying them. But also the basic ansatz was applauded, to geometrize matter by modeling elementary particles as strings in place of points and their world lines as Riemann surfaces, thus reviving an almost forgotten but still largely unexplored theory. The famous triad "Space · Time · Matter," the title of Hermann Weyl's marvellous book on Einstein's theory of General Relativity discussed above, suggested itself effortlessly as overall title of our enterprise, and it was embraced unanimously. On the mathematics side, important topics of interest were nonlinear partial differential equations, notably of geometric interest – like the minimal surface equation – and geometric flows – like the Ricci or the mean curvature flow – that were of high interest at the time, but also field equations – like the Strominger equations in heterotic String Theory – or the dynamical systems arising in cosmology. Naturally, also the Dirac operators, the Bochner Laplacians, and the Schrödinger operators played an important role, in particular in the presence of singularities, as well as genuinely geometric questions like the classification of (semi-)Riemannian manifolds with special holonomy. All of these themes had natural connections with mathematical physics and string theory in particular. In this fertile climate, it was not difficult to invite scientists with corresponding areas of interest among the overall topics just mentioned, to cooperate in what was already envisaged as a new CRC. We had a favorable response from most scientists we approached, but the formation of the individual projects and their interconnections turned out to be a time-consuming enterprise, involving a fair number of daylong meetings. The overall strategic goal of reinstalling mathematical physics in the Berlin area and hopefully following the approach of Weyl, mutatis mutandis, was commonly agreed on from the beginning, though, in particular since both FU and TU did not adequately refill several professorships in the field. Also, the incorporation of discrete mathematics in the program was seen as an important asset, to be strengthened in the future if possible. What appeared as a difficult problem, though, was a natural ordering of the scientific interests according to the key words space, time, and matter.

Eventually, it seemed more logical to organize the proposed projects in two groups: the "Geometry of Matter" group, comprising projects A1 to A6 that should deal with model spaces of interest in String Theory; and the "Evolution of Geometric Structures" group, comprising projects B1 to B6 that were supposed to analyze the partial differential equations that describe properties of the models from group A and related objects as well. The protagonists of the enterprise consulted many specialists, either informally or by invitation to Berlin, to give a talk with subsequent discussion. A very impressive and stimulating event was the visit of *Grigory Perelman* to the AEI who gave a talk on January 28, 2003, at FU that attracted people from all over Germany: At this occasion, he announced for the first time that he had proved the Poincaré conjecture.

The other topic of great interest in our discussions concerned the modes of cooperation. The participating institutions, HU (with its Adlershof campus), FU (with its campus in Dahlem), UP (with its campus near the Neue Palais and Park Sanssouci), and the AEI (on the new campus of the sciences of Max Planck Society and UP in Golm) are situated at significant distances which lead to traveling times up to one hour each way. Therefore, we agreed on at least one regular seminar day each month during the semester, featuring two related one-hour talks and leaving enough time for discussions. Each seminar had to be arranged responsibly by one or two of the projects in turn, thus also determining the location for the seminar. In this way, the work of the projects could be explained and discussed. Since mathematics and physics are "separated by a common language" (Michael Atiyah), we also hoped that the seminar speakers would contribute to the development of a common language eventually spoken by all members of the CRC. Since the young researchers could be expected to learn such a new language much faster, we planned to also apply for an Integrated Research Training Group which should systematically further this purpose, as a byproduct of presenting mathematical and physical content at the same time. This, however, was realized only in the second period. Among the scientists willing to contribute to the project, an atmosphere developed that was characterized by mutual interest, curiosity, and engagement. Thus, the ideas just described were agreed on and a sketch of our project was submitted to DFG in the spring of 2004. Unfortunately, it was decided at this time that Dieter Lüst left Berlin for München, while Albrecht Klemm, another string theorist very interested in participating, had already accepted an offer of the University of Wisconsin-Madison. This weakened the physics group of principal investigators considerably, reducing it to one-third of the expected number. Nevertheless, and much to our delight, we were invited to submit the full application, and the CRC 647 "Space · Time · Matter" was established for the first four years on January 1, 2005. However, our proposal was not accepted in full: the projects A5 – "Minimal and Monotone Lagrangians in Kähler-Einstein spaces" – and B2 – "Geometry in 4-Space and Integrable Systems" – were not accepted by the reviewing committee, thus cutting a viable link to the heritage of CRC 288 and, more significantly, to the needed expertise in discrete geometry.

3 The program and its implementation in the third period

With the changes in projects and project leaders necessary in the second period, the steering committee of the CRC discussed intensely the question of restructuring the projects in order to allow a more concentrated work in neighboring fields, and to direct it towards summarizing – as far as possible – the work done within the full time span granted to the CRC, and to collect the main results in a book. This discussion was not easy but furnished a satisfying result eventually. We agreed on implementing radical changes: We reduced the number of projects from 13 to 9 (counted now as C1 through C9), such that the old balance of six projects dominated by mathematical and three by physical content, was reinstalled (though a slight distortion was caused by the reviewing committee in rejecting project C9 and leaving us with eight projects). More importantly, the projects would comprise all activities within the CRC pertaining to a certain field of importance, in this way becoming competence centers inside and to some extent also outside the CRC.

We now turn to the thematic introduction to the chapters collected in this book. We will discuss the work done in the last period of the CRC, highlighting the intentions, the interactions, and the major achievements and their impact. We organize the presentation along the project structure of the last period.

3.1 C1 Algebraic topology: Rigidity and dynamics

Chapter: **1** Holger Reich and Marco Varisco: *Algebraic K-theory, assembly maps, controlled algebra, and trace methods*

Project C1 was directed by Holger Reich who joined the CRC essentially only for the last period. His arrival was highly welcomed since topological expertise was scarce in the Berlin area, and especially expertise in geometric topology, the part of topology dealing with the topological properties of manifolds. An illustrative example is given by the Borel Conjecture, stating that for any closed, aspherical[6] topological manifold the homotopy class is determined by its homeomorphism class. For example, all closed and connected surfaces except the 2-sphere, and all closed and connected manifolds of non positive curvature satisfy this conjecture, but there are counterexamples in the smooth category (like the 5-torus). One motivation for the Borel Conjecture is Mostow's (differential geometric) rigidity theorem, stating that two homotopy equivalent closed hyperbolic manifolds are already isometric. A proof of the conjecture then obviously hinges only on the group $\pi_1(M)$ and its properties, and another conjecture concerning groups, due to Farrell and Jones, provides an affirmative answer to Borel if satisfied. This latter conjecture is quite algebraic and states that, for any group and any coefficient ring, the algebraic K-theory can be constructed from the algebraic K-theory of all virtually cyclic[7] subgroups of the group. Together with Wolfgang Lück and Arthur Bartels, Holger Reich has verified quite a few difficult cases of the Farrell–Jones Conjecture (e.g., for Gromov hyperbolic groups or the general linear groups over the integers, [5, 6]) such that this research group is among the world leaders in this area of topology. Topological information was needed and used in projects C2 (classification of closed Lorentzian manifolds), C3, C7 (topological structure of Thom–Mather spaces), and C8 (topological structure of flows with global attractors). At first sight, it is quite surprising that fairly deep topological facts (as stated in the Borel Conjecture) admit a (very complicated) proof that is purely algebraic. This phenomenon led to a common CRC seminar on so-called *Temperley–Lieb algebras*, a type of algebra that arose in statistical mechanics via transfer matrices, but surfaced in many other fields,

6 A manifold is called *aspherical* if it is connected and has trivial homotopy groups in any degree bigger than 1.

7 These are groups that have a cyclic subgroup of finite index.

for example knot theory. Specifically, it figured systematically in the work of C1, and was used also in C8 (study of attractors) as well as in C5 (computations in QFT).

3.2 C2 Differential geometry: Geometric and spectral invariants of Riemannian, Lorentzian, and conformal manifolds

Chapters:
2.1 Helga Baum: *Lorentzian manifolds with special holonomy – Constructions and global properties*
2.2 Sebastian Boldt and Dorothee Schueth: *Contributions to the spectral geometry of locally homogeneous spaces*
2.3 Andreas Juhl: *On conformally covariant differential operators and spectral theory of the holographic Laplacian*

The work of the group in metric differential geometry is very well known in the community; the chapters concern the most important results of its three main areas of research. Right at the end of the CRC, several major questions have been given a definite answer. This pleasant fact is owed to a good part to a very careful choice and intense education of young researchers conducted by Helga Baum and Dorothee Schüth.

In Chapter 2.1, the Lorentzian manifolds with special holonomy are the key object with two major tasks: (1) Classify these manifolds in the spirit of Marcel Berger (who did it in the Riemannian case); (2) give important geometric properties of these spaces; (3) describe convenient methods of constructing Lorentzian manifolds with prescribed special holonomy. The classification problem was first tackled by Bérard-Bergery and Ikemakhen in 1990, when they classified all closed subgroups of O(1,n+1) that act indecomposably but non-irreducibly. Thomas Leistner [17] and Anton Galaev [16] (both working in C2) completed the classification by characterizing the orthogonal parts as holonomy groups of a Riemannian manifold, respectively, showing that all groups in the final list are realized as holonomy groups of a Lorentzian manifold.

Special properties are dealt with in the second part. One important question arises from the fact that compactness does not imply completeness in the Lorentzian case: It was then shown in [18], by Schliebner and Leistner, that completeness follows for Lorentzian manifolds from Abelian holonomy (these are the pp-wave space-times, modeling radiation at the speed of light). Interesting results of a more technical character are also given in the first chapter and are worth to be studied.

In the third part, the construction problem is attacked. The aim is to construct a globally hyperbolic Lorentzian manifold possessing a parallel light-like spinor field. First, the resulting constraints on the initial data for the Cauchy problem on a three-dimensional Riemannian manifold (M, g) are formulated, then it is solved by various methods. Andree Lischewski in his excellent dissertation with Helga Baum (cf. Section 4.3) applied subtle first-order hyperbolic differential equation techniques to construct a Lorentzian manifold with the desired properties into which the initial data embed. It should be said that he also produced a number of subtle and interesting constructions pertaining to string theory or quantum field theory which did not arouse interest among the physicists, unfortunately. Even more unfortunately, after getting his PhD, Lischewski left academia leaving us with a definite loss of talent.

In Chapter 2.2, the spectral geometry group around Dorothee Schüth presents several interesting and subtle results. In the first part, Sebastian Boldt explains his rather deep analysis of the Dirac spectrum (with Emilio Lauret [9]) on Lens spaces, based on a totally new description of the spectrum. As a consequence, neither spin structures nor isometry classes of lens spaces are determined by the Dirac spectrum. On the other hand, there are no non trivial Dirac isospectral pairs of three-dimensional lens spaces with fundamental group of prime order. In the second part, the heat coefficients of the Laplace–Beltrami operator on a closed Riemannian manifold are examined: It is well known that each such coefficient is given as the integral of a universal $O(n)$-invariant polynomial in the curvature tensor and its covariant derivatives. While this quantity is spectrally determined, it is not clear what happens to the individual curvature terms that add up to the heat coefficient. Here, Dorothee Schüth shows for the fourth coefficient that no individual term is spectrally determined [2]. Finally, a curious question by Victor Guillemin is (partially) answered in [20]: In analogy with a result of Karen Uhlenbeck, saying that for a generic closed Riemannian manifold the spectrum of the Laplace–Beltrami operator is simple, Guillemin asks whether for a generic left invariant metric on a compact Lie group, the group acts irreducibly on the eigenspaces of the Laplace–Beltrami operator on the group. Here, Dorothee showed that this is true for SU(2) and an array of related groups.

Chapter 2.3 is written by Andreas Juhl. He produced a truely outstanding result by solving the recursion relations for the GJMS (Graham, Jenne, Mason, Sparling) operators explicitly in [15]; knowing these formulas, Fefferman and Graham [15] gave a simpler proof of this result. For many reasons, it is of great interest to extend this machinery to conformally covariant operators acting on differential forms. The purpose of Juhl's chapter is to develop this theory in analogy with [15]. Quite naturally, this analysis is complicated and technical while involving a fair number of new ideas; but certainly, every specialist will enjoy this work.

C3 Algebraic geometry: Deformations, moduli, and vector bundles

Chapters:
3.1 Klaus Altmann and Gavril Farkas: *Moduli and deformations*
3.2 Björn Andreas and Alexander Schmitt: *Vector bundles in algebraic geometry and mathematical physics*

The work of the group in algebraic geometry is very well known in the community and also beyond it. The chapters concern the most important recent work done by the members of the group.

Chapter 3.1 presents a survey of results that have been obtained with support of the CRC, with some spectacular results being obtained only recently. The material presented concerns two large and highly active areas of research, with strong connections to problems posed in string and/or gauge theory. Moduli spaces are of eminent importance in general; the chapter addresses moduli spaces – like the Deligne–Mumford moduli space of complex curves of genus g – that play an important role in string theory and also in the (physics) geometric Langlands program. The said space constitutes an algebraic variety, but to apply useful tools to its analysis like intersection theory, it needs to be compactified. Thus a boundary must be constructed with degenerate but perhaps simpler objects than the elements of the moduli space. This can be achieved by global constructions as well as local ones, notably versal deformations; both aspects are treated. Then one wants to study the complex geometry of the compactified moduli space X, for example its Kodaira dimension, $\kappa(X) \leq \dim X$, which provides an important means of classifying varieties birationally.

The first part deals with the deformation side starting from polyhedra, i.e., combinatorial objects that can be generalized to T-varieties, notably the toric varieties that contain a torus (of full dimension) as an open dense set whose action extends to the full variety. Then the deformation theory is developed by using combinatorial methods only; the theory is fully described in the basic publication on the subject, [1].

The second part deals with the birational classification of certain moduli spaces as above, depending on the genus, which changes quite mysteriously, e.g., in terms of its Kodaira dimension. The work reported on clarifies the situation in quite a few cases; two very important results are [14], on the geometry of the moduli spaces of spin curves, and also [13], on the uniformization of the moduli spaces of abelian six folds.

Chapter 3.2 addresses mainly two topics: first, a comprehensive explanation of the Strominger system of differential equations together with a new technique of solving it under some additional conditions. The second topic is slightly technical and deals with finiteness results of modules under stability or semi stability conditions.

C4 Structure of quantum field theory: Hopf algebras versus integrability

Chapters:
4.1 Dirk Kreimer: *Dyson–Schwinger equations: Fix-point equations for quantum fields*
4.2 Dhritiman Nandan and Gang Yang: *Hidden structure in the form factors of N=4 SYM*

Project C4 was directed by Dirk Kreimer and Matthias Staudacher. They had joined Humboldt University during the second funding period as "bridge professors," both of them holding positions in the Physics as well as the Mathematics Department.

This reestablished Mathematical Physics and in particular the mathematical structure of quantum field theory as a university-funded research field in the Berlin area. Quantum field theory (QFT) plays a curious role in the last, still unfinished chapter of the long history of mutual enrichment as well as periodic alienation between physics and mathematics. Its dominating role and stunning successfulness in the description of all physical interactions, with the notable exception of gravitation, on the one hand, and its stubborn resistance to a rigorous mathematical definability of its foundations on the other, are surely frustrating for all attempts at declaring QFT an overarching common research goal of both disciplines. Yet, in recent years a new point of view is clearly emerging: If one brazenly excludes, for the time being, the "mathematically rigorous definition" of QFT, amazingly beautiful hidden mathematical structures emerge from QFT. And these "one-level-up" structures often *are* well defined mathematically. They typically allow for powerful new, analytically exact results in theoretical physics, and invite deep mathematical analysis. Two examples of such initially hidden structures were a focal point of the activities of project C4: The Hopf algebra structure of perturbative QFT and the quantum integrability of the "simplest" QFT in four spacetime dimensions, $N = 4$ Super Yang–Mills Theory (SYM). Chapters 4.1 and 4.2 were selected accordingly.

In Chapter 4.1, Dirk Kreimer describes recent progress in the derivation of Dyson–Schwinger equations (well known in QFT) from fix-point equations in Hochschild cohomology (originating from the mathematical theory of associative algebras over rings).

In Chapter 4.2, the postdoc hired for this project, Dhritiman Nandan, gives an overview of the hidden, integrable structure of the form factors, quantities of prime importance in high-energy physics, of the $N = 4$ SYM model, in collaboration with Gang Yang, a postdoc originally employed for project C6. The intriguing impetus for project C4 remains an exciting open problem for the future, however: What, if any, is the connection between the Hopf algebra structure and the integrable structure in $N = 4$ SYM?

C5 AdS/CFT correspondence: Integrable structures and observables

Chapters:
5.1 Valentina Forini: *On regulating the AdS superstring*
5.2 Pedro Liendo: *Constraints on CFT observables from the bootstrap program*

Project C5 was directed by Valentina Forini, Jan Plefka, and Matthias Staudacher, all from Humboldt University. Its general focus was on exactly solvable, integrable string and gauge theories, and the intriguing dualities often connecting them. The prime example for such a duality continues to be the Maldacena conjecture of 1997, which holds that a superstring on the five-dimensional Anti-de Sitter space, with locally attached five-dimensional sphere, is exactly equivalent to the "simplest" QFT in four spacetime dimensions, the superconformal $N = 4$ Super Yang–Mills model (SYM) already mentioned above in project C4. Accordingly, this conjectured duality has been termed "AdS/CFT (Anti-de Sitter/Conformal Field Theory) correspondence." The grand goal of the project was to find a proof for this enigmatic duality, at least in the so-called planar limit of infinitely large gauge group rank. Proving that a string theory can be *equal* to a gauge field theory would clearly be of utmost importance for mathematical physics, especially in the light of the mathematical ill-definedness of quantum field theory. However, this goal proved too ambitious so far, so one needs to proceed in steps. Two such important steps are highlighted by the choice of essays for this project.

In Chapter 5.1, Valentina Forini explains her pioneering approach to numerically test the correspondence by applying lattice gauge theory methods directly to the quantization of the AdS superstring. This has never been attempted before, and, as she explains in the chapter, her initial steps for this new research program are very promising. We might soon have a new precision tool for examining the detailed structure of the AdS/CFT correspondence, which might inspire new ideas for the above-mentioned missing proof.

While Chapter 5.1 is concerned with *string theory*, Chapter 5.2, written by postdoc Pedro Liendo, who had been hired for this project, deals with deepening our understanding of superconformal *gauge theory*. It applies the recently reborn and currently intensively researched (on a worldwide scale) "conformal bootstrap approach" to supersymmetric gauge field theories. This also leads to a large number of at least numerically nearly exact results, which can be considered "complementary", in terms of the method applied and many of the results obtained, to the ones obtained from gauge-string dualities.

C6 Scattering amplitudes: Symmetries and interrelations in maximal supergravity and Yang–Mills

Chapters:
6.1 Marco Chiodaroli: *Simplifying amplitudes in Maxwell–Einstein and Yang–Mills–Einstein supergravities*
6.2 Livia Ferro, Jan Plefka, and Matthias Staudacher: *Yangian symmetry in maximally supersymmetric Yang–Mills theory*

The focus of project C6 was on scattering amplitudes in gravitational and gauge theories with and without supersymmetry. Substantial results in Einstein-Maxwell and Einstein–Yang–Mills amplitudes were obtained.

This is reviewed in Chapter 6.1 by Marco Chiodaroli, who was hired for the project at the Max-Planck Institute for Gravitational Physics (Albert-Einstein-Institute) in

Potsdam-Golm. The method applied was the so-called double-copy construction, a still rather mysterious – especially from the mathematical point of view – relation between theories with and without gravity discovered a few years ago by Bern, Carrasco, and Johansson (BCJ). It is a modern incarnation of the Kawai, Lewellen, and Tye (KLT) relations of early-day string theory, and, very roughly, holds that gauge theory is a "kind of square root" of gravity, or, conversely, that quantities in Einsteinian gravity theories can be obtained, both on the classical and on the perturbative quantum level, from "squaring" gauge theory quantities. Generally, it should be considered to be a further but quite separate hint, on top of gauge-string dualities (see project C5), that the mathematical structures underlying both string theories and quantum field theories are incompletely understood.

In Chapter 6.2, written by Livia Ferro, a postdoc at the physics department of Humboldt University from 2010 to 2013, together with C6 project leader Jan Plefka as well as with Matthias Staudacher, the emergence of Yangian symmetry from two copies of superconformal symmetry, one of them rather obvious, the other hidden, is studied in the context of scattering amplitudes for $N = 4$ Super Yang–Mills Theory (SYM). The described novel results go back to the original work of Matthias Staudacher on the integrability of this model in the planar limit, as well as to the seminal 2009 paper by Jan Plefka (with James Drummond and CRC 647 postdoc Johannes Henn) on the Yangian symmetry of these scattering amplitudes. Apart from furnishing a general review of Yangian symmetry in this context, the chapter describes novel ways to directly derive tree-level four-dimensional scattering amplitudes from the two-dimensional "quantum inverse scattering method", i.e., from integrability. Extending the methodology to loop level is an active field of ongoing research.

C7 Differential operators of mathematical physics: Spectral theory and dynamics

Chapters:
7.1 Lars Andersson and Christian Bär: *Wave and Dirac equations on manifolds*
7.2 Francesco Bei, Jochen Brüning, Batu Güneysu, and Matthias Ludewig: *Geometric analysis on singular spaces*

Project C7 comprised a group at HU, directed by Jochen Brüning, and a group at UP, directed by Christian Bär.

In Chapter 7.2, the HU group unifies and extends previous work done in B1 and, partially, in B6. Notably, the notion of "singular space" is extended to the class of "Thom–Mather spaces", as described in [19]. These spaces are locally compact, second countable Hausdorff spaces, W, which admit a locally finite partition by smooth manifolds, called the strata. They satisfy the crucial *condition of frontier*, saying that a stratum intersecting the closure of another stratum is already contained in this closure. The prototypical example of a Thom–Mather space is a manifold with boundary, with two obvious strata. In this case, there is an open and dense stratum, called the *top stratum*, and we assume this in general, and also that all other strata, called *singular*, have codimension at least 2. Almost all singular spaces that play a role in mathematical physics are of this type, for example simplicial complexes, algebraic varieties, in particular moduli spaces (e.g., of curves) and their closures, orbit spaces of proper Lie group actions on manifolds, manifolds with corners, or conifolds. The differential operators of interest are the Dirac operators, notably the de Rham–Hodge operator, the Bochner Laplacians, and the covariant Schrödinger operators. Naturally, these operators should be defined on the top stratum equipped with a suitable metric. The metric shall extend to the whole of W, with a necessary degeneration near a singular stratum. Such metrics do always exist and are known as *Cheeger metrics*, generalizing the well-known conic metrics in the case that all singular strata reduce to points. Unfortunately, there are other canonical metrics – like the *Fubini-Study* metric induced on a projective variety – that are somehow similar to the Cheeger metric but are analytically much more complicated. The project work was aimed at the most important aspects of this setting under the perspective of mathematical physics, and considerable and rather important work was done in the last period. This was possible since, happily, we could form a powerful team of young postdocs with different funding: Batu Güneysu (CRC and HU), Bo Liu, Shu Shen (CRC), Jörn Müller, Francesco Bei (HU), and Nils Waterstraat (BMS), in addition PhD students Juan Orduz and Asilya Suleimanova (both BMS-funded) who have obtained their PhD just after the end of CRC 647.

In Chapter 7.2, the analysis of elliptic operators on Thom–Mather spaces is treated. Batu Güneysu reports on his recent Habilitationsschrift "Covariant Schrödinger Operators on Non-compact Riemannian Manifolds", soon to appear in the Birkhäuser series "Operator Theory: Advances and Applications". This impressive work studies covariant Schrödinger operators on *completely arbitrary* non compact Riemannian manifolds; the author shows that many known statements from the compact case admit generalizations to the general case, under surprisingly similar conditions that are difficult to verify, though.

Francesco Bei, Jochen Brüning, and the two PhD students worked on the differential topology of Thom–Mather spaces and the spectral theory of the de Rham–Hodge operator, respectively, the Hodge Laplacian under various special assumptions, obtaining in particular eigenvalue estimates, spectral invariants and spectral geometry, and index theorems, cf. [8] and [4, 10, 11]. Jörn Müller specialized in scattering theory but left the CRC and Humboldt University for a job in industry in 2014. The two Chinese postdocs spent only 18 months at the CRC but produced very good work: Bo Liu finished an important paper on equivariant differential K-theory while Shu Shen finished his proof of the Fried Conjecture from 1986, stating that on a closed locally symmetric reductive manifold equipped with an acyclic unitarily flat vector bundle, the analytic torsion equals the value at zero of a dynamical zeta function associated with the geodesic flow, a truely outstanding result with a very difficult proof [21]. Summing up, one can say that the project work laid a solid basis for extending the geometric and spectral invariants associated with Dirac and Schrödinger-type operators on closed manifolds to a huge class of singular spaces. This work – except the contributions of Bo Liu and Shu Shen – is described in the chapter.

Also included is a contribution by Matthias Ludewig (UP) (Section 3.4) where he derives the Feynman–Kac formula for Schrödinger operators, proved by Güneysu and others using stochastic calculus, via a finite-dimensional approximation based on more intuitive arguments of differential geometric nature. The corresponding formula for the wave operator admits, so far, no rigorous proof in some generality. It seems, however, that Ludewig and Hanisch made some progress toward this problem recently; further progress would be, of course, of the greatest interest.

Chapter 7.1 presents the work of the Potsdam group, mostly focussing on geometric wave equations, quantum field theory on curved backgrounds, and the mathematical theory of path integrals. Building on work by Christian Bär, Matthias Ludewig wrote a very strong PhD thesis in which he was able to construct path integrals on manifolds with boundary; this resulted in several papers and is described in Chapter 7.2.

Wave equations on Lorentzian manifolds were studied from various points of view. Oliver Lindblad Petersen studied the analytic theory of the linearized Einstein equations that describe gravitational waves. Viktoria Rothe studied the Lorentzian analog of the Yamabe problem; this amounts to studying a nonlinear scalar wave equation. Christian Bär clarified basic properties of such equations together with Roger Tagne Wafo. Together with Alexander Strohmaier he proved an index theorem for the Dirac operator on Lorentzian manifolds [7]. This is surprising because the Fredholm property of differential operators is, generally, strongly linked to ellipticity of the operator rather than to hyperbolicity. As an application they gave a geometric formula for the chiral anomaly in quantum field theory on curved space-times. Higher gauge theory was applied to quantum field theory by Christian Bär, Christian Becker, and coauthors. These results are reviewed in this chapter in the common framework of globally hyperbolic space-times.

C8 Mathematical physics: Dynamics and nonlinear evolution equations in general relativity

Chapter: 8 Klaus Ecker, Bernold Fiedler et al.: *Singularities and long-time behavior in nonlinear evolution equations and general relativity*

Many central problems in geometry, topology, and mathematical physics reduce to questions regarding the behavior of solutions of nonlinear evolution equations. Examples are Thurston's classification of compact 3-manifolds based on Hamilton's

Ricci flow, and the work of Christodoulou and Klainerman on the nonlinear stability of Minkowski spaces. Examples of hyperbolic equations include the Einstein field equations of general relativity as well as semilinear wave equations. Ricci flow and semilinear reaction-advection-diffusion equations are of parabolic type. In all these equations, the global dynamical behavior of bounded solutions for large times is of significant interest. Specific questions concern the convergence to equilibria, the existence of periodic, homoclinic, and heteroclinic solutions, and the existence and geometric structure of global attractors. On the other hand, many solutions develop singularities in finite time. The singularities have to be analyzed in detail before attempting to extend solutions beyond their singularities, or to understand their geometry in conjunction with globally bounded solutions. In this context, the authors have been particularly interested in global qualitative descriptions of blow-up and grow-up profiles. In particular, Bernhard Brehm was able to solve the Belinsky-Khalatnikov-Lifshitz (BKL) conjecture in his dissertation, concerning models for the early universe, that has been open for almost 50 years. The chapter gives a review on the project work of B3, B7, and C8, outlining methods and results in the framework just sketched with many instructive examples.

Bibliography

[1] Altmann K, Ilten NO, Petersen L, Süß H, Vollmert R. The geometry of T-varieties. Contributions to algebraic geometry. EMS Ser Congr Rep Eur Math Soc Zürich, 2012:17–69. MR 2975658

[2] Arias-Marco T, Schueth D. Inaudibility of sixth order curvature invariants. Rev R Acad Cienc Exactas Fís Nat Ser A Math RACSAM 2017;111(2):547–74. MR 3623858

[3] Atiyah MF, Singer IM. The index of elliptic operators on compact manifolds. Bull Am Math Soc 1963;69:422–33. MR 0157392

[4] Ballmann W, Brüning J, Carron G. Index theorems on manifolds with straight ends. Compos Math 2015;148(6):1897–68. MR 2999310

[5] Bartels A, Lück W, Reich H. The K-theoretic Farrell-Jones conjecture for hyperbolic groups, Invent Math 2008;172(1):29–70. MR 2385666

[6] Bartels A, Lück W, Reich H, Rüping H. K and L-theory of group rings over GLn(Z), Publ Math Inst Hautes Études Sci 2014;119:97–125. MR 3210177

[7] Bär C, Strohmaier A. A rigorous geometric derivation of the chiral anomaly in curved backgrounds. Comm Math Phys 2016;347(3):703–21. MR 3551253

[8] Bei F. General perversities and L2 de Rham and Hodge theorems for stratified pseudomanifolds. Bull Sci Math 2014;138(1):2–40. MR 3245491

[9] Boldt S, Lauret EA. An explicit formula for the Dirac multiplicities on lens spaces. J Geom Anal 2017;27(1):689–725. MR 3606566

[10] Brüning J. The signature operator on manifolds with a conical singular stratum. Astérisque 2009;328:1–44. MR 2664466

[11] Brüning J, Ma X. On the gluing formula for the analytic torsion. Math Z 2013;273(3–4):1085–17. MR 3030691

[12] Dyson FJ. Missed opportunities. Bull Am Math Soc 1972;78:635–52. MR 0522147

[13] Farkas G, Ortega A. Higher rank Brill-Noether theory on sections of K3 surfaces. Internat J Math 2012;23(7):1250075. 18 MR 2945654

[14] Farkas G, Verra A. The geometry of the moduli space of odd spin curves. Ann of Math 2014;180(3):927–70. MR 3245010
[15] Fefferman C, Graham CR. Juhl's formulae for GJMS operators and Q-curvatures. J Am Math Soc 2013;26(4):1191–207. MR 3073887
[16] Galaev AS. Metrics that realize all Lorentzian holonomy algebras. Int J Geom Methods Mod Phys 2006;3(5–6):1025–45. MR 2264404
[17] Leistner T. On the classification of Lorentzian holonomy groups. J Differential Geom 2007;76(3):423–84. MR 2331527
[18] Leistner T, Schliebner D. Completeness of compact Lorentzian manifolds with abelian holonomy. Math Ann 2016;364(3–4):1469–503. MR 3466875
[19] Mather J. Notes on topological stability. Bull Amer Math Soc (NS) 2012;49(4):475–506. MR 2958928
[20] Schueth D. Generic irreducibility of Laplace eigenspaces on certain compact Lie groups. Ann Global Anal Geom 2017;52(2):187–200. MR 3690014
[21] Shen S. Analytic torsion, dynamical zeta functions and orbital integrals. C R Math Acad Sci Paris 2016;354(4):433–6. MR 3473562
[22] Weyl H. Raum, Zeit, Materie Vorlesungen über allgemeine Relativitätstheorie. Erste Auflage, Berlin: Julius Springer, 1918.
[23] Weyl H. Gruppentheorie und Qantenmechanik. Erste Auflage, Leipzig, Hirzel, 1928.
[24] Witten E. A new proof of the positive energy theorem. Comm Math Phys 1981;80(3):381–402. MR 626707
[25] Witten E. Index of Dirac operators, Quantum fields and strings: a course for mathematicians. (Princeton: 1996/1997), Am Math Soc 1999;1(2):475–511. Providence. MR 1701605

Holger Reich and Marco Varisco

Algebraic K-theory, assembly maps, controlled algebra, and trace methods

A primer and a survey of the Farrell–Jones Conjecture

Abstract: We give a concise introduction to the Farrell–Jones Conjecture in algebraic K-theory and to some of its applications. We survey the current status of the conjecture and illustrate the two main tools that are used to attack it: controlled algebra and trace methods.

Keywords: algebraic K-theory, Farrell–Jones Conjecture.

Classification: 19-02

1 Introduction

The classification of manifolds and the study of their automorphisms are central problems in mathematics. For manifolds of sufficiently high dimension, these problems can often be successfully solved using algebraic topological invariants in the algebraic K-theory and L-theory of group rings.

In an article published in 1993 [40], Tom Farrell and Lowell Jones formulated a series of *Isomorphism Conjectures* about the K- and L-theories of group rings, which became universally known as the *Farrell–Jones Conjectures*. On the one hand, these conjectures represented the culmination of decades of seminal work by Farrell, Jones, and Wu Chung Hsiang, e.g. [34–36, 38, 39, 56]. On the other hand, they have motivated and continue to motivate an impressive body of research.

In this chapter, we focus only on the Farrell–Jones Conjecture for algebraic K-theory, and mention briefly some of its variants in Section 2.6. We give a concise introduction to this conjecture and to some of its applications, survey its current status, and most importantly we explain the main ideas and tools that are used to attack the conjecture: controlled algebra and trace methods.

Section 2 begins with some fundamental conjectures in algebra and geometric topology, which can be reformulated in terms of K_0 and K_1 of group rings. These conjectures are all implied by the Farrell–Jones Conjecture, but they are more accessible and elementary; moreover, their importance and appeal do not require algebraic K-theory, but may serve as motivation to study it.

In Sections 2.3 and 2.4, we define assembly maps and use them to formulate the Farrell–Jones Conjecture. Then we discuss how the Farrell–Jones Conjecture implies all other conjectures discussed in this chapter.

DOI 10.1515/9783110452150-001

In Section 3, we collect most of what is known today about the Farrell–Jones Conjecture in algebraic K-theory. We invite the reader to compare that section to the corresponding Section 2.6 in the survey article [72] from 2005, to appreciate the tremendous amount of activity and progress that has taken place since then.

The last two sections focus on proofs. In Section 4, we introduce the basic concepts of controlled algebra and see them at work. In particular, we give an almost complete proof of the Farrell–Jones Conjecture in the simplest nontrivial case, that of the free abelian group on two generators. Many ingenious ideas, mainly going back to Farrell and Hsiang, enter the proof already in this seemingly basic case. This section is meant to be an accessible introduction to controlled algebra. We do not even mention the very important flow techniques, and highly recommend Arthur Bartels's survey article [4].

In Section 5, we illustrate how trace methods are used to prove rational injectivity results about assembly maps. We give a complete proof of an elementary but illuminating statement about K_0 in Section 5.1, and then explain how this idea can be generalized using more sophisticated tools like topological Hochschild homology and topological cyclic homology. The complicated technical details underlying the construction of these tools are beyond the scope of this chapter, and we refer the reader to [31, 53, 78] for more information. However, we carefully explain the structure of the proof of the algebraic K-theory Novikov Conjecture due to Marcel Bökstedt, Hsiang, and Ib Madsen [22]. We follow the point of view used by the authors in joint work with Wolfgang Lück and John Rognes [73], leading to a generalization of this theorem for the Farrell–Jones assembly map. In particular, we highlight the importance of a variant of topological cyclic homology, Bökstedt–Hsiang–Madsen's functor C, which has seemingly disappeared from the literature since [22].

We tried to make our exposition accessible to nonexperts, and no deeper knowledge of algebraic K-theory is required. However, we expect our reader to have seen the basic definitions and properties of K_0 and K_1, and to be willing to accept the existence of a spectrum-valued algebraic K-theory functor. Classical and less classical sources for the K-theoretic background include [18, 26, 31, 80, 90, 109].

There are other survey articles about the Farrell–Jones and related conjectures: [4, 72, 78], which we already recommended, and also [71] and the voluminous book project [69]. Our hope is that this contribution may serve as a more concise and accessible starting point, preparing the reader for these other more advanced surveys and for the original articles.

2 Conjectures

In this section, we discuss many conjectures related to group rings and their algebraic K-theory. These conjectures are all implied by the Farrell–Jones Conjecture, which we formulate in Section 2.4. All of these conjectures are known in many cases but open in general, as we review in Section 3.

2.1 Idempotents and projective modules

An element p in a ring is an idempotent if $p^2 = p$. The trivial examples are the elements 0 and 1.

Conjecture 1 (trivial idempotents). *Let $\Bbbk$ be a field of characteristic zero and let G be a torsion-free group. Then every idempotent in the group ring $\Bbbk[G]$ is trivial.*

The assumption that G is torsion-free is necessary: if $g \in G$ is an element of finite order n, then $\frac{1}{n}\sum_{i=0}^{n-1} g^i$ is a nontrivial idempotent in $\mathbb{Q}[G]$.

A counterexample to the conjecture above would be in particular a zero divisor in $\Bbbk[G]$, and hence a counterexample to Problem 6 in Irving Kaplansky's famous problem list [59], which is reproduced in [60].

It is interesting to notice that the analog of Conjecture 1 for the integral group ring is true for all groups, even for groups with torsion. The proof that we give below uses operator algebras, as suggested in [60, page 451], and therefore it is very different from the rest of this chapter, even though the idea of using traces plays a central role in Section 5.

Theorem 2. *For any group G, every idempotent in the integral group ring $\mathbb{Z}[G]$ is trivial.*

Proof. The integral group ring embeds into the reduced complex group C^*-algebra $C_r^* G$, and the map $\mathbb{Z}[G] \longrightarrow \mathbb{Z}$, $\sum a_g g \longmapsto a_e$ extends to a positive faithful trace $\mathrm{tr}\colon C_r^* G \longrightarrow \mathbb{C}$. Let $p \in \mathbb{Z}[G]$ be an idempotent, i.e., $p = p^2$. It is known that in the C^*-algebra $C_r^* G$ every idempotent is similar to a projection, i.e., there exist $q, u \in C_r^* G$ such that $q = q^2 = q^*$, u is invertible, and $p = u^{-1} q u$; see for example [27, Proposition 1.8, Lemma 1.18]. Therefore, $\mathrm{tr}(p) = \mathrm{tr}(q)$. Applying the trace to $1 = q + (1-q) = q^* q + (1-q)^*(1-q)$ and using positivity one sees that the trace of q lies in $[0,1]$. The trace of p is clearly an integer. Therefore, $\mathrm{tr}(q) = 0$ or $\mathrm{tr}(q) = 1$. By faithfulness of the trace this implies that $q = 0$ or $q = 1$, and then the same holds for $p = u^{-1} q u$. □

The module Rp for an idempotent $p = p^2$ in the ring R is an example of a finitely generated projective left R-module. In view of the conjecture and the result above it seems natural to ask whether all finitely generated projective modules over group rings of torsion-free groups are necessarily free. Again, the assumption that G is torsion-free is necessary: if $g \in G$ is an element of finite order n, then for the non-trivial idempotent $p = \frac{1}{n}\sum_{i=0}^{n-1} g^i \in \mathbb{Q}[G]$ the module $\mathbb{Q}[G]p$ is projective but not free.

Example 3.

(i) Over fields and over principal ideal domains, hence in particular over the polynomial and Laurent polynomial rings $\Bbbk[t]$ and $\Bbbk[t^{\pm 1}]$ with coefficients in a field $\Bbbk$, all projective modules are free.

(ii) The question whether finitely generated projective modules over the polynomial ring $\Bbbk[t_1, \dots, t_n]$ for $n \geq 2$ are necessarily free was raised by Jean-Pierre Serre in [96], and was answered affirmatively only 21 years later independently by Dan Quillen and Andrei Suslin. The wonderful book [66] gives a detailed account of this exciting story.

The polynomial ring $R[t_1, \dots, t_n]$ is the monoid algebra $R[A]$ of the free abelian monoid A generated by $\{t_1, \dots, t_n\}$. The statement in (ii) was generalized as follows to monoid algebras.

(iii) If R is a principal ideal domain, then every finitely generated projective module over the monoid algebra $R[A]$ is free provided that A is a semi-normal, abelian, cancellative monoid without nontrivial units [51, 101]. Free abelian groups are examples of monoids satisfying these conditions.
(iv) If R is a principal ideal domain and F a finitely generated free group, then every finitely generated projective module over the group ring $R[F]$ is free [17].

At this point one could overoptimistically conjecture that every finitely generated projective $\mathbb{Q}[G]$-module is free if G is a torsion-free group. However:

(v) Martin Dunwoody constructed in [32] a torsion-free group G and a finitely generated projective $\mathbb{Z}[G]$-module P which is not free but has the property that $P \oplus \mathbb{Z}[G] \cong \mathbb{Z}[G] \oplus \mathbb{Z}[G]$. There are also finitely generated projective modules over $\mathbb{Q}[G]$ with analogous properties.

A weakening of the question above is whether all finitely generated projective $R[G]$-modules are induced from finitely generated projective R-modules when G is torsion-free. Recall that $K_0(R)$ is defined as the group completion of the monoid of isomorphism classes of finitely generated projective R-modules. The surjectivity of the natural map

$$K_0(R) \longrightarrow K_0(R[G])$$

induced by $[M] \longmapsto \left[R[G] \otimes_R M\right]$ studies the stable version of this question: is every finitely generated projective $R[G]$-module P stably induced? That is, is there an $n \geq 0$ such that $P \oplus R[G]^n$ is induced from a finitely generated projective R-module? Notice that this is true for Dunwoody's example (v). The stable version of Serre's Conjecture (ii) is a lot easier to prove and was established much earlier in [97, Proposition 10].

This discussion leads to the following conjecture. In order to formulate it, we need to recall some notions from the theory of rings. A ring is called left Noetherian if submodules of finitely generated left modules are always finitely generated, and it is said to have finite left global dimension if every left module has a projective resolution of finite length. If both properties hold, then R is called left regular. In the sequel we only consider left modules and therefore simply say *regular* instead of left regular. The ring of integers $\mathbb{Z}$, all principal ideal domains, and all fields are examples of regular rings.

Conjecture 4. *Let R be a regular ring, and assume that the orders of all finite subgroups of G are invertible in R. Then the map*

$$\operatorname*{colim}_{H \in \text{obj}\, \mathrm{Sub}G(\mathcal{F}in)} K_0(R[H]) \xrightarrow{\cong} K_0(R[G])$$

is an isomorphism. In particular, if G is torsion-free, then for any regular ring R there is an isomorphism

$$K_0(R) \xrightarrow{\cong} K_0(R[G]) .$$

Here the colimit is taken over the finite subgroup category $\mathrm{Sub}G(\mathcal{F}in)$, whose objects are the finite subgroups H of G and whose morphisms are defined as follows. Given finite subgroups H and K of G, let $\mathrm{conhom}_G(H, K)$ be the set of all group homomorphisms $H \longrightarrow K$ given by conjugation by an element of G. The group $\mathrm{inn}(K)$ of inner automorphisms of K acts on $\mathrm{conhom}_G(H, K)$ by post-composition. The set of morphisms in $\mathrm{Sub}G(\mathcal{F}in)$ from H to K is then defined as the quotient $\mathrm{conhom}_G(H, K)/\,\mathrm{inn}(K)$. Since inner conjugation induces the identity on $K_0(R[-])$, this is indeed a well-defined functor on $\mathrm{Sub}G(\mathcal{F}in)$. In the special case when G is abelian, the category $\mathrm{Sub}G(\mathcal{F}in)$ is just the poset of finite subgroups of G ordered by inclusion.

Proposition 5. *Conjecture 4 implies Conjecture 1.*

Proof. Let $\mathbb{k}$ be a field of characteristic zero and let G be a torsion-free group. Let $\epsilon\colon \mathbb{k}[G] \longrightarrow \mathbb{k}$ denote the augmentation and write $\epsilon_* M = \mathbb{k} \otimes_{\mathbb{k}[G]} M$. If $p \in \mathbb{k}[G]$ is an idempotent, then

$$\mathbb{k}[G] \cong \mathbb{k}[G]p \oplus \mathbb{k}[G](1-p) \qquad \text{and} \qquad \mathbb{k} \cong \epsilon_*\mathbb{k}[G] \cong \epsilon_*\mathbb{k}[G]p \oplus \epsilon_*\mathbb{k}[G](1-p) .$$

Since $\mathbb{k}$ is a field, either $\epsilon_*\mathbb{k}[G]p$ or $\epsilon_*\mathbb{k}[G](1-p)$ is the zero module. Replacing p by $1-p$ if necessary, let us assume that $\epsilon_*\mathbb{k}[G]p$ is zero. The assumption $\mathbb{Z} \cong K_0(\mathbb{k}) \cong K_0(\mathbb{k}[G])$ implies that there exist n and m such that

$$\mathbb{k}[G]p \oplus \mathbb{k}[G]^n \cong \mathbb{k}[G]^m .$$

Applying ϵ_* we see that $n = m$, and from this we conclude that $\mathbb{k}[G]p$ is zero as follows. Recall that a ring R is called *stably finite* if $M \oplus R^n \cong R^n$ always implies that M is zero; see [65, Section 1B]. Kaplansky showed that, if $\mathbb{k}$ is a field of characteristic 0, then any group ring $\mathbb{k}[G]$ is stably finite; compare [82]. □

2.2 *h*-Cobordisms

Recall that a smooth cobordism over a closed n-dimensional smooth manifold M consists of another closed n-dimensional smooth manifold N and an $(n+1)$-dimensional compact smooth manifold W with boundary ∂W together with a diffeomorphism $(f, g)\colon M \amalg N \xrightarrow{\cong} \partial W$. This is called an h-cobordism if both $\mathrm{incl} \circ f$ and $\mathrm{incl} \circ g$ are homotopy equivalences, where incl denotes the inclusion of ∂W in W. Two cobordisms W and W' over M are called isomorphic if there exists a diffeomorphism $F\colon W \xrightarrow{\cong} W'$ such that $F_{|\partial W} \circ f = f'$. A cobordism over M is called trivial if it is isomorphic to the cylinder $M \times [0, 1]$ (and this in particular implies that M and N are diffeomorphic).

Conjecture 6 (trivial h-cobordisms). *Let M be a closed, connected, smooth manifold of dimension at least 5 and with torsion-free fundamental group. Then every h-cobordism over M is trivial.*

Surprisingly, this conjecture can be reinterpreted in terms of algebraic K-theory. In fact, the celebrated s-Cobordism Theorem of Stephen Smale, Barry Mazur, John Stallings, and Dennis Barden (e.g., see [63, 79]), states that there is a bijection

$$\{\, h\text{-cobordisms over } M \,\}/\text{iso} \;\cong\; Wh(\pi_1(M))$$

between the set of isomorphism classes of smooth cobordisms over M and the Whitehead group $Wh(\pi_1(M))$ of the fundamental group of M, whose definition we now review.

Recall that, given a ring R, invertible matrices with coefficients in R represent classes in $K_1(R)$. Given any group G, the elements $\pm g \in \mathbb{Z}[G]$ are invertible for any $g \in G$, and hence represent elements in $K_1(\mathbb{Z}[G])$. By definition, the Whitehead group $Wh(G)$ is the quotient of $K_1(\mathbb{Z}[G])$ by the image of the map that sends $(\pm 1, g)$ to the element represented by $\pm g$ in $K_1(\mathbb{Z}[G])$. This map factors over $\{\pm 1\} \oplus G_{\mathrm{ab}}$, where G_{ab} is the abelianization of G, and the induced map $\{\pm 1\} \oplus G_{\mathrm{ab}} \longrightarrow K_1(\mathbb{Z}[G])$ is in fact injective; see for example [72, Lemma 2]. So there is a short exact sequence

$$0 \longrightarrow \{\pm 1\} \oplus G_{\mathrm{ab}} \longrightarrow K_1(\mathbb{Z}[G]) \longrightarrow Wh(G) \longrightarrow 0 \,. \tag{1}$$

For Whitehead groups there is the following well-known folklore conjecture. The cases of the infinite cyclic group [55], of finitely generated free abelian groups [19], and of finitely generated free groups [99] provided early evidence for this conjecture.

Conjecture 7. *If G is a torsion-free group, then $Wh(G) = 0$.*

By the s-Cobordism Theorem recalled above, the connection between the last two conjectures is as follows.

Proposition 8. *Let M be a closed, connected, smooth manifold of dimension at least 5 and with torsion-free fundamental group. Then Conjecture 6 for M is equivalent to Conjecture 7 for $G = \pi_1(M)$.*

For groups with torsion, the situation is much more complicated. For example, if C_n is a finite cyclic group of order $n \notin \{1, 2, 3, 4, 6\}$, then $Wh(C_n) \neq 0$, and in fact even $Wh(C_n) \otimes_{\mathbb{Z}} \mathbb{Q} \neq 0$. The analog of Conjecture 7 for arbitrary groups is the following.

Conjecture 9. *For any group G the map*

$$\operatorname*{colim}_{H \in \operatorname{obj} \mathrm{Sub}G(\mathcal{F}in)} Wh(H) \underset{\mathbb{Z}}{\otimes} \mathbb{Q} \longrightarrow Wh(G) \underset{\mathbb{Z}}{\otimes} \mathbb{Q} \tag{2}$$

is injective.

We highlight two differences with the corresponding Conjecture 4 for K_0. First, Conjecture 9 is only a rational statement, i.e., after applying $- \otimes_{\mathbb{Z}} \mathbb{Q}$. Second, it is only an injectivity statement. In order to obtain a rational isomorphism conjecture for $Wh(G)$ one needs to enlarge the source of map (2). This requires some additional explanations and is postponed to Conjecture 18.

2.3 Assembly maps

The Farrell–Jones Conjecture, which we formulate in the next section, generalizes Conjectures 4, 7, and 9 from statements about the abelian groups K_0 and Wh to statements about the nonconnective algebraic K-theory spectra $\mathbf{K}(R[G])$ of group rings, for arbitrary coefficient rings and arbitrary groups. In order to formulate the Farrell–Jones Conjecture, we need to first introduce the fundamental concept of assembly maps.

Fixing a ring R, algebraic K-theory defines a functor $\mathbf{K}(R[-])$ from groups to spectra. In fact, it is very easy to promote this to a functor

$$\mathbf{K}(R[-])\colon \mathrm{Groupoids} \longrightarrow \mathrm{Sp}$$

from the category of small groupoids (i.e., small categories whose morphisms are all isomorphisms) to the category of spectra. Moreover, this functor preserves equivalences, in the sense that it sends equivalences of groupoids to π_*-isomorphisms (i.e., weak equivalences) of spectra. For any such functor we now proceed to construct assembly maps, following the approach in [28]. It is not enough to work in the stable homotopy category of spectra, but any point-set level model would work.

Let $\mathbf{T}\colon \mathrm{Groupoids} \longrightarrow \mathrm{Sp}$ be a functor that preserves equivalences. Given a group G, consider the functor $G\int - \colon \mathrm{Sets}^G \longrightarrow \mathrm{Groupoids}$ that sends a G-set S to its action groupoid $G\int S$, with $\operatorname{obj} G\int S = S$ and $\operatorname{mor}_{G\int S}(s, s') = \{g \in G \mid gs = s'\}$. Restricting to the orbit category $\mathrm{Or}G$, i.e., the full subcategory of Sets^G with objects G/H as H varies among the subgroups of G, we obtain the horizontal composition in the following diagram.

$$\begin{array}{ccccccc} \mathrm{Or}G & \hookrightarrow & \mathrm{Sets}^G & \xrightarrow{G\int -} & \mathrm{Groupoids} & \xrightarrow{\mathbf{T}} & \mathrm{Sp} \\ \iota \big\downarrow & & & & & \nearrow_{\mathrm{Lan}_\iota \mathbf{T}(G\int -)} & \\ \mathrm{Top}^G & & & & & & \end{array}$$

Now we take the left Kan extension [77, Section X.3] of $\mathbf{T}(G\int -)$ along the full and faithful inclusion functor $\iota\colon \mathrm{Or}G \hookrightarrow \mathrm{Top}^G$ of $\mathrm{Or}G$ into the category of all G-spaces. The left Kan extension evaluated at a G-space X can be constructed as the coend [77, Sections IX.6 and X.4]

$$\big(\mathrm{Lan}_\iota \mathbf{T}(G\textstyle\int -)\big)(X) = X_+ \underset{\mathrm{Or}G}{\wedge} \mathbf{T}(G\textstyle\int -)$$

of the functor

$$\begin{aligned}(\mathrm{Or}G)^{\mathrm{op}} \times \mathrm{Or}G &\longrightarrow \mathrm{Sp},\\ (G/H, G/K) &\longmapsto \mathrm{map}(G/H, X)_+^G \wedge \mathbf{T}(G\textstyle\int (G/K)) \cong X_+^H \wedge \mathbf{T}(G\textstyle\int (G/K)).\end{aligned}$$

There are natural isomorphisms $G/H_+ \wedge_{\mathrm{Or}G} \mathbf{T}(G\int -) \cong \mathbf{T}(G\int (G/H))$, and the fact that $\mathbf{T}$ preserves equivalences implies that these spectra are π_*-isomorphic to $\mathbf{T}(H)$. Notice that for $\mathrm{pt} = G/G$ we even have an isomorphism $\mathrm{pt}_+ \wedge_{\mathrm{Or}G} \mathbf{T}(G\int -) \cong \mathbf{T}(G)$.

To define the assembly map we apply this construction to the following G-spaces. Consider a family $\mathcal{F}$ of subgroups of G (i.e., a collection of subgroups closed under passage to subgroups and conjugates) and consider a universal G-space $EG(\mathcal{F})$. This is a G-CW complex characterized up to G-homotopy equivalence by the property that, for any subgroup $H \le G$, the H-fixed point space

$$\big(EG(\mathcal{F})\big)^H \text{ is } \begin{cases} \text{empty} & \text{if } H \notin \mathcal{F};\\ \text{contractible} & \text{if } H \in \mathcal{F}.\end{cases}$$

The *assembly map* is by definition the map

$$\mathrm{asbl}_{\mathcal{F}}\colon EG(\mathcal{F})_+ \underset{\mathrm{Or}G}{\wedge} \mathbf{T}(G\textstyle\int -) \longrightarrow \mathbf{T}(G)$$

induced by the projection $EG(\mathcal{F}) \longrightarrow \mathrm{pt}$ (where, in the target, we use the isomorphism $\mathrm{pt}_+ \wedge_{\mathrm{Or}G} \mathbf{T}(G\int -) \cong \mathbf{T}(G)$).

Remark 10.

(i) In the special case of the trivial family $\mathcal{F} = 1$, a universal space $EG(1)$ is by definition a free and nonequivariantly contractible G-CW complex, i.e., the universal cover of a classifying space BG. In this case, there is an identification

$$EG(1)_+ \underset{\mathrm{Or}G}{\wedge} \mathbf{T}(G\textstyle\int -) \cong BG_+ \wedge \mathbf{T}(1)$$

and therefore we obtain the so-called *classical* assembly map

$$\mathrm{asbl}_1\colon BG_+ \wedge \mathbf{T}(1) \longrightarrow \mathbf{T}(G).$$

(ii) Any G-CW complex whose isotropy groups all lie in the family $\mathcal{F}$ has a map to $EG(\mathcal{F})$, and this map is unique up to G-homotopy. This applies in particular to $EG(\mathcal{F}')$ when $\mathcal{F}' \subseteq \mathcal{F}$, and we refer to the induced map

$$\mathrm{asbl}_{\mathcal{F}' \subseteq \mathcal{F}} : EG(\mathcal{F}')_+ \underset{\mathrm{Or}G}{\wedge} \mathbf{K}(R[G\int -]) \longrightarrow EG(\mathcal{F})_+ \underset{\mathrm{Or}G}{\wedge} \mathbf{K}(R[G\int -])$$

as the *relative* assembly map.

(iii) The source of the assembly map is a model for

$$\underset{\substack{G/H \in \mathrm{obj}\, \mathrm{Or}G \\ \mathrm{s.t.}\, H \in \mathcal{F}}}{\mathrm{hocolim}} \mathbf{T}(G\int (G/H)),$$

the homotopy colimit of the restriction of $\mathbf{T}(G\int -)$ to the full subcategory of $\mathrm{Or}G$ of objects G/H with $H \in \mathcal{F}$; compare [28, Section 5.2].

(iv) Taking the homotopy groups of $X_+ \wedge_{\mathrm{Or}G} \mathbf{T}(G\int -)$ defines a G-equivariant homology theory for G-CW complexes X. This is an equivariant generalization of the well-known statement that $\pi_*(X_+ \wedge \mathbf{E})$ gives a nonequivariant homology theory for any spectrum $\mathbf{E}$. The Atiyah–Hirzebruch spectral sequence converging to $\pi_{s+t}(X_+ \wedge \mathbf{E})$ with $E^2_{s,t} = H_s(X; \pi_t \mathbf{E})$ also generalizes to a spectral sequence converging to $\pi_{s+t}(X_+ \wedge_{\mathrm{Or}G} \mathbf{T}(G\int -))$ with

$$E^2_{s,t} = H^G_s(X; \pi_t \mathbf{T}(G\int -)),$$

the Bredon homology of X with coefficients in $\pi_t \mathbf{T}(G\int -)\colon \mathrm{Or}G \longrightarrow \mathrm{Ab}$; compare [28, Theorem 4.7]. Using this we see that, if $\mathrm{asbl}_{\mathcal{F}}$ is a π_*-isomorphism, then in general all $\pi_t(\mathbf{T}(H))$ with $H \in \mathcal{F}$ and $-\infty < t \le n$ contribute to $\pi_n(\mathbf{T}(G))$.

We conclude with a historical comment. The classical assembly map asbl_1 from Remark 10(i) for algebraic K-theory was originally introduced in Jean-Louis Loday's thesis [68, Chapitre IV] using pairings in algebraic K-theory and the multiplication map

$$G \times GL(R) \longrightarrow GL(R[G]).$$

Friedhelm Waldhausen [104, Section 15] characterized this map as a universal approximation by a homology theory evaluated on a classifying space. This point of view was nicely explained by Michael Weiss and Bruce Williams in [110]. In their original work [40], Farrell and Jones used the language developed by Frank Quinn [86, Appendix]. Later, Jim Davis and Wolfgang Lück [28] gave an equivariant version of the point of view of [110], clarifying and unifying the underlying principles. Their approach leads to the concise description of the assembly map given above. The different approaches are compared and shown to agree in [52].

2.4 The Farrell–Jones Conjecture

We begin by formulating the Farrell–Jones Conjecture in the special case of torsion-free groups and regular rings.

Farrell–Jones Conjecture 11 (special case). *For any* torsion-free *group G and for any* regular *ring R the classical assembly map*

$$\mathrm{asbl}_1 \colon BG_+ \wedge \mathbf{K}(R) \longrightarrow \mathbf{K}(R[G])$$

is a π_*-isomorphism.

On π_0 the classical assembly map produces the map $K_0(R) \longrightarrow K_0(R[G])$ induced by the inclusion $R \longrightarrow R[G]$. So we see that the Farrell–Jones Conjecture 11 implies the torsion-free case of Conjecture 4.

On π_1, in the special case when $R = \mathbb{Z}$, we have

$$\pi_1\big(BG_+ \wedge \mathbf{K}(\mathbb{Z})\big) \cong H_0\big(BG; K_1(\mathbb{Z})\big) \oplus H_1\big(BG; K_0(\mathbb{Z})\big) \cong \{\pm 1\} \oplus G_{\mathrm{ab}} \,. \tag{3}$$

The first isomorphism comes from the Atiyah–Hirzebruch spectral sequence, which is concentrated in the first quadrant because regular rings have vanishing negative K-theory. The second isomorphism comes from the computations $K_1(\mathbb{Z}) \cong \{\pm 1\}$ and $K_0(\mathbb{Z}) \cong \mathbb{Z}$. Under isomorphism (3), it can be shown [104, Assertion 15.8] that the classical assembly map produces on π_1 the left-hand map in (1), whose cokernel is by definition the Whitehead group $Wh(G)$. So we see that the Farrell–Jones Conjecture 11 implies Conjecture 7.

From these identifications and computations of $K_0(\mathbb{Z}[G])$ and $Wh(G)$ for finite groups we see that $\pi_0(\mathrm{asbl}_1)$ and $\pi_1(\mathrm{asbl}_1)$ may not be surjective for groups with torsion, even when $R = \mathbb{Z}$. The classical assembly map may also fail to be injective on homotopy groups if we drop the assumption torsion-free. This happens, for example, for $\pi_2(\mathrm{asbl}_1)$ if $R = \mathbb{F}$ is a finite field of characteristic prime to 2 and G is the noncyclic group with four elements [103].

The regularity assumption cannot be dropped either. For example, consider the case when $G = C_\infty$ is the infinite cyclic group. Then of course $BC_\infty = S^1$ and $R[C_\infty] = R[t, t^{-1}]$, and it can be shown that on π_n the classical assembly map produces the left-hand map in the short exact sequence

$$0 \longrightarrow K_n(R) \oplus K_{n-1}(R) \longrightarrow K_n(R[t, t^{-1}]) \longrightarrow NK_n(R) \oplus NK_n(R) \longrightarrow 0$$

given by the Fundamental Theorem of algebraic K-theory; see for example [19] in low dimensions, [102, Section 10], and [104, Theorem 18.1]. Recall that the groups $NK_n(R)$ are defined as the cokernel of the split injection $K_n(R) \longrightarrow K_n(R[t])$ induced by the natural map $R \longrightarrow R[t]$. It is known that $NK_n(R) = 0$ for each n if R is regular [102,

Theorems 10.1(1) and 10.3], but $NK_n(R)$ can be nontrivial for arbitrary rings. So we see that the classical assembly map for the infinite cyclic group is a π_*-isomorphism if the ring R is regular, but otherwise it may fail to be surjective on homotopy groups.

For arbitrary groups and rings, the generalization of Conjecture 11 is the following.

Farrell–Jones Conjecture 12. *For any group G and for any ring R the Farrell–Jones assembly map*

$$\mathrm{asbl}_{\mathcal{V}\mathcal{C}yc}\colon EG(\mathcal{V}\mathcal{C}yc)_+ \underset{\mathrm{Or}G}{\wedge} \mathbf{K}(R[G\textstyle\int -]) \longrightarrow \mathbf{K}(R[G])$$

is a π_*-isomorphism.

Here $\mathcal{V}\mathcal{C}yc$ denotes the family of virtually cyclic subgroups of G. A group is called virtually cyclic if it contains a cyclic subgroup of finite index.

Farrell–Jones Conjectures 11 and 12 are related as follows.

Proposition 13. *If G is a torsion-free group and R is a regular ring, then the Farrell–Jones Conjectures* 11 *and* 12 *are equivalent.*

Proof. This is an application of the following principle, which is proved in [72, Theorem 65].

Transitivity Principle 14. *Let $\mathcal{F}$ and $\mathcal{F}'$ be families of subgroups of G with $\mathcal{F} \subseteq \mathcal{F}'$. Assume that for each $H \in \mathcal{F}'$ the assembly map*

$$EH(\mathcal{F}|_H)_+ \underset{\mathrm{Or}H}{\wedge} \mathbf{K}(R[H\textstyle\int -]) \longrightarrow \mathbf{K}(R[H])$$

is a π_*-isomorphism, where $\mathcal{F}|_H = \{K \le H \mid K \in \mathcal{F}\}$. Then the relative assembly explained in Remark 10(ii)*, i.e., the left vertical map in the following commutative triangle, is a π_*-isomorphism:*

$$\begin{array}{ccc} EG(\mathcal{F})_+ \underset{\mathrm{Or}G}{\wedge} \mathbf{K}(R[G\textstyle\int -]) & & \\ \Big\downarrow \mathrm{asbl}_{\mathcal{F}\subseteq\mathcal{F}'} & \overset{\mathrm{asbl}_{\mathcal{F}}}{\searrow} & \mathbf{K}(R[G]) \\ EG(\mathcal{F}')_+ \underset{\mathrm{Or}G}{\wedge} \mathbf{K}(R[G\textstyle\int -]) & \underset{\mathrm{asbl}_{\mathcal{F}'}}{\nearrow} & \end{array}$$

Therefore, $\mathrm{asbl}_{\mathcal{F}}$ *is a π_*-isomorphism if and only if* $\mathrm{asbl}_{\mathcal{F}'}$ *is a π_*-isomorphism.*

We now apply the transitivity principle in the case $\mathcal{F} = 1$ and $\mathcal{F}' = \mathcal{V}\mathcal{C}yc$. Any nontrivial torsion-free virtually cyclic group is infinite cyclic. Recall that asbl_1 can be identified with the classical assembly map in Conjecture 11. So it is enough to show that the classical assembly map is a π_*-isomorphism for the infinite cyclic group C. The fact that this is true in the case of regular rings is explained above, before the statement of Conjecture 12. □

The next result shows, as promised, that the Farrell–Jones Conjecture implies all the other conjectures introduced in the first two sections; the case of Conjecture 9 is considered right after Conjecture 18.

Proposition 15. *The Farrell–Jones Conjecture* 12 *implies Conjectures* 4 *and* 7*, and so also Conjectures* 1 *and* 6 *by Propositions* 5 *and* 8.

Proof. The case of Conjecture 7 and the torsion-free case of Conjecture 4 are explained above, directly after the statement of Conjecture 11. The general case of Conjecture 4 follows from the following isomorphisms:

$$\begin{aligned}
\pi_0\left(EG(\mathcal{VC}yc)_+ \underset{\mathrm{Or}G}{\wedge} \mathbf{K}(R[G\textstyle\int -])\right) &\overset{①}{\cong} \pi_0\left(EG(\mathcal{F}in)_+ \underset{\mathrm{Or}G}{\wedge} \mathbf{K}(R[G\textstyle\int -])\right) \\
&\overset{②}{\cong} \pi_0\left(EG(\mathcal{F}in)_+ \underset{\mathrm{Or}G(\mathcal{F}in)}{\wedge} \mathbf{K}(R[G\textstyle\int -])\right) \\
&\overset{③}{\cong} H_0\left(C(EG(\mathcal{F}in)) \underset{\mathbb{Z}\mathrm{Or}G(\mathcal{F}in)}{\otimes} K_0(R[G\textstyle\int -])\right) \\
&\overset{④}{\cong} \mathbb{Z} \underset{\mathbb{Z}\mathrm{Or}G(\mathcal{F}in)}{\otimes} K_0(R[G\textstyle\int -]) \\
&\overset{⑤}{\cong} \operatorname*{colim}_{\mathrm{Or}G(\mathcal{F}in)} K_0(R[G\textstyle\int -]) \\
&\overset{⑥}{\cong} \operatorname*{colim}_{\mathrm{Sub}G(\mathcal{F}in)} K_0(R[-]).
\end{aligned}$$

Theorem 16(ii) yields isomorphism ①. Since $EG(\mathcal{F}in)^H = \emptyset$ if H is not finite, isomorphism ② follows immediately by inspecting the construction of the coend. The assumptions that R is regular and that the order of every finite subgroup H of G is invertible in R imply that also $R[H]$ is regular. For regular rings the negative K-groups vanish [90, 3.3.1], and therefore the equivariant Atiyah–Hirzebruch spectral sequence explained in Remark 10(iv) is concentrated in the first quadrant. This gives isomorphism ③. The singular or cellular chain complex $C(EG(\mathcal{F}in))$, considered as a contravariant functor $G/H \longmapsto C(EG(\mathcal{F}in)^H)$, resolves the constant functor $\mathbb{Z}$, therefore ④ follows from right exactness of $- \otimes_{\mathbb{Z}\mathrm{Or}G(\mathcal{F}in)} M$ for any fixed $M\colon \mathrm{Or}G(\mathcal{F}in) \longrightarrow \mathrm{Ab}$. The coend with the constant functor $\mathbb{Z}$ is one possible construction of the colimit in abelian groups, hence ⑤. Since $K_n(R[G\int G/H]) \cong K_n(R[H])$ and since inner automorphisms induce the identity on K-theory, the functor $K_n(R[G\int -])$ factors over $\mathrm{Or}G(\mathcal{F}in) \longrightarrow \mathrm{Sub}G(\mathcal{F}in)$, the functor sending $G/H \to G/K$, $gH \mapsto gaH$ to the class of $H \to K$, $h \mapsto a^{-1}ha$. Isomorphism ⑥ then follows by standard properties of colimits. □

The next result deals with the passage from finite to virtually cyclic subgroups in the source of the Farrell–Jones assembly map.

Theorem 16 (finite to virtually cyclic).

(i) *The relative assembly map*

$$\mathrm{asbl}_{\mathcal{F}in\subseteq\mathcal{V}\mathcal{C}yc}\colon EG(\mathcal{F}in)_+ \underset{\mathrm{Or}G}{\wedge} \mathbf{K}(R[G\int -]) \longrightarrow EG(\mathcal{V}\mathcal{C}yc)_+ \underset{\mathrm{Or}G}{\wedge} \mathbf{K}(R[G\int -])$$

is always split injective.

(ii) *If R is regular and the order of every finite subgroup of G is invertible in R, then* $\mathrm{asbl}_{\mathcal{F}in\subseteq\mathcal{V}\mathcal{C}yc}$ *is a π_*-isomorphism.*

(iii) *If R is regular then* $\mathrm{asbl}_{\mathcal{F}in\subseteq\mathcal{V}\mathcal{C}yc}$ *is a $\pi_*^{\mathbb{Q}}$-isomorphism, i.e., it induces isomorphisms on $\pi_n(-)\otimes_{\mathbb{Z}}\mathbb{Q}$ for all $n\in\mathbb{Z}$.*

Proof. Part (i) is the main result of [2]. Part (ii) is shown in [72, Proposition 70]. Part (iii) is proved in [76, Theorem 0.2] and generalizes [50, Corollary on page 165]. □

2.5 Rational computations

After tensoring with the rational numbers, the Farrell–Jones Conjecture 12 for regular rings can be reformulated in a more concrete and computational fashion as follows.

Assume that R is a regular ring. Recall from Theorem 16(iii) that the relative assembly map $\mathrm{asbl}_{\mathcal{F}in\subseteq\mathcal{V}\mathcal{C}yc}$ induces isomorphisms

$$\pi_n\Big(EG(\mathcal{F}in)_+ \underset{\mathrm{Or}G}{\wedge} \mathbf{K}(R[G\int -])\Big)\underset{\mathbb{Z}}{\otimes}\mathbb{Q} \xrightarrow{\cong} \pi_n\Big(EG(\mathcal{V}\mathcal{C}yc)_+ \underset{\mathrm{Or}G}{\wedge} \mathbf{K}(R[G\int -])\Big)\underset{\mathbb{Z}}{\otimes}\mathbb{Q}. \tag{4}$$

The theory of equivariant Chern characters developed by Lück in [70] yields the following isomorphisms:

$$\begin{gathered}\bigoplus_{(C)\in(\mathcal{F}\mathcal{C}yc)}\bigoplus_{s+t=n} H_s(BZ_GC;\mathbb{Q}) \underset{\mathbb{Q}[W_GC]}{\otimes} \Theta_C\Big(K_t(R[C])\underset{\mathbb{Z}}{\otimes}\mathbb{Q}\Big)\\ \big\downarrow\cong\\ \pi_n\Big(EG(\mathcal{F}\mathcal{C}yc)_+ \underset{\mathrm{Or}G}{\wedge} \mathbf{K}(R[G\int -])\Big)\underset{\mathbb{Z}}{\otimes}\mathbb{Q}\\ \big\downarrow\cong\\ \pi_n\Big(EG(\mathcal{F}in)_+ \underset{\mathrm{Or}G}{\wedge} \mathbf{K}(R[G\int -])\Big)\underset{\mathbb{Z}}{\otimes}\mathbb{Q}.\end{gathered} \tag{5}$$

Before we explain the notation, notice the analogy with the well-known isomorphism

$$\bigoplus_{s+t=n} H_s(BG;\mathbb{Q})\underset{\mathbb{Z}}{\otimes}\Big(K_t(R)\underset{\mathbb{Z}}{\otimes}\mathbb{Q}\Big) \xrightarrow{\cong} \pi_n\big(BG_+\wedge\mathbf{K}(R)\big)\underset{\mathbb{Z}}{\otimes}\mathbb{Q},$$

whose source corresponds to the summand in (5) indexed by $C = 1$.

Given a subgroup H of G, we denote by N_GH the normalizer and by Z_GH the centralizer of H in G, and we define the Weyl group as the quotient $W_GH = N_GH/(Z_GH\cdot H)$.

Notice that the Weyl group $W_G H$ of a finite subgroup H is always finite, since it embeds into the outer automorphism group of H. We write $\mathcal{FC}yc$ for the family of finite cyclic subgroups of G, and $(\mathcal{FC}yc)$ for the set of conjugacy classes of finite cyclic subgroups. Furthermore, Θ_C is an idempotent endomorphism of $K_t(R[C]) \otimes_{\mathbb{Z}} \mathbb{Q}$, which corresponds to a specific idempotent in the rationalized Burnside ring of C, and whose image is a direct summand of $K_t(R[C]) \otimes_{\mathbb{Z}} \mathbb{Q}$ isomorphic to

$$\operatorname{coker}\left(\bigoplus_{D \subsetneq C} \operatorname{ind}_D^C \colon \bigoplus_{D \subsetneq C} K_t(R[D]) \otimes_{\mathbb{Z}} \mathbb{Q} \longrightarrow K_t(R[C]) \otimes_{\mathbb{Z}} \mathbb{Q} \right). \tag{6}$$

The Weyl group acts via conjugation on C and hence on $\Theta_C(K_t(R[C]) \otimes_{\mathbb{Z}} \mathbb{Q})$. The Weyl group action on the homology groups in the source of (5) comes from the fact that $EN_G C/Z_G C$ is a model for $BZ_G C$.

Farrell–Jones Conjecture 17 (rationalized version). *For any group G and for any* regular *ring R the composition of the Farrell–Jones assembly map and isomorphisms* (5) *and* (4)

$$\bigoplus_{(C) \in (\mathcal{FC}yc)} \bigoplus_{s+t=n} H_s(BZ_G C; \mathbb{Q}) \underset{\mathbb{Q}[W_G C]}{\otimes} \Theta_C\left(K_t(R[C]) \underset{\mathbb{Z}}{\otimes} \mathbb{Q}\right) \longrightarrow K_n(R[G]) \underset{\mathbb{Z}}{\otimes} \mathbb{Q}$$

is an isomorphism for each $n \in \mathbb{Z}$.

Analogously one obtains the following conjecture for Whitehead groups, which is the correct generalization of Conjecture 9 mentioned at the end of Section 2.2.

Conjecture 18. *For any group G there is an isomorphism*

$$\begin{gathered}\bigoplus_{(C) \in (\mathcal{FC}yc)} \left(\mathbb{Q} \underset{\mathbb{Q}[W_G C]}{\otimes} \Theta_C\left(Wh(C) \underset{\mathbb{Z}}{\otimes} \mathbb{Q}\right) \oplus H_2(BZ_G C; \mathbb{Q}) \underset{\mathbb{Q}[W_G C]}{\otimes} \Theta_C\left(K_{-1}(\mathbb{Z}[C]) \underset{\mathbb{Z}}{\otimes} \mathbb{Q}\right)\right) \\ \downarrow \cong \\ Wh(G) \underset{\mathbb{Z}}{\otimes} \mathbb{Q}\,.\end{gathered}$$

Conjecture 18 implies Conjecture 9, because in fact

$$\operatorname*{colim}_{H \in \operatorname{obj} \mathrm{Sub}G(\mathcal{F}in)} Wh(H) \underset{\mathbb{Z}}{\otimes} \mathbb{Q} \quad \cong \quad \bigoplus_{(C) \in (\mathcal{FC}yc)} \mathbb{Q} \underset{\mathbb{Q}[W_G C]}{\otimes} \Theta_C\left(Wh(C) \underset{\mathbb{Z}}{\otimes} \mathbb{Q}\right)$$

and map (2) coincides with the restriction to this summand of the map in Conjecture 18.

Remark 19. For finite groups H we have that $Wh(H) \otimes_{\mathbb{Z}} \mathbb{Q} \cong K_1(\mathbb{Z}[H]) \otimes_{\mathbb{Z}} \mathbb{Q}$ by the exact sequence (1). The only difference between the sources of the maps in Conjectures 17 and 18 is the absence from 18 of the summands with $(s,t) = (1,0)$. For finite groups H the natural map $\mathbb{Q} \cong K_0(\mathbb{Z}) \otimes_{\mathbb{Z}} \mathbb{Q} \longrightarrow K_0(\mathbb{Z}[H]) \otimes_{\mathbb{Z}} \mathbb{Q}$ is an isomorphism, and hence it follows from (6) that the only nonvanishing summand among these is $H_1(BG;\mathbb{Q}) \cong G_{ab} \otimes_{\mathbb{Z}} \mathbb{Q}$ corresponding to $C = 1$. This is consistent with the exact sequence (1).

Finally, we note that in the special case when $R = \mathbb{Z}$ the dimensions of the $\mathbb{Q}$-vector spaces in (6) for any t and any finite cyclic group C can be explicitly computed as follows.

Theorem 20. *Let C be a cyclic group of order c. Then*

$$\dim_{\mathbb{Q}} \Theta_C\Big(K_t(\mathbb{Z}[C]) \underset{\mathbb{Z}}{\otimes} \mathbb{Q}\Big) = \begin{cases} s(c) - 1 & \text{if } t = -1; \\ \varphi(c)/2 - 1 & \text{if } t = 1 \text{ and } c > 2; \\ 1 & \text{if } t > 1,\ t \equiv 1 \mod 4, \text{ and } c = 2; \\ \varphi(c)/2 & \text{if } t > 1,\ t \equiv 1 \mod 2, \text{ and } c > 2; \\ 0 & \text{otherwise.} \end{cases}$$

Here $\varphi(c) = \#\{x \in C \mid x \text{ generates } C\}$ is Euler's φ-function, $c = \prod_{i=1}^{s} p_i^{e_i}$ is the prime factorization of c, and $s(c) = \sum_{i=1}^{s} \varphi(n/p_i^{e_i})/f_{p_i}$, where f_{p_i} is the smallest number such that $p_i^{f_{p_i}} \equiv 1 \bmod n/p^{e_i}$.

This result is proved in [83, Theorem on page 9], and more details will appear in [84].

2.6 Some related conjectures

We now survey very briefly some other conjectures that are analogous to Conjecture 12. For details and further explanations we recommend [45, 63, 69, 72, 81].

In [40], Farrell and Jones formulated Conjecture 12 not only for algebraic K-theory but also for L-theory; more precisely, for $\mathbf{L}^{\langle -\infty \rangle}(R[G])$, the quadratic algebraic L-theory spectrum of $R[G]$ with decoration $-\infty$, for any ring with involution R. The corresponding assembly map is constructed completely analogously, by applying the machinery of Section 2.3 to the functor $\mathbf{L}^{\langle -\infty \rangle}(R[-])$. In the special case of torsion-free groups G, this conjecture is equivalent to the statement that the classical assembly map $BG_+ \wedge \mathbf{L}^{\langle -\infty \rangle}(R) \longrightarrow \mathbf{L}^{\langle -\infty \rangle}(R[G])$ is a π_*-isomorphism, for any ring R, not necessarily regular.

If G is a torsion-free group and the Farrell–Jones Conjectures hold for both $\mathbf{K}(\mathbb{Z}[G])$ and $\mathbf{L}^{\langle -\infty \rangle}(\mathbb{Z}[G])$, then the Borel Conjecture is true for manifolds with fundamental group G and dimension at least 5. The Borel Conjecture states that, if M and N are closed connected aspherical manifolds with isomorphic fundamental groups, then M and N are homeomorphic, and every homotopy equivalence between M and N is homotopic to a homeomorphism. In short, the Borel Conjecture says that closed aspherical manifolds are topologically rigid. Recall that a connected CW complex X is aspherical if its universal cover is contractible, or equivalently if $\pi_n(X) = 0$ for all $n > 1$.

We also mention that the Farrell–Jones Conjecture in algebraic L-theory implies the Novikov Conjecture about the homotopy invariance of higher signatures.

Furthermore, Farrell and Jones also formulated an analog of Conjecture 12 for the stable pseudo-isotopy functor, or equivalently for Waldhausen's A-theory, also known as algebraic K-theory of spaces. We refer to [66] for a modern approach to this conjecture and in particular for its many applications to automorphisms of manifolds.

Finally, the analog of the Farrell–Jones Conjecture 12 for the complex topological K-theory of the reduced complex group C^*-algebra of G is equivalent to the famous Baum–Connes Conjecture, formulated by Paul Baum, Alain Connes, and Nigel Higson in [20]. For the Baum–Connes Conjecture, the relative assembly map $\mathrm{asbl}_{\mathcal{F}in \subseteq \mathcal{V}\mathcal{C}yc}$ is always a π_*-isomorphism; compare and contrast with Theorem 16. Also the Baum–Connes Conjecture implies the Novikov Conjecture. For more information on the relation between the Baum–Connes Conjecture and the Farrell–Jones Conjecture in L-theory we refer to [67, 91].

3 State of the art

We now overview what we know and don't know about the Farrell–Jones Conjecture 12, to the best of our knowledge in January 2017. We aim to give immediately accessible statements, which may not always reflect the most general available results. We restrict our attention to algebraic K-theory and ignore the related conjectures mentioned in the previous section.

3.1 What we know already

The following theorem is the result of the effort of many mathematicians over a long period of time. The methods of controlled algebra and topology that underlie this theorem (and that we illustrate in the next section) were pioneered by Steve Ferry [44] and Frank Quinn [85], and were then applied with enormous success by Farrell–Hsiang [34, 36, 37] and Farrell–Jones [38–41]. Many ideas in the proofs of the following results originate in these articles. The formulation of the theorem below is meant to be a snapshot of the best results available today, as opposed to a comprehensive historical overview of the many important intermediate results predating the works quoted here.

Theorem 21. *Let $\mathfrak{G}$ be the smallest class of groups that satisfies the following two conditions.*

(1) The class $\mathfrak{G}$ contains:
- *(a) hyperbolic groups [12];*
- *(b) finite-dimensional CAT(0)-groups [10, 106];*
- *(c) virtually solvable groups [43, 107];*
- *(d) Baumslag–Solitar groups and graphs of abelian groups [43, 47];*
- *(e) lattices in virtually connected Lie groups [9, 58];*
- *(f) arithmetic and S-arithmetic groups [13, 94];*
- *(g) fundamental groups of connected manifolds of dimension at most 3 [93];*
- *(h) Coxeter groups;*
- *(i) Artin braid groups [1];*
- *(j) mapping class groups of oriented surfaces of finite type [6].*

(2) The class $\mathfrak{G}$ is closed under:
- *(A) subgroups [14];*
- *(B) overgroups of finite index [13, Section 6];*
- *(C) finite products;*
- *(D) finite coproducts;*
- *(E) directed colimits [7];*
- *(F) graph products [48];*
- *(G) if $1 \longrightarrow N \longrightarrow G \xrightarrow{p} Q \longrightarrow 1$ is a group extension such that $Q \in \mathfrak{G}$ and $p^{-1}(C) \in \mathfrak{G}$ for each infinite cyclic subgroup $C \le Q$, then $G \in \mathfrak{G}$;*
- *(H) if G is a countable group that is relatively hyperbolic to subgroups $P_1, \dots, P_n$ and each $P_i \in \mathfrak{G}$, then $G \in \mathfrak{G}$ [5].*

Then the Farrell–Jones Conjecture 12 holds for any ring R and for any group $G \in \mathfrak{G}$.

Proof. In order to have the inheritance properties formulated in (2) one needs to work with a slight generalization of the Farrell–Jones Conjecture. First, one needs to allow coefficients in arbitrary additive categories with G-actions [14]; then, one says that the conjecture with finite wreath products is true for G if the conjecture holds not only for G but also for all wreath products $G \wr F$ of G with finite groups F [92], [13, Section 6]. The Farrell–Jones Conjecture with coefficients and finite wreath products is true for all groups listed under (1) and has all the inheritance properties listed under (2).

Some of the earlier references given above omit the discussion of the version with finite wreath products; consult [13, Section 6] and [48, Proposition 1.1] for the corresponding extensions.

We discuss the statements (h), (C), (D), and (G), for which no reference was provided. Coxeter groups (h) are known to fall under (b) by a result of Moussong; compare [29, Theorem 12.3.3]. For (C) use [14, Corollary 4.3] applied to the projection to the factors, the Transitivity Principle 14, and the fact that the Farrell–Jones Conjecture is known for finite products of virtually cyclic groups. The extension to the version with finite wreath products uses the fact that $(G_1 \times G_2) \wr F$ is a subgroup of $(G_1 \wr F) \times (G_2 \wr F)$.

Finite coproducts are treated similarly using property (G) and the natural map from the coproduct to the product; compare [48, Proposition 1.1]. Statement (G) itself is simply a combination of [14, Corollary 4.3] and the Transitivity Principle 14. □

3.2 What we don't know yet

At the time of writing the Farrell–Jones Conjecture 12 seems to be open for the following classes of groups:

(i) Thompson's groups;
(ii) outer automorphism groups of free groups;
(iii) linear groups;
(iv) (elementary) amenable groups;
(v) infinite products of groups (satisfying the Farrell–Jones Conjecture).

However, for some of these groups there are partial injectivity results, as we explain in Remark 23.

3.3 Injectivity results

The next theorem gives two examples of injectivity results for assembly maps in algebraic K-theory. Part (i) is proved using the trace methods explained in Section 5, where more rational injectivity results are described. Part (ii) is based on a completely different approach using controlled algebra, the descent method due to Gunnar Carlsson and Erik Pedersen [25]. For this method to work, the group has to satisfy some mild metric conditions, which are not needed for the weaker statement in part (i). One such condition goes back to [35]. The condition of finite asymptotic dimension appeared in the context of algebraic K-theory in [3, 23, 24], and was later generalized to finite decomposition complexity in [88]. The extension to nonclassical assembly maps appeared in [15, 16, 61]. The statement in part (ii) is from [62] and further improves and combines these developments. We also mention [46] for yet another approach to injectivity results.

Recall from Theorem 16 that the relative assembly map

$$\pi_n\Big(EG(\mathcal{F}in)_+ \underset{\mathrm{Or}G}{\wedge} \mathbf{K}(\mathbb{Z}[G\int -])\Big) \longrightarrow \pi_n\Big(EG(\mathcal{V}\mathcal{C}yc)_+ \underset{\mathrm{Or}G}{\wedge} \mathbf{K}(\mathbb{Z}[G\int -])\Big)$$

is always split injective, and it becomes an isomorphism after applying $- \otimes_{\mathbb{Z}} \mathbb{Q}$ if R is regular, e.g., if $R = \mathbb{Z}$. Therefore, the results below would follow if we knew the Farrell–Jones Conjecture 12.

Theorem 22. *Assume that there exists a finite-dimensional $EG(\mathcal{F}in)$, and that there exists an upper bound on the orders of the finite subgroups of G.*

(i) *If $R = \mathbb{Z}$, then there exists an integer $L > 0$ such that for every $n \geq L$ the rationalized assembly map*

$$\pi_n\Big(EG(\mathcal{F}in)_+ \underset{\mathrm{Or}G}{\wedge} \mathbf{K}(\mathbb{Z}[G\int -])\Big) \underset{\mathbb{Z}}{\otimes} \mathbb{Q} \longrightarrow K_n(\mathbb{Z}[G]) \underset{\mathbb{Z}}{\otimes} \mathbb{Q}$$

is injective.

(ii) *Assume furthermore that G has regular finite decomposition complexity. Then for any ring R the assembly map*

$$EG(\mathcal{F}in)_+ \underset{\mathrm{Or}G}{\wedge} \mathbf{K}(R[G\int -]) \longrightarrow \mathbf{K}(R[G])$$

is split injective on π_*.

Proof. (i) is a consequence of Theorem 44, or rather of its more general version in [73, Main Technical Theorem 1.16]; see Remark 45(iv) and [73, Theorem 1.15], where the result is only stated for cocompact $EG(\mathcal{F}in)$, but the proof given on page 1015 only uses finite dimensionality and the existence of a bound on the order of the finite cyclic subgroups. (ii) is [62, Theorem 1.3]. □

Remark 23. Theorem 22 applies to groups for which no isomorphism results were known at the time of writing:

(i) The existence of an upper bound on the orders of the finite subgroups of G follows from the existence of a cocompact $EG(\mathcal{F}in)$. For example, this is the case for outer automorphism groups of free groups, to which Theorem 22(i) then applies.

(ii) Regular finite decomposition complexity is a property shared by all groups that are either (a) of finite asymptotic dimension, (b) elementary amenable, (c) linear, or (d) subgroups of virtually connected Lie groups.

4 Controlled algebra methods

As noted in the previous section, most proofs of the Farrell–Jones Conjecture 12 use the ideas and technology of controlled algebra, which are the focus of this section. The ultimate goal is to explain the Farrell–Hsiang Criterion for assembly maps to be π_*-isomorphisms. The criterion goes back to [34] and has been successfully applied in many cases, e.g., [36, 37, 87], and plays an important role in the proof of Theorem 21(1)(e) [9]. The formulation that we give here in Theorem 39 is due to [11].

Our goal is to keep the exposition as concrete as possible, and to work out the main details of the proof of the following result, establishing the first nontrivial case of the Farrell–Jones Conjecture.

Theorem 24. *The Farrell–Jones Conjecture 12 holds for finitely generated free abelian groups, i.e., for any $n \geq 2$ and for any ring R, the assembly map*

$$E\mathbb{Z}^n(\mathcal{C}yc)_+ \underset{\mathrm{Or}\mathbb{Z}^n}{\wedge} \mathbf{K}(R[\mathbb{Z}^n \int -]) \longrightarrow \mathbf{K}(R[\mathbb{Z}^n])$$

is a π_*-isomorphism.

Before we get to the proof, we want to show how Theorem 24 leads to a simple formula for the Whitehead groups of $\mathbb{Z}^n$; the article [75] contains many similar but way more general explicit computations. The Whitehead groups of G over R are defined as $Wh_k^R(G) = \pi_k(\mathbf{Wh}^R(G))$, where $\mathbf{Wh}^R(G)$ is the homotopy cofiber of the classical assembly map $\mathrm{asbl}_1 \colon BG_+ \wedge \mathbf{K}(R) \longrightarrow \mathbf{K}(R[G])$ appearing in Conjecture 11. Of course, $Wh_1^{\mathbb{Z}}(G) = Wh(G)$.

Corollary 25. *For any $n \geq 2$ and $k \in \mathbb{Z}$ there are isomorphisms*

$$Wh_k^R(\mathbb{Z}^n) \cong \bigoplus_{C \in \mathcal{M}ax\mathcal{C}yc} Wh_k^R(C) \cong \bigoplus_{C \in \mathcal{M}ax\mathcal{C}yc} NK_k(R) \oplus NK_k(R)\,,$$

where $\mathcal{M}ax\mathcal{C}yc$ denotes the set of maximal cyclic subgroups of $\mathbb{Z}^n$.

Observe that the set of maximal cyclic subgroups of $\mathbb{Z}^n$ can be identified with $\mathbb{P}^{n-1}(\mathbb{Q})$, the set of all one-dimensional subspaces of the $\mathbb{Q}$-vector space $\mathbb{Q}^n$.

Proof. There is a $\mathbb{Z}^n$-equivariant homotopy pushout square

$$\begin{array}{ccc} \coprod_{C \in \mathcal{M}ax\mathcal{C}yc} \mathbb{Z}^n \times_C EC & \longrightarrow & E\mathbb{Z}^n \\ \downarrow & & \downarrow \\ \coprod_{C \in \mathcal{M}ax\mathcal{C}yc} \mathbb{Z}^n \times_C \mathrm{pt} & \longrightarrow & E\mathbb{Z}^n(\mathcal{C}yc)\,. \end{array}$$

Applying $(\,?\,)_+ \wedge_{\mathrm{Or}(\mathbb{Z}^n)} \mathbf{K}(R[\mathbb{Z}^n \int -])$ preserves homotopy pushout squares, and the induced left vertical map can be identified with a wedge sum of copies of the classical assembly map asbl_1 for C, using induction isomorphisms. The homotopy cofibration sequence

$$BC_+ \wedge \mathbf{K}(R) \cong EC_+ \underset{\mathrm{Or}(C)}{\wedge} \mathbf{K}(R[C \int -]) \xrightarrow{\mathrm{asbl}_1} \mathbf{K}(R[C]) \longrightarrow \mathbf{Wh}^R(C)$$

is known to split, and $Wh_n^R(C) \cong NK_n(R) \oplus NK_n(R)$; compare [102, Section 10] and [104, Theorem 18.1]. Therefore, we obtain the following homotopy pushout square.

$$\begin{array}{ccc} \mathrm{pt} & \longrightarrow & E\mathbb{Z}^n{}_+ \underset{\mathrm{Or}\mathbb{Z}^n}{\wedge} \mathbf{K}(R[\mathbb{Z}^n \int -]) \\ \downarrow & & \downarrow \\ \bigvee_{C\in \mathcal{M}ax\mathcal{C}yc} \mathbf{Wh}^R(C) & \longrightarrow & E\mathbb{Z}^n(\mathcal{C}yc)_+ \underset{\mathrm{Or}\mathbb{Z}^n}{\wedge} \mathbf{K}(R[\mathbb{Z}^n \int -]) \end{array}$$

Theorem 24 identifies the bottom right corner with $\mathbf{K}(R[\mathbb{Z}^n])$, and therefore the homotopy cofiber of the right vertical map agrees with the homotopy cofiber of the classical assembly map for $\mathbb{Z}^n$, completing the proof. □

Working with the Farrell–Jones Conjecture with coefficients mentioned in the proof of Theorem 21, we can use induction and reduce the proof of Theorem 24 to the case $n = 2$, by applying the inheritance property formulated in Theorem 21(2)(G) to a surjective homomorphism $\mathbb{Z}^n \longrightarrow \mathbb{Z}^2$. Notice that for $\mathbb{Z}^2$ itself Theorem 21(2)(G) is useless.

However, even in the case $n = 2$ the full proof of Theorem 24 involves many technicalities that obscure the underlying ideas. For this reason, we concentrate on the following partial result.

Proposition 26. *The assembly map*

$$\pi_1\Big(E\mathbb{Z}^2(\mathcal{C}yc)_+ \underset{\mathrm{Or}G}{\wedge} \mathbf{K}(R[\mathbb{Z}^2 \int -])\Big) \longrightarrow K_1(R[\mathbb{Z}^2]) \tag{7}$$

is surjective for any ring R.

In the rest of this section we give a complete proof of this proposition modulo Theorem 29, which we use as a black box. The proof is completed right after the statement of Claim 36.

4.1 Geometric modules

The main characters of controlled algebra are defined next.

Definition 27 (geometric modules). Given a ring R and G-space X, the category $\mathcal{C}(X) = \mathcal{C}^G(X;R)$ of *geometric* $R[G]$*-modules over* X is defined as follows. The objects of $\mathcal{C}(X)$ are cofinite free G-sets S together with a G-map $\varphi\colon S \longrightarrow X$. Notice that, given a cofinite free G-set S, the R-module $R[S]$ is in a natural way a finitely generated free $R[G]$-module. The morphisms in $\mathcal{C}(X)$ from $\varphi\colon S \longrightarrow X$ to $\varphi'\colon S' \longrightarrow X$ are simply the $R[G]$-linear maps $R[S] \longrightarrow R[S']$.

The category $\mathcal{C}(X)$ is additive and depends functorially on X, in the sense that a G-map $f\colon X \longrightarrow X'$ induces an additive functor $f_*\colon \mathcal{C}(X) \longrightarrow \mathcal{C}(X')$ which sends the object φ to $f \circ \varphi$. Let $\mathcal{F}(R[G])$ be the category of finitely generated free $R[G]$-modules. The functor $\mathbb{U}\colon \mathcal{C}(X) \longrightarrow \mathcal{F}(R[G])$ (where $\mathbb{U}$ stands for underlying) that sends $\varphi\colon S \longrightarrow X$ to $R[S]$ is obviously an equivalence of additive categories, since φ does not enter the definition of the morphisms in $\mathcal{C}(X)$. Therefore, we obtain a π_*-isomorphism

$$\mathbf{K}(\mathcal{C}(X)) \xrightarrow[\mathbb{U}]{\cong} \mathbf{K}(R[G])\,. \tag{8}$$

However, the advantage of $\mathcal{C}(X)$ is that morphisms have a geometric shadow in X, and if X is equipped with a metric we can talk about their size.

Definition 28 (support and size). Let $\alpha\colon R[S] \longrightarrow R[S']$ be a morphism in $\mathcal{C}(X)$ from $\varphi\colon S \longrightarrow X$ to $\varphi'\colon S' \longrightarrow X$. Let $(\alpha_{s's})_{(s',s)\in S'\times S}$ be the associated matrix. Define the *support* of α to be

$$\operatorname{supp}\alpha = \left\{ \left(\varphi'(s'), \varphi(s)\right) \in X \times X \;\middle|\; \alpha_{s's} \neq 0 \right\} \subseteq X \times X\,.$$

If X is equipped with a G-invariant metric d, define the *size* of α to be

$$\operatorname{size}\alpha = \sup\left\{ d\left(\varphi'(s'), \varphi(s)\right) \;\middle|\; \alpha_{s's} \neq 0 \right\}.$$

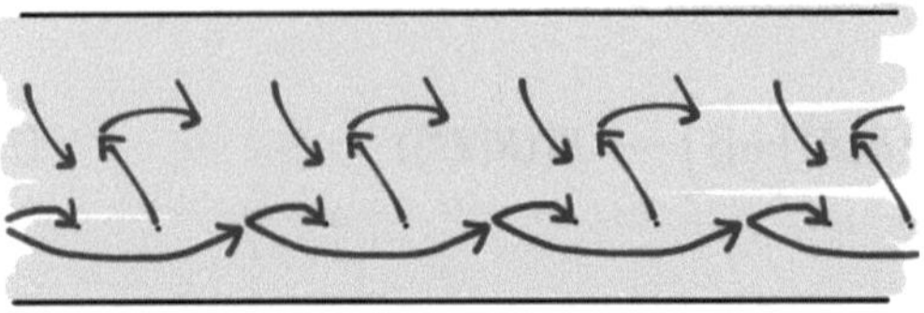

Figure 1: Support of a morphism with $G = \mathbb{Z}$ acting via shift on a band.

Note that the supremum is really a maximum, since α is G-equivariant, d is G-invariant, and S is cofinite. As we will see, sometimes it is convenient to work with extended metrics, i.e., metrics for which $d(x,x') = \infty$ is allowed. Being of finite size is then a severe restriction on α. In the support picture no arrow is allowed between points at distance ∞; compare Figure 2 on page 27.

The main idea now is that assembly maps can be described as *forget control* maps. Proving that an element is in the image of an assembly map can be achieved by proving that it has a representative of small size. Before making this precise we introduce some more conventions and definitions.

Recall that a point a in a simplicial complex Z can be written uniquely in the form

$$a = \sum_{v\in V} a_v v\,,$$

where V is the set of vertices of the underlying abstract simplicial complex, $a_v \in [0,1]$, and $\sum_{v\in V} a_v = 1$. The point a lies in the interior of the realization Δ_v of the unique abstract simplex given by $\{v \mid a_v \neq 0\}$. The l^1-metric on Z is defined as

$$d^1(a,b) = \sum_{v\in V} |a_v - b_v|\,.$$

Observe that the distance between points is always ≤ 2, and that every simplicial automorphism is an isometry with respect to the l^1-metric.

Theorem 29 (small elements are in the image). *For any integer $n > 0$ there is an $\varepsilon = \varepsilon(n) > 0$ such that for every G-simplicial complex Z of dimension n the following is true. Let $x \in K_1(R[G])$ and consider the assembly map* asbl_Z *induced by* $Z \longrightarrow \mathrm{pt}$.

$$\begin{array}{ccccc} & & K_1(\mathscr{C}(Z)) & \ni & [\alpha] \\ & & \Big\downarrow{\scriptstyle \mathbb{U}\ \cong} & & \big\downarrow \\ \pi_1\Big(Z_+ \underset{\mathrm{Or}G}{\wedge} \mathbf{K}(R[G\int -])\Big) & \xrightarrow{\mathrm{asbl}_Z} & K_1(R[G]) & \ni & x \end{array}$$

Then $x \in \mathrm{im}(\mathrm{asbl}_Z)$ if there exists an automorphism α in $\mathscr{C}(Z)$ with $\mathbb{U}([\alpha]) = x$ and

$$\mathrm{size}(\alpha) \leq \varepsilon \qquad \text{and} \qquad \mathrm{size}(\alpha^{-1}) \leq \varepsilon\,.$$

Corollary 30. *Retain the notation and assumptions of Theorem 29. If all isotropy groups of Z belong to the family $\mathscr{F}$, then x is also in the image of the assembly map*

$$\pi_1\Big(EG(\mathscr{F}) \underset{\mathrm{Or}G}{\wedge} \mathbf{K}(R[G\int -])\Big) \xrightarrow{\mathrm{asbl}_{\mathscr{F}}} K_1(R[G])\,. \tag{9}$$

Proof. The universal property of $EG(\mathscr{F})$ in Remark 10(ii) gives a G-equivariant map $Z \longrightarrow EG(\mathscr{F})$. Hence the assembly map asbl_Z, which is induced by $Z \longrightarrow \mathrm{pt}$, factors over the assembly map $\mathrm{asbl}_{\mathscr{F}}$, which is induced by $EG(\mathscr{F}) \longrightarrow \mathrm{pt}$. □

The sufficient condition for surjectivity on π_1 from the preceding two results is generalized in Theorem 37 to a necessary and sufficient condition for assembly maps to be π_*-isomorphisms. In Remark 38 we explain how and where in the literature Theorem 29 is proved.

4.2 Contracting maps

In view of Theorem 29 and Corollary 30, a possible strategy to prove surjectivity of $\mathrm{asbl}_{\mathscr{F}}$ is to look for contracting maps. This leads to the following criterion.

Criterion 31. *Fix G, R, $\mathscr{F}$, and a word metric d^G for G. Suppose that there is an $N > 0$ such that for any arbitrarily large $D > 0$ there exists a simplicial complex Z_D with a simplicial G-action and a G-equivariant map $f_D\colon G/1 \longrightarrow Z_D$ satisfying the following conditions:*

(i) *$\dim Z_D \leq N$;*
(ii) *all isotropy groups of Z_D lie in $\mathscr{F}$;*

(iii) *the map f_D is D-contracting with respect to the l^1-metric in the target and the word metric in the source, i.e., for all $g, g' \in G$ we have*

$$d^1(f_D(g), f_D(g')) \leq \frac{1}{D} d^G(g, g').$$

Then map (9) is surjective.

The projection map to a point always satisfies (i) and (iii) but not (ii). The N-skeleton of a simplicial model for $EG(\mathcal{F})$ always satisfies (i) and (ii). But how can we produce contracting maps f_D that satisfy all three conditions? In Remark 32 we explain why the assumptions of the criterion are too strong to be useful. Nevertheless, we spell out the proof of the criterion as a warm-up exercise.

Proof. Set $\epsilon = \min\{\epsilon(n) \mid n \leq N\}$, where $\epsilon(n)$ comes from Theorem 29. Given any $x \in K_1(R[G])$ consider the following diagram.

$$\begin{array}{ccccc} K_1(\mathcal{C}(Z_D)) & \xleftarrow{f_{D*}} & K_1(\mathcal{C}(G/1)) & \ni & [\alpha] \\ & \underset{\mathbb{U}\ \cong}{\searrow} & \downarrow{\scriptstyle \cong\ \mathbb{U}} & & \downarrow \\ & & K_1(R[G]) & \ni & x \end{array}$$

Choose an automorphism α in $\mathcal{C}(G/1)$ whose class $[\alpha] \in K_1(\mathcal{C}(G/1))$ maps to x under $\mathbb{U}$. Determine the sizes of α and α^{-1}, and then choose D so large that

$$\operatorname{size} f_{D*}(\alpha) \leq \frac{1}{D} \operatorname{size} \alpha < \epsilon$$

and analogously for α^{-1}. Then Corollary 30 implies that x is in the image of the assembly map $\operatorname{asbl}_{\mathcal{F}}$ in (9). □

Remark 32. The case of Proposition 26 is when $G = \mathbb{Z}^2$ and $\mathcal{F} = \mathcal{C}yc$. Unfortunately, the conditions of Criterion 31 cannot possibly be satisfied in this case. To explain why, we need the following lemma.

Lemma 33. *Let s be a simplicial automorphism of a simplicial complex Z with $\dim Z \leq N$. If $x = \sum x_v v \in Z$ is such that*

$$d^1(x, sx) < \frac{1}{(N+1)^2},$$

then a barycenter of a face of the simplex Δ_x spanned by $V_x = \{v \mid x_v \neq 0\}$ is fixed under s.

Proof. For a vertex v with $x_v \neq 0$ set $A_v = \{v, sv, s^2v, \dots\}$. If $A_v \subset V_x$ then s permutes the finitely many elements in A_v and in particular fixes the barycenter of the face spanned by A_v.

Suppose that for no vertex v with $x_v \neq 0$ we have $A_v \subset V_x$. Then for all $v \in V_x$ there exists a smallest $n(v) \geq 1$ such that $s^{n(v)}v \notin V_x$ and hence $x_{s^{n(v)}v} = 0$. Since V_x contains at most $N+1$ vertices we know that $n(v) \leq N+1$ for all $v \in V_x$. Write $\epsilon = \frac{1}{(N+1)^2}$; then from

$$d^1(x, s^{-1}x) = d^1(s^{-1}x, s^{-2}x) = \cdots = d^1(s^{-N}x, s^{-(N+1)}x) < \epsilon$$

and $x_{s^{n(v)}v} = 0$ we conclude that

$$x_{s^{n(v)-1}v} < \epsilon, \quad x_{s^{n(v)-2}v} < 2\epsilon, \quad \dots, \quad x_v < n(v)\epsilon \leq (N+1)\epsilon = \frac{1}{N+1}.$$

However, since $\sum_{v \in V_x} x_v = 1$, the last inequality cannot be true for all vertices in V_x. □

Now suppose we can arrange (i) and (iii) from Criterion 31. If S is a (very large) finite subset of G, then by (iii) there exists a G-equivariant map $f\colon G/1 \longrightarrow Z$ to a G-simplicial complex that is contracting enough in order to have $d^1(f(1), f(s)) < \frac{1}{(N+1)^2}$ for all $s \in S$. The lemma implies that for each $s \in S$ a barycenter of a face of the simplex $\Delta_{f(1)}$ determined by $f(1)$ is fixed under s. Let $b(N)$ be the number of vertices in the barycentric subdivision of an N-simplex. Then there exists a subset $T \subset S$ with cardinality $\#T \geq \frac{\#S}{b(N)}$ and a point in $\Delta_{f(1)}$ fixed by all elements of T. The subgroup generated by T must lie in $\mathcal{F}$ if we require (ii). Since S can be arbitrarily large it seems difficult to keep $\mathcal{F}$ small.

In the case $G = \mathbb{Z}^2$ we can choose

$$S = S_l = \left\{ (x_1, x_2) \;\middle|\; |x_1| \leq \frac{l}{2} \text{ and } |x_2| \leq \frac{l}{2} \right\}.$$

Then if $l > b(N)$ we have $\#T \geq \frac{l^2}{b(N)} > l$, and since a subset of S_l with more than l elements cannot be contained in a line, the set T generates a finite index subgroup. Hence we can never arrange $\mathcal{F} = \mathcal{C}yc$ as desired, proving the claim in Remark 32.

4.3 The Farrell–Hsiang Criterion

The trick to obtain sufficiently contracting maps is to relax the requirement that the maps are G-equivariant, and instead only ask for equivariance with respect to (finite index) subgroups. We first illustrate this phenomenon in an example that is too simple to be useful.

Example 34. Consider the standard shift action of the infinite cyclic group $G = \mathbb{Z}$ on the real line:

$$\mathbb{Z} \times \mathbb{R} \longrightarrow \mathbb{R}, \quad (z, x) \longmapsto z + x.$$

This is a simplicial action if we consider $\mathbb{R}$ as one-dimensional simplicial complex with set of vertices $\mathbb{Z} \subset \mathbb{R}$. The map

$$f_D : \mathbb{Z} \longrightarrow \mathbb{R}, \quad z \longmapsto \frac{1}{D} z$$

is D-contracting but not $\mathbb{Z}$-equivariant. It becomes $\mathbb{Z}$-equivariant if we change the action on $\mathbb{R}$ to the action given by

$$\mathbb{Z} \times \mathbb{R} \longrightarrow \mathbb{R}, \quad (z, x) \longmapsto \frac{1}{D} z + x.$$

However, this action is no longer simplicial. If we restrict the action to the subgroup $D\mathbb{Z} < \mathbb{Z}$ or to any subgroup H with $H \leq D\mathbb{Z}$, then the H-action on $\operatorname{res}_H^{\mathbb{Z}} \mathbb{R}$ is simplicial, and

$$f_D : \operatorname{res}_H^{\mathbb{Z}} \mathbb{Z} \longrightarrow \operatorname{res}_H^{\mathbb{Z}} \mathbb{R}$$

is a D-contracting H-equivariant map.

Assume for a moment that for a subgroup $H \leq G$ of finite index we have an H-equivariant map

$$f_D \colon \operatorname{res}_H^G G/1 \longrightarrow E_H$$

to an H-simplicial complex E_H that is D-contracting with respect to a word metric in the source and the l^1-metric in the target. Let us see what happens when we induce up to G.

If (X, d) is a metric space with an isometric H-action, then $\operatorname{ind}_H^G X = G \times_H X$ has an isometric G-action with respect to the extended metric

$$d([g, x], [g', x']) = \begin{cases} d_X(x, g^{-1} g' x) & \text{if } g^{-1} g' \in H; \\ \infty & \text{if } g^{-1} g' \notin H. \end{cases}$$

Applying this to f_D we obtain a map

$$\operatorname{ind}_H^G f_D : \operatorname{ind}_H^G \operatorname{res}_H^G G/1 \longrightarrow \operatorname{ind}_H^G E_H$$

which is still D-contracting. However, observe that $\frac{1}{D}\infty = \infty$, and that a pair of points at distance ∞ in the source is mapped to a pair of points still at distance ∞ in the target. Hence the map can be used to diminish the size of a morphism between geometric modules only if the morphism over $\operatorname{ind}_H^G \operatorname{res}_H^G G/1$ is of finite size, i.e., only if it has no components that connect points at distance ∞.

The usual induction homomorphism $\operatorname{ind}_H^G: K_1(R[H]) \longrightarrow K_1(R[G])$ given by the functor $R[G] \otimes_{R[H]} -$ can be easily lifted to the categories of geometric modules, i.e., for any metric space X the functor

$$\operatorname{ind}_H^G : \mathscr{C}(X) \longrightarrow \mathscr{C}(\operatorname{ind}_H^G X), \quad (\varphi: S \longrightarrow X) \longmapsto (\operatorname{ind}_H^G \varphi : \operatorname{ind}_H^G S \longrightarrow \operatorname{ind}_H^G X)$$

induces the upper horizontal map in the following commutative diagram:

$$\begin{array}{ccc} K_1(\mathscr{C}(X)) & \xrightarrow{\operatorname{ind}_H^G} & K_1(\mathscr{C}(\operatorname{ind}_H^G X)) \\ \mathbb{U}\downarrow\cong & & \mathbb{U}\downarrow\cong \\ K_1(R[H]) & \xrightarrow{\operatorname{ind}_H^G} & K_1(R[G]) \end{array}$$

If X is a metric space in the usual sense (where ∞ is not allowed), then morphisms in the image of ind_H^G have the desired property: the size of $\operatorname{ind}_H^G \alpha$ is finite even though $\operatorname{ind}_H^G X$ is a metric space in the extended sense. Moreover,

$$\operatorname{size} \operatorname{ind}_H^G \alpha = \operatorname{size} \alpha .$$

Therefore, using the map $\operatorname{ind}_H^G f_D$ we can hope to show that, maybe not arbitrary elements, but at least elements of the form $\operatorname{ind}_H^G[\beta]$ belong to the image of $\operatorname{asbl}_{\mathscr{F}}$.

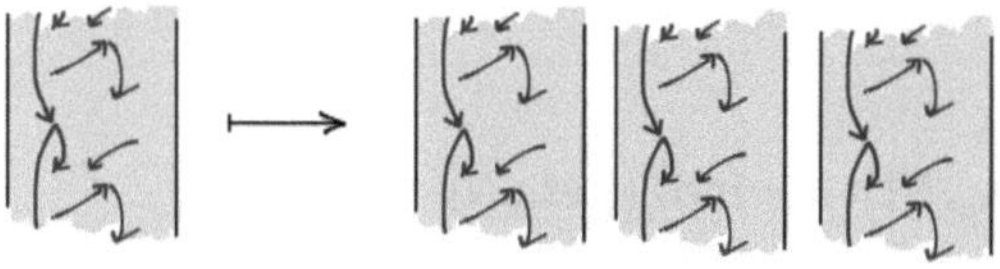

Figure 2: Support of α and $\operatorname{ind}_H^G \alpha$ in an index 3 situation.

The reason why this is useful is the following theorem of Swan. Recall that a finite group E is called hyperelementary if it fits into a short exact sequence $1 \longrightarrow C \longrightarrow E \longrightarrow P \longrightarrow 1$ where C is cyclic and the order of P is a prime power.

Theorem 35 (Swan induction). *Let F be a finite group,* $\operatorname{pr}: G \longrightarrow F$ *a surjective homomorphism, and*

$$\mathscr{H}_{\operatorname{pr}} = \left\{ \operatorname{pr}^{-1}(E) \;\middle|\; E \text{ is a hyperelementary subgroup of } F \right\} .$$

Then for every $H \in \mathcal{H}_{\mathrm{pr}}$ there exist $\mathbb{Z}[H]$-modules M_H^+ and M_H^- that are finitely generated free as $\mathbb{Z}$-modules, and such that, for each $n \in \mathbb{Z}$ and each $x \in K_n(R[G])$, we have

$$x = \sum_{H \in \mathcal{H}_p} \operatorname{ind}_H^G \left([M_H^+] \cdot \operatorname{res}_H^G x\right) - \operatorname{ind}_H^G \left([M_H^-] \cdot \operatorname{res}_H^G x\right). \tag{10}$$

Here, for $y \in K_n(R[H])$ and a $\mathbb{Z}[H]$-module M which is finitely generated free as a $\mathbb{Z}$-module, we write $[M] \cdot y = l_M(y)$ for the image of y under the map induced in K-theory by the functor l_M that sends the $R[H]$-module P to $M \otimes_{\mathbb{Z}} P$ equipped with the diagonal H-action.

Proof. The Swan group $Sw(H;\mathbb{Z})$ is by definition the K_0-group of $\mathbb{Z}H$-modules that are finitely generated free as $\mathbb{Z}$-modules. The relation is the usual additivity relation for (not necessarily split) short exact sequences. Tensor products over $\mathbb{Z}$ equipped with the diagonal H-actions induce the structure of a unital commutative ring on $Sw(H;\mathbb{Z})$, and also define an action of $Sw(H;\mathbb{Z})$ on $K_n(R[H])$. Swan [100] showed that for a finite group F there exist $\mathbb{Z}[E]$-modules N_E^+ and N_E^-, where E runs through all hyperelementary subgroups of F, such that in $Sw(F;\mathbb{Z})$ we have

$$1 = [\mathbb{Z}] = \sum_E \operatorname{ind}_E^F [N_E^+] - \operatorname{ind}_E^F [N_E^-]. \tag{11}$$

The natural isomorphisms

$$\operatorname{ind}_H^G \left(M \underset{\mathbb{Z}}{\otimes} \operatorname{res}_H^G P\right) \xrightarrow{\cong} (\operatorname{ind}_H^G M) \underset{\mathbb{Z}}{\otimes} P \quad \text{and} \quad \operatorname{ind}_{\mathrm{pr}^{-1}(E)}^G \operatorname{res}_{\mathrm{pr}} N \xrightarrow{\cong} \operatorname{res}_{\mathrm{pr}} \operatorname{ind}_E^F N,$$

given by $g \otimes m \otimes p \mapsto g \otimes m \otimes gp$ and $g \otimes n \mapsto \mathrm{pr}(g) \otimes n$, respectively, yield the following identity in $K_n(R[G])$ for $H = \mathrm{pr}^{-1}(E)$:

$$\left(\operatorname{res}_{\mathrm{pr}} \operatorname{ind}_E^F [N]\right) \cdot x = \left(\operatorname{ind}_H^G [\operatorname{res}_{\mathrm{pr}} N]\right) \cdot x = \operatorname{ind}_H^G \left([\operatorname{res}_{\mathrm{pr}} N] \cdot \operatorname{res}_H^G x\right).$$

Using this and $\operatorname{res}_{\mathrm{pr}} 1 = [\operatorname{res}_{\mathrm{pr}} \mathbb{Z}] = [\mathbb{Z}] = 1$ one derives the statement in the theorem with $M_{\mathrm{pr}^{-1}(E)} = \operatorname{res}_{\mathrm{pr}} N_E$ from (11). □

If we want to use H-equivariant contracting maps, as explained above, to show that each of the summands in (10) is in the image of $\mathrm{asbl}_{\mathcal{F}}$, we need to control the size of a geometric representative of $[M_H^\pm] \cdot \operatorname{res}_H^G x$ in terms of the size of a representative of x.

This is indeed easy. Similarly to induction, also the functors restriction res_H^G and $l_M = M \otimes_{\mathbb{Z}}$ – can be lifted to categories of geometric modules. For restriction simply send the object given by $\phi \colon S \longrightarrow Z$ to $\operatorname{res}_H^G \phi \colon \operatorname{res}_H^G S \longrightarrow \operatorname{res}_H^G Z$. For l_M observe that if B is a finite $\mathbb{Z}$-basis for the $\mathbb{Z}[H]$-module M, then there are isomorphisms of $\mathbb{Z}[H]$-modules

$$\mathbb{Z}[B] \underset{\mathbb{Z}}{\otimes} R[\coprod H/1] \xrightarrow{\cong} \mathbb{Z}[B] \underset{\mathbb{Z}}{\otimes} R[\coprod H/1] \xrightarrow{\cong} R[B \times \coprod H/1]\,. \tag{12}$$

Here the first isomorphism is given by $m \otimes h \mapsto h^{-1}m \otimes h$, where in the source one uses the diagonal H-action, and in the target the H-action on the right tensor factor. The second isomorphism is the obvious one. One constructs the desired functor by working only with objects of the form $\phi\colon \coprod H/1 \longrightarrow Z$ and sending such a ϕ to $\phi \circ \mathrm{pr}$, where $\mathrm{pr}\colon B \times \coprod H/1 \longrightarrow \coprod H/1$ is the projection onto the second factor. The behavior on morphisms is determined by isomorphism (12): one defines $l_M\alpha$ between the objects on the right in (12) in such a way that on the left it corresponds to $\mathrm{id} \otimes \alpha$. One then checks easily that

$$\operatorname{size} \operatorname{res}_H^G \alpha = \operatorname{size} \alpha \quad \text{and} \quad \operatorname{size} l_M \alpha = \operatorname{size} \alpha\,.$$

In summary, given a finite index subgroup $H \le G$, a $\mathbb{Z}[H]$-module M that is finitely generated free as a $\mathbb{Z}$-module, and an H-equivariant D-contracting map $f_D\colon \operatorname{res}_H^G G/1 \longrightarrow Z$ to an H-simplicial complex, we have a commutative diagram

$$\begin{array}{ccccccc}
[\alpha] & \longmapsto & & & \big[(\operatorname{ind}_H^G f_D)_*(\operatorname{ind}_H^G l_M \operatorname{res}_H^G \alpha)\big] \in & & K_1(\mathscr{C}(\operatorname{ind}_H^G Z)) \\
\in & & & & & & \big\uparrow (\operatorname{ind}_H^G f_D)_* \\
K_1(\mathscr{C}(G/1)) & \xrightarrow{\operatorname{res}_H^G} & K_1(\mathscr{C}(\operatorname{res}_H^G G/1)) & \xrightarrow{l_M} & K_1(\mathscr{C}(\operatorname{res}_H^G G/1)) & \xrightarrow{\operatorname{ind}_H^G} & K_1(\mathscr{C}(\operatorname{ind}_H^G \operatorname{res}_H^G G/1)) \\
\mathrm{U}\big\downarrow\cong & & \mathrm{U}\big\downarrow\cong & & \mathrm{U}\big\downarrow\cong & & \mathrm{U}\big\downarrow\cong \\
K_1(R[G]) & \xrightarrow{\operatorname{res}_H^G} & K_1(R[H]) & \xrightarrow{l_M} & K_1(R[H]) & \xrightarrow{\operatorname{ind}_H^G} & K_1(R[G])
\end{array} \tag{13}$$

and the estimate

$$\operatorname{size}\left((\operatorname{ind}_H^G f_D)_*(\operatorname{ind}_H^G l_M \operatorname{res}_H^G \alpha)\right) \le \frac{1}{D} \operatorname{size}\left(\operatorname{ind}_H^G l_M \operatorname{res}_H^G \alpha\right) = \frac{1}{D} \operatorname{size} \alpha < \infty\,. \tag{14}$$

In order to prove surjectivity of $\mathrm{asbl}_{\mathcal{F}}$ it remains to find suitable finite quotients $\mathrm{pr}\colon G \longrightarrow F$ and suitable H-equivariant contracting maps for each $H \in \mathcal{H}_{\mathrm{pr}}$. This leads to the criterion formulated in Theorem 39 for arbitrary groups G. Groups that meet this criterion have been named Farrell–Hsiang groups in [11].

4.4 $\mathbb{Z}^2$ is a Farrell–Hsiang group

Now we concentrate on the concrete situation where $G = \mathbb{Z}^2$, and explain how the criterion is met in this special case.

Claim 36 ($\mathbb{Z}^2$ is a Farrell–Hsiang group with respect to $\mathcal{C}yc$). *Fix a word metric $d^{\mathbb{Z}\times\mathbb{Z}}$ on $\mathbb{Z} \times \mathbb{Z}$. Consider $\mathbb{R}$ as a simplicial complex with vertices $\mathbb{Z} \subset \mathbb{R}$ and with the corresponding ℓ^1-metric d^1. For any arbitrarily large $D > 0$ there exists a surjective homomorphism $\mathrm{pr}_D \colon \mathbb{Z} \times \mathbb{Z} \longrightarrow F$ to a finite group F with the following property. For each*

$$H \in \mathcal{H}_{\mathrm{pr}_D} = \left\{ \mathrm{pr}_D^{-1}(E) \;\middle|\; E \text{ is a hyperelementary subgroup of } F \right\}$$

there exist:

(i) *a simplicial H-action on $\mathbb{R}$ with only cyclic isotropy,*
(ii) *a map $f_H \colon \mathrm{res}_H(\mathbb{Z} \times \mathbb{Z}) \longrightarrow \mathbb{R}$ that is H-equivariant and D-contracting, i.e.,*

$$d^1(f_H(g), f_H(g')) \le \tfrac{1}{D}\, d^{\mathbb{Z}\times\mathbb{Z}}(g, g') \tag{15}$$

for all $g, g' \in \mathbb{Z} \times \mathbb{Z}$.

We first show that this implies Proposition 26.

Proof of Proposition 26. The simplicial complex $\mathbb{R}$ is one-dimensional. Let $\epsilon = \epsilon(1)$ be as in Theorem 29. Given $x \in K_1(R[G])$ choose an automorphism α in $\mathcal{C}(G/1)$ such that $[\alpha]$ maps to x under the forgetful map $\mathbb{U} \colon K_1(\mathcal{C}(G/1)) \longrightarrow K_1(R[G])$. Choose $D > 0$ so large that $\frac{1}{D} \max\{\mathrm{size}(\alpha), \mathrm{size}(\alpha^{-1})\} \le \epsilon$. Use Claim 36 in order to find a finite quotient $\mathrm{pr}_D \colon \mathbb{Z} \times \mathbb{Z} \longrightarrow F$ and H-equivariant D-contracting maps $f_H \colon \mathrm{res}_H^G G/1 \longrightarrow \mathbb{R}$ for every $H \in \mathcal{H}_{\mathrm{pr}_D}$.

For each $H \in \mathcal{H}_{\mathrm{pr}_D}$, let $M = M_H^{\pm}$ be as in Theorem 35, and send $[\alpha]$ through the upper row in diagram (13). Use estimate (14) to conclude that

$$\mathrm{size}\left((\mathrm{ind}_H^G f_H)_* (\mathrm{ind}_H^G l_M \mathrm{res}_H^G \alpha) \right) \le \epsilon .$$

By Corollary 30 and the commutativity of diagram (13), we see that $\mathrm{ind}_H^G\, l_M\, \mathrm{res}_H^G\, x$ is in the image of map (7). Because of decomposition (10) in Theorem 35, also x is in the image. □

Proof of Claim 36. We begin with some simplifications. With respect to the standard generating set $\{(\pm 1, 0), (0, \pm 1)\}$, the word metric is Lipschitz equivalent to the Euclidean metric on $\mathbb{Z} \times \mathbb{Z} \subset \mathbb{R} \times \mathbb{R}$. On $\mathbb{R}$ the simplicial ℓ^1 metric and the Euclidean metric satisfy

$$d^1(x, y) \le C\, d^{\mathrm{Eucl}}(x, y)$$

for some fixed constant C. Therefore, it is enough to establish (15) with respect to the Euclidean metrics on $\mathbb{Z} \times \mathbb{Z} \subset \mathbb{R} \times \mathbb{R}$ and on $\mathbb{R}$, instead of the word and

ℓ^1-metrics. Moreover, it is enough to consider only maximal hyperelementary subgroups of F, because then for any $H' < H$ we can take $f_{H'} = \mathrm{res}_{H'} f_H$.

Let us start to look for suitable finite quotients F of $\mathbb{Z} \times \mathbb{Z}$. If F itself were hyperelementary, then we would have to find a contracting map $f_{\mathbb{Z}\times\mathbb{Z}}$ to a $(\mathbb{Z} \times \mathbb{Z})$-simplicial complex with cyclic isotropy that is $(\mathbb{Z} \times \mathbb{Z})$-equivariant. But in Remark 32 we saw that this is impossible.

Every finite quotient F of $\mathbb{Z} \times \mathbb{Z}$ is isomorphic to $\mathbb{Z}/a \times \mathbb{Z}/ab$, which is hyperelementary if and only if a is a prime power. Hence a simple choice of F which is not itself hyperelementary is $\mathbb{Z}/pq \times \mathbb{Z}/pq$ for distinct primes p and q. In order to achieve the contracting property we will later choose the primes to be very large.

Let $\mathrm{pr}_{pq}\colon \mathbb{Z} \times \mathbb{Z} \longrightarrow \mathbb{Z}/pq \times \mathbb{Z}/pq$ be the projection. A maximal hyperelementary subgroup E of $\mathbb{Z}/pq \times \mathbb{Z}/pq$ has order pq^2 or p^2q. By symmetry it is enough to consider the case where the order of E is pq^2. Let $H = \mathrm{pr}_{pq}^{-1}(E)$. Now we need to construct f_H.

For every $v \in \mathbb{Z} \times \mathbb{Z}$ with $v \neq 0$, consider the map

$$f_v\colon \mathbb{Z} \times \mathbb{Z} \xrightarrow{\ell_v = \langle v, -\rangle} \mathbb{Z} \xrightarrow{-/p} \mathbb{R}, \quad w \longmapsto \frac{1}{p}\langle v, w\rangle,$$

where $\langle -, -\rangle$ is the standard inner product on $\mathbb{R}^2$. If we equip $\mathbb{R}$ with the $(\mathbb{Z} \times \mathbb{Z})$-action given by

$$(\mathbb{Z} \times \mathbb{Z}) \times \mathbb{R} \longrightarrow \mathbb{R}, \qquad (w, x) \longmapsto x + \tfrac{1}{p}\langle v, w\rangle,$$

then f_v is $(\mathbb{Z} \times \mathbb{Z})$-equivariant. More importantly, we have that:

(A) f_v is $p/\|v\|$-contracting, i.e., $|f_v(w) - f_v(w')| \leq \frac{\|v\|}{p}\|w - w'\|$. This follows immediately from the linearity of f_v and the Cauchy–Schwarz inequality.

(B) The isotropy group at every point of $\mathbb{R}$ is $\ker(\ell_v) = \{w \in \mathbb{Z} \times \mathbb{Z} \mid \langle v, w\rangle = 0\}$, and hence cyclic since we assumed that $v \neq 0$.

(C) The action restricts to a simplicial H-action if $\ell_v(H) \subseteq p\mathbb{Z}$.

Let us reformulate the last condition. Consider the following commutative diagram.

$$\begin{array}{ccccc}
H = \mathrm{pr}_{pq}^{-1}(E) & \hookrightarrow & \mathbb{Z} \times \mathbb{Z} & \xrightarrow{\ \ell_v\ } & \mathbb{Z} \\
\downarrow & & \downarrow{\scriptstyle \mathrm{pr}_{pq}} & & \downarrow \\
E & \hookrightarrow & \mathbb{Z}/pq \times \mathbb{Z}/pq & & \\
\downarrow & & \downarrow{\scriptstyle \mathrm{pr}} & & \downarrow \\
\mathrm{pr}(E) = \mathbb{F}_p \cdot \overline{u} & \hookrightarrow & \mathbb{F}_p \times \mathbb{F}_p & \xrightarrow[\ \overline{\ell_v} = \ell_{\overline{v}}\ ]{} & \mathbb{F}_p
\end{array} \tag{16}$$

Here $\overline{u} \in \mathbb{F}_p \times \mathbb{F}_p$ is a generator of the $\mathbb{F}_p$-vector space $\mathrm{pr}(E) < \mathbb{F}_p \times \mathbb{F}_p$. Observe that $\mathrm{pr}(E) \neq \mathbb{F}_p \times \mathbb{F}_p$ because the order of E is pq^2. Then the last condition above is equivalent to saying that the composition in diagram (16) from H to $\mathbb{F}_p$ is trivial, i.e., that $\ell_{\overline{v}}(\overline{u}) = 0$.

Hence, if we can find a vector $v \in \mathbb{Z} \times \mathbb{Z}$ such that

$$0 < \|v\| \leq 4\sqrt{p} \qquad \text{and} \qquad \ell_{\overline{v}}(\overline{u}) = 0\,, \tag{17}$$

then from (A) we get that f_v is a $(p/4\sqrt{p} = \sqrt{p}/4)$-contracting H-equivariant map to $\mathbb{R}$, where $\mathbb{R}$ is equipped with a simplicial H-action by (C) and has cyclic isotropy by (B).

The existence of such a vector v is established by the following counting argument. Consider the set

$$S = \left\{ v = (x_1, x_2) \in \mathbb{Z} \times \mathbb{Z} \;\middle|\; |x_1| \leq \sqrt{2p} \text{ and } |x_2| \leq \sqrt{2p} \right\}.$$

This set has more than p elements, and therefore the map

$$S \longrightarrow \mathbb{F}_p\,, \qquad v \longmapsto \ell_{\overline{v}}(\overline{u})$$

is not injective, where $\overline{u}$ was defined right after diagram (16). If v_0 and v_1 are two distinct vectors in S with $\ell_{\overline{v_0}}(\overline{u}) = \ell_{\overline{v_1}}(\overline{u})$, then $v = v_0 - v_1$ is a vector which satisfies the equality in (17). For the inequality in (17) we estimate

$$\|v\| \leq \|v_0\| + \|v_1\| \leq 2\sqrt{2}\sqrt{2p} = 4\sqrt{p}\,.$$

So we define $f_H = f_v$ for such a v and finish the argument using Euclid's Theorem: since there are infinitely many primes, for any given $D > 0$ we can find distinct primes p and q such that both $\sqrt{p}/4 \geq D$ and $\sqrt{q}/4 \geq D$, and hence for every

$$H \in \mathscr{H}_{\mathrm{pr}_{pq}} = \left\{ \mathrm{pr}_{pq}^{-1}(E) \;\middle|\; \text{hyperelementary } E < \mathbb{Z}/pq \times \mathbb{Z}/pq \right\}$$

the map f_H is D-contracting. □

4.5 The Farrell–Hsiang Criterion (continued)

We now indicate how the ideas developed in this section can be used to prove isomorphism results in all dimensions instead of just surjectivity results for K_1. In [12] the authors introduce, for an arbitrary G-space X, the additive categories $\mathscr{T}^G(X)$, $\mathscr{O}^G(X)$, and $\mathscr{D}^G(X)$, and establish in [12, Lemma 3.6] a homotopy fibration sequence

$$\mathbf{K}(\mathscr{T}^G(X)) \longrightarrow \mathbf{K}(\mathscr{O}^G(X)) \longrightarrow \mathbf{K}(\mathscr{D}^G(X))\,. \tag{18}$$

The category $\mathscr{T}^G(X)$ is a variant of the category denoted $\mathscr{C}(X)$ in this section. The functor $X \longmapsto \mathbf{K}(\mathscr{D}^G(X))$ is a G-equivariant homology theory on G-CW complexes

[8, Section 5], and the value at G/H is π_*-isomorphic to $\Sigma\mathbf{K}(R[H])$ [8, Section 6]. Therefore, the general principles in [28, 110] identify the map

$$\mathbf{K}(\mathcal{D}^G(EG(\mathcal{F}))) \longrightarrow \mathbf{K}(\mathcal{D}^G(\mathrm{pt}))$$

with the (suspended) assembly map $\mathrm{asbl}_{\mathcal{F}}$.

A variant of the category $\mathcal{O}^G(X)$ can be defined as follows. Objects are G-equivariant maps $\varphi\colon S \longrightarrow X \times [1,\infty)$, where now the free G-set S is allowed to be cocountable instead of only cofinite. Moreover, we require that $\varphi^{-1}(X\times[1,N])$ is cofinite for every N.

A morphism α from φ to φ' is again an $R[G]$-linear map $\alpha\colon R[S] \longrightarrow R[S']$, but now there is a severe restriction on the support of a morphism: toward ∞ the arrows representing nonvanishing components must become smaller and smaller. Notice though that X is only a topological and not a metric space, and "small" has no immediate meaning. We refer to [8, Definition 2.7] for the precise definition of this condition, which is known as *equivariant continuous control at infinity*.

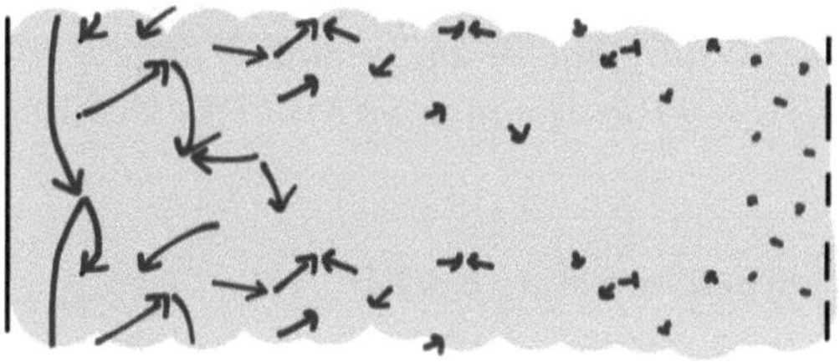

Figure 3: A morphism in the obstruction category $\mathcal{O}^G(X)$.

The following result explains the choice of notation: the category $\mathcal{O}^G(X)$ is the *obstruction category*.

Theorem 37. *The assembly map*

$$EG(\mathcal{F})_+ \underset{\mathrm{Or}G}{\wedge} \mathbf{K}(R[G\int -]) \longrightarrow \mathbf{K}(R[G])$$

is a π_*-isomorphism if and only if $K_*(\mathcal{O}^G(EG(\mathcal{F}))) = 0$.

Proof. The map $EG(\mathcal{F}) \longrightarrow \mathrm{pt}$ and the homotopy fibration sequence (18) induce the following commutative diagram with exact rows:

$$\begin{array}{ccccccc}
\cdots \to K_n(\mathcal{O}^G(EG(\mathcal{F}))) & \to & K_n(\mathcal{D}^G(EG(\mathcal{F}))) & \overset{④}{\to} & K_{n-1}(\mathcal{T}^G(EG(\mathcal{F}))) & \to & \cdots \\
\downarrow & & ②\downarrow & & \cong\downarrow ① & & \\
\cdots \to K_n(\mathcal{O}^G(\mathrm{pt})) = 0 & \longrightarrow & K_n(\mathcal{D}^G(\mathrm{pt})) & \underset{③}{\overset{\cong}{\longrightarrow}} & K_{n-1}(\mathcal{T}^G(\mathrm{pt})) & \longrightarrow & \cdots
\end{array}$$

The map ① is an isomorphism, because source and target are both isomorphic to $K_{n-1}(R[G])$ via the forgetful map (8). Using the shift map $[1,\infty) \longrightarrow [1,\infty)$, $x \longmapsto x+1$, it is not difficult to prove that $\mathscr{O}^G(\mathrm{pt})$ admits an Eilenberg swindle, and so $K_*(\mathscr{O}^G(\mathrm{pt})) = 0$. Therefore also the map ③ is an isomorphism. Since the map ② is identified with the assembly map, the result follows. □

Remark 38 (Proof of Theorem 29). Consider the ladder diagram in the previous proof, but replace $EG(\mathscr{F})$ with a simplicial complex Z. Maps ① and ③ are still isomorphisms. Maps ② and ④ for $n = 2$ are both models for the assembly map asbl_Z in Theorem 29. Exactness implies that $[\alpha] \in K_1(\mathscr{T}^G(Z))$ is in the image of the assembly map if it maps to $0 \in K_1(\mathscr{O}^G(Z))$. The statement of Theorem 29 is now a special case of [10, Theorem 5.3(i)].

With some additional work, the program carried out above to decompose an arbitrary K_1-element into summands with sufficiently small representatives can be generalized to show that the K-theory of the obstruction category in Theorem 37 vanishes. This leads to the following theorem, which is the main result of [11].

Theorem 39 (Farrell–Hsiang Criterion). *Let $\mathscr{F}$ be a family of subgroups of G. Fix a word metric on G. Assume that there exists an $N > 0$ such that for any arbitrarily large $D > 0$ there exists a surjective homomorphism $\mathrm{pr}_D\colon G \longrightarrow F$ to a finite group F with the following property. For each*

$$H \in \mathscr{H}_{\mathrm{pr}_D} = \left\{ \mathrm{pr}_D^{-1}(E) \;\middle|\; \textit{hyperelementary } E \le F \right\}$$

there exist:

(i) *an H-simplicial complex Z_H of dimension at most N and whose isotropy groups are all contained in $\mathscr{F}$;*

(ii) *a map $f_H\colon \mathrm{res}_H G \longrightarrow Z_H$ that is H-equivariant and D-contracting, i.e., $d^1(f_H(g), f_H(g')) \le \frac{1}{D}\, d^G(g, g')$ for all $g, g' \in G$.*

Then the assembly map

$$EG(\mathscr{F})_+ \underset{\mathrm{Or}G}{\wedge} \mathbf{K}(R[G\!\int -]) \longrightarrow \mathbf{K}(R[G])$$

is a π_*-isomorphism.

5 Trace methods

Trace maps are maps from algebraic K-theory to other theories like Hochschild homology, topological Hochschild homology, and their variants, which are usually easier to compute than K-theory. These trace maps have been used successfully to prove injectivity results about assembly maps in algebraic K-theory. In fact, the most sophisticated trace invariant, topological cyclic homology, was invented by Bökstedt, Hsiang, and Madsen specifically to attack the rational injectivity of the classical assembly map for $\mathbf{K}(\mathbb{Z}[G])$, as explained in Section 5.2. In joint work with Lück and Rognes, we applied similar techniques to the Farrell–Jones assembly map, and in particular we obtained the following partial verification of Conjecture 9; see [73, Theorem 1.1].

Theorem 40. *Assume that, for every finite cyclic subgroup C of a group G, the first and second integral group homologies $H_1(BZ_GC;\mathbb{Z})$ and $H_2(BZ_GC;\mathbb{Z})$ of the centralizer Z_GC of C in G are finitely generated abelian groups. Then G satisfies Conjecture 9, i.e., the map*

$$\operatorname*{colim}_{H\in\mathrm{obj}\,\mathrm{Sub}G(\mathcal{F}in)} Wh(H)\otimes_{\mathbb{Z}}\mathbb{Q} \longrightarrow Wh(G)\otimes_{\mathbb{Z}}\mathbb{Q}$$

is injective.

In this section we want to explain the ideas and the structure of the proofs of Bökstedt–Hsiang–Madsen's Theorem 43 and its generalization, suppressing some of the technical details. We first consider a K_0-analog of Theorem 40 and explain in full detail its proof, which is an illuminating example of the trace methods.

5.1 A warm-up example

Proposition 41. *Let $\Bbbk$ be any field of characteristic zero. Then for any group G the map*

$$\operatorname*{colim}_{H\in\mathrm{obj}\,\mathrm{Sub}G(\mathcal{F}in)} K_0(\Bbbk[H])\otimes_{\mathbb{Z}}\mathbb{Q} \longrightarrow K_0(\Bbbk[G])\otimes_{\mathbb{Z}}\mathbb{Q}$$

is injective.

This is closely related to Conjecture 4 for $R = \Bbbk$, but observe that, even though $K_0(\Bbbk[H])$ is a finitely generated free abelian group for each finite group H, the colimit in the source of the map in Conjecture 4 may contain torsion [64]. Therefore, Proposition 41 does not imply the injectivity of the map in Conjecture 4.

The key ingredient in the proof of Proposition 41 is the trace map

$$\mathrm{tr}\colon K_0(R) \longrightarrow R/[R,R]\,,$$

where $[R,R]$ denotes the subgroup of the additive group of R generated by commutators. The trace map is defined as follows. The projection $R \longrightarrow R/[R,R]$ extends to a map

$$\mathrm{tr}\colon M_n(R) \longrightarrow R/[R,R]\,, \qquad a = (a_{ji}) \longmapsto \mathrm{tr}(A) = \sum_{i=1}^{n}[a_{ii}]\,,$$

which is easily seen to be the universal additive map out of $M_n(R)$ with the trace property: $\mathrm{tr}(ab) = \mathrm{tr}(ba)$. If p is an idempotent matrix in $M_n(R)$, then $\mathrm{tr}(p)$ only depends on the isomorphism class of the projective R-module $R^n p$. Since the trace sends the block sum of matrices to the sum of the traces, it induces a group homomorphism

$$\mathrm{tr}\colon K_0(R) \longrightarrow R/[R,R]\,, \qquad [(p_{ji})] \longmapsto \sum_i [p_{ii}]\,. \tag{19}$$

Now consider the case of group algebras. We denote by $\operatorname{conj} G$ the set of conjugacy classes of elements of G. The map $R[G] \longrightarrow R[\operatorname{conj} G]$ induced by the projection sends $\big[R[G], R[G]\big]$ to zero, and it induces an isomorphism

$$R[G]/\big[R[G], R[G]\big] \cong R[\operatorname{conj} G]\,.$$

The composition of the trace map tr from (19) with this isomorphism gives a map

$$\mathrm{tr}\colon K_0(R[G]) \longrightarrow R[\operatorname{conj} G]\,,$$

which is known as the Hattori–Stallings rank. In the special case of group algebras of finite groups with coefficients in fields of characteristic zero we have the following result.

Lemma 42. *Suppose that the group G is finite and that $R = \Bbbk$ is a field of characteristic zero. Let $R_{\Bbbk}(G)$ be the representation ring of G over $\Bbbk$, and consider the map*

$$\chi\colon R_{\Bbbk}(G) \longrightarrow \Bbbk[\operatorname{conj} G]\,, \qquad \rho \longmapsto \big(\chi_\rho\colon g \mapsto \mathrm{tr}_{\Bbbk}(\rho(g))\big)$$

that sends each representation to its character. Then there is a commutative diagram

$$\begin{array}{ccc} K_0(\Bbbk[G]) & \xrightarrow{\ \mathrm{tr}\ } & \Bbbk[\operatorname{conj} G] \\ \cong\big\downarrow & & \big\downarrow\cong \\ R_{\Bbbk}(G) & \xrightarrow[\ \chi\ ]{} & \Bbbk[\operatorname{conj} G] \end{array} \qquad\qquad \begin{array}{c} [g] \\ \big\downarrow \\ \#\big(Z_G\langle g\rangle\big)[g^{-1}] \end{array}$$

whose vertical maps are isomorphisms.

In other words, the Hattori–Stallings rank can be identified up to isomorphism with the character map χ. Notice, though, that unlike χ the Hattori–Stallings rank is natural in G.

Proof of Lemma 42. Since G is finite and $\Bbbk$ has characteristic zero, a finitely generated projective $\Bbbk[G]$-module V is the same as a finite-dimensional $\Bbbk$-vector space V equipped with a linear G-action $\rho\colon G \longrightarrow GL(V)$. This explains the left vertical isomorphism in the diagram above. It is well known that every irreducible representation is contained as a direct summand in the regular representation $\Bbbk[G]$. Therefore, we can assume that the idempotent $p = p^2 = \sum_{k\in G} p_k k$ lies in $\Bbbk[G]$. Let $\langle -,-\rangle$ be the $\Bbbk$-bilinear form on $\Bbbk[G]$ that is determined on group elements by $\langle g,h\rangle = \delta_{gh}$. Then

$$\chi_p(g) = \mathrm{tr}_{\Bbbk}(\Bbbk[G]p \to \Bbbk[G]p,\ x \mapsto gx) = \mathrm{tr}_{\Bbbk}(\Bbbk[G] \to \Bbbk[G],\ x \mapsto gxp) =$$

$$= \sum_{h\in G} \langle h, ghp\rangle = \sum_{h\in G}\sum_{k\in G} p_k \langle h, ghk\rangle = \sum_{h\in G} p_{h^{-1}g^{-1}h} = \sum_{x\in[g^{-1}]} \#\big(Z_G\langle g^{-1}\rangle\big)p_x .$$

For the last equality observe that the stabilizer of $g \in G$ under the action of G on itself via conjugation is the centralizer $Z_G\langle g\rangle$. For the Hattori–Stallings rank we have $\mathrm{tr}(p)([g]) = \sum_{x\in[g]} p_x$. □

We are now ready to prove Proposition 41.

Proof of Proposition 41. It suffices to prove the injectivity of the map in Proposition 41 with $-\otimes_{\mathbb{Z}} \mathbb{Q}$ replaced by $-\otimes_{\mathbb{Z}} \Bbbk$. We explain the proof in the case $\Bbbk = \mathbb{C}$. Consider the following commutative diagram.

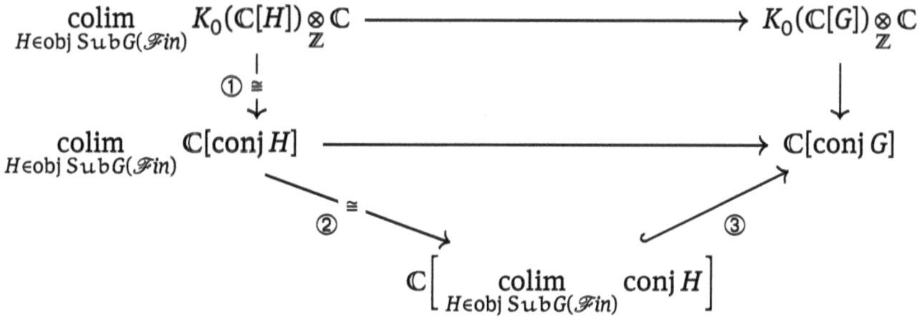

The vertical maps are induced by the $\mathbb{C}$-linear extension of the Hattori–Stallings rank. For each finite group H this extension is an isomorphism by Lemma 42 and [98, Corollary 1 in §12.4], and so map ① is an isomorphism. Map ② is an isomorphism because the functor $\mathbb{C}[-]$ is left adjoint and hence preserves colimits. Since conjugation with elements in G represents morphisms in $\mathrm{Sub}G(\mathcal{F}in)$, map ③ is easily seen to be injective already before applying $\mathbb{C}[-]$.

The proof for an arbitrary field $\Bbbk$ of characteristic zero is completely analogous, but the set $\mathrm{conj}\, G$ needs to be replaced by the set $\mathrm{conj}_{\Bbbk}\, G$ of $\Bbbk$-conjugacy classes, a certain quotient of $\mathrm{conj}\, G$. □

Notice that for each finite group H the Hattori–Stallings rank itself (before $\Bbbk$-linear extension) is always injective. But we cannot leverage this fact to prove integral injectivity results because colimits need not preserve injectivity.

5.2 Bökstedt–Hsiang–Madsen's theorem

The map tr in (19) is just the first (or rather the zeroth) and the easiest trace invariant of the algebraic K-theory of R. We now briefly overview how it can be generalized, starting with the Dennis trace with values in Hochschild homology.

Consider the simplicial abelian group

$$\cdots \quad R \otimes R \otimes R \;\substack{\rightarrow\\[-2pt]\rightarrow\\[-2pt]\rightarrow}\; R \otimes R \;\substack{\rightarrow\\[-2pt]\rightarrow}\; R \tag{20}$$

whose face maps are

$$d_i(r_0 \otimes \cdots \otimes r_n) = \begin{cases} r_0 \otimes \cdots \otimes r_i r_{i+1} \otimes \cdots \otimes r_n & \text{if } i < n; \\ r_n r_0 \otimes r_1 \otimes \cdots \otimes r_{n-1} & \text{if } i = n. \end{cases}$$

The geometric realization of the simplicial abelian group (20) is the zeroth space of an Ω-spectrum denoted $\mathbf{HH}(R) = \mathbf{HH}(R \mid \mathbb{Z})$, whose homotopy groups

$$HH_*(R) = \pi_* \mathbf{HH}(R) \quad \text{are the } \textit{Hochschild homology} \text{ groups of } R.$$

In particular, we see that $HH_0(R)$ is the cokernel of the map $r \otimes s \longmapsto rs - sr$, and hence

$$HH_0(R) \cong R/[R, R]\,.$$

The trace map $\mathrm{tr}\colon K_0(R) \longrightarrow HH_0(R)$ in (19) lifts to a map of spectra

$$\mathrm{trd}\colon \mathbf{K}^{\geq 0}(R) \longrightarrow \mathbf{HH}(R)$$

called the *Dennis trace*, such that $\pi_0 \,\mathrm{trd} = \mathrm{tr}$. We use $\mathbf{K}^{\geq 0}$ to denote connective algebraic K-theory, the (-1)-connected cover of the functor $\mathbf{K}$ we used throughout.

Following ideas of Goodwillie and Waldhausen, Bökstedt [21] introduced a far-reaching generalization of $\mathbf{HH}(R)$, called *topological Hochschild homology* and denoted $\mathbf{THH}(R)$. We omit the technical details of the definitions, and we rather explain the underlying ideas and structures.

The key idea in the definition of topological Hochschild homology is to pass from the ring R to its Eilenberg–Mac Lane ring spectrum $\mathbb{H}R$, and to replace the tensor products (over the initial ring $\mathbb{Z}$) with smash products (over the initial ring spectrum $\mathbb{S}$). In order to make this precise, one needs to work within a symmetric monoidal model category of spectra (e.g., symmetric spectra), or with ad hoc point-set level constructions (as Bökstedt did, long before symmetric spectra and the like were discovered). Once these technical difficulties are overcome, one obtains a simplicial spectrum

$$\cdots \quad \mathbb{H}R \wedge \mathbb{H}R \wedge \mathbb{H}R \;\substack{\rightarrow\\[-2pt]\rightarrow\\[-2pt]\rightarrow}\; \mathbb{H}R \wedge \mathbb{H}R \;\substack{\rightarrow\\[-2pt]\rightarrow}\; \mathbb{H}R\,, \tag{21}$$

whose geometric realization is $\mathbf{THH}(R) = \mathbf{HH}(\mathbb{H}R \mid \mathbb{S})$. Notice that of course this definition applies not only to Eilenberg–Mac Lane ring spectra $\mathbb{H}R$ but to arbitrary ring spectra $\mathbb{A}$.

Bökstedt also lifted the Dennis trace to topological Hochschild homology for any connective ring spectrum $\mathbb{A}$:

$$\begin{array}{ccc} & \overset{\mathrm{trb}}{\nearrow} & \mathbf{THH}(\mathbb{A}) \\ & & \downarrow \\ \mathbf{K}^{\geq 0}(\mathbb{A}) & \xrightarrow[\mathrm{trd}]{\qquad} & \mathbf{HH}(\pi_0\mathbb{A})\,. \end{array}$$

Cyclic permutation of the tensor factors in (20) or smash factors in (21) makes those simplicial objects into *cyclic* objects, thus inducing a natural S^1-action on their geometric realizations; see for example [57, Section 3] and [30]. Bökstedt, Hsiang, and Madsen [22] discovered that topological Hochschild homology has even more structure, which Hochschild homology lacks. Fix a prime p. As n varies, the fixed points of the induced C_{p^n}-actions are related by maps

$$\mathbf{THH}(\mathbb{A})^{C_{p^n}} \underset{F}{\overset{R}{\rightrightarrows}} \mathbf{THH}(\mathbb{A})^{C_{p^{n-1}}}\,, \tag{22}$$

called Restriction and Frobenius. The map F is simply the inclusion of fixed points, whereas the definition of the map R is much more delicate and specific to the construction of $\mathbf{THH}$. The homotopy equalizer of (22) is denoted $\mathbf{TC}^{n+1}(\mathbb{A};p)$. One important property of the maps R and F is that they commute, and therefore they induce a map $\mathbf{TC}^{n+1}(\mathbb{A};p) \longrightarrow \mathbf{TC}^{n}(\mathbb{A};p)$. The *topological cyclic homology* of $\mathbb{A}$ at the prime p is then defined as the homotopy limit

$$\mathbf{TC}(\mathbb{A};p) = \operatorname*{holim}_{n} \mathbf{TC}^{n}(\mathbb{A};p)\,.$$

Bökstedt, Hsiang, and Madsen lifted the Bökstedt trace to topological cyclic homology, thus obtaining the following commutative diagram for any connective ring spectrum $\mathbb{A}$:

$$\begin{array}{ccc} & \overset{\mathrm{trc}}{\nearrow} & \mathbf{TC}(\mathbb{A};p) \\ & & \downarrow \\ & \overset{\mathrm{trb}}{\nearrow} & \mathbf{THH}(\mathbb{A}) \\ & & \downarrow \\ \mathbf{K}^{\geq 0}(\mathbb{A}) & \xrightarrow[\mathrm{trd}]{\qquad} & \mathbf{HH}(\pi_0\mathbb{A})\,. \end{array} \tag{23}$$

The map trc is called the *cyclotomic trace map*.

They then used this technology to prove the following striking theorem, which is often referred to as the algebraic K-theory Novikov Conjecture; see [22, Theorem 9.13] and [78, Theorem 4.5.4].

Theorem 43 (Bökstedt–Hsiang–Madsen). *Let G be a group. Assume that the following condition holds.*

[A_1] *For every $s \geq 1$, the integral group homology $H_s(BG; \mathbb{Z})$ is a finitely generated abelian group.*

Then the classical assembly map

$$\mathrm{asbl}_1 \colon BG_+ \wedge \mathbf{K}(\mathbb{Z}) \longrightarrow \mathbf{K}(\mathbb{Z}[G])$$

is $\pi_*^{\mathbb{Q}}$-injective, i.e., $\pi_n(\mathrm{asbl}_1) \otimes_{\mathbb{Z}} \mathbb{Q}$ is injective for all $n \in \mathbb{Z}$.

We now explain the structure of the proof of Theorem 43, following the approach of [73]. As mentioned above, the idea is to use the cyclotomic trace map. However, it is not enough to work with topological cyclic homology, and one needs a variant of it that we proceed to explain. Instead of taking the homotopy equalizer of R and F in (22), we may consider just the homotopy fiber of R and define

$$\mathbf{C}^{n+1}(\mathbb{A}; p) = \mathrm{hofib}\Big(\mathbf{THH}(\mathbb{A})^{C_{p^n}} \xrightarrow{R} \mathbf{THH}(\mathbb{A})^{C_{p^{n-1}}}\Big).$$

The map F induces a map $\mathbf{C}^{n+1}(\mathbb{A}; p) \longrightarrow \mathbf{C}^n(\mathbb{A}; p)$, and we define

$$\mathbf{C}(\mathbb{A}; p) = \operatorname*{holim}_n \mathbf{C}^n(\mathbb{A}; p).$$

A fundamental property, also established in [22], is that $\mathbf{C}^{n+1}(\mathbb{A}; p)$ can be identified with $\mathbf{THH}(A)_{hC_{p^n}}$, up to a zigzag of π_*-isomorphisms. In [73, Section 8] we provided a *natural* zigzag of π_*-isomorphisms between $\mathbf{THH}(A)_{hC_{p^n}}$ and $\mathbf{C}^{n+1}(\mathbb{A}; p)$, natural even before passing to the stable homotopy category of spectra. The key tool here is the natural Adams isomorphism for equivariant orthogonal spectra developed in [89].

In the special case when $\mathbb{A} = \mathbb{S}[G]$ is a spherical group ring, then the maps R split, and these splittings can be used to construct a map

$$\mathbf{TC}(\mathbb{S}[G]; p) \longrightarrow \mathbf{C}(\mathbb{S}[G]; p). \tag{24}$$

The crucial advantage of using $\mathbf{C}$ instead of $\mathbf{TC}$ is that more general rational injectivity statements can be proved for the assembly maps for $\mathbf{C}$; compare Remark 48.

In order to prove Theorem 43 one studies the following commutative diagram:

$$\begin{array}{ccc}
BG_+ \wedge \mathbf{K}(\mathbb{Z}) & \xrightarrow{\ \text{asbl}_1\ } & \mathbf{K}(\mathbb{Z}[G]) \\
\text{①}\uparrow & & \uparrow\text{❶} \\
BG_+ \wedge \mathbf{K}^{\geq 0}(\mathbb{Z}) & \longrightarrow & \mathbf{K}^{\geq 0}(\mathbb{Z}[G]) \\
\text{②}\uparrow & & \uparrow\text{❷} \\
BG_+ \wedge \mathbf{K}^{\geq 0}(\mathbb{S}) & \longrightarrow & \mathbf{K}^{\geq 0}(\mathbb{S}[G]) \\
\text{③}\downarrow & & \downarrow\text{❸} \\
BG_+ \wedge \mathbf{TC}(\mathbb{S};p) & \longrightarrow & \mathbf{TC}(\mathbb{S}[G];p) \\
\text{④}\downarrow & & \downarrow\text{❹} \\
BG_+ \wedge \big(\mathbf{THH}(\mathbb{S}) \times \mathbf{C}(\mathbb{S};p)\big) & \xrightarrow[\text{⑤}]{} & \mathbf{THH}(\mathbb{S}[G]) \times \mathbf{C}(\mathbb{S}[G];p)
\end{array} \tag{25}$$

The horizontal maps are all classical assembly maps, and we want to prove that the one at the top of the diagram is $\pi_*^{\mathbb{Q}}$-injective. Maps ① and ❶ are induced by the natural maps from connective to nonconnective algebraic K-theory. Since $\mathbb{Z}$ is regular, ① is a π_*-isomorphism. Maps ② and ❷ come from the linearization (or Hurewicz) map $\mathbb{S} \longrightarrow \mathbb{Z}$, and they are both $\pi_*^{\mathbb{Q}}$-isomorphisms by a result of Waldhausen [105, Proposition 2.2]. Maps ③ and ❸ are given by the cyclotomic trace map, and ④ and ❹ by the natural maps in eqs. (23) and (24).

So, in order to prove that the top horizontal map in diagram (25) is $\pi_*^{\mathbb{Q}}$-injective, it is enough to show that

(a) The assembly map ⑤ is $\pi_*^{\mathbb{Q}}$-injective.
(b) The composition ④ ∘ ③ is $\pi_*^{\mathbb{Q}}$-injective.

The assumption $[A_1]$ is then shown to imply (a), and in fact not just for $\mathbb{S}$ but for arbitrary connective ring spectra $\mathbb{A}$. This is the special case $\mathcal{F} = 1$ of Theorems 46 and 47. The difficult part in proving (b) is the analysis of map ③. The Atiyah–Hirzebruch spectral sequences collapse rationally, and therefore it is enough to study the rational injectivity of trc: $\mathbf{K}^{\geq 0}(\mathbb{S}) \longrightarrow \mathbf{TC}(\mathbb{S};p)$. To this end, consider the following commutative diagram:

$$\begin{array}{ccccccc}
 & & & \nearrow^{\text{⑥}} & \mathbf{K}^{\geq 0}(\mathbb{Z}_p) & \searrow^{\text{⑦}} & \\
\mathbf{K}^{\geq 0}(\mathbb{S}) & \xrightarrow{\ \text{②}\ } & \mathbf{K}^{\geq 0}(\mathbb{Z}) & & & & \mathbf{K}^{\geq 0}(\mathbb{Z}_p)^{\wedge}_{p} \\
 & & & \searrow_{\text{⑧}} & \mathbf{K}^{\geq 0}(\mathbb{Z})^{\wedge}_{p} & \nearrow_{\text{⑨}} & \\
\text{trc}\downarrow & & & & & & \downarrow\text{trc}^{\wedge}_{p} \\
\mathbf{TC}(\mathbb{S};p) & & & \longrightarrow & & & \mathbf{TC}(\mathbb{Z}_p;p)^{\wedge}_{p}
\end{array}$$

Here $(-)^{\wedge}_{p}$ denotes the p-completion of spectra and $\mathbb{Z}_p$ are the p-adic numbers. The map $\mathrm{trc}^{\wedge}_{p}$ is a π_n-isomorphism for each $n \geq 0$ by a result of Hesselholt and Madsen [54, Theorem D]. We already mentioned above that ② is a $\pi_*^{\mathbb{Q}}$-isomorphism.

It remains to discuss the diamond. Since the groups $K_n(\mathbb{Z})$ are known to be finitely generated, ⑧ is $\pi_*^{\mathbb{Q}}$-injective. The question whether ⑨ is $\pi_*^{\mathbb{Q}}$-injective is open in general. It can be reformulated in terms of similar maps in étale K-theory, étale cohomology, or Galois cohomology, as surveyed in [73, Section 18]. Luckily the equivalent conjecture in Galois cohomology is known to be true if p is a regular prime by results in [95]; see [73, Proposition 2.9]. Recall that a prime p is regular if it does not divide the order of the ideal class group of $\mathbb{Q}(\zeta_p)$. Since regular primes exist we obtain the following statement and we are done.

[B_1] There exists a prime p such that ⑨ ∘ ⑧ is $\pi_*^{\mathbb{Q}}$-injective.

We remark that little is known about the rationalized homotopy groups of $\mathbf{K}^{\geq 0}(\mathbb{Z}_p)$ without p-completion; compare [108, Warning 60].

This concludes our explanation of the proof of Theorem 43.

5.3 Generalizations

The following result generalizes Theorem 43 from the classical to the Farrell–Jones assembly map and is a special case of [73, Main Technical Theorem 1.16].

Theorem 44. *Let G be a group and let $\mathcal{F} \subseteq \mathcal{FC}yc$ be a family of finite cyclic subgroups of G. Assume that the following two conditions hold:*

[$A_{\mathcal{F}}$] *For every $C \in \mathcal{F}$ and every $s \geq 1$, the integral group homology $H_s(BZ_GC;\mathbb{Z})$ of the centralizer of C in G is a finitely generated abelian group.*

[$B_{\mathcal{F}}$] *For every $C \in \mathcal{F}$ and every $t \geq 0$, the natural homomorphism*

$$K_t(\mathbb{Z}[\zeta_c]) \underset{\mathbb{Z}}{\otimes} \mathbb{Q} \longrightarrow \prod_{p \text{ prime}} K_t\Big(\mathbb{Z}_p \underset{\mathbb{Z}}{\otimes} \mathbb{Z}[\zeta_c]; \mathbb{Z}_p\Big) \underset{\mathbb{Z}}{\otimes} \mathbb{Q}$$

is injective, where c is the order of C, ζ_c is any primitive cth root of unity, and $K_t(R;\mathbb{Z}_p) = \pi_t(\mathbf{K}(R)^{\wedge}_{p})$.

Then the assembly map

$$\mathrm{asbl}_{\mathcal{F}} \colon EG(\mathcal{F})_+ \underset{\mathrm{Or}G}{\wedge} \mathbf{K}^{\geq 0}(\mathbb{Z}[G\!\int -]) \longrightarrow \mathbf{K}^{\geq 0}(\mathbb{Z}[G])$$

is $\pi_*^{\mathbb{Q}}$-injective.

Several comments are in order.

Remark 45.

(i) When $\mathcal{F} = 1$ is the trivial family, Theorems 43 and 44 coincide. This is because assumption $[A_1]$ of Theorem 44 is literally the same as assumption $[A_1]$ of Theorem 43, and assumption $[B_1]$ follows at once from the corresponding true statement explained at the end of the previous section.

(ii) When $\mathcal{F} = \mathcal{FCyc}$, then the rationalized assembly map for *connective* algebraic K-theory studied in Theorem 44 can be rewritten as in Conjecture 17, because isomorphisms (4) and (5) hold for both connective and nonconnective algebraic K-theory. The only difference is that the summands indexed by $t = -1$ in the source of the map in Conjecture 17 are now missing. Notice that the negative K-groups $K_t(\mathbb{Z}[C])$ are known to vanish for any $t < -1$ if C is finite or even virtually cyclic [42].

(iii) As noted above, assumption $[A_{\mathcal{F}}]$ implies and is the obvious generalization of assumption $[A_1]$. For any $\mathcal{F} \subseteq \mathcal{FCyc}$, assumption $[A_{\mathcal{F}}]$ is satisfied if there is a universal space $E G(\mathcal{Fin})$ of finite type, i.e., whose skeleta are all cocompact. Hyperbolic groups, finite-dimensional CAT(0)-groups, cocompact lattices in virtually connected Lie groups, arithmetic groups in semisimple connected linear $\mathbb{Q}$-algebraic groups, mapping class groups, and outer automorphism groups of free groups are all examples of groups that even have a finite-dimensional and cocompact $E G(\mathcal{Fin})$. Among these groups, outer automorphism groups of free groups do not appear in Theorem 21, and for them Theorem 44 gives the first result about the Farrell–Jones Conjecture. An interesting example of a group that satisfies $[A_{\mathcal{FCyc}}]$ without having an $E G(\mathcal{Fin})$ of finite type is given by Thompson's group T of orientation-preserving, piecewise-linear, dyadic homeomorphisms of the circle; see [49].

(iv) Conjecturally assumption $[B_{\mathcal{F}}]$ of Theorem 44 is always satisfied; in fact, it is implied by a weak version of the Leopoldt–Schneider Conjecture for cyclotomic fields, as explained carefully in [73, Sections 2 and 18]. When $t = 0$ or $t = 1$, i.e., for K_0 and K_1, the map in $[B_{\mathcal{F}}]$ is injective for arbitrary c by direct computation; compare [73, Proposition 2.4]. For any fixed c it is known that injectivity may fail for at most finitely many values of t. These two facts allow to deduce Theorems 40 and 22(i) from Theorem 44, or rather from its more general version in [73, Main Technical Theorem 1.16], as explained in loc. cit., Section 17 and page 1015. Notice that, on the other hand, Theorem 43 cannot be used to deduce information about the Whitehead group $Wh(G)$, which is the cokernel of the map induced on π_1 by the classical assembly map asbl_1.

The proof of Theorem 44 follows the same strategy as the proof of Theorem 43 outlined above. We consider the analog of diagram (25) for the generalized assembly map $\mathrm{asbl}_{\mathcal{F}}$; compare [73, "main diagram" (3.1)]. The key results about assembly maps are summarized in the following two theorems [73, Theorem 1.19, parts (i) and (ii)]. We point out that all the following results hold for arbitrary connective ring spectra $\mathbb{A}$.

Theorem 46. *For any group G and for any family $\mathcal{F}$ of subgroups of G, the assembly map*

$$\mathrm{asbl}_{\mathcal{F}}\colon EG(\mathcal{F})_+ \underset{\mathrm{Or}G}{\wedge} \mathbf{THH}(\mathbb{A}[G\int -]) \longrightarrow \mathbf{THH}(\mathbb{A}[G])$$

induces split monomorphisms on π_*, and it is a π_*-isomorphism if and only if $\mathcal{F}$ contains all cyclic subgroups of G, i.e., $\mathcal{F} \supseteq \mathcal{C}yc$.

Theorem 47. *Let G be a group and let $\mathcal{F} \subseteq \mathcal{F}\mathcal{C}yc$ be a family of finite cyclic subgroups of G. Assume that the following condition holds:*

[$A_{\mathcal{F}}$] *For every $C \in \mathcal{F}$ and every $s \geq 1$, the integral group homology $H_s(BZ_GC;\mathbb{Z})$ of the centralizer of C in G is a finitely generated abelian group.*

Then the assembly map

$$\mathrm{asbl}_{\mathcal{F}}\colon EG(\mathcal{F})_+ \underset{\mathrm{Or}G}{\wedge} \mathbf{C}(\mathbb{A}[G\int -;p]) \longrightarrow \mathbf{C}(\mathbb{A}[G];p)$$

is $\pi_*^{\mathbb{Q}}$-injective.

Remark 48. In order to establish an analog of Theorem 47 for the assembly map

$$\mathrm{asbl}_{\mathcal{F}}\colon EG(\mathcal{F})_+ \underset{\mathrm{Or}G}{\wedge} \mathbf{TC}(\mathbb{A}[G\int -;p]) \longrightarrow \mathbf{TC}(\mathbb{A}[G];p) \tag{26}$$

in topological cyclic homology, we need to assume not only condition [$A_{\mathcal{F}}$] but also the following two conditions:

[$A'_{\mathcal{F}}$] the family $\mathcal{F}$ contains only finitely many conjugacy classes of subgroups;
[$A''_{\mathcal{F}}$] for every $g \in G$, $\langle g\rangle \in \mathcal{F}$ if and only if $\langle g^p\rangle \in \mathcal{F}$.

The fact that the assembly map (26) is $\pi_*^{\mathbb{Q}}$-injective under assumptions [$A_{\mathcal{F}}$], [$A'_{\mathcal{F}}$], and [$A''_{\mathcal{F}}$] is a special case of [74, Theorem 1.8]. Notice the following facts.

(i) When $\mathcal{F} = 1$, assumption [A'_1] is vacuously true, but [A''_1] is *not* satisfied if G has p-torsion. This is the reason why, in the proof of Bökstedt–Hsiang–Madsen's Theorem 43, we need to work with **C** and not just **TC**.

(ii) As pointed out in Remark 45(iii), Thompson's group T satisfies [$A_{\mathcal{F}\mathcal{C}yc}$] and obviously also [$A''_{\mathcal{F}\mathcal{C}yc}$]. However, T contains finite cyclic subgroups of any given order, and therefore does not satisfy [$A'_{\mathcal{F}\mathcal{C}yc}$]. It is an interesting open question whether map (26) is $\pi_*^{\mathbb{Q}}$-injective for $G = T$.

(iii) Without homological finiteness assumptions on G, the assembly map (26) is not rationally injective in general. For example, if $G = \mathbb{Q}$ and $\mathcal{F} = 1 = \mathcal{F}\mathcal{C}yc$, then map (26) is essentially trivial after applying $\pi_*(-) \otimes_{\mathbb{Z}} \mathbb{Q}$. This is explained in [73, Remark 3.7]. Of course, the group $G = \mathbb{Q}$ does not satisfy [A_1].

Finally, we mention the following two additional results about assembly maps for topological cyclic homology, which we proved in [74, Theorems 1.1, 1.4(ii), and 1.5].

One should view Theorem 49 as a cyclic induction theorem for the topological cyclic homology of any finite group, with coefficients in any connective ring spectrum. It allows to reduce the computation of **TC** of any finite group to the case of the finite cyclic subgroups; this is carried out explicitly in [74, Proposition 1.2] for the basic case of the symmetric group on three elements.

Theorem 50 studies the analog for **TC** of the Farrell–Jones Conjecture 12. For a large class of groups (for which Conjecture 12 is already known; see Theorem 21), we prove that $\mathrm{asbl}_{\mathcal{V}\mathcal{C}yc}$ is injective, but surprisingly not surjective.

Theorem 49. *For any finite group G the assembly map*

$$\mathrm{asbl}_{\mathcal{C}yc}\colon EG(\mathcal{C}yc)_+ \underset{\mathrm{Or}G}{\wedge} \mathbf{TC}(\mathbb{A}[G\int -];p) \longrightarrow \mathbf{TC}(\mathbb{A}[G];p)$$

is a π_*-isomorphism.

Theorem 50. *Assume that G is either hyperbolic or virtually finitely generated abelian. Then the assembly map*

$$\mathrm{asbl}_{\mathcal{V}\mathcal{C}yc}\colon EG(\mathcal{V}\mathcal{C}yc)_+ \underset{\mathrm{Or}G}{\wedge} \mathbf{TC}(\mathbb{A}[G\int -];p) \longrightarrow \mathbf{TC}(\mathbb{A}[G];p)$$

is always injective but in general not surjective on homotopy groups. For example, it is not surjective on π_{-1} if $\mathbb{A} = \mathbb{Z}_{(p)}$ and G is either finitely generated free abelian or torsion-free hyperbolic, but not cyclic.

Acknowledgments: The Collaborative Research Center 647 *Space–Time–Matter* in Berlin and Potsdam, which involved many scientists from Freie Universität and Humboldt Universität, provided a fruitful research environment to the first author, supported mutual visits of the two authors, and substantially enabled progress in the area of mathematics surveyed here. Among the results in this chapter, the following ones are a direct output of the research within subprojects A12 and C1 of the CRC 647: Theorem 20, Theorem 21(1)(d), Theorem 21(1)(f), Theorem 22(i), Theorem 40, Theorem 44, Theorem 46, Theorem 47, Theorem 49, and Theorem 50.

The second author was also supported by a grant from the Simons Foundation (#419561, Marco Varisco).

Bibliography

[1] Aravinda CS, Farrell FT, Roushon SK. Algebraic K-theory of pure braid groups. Asian J Math 2000;4(2):337–43. DOI:10.4310/AJM.2000.v4.n2.a4. MR 1797585.

[2] Bartels A. On the domain of the assembly map in algebraic K-theory. Algebr Geom Topol 2003;3:1037–50. DOI:10.2140/agt.2003.3.1037. MR 2012963.

[3] Bartels A. Squeezing and higher algebraic K-theory. K-Theory 2003;28(1):19–37. DOI:10.1023/A:1024166521174. MR 1988817.

[4] Bartels A. On proofs of the Farrell–Jones Conjecture. In: Topology and geometric group theory, volume 184 of Springer Proc Math Stat. Cham, Springer: 2016:1–31. DOI:10.1007/978-3-319-43674-6_1. MR 3598160.

[5] Bartels A. Coarse flow spaces for relatively hyperbolic groups. Compos Math 2017;153(4):745–79. DOI:10.1112/S0010437X16008216. MR 3631229.

[6] Bartels A, Bestvina M. The Farrell–Jones Conjecture for mapping class groups. Preprint. arXiv:1606.02844, 2016.

[7] Bartels A, Echterhoff S, Lück W. Inheritance of Isomorphism Conjectures under colimits. In: K-theory and noncommutative geometry, EMS Ser. Congr Rep. Eur Math Soc, Zürich, 2008:41–70. DOI:10.4171/060-1/2. MR 2513332.

[8] Bartels A, Farrell FT, Jones LE, Reich H. On the Isomorphism Conjecture in algebraic K-theory. Topology 2004;43(1):157–213. DOI:10.1016/S0040-9383(03)00032-6. MR 2030590.

[9] Bartels A, Farrell FT, Lück W. The Farrell–Jones Conjecture for cocompact lattices in virtually connected Lie groups. J Am Math Soc 2014:27(2):339–88. DOI:10.1090/S0894-0347-2014-00782-7. MR 3164984.

[10] Bartels A, Lück W. The Borel Conjecture for hyperbolic and CAT(0)-groups. Ann Math (2) 2012;175(2):631–89. DOI:10.4007/annals.2012.175.2.5. MR 2993750.

[11] Bartels A, Lück W. The Farrell–Hsiang method revisited. Math Ann 2012;354(1):209–26. DOI:10.1007/s00208-011-0727-3. MR 2957625.

[12] Bartels A, Lück W, Reich H. The K-theoretic Farrell–Jones Conjecture for hyperbolic groups. Invent Math 2008;172(1):29–70. DOI:10.1007/s00222-007-0093-7. MR 2385666.

[13] Bartels A, Lück W, Reich H, Rüping H. K- and L-theory of group rings over $GL_n(\mathbb{Z})$. Publ Math Inst Hautes Études Sci 2014;119:97–125. DOI:10.1007/s10240-013-0055-0. MR 3210177.

[14] Bartels A, Reich H. Coefficients for the Farrell–Jones Conjecture. Adv Math 2007;209(1):337–62. DOI:10.1016/j.aim.2006.05.005. MR 2294225.

[15] Bartels A, Rosenthal D. On the K-theory of groups with finite asymptotic dimension. J Reine Angew Math 2007;612:35–57. DOI:10.1515/CRELLE.2007.083. MR 2364073.

[16] Bartels A, Rosenthal D. Erratum to "On the K-theory of groups with finite asymptotic dimension". J Reine Angew Math 2017;726:291–92. DOI:10.1515/crelle-2016-0040. MR 3641660.

[17] Bass H. Projective modules over free groups are free. J Algebra 1964;1:367–73. DOI:10.1016/0021-8693(64)90016-X. MR 0178032.

[18] Bass H. Algebraic K-theory. New York–Amsterdam: WA Benjamin, Inc., 1968. MR 0249491.

[19] Bass H, Heller A, Swan RG. The Whitehead group of a polynomial extension. Publ Math Inst Hautes Études Sci 1964;22:61–79. MR 0174605.

[20] Baum P, Connes A, Higson N. Classifying space for proper actions and K-theory of group C^*-algebras. In: C^*-algebras: 1943–1993 (San Antonio, TX, 1993), volume 167 of Contemp Math. Am Math Soc. Providence, RI, 1994:240–91. DOI:10.1090/conm/167/1292018. MR 1292018.

[21] Bökstedt M. Topological Hochschild homology. Preprint, 1986.

[22] Bökstedt M, Hsiang WC, Madsen I. The cyclotomic trace and algebraic K-theory of spaces. Invent Math 1993;111(3):465–539. DOI:10.1007/BF01231296. MR 1202133.

[23] Carlsson G, Goldfarb B. The integral K-theoretic Novikov Conjecture for groups with finite asymptotic dimension. Invent Math 2004;157(2):405–18. DOI:10.1007/s00222-003-0356-x. MR 2076928.

[24] Carlsson G, Goldfarb B. Addendum: "The integral K-theoretic Novikov Conjecture for groups with finite asymptotic dimension". Invent Math 2005;159(3):677. DOI:10.1007/s00222-004-0401-4. MR 2125739.

[25] Carlsson G, Pedersen EK. Controlled algebra and the Novikov Conjectures for K- and L-theory. Topology 1995;34(3):731–58. DOI:10.1016/0040-9383(94)00033-H. MR 1341817.

[26] Cortiñas G. Algebraic v. topological K-theory: a friendly match. In: Topics in algebraic and topological K-theory, volume 2008 of Lecture Notes in Math. Berlin: Springer, 2011: 103–65. DOI:10.1007/978-3-642-15708-0_3. MR 2762555.

[27] Cuntz J, Meyer R, Rosenberg, JM. Topological and bivariant K-theory, volume 36 of Oberwolfach Seminars. Basel: Birkhäuser, 2007. DOI:10.1007/978-3-7643-8399-2. MR 2340673.

[28] Davis JF, Lück W. Spaces over a category and assembly maps in Isomorphism Conjectures in K- and L-theory. K-Theory 1998;15(3):201–52. DOI:10.1023/A:1007784106877. MR 1659969.

[29] Davis MW. The geometry and topology of Coxeter groups, volume 32 of London Math Soc Monographs Ser. Princeton, NJ: Princeton University Press, 2008. MR 2360474.

[30] Drinfeld V. On the notion of geometric realization. Mosc Math J 2004;4(3):619–26, 782. MR 2119142.

[31] Dundas BI, Goodwillie TG, McCarthy R. The local structure of algebraic K-theory, volume 18 of Algebra and Applications. London: Springer, 2013. DOI:10.1007/978-1-4471-4393-2. MR 3013261.

[32] Dunwoody MJ. Relation modules. Bull London Math Soc 1972;4:151–5. DOI:10.1112/blms/4.2.151. MR 0327915.

[33] Enkelmann N, Lück W, Pieper M, Ullmann M, Winges C. On the Farrell–Jones Conjecture for Waldhausen's A-theory. Preprint. arXiv:1607.06395, 2016.

[34] Farrell FT, Hsiang, WC. The topological-Euclidean space form problem. Invent Math 1978;45(2):181–92. DOI:10.1007/BF01390272. MR 0482771.

[35] Farrell FT, Hsiang WC. On Novikov's Conjecture for nonpositively curved manifolds. I. Ann. Math. (2) 1981;113(1):199–209. DOI:10.2307/1971138. MR 604047.

[36] Farrell FT, Hsiang WC. The Whitehead group of poly-(finite or cyclic) groups. J London Math Soc (2) 1981;24(2):308–24. DOI:10.1112/jlms/s2-24.2.308. MR 631942.

[37] Farrell FT, Hsiang WC. Topological characterization of flat and almost flat Riemannian manifolds M^n ($n \neq 3, 4$). Am J Math 1983;105(3):641–72. DOI:10.2307/2374318. MR 704219.

[38] Farrell FT, Jones LE. K-theory and dynamics. I. Ann Math (2) 1986;124(3):531–69. DOI:10.2307/2007092. MR 866708.

[39] Farrell FT, Jones LE. A topological analogue of Mostow's rigidity theorem. J Am Math Soc 1989;2(2):257–370. DOI:10.2307/1990978. MR 973309.

[40] Farrell FT, Jones LE. Isomorphism Conjectures in algebraic K-theory. J Am Math Soc 1993;6(2):249–97. DOI:10.2307/2152801. MR 1179537.

[41] Farrell FT, Jones LE. Topological rigidity for compact nonpositively curved manifolds. In: Differential geometry: Riemannian geometry (Los Angeles, CA, 1990), volume 54 of Proc Sympos Pure Math. Am Math Soc, Providence, RI, 1993: 229–74. MR 1216623.

[42] Farrell FT, Jones LE. The lower algebraic K-theory of virtually infinite cyclic groups. K-Theory 1995;9(1):13–30. DOI:10.1007/BF00965457. MR 1340838.

[43] Farrell FT, Wu X. The Farrell–Jones Conjecture for the solvable Baumslag-Solitar groups. Math Ann 2014;359(3–4):839–62. DOI:10.1007/s00208-014-1021-y. MR 3231018.

[44] Ferry SC. The homeomorphism group of a compact Hilbert cube manifold is an ANR. Ann Math (2) 1977;106(1):101–19. DOI:10.2307/1971161. MR 0461536.

[45] Ferry SC, Ranicki A, Rosenberg J, editors. Novikov Conjectures, index theorems and rigidity, vols. 1 and 2, volumes 226 and 227 of London Math Soc Lecture Note Ser. Cambridge: Cambridge University Press, 1995. DOI:10.1017/CBO9780511662676, DOI:10.1017/CBO9780511629365. MR 1388294, MR 1388306.

[46] Ferry SC, Weinberger S. Curvature, tangentiality, and controlled topology. Invent Math 1991;105(2):401–14. DOI:10.1007/BF01232272. MR 1115548.

[47] Gandini G, Meinert S, Rüping R. The Farrell–Jones Conjecture for fundamental groups of graphs of abelian groups. Groups Geom Dyn 2015;9(3):783–92. DOI:10.4171/GGD/327. MR 3420543.

[48] Gandini G, Rüping H. The Farrell–Jones Conjecture for graph products. Algebr Geom Topol 2013;13(6):3651–60. DOI:10.2140/agt.2013.13. 3651. MR 3248744.

[49] Geoghegan R, Varisco M. On Thompson's group T and algebraic K-theory. In: Geometric and Cohomological Group Theory, volume 444 of London Math Soc Lecture Note Ser. Cambridge: Cambridge Univ Press, 2017:34–45. DOI:10.1017/9781316771327.005.

[50] Grunewald, J. The behavior of Nil-groups under localization and the relative assembly map. Topology 2008;47(3):160–202. DOI:10.1016/j.top.2007.03.007. MR 2414359.

[51] Gubeladze J. The Anderson Conjecture and a maximal class of monoids over which projective modules are free. Mat Sb (N.S.) 1988;135(177)(2):169–85, 271. MR 937805.

[52] Hambleton I, Pedersen EK. Identifying assembly maps in K- and L-theory. Math Ann 2004;328(1–2):27–57. DOI:10.1007/s00208-003-0454-5. MR 2030369.

[53] Hesselholt L. K-theory of truncated polynomial algebras. In: Handbook of K-theory, vol. 1. Berlin: Springer, 2005:71–110. DOI:10.1007/3-540-27855-9_3. MR 2181821.

[54] Hesselholt L, Madsen I. On the K-theory of finite algebras over Witt vectors of perfect fields. Topology 1997;36(1):29–101. DOI:10.1016/0040-9383(96)00003-1. MR 1410465.

[55] Higman G. The units of group-rings. Proc London Math Soc (2) 1940;46:231–48. DOI:10.1112/plms/s2-46.1.231. MR 0002137.

[56] Hsiang WC. Geometric applications of algebraic K-theory. In: Proc Int Congr Math, vol. 1, 2 (Warsaw, 1983). Warsaw: PWN, 1984:99–118. MR 804679.

[57] Jones JDS. Cyclic homology and equivariant homology. Invent Math 1987;87(2):403–23. DOI:10.1007/BF01389424. MR 870737.

[58] Kammeyer H, Lück W, Rüping H. The Farrell–Jones Conjecture for arbitrary lattices in virtually connected Lie groups. Geom Topol 2016;20(3):1275–87. DOI:10.2140/gt.2016.20.1275. MR 3523058.

[59] Kaplansky I. Problems in the theory of rings. Report of a conference on linear algebras. Washington: National Academy of Sciences-National Research Council, June 1956:1–3. Publ. 502, 1957. MR 0096696.

[60] Kaplansky I. "Problems in the theory of rings" revisited. Am Math Monthly 1970:77:445–54. DOI:10.2307/2317376. MR 0258865.

[61] Kasprowski D. On the K-theory of groups with finite decomposition complexity. Proc London Math Soc (3) 2015;110(3):565–92. DOI:10.1112/plms/pdu062. MR 3342098.

[62] Kasprowski D, Nicas A, Rosenthal D. Regular finite decomposition complexity. J Topol Anal 2018. DOI:10.1142/S1793525319500286.

[63] Kreck M, Lück W. The Novikov Conjecture. Geometry and algebra, volume 33 of Oberwolfach Seminars. Basel: Birkhäuser, 2005. DOI:10.1007/b137100. MR 2117411.

[64] Kropholler PH, Moselle B. A family of crystallographic groups with 2-torsion in K_0 of the rational group algebra. Proc Edinburgh Math Soc (2) 1991;34(2):325–31. DOI:10.1017/S0013091500007215. MR 1115651.

[65] Lam T-Y. Lectures on modules and rings, volume 189 of Graduate Texts in Mathematics. New York: Springer, 1999. DOI:10.1007/978-1-4612-0525-8. MR 1653294.

[66] Lam T-Y. Serre's problem on projective modules. Springer Monographs in Mathematics. Berlin: Springer, 2006. DOI:10.1007/978-3-540-34575-6. MR 2235330.

[67] Land M, Nikolaus T. On the relation between K- and L-theory of C^*-algebras. Math Ann 2017. DOI:10.1007/s00208-017-1617-0.

[68] Loday J-L. K-théorie algébrique et représentations de groupes. Ann Sci École Norm Sup (4) 1976;9(3):309–77. DOI:10.24033/asens.1312. MR 0447373.

[69] Lück W. Isomorphism Conjectures in K- and L-theory. In preparation, preliminary version. Available at: him.uni-bonn.de/lueck/.

[70] Lück W. Chern characters for proper equivariant homology theories and applications to K- and L-theory. J Reine Angew Math 2002;543:193–234. DOI:10.1515/crll.2002.015. MR 1887884.

[71] Lück W. K- and L-theory of group rings. In: Proc Int Cong Math, vol II. New Delhi: Hindustan Book Agency, 2010:1071–98. MR 2827832.

[72] Lück W, Reich H. The Baum-Connes and the Farrell–Jones Conjectures in K- and L-theory. In: Handbook of K-theory, vol. 2. Berlin: Springer, 2005:703–842. DOI:10.1007/978-3-540-27855-9_15. MR 2181833.

[73] Lück W, Reich H, Rognes J, Varisco M. Algebraic K-theory of group rings and the cyclotomic trace map. Adv Math 2017;304:930–1020. DOI:10.1016/j.aim.2016.09.004. MR 3558224.

[74] Lück W, Reich H, Rognes J, Varisco M. Assembly maps for topological cyclic homology of group algebras. J Reine Angew Math 2017. DOI:10.1515/crelle-2017-0023.

[75] Lück W, Rosenthal D. On the K- and L-theory of hyperbolic and virtually finitely generated abelian groups. Forum Math 2014;26(5):1565–609. DOI:10.1515/forum-2011-0146. MR 3334038.

[76] Lück W, Steimle W. Splitting the relative assembly map, Nil-terms and involutions. Ann K-Theory 2016;1(4):339–77. DOI:10.2140/akt.2016.1.339. MR 3536432.

[77] MacLane S. Categories for the working mathematician, volume 5 of Graduate Texts in Mathematics. New York: Springer, 1971. DOI:10.1007/978-1-4612-9839-7. MR 0354798.

[78] Madsen I. Algebraic K-theory and traces. In: Current developments in mathematics, 1995 (Cambridge, MA). Cambridge, MA: Int. Press, 1994:191–321. MR 1474979.

[79] Milnor J. Lectures on the h-cobordism theorem. Princeton, NJ: Princeton University Press, 1965. MR 0190942.

[80] Milnor J. Introduction to algebraic K-theory. Annals of Mathematics Studies, No. 72. Princeton, NJ: Princeton University Press, 1971. MR 0349811.

[81] Mislin G, Valette A. Proper group actions and the Baum-Connes Conjecture. Advanced Courses in Mathematics. CRM Barcelona. Basel: Birkhäuser, 2003. DOI:10.1007/978-3-0348-8089-3. MR 2027168.

[82] Montgomery MS. Left and right inverses in group algebras. Bull Am Math Soc 1969;75:539–40. DOI:10.1090/S0002-9904-1969-12234-2. MR 0238967.

[83] Patronas D. The Artin defect in algebraic K-theory. PhD thesis, Freie Universität Berlin, 2014. Available at: diss.fu-berlin.de/diss/receive/FUDISS_thesis_ 000000097443.

[84] Patronas D, Reich H, Varisco M. In preparation.

[85] Quinn F. Ends of maps. I. Ann Math (2) 1979;110(2):275–331. DOI:10.2307/1971262. MR 549490.

[86] Quinn F. Ends of maps. II. Invent Math 1982;68(3):353–424. DOI:10.1007/BF01389410. MR 669423.

[87] Quinn F. Algebraic K-theory over virtually abelian groups. J Pure Appl Algebra 2012;216(1):170–83. DOI:10.1016/j.jpaa.2011.06.001. MR 2826431.

[88] Ramras DA, Tessera R, Yu G. Finite decomposition complexity and the integral Novikov Conjecture for higher algebraic K-theory. J Reine Angew Math 2014;694:129–78. DOI:10.1515/crelle-2012-0112. MR 3259041.

[89] Reich H, Varisco M. On the Adams isomorphism for equivariant orthogonal spectra. Algebr Geom Topol 2016;16(3):1493–566. DOI:10.2140/agt.2016.16.1493. MR 3523048.

[90] Rosenberg J. Algebraic K-theory and its applications, volume 147 of Graduate Texts in Mathematics. New York: Springer, 1994. DOI:10.1007/978-1-4612-4314-4. MR 1282290.

[91] Rosenberg J. Analytic Novikov for topologists. In: Novikov Conjectures, index theorems and rigidity, vol. 1 (Oberwolfach, 1993), volumes 226 of London Math Soc Lecture Note Ser. Cambridge: Cambridge University Press, 1995:338–72. DOI:10.1017/CBO9780511662676.013. MR 1388305.

[92] Roushon SK. Algebraic K-theory of groups wreath product with finite groups. Topol Appl 2007;154(9):1921–30. DOI:10.1016/j.topol.2007.01.019. MR 2319263.

[93] Roushon SK. The Farrell–Jones Isomorphism Conjecture for 3-manifold groups. J K-Theory 2008;1(1):49–82. DOI:10.1017/is007011012jkt005. MR 2424566.

[94] Rüping H. The Farrell–Jones Conjecture for S-arithmetic groups. J Topol 2016;9(1):51–90. DOI:10.1112/jtopol/jtv034. MR 3465840.

[95] Schneider P. Über gewisse Galoiscohomologiegruppen. Math Z 1979;168(2):181–205. DOI:10.1007/BF01214195. MR 544704.

[96] Serre J-P. Faisceaux algébriques cohérents. Ann Math (2) 1955;61:197–278. DOI:10.2307/1969915. MR 0068874.

[97] Serre J-P. Modules projectifs et espaces fibrés à fibre vectorielle. In: Séminaire P. Dubreil, M.-L. Dubreil-Jacotin et C. Pisot, 1957/58, Fasc. 2, Exposé 23. Secrétariat mathématique, Paris, 1958:18. MR 0177011.

[98] Serre J-P. Représentations linéaires des groupes finis. Paris: Hermann, revised edition, 1978. MR 543841.

[99] Stallings J. Whitehead torsion of free products. Ann Math (2) 1965;82:354–63. DOI:10.2307/1970647. MR 0179270.

[100] Swan RG. Induced representations and projective modules. Ann Math (2) 1960;71:552–78. DOI:10.2307/1969944. MR 0138688.

[101] Swan RG. Gubeladze's proof of Anderson's conjecture. In: Azumaya algebras, actions, and modules (Bloomington, IN, 1990), volume 124 of Contemp Math. Am Math Soc. Providence, RI, 1992:215–50. DOI:10.1090/conm/124/1144038. MR 1144038.

[102] Swan RG. Higher algebraic K-theory. In: K-theory and algebraic geometry: connections with quadratic forms and division algebras (Santa Barbara, CA, 1992), volume 58 of Proc Sympos Pure Math. Am Math Soc. Providence, RI, 1995:247–93. MR 1327284.

[103] Ullmann M, Wu X. Note on the injectivity of the Loday assembly map. J Algebra 2017;489:460–462. DOI:10.1016/j.jalgebra.2017.06.037. MR 3686987.

[104] Waldhausen F. Algebraic K-theory of generalized free products. III, IV. AnnMath (2) 1978;108(2):205–56. DOI:10.2307/1971166. MR 0498808.

[105] Waldhausen F. Algebraic K-theory of topological spaces. I. In: Algebraic and geometric topology (Proc Sympos Pure Math, Stanford, CA: Stanford University, 1976), Part 1, Proc Sympos Pure Math, XXXII. Am Math Soc. Providence, RI, 1978:35–60. MR 520492.

[106] Wegner C. The K-theoretic Farrell–Jones Conjecture for CAT(0)-groups. Proc Am Math Soc 2012;140(3):779–93. DOI:10.1090/S0002-9939-2011-11150-X. MR 2869063.

[107] Wegner C. The Farrell–Jones Conjecture for virtually solvable groups. J Topol 2015;8(4):975–1016. DOI:10.1112/jtopol/jtv026. MR 3431666.

[108] Weibel CA. Algebraic K-theory of rings of integers in local and global fields. In: Handbook of K-theory, vol. 1. Berlin: Springer, 2005:139–90. DOI:10.1007/3-540-27855-9_5. MR 2181823.

[109] Weibel CA. The K-book. An introduction to algebraic K-theory, volume 145 of Graduate Studies in Mathematics. Am Math Soc. Providence, RI, 2013. MR 3076731.

[110] Weiss M, Williams B. Assembly. In: Novikov Conjectures, index theorems and rigidity, vol. 2 (Oberwolfach, 1993), volume 227 of London Math Soc Lecture Note Ser. Cambridge: Cambridge University Press, 1995:332–52. DOI:10.1017/CBO9780511629365.014. MR 1388318. June 29, 2017.

Helga Baum

Lorentzian manifolds with special holonomy – Constructions and global properties

Abstract: We report on several new results concerning global properties of Lorentzian manifolds with special holonomy. The first states that compact Lorentzian manifolds with abelian holonomy are geodesically complete. The second describes a Bochner-type estimate for the first Betti number of Lorentzian manifolds with a parallel light-like vector field V and nonnegative Ricci tensor along $V^{\perp}$. In the third part, we discuss Cauchy problems that allow to construct Lorentzian manifolds with special holonomy using appropriate evolution equations for initial data on Riemannian manifolds.

Keywords: Lorentzian manifolds, holonomy groups, pp-waves, completeness, globally hyperbolic manifolds, Cauchy problem, parallel light-like vector fields, parallel spinor fields, generalized Killing spinors

Mathematics Subject Classification 2010: 53C50, 53C27, 53C29, 53C12, 53C22, 53C44, 53A10, 83C05

1 Introduction

Besides the isotropy representations of irreducible Riemannian symmetric spaces, the well-known *Berger list* describes all connected *irreducible* proper subgroups of O(n), which can appear as holonomy group of an n-dimensional connected Riemannian manifold, namely $U(n/2)$, $SU(n/2)$, $Sp(n/4)$, $Sp(n/4) \cdot Sp(1)$, G_2 and $Spin(7)$. Any of these special holonomy groups is related to a rich, interesting and widely studied Riemannian geometry, described by the corresponding parallel geometric object. Contrary to this situation, if $H \subset O(1, n+1)$ is a connected Lie subgroup of the Lorentz group $O(1, n+1)$, acting *irreducibly* on the Minkowski space $\mathbb{R}^{1,n+1}$, then $H = SO^0(1, n+1)$ [10].

Let us shortly describe the structure of connected Lorentzian holonomy groups. All Lorentzian manifolds we consider in this chapter are connected and of dimension $n+2 > 2$. By $\mathrm{Hol}_x(M, g) \subset O(T_xM, g_x)$ we denote the holonomy group of (M, g) with respect to $x \in M$ and by $\mathrm{Hol}_x^0(M, g)$ its connected component. If there is a nondegenerate $\mathrm{Hol}_x^0(M, g)$-invariant subspace $E \subset T_xM$, then (M, g) is locally isometric to a product of a Riemannian and a Lorentzian manifold. Moreover, $\mathrm{Hol}_x^0(M, g)$ decomposes into a product of two orthogonal groups. Hence, for the classification of the connected Lorentzian holonomy groups one can suppose that $\mathrm{Hol}_x^0(M, g)$ acts indecomposable, i.e., that there is no proper nondegenerate $\mathrm{Hol}_x^0(M, g)$-invariant subspace in T_xM.

DOI 10.1515/9783110452150-002_s_001

Then there are two cases: $\mathrm{Hol}^0_x(M,g)$ acts irreducible, hence $\mathrm{Hol}^0_x(M,g) = SO^0(1,n+1)$, or there is a *degenerate* $\mathrm{Hol}^0_x(M,g)$-invariant subspace $W \subset T_xM$, which defines an invariant light-like line $L := W \cap W^\perp \subset T_xM$ and the connected holonomy group is contained in the stabilizer of this line L

$$\mathrm{Hol}^0(M,g) \subset \mathrm{Stab}_{SO^0(1,n+1)}(L) = (\mathbb{R}^+ \times SO(n)) \ltimes \mathbb{R}^n.$$

In the latter case we say that (M,g) has a special holonomy. Such manifolds admit a filtration

$$\mathbb{V} \subset \mathbb{V}^\perp \subset TM, \tag{1}$$

where $\mathbb{V}$ is the parallel light-like line bundle arising by parallel displacement of L. Its orthogonal complement $\mathbb{V}^\perp$ is light-like and parallel as well and the rank n vector bundle $\mathbb{E} := \mathbb{V}^\perp/\mathbb{V}$ on M, called *screen bundle*, admits a positive definite bundle metric $g^{\mathbb{E}}$ induced by g and a connection $\nabla^{\mathbb{E}}$ induced by the Levi–Civita connection ∇^g. It is an old result of Walker [28], that a Lorentzian manifold with a parallel light-like line bundle $\mathbb{V} \subset TM$ admits local coordinates $(U,(u,v,x^1,\dots,x^n))$ such that the metric is given by

$$g|_U = 2dvdu + 2A_i dx^i du + Hdu^2 + h_{ij}dx^i dx^j, \tag{2}$$

where H, A_i, h_{ij} are smooth functions on U, A_i and h_{ij} not depending on v.

The connected holonomy groups of four-dimensional space-times were classified by physicists working in general relativity [25, 27]. The first step of the classification in arbitrary dimension was done by L. Berard-Bergery and A. Ikemakhen, who classified all connected subgroups $H \subset O(1,n+1)$, which act indecomposable but nonirreducible [5]. T. Leistner completed the classification of connected Lorentzian holonomy groups by proving that the orthogonal part $G^0 := \mathrm{proj}_{SO(n)}(Hol^0(M,g)) \subset SO(n)$ is the holonomy representation of a Riemannian manifold [18]. Finally, A. Galaev proved that *all* groups in the list of Berard-Bergery/Ikemakhen and Leistner can be realized by local metrics [13]. The final classification results are the following:

Theorem 1 ([18]). *Let $(M^{1,n+1},g)$ be a connected locally indecomposable Lorentzian manifold of dim $n+2>2$ and let $H := Hol(M,g)$ be its holonomy group.*
Then either H^0 is the full Lorentzian group $SO^0(1,n+1)$ or H is a subgroup of the stabilizer of a light-like line L, i.e.,

$$H \subset Stab_{O(1,n+1)}(L) = (\mathbb{R}^* \times O(n)) \ltimes \mathbb{R}^n.$$

In the second case, let us denote by G^0 the projection of H^0 onto the orthogonal part, i.e., $G^0 := pr_{SO(n)}(H^0)$, by $Z(G^0)$ the center of G^0 and by S its semi-simple part. Then G^0

is a Riemannian holonomy group and H^0 *equals to one of the following subgroups of* $(\mathbb{R}^+ \times SO(n)) \ltimes \mathbb{R}^n$*:*
1. $G^0 \ltimes \mathbb{R}^n$.
2. $(\mathbb{R}^+ \times G^0) \ltimes \mathbb{R}^n$.
3. $(A \times S) \ltimes \mathbb{R}^n$, *where* $A = graph(\Psi)$ *for* $\Psi \in Hom(Z(G^0), \mathbb{R}^+)$.
4. $(B \times S) \ltimes \mathbb{R}^{n-k}$, *where* $B = graph(\Phi)$ *for* $\Phi \in Hom(Z(G^0), \mathbb{R}^k)$.

Some properties of the full holonomy group $\mathrm{Hol}(M, g)$ can be found in [3].

After the classification of the connected holonomy groups is completed, one is interested in local and global properties of Lorentzian manifold with special holonomy as well as in the construction of such manifolds. The state of art in this field until 2011 is described [2]. We present here several new results that were obtained during the third funding period of the CRC 647 in cooperation with T. Leistner. All manifolds we discuss below are so-called *Brinkmann spaces*, i.e., Lorentzian manifolds with a parallel light-like vector field, which we will denote in the following always by V. Then filtration (1) is given by $\mathbb{V} := \mathbb{R}V$. In particular, Brinkman spaces admit a time orientation and (in the locally indecomposable case) the connected holonomy group $\mathrm{Hol}^0(M, g)$ equals to one of the groups $G^0 \ltimes \mathbb{R}^n$ or $(B \times S) \ltimes \mathbb{R}^{n-k}$ in Theorem 1. Moreover, we can always choose a *screen distribution* $\mathbb{S}$ on M, which is a rank n subbundle of $\mathbb{V}^\perp$ on which the metric g is nondegenerate, or equivalently, a *screen vector field* Z on M, which is a light-like vector field such that $g(V, Z) = 1$, related to each other by $\mathbb{S} = V^\perp \cap Z^\perp$ [3, Prop. 2]. The choice of a screen distribution $\mathbb{S}$ allows to change the Lorentzian metric g to a Riemannian metric $h = h^{\mathbb{S}}$ via

$$h(V, \cdot) := g(Z, \cdot), \quad h(Z, \cdot) := g(V, \cdot), \quad h(X, \cdot) := g(X, \cdot) \ \text{for } X \in \mathbb{S} \tag{3}$$

and extension by linearity. This construction allows to apply methods of Riemannian geometry in order to derive properties of Lorentzian manifolds with special holonomy.

The first result which we will explain in this report concerns a classical question in global Lorentzian geometry, namely under which conditions a compact Lorentzian manifold is geodesically complete – a property, which always holds for compact Riemannian manifolds. In Lorentzian signature, only under strong additional conditions compactness implies completeness. For example, a compact Lorentzian manifold is complete if it is flat [9], if it has constant curvature [16], if it admits a time-like conformal Killing field [24] or if it is homogeneous [21]. Moreover, any compact, locally homogeneous three-dimensional Lorentzian manifold is complete [11]. In a joint paper, D. Schliebner and T. Leistner studied this question for Lorentzian manifolds with abelian holonomy, the so-called pp-waves (cf [19]). In general, these manifolds do not satisfy any of the above conditions. The authors prove that any compact Lorentzian manifold with abelian holonomy is geodesically complete. As application follows that any compact, indecomposable, locally symmetric Lorentzian manifold is geodesically complete.

The second result concerns a Bochner-type theorem for Brinkman spaces which was obtained by K. Lärz and D. Schliebner [17, 26]. The subbundle $\mathbb{V}^\perp \subset TM$ induces a light-like codimension one foliation on M. Assuming compactness of the leaves and nonnegative Ricci curvature on the leaves it can be shown that the first Betti number is bounded by the dimension of the manifold or the leaves if the manifold is compact or noncompact, respectively. Moreover, in the case of the maximality of the first Betti number every such Lorentzian manifold is (up to finite cover) diffeomorphic to the torus (in the compact case) or to the product of the real line with a torus (in the noncompact case) and has very degenerate curvature, i.e., the curvature tensor induced on the leaves is light-like.

In the third part, we will explain joint results of H. Baum, A. Lischewski and T. Leistner concerning the Cauchy problem for Lorentzian manifolds with special holonomy. Most of the Lorentzian metrics with special holonomy are given in a form which is adapted to the canonical filtration $\mathbb{V} \subset \mathbb{V}^\perp \subset TM$ (see [2]). Sánchez and Flores proved in 2005 that any globally hyperbolic Lorentzian manifold (M, g) is isometric to $(\mathbb{R} \times \Sigma, -\lambda^2 dt^2 + g_t)$, where $\{t\} \times \Sigma$ is a smooth space-like Cauchy surface for any $t \in \mathbb{R}$ and λ is a smooth positive function on M. Our aim is to describe special holonomy metrics in this canonical form by solving a Cauchy problem for appropriate data (constraint conditions) on a space-like initial hypersurface. In [4] we found such a Cauchy problem in a Cauchy–Kowalewki-type form which can be solved in the real-analytic setting. In the smooth category, A. Lischewski [20] was able to state and solve a well-posed Cauchy problem for Lorentzian holonomies which allow parallel spinors by hyperbolic reduction and first-order symmetric hyperbolic partial differential equation (PDE) methods.

2 Completeness of compact Lorentzian manifolds with abelian holonomy

In this section we describe properties of Lorentzian manifolds with the simplest form of the holonomy groups in Theorem 1, i.e., with $\mathrm{Hol}^0(M, g) \subset \mathbb{R}^n$. First, let us note the following equivalent characterizations.

Lemma 2 ([19]). *Let (M, g) be a Lorentzian manifold with a parallel light-like vector field V and curvature tensor R^g. Then the following conditions are equivalent:*

a) *$Hol^0(M, g) \subset \mathbb{R}^n$.*
b) *$R^g(U, W) = 0$ for all $U, W \in \mathbb{V}^\perp$.*
c) *$R^g(X, Y)W \in \mathbb{V}$ for all $X, Y \in TM$, $W \in \mathbb{V}^\perp$.*
d) *The screen bundle $(\mathbb{E}, \nabla^{\mathbb{E}})$ is flat, i.e., the curvature of $\nabla^{\mathbb{E}}$ vanishes.*

Lorentzian manifolds with curvature conditions as in Lemma 2 are well known in physics.

Definition 3. A Lorentzian manifold $(M^{1,n+1}, g)$ is called *pp-wave*, if it admits a parallel light-like vector field V and if its curvature tensor R^g satisfies

$$R^g(U, W) = 0, \qquad \text{for all } U, W \in \mathbb{V}^{\perp}.$$

A pp-wave metric locally depends only on one function. There are local coordinates $(U, (u, v, x^1, \dots, x^n))$, such that

$$g|_U = 2du(dv + Hdu) + \delta_{ij}dx^i dx^j, \tag{4}$$

where $H = H(u, x^1, \dots, x^n)$ is a smooth function on U not depending on v.
If $M = \mathbb{R}^{n+2}$ and g is globally of the form (4), (M, g) is called *standard pp-wave*.

Four-dimensional standard pp-waves were discovered by Brinkmann in the context of conformal geometry [6], and then played an important role in general relativity (e.g., see [12]). More recently, as manifolds with a maximal number of parallel spinors, higher dimensional pp-waves appeared in supergravity theories, e.g. in [14], and there is now a vast physics literature on them. pp-waves appear in geometry among Lorentzian symmetric spaces. There is the following classification theorem for Lorentzian symmetric spaces:

Theorem 4 (Cahen–Wallach, [7]). *Let (M, g) be an 1-connected Lorentzian symmetric space. Then (M, g) is isometric to a product of an 1-connected Riemannian symmetric space and one of the following spaces:*
1. $\mathbb{R}$ *with the metric* $-dt^2$;
2. *the universal cover of an anti-de Sitter space AdS^n or a de Sitter space dS^n, where $n \geq 2$;*
3. *a Cahen–Wallach space $CW_n(\lambda_1, \dots, \lambda_{n-2})$, with $n \geq 3$, where*

$$CW_n(\underline{\lambda}) := \left(\mathbb{R}^n, g_{\underline{\lambda}} := 2dvdu + \sum_{i=1}^{n-2} \lambda_i x_i^2 \, du^2 + \sum_{i=1}^{n-2} dx_i^2 \right),$$

and $\underline{\lambda} := (\lambda_1, \dots, \lambda_{n-2})$, $\lambda_i \in \mathbb{R}$, $\lambda_i \neq 0$.

The first step in proving the completeness of compact pp-waves is the following observation:

Theorem 5 ([19, Theorem 1]). *The universal cover of an $(n + 2)$-dimensional compact pp-wave (M, g) is globally isometric to a standard pp-wave*

$$\left(\mathbb{R}^{n+2}, g^H = 2dudv + 2H(u, x_1, \dots, x_n)du^2 + \sum_{i=1}^{n} dx_i^2 \right).$$

The proof uses a detailed study of screen distributions on the universal cover $\widetilde{M}$ of M. First, let $\mathbb{S}$ be an arbitrary chosen screen distribution on M and $h = h^{\mathbb{S}}$ the Riemannian metric defined in eq. (3). We mark all induced objects on the universal cover by a tilde. Since Z is a complete vector field on M, $\tilde{Z}$ is complete on $\widetilde{M}$ as well, hence $\widetilde{M}$ is diffeomorphic to $\mathbb{R} \times \widetilde{N}$, where $\widetilde{N}$ is a leaf of the integrable distribution $\widetilde{\mathbb{V}}^{\perp}$. Moreover, $\tilde{h}|_{\widetilde{N}}$ is complete. Since the connected holonomy group of the pp-wave (M, g) is abelian by Lemma 2, the screen bundle $(\mathbb{E}, \nabla^{\mathbb{E}})$ is flat. This implies that there are global orthonormal frame fields $S_1, \dots, S_n$ of $\widetilde{\mathbb{S}}$ such that $\nabla^{\tilde{g}} S_i = \alpha^i \otimes \widetilde{V}$ with $d\alpha^i|_{\widetilde{V}^{\perp} \times \widetilde{V}^{\perp}} = 0$. Now, Leistner and Schliebner proved that it is possible to change the screen distribution $\widetilde{\mathbb{S}}$ to a screen distribution $\mathbb{S}'$ on $\widetilde{M}$ which is spanned by orthonormal vector fields $S_1', \dots, S_n'$ such that $\nabla^{\tilde{g}}_U S_i' = 0$ for all $U \in \widetilde{\mathbb{V}}^{\perp}$. Let h' be the metric on $\widetilde{N}$ induced by $h^{\mathbb{S}'}$. Then $(\widetilde{V}, S_1', \dots, S_n')$ yield an orthonormal basis of complete and $\nabla^{h'}$-parallel vector fields on $(\widetilde{N}, h')$. Hence, by the result of Palais, $\widetilde{N}$ is diffeomorphic to $\mathbb{R}^{n+1}$. Using these constructions, it can be shown that the smooth map

$$\Phi : \mathbb{R}^{n+2} \ni (u, v, x_1, \dots, x_n) \mapsto \exp^{\tilde{g}}_{\gamma(u)} \Big(v\widetilde{V}(\gamma(u)) + \sum_{i=1}^{n} x_i S_i(\gamma(u)) \Big) \in \widetilde{M},$$

where γ is the integral curve of the complete screen vector field $\tilde{Z}$ through a fixed point $p_0 \in \widetilde{M}$, is a diffeomorphism with

$$\Phi^* \tilde{g} = 2dudv + 2H(u, x_1, \dots, x_n)du^2 + \sum_{i=1}^{n} x_i^2,$$

where $2H := (\Phi^* \tilde{g})(\partial_u \partial_u)$. Moreover, it can be proved that all second derivatives of H in x_i-directions are bounded,

$$0 \leq \partial_i \partial_j H \leq c \qquad \text{for all } i, j = 1, \dots, n.$$

This condition on the defining function H yields geodesic completeness of the standard pp-wave, cf. [8]. This implies

Theorem 6 ([19, Theorem 2]). *Every compact pp-wave (M, g) is geodesically complete.*

Theorem 6 has an interesting consequence for locally symmetric spaces. Theorem 4 shows that an indecomposable simply connected Lorentzian symmetric space is either a Cahen–Wallach space or has constant sectional curvature. Neither de Sitter spaces nor even-dimensional anti-de Sitter spaces have compact quotients. Compact quotient of Cahen–Wallach spaces can be found in [15]. Then, for compact locally symmetric spaces, i.e., with $\nabla^g R^g = 0$, Theorem 6 and Klingler's result on completeness of compact Lorentzian manifolds of constant sectional curvature imply

Proposition 7 ([19, Corollary 2]). *An indecomposable, compact locally symmetric Lorentzian manifold is geodesically complete. Moreover, it is a quotient by a lattice of either an odd-dimensional simply connected anti-de Sitter space or of a Cahen–Wallach space.*

3 A Bochner-type theorem for Lorentzian manifolds with special holonomy

Motivated by the classical Bochner result for compact oriented Riemannian manifolds (N, h) with nonnegative Ricci curvature saying that the first Betti number is bounded by the dimension, $b_1(N) \leq \dim N$, and equality holds if and only if (N, h) is isometric to the flat torus, we ask for a similar result for Lorentzian manifolds with lower Ricci curvature bound. An answer for Brinkman spaces (M, g) can be given by studying properties of the Riemannian metric $h^{\mathbb{S}}$ which is associated with g and a chosen screen distribution $\mathbb{S}$ on (M, g) by eq. (3). In this Lorentzian context instead of flat spaces in the limit case, a generalization of pp-waves occurs, which we will define now.

Definition 8. A Lorentzian manifold (M, g) with a parallel light-like vector field V has *light-like hypersurface curvature*, iff the curvature tensor R^g satisfies

$$R^g(X, Y)W \in \Gamma(\mathbb{V}) \qquad \text{for all } X, Y, W \in \Gamma(\mathbb{V}^\perp).$$

We now consider Brinkman spaces with nonnegative Ricci tensor along the distribution $\mathbb{V}^\perp$. Then the following analog of the Riemannian Bochner theorem holds.

Theorem 9 ([26, Main Theorem]). *Let (M, g) be an oriented $(n + 2)$-dimensional Lorentzian manifold with a parallel light-like vector field V. Assume that the leaves of the codimension one foliation induced by the distribution $\mathbb{V}^\perp$ are compact and $Ric^g|_{\mathbb{V}^\perp \times \mathbb{V}^\perp} \geq 0$. Then:*
1. *If M is compact, then $b_1(M) \leq n + 2$ and $b_1(M) = n + 2$ iff M is – up to finite cover – diffeomorphic (homeomorphic if $\dim M = 4$) to the torus and g has light-like hypersurface curvature.*
2. *If M is noncompact, then $b_1(M) \leq n+1$ and $b_1(M) = n+1$ iff M is isometric to $\mathbb{R}\times\mathbb{T}^{n+1}$ and g has light-like hypersurface curvature.*

In both cases, if the first Betti number is maximal, the leaves of $\mathbb{V}^\perp$ are all diffeomorphic to the torus $\mathbb{T}^{n+1}$.

The proof makes use of the properties of the Riemannian foliation which arises by the light-like parallel vector field V and the Riemannian metric $h^{\mathbb{S}}$.

Lemma 10 ([17, 26]). *Let (M, g) be a Lorentzian manifold with a parallel light-like vector field V, $\mathbb{S}$ a screen distribution, N a leaf of the distribution $\mathbb{V}^\perp$ and h the metric induced on N by $h^{\mathbb{S}}$. Then:*

a) *$(N, \mathcal{F}, h)$ is a Riemannian flow, where the foliation $\mathcal{F}$ is given by the flow of $V_{|N}$. If M is oriented, then N is orientable and the Riemannian flow $(N, \mathcal{F}, h)$ is transversally orientable.*
b) *If $\mathbb{S}_{|N} \subset TN$ is horizontal, i.e., $[\Gamma(\mathbb{S}_{|N}), \Gamma(\mathbb{V}_{|N})] \subset \Gamma(\mathbb{S}_{|N})$, and integrable, then $\nabla^h V_{|N} = 0$ and $Ric^g|_{TN\times TN} = Ric^h$.*

Then, by studying the basis cohomology of the Riemannian flow $(N, \mathcal{F}, h)$ and an appropriate Weitzenböck formula for the transversal Laplacian, one obtains

Lemma 11 ([26]). *Let N be a compact $(n + 1)$-dimensional leaf of $\mathbb{V}^{\perp}$. Then:*
a) $H^1_{dR}(N) = H^1_B(\mathcal{F}) \oplus H$ *for H a subgroup of* $H^n_B(\mathcal{F}) \in \{0, \mathbb{R}\}$.
b) *If in addition* $Ric^g|_{TN\times TN} \geq 0$, *then* $b_1(N) \leq H^1_B(\mathcal{F}) + 1 \leq \dim N$.
c) $b_1(M) \leq b_1(N) + 1$ *and* $\pi_1(M) \cong \mathbb{Z} \ltimes \pi_1(N)$ *for some homomorphism* $\varphi \in Hom(\mathbb{Z}, Aut(\pi_1(N))$.

These lemmata yield the estimate for $b_1(M)$ in Theorem 9. To derive the properties in the limit case, one shows that under this condition one can change the screen distribution $\mathbb{S}$ to a screen distribution $\mathbb{S}'$ such that $\mathbb{S}'_{|N}$ is integrable and horizontal along the leaf N. Then the Riemannian metric h' induced on N by $\mathbb{S}'_{|N}$ has nonnegative Ricci curvature by Lemma 10, hence (N, h') is the flat torus. Moreover, the connection $\nabla^g_{|N}$ on each leaf N of $\mathbb{V}^{\perp}$ induced by the Levi–Civita connection of g has light-like curvature. This is equivalent for (M, g) to have light-like hypersurface curvature.

4 Cauchy problems for Lorentzian manifolds with special holonomy and parallel spinor field

In this part we consider another typical class of Lorentzian manifolds with special holonomy, namely Lorentzian spin manifolds with parallel spinor fields. Let (M, g) be a Lorentzian spin manifold with a parallel spinor field $\phi \in \Gamma(S)$, i.e., $\phi \neq 0$ and $\nabla^S \phi = 0$, where ∇^S is the covariant derivative on the spinor bundle S induced by the Levi–Civita connection of g. Each parallel spinor field ϕ defines a nonvanishing vector field V_ϕ, its Dirac current, by

$$g(V_\phi, X) = -\langle X \cdot \phi, \phi\rangle \quad \text{for all } X \in \Gamma(TM).$$

V_ϕ is parallel as well and either time-like or light-like. In the first case, (M, g) is decomposable, i.e., locally isometric to a product $(\mathbb{R}\times\Sigma, -dt^2+h)$, where (Σ, h) is a Riemannian spin manifold with parallel spinors. In the latter case, (M, g) is a Brinkmann space, hence a Lorentzian manifold of special holonomy. If V_ϕ is light-like, we call ϕ *light-like spinor field* for short. The holonomy group of a $(n+2)$-dimensional Lorentzian spin manifold with a parallel light-like spinor field is contained in $G \ltimes \mathbb{R}^n \subset SO(n) \ltimes \mathbb{R}^n$, where the connected component G^0 of G is a product of groups of the form $\{1\}$, $SU(k)$, $Sp(m)$, G_2 or $Spin(7)$. The possible nonconnected groups G are described in [3, Theorem 2].

We are interested in globally hyperbolic Lorentzian spin manifolds with a parallel light-like spinor field of the form

$$(M = \mathbb{R} \times \Sigma, \; g = -\lambda^2 dt^2 + g_t), \tag{5}$$

where (Σ, g_t) is a geodesically complete Riemannian manifold and $\{t\} \times \Sigma$ a Cauchy surface of (M, g) for any t. To construct such manifolds we will answer the following two questions:

(A) *Constraint conditions*: What are the *constraint conditions* that are imposed on a space-like hypersurface Σ in (M, g) by the existence of a parallel light-like spinor field?

(B) *Cauchy problem*: Can we extend a given Riemannian manifold satisfying these constraint conditions to a Lorentzian manifold with a metric as in eq. (5) with a parallel light-like spinor field?

Let us note that this approach is used in general relativity to solve the vacuum Einstein field equation $Ric^g = 0$. Let $\Sigma \subset M$ be a space-like hypersurface in (M, g) with the induced Riemannian metric g^Σ and the Weingarten tensor W. It is well known that $Ric^g = 0$ imposes on Σ the constraint equations

$$d\,\mathrm{tr}_{g^\Sigma} W + \delta^\Sigma W = 0, \tag{6}$$

$$\mathrm{scal}^\Sigma - \mathrm{tr}_{g^\Sigma}(W^2) + (\mathrm{tr}_{g^\Sigma} W)^2 = 0. \tag{7}$$

Conversely, given (Σ, g^Σ, W) solving systems (6) and (7), we can find a Ricci-flat and globally hyperbolic Lorentzian metric on a neighborhood of Σ in $\mathbb{R} \times \Sigma$ in which (Σ, g^Σ) embeds with W; see [23] and references therein.

4.1 The constraint condition

First, let us consider the constraint condition for space-like hypersurfaces in Lorentzian manifolds with parallel light-like spinor field. Note that contrary to the Riemannian case, a Lorentzian manifold with a parallel spinor field need not be Ricci flat, but the Ricci tensor is very degenerate.

Lemma 12. *Let (M, g) be a Lorentzian spin manifold with a light-like parallel spinor field ϕ and denote by $V := V_\phi$ its Dirac current. Then:*

1. $V \cdot \phi = 0$.
2. *The scalar curvature of (M, g) vanishes.*

3. *There is a smooth function f on M such that*

$$Ric^g = f \cdot V^\flat \otimes V^\flat \quad \textit{and} \quad V(f) = 0.$$

Let us assume that $\Sigma \subset M$ is a space-like hypersurface with induced Riemannian metric g^Σ, Levi–Civita connection ∇^Σ and future-directed unit normal vector field T along Σ. Furthermore, let $W^\Sigma := -\nabla^g T|_{T\Sigma}$ denote the Weingarten operator of Σ and $(S^\Sigma, \nabla^\Sigma)$ the spinor bundle of (Σ, g^Σ) with respect to the spin structure canonically induced on Σ with its spin derivative. Then there is a canonical identification of S^Σ with $S_{|\Sigma}$ if n is even and of S^Σ with the half-spinors $S^+_{|\Sigma}$ if n is odd. In this identification, the Clifford product with a vector field X on Σ in both bundles is related via

$$X \cdot \varphi = i\,T \cdot X \cdot \phi_{|\Sigma},$$

where $\varphi \in \Gamma(S^\Sigma)$ is identified with $\phi_{|\Sigma} \in \Gamma(S^{(+)}_{|\Sigma})$ and the $\cdot$ on both sides denotes the Clifford product in the spinor bundle in question. The Dirac current U_ψ of a spinor field ψ on a Riemannian spin manifold (N, h) is given by

$$h(U_\psi, X) := -i\,(X \cdot \psi, \psi), \qquad X \in \Gamma(TN).$$

If $\phi \in \Gamma(S^{(+)})$ is a spinor field on (M, g) and $\varphi := \phi_{|\Sigma} \in \Gamma(S^\Sigma)$ its restriction to the hypersurface Σ, the Dirac currents satisfy

$$(V_\phi)_{|\Sigma} = \|\varphi\|^2 T_{|\Sigma} - U_\varphi.$$

Using the above identification of the spinor bundles, the conditions $\nabla^S \phi = 0$ and $V_\phi \cdot \phi = 0$ translate into the following *constraint conditions* for the spinor field $\varphi = \phi_{|\Sigma}$:

Proposition 13 (Spin constraint conditions, [4, Prop.5.1]).
Let (M, g) be a Lorentzian spin manifold with a parallel light-like spinor field ϕ and $\Sigma \subset M$ a space-like hypersurface. Then the spinor field $\varphi := \phi_{|\Sigma}$ on Σ satisfies

$$\nabla^\Sigma_X \varphi = \frac{i}{2}\, W^\Sigma(X) \cdot \varphi \qquad \textit{for all } X \in \Gamma(T\Sigma), \tag{8}$$

$$U_\varphi \cdot \varphi = i\, u_\varphi\, \varphi, \tag{9}$$

where U_φ is the Dirac current of φ and $u_\varphi := \sqrt{g^\Sigma(U_\varphi, U_\varphi)} = \|\varphi\|^2$ is a smooth function.

Remark 14. A spinor field φ on Riemannian spin manifold (Σ, g^Σ) with a symmetric endomorphism field W satisfying eqs. (8) and (9) is called *imaginary W-Killing spinor.* Note that the Dirac current U_φ of an imaginary W-Killing spinor satisfies

$$\nabla^\Sigma_X U_\varphi = -u_\varphi W(X). \tag{10}$$

Hence, $U_\varphi^\flat$ is a closed 1-form on Σ, and the integral manifolds of the distribution $U_\varphi^\perp \subset T\Sigma$ are Riemannian spin manifolds with a parallel spinor field. Examples of geodesically complete Riemannian spin manifolds with imaginary W-Killing spinors can be found in [4, examples 6.2–6.4].

Remark 15. Let us relax our conditions on (M, g) and require only that (M, g) is time-oriented and admits a parallel light-like vector field V. Let $\Sigma \subset M$ be a space-like hypersurface with induced Riemannian metric g^Σ and future-directed unit normal vector field T along Σ. Then, denote by U the orthogonal projection of $V|_\Sigma$ to $T\Sigma$. Then

$$\nabla_X^\Sigma U = -u\, W^\Sigma(X), \tag{11}$$

where u is the smooth function $u := \sqrt{g^\Sigma(U, U)}$. This is just the condition which holds for the Dirac current U_φ of $\varphi := \phi|_\Sigma$ above.

4.2 A Cauchy problem for the real-analytic case

Since a Lorentzian manifold with a parallel spinor field need not be Ricci flat, we cannot use the solution of the Cauchy problem for the vacuum Einstein equation (as it was done in the Riemannian setting, cf. [1]). In order to construct Lorentzian manifolds of form (11) with a parallel light-like spinor field we use the existence of the parallel light-like Dirac current instead of the condition on the Ricci curvature and proceed as follows:

Step 1: We start with a Riemannian manifold (Σ, g^Σ), equipped with a symmetric endomorphism field W and a vector field U satisfying the constraint condition (11) and ask whether this can be extended to a Lorentzian manifold (M, g) of the form (5) with a parallel light-like vector field V that projects on U along Σ. The idea is to consider the light-like vector field $V := uT - U$ with $\nabla^g V = 0$ *along* Σ and then transport V parallel along the lines $t \mapsto (t, x)$ to get a vector field on $\mathbb{R} \times \Sigma$ with $\nabla_T^g V = 0$. The integrability conditions for $\nabla^g V|_{T\Sigma} = 0$ then imply second-order *evolution equations* for a triple $\Psi(t, x^i) = (g(t, x^i), U(t, x^i), u(t, x^i))$ of symmetric bilinear forms, vector fields and functions depending on t and x^i of the form

$$\partial_t^2 \Psi = F(\Psi, \partial_i \Psi, \partial_t \Psi, \partial_i \partial_j \Psi, \partial_i \partial_t \Psi).$$

This evolution equation is of Cauchy–Kowalevski-type form for which the Cauchy–Kowalevski theorem can be applied. Hence, if all data are real analytic, it can be uniquely solved on an open neighborhood M of Σ in $\mathbb{R} \times \Sigma$ with the given initial conditions on Σ.

Step 2: We start with a Riemannian manifold (Σ, g^Σ), equipped with a symmetric endomorphism field W and an imaginary W-Killing spinor φ satisfying the constraint equations (8) and (9) and consider the Dirac current $U = U_\varphi$. If we suppose that all data are real analytic we can solve the evolution equation for the system (g_t, U_t, u_t) as in step 1. Then it can be shown that the spinor field φ can be extended to a parallel light-like vector field ϕ on M with Dirac current V.

Now, let (M, g) be of the form (5). Then $T = \frac{1}{\lambda}\partial_t$ is a time-like unit vector field.

Lemma 16 ([4], Lemma 3.3). *Let (M, g) be a Lorentzian manifold of the form (5) and let V be a light-like vector field on (M, g). Then $\nabla^g V = 0$, if and only if:*

$$R^g(\partial_t, X)V = 0 \quad \text{for all } X \in T\Sigma, \tag{12}$$

$$\nabla^g_{\partial_t}\nabla^g_{\partial_t} V = 0, \tag{13}$$

$$\nabla^g_X V|_{\{0\}\times\Sigma} = 0 \quad \text{for all } X \in T\Sigma\,, \tag{14}$$

$$\nabla^g_{\partial_t} V|_{\{0\}\times\Sigma} = 0. \tag{15}$$

Now, let V be a future directed light-like vector field on (M, g), $T := \frac{1}{\lambda}\partial_t$, $u := -g(T, V) > 0$ and $U := uT - V$. Then $g(U, U) = u^2$ and U is tangent to Σ in any point. We consider U and u as t-parameter family of vector fields and functions, $U_t = U(t, \cdot)$, $u_t := u(t, \cdot)$. Rewriting eqs. (12) and (13) in terms of the metric g_t we obtain evolution equations for the triple (g_t, U_t, u_t). Equations (14) and (15) imply the initial conditions for u_0 and U_0 as well as for $\dot{U}_0$ and $\dot{u}_0$, where $\dot{}$ denotes the t-derivative. We obtain

Theorem 17 ([4], Theorem 3.7). *Let (Σ, g^Σ) be a Riemannian manifold, W a field of g^Σ-symmetric endomorphisms on $T\Sigma$ and U a vector field on Σ satisfying the constraint equation*

$$\nabla^\Sigma U + u\, W = 0,$$

where $u^2 := g^\Sigma(U, U)$. Then, for any positive smooth function λ on $\mathbb{R}\times\Sigma$, a triple (g_t, U_t, u_t) of smooth one-parameter families of Riemannian metrics, vector fields and functions on Σ defines a Lorentzian metric

$$g = -\lambda^2 dt^2 + g_t$$

on an open neighborhood M of Σ in $\mathbb{R}\times\Sigma$ with parallel light-like vector field

$$V = \frac{u_t}{\lambda}\partial_t - U_t,$$

if and only if g_t, U_t and u_t satisfy the following system of PDEs on M:

$$\begin{aligned}\ddot{g}_t(X, Y) = {} & \frac{\lambda^2}{u_t} d^{\nabla^{g_t}}\Big(\frac{\dot{g}_t}{\lambda}\Big)(U_t, Y, X) + \frac{1}{2}\dot{g}_t(X, \dot{g}_t^\sharp(Y)) + (\dot{\log\lambda})\dot{g}_t(X, Y) \\ & + 2\lambda\, \mathrm{Hess}^{g_t}(\lambda)(X, Y),\end{aligned} \tag{16}$$

$$\begin{aligned} g_t(\ddot{U}_t, X) = &-\frac{\lambda^2}{2u_t} d^{\nabla^t}\Big(\frac{\dot{g}_t}{\lambda}\Big)(U_t, X, U_t) - \dot{g}_t(\dot{U}_t, X) - \frac{\dot{\overline{\log\lambda}}}{2}\dot{g}_t(U_t, X) \\ &- \lambda\,\mathrm{Hess}^{g_t}(\lambda)(U_t, X) + u_t g_t([\partial_t, \mathrm{grad}^{g_t}\lambda], X) \\ &+ \frac{u_t}{2}\dot{g}_t(\mathrm{grad}^{g_t}\lambda, X) + (2\dot{u}_t - d\lambda(U_t))\, d\lambda(X), \\ \ddot{u}_t = &\; g_t([\partial_t, \mathrm{grad}^{g_t}\lambda], U_t) + 2d\lambda(\dot{U}_t) + \frac{3}{2}\dot{g}_t(\mathrm{grad}^{g_t}(\lambda), U_t) - u_t \|\mathrm{grad}^{g_t}\lambda\|^2_{g_t}, \end{aligned}$$

with the initial conditions

$$\begin{aligned} g_0 &= g^{\Sigma}, & U_0 &= U, & u_0 &= u, \\ \dot{g}_0 &= -2\lambda_0 g^{\Sigma}(W\cdot, \cdot), & \dot{U}_0 &= u\,\mathrm{grad}^{\Sigma}(\lambda_0) + \lambda_0 W(U), & \dot{u}_0 &= d\lambda_0(U). \end{aligned}$$

Here, $\dot{g}_t^{\sharp}$ denotes the metric dual of $\dot{g}_t$, i.e., $g_t(X, \dot{g}_t^{\sharp}(Y)) = \dot{g}_t(X, Y)$.

Remark 18. We observe that the Cauchy–Kowalevski theorem can be applied to the PDE system in Theorem 17, provided that all initial data are assumed as real analytic. It guarantees existence and uniqueness of solutions to the given evolution equations for the described constraints and initial conditions for data on Σ in an open neighborhood of Σ in $\mathbb{R} \times \Sigma$. However, it is not clear that the solution g_t defines a family of *symmetric* bilinear forms, and hence Riemannian metrics for small t. The issue here is that the right-hand side of eq. (16) in general does not map symmetric bilinear forms to symmetric bilinear forms. Nevertheless, in Proposition 22 we will give a class of examples where the solution g_t is symmetric.

The key observation to overcome this problem is to replace condition (12) by a weaker one. The prize one has to pay is to assume real analyticity of all data.

Lemma 19 ([4], Lemma 4.1). *Let (M, g) be a real-analytic Lorentzian manifold of the form (5) and let V be a real-analytic light-like vector field on (M, g). Then $\nabla^g V = 0$, if and only if:*

$$R^g(X, V, V, Y) = 0 \quad \text{for all } X, Y \in T\Sigma, \tag{17}$$

$$\nabla^g_{\partial_t}\nabla^g_{\partial_t} V = 0, \tag{18}$$

$$\nabla^g_X V|_{\{0\}\times\Sigma} = 0 \quad \text{for all } X \in T\Sigma, \tag{19}$$

$$\nabla^g_{\partial_t} V|_{\{0\}\times\Sigma} = 0. \tag{20}$$

Then rewriting eq. (17) in terms of the metric g_t instead of eq. (12) we obtain evolution equations for the triple (g_t, U_t, u_t) where g_t are in the class of *symmetric* bilinear forms. We obtain

Theorem 20 ([4], Theorem 4.3). *Let (Σ, g^{Σ}) be a real-analytic Riemannian manifold, W a field of g^{Σ}-symmetric and real-analytic endomorphisms on $T\Sigma$ and U a real-analytic vector field satisfying the constraint equation*

$$\nabla^{\Sigma} U + u\, W = 0,$$

where $u^2 = g^{\Sigma}(U, U)$. Then, for any positive real-analytic function λ on $\mathbb{R} \times \Sigma$, the triple (g_t, U_t, u_t) of real-analytic one-parameter families of Riemannian metrics, vector fields and functions on Σ define a real-analytic Lorentzian metric

$$g = -\lambda^2\, dt^2 + g_t$$

on an open neighborhood M of Σ in $\mathbb{R} \times \Sigma$ with real-analytic parallel vector field

$$V(t, \cdot) = \frac{u_t}{\lambda} \partial_t - U_t,$$

if and only if g_t, U_t and u_t satisfy the following system of PDEs on M:

$$\begin{aligned}
\ddot{g}_t(X, Y) &= \frac{\lambda^2}{u_t} \left(d^{\nabla^t}\left(\frac{\dot{g}_t}{\lambda}\right)(U_t, X, Y) + d^{\nabla^t}\left(\frac{\dot{g}_t}{\lambda}\right)(U_t, Y, X) \right) + \frac{1}{2}\dot{g}_t(X, \dot{g}_t^{\sharp}(Y)) \\
&\quad + (\dot{\log \lambda})\dot{g}_t(X, Y) + 2\lambda\, \mathrm{Hess}^{g_t}(\lambda)(X, Y) + 2\frac{\lambda^2}{u_t^2} R^{g_t}(X, U_t, U_t, Y) \\
&\quad + \frac{1}{2u_t^2}\Big(\dot{g}_t(X, Y)\dot{g}_t(U_t, U_t) - \dot{g}_t(X, U_t)\dot{g}_t(Y, U_t) \Big), \qquad (21) \\
g_t(\ddot{U}_t, X) &= -\frac{\lambda^2}{2u_t} d^{\nabla^t}\left(\frac{\dot{g}_t}{\lambda}\right)(U_t, X, U_t) - \frac{1}{2}(\dot{\log \lambda})\dot{g}_t(U_t, X) - \lambda\, \mathrm{Hess}^{g_t}(\lambda)(U_t, X) \\
&\quad - \dot{g}_t(\dot{U}_t, X) + u_t\, g_t([\partial_t, \mathrm{grad}^{g_t}\lambda], X) + \frac{u}{2}\dot{g}_t(\mathrm{grad}^{g_t}\lambda, X) \\
&\quad + (2\dot{u}_t - d\lambda(U_t))\, d\lambda(X), \\
\ddot{u}_t &= g_t([\partial_t, \mathrm{grad}^{g_t}\lambda], U_t) + 2d\lambda(\dot{U}_t) + \frac{3}{2}\dot{g}_t(\mathrm{grad}^{g_t}(\lambda), U_t) - u_t \|\mathrm{grad}^{g_t}\lambda\|_{g_t}^2,
\end{aligned}$$

with the initial conditions

$$\begin{aligned}
g_0 &= g^{\Sigma}, & U_0 &= U, & u_0 &= u, \\
\dot{g}_0 &= -2\lambda_0 g^{\Sigma}(W\cdot, \cdot), & \dot{U}_0 &= u\, \mathrm{grad}^{g^{\Sigma}}(\lambda_0) + \lambda_0 W(U), & \dot{u}_0 &= d\lambda_0(U).
\end{aligned}$$

In contrast to eq. (16), the g_t-evolution equation (21) is manifestly an equation in the bundle of symmetric bilinear forms on M, i.e., at least for small t the solutions g_t are Riemannian metrics on Σ. By the Cauchy–Kowalevski theorem we obtain the following corollary.

Corollary 21 ([4], Corollary 4.5). *Let (Σ, g^Σ) be a real-analytic Riemannian manifold, W a field of g^Σ-symmetric, real-analytic endomorphisms and U a real-analytic vector field U satisfying the constraint equation*

$$\nabla^\Sigma U + uW = 0,$$

where $u^2 = g^\Sigma(U, U)$. Then, for any positive real-analytic function λ on $\mathbb{R} \times \Sigma$ there exists an open neighborhood M of Σ in $\mathbb{R} \times \Sigma$ and a unique real-analytic Lorentzian metric

$$g = -\lambda^2 dt^2 + g_t$$

on M which admits a real-analytic light-like parallel vector field $V = \frac{u_t}{\lambda}\partial_t - U_t$, where (g_t, U_t, u_t) are solutions of the evolution equations of Theorem 20 with the given initial conditions. Moreover, M can be chosen such that $\Sigma \subset M$ is a space-like Cauchy hypersurface.

Let us consider an explicit example where we find a solution for the evolution equations in both Theorems 17 and 20.

Proposition 22. *Let (Σ, g^Σ) be a Riemannian manifold, W a symmetric endomorphism field on $T\Sigma$, U a vector field and u a function on Σ satisfying the constraint equations*

$$\nabla^\Sigma U = -uW, \qquad g^\Sigma(U, U) = u^2 > 0.$$

Let, in addition, W be a Codazzi tensor, i.e., $d^{\nabla^\Sigma} W = 0$, and $\lambda = 1$. Then

$$\begin{aligned} g_t &:= g^\Sigma - 2t g^\Sigma(W\cdot, \cdot) + t^2 g^\Sigma(W^2\cdot, \cdot) = g^\Sigma((1 - tW)^2\cdot, \cdot), \\ U(t, x) &:= \frac{1}{(1 - t\,W_x)} U(x) = \sum_{k=0}^{\infty} W_x^k(U(x))\, t^k, \\ u(t, x) &:= u(x). \end{aligned}$$

are solutions to the evolution equations in both Theorems 17 and 20, defined on

$$M := \{(t, x) \in \mathbb{R} \times \Sigma \mid t\|W_x\|_{g_x} < 1\}.$$

In particular, the above solution g_t to the evolution equation (16) is a symmetric bilinear form.

This can now be applied to solve the Cauchy problem for parallel light-like spinors in the real-analytic setting.

Theorem 23 ([4, Theorem 5.3]). *Let (Σ, g^Σ) be a real-analytic Riemannian spin manifold, W a real-analytic g^Σ-symmetric endomorphism field on $T\Sigma$ and φ a real-analytic imaginary W-Killing spinor on Σ. Let $(M, g := -\lambda^2 dt^2 + g_t)$ be the Lorentzian manifold* *with parallel light-like vector field V arising as the solution of the evolution equations in Theorem 20 with the initial conditions given by $(\Sigma, g^\Sigma, W, U_\varphi)$. Then φ extends on (M, g) to a parallel spinor ϕ with Dirac current $V_\phi = V$.*

4.3 A Cauchy problem for the smooth case

A. Lischewski [20] used another approach to solve the Cauchy problem for parallel light-like spinors on *smooth* Lorentzian manifolds, based on first-order symmetric hyperbolic PDE methods that generalize the well-known PDE system appearing in the Cauchy problem for the vacuum Einstein equation in classical general relativity.

He proved in the smooth category:

Theorem 24 ([20], Theorem 1). *Let (Σ, g^Σ, W) be a smooth Riemannian spin manifold with a symmetric endomorphism field W admitting a nontrivial imaginary W-Killing spinor φ and let $h := -\lambda^2 dt^2 + h_t$ be an arbitrary chosen smooth (background) metric on $\mathbb{R} \times \Sigma$, where λ is a smooth positive function on $\mathbb{R} \times \Sigma$ and h_t a family of Riemannian metrics on Σ with $h_0 = g^\Sigma$.*

Then there exist an open neighborhood M of Σ in $\mathbb{R} \times \Sigma$ and a unique smooth Lorentzian metric $g = g_h$ on M such that

a) *(M, g) is spin and admits a parallel light-like spinor field ϕ which extends φ.*
b) *$g|_\Sigma = \lambda^2|_\Sigma dt^2 + g^\Sigma$.*
c) *g_h depends on h in terms of the following PDE system: the contracted difference tensor of the Levi–Civita connection of g and h vanishes, i.e.,*

$$E(X) := -\mathrm{tr}_g\big(g(A(\cdot,\cdot), X)\big) = 0 \quad \text{for all} \quad X \in TM,$$

where $A(Y, Z) := \nabla^g_Y Z - \nabla^h_Y Z$.

In particular, (Σ, g^Σ) embeds into (M, g) with Weingarten tensor W. Moreover, for any given metric h, (M, g_h) can be chosen to be globally hyperbolic with space-like Cauchy hypersurface Σ.

The proof consists of three steps. At first, having Lemma 12 in mind, one again prolongs the overdetermined PDE $\nabla\phi = 0$ and considers the system

$$\begin{aligned} Ric^g &= fV^\flat \otimes V^\flat, \\ D\phi &= 0, \\ V(f) &= 0, \end{aligned}$$

where D is the Dirac operator of (M, g), as evolution equation for the triple (g, ϕ, f). With the help of the background metric h one can rewrite this evolution equation by hyperbolic reduction as a first-order quasilinear symmetric hyperbolic PDE of the form

$$A_0(t,x,u)\partial_t u = \sum_{\mu>0} A_\mu(t,x,u)\partial_\mu u + b(t,x,u),$$

where u collects the data $(g, \partial g, \phi, f)$. As initial data, one chooses $(g^\Sigma, W, \varphi, f|\Sigma)$. For such PDEs a local existence and uniqueness result is available in the smooth setting and gives locally defined (g, ϕ, f).

In the second step, one shows that the locally given spinors ϕ are in fact parallel. To this end one shows that $\nabla_X\phi$ lies in the kernel of a locally defined second-order normally hyperbolic operator P (with respect to the metric g). Since initially $\nabla\phi = 0$ by the Killing spinor condition for φ, the uniqueness result for the Cauchy problem for operators of type P yield $\nabla\phi = 0$. In the third step, because of the uniqueness condition of the local solutions (g, ϕ, f), one can patch their domains together and obtain an open neighborhood M of Σ in $\mathbb{R} \times \Sigma$ on which ϕ is parallel. Furthermore, as in the real-analytic case, one can show that M is globally hyperbolic.

Remark 25. If all initial data $(\Sigma, g^\Sigma, W, \varphi)$ are real analytic, then one can use the real-analytic metric g_{analytic} which exists by Theorem 23 as background metric h in Theorem 24. In this case, the constructed metrics g_h and g_{analytic} coincide.

Acknowledgments: The results described in the chapter were obtained during the third funding period of the Collaborative Research Center 647 *Space – Time – Matter* within the project C2 of CRC 647. The CRC allowed a fruitful cooperation between Th. Leistner (School of Mathematics, University of Adelaide) and Helga Baum, Kordian Lärz, Daniel Schliebner and Andree Lischewski (HU Berlin). We were also supported by the Group of Eight Australia and the German Academic Exchange Service through Go8-DAAD Joint Research Co-operation Scheme.

Bibliography

[1] Ammann B, Moroianu A, Moroianu S. The Cauchy problems for Einstein metrics and parallel spinors. Comm Math Phys 2013;320:173–98.

[2] Baum H. Holonomy groups of Lorentzian manifolds. A status report. In: Bär, C, Lohkamp, J, Schwarz, M, editors. Global differential geometry. Springer Proceedings in Mathematics, vol. 17. Berlin: Springer Verlag, 2012:163–200.

[3] Baum H, Laerz K, Leistner T. On the full holonomy group of special Lorentzian manifolds. Math Zeitschrift 2014;277:797–828.

[4] Baum H, Leistner T, Lischewski A. Cauchy problems for Lorentzian manifolds with special holonomy. Diff Geom Appl 2016;45:43–66.

[5] Bérard-Bergery L, Ikemakhen A. On the holonomy of Lorentzian manifolds. In: Differential geometry: geometry in mathematical physics and related topics (Los Angeles, CA, 1990), volume 54 of Proc Sympos Pure Math. Am Math Soc. Providence, RI, 1993:27–40.

[6] Brinkmann HW. Einstein spaces which are mapped conformally on each other. Math Ann 1925;94(1):119–45.

[7] Cahen M, Wallach N. Lorentzian symmetric spaces. Bull Am Math Soc 1970;79:585–91.

[8] Candela A M, Romero A. Completeness of the trajectories of particles coupled to a general force field. Mech Anal 2013;208(1):255–74.
[9] Carrière Y. Autour de la conjecture de L. Markus sur les variétés affines. Invent Math 1989;95(3):615–28.
[10] Di Sacala AJ, Olmos C. The geometry of homogenous submanifolds of hyperbolic space. Math Z 2001;237(1):199–209.
[11] Dumitescu S Zeghib A. Géométries lorenziennes de dimension 3: classification et complétude. Geom Dedicata 2010;149:243–73.
[12] Ehlers J, Kundt W. Exact solutions of the gravitational field equations. In: Witten, L, editor. Gravitation: an introduction of current research. New York: Wiley, 1962:49–101.
[13] Galaev AS. Metrics that realize all Lorentzian holonomy algebras. Int J Geom Meth Mod Phys 2006;3(5–6):1025–45.
[14] Hull CM. Exact pp-wave solutions in 11-dimensional supergravity. Phys Lett B 1984;139(1–2):39–41.
[15] Kath I, Olbrich M. Compact quotients of Cahen-Walacah spaces. Amer Math Soc 2017. arXiv: 1501.01474.
[16] Klingler B. Complétude des variétés lorentziennes à courbure constante. Math Ann 1996;306(2):353–70.
[17] Lärz K. Global aspects of holonomy in Pseudo-Riemannian geometry. Ph.D. thesis, Humboldt-Universität Berlin, 2011.
[18] Leistner T. On the classification of Lorentzian holonomy groups. J Diff Geom 2007;76(3):423–84.
[19] Leistner T, Schliebner D. Completeness of compact Lorentzian manifolds with Abelian holonomy. Math Ann 2016;364(3):1469–503.
[20] Lischewski A. The Cauchy problem for parallel spinors as first order symmetric hyperbolic system. arXiv:1503.04946, 2015.
[21] Marsden J. On completeness of homogeneous pseudo-Riemannian manifolds. Indiana Univ J 1972/3;22:1065–6.
[22] O'Neill B. Semi-Riemannian geometry. London: Academic Press, 1983.
[23] Ringström H. The Cauchy problem in general relativity. ESI Lectures in Mathematics and Physics. Zürich: EMS Publishing House, 2009.
[24] Romero A, Sánchez M. Completeness of compact Lorentzian manifolds admitting a time like conformal vector field. Proc Am Math Soc 1995;123(9):2831–33.
[25] Schell JF. Classification of 4-dimensional Riemannian spaces. J Math Phys 1960;2:202–6.
[26] Schliebner D. On Lorentzian manifolds with highest first Betti number. Ann Institut Fourier 2015;65(4):1423–36.
[27] Shaw R. The subgroup structure of the homogeneous Lorentz group. Quart J Math 1970;21:101–24.
[28] Walker AG. Canonical form for a Riemannian metric with a parallel field of null planes. Quart J Math Oxford 1950;1(2):69–70.

Sebastian Boldt and Dorothee Schueth

Contributions to the spectral geometry of locally homogeneous spaces

Abstract: We report on several new results concerning the spectral geometry of locally homogeneous spaces. The first is a systematic method for constructing Dirac isospectral lens spaces, including many examples. The second is the result that many sixth-order curvature invariants – in particular, the integral of $|\nabla R|^2$ – are not determined by the Laplace spectrum of a closed Riemannian manifold. The third establishes irreducibility of Laplace eigenspaces associated with generic left invariant metrics on certain compact Lie groups.

Keywords: Dirac spectrum, lens spaces, isospectrality, affine lattices, Laplace spectrum, Lie groups, left invariant metrics, heat invariants, curvature invariants, two-step nilmanifolds, Clifford modules, eigenvalue multiplicities

Mathematics Subject Classification 2010: 58J50, 58J53, 53C20, 53C25, 53C27, 53C30, 22E25, 22E46

1 Introduction

We present several contributions to the spectral geometry of locally homogeneous spaces that were obtained during the third funding period of the SFB 647.

The first concerns the Dirac spectrum of lens spaces and was obtained in joint work of the first author with Emilio Lauret [4]. We give a new description of the spectrum of the (spin-)Dirac operator on lens spaces. A main tool is the introduction of so-called affine congruence lattices $\mathscr{L}_\tau$ associated with lens spaces L and spin structures τ on L. Not only does existence of isometries between such pairs (L, τ) turn out to be equivalent to certain one-norm-preserving isometries between the associated affine congruence lattices (Proposition 4), but also the multiplicities in the Dirac eigenvalue spectrum of (L, τ) (which is always contained in that of the covering sphere) can be expressed in terms of the associated lattice; see Theorem 6. The proof of this theorem, which we will not present here, makes use of the representation theory of the spin groups Spin$(2m)$. From Theorem 6 we obtain explicit conditions for lens spaces to be Dirac isospectral. Using these, we give several infinite families of pairs of isospectral examples which show, in particular, that neither spin structures nor isometry classes of lens spaces are spectrally determined. In contrast, the first author proved in [3] that there are no nontrivial Dirac isospectral pairs of three-dimensional lens spaces with fundamental group of prime order.

The second topic concerns inaudibility of certain curvature invariants, which we show using Laplace isospectral nilmanifolds. It is known from earlier work of the

DOI 10.1515/9783110452150-002_s_002

second author [18, 19] that the spectrum of the Laplace operator on functions of a closed Riemannian manifold does not determine the integrals of the individual fourth-order curvature invariants scal^2, $|\mathrm{ric}|^2$, $|R|^2$ which appear as summands in the heat invariant a_2. In joint work with Teresa Arias-Marco [1], the second author has studied the analogous question for the integrals of the sixth-order curvature invariants appearing as summands in a_3. Our result is that none of them is determined individually by the spectrum, which can be shown using various examples. In particular, we prove that two isospectral nilmanifolds of Heisenberg type with three-dimensional center are locally isometric if and only if they have the same value of $|\nabla R|^2$. This is interesting in view of the (still) open question whether the Einstein property ($\nabla R = 0$) is determined by the Laplace spectrum. Isospectral pairs of nilmanifolds of Heisenberg type are closely linked to the existence of nonisomorphic Clifford modules over the Clifford algebra associated with the central subalgebra $\mathfrak{z} \cong \mathbb{R}^r$ for certain dimensions r. The crucial ingredient in the curvature invariant $|\nabla R|^2$ for $r = 3$ turns out to be given by the squared trace of the action of the Clifford volume element.

Concerning our last topic, recall that in a famous result by K. Uhlenbeck [22], for a generic Riemannian metric on a compact manifold, all eigenvalues of the Laplace operator on functions have multiplicity one. In contrast, if the metric is homogeneous then every nonzero eigenvalue is multiple since the isometry group acts on each eigenspace. A natural question in this context, raised by V. Guillemin, is whether for generic left invariant metrics g on a compact Lie group G, the group G acts irreducibly on each eigenspace of the Laplacian; in other words, whether the eigenvalues have no higher multiplicities than necessitated by the prescribed symmetries of the metric. We give a representation theoretic reformulation of this problem (Proposition 21) and explain why generic left invariant metrics on SU(2) do have the mentioned property. These are results from [21], where the second author proved the same for groups of the form $G = SU(2) \times \cdots \times SU(2) \times T$, where T is a torus, and for quotients of such groups G by discrete central subgroups.

2 An explicit formula for the Dirac multiplicities on lens spaces

Our first topic is the explicit description of the spectrum of the (spin-)Dirac operator on quotients of sphere S^n by cyclic groups of isometries $\Gamma = \langle\gamma\rangle \subset \mathrm{O}(n)$. We presuppose familiarity of the reader with the concept of spin structures and the Dirac operator.

2.1 Lens spaces and their spin structures

Odd-dimensional manifolds with cyclic fundamental group which are covered by S^{2m-1} are called *lens spaces*. Restricting to odd dimensions is no loss of generality since the only manifolds which are covered by S^{2m} are the sphere itself and $\mathbb{RP}^{2m}$.

The latter is not orientable and so, in particular, not spin. We describe lens spaces, more explicitly, as follows:

Definition 1. Let $m \in \mathbb{N}$, $m \geq 2$, $q \in \mathbb{N}$, and $s = (s_1, \dots, s_m) \in \mathbb{Z}^m$ such that $(s_j, q) = 1$ for every $1 \leq j \leq m$. The lens space $L(q; s)$ is defined as

$$L(q; s) := \langle \gamma \rangle \backslash S^{2m-1},$$

where

$$\gamma := \operatorname{diag}\left(\begin{bmatrix} \cos(2\pi s_1/q) & -\sin(2\pi s_1/q) \\ \sin(2\pi s_1/q) & \cos(2\pi s_1/q) \end{bmatrix}, \dots, \begin{bmatrix} \cos(2\pi s_m/q) & -\sin(2\pi s_m/q) \\ \sin(2\pi s_m/q) & \cos(2\pi s_m/q) \end{bmatrix} \right) \in \mathrm{SO}(2m).$$

We equip the manifold $L(q; s)$ with the unique orientation and metric such that the canonical projection $\pi : S^{2m-1} \to L(q; s)$ is an orientation-preserving Riemannian covering. Here, S^{2m-1} is equipped with the standard metric and is oriented such that its oriented orthonormal frame bundle is $\mathrm{SO}(2m)$ with projection to the last column vector.

Note that $L(q; s)$ depends only on $[s] \in \mathbb{Z}_q^m$. In particular, for given m and q there are only finitely many isometry classes of lens spaces $L(q; s)$.

In order to be able to speak of the Dirac operator and its spectrum on a lens space $L = L(q; s)$, we have to know about the spin structures τ on L. The following result classifies the spin structures on lens spaces.

Theorem 2 ([8, Theorem 1]). *Let $L = L(q; s_1, \dots, s_m)$. If q is odd then L admits (up to equivalence) precisely one spin structure τ. If q is even then L does not admit any spin structure when m is odd, and admits precisely two inequivalent spin structures τ_0, τ_1 when m is even.*

We will regard subscripts of spin structures modulo 2 (e.g., $\tau_2 = \tau_0$). To fix the roles of τ_0 and τ_1 for future reference, we note that spin structures on $L(q; s_1, \dots, s_m)$ here correspond to homomorphisms from $\langle \gamma \rangle$ to $\mathrm{Spin}(2m)$, and we define

$$\tau_h(\gamma^k) := (-1)^{k(h+h_{q;s})} \prod_{j=1}^{m} \left(\cos\left(\tfrac{k s_j \pi}{q}\right) + \sin\left(\tfrac{k s_j \pi}{q}\right) e_{2j-1} e_{2j} \right),$$

where $h_{q;s} := \sum_j \left\lfloor \frac{s_j}{q} \right\rfloor$.

Remark 3. Let q and m be even, and let $s, s' \in \mathbb{Z}^m$ with $(q, s_j) = (q, s'_j) = 1$ for every j. Assume that $\psi : L(q; s) \to L(q; s')$ is an isometry. If $h' \in \mathbb{Z}$, then the pullback $\psi^* \tau'_{h'}$ of the spin structure $\tau'_{h'}$ is again a spin structure. By the last theorem, there is $h \in \mathbb{Z}$ such that $\psi^* \tau'_{h'} = \tau_h$ (where "=" means equivalence of spin structures). Since ψ is a diffeomorphism, we must then have $\psi^*\tau'_{h'+1} = \tau_{h+1}$. Therefore, for each isometry $\psi : L(q;s) \to L(q;s')$ there exists $\rho \in \{0,1\}$ such that ψ takes τ_h to $\tau'_{h+\rho}$ for every $h \in \{0,1\}$.

2.2 Associated affine congruence lattices

With every lens space L and spin structure τ on L we associate an *affine congruence lattice* $\mathscr{L}$. This affine congruence lattice will fully characterize the isometry class of L. The multiplicities of the eigenvalues of the Dirac operator on L will turn out to be determined by the number of points in $\mathscr{L}$ with a certain one-norm.

We fix some notation. Let

$$\mathbb{E}^m := (1/2+\mathbb{Z})^m = \left\{\frac{1}{2}(a_1,\dots,a_m) \in \mathbb{Q}^m : a_j \in \mathbb{Z} \text{ odd } \forall 1 \le j \le m\right\},$$

the image of $\mathbb{Z}^m$ under translation by $(1/2,\dots,1/2)$. For $q \in \mathbb{N}$ and $s = (s_1,\dots,s_m) \in \mathbb{Z}^m$ such that $(q,s_j) = 1$ for every $1 \le j \le m$, we set

$$\mathscr{L}(q;s) = \left\{\frac{1}{2}(a_1,\dots,a_m) \in \mathbb{E}^m : \sum_j a_j s_j \equiv 0 \pmod q\right\}.$$

Furthermore, if q is even and $h \in \mathbb{Z}$, we partition $\mathscr{L}(q;s)$ into the subsets

$$\mathscr{L}(q;s;h) = \left\{\frac{1}{2}(a_1,\dots,a_m) \in \mathbb{E}^m : \sum_j a_j s_j \equiv \left(h + h_{q;s}\right) q \pmod{2q}\right\}.$$

Note that $\mathscr{L}(q;s;h)$ depends only on the parity of h and that $\mathscr{L}(q;s)$ is the disjoint union of $\mathscr{L}(q;s;0)$ and $\mathscr{L}(q;s;1)$.

By $\|\mu\|_1$ we denote the *one-norm* of $\mu = \frac{1}{2}(a_1,\dots,a_m) \in \mathbb{E}^m$, that is, $\|\mu\|_1 = \frac{1}{2}\sum_{j=1}^m |a_j|$. A linear bijection $\varphi : \mathbb{E}^m \to \mathbb{E}^m$ is a $\|\,.\,\|_1$-isometry if and only if there is a permutation $\sigma \in \mathfrak{S}_m$ and $\epsilon_j \in \{\pm 1\}$, $1 \le j \le m$ such that

$$\varphi(\tfrac{1}{2}(a_1,\dots,a_m)) = \tfrac{1}{2}(\epsilon_{\sigma^{-1}(1)} a_{\sigma^{-1}(1)},\dots,\epsilon_{\sigma^{-1}(m)} a_{\sigma^{-1}(m)}).$$

The $\|\,.\,\|_1$-isometry φ is said to be *orientation preserving* if $\epsilon := \prod_{j=1}^m \epsilon_j = 1$ and *orientation reversing* if $\epsilon = -1$. Note that this definition of orientation preservation does not agree with the usual one when φ is viewed as $\varphi : \mathbb{R}^m \supset \mathbb{E}^m \to \mathbb{E}^m \subset \mathbb{R}^m$. For $\epsilon = \pm 1$, we call a $\|\,.\,\|_1$-isometry φ or an isometry $\psi : L \to L'$ between lens spaces ϵ*-oriented* if it is orientation preserving in the case $\epsilon = 1$, resp. orientation reversing in the case $\epsilon = -1$.

Proposition 4 ([4, Corollary 3.3]). *Let $q \in \mathbb{N}$, and let s and s' be in $\mathbb{Z}^m$ with all coordinates coprime to q. Then, if q is odd, there is an ϵ-oriented isometry ψ between $L(q;s)$ and $L(q;s')$ if and only if there is an ϵ-oriented $\|\,.\,\|_1$-isometry φ between $\mathscr{L}(q;s)$ and*

$\mathscr{L}(q;s')$. If q is even and $\rho \in \{0,1\}$, there is an ϵ-oriented isometry ψ between $L(q;s)$ and $L(q;s')$ taking the spin structures τ_h to $\tau'_{h+\rho}$ for $h \in \{0,1\}$ if and only if there is an ϵ-oriented $\|\,.\,\|_1$-isometry $\varphi : \mathbb{E} \to \mathbb{E}$ with $\varphi(\mathscr{L}(q;s;h)) = \mathscr{L}(q;s;h+\rho)$ for $h \in \{0,1\}$.

Remark 5. The Dirac operator depends on a spin structure which in turn depends on an orientation. A change of orientation causes the Dirac operator to change sign. Accordingly, its eigenvalue spectrum is reflected about zero. In general, there is no connection between the spectra of the Dirac operators associated with the inequivalent spin structures τ_0 and τ_1 on a lens space $L(q;s)$ with q even. If, however, there exists an isometry $\psi : L(q;s) \to L(q;s')$ and $\rho \in \{0,1\}$ such that $\psi^*\tau'_h = \tau_{h+\rho}$, $h \in \{0,1\}$, then the spectra of $(L(q;s), \tau_{h+\rho})$ and $(L(q;s'), \tau_h)$ coincide if ψ is orientation preserving and are images of one another under the reflection about zero if ψ is orientation reversing. Consider, for example, the lens space $L = L(16;1,3,5,7)$ and the map $\varphi : \mathbb{E}^4 \to \mathbb{E}^4$ given by

$$1/2\,(a_1,a_2,a_3,a_4) \mapsto 1/2\,(a_3,-a_1,-a_4,a_2)\ .$$

An easy calculation shows that $\varphi\,(\mathscr{L}(16;1,3,5,7;0)) = \mathscr{L}(16;1,3,5,7;1)$. Since φ is orientation preserving (recall our above definition), the lens spaces (L,τ_0) and (L,τ_1) are Dirac isospectral. However, we want to regard such an isospectrality as "trivial." One has to be careful about such trivial isospectralities when in search of Dirac isospectral examples.

If $L = L(q;s)$ is a lens space with spin structure τ then we define the *associated affine congruence lattice* $\mathscr{L}$ as

$$\mathscr{L} := \mathscr{L}_\tau := \begin{cases} \mathscr{L}(q;s) & \text{if } q \text{ is odd}\,, \\ \mathscr{L}(q;s;h) & \text{if } q \text{ is even and } \tau = \tau_h\,. \end{cases}$$

Furthermore, for any $r \geq 0$ and $\epsilon \in \mathbb{Z}$, we let

$$N_{\mathscr{L}}(\epsilon, r) := \#\{\mu \in \mathscr{L} : \|\mu\|_1 = r + \tfrac{m}{2},\ R(\mu) \equiv \epsilon \pmod 2\}\,,$$

where $R(\mu) := \#\{j : 1 \leq j \leq m,\ a_j < 0\}$. Using the representation theory of Spin$(2m)$, E. Lauret and the first author obtained the following explicit description of the Dirac spectrum on lens spaces:

Theorem 6 ([4, Theorem 4.3]). *Let $L = L(q;s)$ be a lens space with spin structure τ, and let $\mathscr{L} = \mathscr{L}_\tau$ be its associated affine congruence lattice. For $k \geq 0$, let $\lambda_k := k + \frac{2m-1}{2}$. Then the eigenvalues of the Dirac operator on L are $\pm\lambda_k$ with multiplicity*

$$\mathrm{mult}_{(L,\tau)}(-\lambda_k) = \sum_{r=0}^{k} \binom{r+m-2}{m-2} N_{\mathscr{L}}(r, k-r),$$

$$\mathrm{mult}_{(L,\tau)}(+\lambda_k) = \sum_{r=0}^{k} \binom{r+m-2}{m-2} N_{\mathscr{L}}(r+1, k-r).$$

Definition 7. Two subsets $\mathscr{L}$ and $\mathscr{L}'$ of $\mathbb{E}^m$ are called *oriented* $\|\,.\,\|_1$*-isospectral* if $N_{\mathscr{L}}(\epsilon, k) = N_{\mathscr{L}'}(\epsilon, k)$ for every nonnegative integer k and every $\epsilon \in \{0, 1\}$.

Corollary 8 ([4, Corollary 4.5]). *Let L and L' be lens spaces with spin structures τ, τ' and associated affine congruence lattices $\mathscr{L}$ and $\mathscr{L}'$, respectively. Then L and L' are Dirac isospectral if and only if $\mathscr{L}$ and $\mathscr{L}'$ are oriented $\|\,.\,\|_1$-isospectral.*

2.3 Dirac isospectral lens spaces

Corollary 8 is not only a characterization of Dirac isospectrality of lens spaces but enables one to construct Dirac isospectral lens spaces by constructing oriented $\|\,.\,\|_1$-isospectral affine congruence lattices. We present three infinite examples of finite families of Dirac isospectral lens spaces that were found by using this method. Each of these families of examples exhibits different features. The most striking ones are that neither the spin structure nor the isometry class of a lens space is spectrally determined.

Our first example shows that there exist arbitrarily large finite families of pairwise Dirac isospectral lens spaces with a fixed order of the fundamental group.

Theorem 9 ([4, Theorem 5.2]). *Let $q = 40$. For each $r \geq 1$ we set $m = 4r + 2$, and for each $0 \leq p \leq \frac{m-2}{4} = r$ we let*

$$s^{(p)} = (\underbrace{1, 11, \ldots, 1, 11}_{m-2p}, \underbrace{21, 31, \ldots, 21, 31}_{2p}).$$

Then the lens spaces in the family $\{L(q; s^{(p)}) : 0 \leq p \leq r\}$ are pairwise nonisometric and, when each is endowed with the spin structure $\tau_0^{(p)}$, they are pairwise Dirac isospectral. The dimension of the lens spaces is $8r + 3$ and the cardinality of the family is $r + 1$.

The next example is a family of pairs of Dirac isospectral lens spaces that have the same underlying manifold but different spin structures.

Theorem 10 ([4, Theorem 5.3]). *For any $r \geq 1$, we consider the lens space*

$$L = L(32r; 1, 1 + 4r, 1 + 16r, 1 + 28r).$$

Then L does not carry an isometry which takes the spin structure τ_0 to τ_1, but (L, τ_0) and (L, τ_1) are Dirac isospectral.

Our last example is an infinite sequence of pairs of Dirac isospectral lens spaces. In this case, the order of the fundamental group of each lens space is odd and thus the lens spaces admit only one spin structure.

Theorem 11 ([4, Theorem 5.4]). *For any odd positive integer $r \geq 7$, we consider the lens spaces*

$$\begin{aligned} L &= L(r^2; 1, 1+r, 1+2r, 1+4r), \\ L' &= L(r^2; 1, 1-r, 1-2r, 1-4r), \end{aligned}$$

endowed with their unique spin structure. Then L and L' are nonisometric and Dirac isospectral.

Of course, the families in the last three theorems do not exhaust all examples of Dirac isospectral lens spaces. We now present a finiteness result that makes the search for Dirac isospectral lens spaces with a computer feasible.

Let q be a positive integer, $\mathscr{L} \subseteq \mathbb{E}^m$ and

$$\begin{aligned} C(q) &:= \left\{ \tfrac{1}{2}(a_1, \ldots, a_m) \in \mathbb{E}^m : |\tfrac{a_j}{2}| < q \, \forall j \right\}, \\ N_{\mathscr{L}}^{\mathrm{red}}(\epsilon, k) &:= \#\{\mu \in \mathscr{L} \cap C(q) : \|\mu\|_1 = k + \tfrac{m}{2},\ R(\mu) \equiv \epsilon \pmod 2\}. \end{aligned}$$

Note that $N_{\mathscr{L}}^{\mathrm{red}}(\epsilon, k) = 0$ for every $k \geq mq - \frac{m}{2}$; thus, only finitely many of these numbers are nonzero.

Proposition 12 ([4, Proposition 6.1]). *Let $\mathscr{L}$ be an affine congruence lattice associated with a spin lens space of dimension $2m-1$ with fundamental group of order q. Then*

$$N_{\mathscr{L}}(\epsilon, k) = \sum_{\beta=0}^{\lfloor k/q \rfloor} \binom{\beta+m-1}{m-1} N_{\mathscr{L}}^{\mathrm{red}}(\epsilon, k - \beta q).$$

Corollary 13 ([4, Corollary 6.2]). *Two $(2m-1)$-dimensional spin lens spaces (L, τ) and (L', τ') with fundamental group of order q are Dirac isospectral if and only if the multiplicities $\mathrm{mult}_{(L,\tau)}(\pm\lambda_k)$ and $\mathrm{mult}_{(L',\tau')}(\pm\lambda_k)$ of the Dirac eigenvalues $\pm\lambda_k = \pm(k + \frac{2m-1}{2})$ of (L, τ) and (L', τ'), respectively, coincide for all $0 \leq k < mq$.*

Proposition 12 opens the door to a systematic search for Dirac isospectral lens spaces. For any given m and q, representatives $(L(q; s), \tau)$ of isometry classes of lens spaces $L(q; s)$ and spin structures τ are easy to enumerate (see [4, Propositions 2.2 and 2.5]). For every such lens space (L, τ), one computes the numbers $N_{\mathscr{L}}^{\mathrm{red}}(\epsilon, r)$, $\epsilon \in \{0, 1\}$, $0 \leq r < mq$. According to these numbers, one then groups the lens spaces into Dirac isospectral families. In Table 1 we show all of the examples for $m = 4$, $q \leq 100$, for $m = 6$, $q \leq 50$, for $m = 8$, $q \leq 60$, and for $m = 10$, $q \leq 40$.

We conclude with some observations about the Dirac isospectral examples.

Table 1: Examples of Dirac isospectral spin lens spaces in dimensions $2m - 1 = 7, 11, 15, 19$ for low values of q.

	Dimension $n = 7$	
$L(32; 1, 3, 5,15)_{\tau_0}$ $L(32; 1, 3, 5,15)_{\tau_1}$	$L(49; 1, 6, 8,22)$ $L(49; 1, 6, 8,20)$	$L(64; 1, 7, 9,31)_{\tau_0}$ $L(64; 1, 7, 9,31)_{\tau_1}$
$L(75; 1, 4, 14,16)$ $L(75; 1, 4, 11,19)$	$L(75; 1, 4, 11,34)$ $L(75; 1, 4, 14,31)$	$L(80; 1, 3, 9,27)_{\tau_0}$ $L(80; 1, 9, 13,37)_{\tau_0}$
$L(81; 1, 8, 19,37)$ $L(81; 1, 8, 26,37)$	$L(81; 1, 8, 10,28)$ $L(81; 1, 8, 10,26)$	$L(81; 1, 8, 10,37)$ $L(81; 1, 8, 10,35)$
$L(96; 1,11, 13,47)_{\tau_0}$ $L(96; 1,11, 13,47)_{\tau_1}$	$L(98; 1,13, 15,43)_{\tau_0}$ $L(98; 1,13, 15,41)_{\tau_1}$	$L(98; 1,13, 15,41)_{\tau_0}$ $L(98; 1,13, 15,43)_{\tau_1}$
	Dimension $n = 11$	
$L(40; 1,1, 1,11,11,11)_{\tau_0}$ $L(40; 1,1, 9,11,11,19)_{\tau_0}$	$L(40; 1,1, 11,11,13,17)_{\tau_0}$ $L(40; 1,1, 3, 7,11,11)_{\tau_0}$ $L(40; 1,3, 7, 9,11,19)_{\tau_0}$	$L(44; 1,3,5,7, 9,19)_{\tau_0}$ $L(44; 1,3,5,7, 13,15)_{\tau_0}$
$L(44; 1,3, 5, 7, 9,19)_{\tau_1}$ $L(44; 1,3, 5, 7,13,15)_{\tau_1}$	$L(48; 1,1, 5, 7, 7,13)_{\tau_0}$ $L(48; 1,5, 7,11,13,19)_{\tau_0}$ $L(48; 1,1, 7, 7,11,19)_{\tau_0}$	$L(48; 1,1,7,7, 17,23)_{\tau_0}$ $L(48; 1,1,1,7, 7, 7)_{\tau_0}$
	Dimension $n = 15$	
$L(39; 1,2, 4,5, 7,10,14,16)$ $L(39; 1,2, 4,7, 8,10,16,17)$		$L(52; 1,3, 5,7, 9,11,17,25)_{\tau_0}$ $L(52; 1,3, 5,7, 9,15,23,25)_{\tau_1}$
$L(52; 1,3, 5,7, 9,11,19,21)_{\tau_1}$ $L(52; 1,3, 5,7, 9,11,17,23)_{\tau_1}$		$L(52; 1,3, 5,7, 9,11,19,21)_{\tau_0}$ $L(52; 1,3, 5,7, 9,11,17,23)_{\tau_0}$
$L(52; 1,3, 5,7, 9,15,23,25)_{\tau_0}$ $L(52; 1,3, 5,7, 9,11,17,25)_{\tau_1}$		$L(56; 1,3, 5,9, 11,13,19,23)_{\tau_0}$ $L(56; 1,3, 5,9, 11,13,15,27)_{\tau_0}$
$L(56; 1,3, 5,9, 11,13,15,27)_{\tau_1}$ $L(56; 1,3, 5,9, 11,13,19,23)_{\tau_1}$		
	Dimension $n = 19$	
$L(24; 1,1, 1,1, 1, 5, 5, 5, 5, 5)_{\tau_0}$ $L(24; 1,1, 1,5, 5, 5, 7, 7,11,11)_{\tau_0}$ $L(24; 1,1, 1,1, 5, 5, 5, 5, 7,11)_{\tau_0}$		$L(40; 1,1, 1, 9, 9,11,11,11,19,19)_{\tau_0}$ $L(40; 1,1, 1, 1, 1,11,11,11,11,11)_{\tau_0}$ $L(40; 1,1, 1, 1, 9,11,11,11,11,19)_{\tau_0}$
$L(40; 1,1, 1,3, 7, 9,11,11,11,19)_{\tau_0}$ $L(40; 1,1, 1,1, 3, 7,11,11,11,11)_{\tau_0}$ $L(40; 1,1, 3,7, 9, 9,11,11,19,19)_{\tau_0}$ $L(40; 1,1, 1,9,11,11,11,13,17,19)_{\tau_0}$ $L(40; 1,1, 1,1,11,11,11,11,13,17)_{\tau_0}$		$L(40; 1,1, 3, 3, 7, 7, 9,11,11,19)_{\tau_0}$ $L(40; 1,1, 3, 7, 9,11,11,13,17,19)_{\tau_0}$ $L(40; 1,1, 3, 3, 7, 7,11,11,13,17)_{\tau_0}$ $L(40; 1,1, 1, 3, 3, 7, 7,11,11,11)_{\tau_0}$ $L(40; 1,1, 1, 3, 7,11,11,11,13,17)_{\tau_0}$ $L(40; 1,1, 1,11,11,11,13,13,17,17)_{\tau_0}$

The most intriguing observation is the absence of Dirac isospectral lens spaces in dimensions $2m-1 \equiv 1 \pmod 4$. Note that the order q of the fundamental group of a lens space $L(q;s)$ which admits a spin structure and is of dimension $2m-1 \equiv 1 \pmod 4$ is necessarily odd. We have searched in the following ranges of the parameters m and q: $m = 3$ and $q \leq 1,001$; $m = 5$ and $q \leq 501$; $m = 7$ and $q \leq 251$; $m = 9$ and $q \leq 125$. Since this search returned no isospectral pairs, while the search for Dirac isospectral lens spaces in dimensions $2m-1 \equiv 3 \pmod 4$ yielded a wealth of examples, it seems reasonable to conjecture the nonexistence of Dirac isospectral lens spaces in dimension $n \equiv 1 \pmod 4$ in general:

Conjecture 14 ([4, Conjecture 6.3]). *Two Dirac isospectral spin lens spaces of dimension $n \equiv 1 \pmod 4$ are necessarily isometric.*

This conjecture cannot be generalized to all spin manifolds due to the existence of a large number of examples in the case of flat manifolds: Miatello and Podestá [16, §4], found families of pairwise Dirac isospectral compact flat manifolds in every dimension $n \geq 4$.

The next observation concerns the topology of the Dirac isospectral examples. In all examples that we have found, the lens spaces are pairwise homotopy equivalent (see [6, 29.6] for criteria of when two lens spaces are homotopy equivalent). This situation is different from the case of the Laplace operator: Ikeda [13] found isospectral, nonhomotopy equivalent lens spaces in dimension 7 with $q = 13$.

The last observation concerns isospectrality of lens spaces with respect to the Hodge–Laplace operator acting on p-forms. In [14], nonisometric lens spaces were found that are p-isospectral for every p. Looking at the examples in [14], it is noticeable that most of the Dirac isospectral seven-dimensional lens spaces above are also p-isospectral for every p; e.g., those from Theorem 11. There are, however, counterexamples in both directions. The seven-dimensional lens spaces in Table 1 with $q = 75$ are Dirac isospectral but not p-isospectral for every p, while the lens spaces $L(100;1,9,11,29)$ and $L(100;1,9,11,31)$ are p-isospectral for every p but not Dirac isospectral.

3 Inaudibility of sixth-order curvature invariants

We now shift attention from the Dirac operator to the Laplace operator. Let (M,g) be a compact, connected Riemannian manifold without boundary, and let $\Delta_g = -\operatorname{div}_g \operatorname{grad}_g$ the associated Laplace operator on functions. By the *spectrum* $\operatorname{spec}(M,g)$ we now mean the eigenvalue spectrum

$$0 = \lambda_0 < \lambda_1 \leq \lambda_2 \leq \cdots \to \infty$$

of Δ_g, taking multiplicities into account.

A central question in spectral geometry is (as already in the previous section) in how far the geometry of (M, g) is determined by the spectrum – this time, the Laplace spectrum. A classical result by Minakshisundaram–Pleijel says that there is an asymptotic expansion of the form

$$\sum_{k=0}^{\infty} e^{-t\lambda_k} \sim (4\pi t)^{-\dim M/2} \sum_{q=0}^{\infty} a_q(g)t^q \quad \text{for } t \searrow 0.$$

Thus, the dimension of M, as well as the coefficients $a_q(g)$ which are called the *heat invariants* of (M, g), are determined by spec(M, g); that is, they are *"audible."* The first few heat invariants are

$$\begin{aligned} a_0(g) &= \mathrm{vol}(M, g), \\ a_1(g) &= \tfrac{1}{6} \int_M \mathrm{scal}\, d\mathrm{vol}_g, \\ a_2(g) &= \tfrac{1}{360} \int_M (5\,\mathrm{scal}^2 - 2|\,\mathrm{ric}\,|^2 + 2|R|^2)\, d\mathrm{vol}_g, \end{aligned}$$

where scal, ric, and R denote the scalar curvature, the Ricci tensor, and the Riemannian curvature tensor of (M, g), respectively. So, for example, the integral of $5\,\mathrm{scal}^2 - 2|\,\mathrm{ric}\,|^2 + 2|R|^2$ is an audible geometric invariant.

Is this the case for the integrals of the individual terms scal^2, $|\,\mathrm{ric}\,|^2$, $|R|^2$ as well? From earlier work of the second author [18, 19], the answer to this question is known to be *no*: For each of these three integrals (and in fact, for any linear combination of them which is not a multiple of the one appearing in $a_2(g)$) there exist pairs of Laplace isospectral manifolds differing in the respective integral.

Note that the terms appearing in $a_2(g)$ are curvature invariants of order four. A *curvature invariant* is a polynomial in the coefficients of the Riemannian curvature tensor R and its covariant derivatives ∇R, $\nabla^2 R$, ..., where the coefficients are taken with respect to some orthonormal basis of the tangent space at the point under consideration, and the polynomial is required to be invariant under changes of the orthonormal basis. A curvature invariant is called *of order k* if it is a sum of terms each of which involves a total of k derivatives of the metric tensor. It is known that each heat invariant $a_q(g)$ is the integral of a certain curvature invariant of order $2q$.

Some examples of curvature invariants of order six are scal^3, $\mathrm{scal}\,|\,\mathrm{ric}\,|^2$, $\mathrm{scal}\,|R|^2$, $|\nabla\,\mathrm{scal}\,|^2$, $|\nabla\,\mathrm{ric}\,|^2$, $|\nabla R|^2$. Among these, $|\nabla R|^2$ is of special interest: Its integral over (M, g) vanishes if and only if (M, g) is an Einstein manifold. It is not known whether the Einstein property is audible in the above sense. When the integrand of $a_3(g)$ is expressed as a linear combination of the elements of a common basis of sixth-order curvature invariants (see [17] for an explicit formula), then $|\nabla R|^2$ does appear with a nonzero coefficient; however, since there are additional terms, this does not imply audibility of $\int |\nabla R|^2$.

Nevertheless, no examples had been known of pairs of isospectral manifolds differing in $\int |\nabla R|^2$ until Arias-Marco and the second author [1] gave the first such examples, consisting of certain pairs of isospectral two-step nilmanifolds. We are going to describe these in the following. Before doing so, we give some necessary framework concerning the structure of curvature invariants on two-step nilmanifolds and, more specifically, on nilmanifolds of Heisenberg type.

3.1 Riemannian two-step nilmanifolds and their curvature invariants

A *Riemannian two-step nilmanifold* is a quotient $\Gamma\backslash G$ of a simply connected two-step nilpotent Lie group G by some discrete subgroup Γ, endowed with a Riemannian metric which is induced by some left invariant Riemannian metric g on G; in this situation, one usually denotes the induced metric on $\Gamma\backslash G$ by g again. Since $(\Gamma\backslash G, g)$ is locally homogeneous and locally isometric to (G, g), all of its curvature invariants are constant functions and are the same as the corresponding invariants of (G, g).

It is well known that each (G, g) as above is of the form $(G(j), g(j))$, where $j : \mathfrak{z} \to \mathfrak{so}(\mathfrak{v})$ is a linear map, $\mathfrak{v} := \mathbb{R}^m$, $\mathfrak{z} := \mathbb{R}^r$ (for some $m, r \in \mathbb{N}_0$) and $G(j), g(j)$ are associated with j as follows: First, let $(\mathfrak{g}(j), \langle\, , \rangle)$ denote the two-step nilpotent metric Lie algebra with underlying vector space $\mathfrak{v} \oplus \mathfrak{z} = \mathbb{R}^{m+r}$, endowed with the standard euclidean inner product $\langle\, , \rangle$ and with Lie bracket $[\, ,]^j$ defined by letting $\mathfrak{z}$ be central, $[\mathfrak{v}, \mathfrak{v}]^j \subseteq \mathfrak{z}$ and

$$\langle j_Z X, Y\rangle = \langle Z, [X, Y]^j\rangle \text{ for all } X, Y \in \mathfrak{v} \text{ and } Z \in \mathfrak{z}.$$

Then define $G(j)$ as the simply connected Lie group with Lie algebra $\mathfrak{g}(j)$, and $g(j)$ as the left invariant metric on $G(j)$ which coincides with the given inner product $\langle\, , \rangle$ on $\mathfrak{g}(j) = T_e G(j)$.

Each curvature invariant of $(G(j), g(j))$ can of course be expressed in terms of the map j. More precisely, we have the following qualitative observation, where here – and later – $\{Z_1, \dots, Z_r\}$ denotes an orthonormal basis of $\mathfrak{z} = \mathbb{R}^m$:

Proposition 15 ([1, Proposition 4.12]). *Let $q \in \mathbb{N}$. On a two-step nilpotent Lie group $G(j)$, endowed with the left invariant metric $g(j)$, each curvature invariant of order $2q$ can be expressed as a linear combination of certain polynomial invariants of j of the form $I_{k_1 \dots k_\lambda | \dots | k_\mu \dots k_{2q}}(j)$, where $(k_1, \dots, k_{2q})$ in $\{1, \dots, q\}^{2q}$ arises as a permutation of $(1, 1, 2, 2, \dots, q, q)$ (i.e., contains each entry exactly twice), and*

$$I_{k_1 \dots k_\lambda | \dots | k_\mu \dots k_{2q}}(j) := \sum \operatorname{Tr}\left(j_{Z_{\alpha_{k_1}}} \cdots j_{Z_{\alpha_{k_\lambda}}} \right) \cdot \dots \cdot \operatorname{Tr}\left(j_{Z_{\alpha_{k_\mu}}} \cdots j_{Z_{\alpha_{k_{2q}}}} \right),$$

where the sum is taken according to the Einstein summation convention: For each pair $k_i = k_j$ the sum runs over α_{k_i} once from 1 to r. (So the sum has precisely r^q summands.)

When the context is clear, we also write $I_{k_1\ldots k_\lambda|\ldots|k_\mu\ldots k_{2q}}$ for $I_{k_1\ldots k_\lambda|\ldots|k_\mu\ldots k_{2q}}(j)$. In order to illustrate both the above statement and its notation, we state the formulas for the fourth-order curvature invariants (see [1, Lemma 4.6]). Here, $q = 2$. For better readability, we use the symbols α, β instead of the numbers k_i on the left-hand side, and also instead of the corresponding summation indices α_{k_i} on the right-hand sides:

$$\operatorname{scal}^2 = \tfrac{1}{16} I_{\alpha\alpha|\beta\beta} = \tfrac{1}{16} \sum_{\alpha,\beta=1}^{r} \operatorname{Tr}(j_{Z_\alpha}^2)\operatorname{Tr}(j_{Z_\beta}^2),$$

$$|\operatorname{ric}|^2 = \tfrac{1}{4} I_{\alpha\alpha\beta\beta} + \tfrac{1}{16} I_{\alpha\beta|\alpha\beta} = \tfrac{1}{4} \sum_{\alpha,\beta=1}^{r} \operatorname{Tr}(j_{Z_\alpha}^2 j_{Z_\beta}^2) + \tfrac{1}{16} \sum_{\alpha,\beta=1}^{r} (\operatorname{Tr}(j_{Z_\alpha} j_{Z_\beta}))^2,$$

$$|R|^2 = \tfrac{1}{2} I_{\alpha\alpha\beta\beta} + \tfrac{3}{8} I_{\alpha\beta|\alpha\beta} + \tfrac{1}{8} I_{\alpha\beta\alpha\beta} = \tfrac{1}{2} \sum_{\alpha,\beta=1}^{r} \operatorname{Tr}(j_{Z_\alpha}^2 j_{Z_\beta}^2)$$
$$+ \tfrac{3}{8} \sum_{\alpha,\beta=1}^{r} (\operatorname{Tr}(j_{Z_\alpha} j_{Z_\beta}))^2 + \tfrac{1}{8} \sum_{\alpha,\beta=1}^{r} \operatorname{Tr}(j_{Z_\alpha} j_{Z_\beta} j_{Z_\alpha} j_{Z_\beta}).$$

Similarly, $|\operatorname{ric}|^2 \operatorname{scal} = \frac{1}{16} I_{\alpha\alpha\beta\beta|\gamma\gamma} + \frac{1}{64} I_{\alpha\beta|\alpha\beta|\gamma\gamma}$ is an example of a curvature invariant of order six.

It is quite tedious to compute the curvature invariant $|\nabla R|^2$ which is of particular interest to us; however, for our purposes the following partial information [1, Lemma 4.13(i)] will be sufficient:

$$|\nabla R|^2 = -\tfrac{3}{2} I_{\alpha\beta\gamma|\alpha\beta\gamma} + L = -\tfrac{3}{2} \textstyle\sum_{\alpha,\beta,\gamma=1}^{r} (\operatorname{Tr}(j_{Z_\alpha} j_{Z_\beta} j_{Z_\gamma}))^2 + L, \tag{1}$$

where L is a linear combination of other sixth-order invariants $I_{k_1\ldots k_\lambda|\ldots|k_\mu\ldots k_6}$ in which, however, the separations $\ldots|\ldots$ occur (if they occur) only after evenly many indices – unlike in $I_{\alpha\beta\gamma|\alpha\beta\gamma}$, where such a separation occurs after three indices.

We will now show that eq. (1) implies that in the case of two-step nilmanifolds of Heisenberg type, the invariant $I_{\alpha\beta\gamma|\alpha\beta\gamma}$ together with the dimensions m and r already determines the value of $|\nabla R|^2$ (see Lemma 16). A linear map $j : \mathfrak{z} \to \mathfrak{so}(\mathfrak{v})$ (or a Riemannian two-step nilmanifold associated with it) is called *of Heisenberg type* if

$$j_Z j_W + j_W j_Z = -2\langle Z, W\rangle \operatorname{Id}_{\mathfrak{v}} \quad \text{for all } Z, W \in \mathfrak{z}, \tag{2}$$

or equivalently, $j_Z^2 = -|Z|^2 \operatorname{Id}_{\mathfrak{v}}$ for all $Z \in \mathfrak{z}$. By this condition, j extends to a representation of the real Clifford algebra C_r over $\mathfrak{z} = \mathbb{R}^r$ and turns $\mathfrak{v}$ into a module over C_r; Clifford multiplication by Z is given by $j_Z : \mathfrak{v} \to \mathfrak{v}$.

In this situation, those invariants from Proposition 15 which contain only separations after evenly many indices simplify as follows:

Lemma 16 ([1, Corollary 5.2(i)]). *In the Heisenberg-type case, any $I_{k_1\ldots k_\lambda|\ldots|k_\mu\ldots k_{2q}}$ as in Proposition 15 in which all the subtuples $(k_1, \ldots, k_\lambda), \ldots, (k_\mu, \ldots, k_{2q})$ are of even length can be expressed as a universal polynomial in $m = \dim\mathfrak{v}$ and $r = \dim\mathfrak{z}$ which does not depend on j.*

To see this, let $(\alpha_1, \dots, \alpha_k)$ be any tuple in $\{1, \dots, r\}^k$. By the Clifford relation (2), $j_{Z_{\alpha_1}}$ and $j_{Z_{\alpha_2}}$ anticommute for $\alpha_1 \neq \alpha_2$. Using this, one first rearranges the factors of $j_{Z_{\alpha_1}} \cdots j_{Z_{\alpha_k}}$ in nondecreasing order w.r.t. the values of the α_i. From this and $(j_{Z_{\alpha_i}})^2 = -\mathrm{Id}_{\mathfrak{v}}$ it follows that there exists $c \in \{0, 1\}$ (depending on the k-tuple, but not on j) such that $j_{Z_{\alpha_1}} \cdots j_{Z_{\alpha_k}} = (-1)^c j_{Z_{\beta_1}} \cdots j_{Z_{\beta_\ell}}$, where $\beta_1 < \cdots < \beta_\ell$ and where $\{\beta_1, \dots, \beta_\ell\}$ consists precisely of those α_i which occur an odd number of times in $(\alpha_1, \dots, \alpha_k)$. Now assume k is even; then ℓ, too, is even. If $\ell = 0$ then $\mathrm{Tr}(j_{Z_{\alpha_1}} \cdots j_{Z_{\alpha_k}}) = (-1)^c \mathrm{Tr}(\mathrm{Id}_{\mathfrak{v}}) = (-1)^c m$. If $\ell > 0$ then $\mathrm{Tr}(j_{Z_{\alpha_1}} \cdots j_{Z_{\alpha_k}}) = (-1)^c \mathrm{Tr}(j_{Z_{\beta_1}} \cdots j_{Z_{\beta_\ell}}) = 0$ by eq. (2) and by the cyclicity of the trace. Lemma 16 now follows by applying the above to each factor in each of the r^q summands of the sum as which $I_{k_1 \dots k_\lambda | \dots | k_\mu \dots k_{2q}}(j)$ is defined.

Let $j : \mathfrak{z} \to \mathfrak{so}(\mathfrak{v})$ be of Heisenberg type. The *Clifford volume element* associated with the orthonormal basis $\{Z_1, \dots, Z_r\}$ is defined as

$$\omega_r := Z_1 \cdot \cdots \cdot Z_r \in C_r.$$

For the value of $|\nabla R|^2$, the Clifford action of ω_r now plays a special role in the case $r = 3$. Since $\mathrm{Tr}(j_{Z_\alpha} j_{Z_\beta} j_{Z_\gamma})$ obviously vanishes whenever at least two of the indices α, β, γ are equal, we have in the case $r = 3$:

$$I_{\alpha\beta\gamma|\alpha\beta\gamma}(j) = 3!\,(\mathrm{Tr}(j_{Z_1} j_{Z_2} j_{Z_3}))^2 = 6(\mathrm{Tr}(j_{\omega_3}))^2.$$

(Recall that j extends to a representation of the real Clifford algebra C_r on $\mathfrak{v}$.) Together with eq. (1) and Lemma 16 this equation immediately implies:

Corollary 17. *Let $j, j' : \mathfrak{z} = \mathbb{R}^3 \to \mathfrak{so}(m) = \mathfrak{so}(\mathfrak{v})$ be two linear maps of Heisenberg type. The Riemannian two-step nilmanifolds associated with j, resp. with j', have the same value of $|\nabla R|^2$ if and only if $(\mathrm{Tr}(j_{\omega_3}))^2 = (\mathrm{Tr}(j'_{\omega_3}))^2$.*

In the next section, we will discuss isospectral examples of Heisenberg-type nilmanifolds. As it will turn out, the criterion for local nonisometry of these isospectral pairs will precisely be (this time regardless of r) equalness of the squared traces of the associated Clifford volume elements.

3.2 Isospectral examples

Laplace isospectrality of Riemannian two-step nilmanifolds is closely linked to "isospectrality" of the associated j-maps. Two linear maps $j, j' : \mathfrak{z} \to \mathfrak{so}(\mathfrak{v})$ are called *isospectral* if for each $Z \in \mathfrak{z}$, the maps $j_Z, j'_Z \in \mathfrak{so}(\mathfrak{v})$ are similar, that is, have the same eigenvalues (with multiplicities) in $\mathbb{C}$. By skew symmetry of the j_Z, this is equivalent to the existence, for each $Z \in \mathfrak{z}$, of some $A_Z \in \mathrm{O}(\mathfrak{z})$ such that $j'_Z = A_Z j_Z A_Z^{-1}$. Note that A_Z may depend on Z.

If $j : \mathfrak{z} = \mathbb{R}^r \to \mathfrak{so}(\mathfrak{v}) = \mathfrak{so}(m)$ satisfies $[\mathbb{Z}, \mathbb{Z}]^j \subseteq \mathbb{Z}^r$ then, by the Campbell–Baker–Hausdorff formula,

$$\Gamma(j) := \exp\left(\mathbb{Z}^m \oplus \tfrac{1}{2}\mathbb{Z}^r\right) \tag{3}$$

is a cocompact discrete subgroup of $G(j)$; here, exp denotes the Lie group exponential map from $\mathfrak{g}(j)$ to $G(j)$. The following is a specialized version of a result by Gordon and Wilson [11] (see [20] for an explanation about how to derive it from the original, more general version):

Proposition 18 ([11, 3.2, 3.7, 3.8]). *Let $j, j' : \mathfrak{z} = \mathbb{R}^r \to \mathfrak{so}(\mathfrak{v}) = \mathfrak{so}(m)$ be isospectral. Assume that both $[\mathbb{Z}^m, \mathbb{Z}^m]^j$ and $[\mathbb{Z}^m, \mathbb{Z}^m]^{j'}$ are contained in $\mathbb{Z}^r$. For each $Z \in \mathbb{Z}^r$ assume that the lattices $\ker(j_Z) \cap \mathbb{Z}^m$ and $\ker(j'_Z) \cap \mathbb{Z}^m$ have the same length spectrum. Then the compact Riemannian manifolds $(\Gamma(j)\backslash G(j), g(j))$ and $(\Gamma(j')\backslash G(j'), g(j'))$ are isospectral for the Laplace operator on functions.*

In the special case that $j, j' : \mathfrak{z} \to \mathfrak{so}(\mathfrak{v})$ are both of Heisenberg type, j and j' are obviously isospectral because the eigenvalues of both of j_Z and j'_Z then are $\pm i|Z|$, each with multiplicity $(\dim \mathfrak{v})/2$. Moreover, $\ker(j_Z) = \ker(j'_Z) = \{0\}$ for all $Z \neq 0$. Therefore, if the matrix entries of each j_{Z_α} with respect to $\{X_1, \dots, X_m\}$ are integer, then all conditions of Proposition 18 are satisfied and $(\Gamma(j)\backslash G(j), g(j))$, $(\Gamma(j')\backslash G(j'), g(j'))$ are isospectral. Note that it was just such a pair of manifolds which Gordon had constructed in [10] as the very first example of isospectral, locally nonisometric manifolds (associated with $j^{3;2,0}$ and $j^{3;1,1}$ in the notation below).

Why do there exist nontrivial isospectral pairs j, j' of Heisenberg type? The reason is that for each $r \in \{3, 7, 11, 15, \dots\}$ there are precisely two nonisomorphic simple real modules $\mathfrak{m}^r_+$ and $\mathfrak{m}^r_-$ over C_r (see [15, p. 32] or [2]; for other values of r, there is only one simple C_r-module up to isometry). The modules $\mathfrak{m}^r_+$ and $\mathfrak{m}^r_-$ have the same dimension d_r and can be distinguished by the action of the Clifford volume element ω_r: After possibly switching names, ω_r acts on $\mathfrak{m}^r_+$ as Id and on $\mathfrak{m}^r_-$ as $-$ Id. By results from [7] we can identify $\mathfrak{m}^r_\pm$ with $\mathbb{R}^{d_r}$ in such a way that for both modules, all matrix entries of the Clifford multiplications with the elements of the given orthonormal basis $\{Z_1, \dots, Z_r\}$ of $\mathbb{R}^r$ are in $\{0, 1, -1\}$.

For $a, b \in \mathbb{N}_0$ let $j^{r;a,b}$ denote the representation of C_r on $\mathfrak{v} := (\mathbb{R}^{d_r})^{\oplus(a+b)}$ viewed as $(\mathfrak{m}^r_+)^{\oplus a} \oplus (\mathfrak{m}^r_-)^{\oplus b}$. For any pair (a, b), $(a', b') \in \mathbb{N}_0 \times \mathbb{N}_0$ with $a + b = a' + b'$ but $\{a, b\} \neq \{a', b'\}$, consider the maps $j^{r;a,b}, j^{r;a',b'} : \mathbb{R}^r \subset C_r \to \mathfrak{so}(m)$, where $m := (a+b)d_r$. We obviously have

$$\left(\mathrm{Tr}(j^{r;a,b}_{\omega_r})\right)^2 = ((a-b)d_r)^2 \neq \left((a'-b')\,d_r\right)^2 = \left(\mathrm{Tr}\left(j^{r;a',b'}_{\omega_r}\right)\right)^2. \tag{4}$$

Moreover, the associated $\Gamma(j^{r;a,b})$, $\Gamma(j^{r;a',b'})$ from eq. (3) are cocompact discrete subgroups. So one indeed has a pair of isospectral compact nilmanifolds of Heisenberg type for each pair $j^{r;a,b}$, $j^{r;a',b'}$ as above. The following lemma, together with eq. (4), implies that the two manifolds are not locally isometric:

Lemma 19. *Let $j, j' : \mathfrak{z} = \mathbb{R}^r \to \mathfrak{so}(m) = \mathfrak{so}(\mathfrak{v})$ be two linear maps of Heisenberg type. Then associated Riemannian metrics $g(j)$ and $g(j')$ are locally isometric if and only if*

$$(\mathrm{Tr}(j_{\omega_r}))^2 = (\mathrm{Tr}(j'_{\omega_r}))^2. \tag{5}$$

In fact, eq. (5) is necessary since by general results of Gordon and Wilson [11], local isometry of $g(j)$ and $g(j')$ always implies the existence of $(A, B) \in \mathrm{O}(\mathfrak{v}) \times \mathrm{O}(\mathfrak{z})$ such that $j'_Z = A j_{B^{-1}(Z)} A^{-1}$ for all $Z \in \mathfrak{z}$; but then $B^{-1}(Z_1) \cdot \dots \cdot B^{-1}(Z_r) = \det(B^{-1})\omega_r = \pm\omega_r$ (see [15, p. 34]) and thus $j'_{\omega_r} = \pm A j_{\omega_r} A^{-1}$, which obviously implies eq. (5). That condition (5) is also sufficient for local isometry basically follows from eq. (4) and the classification of modules over C_r.

Recalling Corollary 17, we conclude that in the case $r = 3$, the isospectral pairs which we just discussed have different values (and different integrals) of $|\nabla R|^2$.

3.3 Additional results

Note that the space of curvature invariants of order six is 17-dimensional (see, e.g., [12]), and there is a seven-dimensional subspace consisting of invariants whose integral over any closed Riemannian manifold is zero. Choosing a commonly used basis for a certain 10-dimensional complement of that subspace, T. Arias Marco and the second author gave, for each of those ten basis elements (of which $|\nabla R|^2$ is only one), isospectral Riemannian two-step nilmanifolds differing in that invariant (see [1]). All those examples arise from Proposition 18. Some more results from [1] are the following:

- If $j, j' : \mathbb{R}^r \to \mathfrak{so}(m)$ are of Heisenberg type and $2q < 2r$ then for each of the invariants from Proposition 15 we have $I_{k_1 \dots k_\lambda | \dots | k_\mu \dots k_{2q}}(j) = I_{k_1 \dots k_\lambda | \dots | k_\mu \dots k_{2q}}(j')$ [1, Proposition 5.5(i)]. In particular, any two isospectral nilmanifolds of Heisenberg type with $\dim \mathfrak{z} = r$ cannot differ in any curvature invariant of order $2q < 2r$ [1, Theorem 5.6(i)].
- Any two nilmanifolds N, N' of Heisenberg type which are not locally isometric are not curvature equivalent [1, Proposition 5.9]; that is, there does not exist a vector space isomorphism from T_pN to $T_{p'}N'$ intertwining the Riemannian curvature tensors. Together with the previous result this implies, using the isospectral pairs associated with $j^{r;a,b}$, $j^{r;a',b'}$ for $r \in \{3, 7, 11, 15, \dots\}$ as in Section 3.2: For any $k \in \mathbb{N}$, there exist pairs of locally homogeneous Riemannian manifolds which are not curvature equivalent, but do not differ in any curvature invariant of order up to $2k$ [1, Theorem 5.11].

4 Generic irreducibility of Laplace eigenspaces on certain compact Lie groups

We now turn to a problem which might be called "minimizing eigenvalue multiplicities under restrictions." Let G be an n-dimensional compact, connected Lie group and g a left invariant metric on G. Then G acts from the left by isometries on (G, g) and, hence, on each eigenspace of the associated Laplacian Δ_g. This implies that each nonzero eigenvalue of Δ_g has multiplicity greater than one. If the action of G on each eigenspace is irreducible, then these multiplicities still are "as small as possible" (or "no greater than necessary"). Victor Guillemin raised the question whether this can always be achieved by choosing g as a generic element within the (finite-dimensional) set of left invariant metrics on G. We will first explain how to translate this into an algebraic problem.

4.1 Algebraic conditions for irreducibility of eigenspaces

Investigating the above question immediately leads to representation theoretic considerations. In fact, letting ρ denote the right regular representation of G on $L^2(G, \mathbb{C})$, given by $\rho(x) : f \mapsto f(\,.\,x)$ for $x \in G$, one has

$$\Delta_g f = -\sum_{k=1}^{n}(\rho_*(Y_k))^2 f, \tag{6}$$

where $\{Y_1, \ldots, Y_n\}$ is a g-orthonormal basis of the associated Lie algebra $\mathfrak{g} = T_eG$. In particular, if $V \subset L^2(G, \mathbb{C})$ is invariant under ρ then it is invariant under Δ_g, too. Recall the famous Peter–Weyl theorem (see, e.g., [5, III.1–III.3]): For any complex irreducible representation $\rho_V : G \to \mathrm{GL}(V)$, the isotypical component $I(V)$ of V as a subrepresentation of ρ is invariant not only under ρ but also under the left regular representation λ of G on $L^2(G, \mathbb{C})$, given by $\lambda(x) : f \mapsto f(x^{-1}\,.\,)$; moreover, there is a vector space isomorphism from $I(V) \subset L^2(G, \mathbb{C})$ to $V^* \otimes V$ under which

$$\rho(x) \text{ corresponds to } \mathrm{Id} \otimes \rho_V(x), \text{ and } \lambda(x) \text{ corresponds to } (\rho_V(x)^{-1})^* \otimes \mathrm{Id}\,.$$

So, if μ is a simple eigenvalue of

$$\Delta_g^V := -\sum_{k=1}^{n}((\rho_V)_*(Y_k))^2$$

which, moreover, happens to be different from the eigenvalues of Δ_g^W for each other complex irreducible representation $W \not\simeq V$, then the multiplicity of μ as an eigenvalue of Δ_g has the minimal possible value, namely, $\dim(V)$: If $v \in V$ is a corresponding eigenvector, then $V^* \otimes \mathbb{C}v \subset V^* \otimes V$ corresponds to the μ-eigenspace of Δ_g on $L^2(G, \mathbb{C})$, and is irreducible with respect to the left action λ of G.

Thus, irreducibility of all complex(!) eigenspaces of Δ_g is obviously equivalent to the following two conditions being jointly satisfied: Simplicity of each eigenvalue of $\Delta_{g|V}$ for every $V \in \mathrm{Irr}(G, \mathbb{C})$, and nonexistence of common eigenvalues of $\Delta_{g|V}$ and $\Delta_{g|W}$ whenever $V \ncong W$, for all $V, W \in \mathrm{Irr}(G, \mathbb{C})$, where the latter denotes a set of representatives of the irreducible complex representations of G.

However, neither of these two conditions can, in general, be achieved. For example, consider $G = \mathrm{SU}(2)$ and let $V = \mathbb{C}^2$ denote its standard representation. For any $Y \in \mathfrak{su}(2)$, the endomorphism $((\rho_V)_*(Y))^2$ of V is a multiple of the identity. So Δ_g^V, too, is a multiple of the identity for each left invariant metric g on G; in particular, the eigenvalues of Δ_g^V are never simple here. On the bright side, we note that this does not prevent the corresponding real(!) eigenspace of Δ_g, for suitable g, from being irreducible nevertheless: The isotypical component $I(V)$ of the standard representation is spanned by the four functions assigning to each $x = \begin{pmatrix} a & -b \\ \bar{b} & \bar{a} \end{pmatrix} \in \mathrm{SU}(2)$ its matrix entries $a, -b, \bar{b}, \bar{a}$. So, $I(V) \cap C^\infty(\mathrm{SU}(2), \mathbb{R})$ has real dimension four and is spanned by the real-valued functions $\mathrm{Re}(a), \mathrm{Im}(a), \mathrm{Re}(b), \mathrm{Im}(b)$. By elementary arguments one checks that this real vector space is irreducible with respect to λ. So, if $\mu(g)$ denotes the eigenvalue of Δ_g^V, then the real $\mu(g)$-eigenspace of Δ_g will indeed be irreducible as long as g is such that $\mu(g)$ differs from all eigenvalues of Δ_g^W for each $W \in \mathrm{Irr}(\mathrm{SU}(2), \mathbb{C})$ with $W \ncong V$ (and this is the case here for generic left invariant metrics g, as can easily be shown).

Similarly, the example $G = U(1)$ shows that one cannot always achieve disjointness of the eigenvalue sets of Δ_g^V and Δ_g^W for $V, W \in \mathrm{Irr}(G, \mathbb{C})$ with $V \ncong W$: For any left invariant metric g on $U(1)$, the eigenvalue of Δ_g^V on the standard representation $V = \mathbb{C}$ of $U(1)$ will be the same as the eigenvalue of $\Delta_g^{V^*}$ on the dual representation V^*, although V and V^* are not isomorphic here. Again, this does not prevent irreducibility of the corresponding real(!) eigenspace of Δ_g, which is spanned by the two functions $U(1) \ni (e^{it}) \mapsto \cos t, \sin t \in \mathbb{R}$ and whose complexification is the complex vector space $I(V) \oplus I(V^*) = V \oplus V^*$.

The reason for the phenomena occurring in the above two examples is that the standard representation of SU(2) is of quaternionic type, and the standard representation of $U(1)$ is of complex type. These notions can be characterized as follows (see, e.g., [5, II.6]): A complex irreducible representation V of G is

- of *quaternionic type* if $V \cong V^*$ and $V \oplus V \cong U \otimes \mathbb{C}$ for some $U \in \mathrm{Irr}(G, \mathbb{R})$,
- of *complex type* if $V \ncong V^*$; in this case, $V \oplus V^* \cong U \otimes \mathbb{C}$ for some $U \in \mathrm{Irr}(G, \mathbb{R})$,
- of *real type* if $V \cong V^*$ and $V \cong U \otimes \mathbb{C}$ for some $U \in \mathrm{Irr}(G, \mathbb{R})$.

(Here, $\mathrm{Irr}(G, \mathbb{R})$ denotes a system of representatives of irreducible real representations of G.) Any $V \in \mathrm{Irr}(G, \mathbb{C})$ is of exactly one of the above three types. And, as it turns out, the correct criterion for irreducibility of each real(!) eigenspace of Δ_g on G takes these different types into account:

Proposition 20 ([21, Corollary 3.3]). *Let g be a left invariant metric on G. Then all real eigenspaces of Δ_g are irreducible if and only if each of the following conditions is satisfied:*

(i) *For any pair $V, W \in \mathrm{Irr}(G, \mathbb{C})$ with $V \not\cong W$ and $V^* \not\cong W$, Δ_g^V and Δ_g^W have no common eigenvalues.*

(ii) *For each $V \in \mathrm{Irr}(G, \mathbb{C})$ of real or complex type, all eigenvalues of Δ_g^V are simple.*

(iii) *For each $V \in \mathrm{Irr}(G, \mathbb{C})$ of quaternionic type, all eigenvalues of Δ_g^V are of multiplicity precisely two.*

Simplicity of the eigenvalues of Δ_g^V, resp. disjointness of the eigenvalue sets of Δ_g^V and Δ_g^W, can be expressed using the discriminant, resp. the resultant, of the associated characteristic polynomials. For any $V \in \mathrm{Irr}(G, \mathbb{C})$ and g-orthonormal basis $\{Y_1, \dots, Y_n\}$ of $\mathfrak{g}$, note that

$$\Delta_g^V = D_V(Y_1^2 + \cdots + Y_n^2),$$

where the linear map $D_V : \mathrm{Sym}^2(\mathfrak{g}) \to \mathrm{End}(V)$ is defined by

$$D_V(Y \cdot Z) := -\tfrac{1}{2}\big((\rho_V)_*(Y) \circ (\rho_V)_*(Z) + (\rho_V)_*(Z) \circ (\rho_V)_*(Y)\big).$$

Here, $\mathrm{Sym}^2(\mathfrak{g}) \subset \mathfrak{g} \otimes \mathfrak{g}$ denotes the second symmetric tensor power of $\mathfrak{g}$. The subset

$$\mathrm{Sym}^2_+(\mathfrak{g}) := \big\{Y_1^2 + \ldots + Y_n^2 : \{Y_1, \ldots, Y_n\} \text{ a basis of } \mathfrak{g}\big\}$$

is open in $\mathrm{Sym}^2(\mathfrak{g})$. We will be interested in showing that certain polynomials on $\mathrm{Sym}^2_+(\mathfrak{g})$ do not vanish identically, and it will be more practical (and not change the problem) if we regard those polynomials as defined on the entire space $\mathrm{Sym}^2(\mathfrak{g})$.

First, let $p_V := \det(D_V(.) - X \cdot \mathrm{Id}) : \mathrm{Sym}^2(\mathfrak{g}) \to \mathbb{C}(X)$ denote the map sending $s \in \mathrm{Sym}^2(\mathfrak{g})$ to the characteristic polynomial of $D_V(s)$. By $\mathrm{res} : \mathbb{C}[X] \times \mathbb{C}[X] \to \mathbb{C}$ we denote the resultant (see, e.g., [9]); for two polynomials p, q, the number $\mathrm{res}(p, q)$ is given by a certain polynomial in the coefficients of p and q which vanishes if and only if p and q have a common zero. For $V, W \in \mathrm{Irr}(G, \mathbb{C})$ we now define:

$$\begin{aligned} a_{V,W} &:= \mathrm{res} \circ (p_V, p_W) : \mathrm{Sym}^2(\mathfrak{g}) \to \mathbb{C}, \\ b_V &:= \mathrm{res} \circ (p_V, p_V') : \mathrm{Sym}^2(\mathfrak{g}) \to \mathbb{C}, \\ c_V &:= \mathrm{res} \circ (p_V, p_V'') : \mathrm{Sym}^2(\mathfrak{g}) \to \mathbb{C}, \end{aligned}$$

where $p_V', p_V'' \in \mathbb{C}[X]$ denote the formal derivatives of $p_V \in \mathbb{C}[X]$ with respect to X. Note that $a_{V,W}$, b_V, c_V are $\mathbb{C}$-valued polynomials on $\mathrm{Sym}^2(\mathfrak{g})$ and have the following crucial properties: The polynomial $a_{V,W}$ is not identically zero if and only if there exists $s \in \mathrm{Sym}^2(\mathfrak{g})$ such that $D_V(s)$ and $D_W(s)$ have no common zeros; by the above considerations, this is equivalent to the existence of a left invariant metric g on G such that Δ_g^V and Δ_g^W have no common eigenvalues. Similarly, $b_V \neq 0$ is equivalent to the existence of a left invariant metric g on G such that all eigenvalues of Δ_g^V are simple. Finally, in the special case that V is of quaternionic type, all eigenvalues of Δ_g^V have even multiplicity; in this case, $c_V \neq 0$ is equivalent to the existence of a left invariant metric g on G such that each eigenvalue of Δ_g^V is of multiplicity precisely two.

Using these observations and Proposition 20, together with the fact that the complement of the union of the zero sets of countable many nontrivial polynomials is a residual set, we conclude:

Proposition 21 ([21, Proposition 3.7]). *Existence of a left invariant metric g on G such that all real eigenspaces of Δ_g are irreducible is equivalent to the following conditions being jointly satisfied:*

(a) $a_{V,W} \neq 0$ *for each pair* $V, W \in \mathrm{Irr}(G, \mathbb{C})$ *with* $V \not\cong W$ *and* $V^* \not\cong W$,
(b) $b_V \neq 0$ *for each* $V \in \mathrm{Irr}(G, \mathbb{C})$ *of real or complex type,*
(c) $c_V \neq 0$ *for each* $V \in \mathrm{Irr}(G, \mathbb{C})$ *of quaternionic type.*

Moreover, existence of one g with the above property implies that generic left invariant metrics g on G have the same property.

Remark 22. Generic irreducibility of eigenspaces is certainly the case if the Lie group is a torus: It is classically known that generic left invariant metrics on $\mathbb{R}^n/\mathbb{Z}^n$ have simple length spectrum. Then for every nonzero eigenvalue, the corresponding eigenspace is spanned by the real and imaginary parts of the associated character, and is certainly irreducible under the group action. It is similarly easy (but instructive) to verify, instead, the conditions of the above proposition in the case of tori.

4.2 The case of SU(2)

We now come to the simplest nontrivial case for our problem: Below, we will explain why the conditions of Proposition 21 are satisfied for SU(2) [21, Theorem 4.1]. Note that the most general class for which this property has been proved so far are the groups of the form $(\mathrm{SU}(2) \times \cdots \times \mathrm{SU}(2) \times T)/\Gamma$, where T is a torus and Γ is a discrete central subgroup [21, Corollary 4.10].

So let $G := \mathrm{SU}(2)$. As is well known, $\mathrm{Irr}(\mathrm{SU}(2), \mathbb{C}) = \{V_0, V_1, V_2, \ldots\}$, where $V_m := \mathrm{span}_{\mathbb{C}}\{v_{m,0}, \ldots, v_{m,m}\}$ with $v_{m,\ell} := z_1^{m-\ell} z_2^{\ell} \in \mathbb{C}[z_1, z_2]$, and where $x = \left(\begin{smallmatrix} a & -b \\ \bar{b} & \bar{a} \end{smallmatrix}\right) \in \mathrm{SU}(2)$ acts on V_m by $\rho_{V_m}(x) : v \mapsto \big((u, w) \mapsto v(ua + w\bar{b}, -ub + w\bar{a})\big)$ (here $v \in V_m$ is viewed as a function on $\mathbb{C}^2$). Moreover, V_m is of real type if of odd dimension $m + 1$, and of quaternionic type for even $m + 1$ [5, VI.4–VI.5]. Consider the basis $\{H, A, B\}$ of $\mathfrak{g} = \mathfrak{su}(2)$ with $H := \left(\begin{smallmatrix} i & 0 \\ 0 & -i \end{smallmatrix}\right)$, $A := \left(\begin{smallmatrix} 0 & i \\ i & 0 \end{smallmatrix}\right)$, $B := \left(\begin{smallmatrix} 0 & -1 \\ 1 & 0 \end{smallmatrix}\right)$.

One easily computes $D_{V_m}(H^2 + A^2 + B^2) = m(m + 2)\,\mathrm{Id}_{V_m}$; in particular, $a_{V_m, V_{m'}}(H^2 + A^2 + B^2) \neq 0$ whenever $m \neq m'$. This shows condition (a) of Proposition 21. For condition (c), note that $\varphi := (\rho_{V_m})_*(H)$ has matrix

$$\mathrm{diag}(im, i(m-2), i(m-4), \ldots, -i(m-2), -im)$$

with respect to the above basis of V_m. In particular, if V_m is quaternionic (i.e., m is odd), then each of the eigenvalues $m^2, (m-2)^2, \dots, 9, 1$ of $D_{V_m}(H^2) = -\varphi^2$ has multiplicity precisely two, so $c_{V_m}(H^2) \neq 0$.

It remains to show condition (b) of Proposition 21 for V_m of real type. The argument we give here is different from that in the proof of [21, Theorem 4.1]; instead, we use an idea from the proof of [21, Lemma 4.8(ii)]. Let $x := \exp(\frac{\pi}{2}B) = \left(\begin{smallmatrix} 0 & -1 \\ 1 & 0 \end{smallmatrix}\right)$ and $\tau := \rho_{V_m}(x)$. Then $\tau^2 = \mathrm{Id}$ since m is even. Let $V^+, V^- \subset V_m$ denote the (+1)- and (−1)-eigenspaces of τ, respectively, and let $\mathscr{R} := \mathrm{span}_{\mathbb{R}}\{v_{m,0}, \dots, v_{m,m}\}$. Since τ has real coefficients, $V^\pm$ is spanned by $V^\pm \cap \mathscr{R}$.

The endomorphism $\psi := (\rho_{V_m})_*(B)$ (which has real coefficients, but the same eigenvalues $im, i(m-2), \dots, -im$ as the above map φ) commutes with τ and preserves both $V^+ \cap \mathscr{R}$ and $V^- \cap \mathscr{R}$. So the nonzero eigenvalues of $\psi_{|V^+}$ come in conjugate pairs, and similarly for $\psi_{|V^-}$. Thus, ${\psi^2}_{|V^+}$ and ${\psi^2}_{|V^-}$ have no eigenvalues in common. In contrast, φ maps V^+ to V^-, and vice versa, because it anticommutes with τ (note that $\mathrm{Ad}_x(H) = -H$). But φ also preserves the eigenspaces of φ^2, and each of those is of dimension two or (in case of the eigenvalue 0) one. Thus, both ${\varphi^2}_{|V^+}$ and ${\varphi^2}_{|V^-}$ have only simple eigenvalues.

Consequently, there is a dense open set $\mathscr{O} \subset \mathbb{R}$ (in fact, the complement of the union of the zeros of two certain polynomials) such that for each $\alpha \in \mathscr{O}$, the endomorphism $D_\alpha := -(\alpha\varphi^2 + (1-\alpha)\psi^2) = D_{V_m}(\alpha H^2 + (1-\alpha)B^2)$ has *both* of the above properties (i.e., no common eigenvalues on V^+ and V^-, and only simple eigenvalues on each of them) and, thus, has only simple eigenvalues. This means $b_{V_m}(\alpha H^2 + (1-\alpha)B^2) \neq 0$ for $\alpha \in \mathscr{O}$; in particular, $b_{V_m} \neq 0$, as desired.

Acknowledgments: The Collaborative Research Center 647 *Space – Time – Matter* in Berlin and Potsdam provided a fruitful research environment and financial support for both authors, including travels to and visits by collaborators. The results in this chapter are a direct output of the research within subproject C2 of the CRC 647.

Bibliography

[1] Arias-Marco T, Schueth D. Inaudibility of sixth order curvature invariants. Rev R Acad Cienc Exactas, Fís Nat 2017;111:547–74.

[2] Berndt J, Tricerri F, Vanhecke L. Generalized Heisenberg groups and Damek-Ricci harmonic spaces. Lecture Notes in Mathematics 1598. New York–Berlin: Springer-Verlag, 1995.

[3] Boldt S. Properties of the Dirac spectrum on three dimensional lens spaces. Osaka J Math 2017;54:749–767.

[4] Boldt S, Lauret E. An explicit formula for the Dirac multiplicities on lens spaces. J Geom Anal 2017;27:689–725.

[5] Bröcker T, Tom Dieck T. Representations of compact Lie groups. Graduate Texts in Mathematics 98. New York–Berlin: Springer-Verlag, 2003.

[6] Cohen M. A course in simple-homotopy theory. Graduate Texts in Mathematics 10. New York–Berlin: Springer-Verlag, 1973.

[7] Crandall G, Dodziuk J. Integral structures on H-type Lie algebras. J Lie Theory 2002;12(1):69–79.
[8] Franc A. Spin structures and Killing spinors on lens spaces. J Geom Phys 1987;4(3):277–87.
[9] Gelfand IM, Kapranov MM, Zelevinsky AV. Discriminants, resultants, and multidimensional determinants. Boston: Birkhäuser, 1994.
[10] Gordon CS. Isospectral closed Riemannian manifolds which are not locally isometric. J Diff Geom 1993;37(3):639–49.
[11] Gordon CS, Wilson EN. Continuous families of isospectral Riemannian manifolds which are not locally isometric. J Diff Geom 1997;47(3):504–29.
[12] Gray A, Vanhecke L. Riemannian geometry as determined by the volumes of small geodesic balls. Acta Math 1979;142(3–4):157–98.
[13] Ikeda A. On lens spaces which are isospectral but not isometric. Ann Sci École Norm Sup (4) 1980;13(3):303–15.
[14] Lauret E, Miatello R, Rossetti JP. Spectra of lens spaces from 1-norm spectra of congruence lattices. Int Math Res Not IMRN 2016:1054–89.
[15] Lawson HB, Michelsohn M-L. Spin geometry. Princeton Mathematical Series 38. Princeton: Princeton University Press, 1989.
[16] Miatello R, Podestá R. The spectrum of twisted Dirac operators on compact flat manifolds. Trans Am Math Soc 2006;358(10):4569–603.
[17] Sakai T. On eigen-values of Laplacian and curvature of Riemannian manifold. Tôhoku Math J (2) 1971;23:589–603.
[18] Schueth D. Continuous families of isospectral metrics on simply connected manifolds. Ann Math 1999;149(1):287–308.
[19] Schueth D. Isospectral manifolds with different local geometries. J Reine Angew Math 2001;534:41–94.
[20] Schueth D. Integrability of geodesic flows and isospectrality of Riemannian manifolds. Math Z 2008;260(3):595–613.
[21] Schueth D. Generic irreducibility of Laplace Eigenspaces on certain compact Lie groups. Ann Glob Anal Geom 2017;52:187–200.
[22] Uhlenbeck K. Generic Properties of Eigenfunctions. Am Math J 1976;98(4):1059–78.

Andreas Juhl

On conformally covariant differential operators and spectral theory of the holographic Laplacian

Abstract: We review recent results on the structure of conformally covariant differential operators on functions and differential forms together with some applications to spectral theory.

Keywords: Conformally invariant operators, Branson–Gover operators, Q-curvature, Q-curvature operators, residue families, heat invariants, generalized Verma modules, F-method

Mathematics Subject Classification 2010: Primary 35J30, 35K08, 53A30, 58J50; Secondary 22E46, 35Q76, 53B20

1 Introduction

In these notes we describe recent results on certain classes of conformally covariant differential operators of high order on functions and differential forms on Riemannian manifolds. The GJMS operators [13] are defined as specific high-order conformally covariant corrections of powers of the Laplacian of a metric g by lower-order curvature terms. They generalize the well-known Yamabe and Paneitz operators. We start with an outline of an approach to the study of these operators which was developed in the monograph [18]. In the later works [19], these methods were used to describe recursive structures in the sequence of these operators. These structures are closely related to similar recursive structures in the sequence of Branson's Q-curvatures [20]. The main tool to unveil these structures is the notion of residue family operators. These are certain conformally covariant differential operators which are naturally associated with the eigenfunctions of the Laplacian of a Poincaré-Einstein metric (in the sense of Fefferman and Graham [7]) associated with the given metric. In Section 3, we describe the consequences of these results for the spectral theory of a geometric deformation of the Yamabe operator which is termed the holographic Laplacian [21]. We recall that the conformal covariance of a geometric elliptic self-adjoint differential operator acting on functions has remarkable consequences for the transformation properties of its heat kernel invariants under conformal changes of the underlying metric [5]. Analogues of these are the conformal transformation laws of Branson's Q-curvatures [3] and also of the so-called renormalized volume coefficients [16]. The theory of the holographic Laplacian sheds new light on the interrelations of these topics. In particular,

DOI 10.1515/9783110452150-002_s_003

it gives a uniform proof for the transformation laws of the heat kernel coefficients of the Yamabe operator and of the renormalized volume coefficients. In the last section, we leave Riemannian geometry and describe the algebraic structure of conformally equivariant differential operators which map differential forms on $\mathbb{R}^n$ to differential forms on a hyperplane $\mathbb{R}^{n-1}$. These operators are not equivariant with respect to the full conformal group of $\mathbb{R}^n$ but only with respect to the subgroup of those maps which leave the hyperplane invariant. The latter lack of equivariance motivates to call them symmetry-breaking operators [23, 26]. These operators are analogues of residue families as introduced in [18]. We fully describe all such operators and discuss some of their basic properties. Of particular significance are their relations to the so-called Branson-Gover operators which constitute a generalization of the GJMS operators to differential forms. The results will have analogues in the curved case and these in turn are expected to shed light on the structure of the still very mysterious Branson-Gover operators.

2 Residue families and the recursive structure of GJMS operators

In the present section, we describe some basic results on the structure of the so-called GJMS operators. These operators are high-order generalizations of the Yamabe operator

$$P_2(g) \stackrel{\text{def}}{=} \Delta_g - \left(\frac{n}{2} - 1\right) \mathsf{J}_g, \quad \mathsf{J}_g = \frac{\operatorname{scal}(g)}{2(n-1)}; \tag{1}$$

here and throughout we use the convention that $-\Delta_g = \delta_g d$ is non-negative. All GJMS operators $P_{2N}(g)$ are natural differential operators acting on functions. They are of the form

$$\Delta_g^N + \text{lower-order terms} \tag{2}$$

and are conformally covariant in the sense that

$$e^{(\frac{n}{2}+N)\varphi} P_{2N}(e^{2\varphi} g)(u) = P_{2N}(g)(e^{(\frac{n}{2}-N)\varphi} u) \tag{3}$$

for all $u, \varphi \in C^\infty(M)$. The operators $P_{2N}(g)$ were constructed by Graham, Jenne, Mason and Sparling in the seminal work [13] in terms of the powers of the Laplacian of the Fefferman–Graham ambient metric [7] on a space of *two* higher dimensions.

The second operator in the sequence $P_2, P_4, P_6, \ldots$ of GJMS operators is known as the Paneitz operator [28]. It is given by the formula

$$P_4 = \Delta^2 + \delta((n-2)\mathsf{J} - 4\mathsf{P})d + \left(\frac{n}{2} - 2\right)\left(\frac{n}{2}\mathsf{J}^2 - 2|\mathsf{P}|^2 - \Delta \mathsf{J}\right). \tag{4}$$

Here P denotes the Schouten tensor, i.e., $(n-2)\mathsf{P} = \mathrm{Ric} - \mathsf{J}g$, the tensor P acts on one-forms using g and $|\mathsf{P}|^2 = \mathsf{P}_{ij}\mathsf{P}^{ij}$.

On manifolds of odd dimension n, there are GJMS operators P_{2N} of any order $2N \geq 2$. However, on manifolds of even dimension n, the construction in [13] does not give a conformally covariant power of Laplacian of order $> n$ for general metrics. In fact, by [14] (for $n = 4$ and $N = 3$) and [12] (in general) no such operator exists for general metrics. However, for locally conformally flat metrics and conformally Einstein metrics, the construction in [13] yields conformally covariant powers of the Laplacian of any even order. In particular, we have an infinite sequence of such operators for all round spheres S^n.

The GJMS operators play a central role in conformal differential geometry and geometric analysis. More information on these aspects can be found in the recent monographs [2, 6, 18] and the references therein.

Since the complexity of the lower-order terms in eq. (2) grows exponentially with N, finding an explicit description of these terms seems to be an almost impossible task. However, the following results will show that the enormous complexity of these operators can be described in terms of a sequence of natural *second-order* operators. For full details we refer to [19].

In order to formulate these results, we introduce some more notation.

First, we shall use the following combinatorial conventions. A sequence $I = (I_1, \ldots, I_r)$ of integers $I_j \geq 1$ will be regarded as a composition of the natural number $|I| = I_1 + I_2 + \cdots + I_r$. In other words, compositions are partitions in which the order of the summands is considered. $|I|$ will be called the size of I. To any composition $I = (I_1, \ldots, I_r)$, we associate the numbers

$$m_I = -(-1)^r |I|!\,(|I|-1)! \prod_{j=1}^{r} \frac{1}{I_j!\,(I_j-1)!} \prod_{j=1}^{r-1} \frac{1}{I_j + I_{j+1}}$$

and

$$n_I = \prod_{j=1}^{r} \binom{\sum_{k\leq j} I_k - 1}{I_j - 1} \binom{\sum_{k\geq j} I_k - 1}{I_j - 1}.$$

Note that $m_{(N)} = n_{(N)} = 1$ for all $N \geq 1$ and $n_{(1,\ldots,1)} = 1$. Moreover, for any composition I, we define the operator

$$P_{2I} = P_{2I_1} \circ \cdots \circ P_{2I_r}.$$

Then we set

$$\mathscr{M}_{2N} = \sum_{|I|=N} m_I P_{2I} = P_{2N} + \text{compositions of lower-order GJMS operators.} \qquad (5)$$

These sums contain 2^{N-1} terms. Similarly, we define the operators

$$\mathcal{M}_{2I} = \mathcal{M}_{2I_1} \circ \cdots \circ \mathcal{M}_{2I_r}.$$

Second, we recall the notion of Poincaré-Einstein metrics in the sense of Fefferman and Graham [7]. Let M be a manifold of dimension n with a given metric g. On the space $X = M \times (0, \varepsilon)$ we consider metrics of the form

$$g_+ = r^{-2}(dr^2 + g(r)), \tag{6}$$

where $g(r)$ is a one-parameter family of metrics on M so that $g_0 = g$. We require that for *odd* $n \geq 3$ the tensor

$$\mathrm{Ric}(g_+) + ng_+$$

vanishes to infinite order along M, and that for *even* $n \geq 4$

$$\mathrm{Ric}(g_+) + ng_+ = O(r^{n-2}) \tag{7}$$

together with a certain vanishing trace condition for $\mathrm{Ric}(g_+) + ng_+$ to the order r^{n-2}. If $g(r)$ is assumed to be even in r and n is odd, these conditions uniquely determine the family $g(r)$ for a general metric g. Similarly, for even n, the conditions uniquely determine the coefficients $g_{(2)}, \ldots, g_{(n-2)}, \tilde{g}_{(n)}$ and the trace of $g_{(n)}$ in the even power series

$$g(r) = g + r^2 g_{(2)} + \cdots + r^{n-2} g_{(n-2)} + r^n (g_{(n)} + \log r \tilde{g}_{(n)}) + \cdots,$$

where $\mathrm{tr}_g(\tilde{g}_{(n)}) = 0$. For even n, all constructions needed here depend only on the terms $g_{(2)}, \ldots, g_{(n-2)}$ and $\mathrm{tr}_g(g_{(n)})$ (which are uniquely determined by g).

The volume form of a Poincaré–Einstein metric g_+ associated with g takes the form

$$d\mathrm{vol}(g_+) = r^{-n-1} v(r) d\mathrm{vol}(g) dr$$

with

$$v(r) = 1 + v_2 r^2 + v_4 r^4 + \cdots + v_n r^n + \cdots;$$

here $v_n = 0$ if n is odd. In odd dimensions n, all coefficients $v_2, v_4, \ldots$ are scalar-valued curvature invariants of g. Similarly, in even dimensions n, the coefficients $v_2, \ldots, v_n$ are uniquely determined by g. The functionals $g \mapsto v_{2j}(g)$ are called the *renormalized volume coefficients* of g [15, 16]. In even dimensions n, the quantity v_n is the (infinitesimal) conformal anomaly of the renormalized volume of g_+ [15]. In the physical literature, v_n is called the *holographic anomaly*. Now we set

$$w(r) = \sqrt{v(r)} = 1 + w_2 r^2 + w_4 r^4 + \cdots + w_n r^n + \cdots$$

with $w_n = 0$ for odd n. The coefficients $w_2, w_4, w_6, \ldots$ are polynomials in $v_2, v_4, v_6, \ldots$. In particular,

$$2w_2 = v_2, \quad 8w_4 = 4v_4 - v_2^2 \quad \text{and} \quad 16w_6 = 8v_6 - 4v_4v_2 + v_2^3.$$

Finally, we introduce the generating function

$$\mathscr{H}(r) \stackrel{\text{def}}{=} \sum_{N\geq 1} \mathscr{M}_{2N} \frac{1}{(N-1)!^2} \left(\frac{r^2}{4}\right)^{N-1} \tag{8}$$

of the sequence $\mathscr{M}_2, \mathscr{M}_4, \ldots$. For even n and general metrics, the sum in eq. (8) is defined to run up to r^{n-2}. However, in special cases with well-defined GJMS operators beyond the order n, the sum in eq. (8) may extend appropriately beyond r^{n-2}. For odd n, GJMS operators are well defined for any order $2N \geq 2$, and the sum in eq. (8) is a formal infinite power series. In particular, the sum is an infinite power series for locally conformally flat metrics in all dimensions $n \geq 3$.

The following result gives formulas for all GJMS operators [19].

Theorem 1. *The GJMS operators of a Riemannian manifold (M, g) of dimension $n \geq 3$ can be written in the form*

$$P_{2N} = \sum_{|I|=N} n_I \mathscr{M}_{2I} = \mathscr{M}_2^N + \textit{compositions with fewer factors}. \tag{9}$$

Moreover, all differential operators $\mathscr{M}_{2N}(g)$ are second order and are given by the formula

$$\mathscr{H}(g)(r) = -\delta(g(r)^{-1}d) - r^{-2}\left(\Delta_{g_+}(\log w) - |d \log w|^2_{g_+}\right). \tag{10}$$

Here δ is defined with respect to g, and we regard $g(r)$ as an endomorphism on one-forms using g. Moreover, w is regarded as a function on X.

The fact that the operators $\mathscr{M}_{2N}$ are second order may be rephrased by stating that any GJMS operator P_{2N} is the sum of the linear combination

$$-\sum_{|I|=N,\, I\neq(N)} m_I P_{2I}$$

of compositions of lower-order GJMS operators and the second-order operator given by the coefficient of

$$\frac{1}{(N-1)!^2}\left(\frac{r^2}{4}\right)^{N-1}$$

in the Taylor series of the one-parameter family

$$\mathscr{D}(r;g) \stackrel{\text{def}}{=} -\delta(g(r)^{-1}d) - r^{-2}\left(\Delta_{g_+}(\log w) - |d\log w|^2_{g_+}\right). \tag{11}$$

Such identities may be regarded as *recursive* formulas for GJMS operators. For example, by an evaluation of these formulas we find the recursive formula

$$P_6 = (2P_2P_4 + 2P_4P_2 - 3P_2^3) - 48\delta(\mathsf{P}^2 d) - \frac{16}{n-4}\delta(\mathscr{B}d) + \mu_6 \tag{12}$$

with the scalar quantity

$$\mu_6 = -8\frac{n-6}{n-4}(\mathscr{B},\mathsf{P}) - 8(n-6)\operatorname{tr}(\mathsf{P}^3) - 16\mathsf{J}|\mathsf{P}|^2 - 4|d\mathsf{J}|^2 + 4\Delta\left(|\mathsf{P}|^2\right) - 16\delta\left(\mathsf{P}d\mathsf{J}\right).$$

Here $\mathscr{B}$ denotes the Bach tensor.

The proof of Theorem 1 in [19] rests on the theory of *residue families* (as introduced in [18]). We continue with the definition of residue families

$$D_{2N}^{\text{res}}(g;\lambda) : C^\infty(M\times[0,\varepsilon)) \to C^\infty(M)$$

(see also [2] for an introduction) and describe their main properties. In particular, we describe their recursive structure which finds its natural expression in the form of systems of factorization identities.

Now let $\iota : M \hookrightarrow M\times[0,\varepsilon)$ denote the embedding $\iota(m) = (m,0)$. We consider formal solutions

$$u(\cdot,r) \sim \sum_{j\ge 0} r^{\lambda+2j}\mathscr{T}_{2j}(g;\lambda)(f)(\cdot),\quad \mathscr{T}_0(f) = f \in C^\infty(M) \tag{13}$$

of the eigen-equation

$$-\Delta_{g_+}u = \lambda(n-\lambda)u,\ \lambda\in\mathbb{C}.$$

The coefficients in expansion (13) are given by rational families $\mathscr{T}_{2j}(g;\lambda)$ (in λ) of differential operators acting on the "boundary value" f of u.

Definition 2 (Residue families). For general g, even n and $2N \le n$, we define

$$D_{2N}^{\text{res}}(g;\lambda) = 2^{2N}N!\left[\left(-\frac{n}{2}-\lambda+2N-1\right)\cdots\left(-\frac{n}{2}-\lambda+N\right)\right]\delta_{2N}(g;\lambda+n-2N), \tag{14}$$

where

$$\delta_{2N}(g;\lambda) = \sum_{j=0}^{N}\frac{1}{(2N-2j)!}\left[\mathscr{T}_{2j}^*(g;\lambda)v_0 + \cdots + \mathscr{T}_0^*(g;\lambda)v_{2j}\right]\iota^*(\partial/\partial r)^{2N-2j}. \tag{15}$$

Here the holographic coefficients act as multiplication operators, and $\mathcal{T}_{2j}^*(g;\lambda)$ denotes the formal adjoint of the differential operator $\mathcal{T}_{2j}(g;\lambda)$ on $C^\infty(M)$ with respect to the metric g. Similarly, we define $D_{2N}^{\mathrm{res}}(g;\lambda)$, $N \geq 1$ for general g and odd n.

Note that $D_0^{\mathrm{res}}(g;\lambda) = i^*$. The main role of the polynomial pre-factor in eq. (14) is to remove the poles. It is obvious that the definition of $D_{2N}^{\mathrm{res}}(\lambda)$ only requires to solve the eigen-equation approximately up to the order $r^{\lambda+2N}$.

The residue families $D_{2N}^{\mathrm{res}}(g;\lambda)$ admit an interpretation as obstructions to the extension of eigenfunctions of Δ_{g_+} through the boundary $r = 0$. These obstructions appear as residues of associated meromorphic families of distributions. It is the latter relation which motivates the name. Let $u \in C^\infty(X)$, $X = (0,\varepsilon) \times M$ be an eigenfunction

$$-\Delta_{g_+} u = \mu(n-\mu)u$$

with "boundary value" $f \in C^\infty(M)$. Let $\varphi \in C_0^\infty(\bar{X})$ be a test function. We consider the integral

$$\int_X r^\lambda u\varphi d\mathrm{vol}(r^2 g_+).$$

It is holomorphic for λ with sufficiently large real part and admits a meromorphic continuation with simple poles at $\lambda = -\mu - 1 - 2N$, $N \in \mathbb{N}_0$. Then

$$\mathrm{Res}_{\lambda=-\mu-1-2N}\left(\int_X r^\lambda u\varphi d\mathrm{vol}(r^2 g_+)\right) = \int_M f\delta_{2N}(\lambda+n-2N)(\varphi)d\mathrm{vol}(g). \tag{16}$$

For full details we refer to [2, 18].

Now residue families have the following basic properties. First of all,

(1) $D_{2N}^{\mathrm{res}}(g;\lambda)$ is a polynomial in λ of degree N,
(2) $D_{2N}^{\mathrm{res}}(g;-n/2+N) = P_{2N}(g)i^*$.

Next, residue families $D_N^{\mathrm{res}}(g;\lambda)$ satisfy the following conformal covariance law.

Theorem 3 (Conformal covariance). *For any* $\varphi \in C^\infty(M)$,

$$e^{(-\lambda+N)\varphi} D_N^{res}(e^{2\varphi}g;\lambda) = D_N^{res}(g;\lambda) \circ \kappa_* \circ \left(\frac{\kappa^*(r)}{r}\right)^\lambda, \tag{17}$$

where the diffeomorphism κ *pulls back the corresponding Poincaré-Einstein metrics, i.e.,*

$$\kappa^*\left(r^{-2}(dr^2 + g(r))\right) = r^{-2}(dr^2 + \hat{g}(r)) \tag{18}$$

with $\hat{g} = e^{2\varphi}g$.

Note that κ restricts to the identity map on the boundary. Thus, eq. (18) implies

$$i^* \left(\frac{\kappa^*(r)}{r}\right)^2 \hat{g} = g,$$

i.e.,

$$i^* \left(\frac{\kappa^*(r)}{r}\right) = e^{-\varphi}. \tag{19}$$

Finally, the even-order residue families $D_{2N}^{\mathrm{res}}(g;\lambda)$ satisfy a system of factorization identities. In fact, it turns out that the above property $D_{2N}^{\mathrm{res}}(g;-n/2+N) = P_{2N}(g)i^*$ is contained in a system of N factorization identities which are satisfied by the family $D_{2N}^{\mathrm{res}}(g;\lambda)$.

Theorem 4 (Factorizations I). *Assume that $2N \le n$ for even n and $N \ge 1$ for odd n. Then for any metric g we have the factorization relations*

$$D_{2N}^{res}\left(g;-\frac{n}{2}+2N-j\right) = P_{2j}(g) \circ D_{2N-2j}^{res}\left(g;-\frac{n}{2}+2N-j\right) \tag{20}$$

for $j = 1, \ldots, N$.

Next, we denote by $\bar{P}_{2N}(g) \stackrel{\mathrm{def}}{=} P_{2N}(dr^2 + g(r))$ the GJMS operator of order $2N$ for the conformal compactification $\bar{g} = dr^2 + g(r)$ of g_+ on $X = M \times [0,\varepsilon)$. Then the family $D_{2N}^{\mathrm{res}}(g;\lambda)$ satisfies the following additional relations.

Theorem 5 (Factorizations II). *Assume that $2N \le n$ for even n and $N \ge 1$ for odd n. Then for any metric g we have the factorization relations*

$$D_{2N}^{res}\left(g;-\frac{n+1}{2}+j\right) = D_{2N-2j}^{res}\left(g;-\frac{n+1}{2}-j\right) \circ \bar{P}_{2j}(g) \tag{21}$$

for $j = 1, \ldots, N$.

In particular, Theorems 4 and 5 yield $2N$ factorizations of the residue family $D_{2N}^{\mathrm{res}}(\lambda)$ of order $2N$ into products of lower-order residue families and GJMS operators.

Theorems 4 and 5 have strong consequences for the GJMS operators itself. In particular, they imply formula (10) for the generating function of the operators $\mathcal{M}_{2N}$. We briefly describe the relevant arguments. Assuming that $D_{2N}^{\mathrm{res}}(g;\lambda)$ is a polynomial of degree $2N-1$ in λ one can use identities (20) and (21) recursively to represent the families $D_{2N}^{\mathrm{res}}(g;\lambda)$ in terms of linear combinations of compositions of the GJMS operators $P_{2j}(g)$ and $\bar{P}_{2j}(g)$ for $j = 1, \ldots, N$.

Theorem 6. *The leading coefficient of $D_{2N}^{res}(g;\lambda)$ regarded as a polynomial of degree $2N-1$ in λ is given by*

$$(-1)^{N-1}\frac{2^{(2N-1)}}{(2N-1)!}\left(\mathcal{M}_{2N}(g)\iota^* - \iota^*\bar{\mathcal{M}}_{2N}(g)\right).$$

But since the family $D_{2N}^{\mathrm{res}}(g;\lambda)$ is actually a polynomial of degree N in λ, Theorem 6 implies the restriction property

$$\mathcal{M}_{2N}(g)\iota^* = \iota^*\bar{\mathcal{M}}_{2N}(g) \tag{22}$$

for $N \geq 2$. Furthermore, the structure of the sub-leading coefficients of residue families implies

Theorem 7 (Commutator relation). *For $N \geq 3$, we have the commutator relations*

$$\mathcal{M}_{2N}i^* = (2N-2)\left(i^*\left[\bar{\mathcal{M}}_2, \bar{\mathcal{M}}_{2N-2}\right] - \left[\mathcal{M}_2, \mathcal{M}_{2N-2}\right]i^*\right).$$

The latter result suffices to create an inductive proof of the formula for the leading parts of the operators $\mathcal{M}_{2N}$. For the corresponding details and the additional machinery needed to complete the proof of eq. (10) we refer to [19].

The proof of formula (9) for the GJMS operators is a consequence of an algebraic inversion theorem for the map which sends the GJMS operators P_{2N} into the operators $\mathcal{M}_{2N}$.

An alternative proof of these formulas for the operators P_{2N} was given in [8]. The idea of that proof is the following. Instead of appealing to the inversion formula of [19], now directly *define* the operators $\mathcal{M}_{2N}(g)$ by the identity

$$\mathcal{D}(r;g) = \sum_{N\geq 1}\mathcal{M}_{2N}(g)\frac{1}{(N-1)!^2}\left(\frac{r^2}{4}\right)^{N-1}$$

with $\mathcal{D}(r;g)$ given by eq. (11). The starting point of the proof is a formula which relates a conjugate of the Laplacian $\tilde{\Delta}$ of the ambient metric $\tilde{g}$ associated with g to $\mathcal{D}(r;g)$. We recall that the ambient metrics $\tilde{g}$ associated with g is closely related to the Poincaré–Einstein metrics g_+ associated with g. In normal form, the ambient metric is given by the formula

$$\tilde{g} = 2\rho dt^2 + 2tdtd\rho + t^2\mathbf{g}(\rho)$$

on $\mathbb{R}_+ \times M \times (-\varepsilon, \varepsilon)$ with coordinates (t, x, ρ) and $\mathbf{g}(-r^2/2) = g(r)$. We set

$$\mathbf{v}(\rho) = \sqrt{\det \mathbf{g}(\rho)}/\sqrt{\det \mathbf{g}(0)} \quad \text{and} \quad v(r) = \sqrt{\det g(r)}/\sqrt{\det g(0)}.$$

Then $\mathbf{v}(-r^2/2) = v(r)$. We also define

$$\mathbf{w}(\rho) = \sqrt{\mathbf{v}(\rho)} \quad \text{and} \quad w(r) = \sqrt{v(r)}.$$

Now we consider the conjugation

$$\tilde{\Delta}_{\mathbf{w}} \stackrel{\text{def}}{=} \mathbf{w} \circ \tilde{\Delta} \circ \mathbf{w}^{-1}.$$

Then a calculation shows that, for any $\lambda \in \mathbb{R}$, we have

$$\tilde{\Delta}_{\mathbf{w}}(t^\lambda \tilde{u}) = t^{\lambda-2}\left(-2\rho(\partial/\partial\rho)^2 + (2\lambda + n - 2)\partial/\partial\rho + \tilde{\mathscr{D}}(\rho)\right)\tilde{u},$$

where $\tilde{\mathscr{D}}(-r^2/2) = \mathscr{D}(r)$ and $\tilde{u} \in C^\infty(M \times (-\varepsilon, \varepsilon))$ is an arbitrary extension of u: $\tilde{u}(x, 0) = u(x)$. Now the definition of GJMS operators in terms of the ambient Laplacian $\tilde{\Delta}$ is easily seen to be equivalent to

$$P_{2N}(u) = \tilde{\Delta}_{\mathbf{w}}^N \left(t^{N-\frac{n}{2}}\tilde{u}\right)|_{\rho=0,t=1}. \tag{23}$$

The (non-trivial) evaluation of eq. (23) then yields a formula for P_{2N} in terms of the power series coefficients of $\tilde{\mathscr{D}}(\rho)$. It turns out that

$$P_{2N} = \sum_{|I|=N} n_I \mathscr{M}_{2I},$$

i.e., that both methods yield the same result.

3 Heat kernel coefficients of the holographic Laplacian

In the following, we review some results of [21] on the spectral theory of the family

$$\mathscr{D}(r; g) : C^\infty(M) \to C^\infty(M).$$

As in [21], we shall refer to $\mathscr{D}(r; g)$ as to the *holographic* Laplacian of g. By eq. (24), the family $\mathscr{D}(r; g)$ can be regarded as a perturbation of the conformal Laplacian $P_2(g)$.

Through the heat kernel coefficients of the family $\mathscr{D}(r; g)$, the heat kernel coefficients of P_2 are embedded into one-parameter families of local Riemannian invariants. Now a key result states that, although the operator $\mathscr{D}(r; g)$ for $r \neq 0$ is not conformally covariant itself, its spectral behaviour resembles that of a conformally covariant one. In particular, this allows one to give a *heat equation proof* of the conformal variational formulas for the renormalized volume coefficients proved in [16].

The behaviour of the family $\mathscr{D}(r; g)$ under conformal changes of the metric can be derived from its role as the generating function of the building block operators $\mathscr{M}_{2N}$

of the GJMS operators. We recall that, by Theorem 1, any GJMS operator P_{2N} can be written as a linear combination

$$P_{2N} = \sum_{|I|=N} n_I \mathscr{M}_{2I}$$

of compositions of natural self-adjoint second-order *building-block* operators $\mathscr{M}_{2N}$ with $\mathscr{M}_2 = P_2$. Moreover, the generating function

$$\sum_{N\geq 1} \mathscr{M}_{2N}(g) \frac{1}{(N-1)!^2} \left(\frac{r^2}{4}\right)^{N-1}$$

of the building-block operators coincides with the Schrödinger-type operator $\mathscr{D}(r;g)$. A calculation shows that

$$\mathscr{D}(r;g) = -\delta(g(r)^{-1}d) - \frac{\left(\partial^2/\partial r^2 - (n-1)r^{-1}\partial/\partial r - \delta(g(r)^{-1}d)\right)(w(r))}{w(r)}.$$

In particular, we find

$$\mathscr{D}(0;g) = P_2(g). \tag{24}$$

Now, by combining the equivariance of the GJMS operators with eq. (5), one can prove the following result.

Theorem 8 (Conformal variation of $\mathscr{D}(r;g)$). *For any metric g and any $\varphi \in C^\infty(M)$, we have*

$$\begin{aligned} &(\partial/\partial t)|_{t=0}\left(e^{(\frac{n}{2}+1)t\varphi}\mathscr{D}(r;e^{2t\varphi}g)e^{-(\frac{n}{2}-1)t\varphi}\right) \\ &\quad = -\frac{1}{2}r(\partial/\partial r)(\varphi\mathscr{D}(r;g) + \mathscr{D}(r;g)\varphi) - [\mathscr{D}(r;g),[\mathscr{K}(r;g),\varphi]] \end{aligned} \tag{25}$$

for sufficiently small r. Here

$$\mathscr{K}(r;g) \stackrel{def}{=} \frac{1}{2}\int_0^r s\mathscr{D}(s;g)ds.$$

On the right-hand side of eq. (25), *φ is regarded as a multiplication operator.*

For odd $n \geq 3$, identity (25) is to be interpreted as a relation of formal power series in r. For even $n \geq 4$ and general metrics, it asserts the equality of the power series expansions of both sides of eq. (25) up to r^{n-2}.

Now we consider spectral invariants of the holographic Laplacian on a closed manifold M. First, we note that, for sufficiently small r, $g(r)$ is a Riemannian metric and the holographic Laplacian is elliptic. The following result is a conformal variational formula for the trace of the heat kernel of the holographic Laplacian. It is a consequence of the variational formula in Theorem 8.

Theorem 9 (Conformal variation of trace of heat kernel of $\mathscr{D}(r;g)$). *For any metric g on a closed manifold M, we have*

$$(\partial/\partial\varepsilon)|_{\varepsilon=0}(\operatorname{Tr}(\exp(t\mathscr{D}(r;e^{2\varepsilon\varphi}g)))) \\ = -2t(\partial/\partial t)(\operatorname{Tr}(\varphi e^{t\mathscr{D}(r;g)})) - rt\frac{1}{2}\operatorname{Tr}\left((\varphi\dot{\mathscr{D}}(r;g) + \dot{\mathscr{D}}(r;g)\varphi)e^{t\mathscr{D}(r;g)}\right) \tag{26}$$

for sufficiently small r and all $\varphi \in C^\infty(M)$. Here traces are taken in $L^2(M,g)$ and the dot denotes the derivative with respect to r.

Again, relation (26) is to be interpreted as an identity of formal power series in r (up to r^{n-2} for even n and general metrics). It is one of the basic features of Theorem 9 that the massive double-commutator term on the right-hand side of eq. (25) does *not* contribute to eq. (26). This follows from the observation that for any differential operator $\mathscr{K}$ on M we have

$$\operatorname{Tr}([\mathscr{D},\mathscr{K}]e^{t\mathscr{D}}) = 0$$

by the cyclicity of the trace and the fact that $\mathscr{D}$ commutes with $e^{t\mathscr{D}}$.

By the ellipticity of $\mathscr{D}(r;g)$ (for sufficiently small r), we have an asymptotic expansion

$$\operatorname{Tr}(\exp(t\mathscr{D}(r;g)) \sim (4\pi t)^{-\frac{n}{2}} \sum_{j\geq 0} t^j \int_M a_{2j}(r;g)dv_g, \quad t \to 0. \tag{27}$$

By $a_{2j}(0;g) = a_{2j}(g)$ it generalizes the asymptotic expansion

$$\operatorname{Tr}(\exp(tP_2(g)) \sim (4\pi t)^{-\frac{n}{2}} \sum_{j\geq 0} t^j \int_M a_{2j}(g)dv_g, \quad t \to 0$$

of the trace of the heat kernel of the conformal Laplacian. Let $a_{(2j,2k)}(g)$ be the coefficients in the Taylor expansion of $a_{2k}(r;g)$, i.e.,

$$a_{2j}(r;g) \sim \sum_{k\geq 0} a_{(2j,2k)}(g)r^{2k}, \quad r \to 0.$$

The coefficients $a_{(2j,2k)}(g)$ are local Riemannian invariants.

Next, we describe some consequences of Theorem 9 for the behaviour of the heat kernel coefficients of $\mathscr{D}(r;g)$ under conformal changes of the metric. First, by the overall factor r in the second term on right-hand side of eq. (26), this equation specializes for $r = 0$ to the conformal variational formula

$$(\partial/\partial\varepsilon)|_{\varepsilon=0}(\mathrm{Tr}(\exp(tP_2(e^{2\varepsilon\varphi}g)))) = -2t(\partial/\partial t)(\mathrm{Tr}(\varphi e^{tP_2(g)}))$$

for the trace of the heat kernel of the conformal Laplacian. In the by now classical works [4, 29] it was shown that the latter formula implies the variational formulas

$$\left(\int_{M^n} a_{2j}(g)dv_g\right)^{\bullet}[\varphi] = (n-2j)\int_{M^n} \varphi a_{2j}(g)dv_g. \tag{28}$$

In particular, for even n, the functional

$$g \mapsto \int_{M^n} a_n(g)dv_g \tag{29}$$

is a global conformal invariant. The integral (29) is also known as the *conformal index*. The second term on the right-hand side of eq. (26) contributes only to the conformal transformation laws of the *non-constant* Taylor coefficients of the heat kernel coefficients $a_{2j}(r;g)$ with respect to r. By the cyclicity of the trace, it equals

$$-rt\frac{1}{2}\mathrm{Tr}\left(\varphi\left(\dot{\mathscr{D}}(r;g)e^{t\mathscr{D}(r;g)} + e^{t\mathscr{D}(r;g)}\dot{\mathscr{D}}(r;g)\right)\right). \tag{30}$$

But if the operators $\mathscr{D}(r;g)$ and $\dot{\mathscr{D}}(r;g)$ commute, this sum can be written in the form

$$-r(\partial/\partial r)(\mathrm{Tr}(\varphi e^{t\mathscr{D}(r;g)})). \tag{31}$$

We emphasize the similarity of both contributions

$$-2t(\partial/\partial t)(\mathrm{Tr}(\varphi e^{t\mathscr{D}(r;g)})) \quad \text{and} \quad -r(\partial/\partial r)(\mathrm{Tr}(\varphi e^{t\mathscr{D}(r;g)})) \tag{32}$$

in these variational formulas. In other words, in these conformal variational formulas the holographic dimension variable r^2 and the time variable t play similar roles.

However, $\mathscr{D}(r;g)$ and $\dot{\mathscr{D}}(r;g)$ do not commute in general and the right-hand side of eq. (26) is the sum of the contributions (32) and a contribution which is given by the difference of eqs. (30) and (31). The additional term can be described in terms of the restriction to the diagonal of the kernel of the operator

$$\mathscr{C}(t;r) = \frac{1}{2\pi i}\int_{\Gamma} [\mathscr{R}(r;\lambda), [\mathscr{R}(r;\lambda), \dot{\mathscr{D}}(r)]]e^{-t\lambda}d\lambda, \tag{33}$$

where $\mathscr{R}(r;\lambda)$ denotes the resolvent of $-\mathscr{D}(r)$ and Γ is an appropriate contour in the complex plane. For special metrics, the operator $\mathscr{C}(t;r)$ may vanish. For instance, the term (33) vanishes at an Einstein metric. Hence by combining Theorem 9 with the above arguments, we obtain the following result.

Theorem 10. *On closed Einstein manifolds (M^n, g) of even dimension n, we have*

$$\left(\int_{M^n} a_{(2j,2k)}(g)dv_g\right)^{\bullet}[\varphi] = (n-2j-2k)\int_{M^n} \varphi a_{(2j,2k)}(g)dv_g.$$

In particular, the total integrals

$$\int_{M^n} a_{(2j,n-2j)}dv$$

are critical at Einstein metrics in their conformal class.

For general metrics, the term (33) does *not* contribute to the leading heat kernel coefficient $a_0(r)$ of $\mathcal{D}(r)$. In fact, we have the following result.

Proposition 11. *The leading heat kernel coefficient $a_0(r;g)$ of $\mathcal{D}(r;g)$ is given by the renormalized volume coefficients of g. More precisely, we have*

$$a_{(0,2k)}(g) = v_{2k}(g).$$

In view of the relations $a_{(2k,0)} = a_{2k}$ and $a_{(0,2k)} = v_{2k}$, the coefficients $a_{(2j,2k-2j)}$ for $j = 0, \ldots, k$ may be regarded as interpolating between the heat kernel coefficients of the Yamabe operator and the renormalized volume coefficients.

Now combining Proposition 11 with the above arguments concerning the variation of the total integrals of $a_0(r;g)$ yields a heat equation proof of the following result.

Theorem 12 ([16]). *For any metric g on a closed manifold M of dimension n (and $2k \leq n$ if n is even), we have*

$$\left(\int_M v_{2k}(g)dv_g\right)^{\bullet}[\varphi] = (n-2k)\int_M \varphi v_{2k}(g)dv_g$$

for all $\varphi \in C^\infty(M)$.

With some more efforts one can also give an analogous heat equation proof of the conformal variational formulas for the coefficients v_{2k} itself.

We finish with an explicit formula for the coefficient $a_2(r)$. The following result shows that this coefficient can be completely described in terms of renormalized volume coefficients.

Theorem 13. *For any metric, the sub-leading heat kernel coefficient $a_2(r;g)$ of the holographic Laplacian $\mathcal{D}(r;g)$ is given by the formula*

$$a_2(r) = -\frac{1}{3}\left(\ddot{v}(r) - \left(\frac{n}{2}-1\right)r^{-1}\dot{v}(r)\right) + (\dot{w}(r))^2. \tag{34}$$

A proof of the latter result follows from a closed formula for the leading term in the asymptotic expansion of the kernel of $\mathscr{C}(t;r)$ for $t \to 0$.

4 Symmetry breaking operators on differential forms

The results in the present section are taken from [9]. They constitute the basis of forthcoming work on residue families acting on differential forms. These families generalize the residue families discussed in Section 2. Here we describe the flat models of those families. They will also be referred to as symmetry breaking operators following the terminology of [23, 27].

The main problem is the description of all differential operators

$$\Omega^p(\mathbb{R}^n) \to \Omega^q(\mathbb{R}^{n-1}) \tag{35}$$

which are equivariant with respect to principal series representations of $\mathfrak{so}(n,1,\mathbb{R})$ and $\mathfrak{so}(n+1,1,\mathbb{R})$ on differential forms on $\mathbb{R}^{n-1}$ and $\mathbb{R}^n$, respectively. Here we regard $\mathfrak{so}(n,1,\mathbb{R})$ and $\mathfrak{so}(n+1,1,\mathbb{R})$ as the conformal Lie algebras of $\mathbb{R}^n$ and $\mathbb{R}^{n-1}$ with the Euclidean metrics and $\mathfrak{so}(n,1,\mathbb{R}) \subset \mathfrak{so}(n+1,1,\mathbb{R})$ is viewed as the subalgebra which leaves the hyperplane $\mathbb{R}^{n-1}$ invariant. The fact that the operators of interest are only required to be equivariant with respect to $\mathfrak{so}(n,1,\mathbb{R})$ motivates the terminology *symmetry breaking operators*.

The operators (35) are the non-compact models of symmetry breaking differential operators $\Omega^p(S^n) \to \Omega^q(S^{n-1})$, where S^{n-1} is an equatorially embedded subsphere of S^n. We shall not go into the details of the structure of these operators (for a discussion of the case $p = q = 0$ we refer to [18]). We also note that symmetry breaking operators which are *not* differential operators remain outside the present discussion. For the corresponding theory of such operators acting on functions we refer to [27].

In order to state the results, we need to fix some notation. We use Euclidean coordinates $x_1,\dots,x_n$ on $\mathbb{R}^n$ and assume that $\mathbb{R}^{n-1} \subset \mathbb{R}^n$ is the hyperplane $x_n = 0$. Let $\iota : \mathbb{R}^{n-1} \to \mathbb{R}^n$ be the corresponding embedding $\iota(x_1,\dots,x_{n-1}) = (x_1,\dots,x_{n-1},0)$. Then ι^* denotes the pull-back induced by ι. Let d and δ be the usual exterior differential and co-differential on forms on $\mathbb{R}^{n-1}$. Their counterparts on $\mathbb{R}^n$ are denoted by $\bar{d}$ and $\bar{\delta}$. Next, we define (principal series) representations $\pi_\lambda^{(p)}$ on $\Omega^p(\mathbb{R}^n)$ by

$$\pi_\lambda^{(p)}(\gamma) = e^{\lambda\Phi_\gamma}\gamma_*$$

for all conformal diffeomorphisms γ of the Euclidean metric g_0 on $\mathbb{R}^n$ so that $\gamma_*(g_0) = e^{2\Phi_\gamma}g_0$ for some $\Phi_\gamma \in C^\infty(\mathbb{R}^n)$. The analogous representations on $\Omega^p(\mathbb{R}^{n-1})$ will be denoted by $\pi_\lambda'^{(p)}$.

It turns out that all equivariant operators (35) are given by two infinite sequences

$$D_N^{(p\to p)}(\lambda) : \Omega^p(\mathbb{R}^n) \to \Omega^p(\mathbb{R}^{n-1}) \quad \text{and} \quad D_N^{(p\to p-1)}(\lambda) : \Omega^p(\mathbb{R}^n) \to \Omega^{p-1}(\mathbb{R}^{n-1})$$

($\mathbb{N} \in \mathbb{N}_0$) of one-parameter families, some additional operators and compositions of these with the Hodge star operator of the Euclidean metric on $\mathbb{R}^{n-1}$. The one-parameter families will be referred to as the first- and second-type families. We shall display explicit formulas for all operators and describe some of their basic mapping properties.

The following theorem is the main result on *even-order* families of the first type.

Theorem 14. *Let $N \in \mathbb{N}$ and $0 \leq p \leq n-1$. Then the one-parameter family*

$$D_{2N}^{(p\to p)}(\lambda) : \Omega^p(\mathbb{R}^n) \to \Omega^p(\mathbb{R}^{n-1}),\ \lambda \in \mathbb{C}$$

of even-order differential operators, which is defined by the formula

$$\begin{aligned} D_{2N}^{(p\to p)}(\lambda) = {} & \sum_{i=1}^{N-1} (\lambda + p - 2i)\alpha_i^{(N)}(\lambda)(d\delta)^{N-i}\iota^*(\bar{\delta}\bar{d})^i \\ & + (\lambda + p) \sum_{i=0}^{N} \alpha_i^{(N)}(\lambda)(d\delta)^{N-i}\iota^*(\bar{d}\bar{\delta})^i \\ & + (\lambda + p - 2N) \sum_{i=0}^{N} \alpha_i^{(N)}(\lambda)(\delta d)^{N-i}\iota^*(\bar{\delta}\bar{d})^i \end{aligned} \tag{36}$$

with the coefficients

$$\alpha_i^{(N)}(\lambda) = (-1)^i 2^N \frac{N!}{(2N)!}\binom{N}{i} \prod_{k=i+1}^{N} (2\lambda + n - 2k) \prod_{k=1}^{i} (2\lambda + n - 2k - 2N + 1),$$

satisfies the intertwining relation

$$d\pi'^{(p)}_{-\lambda+2N-p}(X) D_{2N}^{(p\to p)}(\lambda) = D_{2N}^{(p\to p)}(\lambda) d\pi^{(p)}_{-\lambda-p}(X) \tag{37}$$

for all $X \in \mathfrak{so}(n,1,\mathbb{R})$.

The analogous result for *odd-order* families $\Omega^p(\mathbb{R}^n) \to \Omega^p(\mathbb{R}^{n-1})$ of the first type reads as follows.

Theorem 15. *Let $N \in \mathbb{N}_0$ and $0 \leq p \leq n-1$. Then the one-parameter family*

$$D_{2N+1}^{(p\to p)}(\lambda) : \Omega^p(\mathbb{R}^n) \to \Omega^p(\mathbb{R}^{n-1}),\ \lambda \in \mathbb{C}$$

of odd-order differential operators, which is defined by the formula

$$\begin{aligned} D_{2N+1}^{(p\to p)}(\lambda) = {} & \sum_{i=1}^{N} \gamma_i^{(N)}(\lambda;p)(d\delta)^{N-i} d\iota^* i_{\partial_n}(\bar{\delta}\bar{d})^i \\ & + (\lambda+p)\sum_{i=0}^{N} \beta_i^{(N)}(\lambda)(d\delta)^{N-i} d\iota^* i_{\partial_n}(\bar{d}\bar{\delta})^i \\ & + (\lambda+p-2N-1)\sum_{i=0}^{N} \beta_i^{(N)}(\lambda)(\delta d)^{N-i}\iota^* i_{\partial_n}\bar{d}(\bar{\delta}\bar{d})^i \end{aligned} \tag{38}$$

with the coefficients

$$\beta_i^{(N)}(\lambda) \stackrel{def}{=} (-1)^i 2^N \frac{N!}{(2N+1)!}\binom{N}{i} \prod_{k=i+1}^{N}(2\lambda+n-2k)\prod_{k=1}^{i}(2\lambda+n-2k-2N-1)$$

and

$$\gamma_i^{(N)}(\lambda;p) = (\lambda+p-2i)\beta_i^{(N)}(\lambda) - (\lambda+p-2i+1)\beta_{i-1}^{(N)}(\lambda),$$

satisfies the intertwining relation

$$d\pi'^{(p)}_{-\lambda+2N+1-p}(X) D_{2N+1}^{(p\to p)}(\lambda) = D_{2N+1}^{(p\to p)}(\lambda) d\pi^{(p)}_{-\lambda-p}(X) \tag{39}$$

for all $X \in \mathfrak{so}(n,1,\mathbb{R})$.

We note that the generating functions of the coefficients $\alpha_i^{(N)}(\lambda)$ and $\beta_i^{(N)}(\lambda)$ are given by some Jacobi polynomials. We recall that for $m \in \mathbb{N}$ and $\alpha, \beta \in \mathbb{C}$ such that $(\alpha+1)_m \neq 0$ the corresponding Jacobi polynomial of degree m is defined by

$$P_m^{(\alpha,\beta)}(z) \stackrel{\text{def}}{=} \frac{(\alpha+1)_m}{m!} {}_2F_1\left[\begin{matrix} -m\ ,\ 1+\alpha+\beta+m \\ \alpha+1 \end{matrix} ; \frac{1-z}{2}\right].$$

Now we find

$$\sum_{i=0}^{N} \alpha_i^{(N)}(\lambda)t^i = 4^N \frac{N!}{(2N)!}\left(\lambda+\tfrac{n}{2}-N\right)_N {}_2F_1\left[\begin{matrix} -N, N+\frac{1}{2}-\lambda-\frac{n}{2} \\ 1-\lambda-\frac{n}{2} \end{matrix}; t\right],$$

where the right-hand side is proportional to the Jacobi polynomial $P_N^{(-\lambda-\frac{n}{2},-\frac{1}{2})}(1-2t)$. Similarly, we find

$$\sum_{i=0}^{N} \beta_i^{(N)}(\lambda)t^i = 4^N \frac{N!}{(2N+1)!}\left(\lambda+\tfrac{n}{2}-N\right)_N {}_2F_1\left[\begin{matrix} -N, N+\frac{3}{2}-\lambda-\frac{n}{2} \\ 1-\lambda-\frac{n}{2} \end{matrix}; t\right],$$

where the right-hand side is proportional to the Jacobi polynomial $P_N^{(-\lambda-\frac{n}{2},\frac{1}{2})}(1-2t)$.

Next, the families

$$D_N^{(p\to p)}(\lambda):\Omega^p(\mathbb{R}^n)\to\Omega^p(\mathbb{R}^{n-1})$$

of the first type are related to families

$$D_N^{(p\to p-1)}(\lambda):\Omega^p(\mathbb{R}^n)\to\Omega^{p-1}(\mathbb{R}^{n-1})$$

of the second type through compositions with respective Hodge star operators $\bar{\star}$ and $\star$ on $\mathbb{R}^n$ and $\mathbb{R}^{n-1}$ (Hodge conjugation). This relation yields the following two results.

Theorem 16. *Let $N\in\mathbb{N}$ and $p=1,\dots,n$. Then the one-parameter family*

$$D_{2N}^{(p\to p-1)}(\lambda):\Omega^p(\mathbb{R}^n)\to\Omega^{p-1}(\mathbb{R}^{n-1}),\ \lambda\in\mathbb{C}$$

of even-order differential operators, which is defined by the formula

$$\begin{aligned}D_{2N}^{(p\to p-1)}(\lambda) = &-(\lambda+n-p-2N)\sum_{i=0}^{N}\alpha_i^{(N)}(\lambda)(d\delta)^{N-i}\iota^* i_{\partial_n}(\bar{d}\bar{\delta})^i\\ &-\sum_{i=1}^{N-1}(\lambda+n-p-2i)\alpha_i^{(N)}(\lambda)(\delta d)^{N-i}\iota^* i_{\partial_n}(\bar{d}\bar{\delta})^i\\ &-(\lambda+n-p)\sum_{i=0}^{N}\alpha_i^{(N)}(\lambda)(\delta d)^{N-i}\iota^* i_{\partial_n}(\bar{\delta}\bar{d})^i,\end{aligned}\tag{40}$$

satisfies the intertwining relation

$$d\pi'^{(p)}_{-\lambda+2N-p}(X)D_{2N}^{(p\to p-1)}(\lambda)=D_{2N}^{(p\to p-1)}(\lambda)d\pi^{(p)}_{-\lambda-p}(X)\tag{41}$$

for all $X\in\mathfrak{so}(n,1,\mathbb{R})$.

Here i_{∂_n} denotes the operation of insertion of the normal vector field ∂_n.

Theorem 17. *Let $N\in\mathbb{N}_0$ and $p=1,\dots,n$. Then the one-parameter family*

$$D_{2N+1}^{(p\to p-1)}(\lambda):\Omega^p(\mathbb{R}^n)\to\Omega^{p-1}(\mathbb{R}^{n-1}),\ \lambda\in\mathbb{C}$$

of odd-order differential operators, which is defined by the formula

$$D_{2N+1}^{(p\to p-1)}(\lambda) = -\sum_{i=1}^{N} \gamma_i^{(N)}(\lambda; n-p)(\delta d)^{N-i}\delta\iota^*(\bar{d}\bar{\delta})^i$$
$$+ (\lambda + n - p - 2N - 1)\sum_{i=0}^{N} \beta_i^{(N)}(\lambda)(d\delta)^{N-i}\iota^*\bar{\delta}(\bar{d}\bar{\delta})^i$$
$$- (\lambda + n - p)\sum_{i=0}^{N} \beta_i^{(N)}(\lambda)(\delta d)^{N-i}\delta\iota^*(\bar{\delta}\bar{d})^i, \tag{42}$$

satisfies the intertwining relation

$$d\pi'^{(p)}_{-\lambda+2N+1-(p-1)}(X)D_{2N+1}^{(p\to p-1)}(\lambda) = D_{2N+1}^{(p\to p-1)}(\lambda)d\pi^{(p)}_{-\lambda-p}(X) \tag{43}$$

The families $D_N^{(p\to p)}(\lambda)$ and $D_N^{(p\to p-1)}(\lambda)$ admit an interpretation in terms of residue families

$$D_N^{\mathrm{res},(p\to p)}(g;\lambda) : \Omega^p(M\times[0,\varepsilon)) \to \Omega^p(M)$$

of the first type and residue families

$$D_N^{\mathrm{res},(p\to p-1)}(g;\lambda) : \Omega^p(M\times[0,\varepsilon)) \to \Omega^{p-1}(M)$$

of the second type. These residue families on forms are naturally associated to the asymptotic expansion of eigenforms of the Laplacian for the Poincaré-Einstein metric g_+ on $M\times[0,\varepsilon)$ acting on differential forms. Residue families of the first type generalize the residue families considered in Section 2. They are conformally covariant in the sense that

$$e^{(-\lambda-p+N)\varphi}D_N^{\mathrm{res},(p\to p)}(e^{2\varphi}g;\lambda) = D_N^{\mathrm{res},(p\to p)}(g;\lambda)\circ\kappa_*\circ\left(\frac{\kappa^*(r)}{r}\right)^{\lambda+p} \tag{44}$$

and

$$e^{(-\lambda-(p-1)+N)\varphi}D_N^{\mathrm{res},(p\to p-1)}(e^{2\varphi}g;\lambda)$$
$$= D_N^{\mathrm{res},(p\to p-1)}(g;\lambda)\circ\kappa_*\circ\left(\frac{\kappa^*(r)}{r}\right)^{\lambda+p}, \tag{45}$$

respectively (compare with Theorem 3). In these terms, the families $D_N^{(p\to p)}(\lambda)$ and $D_N^{(p\to p-1)}(\lambda)$ are constant multiples of the respective residue families $D_N^{\mathrm{res},(p\to p)}(g;\lambda)$ and $D_N^{\mathrm{res},(p\to p-1)}(g;\lambda)$ for $M = \mathbb{R}^{n-1}$ with the Euclidean metric g. The details of the theory of residue families on forms will be given in [10].

Example 18. Assume that M^{n-1} is a hypersurface in (X^n, g). Let N be the unit normal vector field, H the corresponding mean curvature and $\mathscr{H}$ the mean curvature vector. Then the first-order residue families of the first and second types can be described in terms of the conformally covariant families

$$D_1^{(p\to p)}(g;\lambda) = \lambda\iota^* i_N d + (\lambda+1)d\iota^* i_N - \lambda(\lambda+1)H\iota^* : \Omega^p(X) \to \Omega^p(M)$$

and

$$\begin{aligned} D_1^{(p\to p-1)}(g;\lambda) = (n-2p+\lambda+1)\iota^*\delta_g - (n-2p+\lambda+2)\delta_{\iota^*(g)}\iota^* \\ + (n-2p+\lambda+1)(n-2p+\lambda+2)i_{\mathscr{H}} : \Omega^p(X) \to \Omega^{p-1}(M). \end{aligned}$$

For the special case $X = M \times [0,\varepsilon)$ with the conformal compactification $r^2 g_+ = dr^2 + g(r)$ of the Poincaré-Einstein metric g_+, we have

$$D_1^{(p\to p)}(r^2 g_+; \lambda + p - 1) = D_1^{\mathrm{res},(p\to p)}(g_0;\lambda)$$

and

$$D_1^{(p\to p-1)}(r^2 g_+; \lambda + p - 2) = D_1^{\mathrm{res},(p\to p-1)}(g_0;\lambda).$$

The conformal covariance properties (44) and (45) of these residue families are consequences of the covariance rules

$$e^{-\lambda\iota^*(\varphi)} D_1^{(p\to p)}(e^{2\varphi}g;\lambda) = D_1^{(p\to p)}(g;\lambda)e^{(-\lambda-1)\varphi}$$

and

$$e^{-\lambda\iota^*(\varphi)} D_1^{(p\to p-1)}(e^{2\varphi}g;\lambda) = D_1^{(p\to p-1)}(g;\lambda)e^{(-\lambda-2)\varphi},$$

which easily follow by direct calculations.

In addition to the one-parameter families $D_N^{(p\to p)}(\lambda)$ and $D_N^{(p\to p-1})$ there are symmetry breaking operators of the type (35) for other choices of the form degrees p and q. The simplest of these is the operator $d\iota^* : \Omega^p(\mathbb{R}^n) \to \Omega^{p+1}(\mathbb{R}^{n-1})$. The following result classifies all such symmetry breaking operators.

Theorem 19 (Classification). *The linear differential operators*

$$D : \Omega^p(\mathbb{R}^n) \to \Omega^q(\mathbb{R}^{n-1})$$

which satisfy the intertwining relation

$$d\pi_\eta'^{(q)}(X)D = Dd\pi_\mu^{(p)}(X) \tag{46}$$

for all $X \in \mathfrak{so}(n,1,\mathbb{R})$ and some $\mu, \eta \in \mathbb{C}$ are generated by the following operators:

(1) *The case $q = p$. The families $D_N^{(p\to p)}(\lambda)$ of the first type. Here $N \in \mathbb{N}_0$, $\lambda \in \mathbb{C}$ and $\mu = -\lambda - p$, $\eta = -\lambda - p + N$.*
(2) *The case $q = p - 1$. The families $D_N^{(p\to p-1)}(\lambda)$ of the second type. Here $N \in \mathbb{N}_0$, $\lambda \in \mathbb{C}$ and $\mu = -\lambda - p$, $\eta = -\lambda - (p-1) + N$.*
(3) *The case $q = 1$ and $p = 0$. The operators $d\dot{D}_N^{(0\to 0)}(N)$ with $N \in \mathbb{N}_0$. Here $\mu = -N$ and $\eta = 0$.*
(4) *The case $q = p + 1$. The operator $d\iota^*$. Here $\mu = \eta = 0$.*
(5) *The case $q = n - 2$ and $p = n$. The operators $\delta\dot{D}_N^{(n\to n-1)}(N)$ with $N \in \mathbb{N}_0$. Here $\mu = -n - N$ and $\eta = -n + 3$.*
(6) *The case $q = p - 2$. The operator $\delta\iota^* i_{\partial_n}$. Here $\mu = n - 2p$ and $\eta = n - 2p + 3$.*
(7) *The compositions of the above operators with $\star$.*

In points (3) and (5), the dot denotes derivatives of the families with respect to the parameter λ.

In dimension $n = 2$, the conformal symmetry breaking operators $D_N^{(1,0)}(\lambda)$ and $D_N^{(1,0)}(\lambda)\bar{\star}$ were found in [24].

We continue with the discussion of analogues of the basic factorization identities for residue families on functions described in Section 2. We recall that the residue families $D_{2N}^{\mathrm{res}}(\lambda) : C^\infty(M\times[0,\varepsilon)) \to C^\infty(M)$ interpolate between the GJMS operators of order $2N$ on $M \times [0,\varepsilon)$ and M. Similarly, the families $D_{2N}^{(p\to p)}(\lambda)$ of the first-type interpolates between

$$D_{2N}^{(p\to p)}\left(N - \tfrac{n-1}{2}\right) = -L_{2N}^{(p)}\iota^* \tag{47}$$

and

$$D_{2N}^{(p\to p)}\left(N - \tfrac{n}{2}\right) = -\iota^* \bar{L}_{2N}, \tag{48}$$

where $L_{2N}^{(p)}$ and $\bar{L}_{2N}^{(p)}$ are the respective Branson–Gover operators of order $2N$ on the Euclidean spaces $\mathbb{R}^{n-1}$ and $\mathbb{R}^n$.

We recall that the Branson–Gover operators $L_{2N}^{(p)}(g)$ on a Riemannian manifold (M^n, g) of dimension $n \geq 3$ are differential operators on $\Omega^p(M)$ of order $2N$ with $2 \leq 2N \leq n-2$ (for n even and $p \neq 0, n$) and order $2 \leq 2N \leq n$ (for n even and $p = 0, n$). They are conformally covariant in the sense that they satisfy the intertwining relation

$$e^{(\frac{n}{2}-p+N)\varphi} L_{2N}^{(p)}(\hat{g})(\omega) = L_{2N}^{(p)}(g)\left(e^{(\frac{n}{2}-p-N)\varphi}\omega\right), \quad \omega \in \Omega^p(M),$$

where $\hat{g} = e^{2\varphi}g$. The operator $L_{2N}^{(0)}(g)$ on $C^\infty(M)$ reduces to a constant multiple of the GJMS operator $P_{2N}(g)$ of order $2N$. On the Euclidean space $\mathbb{R}^{n-1}$, the Branson–Gover operators are given by the explicit formula

$$L_{2N}^{(p)} = \left(\tfrac{n-1}{2} - p + N\right)(\delta d)^N + \left(\tfrac{n-1}{2} - p - N\right)(d\delta)^N. \tag{49}$$

For details on Branson–Gover operators we refer to [1, 4, 11].

Identities (47) and (48) are special cases of the following *main factorizations* of the first-type families $D_{2N}^{(p\to p)}(\lambda)$ (for specific choices of λ) into products of lower-order first-type families and Branson–Gover operators.

Theorem 20 (Main factorizations). *Let $N \in \mathbb{N}$ and $p = 0, \dots, n-1$. Then the even-order families of the first type satisfy the factorization identities*

$$\left(\tfrac{n}{2} - p + k\right) D_{2N}^{(p\to p)}\left(k - \tfrac{n}{2}\right) = D_{2N-2k}^{(p\to p)}\left(-k - \tfrac{n}{2}\right) \circ \bar{L}_{2k}^{(p)}$$

and

$$\left(\tfrac{n-1}{2} - p - k\right) D_{2N}^{(p\to p)}\left(2N - k - \tfrac{n-1}{2}\right) = L_{2k}^{(p)} \circ D_{2N-2k}^{(p\to p)}\left(2N - k - \tfrac{n-1}{2}\right)$$

for $k = 1, \dots, N-1$, $N \geq 2$. Moreover, in the extremal case $k = N$, we have

$$D_{2N}^{(p\to p)}\left(N - \tfrac{n-1}{2}\right) = -L_{2N}^{(p)}\iota^* \quad \text{and} \quad D_{2N}^{(p\to p)}\left(N - \tfrac{n}{2}\right) = -\iota^* \bar{L}_{2N}^{(p)}. \tag{50}$$

It is natural to improve the formulation of Theorem 20 by removing the numerical coefficients on the left-hand sides of the factorization identities. This can be done by introducing renormalized versions of the families $D_{2N}^{(p\to p)}(\lambda)$. In particular, we find the following result.

Theorem 21. *Assume that n is even and $p < \frac{n}{2}$. Then the identities in Theorem 20 are equivalent to the main factorizations*

$$\tilde{D}_{2N}^{(p\to p)}\left(k - \tfrac{n}{2}\right) = \tilde{D}_{2N-2k}^{(p\to p)}\left(-k - \tfrac{n}{2}\right) \circ \tilde{\bar{L}}_{2k}^{(p)},$$
$$\tilde{D}_{2N}^{(p\to p)}\left(2N - k - \tfrac{n-1}{2}\right) = \tilde{L}_{2k}^{(p)} \circ \tilde{D}_{2N-2k}^{(p\to p)}\left(2N - k - \tfrac{n-1}{2}\right)$$

for $k = 1, \dots, N$ with the renormalized families

$$\tilde{D}_{2N}^{(p\to p)}(\lambda) \stackrel{def}{=} \frac{D_{2N}^{(p\to p)}(\lambda)}{\lambda + p - 2N} \tag{51}$$

and the renormalized Branson–Gover operators

$$\tilde{L}_{2N}^{(p)} \stackrel{def}{=} \frac{L_{2N}^{(p)}}{\frac{n-1}{2} - p + N} = (\delta d)^N + \cdots, \quad \tilde{\bar{L}}_{2N}^{(p)} \stackrel{def}{=} \frac{\bar{L}_{2N}^{(p)}}{\frac{n}{2} - p + N} = (\bar{\delta}\bar{d})^N + \cdots. \tag{52}$$

The latter result should be regarded as an analogue of Theorems 4 and 5. We omit the formulation of the analogues of these factorizations for odd-order families of the first type and all families of the second type.

Next, we note that both types of one-parameter families are connected through certain systems of identities which involve the operators d, δ, $\bar{d}$ and $\bar{\delta}$ as factors.

Theorem 22 (Supplementary factorizations). *For $N \in \mathbb{N}$, we have the factorization identities*

$$\begin{aligned} D_{2N}^{(p\to p)}(-p+2N) &= -(2N)dD_{2N-1}^{(p\to p-1)}(-p+2N), \quad 1 \le p \le n-1, \\ D_{2N}^{(p\to p)}(-p) &= (2N)D_{2N-1}^{(p+1\to p)}(-p-1)\bar{d}, \quad 0 \le p \le n-1. \end{aligned} \tag{53}$$

Moreover, for $N \in \mathbb{N}_0$, we have

$$(n-2p-2N-1)D_{2N+1}^{(p\to p-1)}(p-n+2N+1) = \delta D_{2N}^{(p\to p)}(p-n+2N+1) \tag{54}$$

for $1 \le p \le n-1$ and

$$(n-2p+2N)D_{2N+1}^{(p+1\to p)}(-n+p+1) = D_{2N}^{(p\to p)}(-n+p)\bar{\delta}$$

for $0 \le p \le n-1$.

We further omit the formulation of the analogous relations between odd-oder first-type families and even-order second-type families which arise by Hodge conjugation.

As an application of Theorem 22, we describe a formal argument which suggests the double factorization formula

$$L_{n-2p-1}^{(p)}\iota^* = (n-2p-1)\delta\dot{D}_{n-2p-3}^{(p+1\to p+1)}(-p-1)\bar{d} \tag{55}$$

(for odd n and $p < \frac{n-1}{2}$) of the critical Branson–Gover operators $L_{n-2p-1}^{(p)}$. Here the dot denotes the derivative with respect to λ. In fact, identity (54) implies

$$(n-2p-2N-1)D_{2N-1}^{(p+1\to p)}(p-n+2N) = \delta D_{2N-2}^{(p+1\to p+1)}(p-n+2N).$$

In particular, we have

$$\delta D_{2N-2}^{(p+1\to p+1)}(-p-1) = 0$$

if $n-2p-2N-1=0$. Moreover, by a continuation in dimension n argument, we find

$$D_{2N-1}^{(p+1\to p)}(-p-1) = \delta\dot{D}_{2N-2}^{(p+1\to p+1)}(-p-1)$$

if $2N = n - 2p - 1$. In combination with eq. (53) we obtain

$$D_{2N}^{(p\to p)}(-p) = 2N\delta\dot{D}_{2N-2}^{(p+1\to p+1)}(-p-1)\bar{d}$$

for $2N = n-2p-1$. This proves eq. (55) using the first identity in eq. (50). An alternative rigorous proof of eq. (55) follows from Theorem 14 and the explicit formula (49).

In later work, we shall prove analogues for residue families on forms [10] of the factorization identities in Theorems 20 and 22. In particular, there is an analogue of eq. (55) for general metrics. We also recall from Section 2 that the recursive structure of residue families, which manifests itself in form of the (main) factorization identities, is the source of recursive relations among the GJMS operators. We expect that the above results have analogous consequences for the Branson–Gover operators. Finally, we emphasize an important difference between residue families on forms and residue families on functions. In fact, for any fixed order the former are polynomials in λ of exactly one degree higher. This corresponds to the fact that the renormalized families (51) are not polynomial but rational in λ.

We finish with a brief comment on Q-curvature operators. The following result states a formula for the critical Q-curvature operators $Q_{n-1-2p}^{(p)} : \Omega^p(\mathbb{R}^{n-1})|_{\ker(d)} \to \Omega^p(\mathbb{R}^{n-1})$ in terms of the families of the first type (for the definition of the concept of Q-curvature operators we refer to [4]).

Theorem 23. *If $n-1$ is even and $n-2p \geq 3$, we have*

$$\dot{D}_{n-1-2p}^{(p\to p)}(-p)|_{\ker(\bar{d})} = Q_{n-1-2p}^{(p)}\iota^*.$$

Theorem 23 is a consequence of formula (36). The result resembles the holographic formula

$$\dot{D}_n^{\mathrm{res}}(0;g)(1) = Q_n(g)$$

for Q-curvature of general metrics g on a manifold of even dimension n [17]. In later work, we shall also prove an analogue of Theorem 23 for residue families on forms. Combining this result with the analogue of eq. (55) for residue families then reproduces the double factorization property

$$L_{n-2p-1}^{(p)}\iota^* = (n-2p-1)\delta Q_{n-1-2p}^{(p)} d\iota^*$$

of the critical Branson–Gover operators discovered in [4].

The proofs of the results of the present section utilize the so-called F-method. This method has been developed by Kobayashi et al. in recent years in order to determine the homomorphisms of generalized Verma modules [22–26].

Acknowledgments: The results presented in this chapter were obtained within the research environment of the Collaborative Research Center 647 Space-Time-Matter at Humboldt-University Berlin. We are grateful for its support and the stimulating atmosphere.

Bibliography

[1] Aubry E, Guillarmou C. Conformal harmonic forms, Branson-Gover operators and Dirichlet problem at infinity. J. Eur Math Soc 2011;13:911–57. arXiv:0808.0552

[2] Baum H, Juhl A. Conformal differential geometry: Q-curvature and conformal holonomy. Oberwolfach Seminars 40, 2010.

[3] Branson T. Sharp inequalities, the functional determinant, and the complementary series, Trans. Am Math Soc 1995;347:3671–742.

[4] Branson T, Gover R. Conformally invariant operators, differential forms, cohomology and a generalisation of Q-curvature. Comm Part Diff Eq 2005;30(11):1611–69. arXiv:math/0309085.

[5] Branson T, Ørsted B. Conformal indices of Riemannian manifolds. Compos Math 1986;60(3):261–93.

[6] Djadli Z, Guillarmou C, Herzlich M. Opérateurs géométriques, invariants conformes et variétés asymptotiquement hyperboliques. Panoramas et Synthèses, 26, Société Mathématique de France, 2008.

[7] Fefferman C, Graham C R. The ambient metric. Ann Math Stud 78, Princeton University Press, 2011.

[8] Fefferman C, Graham C R, Juhl's formulae for GJMS-operators and Q-curvatures. J Am Math Soc 2013;26(4):1191–207. arXiv:1203.0360.

[9] Fischmann M, Juhl A, Somberg P. Conformal symmetry breaking differential operators on differential forms. Memoirs of AMS (2017) (to appear). 115 pp. arXiv:1605:04517.

[10] Fischmann M, Juhl A, Somberg P. Residue family operators on differential forms and Branson-Gover operators. (in preparation).

[11] Gover R. Conformal de Rham Hodge theory and operators generalizing the Q-curvature. Rend Circ Mat Palermo (2) 2005;Suppl 75:109–37. arXiv:math/0404004.

[12] Gover A R, Hirachi K. Conformally invariant powers of the Laplacian–a complete nonexistence theorem. J Am Math Soc 2004;17(2):389–405. arXiv:math/ 0304082v2

[13] Graham C R, Jenne R, Mason LJ, Sparling GAJ. Conformally invariant powers of the Laplacian. I. Existence. J London Math Soc 1992;46(2):557–565.

[14] Graham C R. Conformally invariant powers of the Laplacian. II Nonexistence, J London Math Soc 1992;46(2):566–76.

[15] Graham C R. Volume and area renormalizations for conformally compact Einstein metrics. Rend Circ Mat Palermo (2) 2000;Suppl 63:31–42. arXiv:math/9909042.

[16] Graham C R. Extended obstruction tensors and renormalized volume coefficients. Adv. Math. 2009;220(6):1956–1985. arXiv:0810.4203

[17] Graham, C R, Juhl A. Holographic formula for Q-curvature. Adv. Math. 2007;216:841–53. arXiv:0704.1673.

[18] Juhl A. Families of conformally covariant differential operators, Q-curvature and holography. Volume 275 of Progress in Mathematics. Birkhäuser, 2009.

[19] Juhl A. Explicit formulas for GJMS-operators and Q-curvatures. Geom Funct Anal 2013;23(4):1278–370. arXiv:1108.0273.

[20] Juhl A. On the recursive structure of Branson's Q-curvature. Math Res Lett 2014;21 (3):1–13. arXiv:1004.1784.v2.

[21] Juhl A. Heat kernels, ambient metrics and conformal invariants. Adv. Math. 2016;286:545–682. arXiv:1411.7851.

[22] Kobayashi T. Restrictions of generalized Verma modules to symmetric pairs. Transf Group 2012;17(2):523–46. arXiv:1008.4544

[23] Kobayashi T. F-method for symmetry breaking operators. Diff Geom Appl 2014;33:272–89. arXiv:1303.3541.

[24] Kobayashi T, Kubo T, Pevzner M. Vector-valued differential operators for the Möbius transformation. Springer Proc Math Stat 2015;101:67–86. arXiv:1406.0674v2.

[25] Kobayashi T, Ørsted B, Somberg P, Soucek V. Branching laws for Verma modules and applications in parabolic geometry. I Adv. Math. 2015;285:1796–852. arXiv:1305. 6040v1.

[26] Kobayashi T, Pevzner M. Differential symmetry breaking operators I. General theory and F-method. Sel Math 2016;22(2):801–45. arXiv:1301.2111v4.

[27] Kobayashi T, Speh B. Symmetry breaking for representations of rank one orthogonal groups. Volume 238 of Memoirs of AMS, Number 1126, 2015. arXiv:1310.3213.

[28] Paneitz S. A quartic conformally covariant differential operator for arbitrary pseudo-Riemannian manifolds (summary), SIGMA Symmetry Int Geom Meth Appl 2008;4, paper 036, 3p. arXiv:0803.4331.

[29] Parker T, Rosenberg S. Invariants of conformal Laplacians, J Diff Geom 1987;25(2):199–222.

Klaus Altmann and Gavril Farkas

Moduli and deformations

Global and local investigations in algebraic geometry

Abstract: Moduli spaces are algebraic objects parametrizing the entire set of algebraic gadgets of a certain type (e.g. curves, vector bundles, singularities). However, to gain enumerative information by using a meaningful intersection theory, it is necessary to consider compactifications of these moduli spaces. Doing so, an essential point is to provide a modular interpretation of the new objects on the boundary – and this can be understood from two, rather opposite viewpoints: On the one hand, we will study degenerations of classical objects, and on the other, we might start with singular, sometimes rather combinatorial objects and look for deformations, e.g. smoothings. In the present survey we are going to explain these general approaches, and we present concrete results within the context of moduli of curves and abelian varieties, toric and spherical varieties.

Keywords: deformations and moduli

Mathematics Subject Classification 2010: 14B07, 14B10, 14H10, 14K10, 14H51, 14M25

1 Introduction

Moduli spaces of algebraic objects like curves, abelian varieties, or sheaves on a fixed variety are parameter spaces that allow the study of these gadgets from a geometric, topological, and arithmetic point of view. These moduli spaces are not just sets, but algebraic varieties, whose geometry and invariants encode fundamental facts about the objects we parametrize. For instance, the fact that a moduli space of a certain class of objects is not unirational implies that the generic object of that kind cannot be written in a family depending on free parameters.

Often, the moduli spaces we encounter are among the most intriguing objects appearing in algebraic geometry. For instance, they have an extremely rich cycle theory, for giving a subvariety in moduli amounts to specifying a geometric property enjoyed by some (but not all) of the classified objects.

However, in most cases, moduli spaces that appear in algebraic geometry are not compact. To allow enumerative investigations, to be able to perform a meaningful intersection theory, one must work with compact spaces, therefore the moduli spaces in question have to be compactified. In doing this, the essential point is to provide

DOI 10.1515/9783110452150-003_s_001

a modular interpretation of the objects being represented by the new points, that is, those sitting in the boundary. This leads to the concept of degeneration.

Understanding degenerations of the classified objects becomes in this way one of the central points in the construction of compactifications in algebraic geometry.

Degenerations provide very often structures that are easier to handle than the original objects. For instance, they allow the usage of combinatorics and representation theory. In the case of moduli spaces of curves, the most degenerate objects become trivalent graphs.

The opposite point of view is to start with easy, rather combinatorial objects having plenty of symmetries and to classify all their deformations. Here, one starts with infinitesimal methods, studies possible obstructions, and tries to construct the so-called versal deformation of the original special objects. This family reflects the local structure of the corresponding moduli space around the points in question. Moreover, a possible splitting into several components might reflect different possibilities and view points of understanding.

In this survey, we are going to explain these general approaches, and we present concrete results within the context of moduli of curves and abelian varieties, toric and spherical varieties, and deformations obeying an additional structure like the compatibility with powers of the canonical sheaf. The first half of the survey is devoted to deformations and questions of local nature, whereas the second half focuses on the study of the geometry of some of the most established moduli spaces in algebraic geometry. The emphasis lies in questions of global nature like Kodaira dimension for various moduli spaces of curves and on uniformization and structure results.

2 Encoding torus actions on algebraic varieties

Torus actions on algebraic varieties allow to translate parts of their information into combinatorics. The link between algebraic geometry and combinatorics is provided by the character lattice of the torus. In Section 2.1, we recall the classical of toric varieties where the torus has the same dimension as the varieties in question. Afterwards, in Section 2.2, we will look at lower-dimensional torus actions and try to save parts of the combinatorial language.

2.1 Toric varieties

Toric varieties are meanwhile a well-established tool for illustrating methods and results from algebraic geometry within a discrete and convex geometric language. Starting at the combinatorial side of this correspondence, we will just recall a few facts, mostly thought to fix notation. For a more detailled information about general notions and concepts, see [16].

Let N and M be two mutually dual free abelian groups of rank n, i.e. $N, M \cong \mathbb{Z}^n$, and we have fixed a perfect pairing

$$\begin{aligned} N \times M &\longrightarrow \mathbb{Z} \\ a, r &\longrightarrow \langle a, r\rangle \end{aligned}$$

We denote by $N_{\mathbb{Q}}$ and $M_{\mathbb{Q}}$ the associated $\mathbb{Q}$-vector spaces. They host strictly convex, polyhedral cones $\sigma \subseteq N_{\mathbb{Q}}$ and their duals $\sigma^{\vee} := \{r \in M_{\mathbb{Q}} \mid \langle \sigma, r\rangle \geq 0\}$. The latter give rise to finitely generated semigroups $\sigma^{\vee} \cap M$, and the associated semigroup algebra leads to the n-dimensional affine toric variety

$$\mathbb{TV}(\sigma, N) := \operatorname{Spec} \mathbb{C}[\sigma^{\vee} \cap M].$$

This construction is functorial, and, moreover, the face relation $\tau \leq \sigma$ among cones from $N_{\mathbb{Q}}$ leads to open embeddings $\mathbb{TV}(\tau, N) \hookrightarrow \mathbb{TV}(\sigma, N)$. This allows to define non-affine toric varieties by gluing, i.e. so-called polyhedral fans Σ in $N_{\mathbb{Q}}$ lead to

$$\mathbb{TV}(\Sigma, N) := \varinjlim_{\sigma \in \Sigma} \mathbb{TV}(\sigma, N).$$

This variety contains the n-dimensional torus $T = \mathbb{TV}(0, N) = \operatorname{Spec} \mathbb{C}[M]$ as the minimal open subset within this direct limit, and M can be recovered as the character lattice of T. Moreover, the group law on T extends to an action of T on all varieties $\mathbb{TV}(\sigma, N)$. This action is reflected by the M-grading of the $\mathbb{C}$-algebras $\mathbb{C}[\sigma^{\vee} \cap M]$, and it is still present on all "naturally defined" modules over this algebra.

The polyhedral cones are rationally spanned by unique systems of extremal rays. By abuse of notation, we will use the same symbols for these rays as well as for the primitive lattice vectors spanning them. For the cones σ and $\sigma^{\vee}$, we denote

$$\sigma = \langle a^1, \ldots, a^k\rangle \quad \text{and} \quad \sigma^{\vee} = \langle w^1, \ldots, w^m\rangle.$$

If $\sigma^{\vee}$ is strictly convex, then the set E of irreducible elements of the semigroup $\sigma^{\vee} \cap M$ is finite, called the Hilbert basis. It contains $\{w^1, \ldots, w^m\}$ as a (usually proper) subset and it provides a minimal embedding

$$\mathbb{TV}(\sigma, N) \hookrightarrow \mathbb{TV}(\mathbb{Q}^E_{\geq 0}, \mathbb{Z}^E) = \mathbb{A}^E_{\mathbb{C}}.$$

2.2 Affine T-varieties via p-divisors

The concept of toric varieties presented in Section 2.1 describes algebro-geometric varieties X by objects from discrete mathematics. This implies a limitation to very special objects from algebraic geometry characterized by a large symmetry group (containing the torus T). In particular, the whole scenario becomes quite rigid. So, it was deformation theory which motivated the investigation of lower-dimensional torus actions in [4, 6].

On the other hand, retaining a torus action at all allows to keep some of the techniques from combinatorics. Compared to Section 2.1, we still have that $\dim X = n$, but now T is just a k-dimensional torus acting effectively on X that cannot be longer assumed to be an open subset of X. It follows that $k \le n$, and we call $c := n - k$ the complexity of the torus action. The case of $c = 0$ corresponds to Section 2.1.

We will now introduce the semi-combinatorial tools describing affine T-varieties X. See [4] or [7] for more details. The presence of T leads directly to (now k-dimensional) mutually dual lattices N and M as in Section 2.1. Let $\Delta \subseteq N_{\mathbb{Q}}$ be a polyhedron. Then,

$$\operatorname{tail}(\Delta) := \{a \in N_{\mathbb{Q}} \mid a + \Delta \subseteq \Delta\}$$

is called the tail cone of Δ. In particular, $\operatorname{tail}(\Delta)$ is the unique polyhedral cone C such that Δ equals the sum $\Delta^c + C$ with some (not unique) bounded polyhedron Δ^c. The minimal choice for Δ^c is the convex hull of all vertices of Δ. Let us, vice versa, fix a polyhedral cone $\sigma \subseteq N_{\mathbb{Q}}$. Then, we consider

$$\operatorname{Pol}^+(\sigma) := \{\text{polyhedra } \Delta \subseteq N_{\mathbb{Q}} \mid \operatorname{tail}\Delta = \sigma\}.$$

This set becomes a semigroup under Minkowski addition $\Delta + \Delta' := \{a + a' \mid a \in \Delta, a' \in \Delta'\}$. The advantage of fixing the tail cone is that Pol^+ gains the cancellation property, i.e. it is naturally embedded into its Grothendieck group $\operatorname{Pol}^+(\sigma) \hookrightarrow \operatorname{Pol}(\sigma)$. The neutral element is the prescribed tail cone σ. While all this has provided the tools on the combinatorial side of the picture, we now turn to the algebro-geometric part.

Let Y be a c-dimensional semi-projective (i.e. projective over something affine) variety over $\mathbb{C}$. We denote by $\operatorname{CaDiv} Y \subseteq \operatorname{Div} Y$ and $\operatorname{CaDiv}_{\mathbb{Q}} Y \subseteq \operatorname{Div}_{\mathbb{Q}} Y$ the Cartier and Weil divisors and their respective $\mathbb{Q}$-versions. Then, if $\mathcal{D} \in \operatorname{Pol}(\sigma) \otimes_{\mathbb{Z}} \operatorname{Div} Y$ is a divisor with polyhedral coefficients, we can define the following notions:

Definition 1.

(1) If $u \in \sigma^\vee$, then $\min\langle \Delta, u\rangle := \min\{\langle a, u\rangle \mid a \in \Delta\} \in \mathbb{Q}$ for $\Delta \in \operatorname{Pol}^+(\sigma)$. This definition naturally extends to a linear form $\min\langle \cdot, u\rangle$ on $\operatorname{Pol}(\sigma)$.

(2) If $u \in \sigma^\vee$ and $\mathcal{D} = \sum_D \Delta_D \otimes D$, then $\mathcal{D}(u) := \sum_D \min\langle \Delta_D, u\rangle \cdot D \in \operatorname{Div}_{\mathbb{Q}} Y$ is called the evaluation of $\mathcal{D}$ on u.

(3) $\mathcal{D}$ is called positive if it allows a representation $\mathcal{D} = \sum_D \Delta_D \otimes D$ with $\Delta_D \in \operatorname{Pol}^+(\sigma)$ and $D \ge 0$, e.g. being prime divisors.

Positive elements $\mathcal{D} \in \operatorname{Pol}(\sigma) \otimes_{\mathbb{Z}} \operatorname{Div} Y$ satisfy the convexity property

$$\mathcal{D}(u) + \mathcal{D}(u') \le \mathcal{D}(u + u') \qquad (u, u' \in \sigma^\vee)$$

meaning that the difference of both sides is an effective divisor. In particular, they give rise to an M-graded sheaf

$$\mathcal{O}_Y(\mathcal{D}) := \oplus_{u \in \sigma^\vee \cap M} \mathcal{O}_Y\big(\mathcal{D}(u)\big)$$

of $\mathcal{O}_Y$-algebras.

Definition 2. A positive $\mathcal{D} \in \operatorname{Pol}(\sigma) \otimes_{\mathbb{Z}} \operatorname{CaDiv} Y$ is called a p-divisor on Y if the evaluations $\mathcal{D}(u)$ are semiample for all $u \in \sigma^\vee \cap M$ and big for $u \in \operatorname{int} \sigma^\vee \cap M$.

For p-divisors $\mathcal{D} \in \operatorname{Pol}(\sigma) \otimes_{\mathbb{Z}} \operatorname{Div} Y$, we may define

$$\mathbb{TV}(\mathcal{D}, N, Y) := \operatorname{Spec}\Gamma(Y, \mathcal{O}_Y(\mathcal{D})).$$

It follows from [4, §3] that this is an affine, normal variety of finite type over $\mathbb{C}$ where the M-grading translates into a T-action, and it is shown that all those varieties can be obtained that way. Moreover, it is possible to define maps among p-divisors and to name the right equivalence relations on both sides such that this construction turns into an equivalence of categories, cf. [4, §8] and [7, §3].

Example 3. Let $\delta \subseteq N \oplus N'$ be a polyhedral cone, where N, N' are lattices as in Section 2.1. Then, we may understand the affine toric variety $X = \mathbb{TV}(\delta, N \oplus N')$ as a T-variety. That is, instead of the big torus $T \times T'$ corresponding to $N \oplus N'$, we focus just on T corresponding to the sublattice $N \subseteq N \oplus N'$, i.e. we forget about the T'-action. It was shown in [7, (4.2)], that, as a T-variety, X equals $\mathbb{TV}(\mathcal{D}, N, Y)$ with $Y = \mathbb{TV}(\pi(\delta), N'))$, where $\pi(\delta)$ denotes the coarsest fan in $N'_{\mathbb{Q}}$ subdividing the images of all faces of δ along the projection $\pi : N \oplus N' \to N'$ and where

$$\mathcal{D} = \sum_{a' \in \pi(\delta)(1)} (\pi^{-1}(a') \cap \delta) \otimes \operatorname{orb}(a').$$

Here we understand the coefficients as polyhedra in $N_{\mathbb{Q}}$, and $\operatorname{orb}(a')$ means the divisor being the 1-codimensional closed orbit in the toric variety Y that corresponds to the ray, i.e. to the 1-dimensional cone a' of the fan $\pi(\delta)(1)$.

Let us look at a very special case: If $N' \cong \mathbb{Z}$, then Example 3 shows how to understand an affine toric variety as a T-variety of complexity 1. The base Y will be $\mathbb{A}^1$ or $\mathbf{P}^1$, and the above sum will involve one or two summands, respectively. Taking the opposite point of view shows that p-divisors on, e.g., $\mathbf{P}^1$ involving at most two summands carry a stronger, namely a toric structure. However, p-divisors with three non-trivial summands or more are true T-varieties.

2.3 Understanding spherical varieties as T-varieties

Spherical varieties is a well-known class of varieties that allow a combinatorial description, namely by the so-called colored fans. In [10], we have studied a canonical torus action on these gadgets turning them into T-varieties. We will explain how the associated p-divisor can be read off the given colored fan. However, we start with recalling basic facts and notation about spherical varieties, cf. [34, 39]. See [10, §3] for detailed references.

Let G be a connected reductive group. A subgroup $H \subseteq G$ is called spherical if there is a Borel subgroup $B \subseteq G$ such that $B \cdot H$ is open and dense in G. In particular, $B \cdot 1$ is an open dense orbit in G/H. The finitely many B-invariant divisors in G/H are called colors; they form a set $\mathscr{C} = \mathscr{C}(G/H)$. The weight lattice $\mathscr{X} = \mathscr{X}(G/H)$ is defined as the sublattice of the character group of B such that for each $\chi \in \mathscr{X}$ there exists a (up to scalars unique) rational function $f \in \mathbb{C}(G/H)$ with

$$f(bg) = \chi(b) \cdot f(g) \quad (\text{for all } b \in B,\ g \in G).$$

Thus, valuations on $\mathbb{C}(G/H)$ with values in $\mathbb{Z}$ induce elements in $\mathscr{X}^* := \operatorname{Hom}(\mathscr{X}, \mathbb{Z})$. If $\mathscr{V}$ denotes the set of all G-invariant valuations, then this even provides an embedding $\varrho : \mathscr{V} \hookrightarrow \mathscr{X}^*$. The rational version $\mathscr{V}_{\mathbb{Q}} \subseteq \mathscr{X}_{\mathbb{Q}}$ of this is the so-called valuation cone. Similarily, since every divisor provides a valuation, we obtain a map $\varrho : \mathscr{C} \to \mathscr{X}^*$ which, however, does not need to be injective.

Note that the pair $(\mathscr{X}^*, \mathscr{X})$ is a generalization of the pair (N, M) in the toric situation of Section 2.1 with $H = \{1\}$ in $G = B = T$ and $\mathscr{C} = \emptyset$, $\mathscr{V} = \mathscr{X}^*$. However, since we are going to consider a special torus action with a different character lattice later, it is useful to use a different notation here. The rank of the lattice $\mathscr{X}$ is called the rank $\operatorname{rk}(G/H)$ of the spherical homogenous space G/H.

The main point about the Luna-Vust theory in [39] is the classification of G-invariant partial compactifications $X \supseteq G/H$ ("embeddings of G/H") by the so-called colored fans Σ^X. They arise in a straightforward manner from the so-called colored cones:

Definition 4. A colored cone is a pair $(C, \mathscr{F})$ consisting of a strictly convex polyhedral cone $C \subseteq \mathscr{X}^*_{\mathbb{Q}}$ with $\operatorname{int}(C) \cap \mathscr{V}_{\mathbb{Q}} \neq \emptyset$ and some $\mathscr{F} \subseteq \mathscr{C}$ with $0 \notin \varrho(\mathscr{F})$ such that C is generated by $\varrho(\mathscr{F})$ and some additional elements from $\mathscr{V}$.

Within spherical varieties, it is the simple ones, i.e. those $X_0 \supseteq G/H$ containing a unique closed G-orbit, which correspond to the colored cones. The respective cone arises from X_0 as follows: $\mathscr{F}$ is the set of colors whose closure in X_0 contain the unique closed G-orbit, and C is spanned by the image under ϱ of the divisors of X_0 outside the dense B-orbit.

Let $H' := H \cdot N_G(H)^\circ$. The associated G/H' is spherical, too. Its structure is much easier than the original G/H. The main difference is that the corresponding valuation cone $\mathcal{V}'_{\mathbb{Q}} \subseteq \mathcal{X}^*(G/H')$ is strictly convex and simplicial. In particular, it may serve as an (un-) colored cone $(\mathcal{V}'_{\mathbb{Q}}, \emptyset)$ providing a simple, toroidal compactification $\overline{Y}$ of G/H'. Every other toroidal compactification $Y \supseteq G/H'$ dominates $\overline{Y}$ and corresponds to a fan Σ_Y refining $\mathcal{V}'_{\mathbb{Q}}$.

There is a natural projection $\Phi : G/H \to G/H'$ called the Tits fibration. Its fiber is the torus $T := H'/H$. This torus acts from the right on all embeddings $X \supseteq G/H$; in particular, this action commutes with the G-action from the left. And it is exactly this action turning X into a T-variety. The lattice of one-parameter subgroups N of T fits into the exact sequence

$$0 \to N \to \mathcal{X}^*(G/H) \xrightarrow{p} \mathcal{X}^*(G/H') \to 0,$$

and $N_{\mathbb{Q}}$ is the fiber of the projection from $\mathcal{V}_{\mathbb{Q}} = p^{-1}(\mathcal{V}'_{\mathbb{Q}})$ onto $\mathcal{V}'_{\mathbb{Q}}$. Using Φ, the set of colors $\mathcal{C}(G/H)$ and $\mathcal{C}(G/H')$ can be naturally identified, and the maps $\varrho : \mathcal{C}(G/H) \to \mathcal{X}^*(G/H)$ and $\varrho' : \mathcal{C}(G/H') \to \mathcal{X}^*(G/H')$ are compatible with p. Moreover, fixing a splitting of the above sequence, the map ϱ splits, too. We obtain a map $\overline{\varrho} : \mathcal{C} \to N$ with $\varrho(D) = \big(\overline{\varrho}(D), \varrho'(D)\big)$.

Assume now that G/H is a spherical homogenous space of minimal rank, i.e. $\mathrm{rk}(G/H) = \mathrm{rk}(G) - \mathrm{rk}(H)$ (in general, only "$\geq$" is satisfied). The main result of [10] is to understand every spherical embedding $X \supseteq G/H$ as a T-variety and to encode a certain T-invariant, affine open covering of X by a collection of p-divisors on open subsets of some spherical embedding $Y \supseteq G/H'$ dominating $\overline{Y}$. This means to split the information about X into a combinatorial part represented by the p-divisors and an algebro geometric one represented by Y. Both parts are constructed out of the colored fan Σ^X as follows:

First, the modification $Y \to \overline{Y} \supseteq G/H'$ is given by the subdivision $\Sigma_Y \leq \mathcal{V}'_{\mathbb{Q}}$ arising as the image fan of $\Sigma^X \cap \mathcal{V}_{\mathbb{Q}}$. The latter means that we forget about the colors of Σ^X and that Σ_Y is supposed to be the coarsest refinement of the union of the projections $p(C \cap \mathcal{V}_{\mathbb{Q}})$, where C runs through all elements of Σ^X. Thus, there are two special types of divisors on Y:

(i) The modification $Y \to \overline{Y}$ is an isomorphism along the common open subset G/H'. And the one-dimensional cones $a \in \Sigma_Y(1)$ are in a one-to-one correspondence to the G-invariant divisors D_a in $Y \setminus (G/H')$.

(ii) The colors $D' \in \mathcal{C}(G/H')$ are B-invariant divisors in G/H'. We denote their closures in Y with $\overline{D'}$.

Similar to the downgrade construction in Example 3, we define for each colored cone $(C, \mathcal{F}) \in \Sigma^X$ the following divisor on Y with polyhedral coefficients:

$$\mathcal{D}(C) := \sum_{a \in \Sigma_Y(1)} \big(p^{-1}(a) \cap C\big) \otimes D_a + \sum_{D' \in \mathcal{C}(G/H')} \big(\overline{\varrho}(D') + (p^{-1}(0) \cap C)\big) \otimes \overline{D'}.$$

Using the splitting of the above exact sequence, the coefficients in the first sum can be understood as polyhedra inside $p^{-1}(0) = N_{\mathbb{Q}}$; their common tail cone is $\sigma(C) := p^{-1}(0) \cap C$. The coefficients in the second sum are just integral shifts of this tail cone. In the theory of p-divisors, those *almost trivial* summands are responsible for creating a certain twist on X understood as a kind of a bundle over Y. So far, $\mathscr{D}(C)$ does neither depend on the colors $\mathscr{F}$, nor do we claim that it is a p-divisor on Y. Instead, we have the following:

Theorem 5. *Let $G/H \subseteq X$ be an embedding of a spherical G/H of minimal rank. Denote by Σ^X the colored fan describing X and by W the Weyl group of G.*
1) *If $(C, \mathscr{F}) \in \Sigma^X$ and $w \in W$, then $\mathscr{D}(C)$ is a p-divisor on $Y \setminus \bigcup_{D' \in \mathscr{C}(G/H')\setminus\mathscr{F}} w\overline{D'}$.*
2) *The corresponding p-divisors $\mathbb{TV}(\mathscr{D}(C), \mathscr{F}, w)$ with $(C, \mathscr{F}) \in \Sigma^X$ and $w \in W$ glue in a natural way, yielding X.*

This is Theorem 1.1 in [10]. We have just changed the original wording slightly to avoid the notion of divisorial fans being introduced in [6]. We have mentioned that $\mathscr{D}(C)$ does not need to be a p-divisor on Y. Nevertheless, we can always build the sheaf $\mathscr{O}_Y(\mathscr{D}(C))$. The difference between the construction of $\mathbb{TV}(\mathscr{D})$ right after Definition 2 and here is that in the theorem one is supposed to take a kind of a relative spectrum with respect to a (usually non-affine) covering of Y. The "degree of relativity" of this spectrum is reflected by the "degree of locality" of this covering – and this is governed by the amount of colors within $\mathscr{F}$. Moreover, for a given $\mathscr{F} \subseteq \mathscr{C}$, one can define the subgroup

$$W_{\mathscr{F}} := \left\{w \in W \mid w\left(\bigcup_{D' \in \mathscr{C}\setminus\mathscr{F}} w\overline{D'}\right) = \bigcup_{D' \in \mathscr{C}\setminus\mathscr{F}} w\overline{D'}\right\} \subseteq W.$$

Then, in (ii) of the previous theorem, the elements w are supposed to run through the cosets of $W/W_{\mathscr{F}}$ only.

Finally, we would like to remark that the assumption of minimal rank is essential for the previous theorem. The easiest counter example is $X = \mathbf{P}^1 \times \mathbf{P}^1$ considered with the diagonal GL_2-action. See [10, (7.5)] for a less trivial example and [10, (3.4)] for a discussion of the property of minimal rank.

3 Infinitesimal deformations of combinatorial singularities

In this section, we would like to treat the infinitesimal deformation theory of singularities given by certain combinatorial data. First, after recalling some general facts about deformation theory in Section 3.1, we focus in Section 3.2 on monomial ideals. In particular, we will demonstrate how this can be used to check rigidity of certain singularities provided by graphs. Afterwards, we return to the toric varieties introduced in Section 2.1. In Section 3.4 we will give a description of the tangent space of the deformation functor, and in Section 3.3 we will impose further requirements about the behavior of the canonical sheaf under deformation. All this can be done without leaving the toric context. Only later, when treating global deformations over reduced parameter spaces in Section 4, it will be useful to use the techniques of T-varieties introduced in Section 2.2.

3.1 General setup of deformation theory

Let X be a $\mathbb{C}$-scheme of finite type or a germ of such an object. A deformation of X over the germ of some $\mathbb{C}$-scheme $S = (S, 0)$ of finite type is defined as a flat map $f : \mathcal{X} \to S$ together with an isomorphism $X \xrightarrow{\sim} f^{-1}(0)$. Restricting this notion to base spaces S being the spectrum of local artinian $\mathbb{C}$-algebras $A \in \mathrm{Art}_{\mathbb{C}}$, this notion yields a functor $\mathrm{Def}_X : \mathrm{Art}_{\mathbb{C}} \to \mathrm{Sets}$ with

$$\mathrm{Def}_X : A \mapsto \{\text{deformations of } X\}/\{\text{isomorphisms}\}.$$

In [44] it was shown that, at least for isolated singularities X, this functor has a pro-representing hull, i.e. there is a unique formal, formally miniversal deformation of X. We denote by T^i_X ($i = 1, 2$) its tangent space and the space containing the obstructions, respectively. The former can be obtained as $\mathrm{Def}_X(\mathbb{C}[\varepsilon]/\varepsilon^2)$, and for normal singularities, this equals $\mathrm{Ext}^1_X(\Omega_X, \mathcal{O}_X)$. See [46] for a detailed treatment of the deformation theory of singularities.

Example 6. The first known example exhibiting a non-smooth versal base space was the cone over the rational normal curve of degree 4, i.e. over the Veronese embedding $\mathbf{P}^1 \hookrightarrow \mathbf{P}^4$, cf. [42]. This is a normal surface singularity with embedding dimension five, and it can be given by six equations encoded via the condition

$$\mathrm{rk}\begin{pmatrix} x_0 & x_1 & x_2 & x_3 \\ x_1 & x_2 & x_3 & x_4 \end{pmatrix} \le 1 \quad \text{or, alternatively,} \quad \mathrm{rk}\begin{pmatrix} x_0 & x_1 & x_2 \\ x_1 & x_2 & x_3 \\ x_2 & x_3 & x_4 \end{pmatrix} \le 1.$$

Here, T^1 is four-dimensional, and the four perturbations can be realized via

$$\mathrm{rk}\begin{pmatrix} x_0 & x_1+\varepsilon_1 & x_2+\varepsilon_2 & x_3+\varepsilon_3 \\ x_1 & x_2 & x_3 & x_4 \end{pmatrix} \le 1 \quad \text{and} \quad \mathrm{rk}\begin{pmatrix} x_0 & x_1 & x_2 \\ x_1 & x_2+\varepsilon_4 & x_3 \\ x_2 & x_3 & x_4 \end{pmatrix} \le 1.$$

In fact, these two deformations provide two irreducible components of dimension three and one of the miniversal deformation.

The higher the codimension of the singular locus the more one should expect that the singularity tends to become rigid, i.e. $T^1 = 0$. The most famous example is the rigidity of isolated quotient singularities in dimension at least three, cf. [45]. A not fully rigid singularity is the cone over the del Pezzo surface of degree eight – in [11, (9.2)] it was shown that the versal base space is the smallest possible non-trivial one, namely $\operatorname{Spec}\mathbb{C}[\varepsilon]/\varepsilon^2$.

3.2 The tangent space T^1 for reduced monomial singularities

In [2] we have investigated (in general non-normal) singularities arising from a simplicial complex Δ on $[n] := \{1, \dots, n\}$ via the associated Stanley–Reisner ring, i.e. the ideal of $X(\Delta) \subseteq \mathbb{A}^n_{\mathbb{C}}$ is given by the squarefree monomials $x^s := \prod_{i \in s} x_i \in \mathbb{C}[x_1, \dots, x_n]$ with $s \in 2^{[n]} \setminus \Delta$. The affine coordinate ring of $X(\Delta)$ is $\mathbb{Z}^n$-graded, and so is the module $T^1_{X(\Delta)}$.

Definition 7 ([25]). Let $I \subseteq \mathbb{C}[x_1, \dots, x_n]$ be a monomial ideal. Then, a monomial ideal $J \subseteq \mathbb{C}[y_i, x_1, \dots, x_n]$ is called a separation of I for x_i if both y_i and x_i divide some minimal generators of J, if $y_i - x_i$ is a non-zero divisor of $\mathbb{C}[y_i, x_1, \dots, x_n]/J$, and if $J/(y_i - x_i) = I$.

Then, denoting the difference $t := y_i - x_i$, the ring $\mathbb{C}[y, x_1, \dots, x_n]/J$ becomes a flat $\mathbb{C}[t]$-algebra providing an (unobstructed) 1-parameter family with special fiber $X = \operatorname{Spec}\mathbb{C}[x_1, \dots, x_n]/I$. Via the Kodaira-Spencer map, this deformation is visible in the homogeneous component $T^1_X(-e_i)$ of multidegree $-e_i \in \mathbb{Z}^n$.

In [2, §1] we have shown that in the case of squarefree monomial ideals, this is the only deformations in this degree: If Δ is a simplicial complex as above, then we attach to each $i \in [n]$ a graph $G_i(\Delta)$ whose vertices are those $f \in \Delta$ for which $f \cup \{i\} \notin \Delta$. Two vertices f, g are connected by an edge if either $f \subsetneq g$ or $g \subsetneq f$.

Theorem 8 (Ref. [2]). *The homogeneous components $T^1_X(-v \cdot e_i)$ $(v > 0)$ vanish unless $v = 1$. In this case it equals the reduced homology of the graph $G_i(\Delta)$. If this dimension is k, then there is a k-dimensional separation whose Kodaira-Spencer map $\mathbb{C}^k \to T^1_X(-e_i)$ is an isomorphism. In particular, $T^1_X(-e_i)$ is unobstructed.*

In [2, §3] we have obtained further results for the special case of so-called edge ideals of graphs. This is the special case where Δ arises as the independence complex of some graph $\mathcal{G}$. In particular, the associated monomial ideal is then generated by all products $x_i x_j$, where $\{i, j\}$ form an edge of the given graph $\mathcal{G}$. In these cases, we have determined in [2, Corollary 3.5] all other strictly negative degrees of T^1:

Theorem 9. *Let $b \in \mathbb{Z}^n_{\geq 0}$ with $\operatorname{supp} b := \{i \in [n] \mid b_i \neq 0\}$ consisting of at least two elements. Then $T^1_X(-b) = 0$ unless $b = \operatorname{supp} b$ is an isolated edge in $\mathcal{G}$. If this is the case, then $T^1_X(-b)$ is one dimensional.*

3.3 qG-deformations of two-dimensional cyclic quotients

Beyond monomial ideals considered in Section 3.2, the next type of ideals governed by combinatorics are toric ideals as introduced in Section 2.1. The easiest ones among them yield the two-dimensional cyclic quotient singularities $X = \mathbb{TV}(\sigma)$, where $\sigma \subseteq N_{\mathbb{Q}}$ has two generators $\alpha, \beta \in N$. One can find an isomorphism $N \cong \mathbb{Z}^2$ such that $\alpha = (1, 0)$ and $\beta = (-q, n)$ for some $n \in \mathbb{Z}_{\geq 2}$ and $q \in (\mathbb{Z}/n\mathbb{Z})^*$, where the latter will be represented by some $q \in \{1, \ldots, n-1\}$. Actually, we will exclude both the smooth case ($n = 1$) and the case of a hypersurface singularity ($q = n - 1$).

A special feature of the two-dimensional case is that the Hilbert basis $E = \{w^1, \ldots, w^e\}$ of $\sigma^\vee \cap M$ is formed by the lattice points of the boundary of the convex hull of $(\sigma^\vee \cap M) \setminus \{0\}$. In particular, it is ordered. It starts with $w^1 = [0, 1] \in \alpha^\perp$, and it ends with $w^e = [n, q]$, where e is the embedding dimension. In between, everything is governed by the rule

$$w^{i-1} + w^{i+1} = a_i \cdot w^i \quad \text{with} \quad \frac{n}{n-q} = [a_2, \ldots, a_{e-1}] := a_2 - \frac{1}{a_3 - \frac{1}{a_4 - \ldots}}, \quad a_i \geq 2.$$

Example 10. The toric description of Example 6 is given by $n = 4$, $q = 1$. In its most symmetric form, it looks like $\sigma = \langle (1, 2), (-1, 2) \rangle$ and $\sigma^\vee = \langle [-2, 1], [2, 1] \rangle$ with $[a_2, a_3, a_4] = [2, 2, 2]$.
(Note that we use (..) and [..] to display elements of N and M, respectively. The latter should not be mistaken with the notation of the continued fraction.)

It follows from [42] that $T^1(-R) = 0$ unless the following few cases for $R \in M$:
(i) $R = w^2$ and w^{e-1} yield $T^1(-R) = N_{\mathbb{C}}/\mathbb{C}\cdot\alpha$ and $N_{\mathbb{C}}/\mathbb{C}\cdot\beta$, respectively.
(ii) $R = w^i$ ($i = 3, \ldots, e-2$) yield two-dimensional $T^1(-R) = N_{\mathbb{C}}$.
(iii) $R = \ell \cdot w^i$ ($i = 2, \ldots, e-1$, $2 \leq \ell \leq a_i - 1$) yield $T^1(-R) = (w^i)^\perp \subset N_{\mathbb{C}}$.

In any case, $T^1(-R)$ is a subquotient of $N_{\mathbb{C}}$, and its elements are represented by derivations $x^{-R}\,\partial_a : x^r \mapsto \langle a, r \rangle \cdot x^{r-R}$ with $a \in N$ and $r \in M$. By definition, they provide an infinitesimal deformation $\widetilde{X} = \widetilde{X}_{R,a}$ of X over the base $B = \mathrm{Spec}\,\mathbb{C}[\varepsilon]/\varepsilon^2$. However, not every flat deformation fits well into moduli theory. We have to impose an additional property:

Definition 11.
(1) For an integer $g \in \mathbb{Z}$, we say that $x^{-R}\,\partial_a$ satisfies the property $(*)_g$ if the reflexive power $\omega^{[g]}_{\widetilde{X}|B}$ on $\widetilde{X} = \widetilde{X}_{R,a}$ is flat over B.
(2) We call $\widetilde{X}$ a V-deformation, if $(*)_g$ is satisfied for $g = \mathrm{index}\,\omega_X$ (implying it for all multiples), and it is called a qG-deformation if $(*)_g$ is satisfied for every $g \in \mathbb{Z}$.

Note that $(*)_0$ encodes the fact that $\widetilde{X}$ is a deformation at all. However, the remaining conditions $(*)_g$ ask for more. Translated into combinatorics, we have shown in [9, §3] that it is characterized by

$$x^{-R}\partial_a \text{ satisfies } (*)_g \iff \left(\tfrac{g}{n}\cdot(w^1+w^e)+Z_R\right)\cap M \subseteq a^\perp + g\cdot R$$

where Z_R denotes the bounded region $Z_R := \sigma^\vee \cap (R - \operatorname{int}\sigma^\vee) \subset M_{\mathbb{Q}}$. As a first consequence, this implies the following explicit description of the V-deformations in [9, Corollary 22]:

Theorem 12. *$x^{-R}\partial_a \in T^1_X(-R)$ is a V-deformation iff $a \in (w^1 + w^e - nR)^\perp$.*

This exposes altogether $(e-4)$ dimensions of V-deformations within the degrees $R = w^i$ of type (ii) of the previous list. Moreover, there is at most one additional series of V-deformations of type (iii). This happens if and only if $w^1 + w^e$ is a multiple of some special $w^\nu \in E$. We call the cone σ grounded in this case; it leads to one-dimensional spaces of V-deformations for each $R = \ell \cdot w^\nu$ with $\ell = 2, \ldots, a_\nu - 1$.

To list the more special qG-deformations, we will switch to an alternative way of naming cyclic quotient singularities. Denoting by $w \in M$ the primitive generator of the ray $\mathbb{Q}_{\geq 0}\cdot(w^1 + w^e)$ (i.e. $w = w^\nu$ in the grounded case), we obtain a one-one correspondence

$$\begin{array}{rcl} \{\text{cones } \sigma\}/\mathrm{SL}(2,\mathbb{Z}) & \longleftrightarrow & \{\text{intervals } I \subseteq \mathbb{Q} \text{ with uniform denominators}\}/\{\mathbb{Z}\text{-shifts}\} \\ \sigma & \longmapsto & \sigma \cap [w = 1] \\ \mathbb{Q}_{\geq 0}\cdot(I, 1) & \longleftarrow & I, \end{array}$$

where we call I to have "*uniform denominators*" (at the end points) if both become equal in the reduced forms. Under this correspondence, the groundedness of cones translates into the property that I contains interior integers. When this is the case, then we may, w.l.o.g., suppose that $0 \in \operatorname{int} I$, i.e. that

$$I = [-A, B] \subseteq \mathbb{Q} \quad \text{with} \quad A, B \in \mathbb{Q}_{>0}$$

having the same denominator in their reduced form. This language allows to express the special ν-th element a_ν of the continued fraction $[a_2, \ldots, a_{e-1}]$ as

$$a_\nu - 2 = \lfloor A \rfloor + \lfloor B \rfloor \leq \lfloor A + B \rfloor \leq A + B = |I|.$$

Theorem 13. *X has no qG-deformations unless it is grounded. If this is the case, and if it is given by $I = [-A, B]$ with $w = w^\nu$, then the homogeneous qG-deformations are formed by the one-dimensional subspaces $w^\perp \subseteq T^1(-\ell \cdot w) \subseteq N_{\mathbb{C}}$ with $\ell = 1, \ldots, \lfloor A + B \rfloor$.*

In particular, X allows a non-trivial qG-deformation at all if and only if $|I| \geq 1$. This should be compared with the characterization of the so-called T-singularities via the property $|I| \in \mathbb{Z}$. Recall from [35] that T-singularities are the cyclic quotients admitting a $\mathbb{Q}$-Gorenstein smoothing. They are the building blocks of the partial "P-resolutions" being in a one-to-one correspondence with the components of the versal base space of cyclic quotient singularities.

3.4 The tangent space T^1 for toric singularities

In Section 3.3, we have treated the easiest among the toric singularities, but somehow it is exactly them causing the main trouble within the general class of toric singularities. Let $\sigma = \langle a^1, \dots, a^k \rangle \subseteq N_{\mathbb{Q}}$ be a strictly convex, polyhedral cone as in Section 2.1. In [12, §2] one can find a formula representing $T^1_X(-R)$ for $R \in M$ as a simple expression along the inclusion-exclusion principle involving the finite sets

$$E_j^R := \{w \in E \mid \langle a^j, w\rangle < \langle a^j, R\rangle\} \subset M.$$

In the special case of $\sigma = \langle \alpha, \beta \rangle$ from Section 3.3, this formula comes down to

$$T^1_X(-R)^* = \left(\mathrm{span}_{\mathbb{C}} E^R_\alpha \cap \mathrm{span}_{\mathbb{C}} E^R_\beta\right) / \mathrm{span}_{\mathbb{C}}(E^R_\alpha \cap E^R_\beta)$$

which is just another version of the characterization of the condition $(*)_0$ right after Definition 11.

In the general toric case, we define $C(R)$ as the polyhedral cone parametrizing the Minkowski summands of $Q(R) := \sigma \cap [R = 1]$, i.e. those polyhedra Q' with $\mathrm{tail}(Q') = \mathrm{tail}Q(R)$ such that there is a polyhedron Q'' with $Q + Q'' = \lambda \cdot Q(R)$ for some $\lambda \geq 0$. To make this more explicit, we denote by $d^1, \dots, d^N \in R^\perp$ the bounded edges of $Q(R)$. Then,

$$C(R) := \{(t_1, \dots, t_N) \in \mathbb{Q}^N_{\geq 0} \mid \textstyle\sum_{i \in f} t_i \cdot (\pm d^i) = 0 \text{ for bounded 2-faces } f \leq Q(R)\}$$

where the signs in front of d^i are chosen in such a way that the edges form an oriented cycle along ∂f. If $V(R) := \mathrm{span}_{\mathbb{C}} C(R)$, then [12, Theorem 2.5] describes $T^1_{\mathbb{TV}(\sigma)}(-R)$ as a certain subspace of $V(R) \oplus W(R) / \mathbb{C} \cdot (\underline{1}, \underline{1})$ with $W(R)$ being the vector space of all formal $\mathbb{C}$-linear combinations of non-integral vertices of $Q(R)$. The particular equations of $T^1(-R)$ depend on the type of cyclic quotient singularities provided by the bounded edges of $Q(R)$. In the special case of their absence, this leads to the following version without using the auxiliary $W(R)$ at all:

Theorem 14. *Assume that $X = \mathbb{TV}(\sigma)$ is smooth in codimension two. Then $T^1_X(-R) \subseteq V(R) / \mathbb{C} \cdot \underline{1}$ is cut out by the equations $t_i = t_j$ arising from bounded edges d^i, d^j sharing a non-integral vertex of $Q(R)$.*

In particular, the Minkowski summands represented by elements of $C(R)$ obeying the $T^1(-R)$-equations are locally, around every non-lattice vertex, just dilations of the original polyhedron $Q(R)$. On the other hand, global dilations of $Q(R)$ are collected in the subspace $\mathbb{C} \cdot \underline{1}$, hence become trivial in $T^1(-R)$.

4 Global deformations of toric singularities

We leave the infinitesimal level and focus on the true structure of the versal deformations. In particular, the component structure of the base space and adjacent fibers will become visible.

4.1 Maximal deformations in a fixed negative degree

Let $X = \mathbb{TV}(\sigma)$ be a toric variety that is smooth in codimension two. Let $R \in \sigma^\vee \cap M$ be a primitive element, i.e. besides its "positivity", we assume that it is not a multiple of some other degree. We have built in Section 3.4 a cone $C(R)$ acting as the "moduli space of Minkowski summands" of $Q(R)$, i.e. every $t = (t_1, \dots, t_N) \in C(R)$ encodes a polyhedron Q_t with bounded edges $t_i \cdot d^i$ instead of just d^i in $Q(R)$. This leads to Definition 15.

Definition 15. We call $\tilde{C}(R) := \{(t, a) \mid t \in C(R),\ a \in Q_t \subset R^\perp \subset N_{\mathbb{Q}}\}$ with its associated projection $\pi : \tilde{C}(R) \to C(R)$ the universal Minkowski summand of $Q(R)$.

Then, $\tilde{C}(R)$ is a polyhedral cone, too, and the projection π gives rise to two interesting subfamilies:

(i) The restriction of π to $\mathbb{Q}_{\geq 0} \cdot \underline{1} \subseteq C(R)$ yields the pre-image $\sigma \subseteq \tilde{C}(R)$.
(ii) $C(R)$ contains the subcone $C'(R) := \{t \in C(R) \mid t_i = t_j \text{ if } d^i \cap d^j \supsetneq d^i \cap d^j \cap N\}$, and we denote $\pi^{-1}C'(R)$ by $\tilde{C}'(R)$. The restriction $\pi : \tilde{C}'(R) \to C'(R)$ contains $\pi : \sigma \to \mathbb{Q}_{\geq 0} \cdot \underline{1}$ from (i).

In [8, Theorem 6.1], we have used these subfamilies to construct the maximal deformation of X in degree $-R$. By this we mean a flat family $f : \tilde{X} \to V(J) \subseteq T^1(-R)$ with minimal ideal J extending the tautological infinitesimal deformation of X over the thick point with tangent space $T^1(-R)$.

Theorem 16 ([8]).

(1) *Denoting by* $\iota : \mathbb{TV}\big(C'(R)\big) \to \mathbb{C}^N$ *the toric map corresponding to* $C'(R) \to \mathbb{Q}^N_{\geq 0}$, *then there is a maximal closed subscheme* $\mathcal{M} \subseteq \iota\big(\mathbb{TV}\big(C'(R)\big)\big) \subseteq \mathbb{C}^N$ *being, at the same time, a full pre-image under the linear projection* $\ell : \mathbb{C}^N \to \mathbb{C}^N/\mathbb{C} \cdot \underline{1}$.
(2) *The projection* π *induces a toric morphism* $\mathbb{TV}(\pi) : \mathbb{TV}\big(\tilde{C}'(R)\big) \to \mathbb{TV}\big(C'(R)\big)$, *and the restriction of* $\big(\ell \circ \iota \circ \mathbb{TV}(\pi)\big) : \mathbb{TV}\big(\tilde{C}'(R)\big) \to \mathbb{C}^N/\mathbb{C} \cdot \underline{1}$ *to* $\ell(\mathcal{M})$ *yields the maximal deformation of* X *in degree* $-R$.

Besides its description by a universal condition, the subscheme $\mathscr{M} \subseteq \mathbb{C}^N$ has a very easy description by equations. They consist of two sets, namely of $t_i = t_j$ arising from bounded edges d^i, d^j sharing a non-integral vertex of $Q(R)$ (just as in the previous definition of $C'(R)$) and of $\sum_{i\in f} t_i^k \cdot (\pm d^i) = 0$ for bounded 2-faces $f \le Q(R)$ and $k \ge 1$ (just as in the definition of $C(R)$ in Section 3.4 with $k = 1$).

4.2 Genuine k-parameter families

We will impose an integrality condition on Minkowski decompositions of polyhedra Q. It will be the natural relaxation of the request for lattice polytopes as summands when Q is not required to be a lattice polytope itself.

Definition 17. Let Q be a polyhedron. A decomposition $Q = Q_0 + \ldots + Q_k$ is called integral if all Q_i have the same tail cone σ, and if for every $u \in \sigma^\vee$ all but (at most) one $\operatorname{face}(Q_i, u) := \{a \in Q_i \mid \langle a, u\rangle \le \langle Q_i, u\rangle\}$ contain lattice points.

If σ and $R \in \sigma^\vee \cap M$ are as in Section 4.1, then every Minkowski decomposition of $Q(R)$ can be observed as a decomposition of $\underline{1}$ in the cone $C(R)$. And it is the absence of two-codimensional singularities in $X = \mathbb{TV}(\sigma)$ that does even ensure that integral decompositions of $Q(R)$ induce integral decompositions of $\underline{1}$ within the smaller cone $C'(R)$. In particular, the summands $Q_i \in C'(R)$ span a $(k+1)$-dimensional subcone isomorphic to $\mathbb{Q}_{\ge 0}^{k+1}$, and the restriction of the family $\pi : \widetilde{C}'(R) \to C'(R)$ to this subcone equals the so-called Cayley construction for $\{Q_0, \ldots, Q_k\}$. It leads directly to a k-parameter family deforming X over the parameter space $\mathbb{C}^{k+1}/\mathbb{C}\cdot\underline{1} \cong \mathbb{A}^k$.

We have seen in [12, §3] that the Cayley construction transforms integral decompositions of $Q(R)$ into k-parameter families even when two-codimensional singularities are present. However, to obtain a maximal deformation in degree $-R$ out of this, i.e. to generalize [8], requires to find the right substitute for the cone $C'(R)$ inside $V(R) \oplus W(R)$, cf. Section 3.4. But even more difficult will be the definition of the appropriate lattice structure within this vector space.

Example 18. Returning to Section 3.3, we assume that the interval $I = [-A, B]$ with uniform denominators represents a grounded cyclic quotient singularity ($A, B \in \mathbb{Q}_{>0}$). If $k := \lfloor A + B \rfloor$, then $I = [-A, B - k] + k \cdot [0, 1]$ is an integral decomposition of I, and the Cayley construction leads to a k-parameter qG-deformation of $X = \mathbb{TV}\big(\mathbb{Q}_{\ge 0} \cdot (I, 1)\big)$.

4.3 Using p-divisors to understand toric deformations

If $R \in \sigma^\vee \cap M$, then the k-parameter families arising from integral Minkowski decompositions of $Q(R)$ mentioned in Section 4.2 have a toric total space. In general, this is to much symmetry to ask for. In [12, §3], we have extended our construction to primitive $R \in M$ outside $\sigma^\vee$, but then the total space ceases to be toric. Both cases allow a unifying treatment when using the language of p-divisors from Section 2.2 as it was done in [33]:

(i) If a primitive $R \in M$ is given, then our new torus T will be the kernel of the corresponding character χ^R : Spec $\mathbb{C}[M] \to \mathbb{C}^*$. The new lattices of one-parameter subgroups and characters are $\overline{N} := R^\perp \subset N$ and $\overline{M} := M/R$, respectively.

(ii) Downgrading $\mathbb{TV}(\sigma)$ along the lines of Example 3 yields $Y = \mathbb{A}^1$ or $\mathbf{P}^1$ depending on whether $R \in \sigma^\vee$ or not. The new tail cone is $\sigma \cap R^\perp$, and the p-divisor becomes

$$\mathscr{D} = Q(R) \otimes \{0\} + Q(-R) \otimes \{\infty\},$$

where the second summand is absent when $R \in \sigma^\vee$, i.e. $Q(-R) = \emptyset$.

(iii) If $Q(R) = Q_0 + Q_1$ is an integral Minkowski decomposition, then we build

$$\mathscr{D}_t := Q_0 \otimes \{0\} + Q_1 \otimes \{t\} + Q(-R) \otimes \{\infty\} \qquad (t \in \mathbb{C}).$$

These gadgets are p-divisors on Y, too. They represent the fiber over $t \in \mathbb{A}^1$ of our one-parameter deformation.

(iv) The total space of the family will be considered as a T-variety, too. It is of complexity two; the base space carrying the p-divisor $\mathscr{D}_{\text{total}}$ is the surface $Y \times \mathbb{A}^1$. The p-divisor looks like

$$\mathscr{D}_{\text{total}} = Q_0 \otimes (0 \times \mathbb{A}^1) + Q_1 \otimes \Delta + Q(-R) \otimes (\infty \times \mathbb{A}^1),$$

where $\Delta \subset Y \times \mathbb{A}^1$ is the diagonal.

(v) If $R \in \sigma^\vee$, then $0 \times \mathbb{A}^1$ and Δ are two divisors on $Y \times \mathbb{A}^1 = \mathbb{A}^2$ which can be brought into a “toric position”. Thus, $\mathscr{D}_{\text{total}}$ represents a (downgraded) toric variety. In contrast, if $R \notin \sigma^\vee$, then $Y \times \mathbb{A}^1 = \mathbf{P}^1 \times \mathbb{A}^1$ is toric, too – but the configuration of the three divisors is not.

4.4 Rational T-varieties of complexity one

Instead of downgrading toric singularities as in Section 4.3, one could start with complexity one T-singularities and investigate their equivariant (i.e. degree 0) deformation theory right away. In dimension three, this was done in [17, Theorem 1.1]: Devyatov considers $X = \mathbb{TV}(\mathscr{D}, \mathbb{Z}^2, \mathbf{P}^1)$ as in Section 2.2, where $\mathscr{D} = \sum_{p \in \mathbf{P}^1} \Delta_p \otimes p$ is a p-divisor with $\Delta_p \subseteq \mathbb{Q}^2$ being lattice polygons sharing a non-trivial tail cone. The latter means that the bounded edges Δ_p^i of the coefficients Δ_p form a (maybe empty) chain $\partial^{\text{bd}}\Delta_p$, but not a cycle.

Theorem 19. *The vector space $T^1_X(0)$ of equivariant, infinitesimal deformations has dimension* $-3+\sum_p \mathrm{length}(\partial^{\mathrm{bd}}\Delta_p)$ (to be corrected by adding $(3-c)$ if there is only $c = 0, 1, 2$ non-trivial terms in the sum).

Moreover, in [17, §6], it was shown that these equivariant deformations are not obstructed. The equivariant versal deformation arises, similar to Section 4.3 (iii, iv), from a complete decomposition of all coefficients Δ_p into polygons built from single edges of length one,

At this point, the assumption of a non-zero tail cone becomes essential. Within the setup of Section 3.4 this corresponds to $R \notin \operatorname{int} \sigma^\vee$, which is almost the opposite of our assumption $R \in \sigma^\vee$ made in Section 4.1. This implies that the polygon $Q(R)$ is not bounded, hence there is no vanishing conditions in the definition of $C(R)$. And it is them causing the obstructions in the general case.

4.5 Examples leading to higher complexity

We proceed with some examples from [47, §2]. Let $2 \le d < m$ be natural numbers and fix some $(m-d)$-dimensional linear subspace $G \subseteq \mathbb{C}^m$. It acts on $\mathbb{C}^{2m}$ with coordinates $\mathbf{x} = (x_1, \dots, x_m)$ and $\mathbf{y} = (y_1, \dots, y_m)$ via

$$x_i \mapsto x_i, \quad y_i \mapsto y_i + g_i x_i \quad \text{for } (g_1, \dots, g_m) \in G \subset \mathbb{C}^m.$$

Denoting by $R^G \subset R := \mathbb{C}[\mathbf{x}, \mathbf{y}]$ the invariants of the G-action, both rings respect the $\mathbb{Z}^{m+1}$-grading given by

$$\deg(x_i) := e_i \quad \text{and} \quad \deg(y_i) := e_0 + e_i \quad \text{for } i = 1, \dots, m,$$

where $e_0, \dots, e_m$ denotes the canonical basis of $M := \mathbb{Z}^{m+1}$. Our goal is to understand the $(m+d)$-dimensional $X = \operatorname{Spec} R^G$ or their toric degenerations obtained from varying G inside $\mathbb{C}^m$ by representing them as T-varieties with complexity $d-1$.

The case $m = d+1$ leads to the affine cone over $\mathrm{Grass}(2, m+1)$; its description as a p-divisor on $\overline{M}_{0,m+1}$ was given in [5]. In contrast, the case $d = 2$ means complexity one, i.e. we are supposed to find p-divisors on $\mathbf{P}^1$. Let us start with the case $d = 2$, $m = 4$. Its toric degenerations were determined in [47, Example 3.3]; there are two types of them – we will call them A and B. Downgraded to T-varieties, they look as follows:

The polyhedral coefficients are contained in $N = \mathbb{Z}^{m+1}$ with basis $e^0, \dots, e^m$. Using the affine embedding,

$$\mathbb{Z}^m \hookrightarrow \mathbb{Z}^{m+1}, \quad (a_1, \dots a_m) \mapsto (1 - \textstyle\sum_{i=1}^m a_i, a_1, \dots, a_m),$$

the tail cone σ of both types is spanned from the m-dimensional hypercube, i.e. from the 2^m points $(0/1, \ldots, 0/1) \in \mathbb{Z}^m$. The polyhedral coefficients are assembled from line segments $s(i) \subseteq N_{\mathbb{Q}}$ connecting e^0 with e^i. In detail, up to some innocent integral shifts, we have

$$\begin{aligned}
&\text{Type A:}\quad \Delta_0^A = s(1) + s(2) + s(3) + \sigma \quad \text{and} \quad \Delta_\infty^A = s(4) + \sigma\\
&\text{Type B:}\quad \Delta_0^B = s(1) + s(2) + \sigma \qquad\quad\ \ \text{and} \quad \Delta_\infty^B = s(3) + s(4) + \sigma.
\end{aligned}$$

The other toric degenerations result from permutations of the summands. Thus, inspired from the shape of the equivariant versal deformation from [17, §6] mentioned in Section 4.4, we obtain, again up to an integral shift, the following.

Conjecture 20. *If $G \subseteq \mathbb{C}^m$ is in general position, then the p-divisor describing Spec R^G is $\mathcal{D} = \sum_{i=1}^m \big(s(i) + \sigma\big) \otimes p_i$ with points $p_i \in \mathbf{P}^1$ depending on G. Moreover, any arrangements of the coefficients into two groups leads to a toric degeneration.*

Finally, we will have a look at an example from the case $d = 3$. Here, R^G equals the Cox ring of $\mathbf{P}^2$ blown up in m points, and varying G means varying the position of these points. The associated p-divisors are now supported on surfaces, namely on the blow up of $\mathbf{P}^2$ in question, cf. [13, (6.2)].

Nevertheless, we will have a look at some of their toric degenerations. The special case $m = 5$ is treated in [47, §4] naming seven different types. Picking just two of them, we obtain p-divisors $\mathcal{D}^{(1)}$ and $\mathcal{D}^{(6)}$ supported on toric surfaces $Y^{(1)} = \mathbb{TV}(\Sigma^{(1)})$ and $Y^{(6)} = \mathbb{TV}(\Sigma^{(6)})$ with

$$\begin{aligned}
\Sigma_1^{(1)} &= \{\pm(1,0), \pm(0,1), \pm(1,1), (2,1), -(1,2)\} \text{ and}\\
\Sigma_1^{(6)} &= \{\pm(1,0), \pm(0,1), \pm(1,1), \pm(1,2)\},
\end{aligned}$$

respectively. In contrast to the previous case of complexity one, there are two new features visible:

(i) While $\mathbb{TV}(\mathcal{D}^{(1)}, Y^{(1)})$ and $\mathbb{TV}(\mathcal{D}^{(6)}, Y^{(6)})$ are supposed to sit in a common family, the base spaces $Y^{(1)}$ and $Y^{(6)}$ are different.

(ii) The polyhedral coefficients on both sides involve specimens with mutually incompatible Minkowski sums.

While the first problem can be overcome by considering a common blowing up, the latter indicates that the theory of Minkowski decompositions does not suffice in complexity two. The reason is that for p-divisors $\mathcal{D} = \sum_i \Delta_i \otimes p_i$ on $\mathbf{P}^1$, the so-called degree polyhedron $\deg \mathcal{D} := \sum_i \Delta_i$ satisfies

$$\langle \deg \mathcal{D}, u\rangle = \deg \mathcal{D}(u) \quad \text{for } u \in (\operatorname{tail}\mathcal{D})^\vee.$$

In particular, it stays constant along flat families. In complexity two, this concept has to be replaced by other invariants obtained from Riemann–Roch. In particular, if $\mathcal{D} = \sum_i \Delta_i \otimes D_i$ is a p-divisor on a surface S, then the polyhedron $(\mathcal{D} \cdot K_S) := \sum_i (D_i \cdot K_S) \cdot \Delta_i$ satisfies

$$\left\langle (\mathcal{D} \cdot K_S),\, u \right\rangle = (\mathcal{D}(u) \cdot K_S) \quad \text{for } u \in (\operatorname{tail}\mathcal{D})^\vee.$$

Applied to the previous example, these weighted sums are indeed constant on the different toric degenerations. However, Riemann–Roch does tell us that we are supposed to look at the self intersections $\left(\mathcal{D}(u) \cdot \mathcal{D}(u)\right)$, too. Thus, we need some new kind of object yielding exactly this number when evaluated on u.

5 Uniformization of moduli spaces of abelian varieties

The moduli space $\mathcal{A}_g$ of principally polarized abelian varieties (ppav) of dimension g is a quasi-projective irreducible variety of dimension $\frac{g(g+1)}{2}$. The study of its geometric and arithmetic properties has been a mainstream topic of research in algebraic geometry since the late nineteenth century, when Riemann, Frobenius, Schottky and others observed that the *theta Nullwerte* give coordinates on the moduli space. Later a more algebraic approach to the study of $\mathcal{A}_g$ took hold. A famous result of Freitag et al. [40] says that $\mathcal{A}_g$ is a variety of general type for $g \geq 7$. Thus, in this range there is no way of writing down the general ppav of dimension g in a family depending on *free* parameters, which are no subject to any algebraic equations.

On the other hand, the general ppav $[A, \Theta] \in \mathcal{A}_g$ of dimension $g \leq 5$ can be realized as a Prym variety corresponding to an unramified double cover between algebraic curves. That is to say, that the *Prym map*

$$P : \mathcal{R}_{g+1} \to \mathcal{A}_g$$

is dominant for $g \leq 5$. Abelian varieties of small dimension can be studied in this way via the rich and concrete theory of curves, without making any reference to the theory of theta functions. In particular, one can establish that $\mathcal{A}_g$ is unirational in the range $g \leq 5$. In the case $g = 5$, Donagi and Smith [43] have proved that the Prym map $P : \mathcal{R}_6 \to \mathcal{A}_5$ is finite of degree 27. Equivalently, a general ppav $[A, \Theta] \in \mathcal{A}_5$ admits 27 Prym realizations, which are related by Donagi's *tetragonal construction*. Three different proofs, due to Donagi, Mori-Mukai and Verra of the unirationality of $\mathcal{R}_6$ are known. The Kodaira dimension of $\mathcal{A}_6$ is unknown and this is one of the outstanding open problems in the field of abelian varieties. Understanding abelian varieties of dimension 6 and their moduli space has been one of our central themes of study in the last few years, and several papers [23, 31] and, above all, [3] emerged out of this research.

The aim of the paper [31] is to establish a structure theorem of the boundary divisor of $\overline{\mathcal{A}}_6$ and derive from this fact important consequences concerning the global geometry of $\mathcal{A}_6$ itself. This is possible because the boundary divisor of any compactification of $\mathcal{A}_g$ is birationally isomorphic to the universal Kummer variety over $\mathcal{A}_{g-1}$. Morally speaking, studying the universal abelian variety of dimension g amounts to studying the boundary divisor of the moduli space of ppav of dimension $g + 1$.

To explain the results of [31], we set some notation. Let $\phi : \mathcal{X}_{g-1} \to \mathcal{A}_{g-1}$ be the universal abelian variety of dimension $g - 1$ (in the sense of stacks). The moduli space of principally polarized abelian varieties of dimension g and their rank one degenerations is a partial compactification of $\mathcal{A}_g$ obtained by blowing-up $\mathcal{A}_{g-1}$ in the Satake compactification. Its boundary is isomorphic to the universal Kummer variety in dimension $g - 1$. The main result of [31] is a simple structure result for the boundary of the moduli space of abelian 6-folds:

Theorem 21. *The universal abelian variety $\mathcal{X}_5$ is unirational.*

This immediately implies that the boundary divisor of $\overline{\mathcal{A}}_6$ is unirational as well. What we prove is actually stronger than Theorem 21. Over the moduli space $\mathcal{R}_g$ of smooth Prym curves of genus g, we consider the universal Prym variety $\varphi : \mathcal{Y}_g \to \mathcal{R}_g$ obtained by pulling-back $\mathcal{X}_{g-1} \to \mathcal{A}_{g-1}$ via the Prym map $P : \mathcal{R}_g \to \mathcal{A}_{g-1}$. If

$$\widetilde{\mathcal{C}}^{g-1} := \widetilde{\mathcal{C}} \times_{\mathcal{R}_g} \dots \times_{\mathcal{R}_g} \widetilde{\mathcal{C}}$$

is the $(g-1)$-fold product of the universal curve $\widetilde{\mathcal{C}}$ of genus $2g-1$ over the Prym moduli space, one has a universal *Abel–Prym* rational map

$$\mathfrak{ap} : \widetilde{\mathcal{C}}^{g-1} \dashrightarrow \mathcal{Y}_g,$$

whose restriction on each individual Prym variety is the usual Abel–Prym map. The rational map $\mathfrak{ap}$ is dominant and generically finite. We prove the following result:

Theorem 22. *The fivefold product $\widetilde{\mathcal{C}}^5$ of the universal Prym curve over $\mathcal{R}_6$ is unirational.*

Theorem 22 is proved by viewing smooth Prym curves of genus 6 as discriminants of conic bundles, via their representation as symmetric determinants of quadratic forms in three variables. The idea of the proof is the following. We fix four general points $u_1, \dots, u_4 \in \mathbf{P}^2$ and set $w_i := (u_i, u_i) \in \mathbf{P}^2 \times \mathbf{P}^2$. We then consider the linear system

$$\mathbf{P}^{15} := \left|\mathcal{I}^2_{\{w_1,\dots,w_4\}}(2,2)\right| \subset \left|\mathcal{O}_{\mathbf{P}^2 \times \mathbf{P}^2}(2,2)\right|$$

of hypersurfaces $Q \subset \mathbf{P}^2 \times \mathbf{P}^2$ of bidegree (2, 2) which are nodal at $w_1, \ldots, w_4$. For a general threefold $Q \in \mathbf{P}^{15}$, the first projection $p : Q \to \mathbf{P}^2$ induces a conic bundle structure with a sextic discriminant curve $\Gamma \subset \mathbf{P}^2$ such that $p(\mathrm{Sing}(Q)) = \mathrm{Sing}(\Gamma)$. The discriminant curve Γ is nodal precisely at the points $u_1, \ldots, u_4$. Furthermore, Γ is equipped with an unramified double cover $p_\Gamma : \widetilde{\Gamma} \to \Gamma$, parametrizing the lines which are components of the singular fibers of the conic bundle $p : Q \to \mathbf{P}^2$.

One can show without much effort that the assignment

$$\mathbf{P}^{15} \ni Q \mapsto [\widetilde{C} \xrightarrow{f} C] \in \mathcal{R}_6$$

is dominant. This offers an alternative, much simpler, proof of the unirationality of $\mathcal{R}_6$. However, much more can be obtained with this construction.

Let $\mathbf{G} := \mathbf{P}^2 \times (\mathbf{P}^2)^\vee = \Big\{(o, \ell) : o \in \mathbf{P}^2, \ell \in \{o\} \times (\mathbf{P}^2)^\vee\Big\}$ be the Hilbert scheme of lines in the fibers of the first projection $p : \mathbf{P}^2 \times \mathbf{P}^2 \to \mathbf{P}^2$. Since containing a given line in a fiber of p imposes *three* linear conditions on the linear system $\mathbf{P}^{15}$ of threefolds $Q \subset \mathbf{P}^2 \times \mathbf{P}^2$ as above, it follows that imposing the condition $\{o_i\} \times \ell_i \subset Q$ for *five* general lines, singles out a *unique* conic bundle $Q \in \mathbf{P}^{15}$. This induces an étale double cover $f : \widetilde{C} \to C$, as above, over a smooth curve of genus 6. Moreover, f comes equipped with five marked points $\ell_1, \ldots, \ell_5 \in \widetilde{C}$. To summarize, we can define a rational map

$$\zeta : \mathbf{G}^5 \dashrightarrow \widetilde{\mathcal{C}}^5, \quad \zeta\Big((o_1, \ell_1), \ldots, (o_5, \ell_5)\Big) := \Big(f : \widetilde{C} \to C, \ell_1, \ldots, \ell_5\Big),$$

between two 20-dimensional varieties, where $\mathbf{G}^5$ denotes the fivefold product of $\mathbf{G}$.

Theorem 23. *The morphism $\zeta : \mathbf{G}^5 \dashrightarrow \widetilde{\mathcal{C}}^5$ is dominant, so that $\widetilde{\mathcal{C}}^5$ is unirational.*

Theorem 26 can be used to give a lower bound for the slope $s(\overline{\mathcal{A}}_6)$ of the effective cone of divisors on the moduli space. The slope is a global invariant which encodes the Kodaira dimension of the moduli space of ppav. For instance, if $s(\overline{\mathcal{A}}_g) < g + 1$, then $\mathcal{A}_g$ is of general type, whereas $s(\overline{\mathcal{A}}_g) > g + 1$ implies that $\mathcal{A}_g$ is uniruled.

It is known that $s(\overline{\mathcal{A}}_4) = 8$ and that the Jacobian locus $\overline{\mathcal{M}}_4 \subset \overline{\mathcal{A}}_4$ achieves the minimal slope. This result is essentially due to Schottky an Jung, who wrote down explicitly the equation of a Siegel modular form of weight 8 (that is, a product of 16 theta constants, each having weight $\frac{1}{2}$), which cuts the equation of the Jacobian locus in genus 4. To explain what is known on the slope of $\overline{\mathcal{A}}_5$, we record the definition of the Andreotti-Mayer divisor.

For a general ppav $[A, \Theta] \in \mathcal{A}_g$, the theta divisor Θ is a smooth variety. The locus N_0 of ppav with a singular theta divisor decomposes into two irreducible components

$$N_0 = \Theta_{\mathrm{null}} + 2N_0',$$

depending whether the singularity appears at a torsion point of order 2 or not. The multiplicity 2 in front of N_0' is explained by the fact that if $x \in \Theta_{\text{sing}}$, then $-x \in \Theta_{\text{sing}}$ as well. The class of the divisor Θ_{null} can be easily calculated; a ppav $[A, \Theta]$ belongs to Θ_{null} if and only if one of the theta-constants used in the classical construction of $\mathcal{A}_g$ vanishes at the point $[A, \Theta]$. The residual divisor N_0' has been first studied by Andreotti–Mayer in their groundbreaking work on linking theta functions to the heat equation, and later by Mumford [40] in the process of showing that $\mathcal{A}_g$ is of general type for $g \geq 7$. Precisely, Mumford computed the following formula over the moduli space of abelian varieties and their rank one degenerations:

$$[N_0'] = \left(\frac{(g+1)!}{4} + \frac{g!}{2} - 2^{g-3}(2^g + 1)\right)\lambda_1 - \left(\frac{(g+1)!}{24} - 2^{2g-6}\right)D \in CH^1(\overline{\mathcal{A}}_g),$$

where λ_1 is the Hodge bundle whose sections are Siegel modular forms and D is the boundary divisor of $\overline{\mathcal{A}}_g$.

In [23], we prove that for $g = 5$ the Andreotti–Mayer divisor N_0' determines the slope of the effective cone of $\overline{\mathcal{A}}_5$:

Theorem 24. *One has that $s(\overline{\mathcal{A}}_5) = \frac{54}{7}$. Furthermore, the only irreducible effective divisor on $\overline{\mathcal{A}}_5$ of minimal slope is the closure of the Andreotti–Mayer divisor N_0'.*

The proof uses in an essential way the degree 27 Prym map $P : \mathcal{R}_6 \to \mathcal{A}_5$ and the uniformization of the ramification divisor of the map P given in terms of Nikulin surfaces in [27]. Concerning the slope of the moduli space $\overline{\mathcal{A}}_6$, using an explicit ruling given by Theorem 26, we establish a lower estimate. This is a revealing instance on how one can deduce information about the *global geometry* of $\mathcal{A}_6$ by having extensive information only about one of its divisors.

Theorem 25. *The following lower bound holds: $s(\overline{\mathcal{A}}_6) \geq \frac{53}{10}$.*

Note that this is the first concrete lower bound on the slope of $\overline{\mathcal{A}}_6$. The canonical class of $\overline{\mathcal{A}}_g$ is given by the formula

$$K_{\overline{\mathcal{A}}_g} = (g+1)\lambda_1 - D \in CH^1(\overline{\mathcal{A}}_g).$$

Theorem 25 offers therefore no conclusive answers as to what is the Kodaira dimension of $\mathcal{A}_6$, but provides an invitation to further research.

5.1 Prym–Tyurin varieties and abelian varieties of dimension 6

We now describe the results of the paper [3] in which a parmetrization of $\mathcal{A}_6$ by means of Prym–Tyurin varieties is obtained. The idea is to realize each ppav of dimension 6 as a particular Prym–Tyurin variety.

We start with a curve C endowed with a correspondence $D \subset C \times C$ that induces an endomorphism $\gamma : JC \to JC$ of the Jacobian variety. If γ satisfies a quadratic equation of the form

$$(\gamma - 1)(\gamma + q - 1) = 0 \in \mathrm{End}(JC),$$

for some integer $q > 0$, we define the *Prym–Tyurin* variety

$$PT(C, D) := \mathrm{Im}(1 - \gamma) \subset JC.$$

This abelian variety comes equipped with a principal polarization Ξ, defined by the equality $\Theta_{C|PT(C,D)} = q \cdot \Xi \subset PT(C, D)$, where Θ_C is the theta divisor on the Jacobian of C and q is the *exponent* of the abelian variety. Since every ppav is a Prym–Tyurin variety (usually of very high exponent), one obtains a stratification of the moduli space $\mathcal{A}_g$, with strata being loci of Prym-Tyurin varieties of fixed exponent. This stratification is still poorly understood.

Constructing Prym–Tyurin varieties has proven to be difficult, for general curves admit no non-trivial correspondences (hence a general Jacobians has only trivial endomorphisms). For a novel realization of classical Jacobian varieties JC of odd genus g as Prym–Tyurin varieties on the Brill-Noether curve $W^1_{\frac{g+3}{2}}(C)$ of pencils of minimal degree on the curve C in question, see the paper [41].

Vassil Kanev has put forward a beautiful proposal for constructing Prym-Tyurin abelian six-folds using the geometry of cubic surfaces, which we now explain. This construction is the starting point of [3].

To a pencil of cubic surfaces $\{X_\lambda\}_{\lambda \in \mathbf{P}^1}$ on a smooth cubic threefold $X \subset \mathbf{P}^4$, we associate the curve C parametrizing the lines on the cubic surfaces in the pencil. This curve is equipped with a cover $C \xrightarrow{27:1} \mathbf{P}^1$ branched over the 24 points corresponding to the singular cubic surfaces in the pencil. Furthermore, the incidence of the lines on each individual cubic surface induces a correspondence $D \subset C \times C$ whose corresponding endomorphism $\gamma : JC \to JC$ verifies the equation

$$(\gamma - 1)(\gamma + 5) = 0.$$

Therefore the induced *Prym–Tyurin variety* $PT(C, D) := \mathrm{Im}(1 - \gamma) \subset JC$ is a ppav of dimension 6 and exponent 6. In fact, $PT(C, D)$ is isomorphic to the product of the intermediate Jacobian of X and the elliptic curve $E \subset X$ which is the base locus of the pencil. This idea can be generalized to coverings over the projective line whose monodromy is the Weyl group $W(E_6)$ of symmetries of the 27 lines on a cubic surface.

We consider the Hurwitz space $\mathcal{H}_{E_6}$ classifying covers $C \to \mathbf{P}^1$ of degree 27 with monodromy group $W(E_6)$, branched over 24 points, and with the ramification over each branch point being given by a root of the E_6-lattice. This is an irreducible variety of dimension $21 = \dim(\mathcal{A}_6)$. Thus we set up a *Prym–Tyurin* map

$$PT : \mathcal{H}_{E_6} \to \mathcal{A}_6, \quad [f : C \to \mathbf{P}^1] \mapsto [P(C, D), \Xi],$$

between varieties of the same dimension. In [3], we establish the following uniformization result:

Theorem 26. *The following results hold:*
1. *The Prym–Tyurin map $PT : \mathcal{H}_{E_6} \to \mathcal{A}_6$ is dominant.*
2. *The Hurwitz space $\mathcal{H}_{E_6}$ is a variety of general type.*

This is a structure theorem for ppav of dimension 6 in terms of curve data and should be regarded as an analogue of Wirtinger's well-known result that a general ppav of dimension at most 5 is a Prym variety. Theorem 26 gives an answer to the riddle: What is the general abelian variety of dimension 6? The answer: a Prym–Tyurin variety of exponent 6 corresponding a $W(E_6)$-cover of the projective line.

Furthermore, we also show in [3] that the ramification divisor of the map PT consists precisely of those E_6-curves $C \to \mathbf{P}^1$ for which the Prym–Tyurin canonical curve

$$\phi_{H^0(C,K_C)^{(-5)}} : C \to \mathbf{P}^5$$

lies on a quadric; equivalently, the multiplication map of sections

$$\mathrm{Sym}^2 H^0(C, K_C)^{(-5)} \longrightarrow H^0(C, K_C^{\otimes 2})$$

is not injective. This result is reminiscent of the description of the ramification of the classical Prym map $P : \mathcal{R}_{g+1} \to \mathcal{A}_g$: The differential of the Prym map P is not injective at a point $[C, \eta] \in \mathcal{R}_{g+1}$, if and only if the Prym-canonical curve $\phi_{H^0(K_C \otimes \eta)} : C \to \mathbf{P}^{g-1}$ lies on a quadric.

The proof of Theorem 26 proceeds by a study of the map PT in the neighborhood of a completely degenerate point, where the corresponding abelian variety is entirely toric (that is, it has no abelian part) and the map PT can be read off as a map of fans. In [3] we establish numerous other results on the geometry of $\mathcal{H}_{E_6}$. For instance, we show that there exist only three boundary divisors of the compactification $\overline{\mathcal{H}}_{E_6}$ by means of admissible coverings, that are not contracted under the Prym-Tyurin map. Finally, we describe the inverse image $PT^{-1}(\mathcal{R}_7)$ of the locus of classical Prym varieties and explain how the new Prym-Tyurin varieties generalize ordinary Prym varieties.

6 Birational classification of moduli space of spin curves

Between 2005 and 2010, a number of results on the slopes of the effective cone of divisors and the Kodaira dimension of $\overline{\mathcal{M}}_g$ were obtained in [19–21]. It was one of the overarching goals of the last years to adapt those techniques to the case of higher level modular varieties. From a physical point of view, the space $\overline{\mathcal{S}}_g$, (corresponding to superstring theory), might be more interesting than $\overline{\mathcal{M}}_g$ itself. Important functions on moduli spaces like the *theta constants*, are defined only at the level of $\overline{\mathcal{S}}_g^+$ and $\overline{\mathcal{S}}_g^-$ rather than that of $\overline{\mathcal{M}}_g$. Furthermore, the fundamental problem of writing down the *genus g superstring scattering amplitudes* can be phrased in terms of divisors of weight 8 on $\overline{\mathcal{S}}_g^+$. Our results in this direction surpassed our expectation. A *complete* birational classification of the spin moduli space has been obtained. The papers [22, 27, 32] are devoted to this topic.

We record the following classification for the spin moduli space:

Theorem 27. *The birational type of the moduli spaces $\overline{\mathcal{S}}_g^+$ and $\overline{\mathcal{S}}_g^-$ of even and odd spin curves of genus g can be summarized as follows:*

$\overline{\mathcal{S}}_g^+$:

$g > 8$	general type
$g = 8$	Calabi–Yau
$g \leq 7$	unirational

$\overline{\mathcal{S}}_g^-$:

$g \geq 12$	general type
$9 \leq g \leq 11$	uniruled
$g \leq 8$	unirational

The remarkable feature of this classification is its completeness, especially since there are significant gaps in the classification of related moduli spaces. The birational nature of $\overline{\mathcal{M}}_g$ is not known for $17 \leq g \leq 21$, see [21]; the Kodaira dimension of the Prym moduli space $\overline{\mathcal{R}}_g$ is not determined for $8 \leq g \leq 12$, see [24]; as already discussed, finding the Kodaira dimension of $\mathcal{A}_6$ is a notorious open problem. Observe that there are genera $g \leq 11$ such that, the odd and even components of the spin space are of different birational nature! The most striking aspect of the above-mentioned results is the case $g = 8$.

We quote from [27]:

Theorem 28. *The even spin moduli space $\overline{\mathcal{S}}_8^+$ has Kodaira dimension zero and the Mukai model of $\overline{\mathcal{S}}_8^+$ is a 21-dimensional Calabi–Yau variety of Picard number one.*

This is the first instance of a moduli space of curves of intermediate Kodaira dimension. To clarify the statement of Theorem 28, we mention that we define in [27] the *Mukai model* $\mathfrak{M}_8$ of $\overline{\mathcal{M}}_8$ as the GIT quotient

$$\mathfrak{M}_8 := G(8, \wedge^2 V)^{ss} // SL(V),$$

where $V = \mathbb{C}^6$. There exists a birational morphism $f : \overline{\mathcal{M}}_8 \dashrightarrow \mathfrak{M}_8$ which corresponds to the fact that the general genus 8 curve is a linear section of the Grassmannian $G(2,6) \subset \mathbf{P}^{14}$ in its Plücker embedding. By taking fiber products, we define the Mukai model $\mathfrak{S}_8$ of the spin space $\overline{\mathcal{S}}^+_8$. We show that $\mathfrak{S}_8$ is a Calabi–Yau variety, birationally equivalent to $\overline{\mathcal{S}}^+_8$. Has this new Calabi–Yau moduli space a (super)string theory meaning?

The general type statements from Theorem 27 are proved by computing the classes of *geometric* divisors. We explain the situation of $\overline{\mathcal{S}}^+_g$, see [22].

Theorem 29. *The closure of the locus*

$$\Theta_{\text{null}} := \left\{[C,\theta] \in \mathcal{S}^+_g : H^0(C,\theta) \neq 0\right\}$$

of vanishing theta nulls has class equal to

$$[\overline{\Theta}_{\text{null}}] = \frac{1}{4}\lambda - \frac{1}{16}\alpha_0 - \frac{1}{2}\sum_{i=1}^{[\frac{g}{2}]}\beta_i \in \operatorname{Pic}(\overline{\mathcal{S}}^+_g).$$

Here λ is the Hodge class and $\{\alpha_i, \beta_i\}_{i=0}^{\lfloor\frac{g}{2}\rfloor}$ are the boundary divisors. Combining the formula of $[\overline{\Theta}_{\text{null}}]$ with the Brill–Noether class from $\overline{\mathcal{M}}_g$, we find that the class $\frac{11g+29}{g+1}\lambda - 2\alpha_0 - 3\beta_0 - \cdots - \operatorname{Pic}(\overline{\mathcal{S}}^+_g)$ is effective. Recalling that $K_{\overline{\mathcal{S}}^+_g} = 13\lambda - 2\alpha_0 - 3\beta_0 - \cdots$, we observe that whenever $\frac{11g+29}{g+1} < 13 \Leftrightarrow g > 8$, the space $\overline{\mathcal{S}}^+_g$ has maximal Kodaira dimension.

In the paper [32], the space $\overline{\mathcal{S}}^-_g$ parametrizing odd spin curves is treated. One can view odd theta characteristics as divisors $x_1 + \cdots + x_{g-1}$ on a curve C, such that the corresponding $(g-2)$-plane $\langle x_1+\cdots+x_{g-1}\rangle \in (\mathbf{P}^{g-1})^\vee$ is everywhere tangent to the canonical curve $C \subset \mathbf{P}^{g-1}$. Computing the class of the locus where two of these tangency points coalesce, we show in [32] that $\overline{\mathcal{S}}^-_g$ is of general type for $g \geq 12$.

The unirationality results from Theorem 27 are proved in [32] using the existence when $g \leq 9$, of *Mukai varieties* V_g such that one-dimensional linear sections of V_g are canonical curves $[C] \in \mathcal{M}_g$ with general moduli. To recover an odd spin curve from the Mukai variety we rely on the concept of *cluster*. This is a zero-dimensional scheme $Z \subset V_g$ of length $2g-2$ such that Z has multiplicity two at each point of its support. Over the parameter space $\mathcal{Z}_{g-1}$ of clusters, we define a correspondence:

$$\mathcal{U}^-_g := \left\{(C,Z) : Z \in \mathcal{Z}_{g-1},\ Z \subset C,\ C \text{ is a curve section of } V_g\right\}.$$

The fibers of the first projection classify odd theta characteristics on genus g curves. For $0 \leq \delta \leq g-1$, we consider the variety of clusters and δ-nodal sections of V_g, that is,

$$\mathcal{U}^-_{g,\delta} := \left\{(\Gamma,Z) \in \mathcal{U}^-_g : \operatorname{sing}(\Gamma) \subset \operatorname{supp}(Z) \ \text{ and } \ |\operatorname{sing}(\Gamma)| = \delta\right\}.$$

Importantly, over $\mathcal{U}^-_{g,\delta}$ one has an incidence correspondence taking into account linear sections of V_g that admit the same cluster. Using this, we reduce the unirationality of $\overline{\mathcal{S}}^-_g$ to a numerical condition on the Mukai variety V_g and to a statement on the geometry of spin moduli spaces of curves of smaller genus. Both conditions are fulfilled when $g \le 8$.

The birational geometry of the moduli space $\mathcal{R}_{g,\ell}$ parametrizing pairs $[C, \eta]$, where C is a smooth curve of genus g and η is an $\ell \ge 3$ torsion point in the Jacobian of C, has been studied using syzygetic methods in [14, 15]. For instance, it is proved that $\mathcal{R}_{g,3}$ is a variety of general type for $g \ge 12$. In [15] a systematic study of the singularities Deligne-Mumford moduli space $\overline{\mathcal{R}}_{g,\ell}$ of level ℓ twisted curves is undertaken. In particular, it is shown that for $\ell \le 5$, singularities of $\overline{\mathcal{R}}_{g,\ell}$ pose no adjunction conditions and pluricanonical forms on the moduli space extend to any resolution of singularities. In a different direction, Lange and Ortega [37, 38] have established beautiful results on the geometry of the moduli spaces of exotic Prym varieties of small dimensions and linked these spaces to the Minimal Model Program for the moduli space of curves. In the interest of brevity, we shall not review these works in this survey.

7 The universal Jacobian variety

The compactified universal Jacobian variety $\overline{\mathcal{J}}_g$ over $\overline{\mathcal{M}}_g$ has been constructed by Caporaso and Pandharipande in the 1990s. Over a point $[C] \in \mathcal{M}_g$ corresponding to a smooth curve, the universal Jacobian parametrizes the Picard variety $\mathrm{Pic}^g(C)$ of line bundles of degree g on the curve C. The cohomology of the universal Jacobian is still little understood, despite interest coming from the Geometric Langlands Program. The paper [28] deals instead with the birational geometry of the universal Jacobian variety $\mathcal{J}_g \to \mathcal{M}_g$.

Since $\mathcal{J}_g \to \mathcal{M}_g$ is a fibration in abelian varieties, using the easy addition formula for the Kodaira dimension, we find that the Kodaira dimension of $\mathcal{J}_g$ can never exceed $3g-3 = \dim(\mathcal{M}_g)$. In particular, $\mathcal{J}_g$ is never of general type. A somewhat naive prediction could be that the Kodaira dimensions of $\overline{\mathcal{J}}_g$ and $\overline{\mathcal{M}}_g$ are always equal. We prove in the paper [28] the following result, completing the birational classification of the universal Jacobian for all genera.

Theorem 30. *The following results hold:*
1. *For $g > 11$, the Kodaira dimension of $\mathcal{J}_g$ equals $3g - 3$.*
2. *The universal Jacobian $\mathcal{J}_{10}$ has Kodaira dimension 0.*
3. *The Kodaira dimension of $\mathcal{J}_{11}$ equals 19.*
4. *$\mathcal{J}_g$ is unirational for $g \le 9$.*

Observe that there exist values of g such that $\mathcal{M}_g$ is uniruled whereas $\mathcal{J}_g$ is of maximal Kodaira dimension $3g - 3$, which at first sight, was surprising to us. The paper [29] is similar in methods and scope to [28]. We achieve a birational classification of the universal theta divisor $\Theta_g \to \mathcal{M}_g$.

For a smooth curve C, the Abel–Jacobi map $C_{g-1} \to \mathrm{Pic}^{g-1}(C)$ provides a resolution of singularities of the theta divisor Θ_C of the Jacobian of C. Thus one may regard the degree $g-1$ universal symmetric product

$$\overline{\mathcal{C}}_{g,g-1} := \overline{\mathcal{M}}_{g,g-1}/\mathfrak{S}_{g-1}$$

as a birational model of Θ_g (having only finite quotient singularities), and ask for the place of Θ_g in the classification of varieties. We provide a complete answer to this question.

Theorem 31. *The following results summarized the birational classification of the universal theta divisor:*
1. Θ_g *is unirational for* $g \le 9$ *and uniruled for* $g \le 11$.
2. Θ_g *is a variety of general type for* $g \ge 12$.

These results are obtained by working directly on the model $\overline{\mathcal{C}}_{g,g-1}$. The proof of Theorem 31 relies on the calculation of the universal *antiramification divisor* class of the Gauss map. For a curve C of genus g, let

$$\gamma : C_{g-1} \dashrightarrow \left(\mathbf{P}^{g-1}\right)^{\vee}$$

be the Gauss map. The branch divisor $\mathrm{Br}(\gamma) \subset (\mathbf{P}^{g-1})^{\vee}$ is the dual of the canonical curve $C \subset \mathbf{P}^{g-1}$. The closure in C_{g-1} of the ramification divisor $\mathrm{Ram}(\gamma)$ is the locus of divisors $D \in C_{g-1}$ such that $\mathrm{supp}(D) \cap \mathrm{supp}(K_C(-D)) \neq \emptyset$. The *antiramification* divisor $\mathrm{Ant}(\gamma)$, defined by the following equality of divisors

$$\gamma^*(\mathrm{Br}(\gamma)) = \mathrm{Ant}(\gamma) + 2 \cdot \mathrm{Ram}(\gamma),$$

splits into the locus of non-reduced divisors $\Delta_C := \{2p + D : p \in C,\ D \in C_{g-3}\}$ and the locus of divisors $D \in C_{g-1}$ such that $K_C(-D)$ has non-reduced support. Globalizing this construction over $\mathcal{M}_g$, we are led to consider the universal antiramification divisor

$$\mathfrak{Ant}_g := \left\{[C, x_1, \dots, x_{g-1}] \in \mathcal{M}_{g,g-1} : \exists p \in C \ \text{ with } H^0\left(K_C\left(-\sum_{i=1}^{g-1} x_i - 2p\right)\right) \neq 0\right\}.$$

We have the following formula for the class of the closure $\overline{\mathfrak{Ant}}_g$ of $\mathfrak{Ant}_g$ in $\overline{\mathcal{M}}_{g,g-1}$:

Theorem 32. *The class in* $\mathrm{Pic}(\overline{\mathcal{M}}_{g,g-1})$ *of the closure of the antiramification locus is*

$$[\overline{\mathfrak{Ant}}_g] = -4(g-7)\lambda - 2\delta_{\mathrm{irr}} -$$

$$\sum_{i=0}^{g}\sum_{s=0}^{i-1}\left(2i^3-5i^2-3i+4g-4i^2s+14si-6gs-s+2s^2g-3s^2+2\right)\delta_{i:s}.$$

This calculation, coupled with known facts on effective divisors on $\mathcal{M}_g$, quickly leads to a proof that Θ_g is of general type for $g \le 12$.

8 Uniformization of moduli spaces of Prym varieties

Some of the most fundamental results in the theory of algebraic curves, like the Lazarsfeld's proof of the Brill–Noether Theorem or Voisin's proof of Green's Conjecture for generic curves use specialization to curves lying on polarized $K3$ surface. The aim of the papers [27, 30] is to show that at the level of the *Prym moduli space* $\mathcal{R}_g$, the locus of curves lying on a *Nikulin* $K3$ *surface* plays a similar role. Nikulin surfaces furnish a unirational parametrization of $\mathcal{R}_g$ in small genus, just like ordinary $K3$ surfaces provide a ruling for $\mathcal{M}_g$, when g is small.

A *Nikulin surface* is a $K3$ surface equipped with a symplectic involution. Precisely, let us begin with a smooth $K3$ surface S endowed with a non-trivial double cover $f : \widetilde{S} \to S$ with a branch divisor $N := N_1 + \cdots + N_8$ consisting of eight disjoint smooth rational curves $N_i \subset S$. Blowing down the (-1)-curves $f^{-1}(N_i) \subset \widetilde{S}$, one obtains another smooth $K3$ surface Y, together with a symplectic involution $\iota \in \mathrm{Aut}(Y)$ with 8 fixed points. Setting

$$e := \frac{1}{2}\mathcal{O}_S(N_1 + \cdots + N_8) \in \mathrm{Pic}(S),$$

we can consider the 11-dimensional moduli space $\mathcal{F}_g^{\mathfrak{N}}$ of polarized Nikulin surfaces $[S, e, \mathcal{O}_S(C)]$, with $C^2 = 2g-2$ and $C \cdot e = 0$. Due to fundamental work of Nikulin, it is known that $\mathcal{F}_g^{\mathfrak{N}}$ is irreducible. Over the moduli space of Nikulin surfaces, we consider the following open set in a projective $\mathbf{P}^g$-bundle

$$\mathcal{P}_g^{\mathfrak{N}} := \left\{(S, e, C) : C \subset S \text{ is a smooth curve such that } [S, e, \mathcal{O}_S(C)] \in \mathcal{F}_g^{\mathfrak{N}}\right\},$$

which is endowed with the two projection maps

$$\begin{array}{ccccc} & & \mathcal{P}_g^{\mathfrak{N}} & & \\ & \swarrow^{p_g} & & \searrow^{\chi_g} & \\ \mathcal{F}_g^{\mathfrak{N}} & & & & \mathcal{R}_g \end{array}$$

defined by $p_g([S, j, C]) := [S, j]$ and $\chi_g([S, j, C]) := [C, e_C := e \otimes \mathcal{O}_C]$, respectively.

Note that $\dim(\mathcal{P}_g^{\mathfrak{N}}) = 11 + g$ and it is natural to ask when is χ_g dominant and induces a uniruled parametrization of $\mathcal{R}_g$. The following result is quoted from [27]:

Theorem 33. *The general Prym curve of genus g lies on a Nikulin surface if and only if $g \leq 7$ and $g \neq 6$. In other words, the morphism $\chi_g : \mathcal{P}_g^{\mathfrak{N}} \to \mathcal{R}_g$ is dominant precisely in this range.*

Observe that $\dim(\mathcal{P}_7^{\mathfrak{N}}) = \dim(\mathcal{R}_7) = 18$. The map $\chi_7 : \mathcal{P}_7^{\mathfrak{N}} \dashrightarrow \mathcal{R}_7$ is a birational isomorphism, precisely $\mathcal{R}_7$ is birational to a Zariski locally trivial $\mathbf{P}^7$-bundle over $\mathcal{F}_7^{\mathfrak{N}}$. This is reminiscent of Mukai's well-known results: A general curve of genus g lies on a $K3$ surface if and only if ≤ 11 and $g \neq 10$. The moduli space $\mathcal{M}_{11}$ of curves of genus 11 is birational to a projective bundle over the moduli space $\mathcal{F}_{11}$ of polarized $K3$ surfaces of genus 11. Note that $\mathcal{M}_{11}$ and $\mathcal{R}_7$ are the only known examples of moduli spaces of curves admitting a non-trivial fiber bundle structure over a moduli space of polarized $K3$ surfaces. In the paper [30], we go deeper into describing the structure of the space $\mathcal{F}_7^{\mathfrak{N}}$:

Theorem 34. *The Nikulin moduli space $\mathcal{F}_7^{\mathfrak{N}}$ is unirational. The Prym moduli space $\mathcal{R}_7$ is birational to a $\mathbf{P}^7$-bundle over $\mathcal{F}_7^{\mathfrak{N}}$. It follows that $\mathcal{R}_7$ is unirational as well.*

To explain the content of this result, we consider the moduli space $\widehat{\mathcal{F}}_g^{\mathfrak{N}}$ of *decorated* Nikulin surfaces consisting of Nikulin surfaces of genus g, together with a distinguished line $N_8 \subset S$ viewed as a component of the branch divisor of the double covering $f : \tilde{S} \to S$. There is an obvious forgetful map $\widehat{\mathcal{F}}_g^{\mathfrak{N}} \to \mathcal{F}_g^{\mathfrak{N}}$ of degree 8 forgetting the marked curve N_8. Having specified $N_8 \subset S$, we can also specify the divisor $N_1 + \cdots + N_7 \subset S$ such that $e^{\otimes 2} = \mathcal{O}_S(N_1 + \cdots + N_7 + N_8)$. We show in [30] that there is a close link between the moduli space of decorated Nikulin surfaces and the moduli space $\mathfrak{Rat}_7$ of rational curves having 7-nodes. The moduli space $\mathfrak{Rat}_7$, which is a quotient of the moduli space $\overline{\mathcal{M}}_{0,14}$ by a finite group, is known to be a rational variety.

We fix a rational quintic curve $R \subset \mathbf{P}^5$ and points $x_1, y_1, \ldots, x_7, y_7 \in R$. Note that $\left[R, (x_1 + y_1) + \cdots + (x_7 + y_7)\right] \in \mathfrak{Rat}_7$. We denote by

$$N_1 := \langle x_1, y_1 \rangle, \ldots, N_7 := \langle x_7, y_7 \rangle \in G(2, 6),$$

the corresponding bisecant lines to R and observe that $C := R \cup N_1 \cup \ldots \cup N_7$ is a nodal curve of genus 7 and degree 12 in $\mathbf{P}^5$. The base locus

$$S := \mathrm{Bs}\left|\mathcal{I}_{C/\mathbf{P}^5}(2)\right|$$

is a smooth $K3$ surface which is a complete intersection of three quadrics in $\mathbf{P}^5$. Obviously, S is equipped with the seven lines $N_1, \ldots, N_7$. In fact, S carries an eight line as well, namely

$$N_8 := 2R + N_1 + \cdots + N_7 - 2H,$$

where H is the hyperplane section. In other words, to a 7-nodal rational curve one can associate a smooth Nikulin surface of genus 7 with a distinguished line N_8. This construction is summarized as follows:

Theorem 35. *The rational map* $\varphi : \mathfrak{Rat}_7 \dashrightarrow \widetilde{\mathcal{F}}_7^{\mathfrak{N}}$ *given by*

$$\varphi\Big(\big[R, (x_1 + y_1) + \cdots + (x_7 + y_7)\big]\Big) := \Big[S, \mathcal{O}_S(R + N_1 + \cdots + N_7), N_8\Big]$$

is a birational isomorphism.

Theorem shows that there is a degree 8 map from a rational variety onto $\mathcal{F}_7^{\mathfrak{N}}$. It would be interesting to know which moduli space of (undecorated) Nikulin surfaces of genus g are rational varieties, as well as determining whether for large g, the space $\mathcal{F}_g^{\mathfrak{N}}$ is of general type.

9 Explicit Brill–Noether–Petri general curves defined over $\mathbb{Q}$

The Petri Theorem asserts that for a general curve C of genus g, the multiplication map

$$\mu_{0,L} : H^0(C, L) \otimes H^0(C, \omega_C \otimes L^{-1}) \to H^0(C, \omega_C)$$

is injective for every line bundle L on C. The theorem is essential in showing the smoothness of the Brill–Noether loci of special linear system on a general curve. While the Petri Theorem, which immediately implies the Brill–Noether Theorem, holds for almost every curve $[C] \in \mathcal{M}_g$, so far no explicitly examples of smooth curves of arbitrary genus satisfying this theorem have been known. Indeed, there are two types of known proofs of the Petri Theorem. The original proofs by degeneration due to Griffiths–Harris, Gieseker, and simplified later by Eisenbud-Harris [18] by their very nature, shed little light on the explicit smooth curves which are Petri general. Then there is the proof by Lazarsfeld [36], asserting that every hyperplane section of a general polarized $K3$ surface (X, H) of degree $2g - 2$ is a Brill–Noether general curve, while a general curve in the linear system $|H|$ is a Petri general curve. However, there are no known concrete examples of polarized $K3$ surfaces of arbitrary degree satisfying the requirement above. For instance, it is a non-trivial recent result of Maulik and Poonen that there exists polarized $K3$ surfaces of degree $2g - 2$ over a number field, having Picard number one.

The aim of the chapter [1] is to show that, by using Nagata's classical results on the effective cone of the blown-up projective plane in 9 points, curves lying on Du Val surfaces provide *computationally explicit* examples of Brill–Noether–Petri general curves of any genus.

We start with nine points $p_1, \dots, p_9 \in \mathbf{P}^2$ which are suitably general, and we let $E_1, \dots, E_9$ be the exceptional curves of this blow-up. We next consider the linear system on the blow-up S of $\mathbf{P}^2$ at these points:

$$L_g := \left|3g\ell - gE_1 - \cdots - gE_8 - (g-1)E_9\right|,$$

where ℓ is a class of a line. This is a g-dimensional system of curves with a base point $p_{10} = p_{10}^{(g)}$. The general element of the linear system L_g is a smooth genus g curve. Denoting by J the unique cubic plane curve passing through $p_1, \dots, p_9$, the point $p_{10}^{(g)}$ can be determined as follows using the group law of J:

$$p_{10} = p_{10}^{(g)} = -gp_1 - \cdots - gp_8 - (g-1)p_9 \in J.$$

A curve in the linear system L_g as above is called a *Du Val curve*. Each Du Val curve of genus g is endowed with a distinguished marked point $p_{10}^{(g)} \in C$.

The main results of [1, 26] can be summarized as follows:

Theorem 36. *A general Du Val curve $C \subset S$ satisfies the Brill–Noether–Petri Theorem. A general pointed Du Val curve $[C, p_{10}]$ verifies the pointed Brill–Noether Theorem.*

The pointed Brill–Noether Theorem refers to the statement that for every linear series $\ell \in G^r_d(C)$, denoting by

$$\rho(\ell, p) := \rho(g, r, d) - w^{\ell}(p)$$

the adjusted Brill–Noether number taking into account the ramification of ℓ at the point p, the inequality $\rho(\ell, p) \geq 0$ holds.

Theorem 36 provides a very concrete example of a Brill–Noether–Petri curve (in both the pointed and the unpointed case) for every value of the genus. Since the points $p_1, \dots, p_9$ can be chosen to have rational coefficients, $p = p_{10}^{(g)} \in \mathbf{P}^2(\mathbb{Q})$ and then $[C, p]$ is also defined over $\mathbb{Q}$. Hence, our results provides examples of Brill-Noether general pointed curves of arbitrary genus g defined over $\mathbb{Q}$. This answers a question of Harris–Morrison concerning the existence of such curves. The generality assumption on the nine points in $\mathbf{P}^2$ is easy to satisfy. It is explained in [1] that the following points lying on the elliptic curve $y^2 = x^3 + 17$ are general: $p_1 = (-2, 3)$, $p_2 = (-1, -4)$, $p_3 = (2, 5)$, $p_4 = (4, 9)$, $p_5 = (52, 375)$, $p_6 = (5234, 37866)$, $p_7 = (8, -23)$, $p_8 = (43, 282)$, and $p_9 = \left(\frac{1}{4}, -\frac{33}{8}\right)$.

Acknowledgments: The CRC 647 brought together many scientists from the Berlin and Potsdam area. It provided a successful environment for doing research and presenting and discussing new results. Moreover, it enabled many collaborators and guests to come to Berlin. The results of the chapter arose from the collaboration in Subproject C3.

Bibliography

[1] Arbarello E, Bruno A, Farkas G, Saccà G. Explicit Brill- Noether general curves. Comm Math Helv 2016;91:477–91.
[2] Altmann K, Bigdeli M, Herzog J, Lu D. Algebraically rigid simplicial complexes and graphs. J Pure Appl Algebra 2016;220(8):2914–935.
[3] Alexeev V, Donagi R, Farkas G, Izadi E, Ortega A. The uniformization of the moduli space of abelian 6-folds, 2015. arXiv:1507.05710.
[4] Altmann K, Hausen J. Polyhedral divisors and algebraic torus actions. Math Ann 2006;334(3):557–607.
[5] Altmann K, Hein G. A fancy divisor on M0;n. J. Pure Appl Algebra 2008; 212(4):840–50.
[6] Altmann K, Hausen J, Süss H. Gluing affine torus actions via divisorial fans. Trans Groups 2008;13(2):215–42.
[7] Altmann K, Ilten NO, Petersen L, Süß H, Vollmert R. The geometry of T-varieties. In: Contributions to algebraic geometry. Impanga lecture notes. Based on the Impanga conference on algebraic geometry, Banach Center, Bedlewo, Poland, 4–10 July 2010. Zürich: European Mathematical Society (EMS), 2012:17–69.
[8] Altmann K, Kastner L. Negative deformations of toric singularities that are smooth in codimension two. In: Deformations of surface singularities. Berlin: Springer; Budapest: János Bolyai Mathematical Society, 2013:13–55.
[9] Altmann K, Kollár J. The dualizing sheaf on first-order deformations of toric surface singularities, 2016. arXiv:1601.07805v1.
[10] Altmann K, Kiritchenko V, Petersen L. Merging divisorial with colored fans. Mich Math J 2015;64(1):3–38.
[11] Altmann K. The versal deformation of an isolated toric Gorenstein singularity. Invent Math 1997;128(3):443–79.
[12] Altmann K. One parameter families containing three-dimensional toric-Gorenstein singularities. In: Explicit birational geometry of 3-folds, volume 281 of London Math Soc Lecture Note Ser. Cambridge: Cambridge University Press, 2000:21–50.
[13] Altmann K, Wisniewski J. Polyhedral divisors of Cox rings. Mich Math J 2011;60(2):463–80.
[14] Chiodo A, Eisenbud D, Farkas G, Schreyer F-O. Syzygies of torsion bundles and the geometry of the level 'modular variety over Mg. Invent Math 2013;194:73–118.
[15] Chiodo A, Farkas G. Singularities of moduli spaces of level curves. J Eu Math Soc, 2017;19: 603–58.
[16] Cox DA, Little JB, Schenck HK. Toric varieties. Providence, RI: American Mathematical Society (AMS), 2011.
[17] Devyatov R. Equivariant deformations of algebraic varieties with an action of an algebraic torus of complexity 1. Available at: www.diss.fuberlin. de/diss/receive/FUDISS_thesis_000000101121. Accessed: 2016.
[18] Eisenbud D, Harris J. A simple proof of the Gieseker-Petri theorem on special divisors. Invent Math 1983;74:269–80.

[19] Farkas G. Syzygies of curves and the effective cone of Mg. Duke Math J 2006;135:53–98.
[20] Farkas G. Koszul divisors on moduli spaces of curves. Am J Math 2009;131:819–67.
[21] Farkas G. Aspects of the birational geometry of Mg. Surv Diff Geom 2010;14:57–111.
[22] Farkas G. The birational type of the moduli space of even spin curves. Adv Math 2010;223:433–43.
[23] Farkas G, Grushevsky S, Manni RS, Verra A. Singularities of theta divisors and the geometry of A5. J. Eu Math Soc 2014;16:1817–48.
[24] Farkas G, Ludwig K. The Kodaira dimension of the moduli space of Prym varieties. J Eu Math Soc 2010;12:755–95.
[25] Fløystad G, Greve BM, Jürgen H. Letterplace and coletterplace ideals of posets, 2015. arXiv:1501.04523.
[26] Farkas G, Tarasca N. Du Val curves and the pointed Brill-Noether theorem, 2016. arXiv:1606.02725.
[27] Farkas G, Verra A. Moduli of theta-characteristics via Nikulin surfaces. Math Annalen 2012;354:465–96.
[28] Farkas G, Verra A. The classification of the universal Jacobian over the moduli space of curves. Comm Math Helv 2013;88:587–611.
[29] Farkas G, Verra A. The universal theta divisor over the moduli space of curves. J Math Pures et Appl 2013;100:591–605.
[30] Farkas G, Verra A. Prym varieties and moduli of Nikulin surfaces. Adv Math 2016;290:314–28.
[31] Farkas G, Verra A. The universal abelian variety over A5. Ann Sci Ecole Normale Sup 2016;49:300–25.
[32] Farkas G, Verra A. The geometry of the moduli space of odd spin curves. Ann Math 2014;180:927–70.
[33] Ilten NO, Vollmert R. Deformations of rational T-varieties. J Algebr Geom 2012;21(3):531–62.
[34] Knop F. The Luna-Vust theory of spherical embeddings. In: Proc Hyd Conf Algebr Group. Manoj-Prakashan, Chennai, 1991:225–49.
[35] Kollár J, Shepherd-Barron, NI. Threefolds and deformations of surface singularities. Invent Math 1988;91(2): 299–338.
[36] Lazarsfeld R. Brill-noether-petri without degenerations. J Diff Geom 1986;23:299–307.
[37] Lange H, Ortega A. Prym varieties of triple coverings. Int Math Res Notices 2011;22:5045–75.
[38] Lange H, Ortega A. Log canonical models of the moduli space of spin curves of genus two. J London Math Soc 2014;90:763–84.
[39] Luna D, Vust Th. Plongements d'espaces homogènes. Comment Math Helv 1983;6:186–245.
[40] Mumford D. On the Kodaira dimension of the Siegel modular variety. Springer Lecture Notes Math 1983;997:348–75.
[41] Ortega A. Prym-Tyurin varieties and the Brill-Noether curve. Math Ann 2013;356:809–17.
[42] Pinkham HC. Deformations of quotient surface singularities. Several complex Variables, Proc Symp Pure Math 30, Part 1, Williamstown 1975:65–7, 1977.
[43] Donagi R, Smith R. The structure of the Prym map. Acta Math 1981;146:25–102.
[44] Schlessinger M. Functors of Artin rings. Trans Am Math Soc 1968;130:208–22.
[45] Schlessinger M. Rigidity of quotient singularities. Invent Math 1971;14: 17–26.
[46] Stevens J. Deformations of singularities. Berlin: Springer, 2003.
[47] Sturmfels B, Xu Z. Sagbi bases of Cox-Nagata rings. J Eu Math Soc (JEMS) 2010;12(2):429–59.

Björn Andreas and Alexander Schmitt

Vector bundles in algebraic geometry and mathematical physics

Abstract: We will present results on vector bundles and sheaves on compact complex manifolds which either directly address problems from string theory or concern fundamental questions on objects arising from gauge theory. The first part discusses the Strominger system of differential equations and strategies to solve it. In the second part, we will deal with quiver sheaves. Such objects occur, for example, as branes. The notion of slope semistability for quiver sheaves depends on two sets of parameters. We will begin the systematic investigation of the set of stability parameters that has been largely ignored up to now.

Keywords: Strominger system, Calabi–Yau manifold, hermitian Yang–Mills connection, quiver, representation, boundedness

Mathematics Subject Classification 2010: 14D20, 16G20, 32L05, 32W50, 53C55

Introduction

In algebraic geometry, vector bundles arise naturally, e.g., as tangent bundles to complex projective manifolds or as normal bundles of submanifolds. Line bundles and their sections describe divisors on a projective algebraic manifold and, thus, govern its projective geometry. In a similar vein, vector bundles and sheaves are related to subvarieties of higher codimension. The relevance of algebraic vector bundles for modern physics is founded on gauge theory. Gauge theory deals with (differentiable) principal bundles on differentiable manifolds and their connections. In physics, a principal bundle is the space of states a particle moving in the base manifold can take on. The potential defines the connection on the principal bundle. For example, the structure group of principal bundles in electrodynamics is U(1), U(1) parameterizing the phase. In 1954, Yang and Mills attempted to formalize particle physics with gauge theory, starting with a non-abelian structure group. Their ideas have been vastly expanded. For example, the standard model of particle physics has been fully understood in this framework, the structure group being $U(1) \times SU(2) \times SU(3)$. It was observed by Penrose and Atiyah/Ward that solutions to the Yang–Mills equations for connections on SU(2)-bundles on the four-dimensional sphere correspond to certain algebraic vector bundles of rank two on the three-dimensional complex projective space. The Kobayashi–Hitchin correspondence between Hermite–Einstein metrics on a given differentiable vector bundle on a complex projective manifold and stable holomorphic

DOI 10.1515/9783110452150-003_s_002

structures on it is a far reaching extension of this. It has been the basis for computing Donaldson's invariants for algebraic surfaces – which are defined in a purely gauge theoretical set-up – on algebro-geometric moduli spaces of stable vector bundles on that surface. Physics entered again the picture when Seiberg and Witten used considerations in physics to define new invariants for differentiable four manifolds which are more easily computable.

This short discussion shows that the theory of algebraic vector bundles is a place of lively exchange between mathematical physics and algebraic geometry with great benefits for both sides. This circle of ideas has influenced research within CRC 647 "Space–Time–Matter" in many ways. In this chapter, we will report on some of the work carried out in CRC-projects A1 "Strings, D-Branes, and Manifolds of Special Holonomy" (2005–2008), A11 "Algebraic Varieties and Principal Bundles: Semistable Objects and their Moduli Spaces" (2009–2012), and C3 "Algebraic Geometry: Deformations, Moduli and Vector Bundles" (2013–2016). More precisely, we will touch upon two questions related to vector bundles and sheaves on projective algebraic manifolds. The first one is the quest for solutions of the Strominger system and stable bundles on Calabi–Yau manifolds. The second one concerns boundedness and variation of moduli spaces of semistable quiver sheaves on projective algebraic varieties.

We would like to thank all members of the CRC, especially the spokesmen Jochen Brüning and Matthias Staudacher, for their efforts to create and energize this unique and successful scientific project. With gratitude, we acknowledge generous support by the German Research Foundation (DFG) of the CRC and, in particular, the projects mentioned above.

1 Stable bundles and the Strominger system

Fu, Li, Tseng, and Yau [27–29, 42] provided first examples of solutions of a system of coupled non-linear differential equations which arise as consistency conditions in heterotic string compactifications. The system of equations is called the Strominger system and has been originally proposed by Strominger in [58]. From a mathematical perspective, it can be considered as a generalization of the kählerian Ricci-flat equation to the case of non-kählerian Calabi–Yau manifolds [27]. Moreover, the Strominger system is expected [1, 27] to play an important role in investigating the geometry of Calabi–Yau threefolds within the Reid conjecture [49] which relates all moduli spaces of smooth Calabi–Yau threefolds by certain birational transformations.

To solve the Strominger system, one has to specify a conformally balanced hermitian form on a compact complex three-dimensional manifold X, a nowhere vanishing holomorphic $(3, 0)$-form, and a hermitian Yang–Mills connection on a bundle E over this manifold. The consistency of the underlying physical theory imposes a constraint on the associated curvature forms of the connection on the bundle E and of a unitary connection on the tangent bundle of X. More precisely, the curvature forms have to satisfy the anomaly equation, a non-linear partial differential equation. The necessary condition for the existence of solutions of this equation is that the second Chern classes of E and X agree. However, although this topological constraint can be fulfilled within the framework of algebraic geometry by finding suitable stable bundles, the problem which remains is to prove that solutions of the Strominger system actually exist.

One difficulty in obtaining smooth solutions of the Strominger system lies in the fact that many theorems of Kähler geometry and thus methods of algebraic geometry do not apply. Therefore, one approach to obtain solutions is to simultaneously perturb Kähler and hermitian Yang–Mills metrics and so to avoid the direct construction of non-kählerian manifolds with stable bundles. This approach has been used in [42] where it is shown that a deformation of the holomorphic structure on the direct sum of the tangent bundle and the trivial bundle and of the Kähler form of a given Calabi–Yau threefold leads to a smooth solution of the Strominger system whereas the original Calabi–Yau space is perturbed to a non-kählerian space. One question which motivated the present paper is whether this method can be used also for any stable bundle (which satisfies the topological constraint imposed by the anomaly equation). A result along these lines has been originally conjectured in a slightly different framework by Witten [61] who gave evidence for this in [63, 67].

We will present solutions of the Strominger system, using a perturbative method which has been inspired by the method developed in [42]. Our starting point will be a solution of the Strominger system in the so-called large radius limit, given by any stable vector bundle on a Calabi–Yau threefold which satisfies the second Chern class constraint. This solution is then perturbed to a solution of the system using the implicit function theorem. Moreover, if the initial Calabi–Yau threefold has strict SU(3)-holonomy, the obtained solutions satisfy also the equations of motion derived from the effective action of the heterotic string theory. For this, we used a result of [23] and extended the original Strominger system by an instanton condition. For the proof of our main result (see Theorem 1), we rely on a previous result in [42]. Previous examples of simultaneous solutions of the Strominger system and the equations of motion on compact nilmanifolds have been obtained in [23].

In Section 1.1, we will briefly review various aspects of the Strominger system and refer to [16, 58] for more details. In Section 1.2, we will state the main result of [8] and explain the method we used to prove it. In Section 1.3, we will point out some open problems.

1.1 Physical background

Strominger [58] proposed to specify a ten-dimensional space-time that is a (warped) product of a maximal symmetric four-dimensional space-time and a compact complex three-dimensional manifold X in order to compactify the heterotic string. On X, one has to specify a hermitian form ω, a holomorphic volume form Ω which is a nowhere vanishing section of the canonical bundle $K_X = \mathcal{O}_X$, and a connection A on a complex vector bundle E over X with vanishing first Chern class $c_1(E) = 0$. To get a supersymmetric theory the gauge field A has to satisfy the hermitian Yang–Mills (HYM) equations

$$F^{2,0} = F^{0,2} = 0, \quad F \wedge \omega^2 = 0, \tag{1}$$

where F denotes the curvature two form of the connection A. If ω is closed, the Donaldson–Uhlenbeck–Yau theorem ([20, 60]) states that, if E is an ω-stable holomorphic vector bundle, then there exists a connection A on E, such that the HYM equations are satisfied. A similar result holds for arbitrary hermitian metrics (i.e., for non closed ω) as proved in [41].

In addition, supersymmetry requires that the hermitian form ω and the holomorphic 3-form Ω have to satisfy the dilatino equation

$$d^* \omega = i(\bar{\partial} - \partial) \log\left(\|\Omega\|_\omega\right), \tag{2}$$

where $\bar{\partial}$ and ∂ denote the Dolbeault operator on X and its conjugate, d^* is the adjoint of the exterior differential d with respect to ω, and $\|\Omega\|_\omega$ denotes the point-wise norm of Ω with respect to ω. In [42, Lemma 3.1], it is shown that this last equation is equivalent to

$$d\left(\|\Omega\|_\omega \omega^2\right) = 0. \tag{3}$$

Note that eq. (3) implies that ω is conformally balanced. Moreover, to get a consistent physical theory, the hermitian form ω and the curvature 2-form F have to satisfy the anomaly equation

$$i\partial\bar{\partial}\omega = \alpha'\left(\operatorname{tr}(R \wedge R) - \operatorname{tr}(F \wedge F)\right), \tag{4}$$

where R is the curvature of an ω-unitary connection on the tangent bundle TX and α' is the slope parameter in string theory. As mentioned in the introduction, the existence of a solution of eq. (4) requires that the topological condition

$$c_2(E) = c_2(X) \tag{5}$$

is satisfied. The coupled system of differential equations (1)–(4) defines what is called the *Strominger system.*

As we are interested in also solving the equation of motion derived from the heterotic string effective action, we apply here a recent result obtained in [23, 36]. This result states that a solution of the Strominger system (the supersymmetry and anomaly equation) implies a solution of the equation of motion if and only if the connection on TX is an SU(3)-instanton, that is, the curvature 2-form R has to satisfy

$$R^{2,0} = R^{0,2} = 0, \quad R \wedge \omega^2 = 0. \tag{6}$$

As pointed out in [36], eq. (6) completely determines the choice of the ω-unitary connection in the anomaly equation (4) up to gauge transformations.

Remark 1. First examples of solutions of the Strominger system have been obtained in [27–29, 42]. One difficulty in obtaining smooth solutions of the Strominger system lies in the fact that many theorems of Kähler geometry and thus methods of algebraic geometry do not apply. Therefore, one approach to obtain solutions is to simultaneously perturb Kähler and hermitian Yang–Mills (HYM) metrics and so to avoid the direct construction of non-kählerian manifolds with stable bundles. This approach has been used in [42] where it is shown that a deformation of the holomorphic structure on the direct sum of the tangent bundle and the trivial bundle and of the Kähler form of a given Calabi–Yau threefold leads to a smooth solution of the Strominger system whereas the original Calabi–Yau space is perturbed to a non-kählerian space.

1.2 Method and results

We now explain the method we adopted in order to obtain simultaneous solutions of the Strominger system (1)–(4) and (6). We first note that the above system is invariant under rescaling of the hermitian form ω, except for the anomaly equation. Given a positive real constant λ, if we change ω to $\lambda\omega$ and define $\varepsilon := \alpha'/\lambda$, we obtain the new system

$$F^{2,0} = F^{0,2} = 0, \quad F \wedge \omega^2 = 0, \tag{7}$$

$$d\big(\|\Omega\|_\omega \omega^2\big) = 0, \tag{8}$$

$$i\partial\bar{\partial}\omega - \varepsilon\big(\mathrm{tr}(R \wedge R) - \mathrm{tr}(F \wedge F)\big) = 0, \tag{9}$$

$$R^{2,0} = R^{0,2} = 0, \quad R \wedge \omega^2 = 0, \tag{10}$$

that will be called in the following the ε-system. Here, we use the equivalence between the dilatino equation (2) and the conformally balanced condition (3), given by [42, Lemma 3.1]. Therefore, any solution of the ε-system with $\varepsilon > 0$ is related to the original system after rescaling. In the limit $\lambda \to \infty$, a solution of the ε-system is given by a degree zero stable holomorphic vector bundle E on a Calabi–Yau threefold. The next step is then to perturb a given solution with $\varepsilon = 0$ to a solution with small $\varepsilon > 0$, i.e., with large λ. For this, we perturb the Kähler form of the given Calabi–Yau threefold into a conformally balanced hermitian form on the fixed complex manifold while also perturbing its Chern connection and the unique HYM connection on the bundle E, whereas we preserve the HYM condition. Moreover, in order for eq. (9) to have a solution, the topological obstruction

$$c_2(E) = c_2(X) \tag{11}$$

must be satisfied. This provides our starting point.

Let (X, ω_0) be a compact Calabi–Yau threefold with nowhere vanishing holomorphic (3, 0)-form Ω. Let E be a stable holomorphic vector bundle over X satisfying eq. (11) with rank r and $c_1(E) = 0$. In this setting, an unknown of the ε-system corresponds to a triple $(h, \omega, \bar\partial_{TX})$, where h is a hermitian metric on E, ω is a hermitian form on X and

$$\bar\partial_{TX} : \Omega^0(TX) \longrightarrow \Omega^{0,1}(TX)$$

is an integrable Dolbeault operator on TX (regarded as a smooth complex vector bundle) satisfying $\bar\partial_{TX}^2 = 0$. Recall that the integrability condition implies that there exist holomorphic coordinates on TX with respect to $\bar\partial_{TX}$. The curvatures R and F correspond to the Chern connections of ω and h on the holomorphic bundles $(TX, \bar\partial_{TX})$ and E, respectively. In holomorphic coordinates for $\bar\partial_{TX}$ and E we can write

$$R = \bar\partial(h_\omega^{-1}\partial h_\omega) \quad \text{and} \quad F = \bar\partial(h^{-1}\partial h), \tag{12}$$

where h_ω denotes the hermitian metric on TX determined by ω. Note that the condition on the (2, 0)- and (0, 2)-part of R and F in eqs. (10) and (7) is always satisfied.

The fixed data (E, X, ω_0) provide a *canonical solution* $(h_0, \omega_0, \bar\partial_0)$ in the limit $\lambda \to \infty$, where h_0 is the unique Hermite–Einstein metric on E, given by the Donaldson–Uhlenbeck–Yau theorem, and $\bar\partial_0$ is the Dolbeault operator on TX determined by the holomorphic structure on X. Our main result in [8] is:

Theorem 1. *Let E be a degree zero holomorphic vector bundle over a compact Calabi–Yau threefold (X, ω_0). If $c_2(E) = c_2(X)$ and E is stable with respect to $[\omega_0]$, then there exist $\lambda_0 \gg 0$ and a C^1-curve*

$$]\lambda_0, +\infty[\ni \lambda \longrightarrow (h_\lambda, \omega_\lambda, \bar\partial_\lambda)$$

of solutions of the Strominger system, such that $\bar\partial_\lambda$ is isomorphic to $\bar\partial_0$ for all λ and $(h_\lambda, \omega_\lambda/\lambda, \bar\partial_\lambda)$ converges uniformly to the canonical solution $(h_0, \omega_0, \bar\partial_0)$ when $\lambda \to \infty$. Moreover, if (X, ω_0) has holonomy equal to SU(3). then $(h_\lambda, \omega_\lambda, \bar\partial_\lambda)$ solves also eq. (6) and so it provides a solution of the equations of motion derived from the effective heterotic string action.

The perturbative process provided by Theorem 1 leaves the holomorphic structure of E unchanged while the one on TX is shifted by a complex gauge transformation and so remains isomorphic to the initial one. Since E admits an irreducible solution of the Hermite–Einstein equations with respect to any element of the family of conformally balanced metrics ω_λ, it is ω_λ-stable for any λ. Recall that the stability condition for a holomorphic vector bundle E over X with respect to ω_λ is defined in terms of slopes of coherent subsheaves as in the Kähler case where the degree is computed by $c_1(E) \cdot [\|\Omega\|_\omega \omega^2]$ (see [43]). The same holds for TX, when (X, ω_0) has holonomy equal to SU(3).

We now sketch the idea underlying the perturbative process in Theorem 1. We consider a 1-parameter family of maps between suitable fixed Hilbert spaces

$$\boldsymbol{L}^{\varepsilon}\colon \mathscr{V}_1 \longrightarrow \mathscr{V}_2,$$

indexed by $\varepsilon \in \mathbb{R}$ which acts on triples $(h, \omega, \bar{\partial}_{TX}) \in \mathscr{V}_1$ and whose zero locus corresponds to solutions of the ε-system. In this way, the equation $\boldsymbol{L}^{\varepsilon} = 0$ with $\varepsilon \neq 0$ corresponding to the ε-system is related to a simpler one at $\varepsilon = 0$ which has a known canonical solution $(h_0, \omega_0, \bar{\partial}_0)$. Proving now that the differential

$$\delta_0 \boldsymbol{L}^0\colon \mathscr{V}_1 \longrightarrow \mathscr{V}_2,$$

at the initial solution is an isomorphism, an implicit function theorem argument shows that there is a solution of $\boldsymbol{L}^{\varepsilon} = 0$ nearby $(h_0, \omega_0, \bar{\partial}_0) \in \mathscr{V}_1$ in an open neighborhood of $\varepsilon = 0 \in \mathbb{R}$.

To give a more precise description of the method, we introduce some notation which we use in the sequel. Let $(h_0, \omega_0, \bar{\partial}_0)$ be the canonical solution of the ε-system at $\varepsilon = 0$ associated with E and (X, ω_0). Let $\mathscr{H}(E)_1$ be the space of hermitian metrics h on E whose induced metric on $\bigwedge^{r} E \cong \mathbb{C}_X$ is the constant metric 1. We identify its elements with determinant one symmetric endomorphisms of (E, h_0) via

$$\langle u, v\rangle_h = \langle hu, v\rangle_{h_0}, \tag{13}$$

for u, v smooth sections on E. Let $\mathscr{H}(X)$ be the cone of positive definite hermitian forms on X, regarded as an open subspace of $\Omega^{1,1}_{\mathbb{R}}(X)$. As in eq. (13), given $\omega \in \mathscr{H}(X)$, we identify the corresponding hermitian metric h_ω with a symmetric endomorphism of (TX, h_{ω_0}). Let $\mathscr{D}(TX)$ be the space of integrable Dolbeault operators on TX, regarded as a smooth complex vector bundle. Given a real vector bundle W over X, we denote by $\Omega^{p,q}_{\mathbb{R}}(W)$ and $\Omega^{m}_{\mathbb{R}}(W)$ the space of smooth real (p, q)-type forms and m-forms on X with values in W. Let $\operatorname{ad}_0 E$ and $\operatorname{ad} TX$ be the smooth real vector bundles of traceless hermitian antisymmetric endomorphisms of (E, h_0) and hermitian antisymmetric endomorphisms of (TX, ω_0), respectively. Then, we define the operator

$$\boldsymbol{L}^{\varepsilon}\colon \mathscr{H}(E)_1 \times \mathscr{H}(X) \times \mathscr{D}(TX) \to \Omega^6_{\mathbb{R}}(\operatorname{ad}_0 E) \oplus \Omega^{2,2}_{\mathbb{R}}(X) \oplus \Omega^1_{\mathbb{R}}(X) \oplus \Omega^6_{\mathbb{R}}(\operatorname{ad} TX) \tag{14}$$

as the direct sum $\boldsymbol{L}^{\varepsilon} = \boldsymbol{L}_1 \oplus \boldsymbol{L}_2^{\varepsilon} \oplus \boldsymbol{L}_3 \oplus \boldsymbol{L}_4$, where

$$\begin{aligned}
\boldsymbol{L}_1(h, \omega) &= (h^{1/2} \cdot F \cdot h^{-1/2}) \wedge \omega^2 && \in \Omega^6_{\mathbb{R}}(\operatorname{ad}_0 E),\\
\boldsymbol{L}_2^{\varepsilon}(h, \omega, \bar{\partial}_{TX}) &= i\partial\bar{\partial}\omega - \varepsilon\big(\operatorname{tr}(R \wedge R) - \operatorname{tr}(F \wedge F)\big) && \in \Omega^{2,2}_{\mathbb{R}}(X),\\
\boldsymbol{L}_3(\omega) &= *d\big(\|\Omega\|_\omega \omega^2\big) && \in \Omega^1_{\mathbb{R}}(X),\\
\boldsymbol{L}_4(\omega, \bar{\partial}_{TX}) &= (h_\omega^{1/2} \cdot R \cdot h_\omega^{-1/2}) \wedge \omega^2 && \in \Omega^6_{\mathbb{R}}(\operatorname{ad} TX).
\end{aligned} \tag{15}$$

Here, $*$ is the Hodge star operator of ω_0, $R = R_{\omega,\bar\partial_{TX}}$ is the curvature of the Chern connection of ω with respect to $\bar\partial_{TX}$ and $F = F_h$ is the curvature of the Chern connection of h on the holomorphic bundle E. Since R and F are of type (1, 1) (see eq. (12)), we have $h^{1/2}Fh^{-1/2} \in \Omega^{1,1}_{\mathbb{R}}(\mathrm{ad}_0 E)$ and $h_\omega^{1/2}Rh_\omega^{-1/2} \in \Omega^{1,1}_{\mathbb{R}}(\mathrm{ad}\, TX)$. Hence, the zero locus of $\boldsymbol{L}^\varepsilon$ corresponds to solutions of the ε-system

$$\boldsymbol{L}^0(h_0, \omega_0, \bar\partial_0) = 0. \tag{16}$$

We then apply the implicit function theorem to perturb the given solution $(h_0, \omega_0, \bar\partial_0)$ of eq. (16) to a solution of the ε-system with $\varepsilon > 0$. In order to do this, we consider suitable Hilbert spaces $\mathscr{V}_1$ and $\mathscr{V}_2$ and $\boldsymbol{L}^\varepsilon$ as a map

$$\boldsymbol{L}^\varepsilon : \mathscr{U} \longrightarrow \mathscr{V}_2 \tag{17}$$

where $\mathscr{U} \subset \mathscr{V}_1$ is an open subset of $\mathscr{V}_1$. The space $\mathscr{V}_1$ is a Sobolev completion of the space of those triples $(h, \mu, \bar\partial_{TX}) \in \mathscr{H}(E)_1 \times \Omega^{1,1}_{\mathbb{R}}(X) \times \mathscr{D}(TX)$, such that $\bar\partial_{TX}$ is isomorphic to $\bar\partial_0$ via a complex gauge transformation on TX, while $\mathscr{U}$ is defined by the locus where μ is positive definite. The space $\mathscr{V}_2$ is simply a Sobolev completion of the codomain in eq. (14). For simplicity, in this section, we identify $(h_0, \omega_0, \bar\partial_0)$ with the origin $0 \in \mathscr{V}_1$. We compute the linearization of eq. (17) at 0 when $\varepsilon = 0$,

$$\delta_0 \boldsymbol{L}^0 = \delta_0 \boldsymbol{L}_1 \oplus \delta_0 \boldsymbol{L}_2^0 \oplus \delta_0 \boldsymbol{L}_3 \oplus \delta_0 \boldsymbol{L}_4 : \mathscr{V}_1 \longrightarrow \mathscr{V}_2,$$

and study its mapping properties. By standard properties of the HYM equation, the higher order derivatives in the linear operators $\delta_0 \boldsymbol{L}_1$ and $\delta_0 \boldsymbol{L}_4$ come from the Laplacian

$$\Delta_A : \Omega^0_{\mathbb{R}}(\mathrm{ad}_0 E) \longrightarrow \Omega^0_{\mathbb{R}}(\mathrm{ad}_0 E) \tag{18}$$

induced by the Chern connection of h_0 on E, and the Laplacian

$$\Delta_0 : \Omega^0_{\mathbb{R}}(\mathrm{ad}\, TX) \longrightarrow \Omega^0_{\mathbb{R}}(\mathrm{ad}\, TX) \tag{19}$$

induced by the Chern connection of ω_0 on TX, respectively. Here, we identify the codomain in eq. (18) (respectively in eq. (19)) with $\Omega^6_{\mathbb{R}}(\mathrm{ad}_0 E)$ (respectively $\Omega^6_{\mathbb{R}}(\mathrm{ad}\, TX)$) in eq. (15) via multiplication by the volume form $\omega_0^3/3!$. A crucial ingredient in our argument is the previous study in [42] of the linear operator

$$\boldsymbol{T} := \delta_0 \boldsymbol{L}_2^0 \oplus \delta_0 \boldsymbol{L}_3. \tag{20}$$

To apply the implicit function theorem, we need that the differential of our operator induces an isomorphism at $0 \in \mathscr{V}_1$, so we redefine eq. (17) as follows: first, we define the domain $\mathscr{W}_1 \subset \mathscr{V}_1$ and codomain $\mathscr{W}_2 \subset \mathscr{V}_2$ of the new operator. Let

$$\mathscr{W}_1 := \mathscr{U} \cap (0 \oplus \operatorname{Ker} \boldsymbol{T} \oplus \operatorname{Ker} \Delta_0)^\perp,$$

where $(0 \oplus \operatorname{Ker} \boldsymbol{T} \oplus \operatorname{Ker} \Delta_0)^{\perp}$ denotes the orthogonal complement in $\mathscr{V}_1$ of the direct sum $0 \oplus \operatorname{Ker} \boldsymbol{T} \oplus \operatorname{Ker} \Delta_0$ of kernels, corresponding to eqs. (18), (19), and (20). Note here that $\operatorname{Ker} \Delta_0$ is given by the covariant constant sections in $\Omega^0_{\mathbb{R}}(\operatorname{ad} TX)$ which correspond to the Lie algebra of the centralizer of the holonomy group of ∇_0 in U(3). Note also that, since E is stable, $\operatorname{Ker} \Delta_A = 0$. By previous analysis in [42], Proposition 3.3, the range $\operatorname{Im} \boldsymbol{T}$ of eq. (20) is closed and so, by the ellipticity of Δ_A and Δ_0,

$$\mathscr{W}_2 := \operatorname{Im} \Delta_A \oplus \operatorname{Im} \boldsymbol{T} \oplus \operatorname{Im} \Delta_0$$

defines a closed subspace of $\mathscr{V}_2$. Then, we consider the orthogonal projection $\Pi_{\mathscr{W}_2} \colon \mathscr{V}_2 \to \mathscr{W}_2$ and define a new operator

$$\boldsymbol{M}^{\varepsilon} \colon \mathscr{W}_1 \longrightarrow \mathscr{W}_2$$

by composition, $\boldsymbol{M}^{\varepsilon} := \Pi_{\mathscr{W}_2} \circ \boldsymbol{L}^{\varepsilon}_{|\mathscr{W}_1}$. In [8], Proposition 4.4, we prove that the differential of $\boldsymbol{M}^0$ induces an isomorphism at $0 \in \mathscr{W}_1$ which allows to apply the implicit function theorem in the proof of Theorem 1. This provides solutions of $\boldsymbol{M}^{\varepsilon} = 0$ with $\varepsilon > 0$ which are smooth by standard elliptic regularity. An explicit description of $\boldsymbol{M}^{\varepsilon}$ shows that a solution $(h, \omega, \bar{\partial}_{TX})$ of $\boldsymbol{M}^{\varepsilon} = 0$ corresponds to a solution of eqs. (7)–(9) (and so of the Strominger system) provided that $c_2(E) = c_2(X)$. Moreover, such a triple satisfies

$$\boldsymbol{L}_4(h, \omega, \bar{\partial}_{TX}) \in \operatorname{Ker} \Delta_0 \cdot \omega_0^3.$$

When (X, ω_0) has strict SU(3)-holonomy, $\operatorname{Ker} \Delta_0$ reduces to the constant endomorphisms $i\mathbb{R}\,\mathrm{Id}$ which allows to prove that actually $\boldsymbol{L}_4(h, \omega, \bar{\partial}_{TX}) = 0$ in the proof of Theorem 1.

Remark 2.

(i) Theorem 1 has been extended to the polystable case [9] and explicit examples of polystable bundles on elliptically fibered Calabi–Yau threefolds where it applies are given. The polystable bundle is given by a spectral cover bundle (for the visible sector) and a suitably chosen bundle (for the hidden sector). This provides a new class of heterotic flux compactifications via non-kählerian deformation of Calabi–Yau geometries with polystable bundles. As an application, we obtained examples of non-kählerian deformations of some three generation GUT models.

(ii) There are presently no general results available concerning the geometry of moduli spaces of stable bundles (or sheaves) on Calabi–Yau threefolds which could provide solutions of the second Chern class constraint (5) and so give a large class of examples where Theorem 1 applies. One way to find examples of stable bundles which satisfy the constraint is to explicitly construct them on a given Calabi–Yau threefold.

The probably most prominent threefold is the quintic Calabi–Yau threefold. A stable rank four vector bundle of vanishing first Chern class on the quintic which satisfies the Chern class constraint has been constructed in [19] using the vector bundle construction of [45]. Another example of a stable rank four bundle on the quintic threefold which satisfies the constraint is given by a smooth deformation of the tangent bundle and the trivial line bundle [34]. We note that the result of [34] combined with Theorem 1 provides also a new proof of [42], Theorem 5.1, which moreover assures that (for the generic quintic) the solutions of the Strominger system satisfy also the equations of motion derived from the effective heterotic string action. An extensive search for vector bundles on the quintic Calabi–Yau threefold which satisfy the second Chern class constraint has been performed in [21] using the monad construction. These bundles have been further investigated in [15] proving stability of some rank three bundles.

Elliptically fibered Calabi–Yau threefolds with section have also been extensively investigated. In [7], a class of stable extension bundles has been constructed which satisfy the second Chern class constraint. More generally, if the Calabi–Yau manifold admits a fibration structure, a natural procedure to describe moduli spaces of sheaves and bundles is to first construct them fiberwise and then to find an appropriate global description. This method has been successfully employed to construct stable vector bundles on elliptic fibrations and on $K3$ fibrations. However, these bundles do not satisfy eq. (5), but only a generalized anomaly constraint [6].

1.3 Some open problems

The study of solving the Strominger system has just begun. There are various directions for future research. One open problem is to understand the moduli space associated with the Strominger system which combines moduli of bundles and of the underlying variety. In more general string backgrounds (i.e., if a number of five-branes contributes to the compactification, cf. [24, 62]), the role of the integrability condition (5) is played by the generalized cohomological condition (which generalizes the above second Chern class condition)

$$c_2(X) = c_2(E) + [W], \tag{21}$$

with $[W]$ an effective curve class. Three approaches to construct holomorphic G-bundles on elliptically fibered Calabi–Yau threefolds X which are capable of satisfying eq. (21) have been introduced in [24–26]. The parabolic bundle approach applies to any simple group G. One considers deformations of certain minimally unstable G-bundles corresponding to special maximal parabolic subgroups of G (cf. also [5]). The spectral cover approach applies to SU(n)- and Sp(n)-bundles and can be essentially understood as a relative Fourier–Mukai transformation. The del Pezzo surface approach applies to E_6-, E_7-, and E_8-bundles and uses the relation between subgroups of G and singularities of del Pezzo surfaces.

Condition (21) can be understood as an integrability condition for a generalized anomaly equation which in the context of the present paper is given by

$$i\partial\bar{\partial}\omega = \alpha' \left(\operatorname{tr}(R \wedge R) - \operatorname{tr}(F \wedge F) - \sum \delta_5\right). \tag{22}$$

Here, the sum is taken over a union of holomorphic curves representing W, and δ_5 can be understood as a current which integrates to one in the direction transverse to a single curve. One way to view the δ_5 contribution to the anomaly equation is to consider the gauge field in eq. (22) as a singular limit of smooth solutions of the HYM equation which degenerate, such that (a part of) the $\operatorname{tr}(F \wedge F)$ term becomes a delta-function source. This limit can be understood as the analog of the small instanton limit in two complex dimensions which leads to the Uhlenbeck compactification of the underlying moduli space. Another way is to consider the fields in eq. (22) as smooth fields in the non-compact manifold given by the complement of a subvariety on X with behavior at 'infinity' fixed by the δ_5 contributions. It is plausible that the implicit function theorem argument can be generalized to find solutions in this last case.

Remark 3. Note that solutions of the Strominger system and of the equations of motion describe supersymmetric configurations in the low energy field theory approximation of heterotic string theory. More generally, these equations include terms of higher order in α'. In [58], it has been argued, heuristically, that solutions of the low energy field theory imply solutions (to all finite orders) of the fully α'-corrected equations. This agrees also with predictions one gets from a non-renormalization theorem for the low energy superpotential. Using a perturbative expansion of the gauge connection and the metric around a Calabi–Yau manifold with fixed complex structure [61], these predictions have been checked ([63, 67], see also [58]) for the first few orders in the expansion in the special case of a deformation of the direct sum of the tangent bundle and the trivial line bundle. The only obstruction encountered in this procedure is the integrability condition imposed by the anomaly equation. However, it is an open problem to prove that the expansion encounters no obstructions at any order and that exact solutions of the α'-corrected equations exist.

Finally, let us point out the problem of solving the Strominger system on group manifolds. Biswas and Mukherjee published a paper [11] in which they claim that they have found an invariant solution to the Strominger system on $\mathrm{SL}_2(\mathbb{C})$. However, in [10], we showed that there was an error in the calculation in [11] and there is actually no solution to the Strominger system in that setting. Furthermore, we proposed to look for solutions to the Strominger system using the Strominger–Bismut connection. Following our proposal, Fei and Yau [22] were able to obtain a few interesting invariant solutions to the Strominger system on complex Lie groups and their quotients.

2 Quivers sheaves

Many classification problems in Algebraic Geometry can be described in terms of principal bundles with extra structures.[1] Examples include the classification of special varieties, such as coverings of a given one [17], representations of fundamental groups which lead to Higgs bundles [33], and objects of arithmetic origin, such as shtukas which are closely related to vector bundles with level structures [39, 46]. Decorated principal bundles are algebro-geometric incarnations of fundamental objects considered in gauge theory (see, e.g., [57, Section 6.9]). There are two classes of decorated principal bundles which have been studied in a very general context: (rational) principal bundles with a section in an associated vector bundle [52] and principal bundles (over curves) with a section in an associated fiber bundle over a point [12]. Decorated principal bundles are closely linked to the theory of quasi-maps [37, 38] and gauged maps [18, 64–66].

The general theory looks as follows: There are notions of semistability and stability, usually depending on many stability parameters, and, for any specific parameter, there is a coarse moduli space for stable objects together with a partial compactification by S-equivalence classes of semistable ones. This is parallel to the usual GIT-setup: Given a quasi-projective variety X, a reductive group G, and an action $\sigma: G \times X \longrightarrow X$, the stability parameter is the choice of a linearization, the stable points possess an orbit space, and this space is (partially) compactified by the set of equivalence classes of semistable points. Here, the equivalence relation is that the orbit closures intersect inside the locus of semistable points. If X is projective or affine, the Hilbert–Mumford criterion characterizes the stable and the semistable points. Once one has a GIT quotient or a moduli space, one would like to study its geometric and topological properties. The most basic question is non-emptiness. Another natural question is how the moduli spaces vary with the stability parameter.

Before we discuss this in more detail, let us remind you of the different stability parameters that exist. In this paper, we will stick to decorated principal bundles in the sense of [52]. Let X be a complex projective manifold and G a complex reductive linear algebraic group. A *rational principal bundle* on X is a pair $(U, \mathscr{P})$ in which $U \subset X$ is a big open subset[2] and $\mathscr{P}$ is a principal G-bundle on U. For example, if $\mathscr{E}$ is a torsion-free coherent $\mathscr{O}_X$-module of rank $n \geq 1$, then the locus U where $\mathscr{E}$ is locally free is a big open subset, and the frame bundle of $\mathscr{E}_{|U}$ is a principal $\mathrm{GL}_n(\mathbb{C})$-bundle, so that $\mathscr{E}$ gives rise to a rational principal $\mathrm{GL}_n(\mathbb{C})$-bundle. We fix a representation $\sigma: G \longrightarrow \mathrm{GL}(H)$ and a line bundle $\mathscr{L}$ on X. Then, an *$\mathscr{L}$-twisted affine σ-bump* is a pair $((U, \mathscr{P}), \varphi)$ which consists of a rational principal G-bundle $(U, \mathscr{P})$ on X and a homomorphism $\sigma: \mathscr{P}_\sigma \longrightarrow \mathscr{L}_{|U}$. Here, $\mathscr{P}_\sigma$ is the vector bundle on U with typical fiber H that is associated with $\mathscr{P}$ and

1 Following a suggestion by Huybrechts and Lehn [35, Section 4.B], we will call such objects *decorated principal bundles*.

2 This means that the closed subset $Z := X \setminus U$ has codimension at least two in X.

the G-action on H induced by σ. We leave it as an easy exercise to the reader to describe the objects in [17] and Higgs bundles on curves as twisted affine bumps. There are three different types of stability parameters. The first one is the choice of a polarization $\mathscr{O}_X(1)$ on X. This is irrelevant, if X is a curve, and is quite tricky to understand, if $\dim(X) \geq 3$, even if one deals "just" with vector bundles or torsion-free coherent sheaves [32, 48, 50]. The second parameter is the choice of a character χ of G. This appeared already in the first examples of decorated vector bundles [13]. In [4], the general formalism for studying this parameter has been laid out. The basic questions and conjectures raised in that paper were answered in [54]. The last parameter is the choice of a faithful representation $\kappa\colon G \longrightarrow \mathrm{GL}(K)$. More generally, one may fix a maximal torus on $T \subset G$ and a Weyl invariant pairing on $X_\star(T) \underset{\mathbb{Z}}{\otimes} \mathbb{C}$ (compare [44]). As far as we know, this parameter has not been systematically studied in the algebro-geometric context. Here, we would like to prepare its detailed investigation.

We will restrict to the structure group $\mathrm{GL}_{\underline{r}}(\mathbb{C}) := \bigtimes_{v \in V} \mathrm{GL}_{r_v}(\mathbb{C})$, V a finite set and $\underline{r} = (r_v, v \in V)$ a tuple of positive integers, referred to as *dimension vector*. Furthermore, we let $Q = (V, A, t, h)$ be a quiver. Finally, let $\underline{M} = (M_a, a \in A)$ be a tuple of locally free coherent $\mathscr{O}_X$-modules. Then, an *$\underline{M}$-twisted Q-sheaf* is a tuple $(\mathscr{E}_v, v \in V, \varphi_a, a \in A)$ in which $\mathscr{E}_v$ is a coherent $\mathscr{O}_X$-module, $v \in V$, and $\varphi_a\colon M_a \otimes \mathscr{E}_{t(a)} \longrightarrow \mathscr{E}_{h(a)}$ is a twisted homomorphism, $a \in A$. We will explain below (Remark 15, ii) how to view twisted Q-sheaves as twisted affine bumps attached to the dual space H of

$$\mathrm{Rep}_{\underline{r}}(Q) := \bigoplus_{a \in A} \mathrm{Hom}_{\mathbb{C}}(\mathbb{C}^{r_{t(a)}}, \mathbb{C}^{r_{h(a)}}).$$

If the quiver Q is not specified, we will speak more loosely of *quiver sheaves*.

The origin of quiver sheaves lies in category and representation theory. In fact, if one forgets about the twisting sheaves, then a Q-sheaf is a functor from Q viewed as a category to the category of coherent $\mathscr{O}_X$-modules. In the foundational paper [3], quiver bundles were used to describe vector bundles on the product of X and a flag manifold. The theory of quasi-maps arising from the $\mathrm{GL}_{\underline{r}}(\mathbb{C})$-action on H has been investigated in [38]. Quiver sheaves naturally arise from quiver gauge theories. In that context, they are used to describe branes (see, e.g., [40, 59]).

The notion of slope stability for quiver sheaves was introduced in [2]. The choice of the character of $\mathrm{GL}_{\underline{r}}(\mathbb{C})$ may be generalized to the choice of a tuple $\underline{\chi} = (\chi_v, v \in V)$ of real numbers. There is a family of faithful representations parameterized by tuples $\underline{\kappa} = (\kappa_v, v \in V)$ of positive integers. In fact, a tuple $\underline{\kappa}$ encodes the embedding

$$\iota_{\underline{\kappa}}\colon \mathrm{GL}_{\underline{r}}(\mathbb{C}) \longrightarrow \mathrm{GL}_R(\mathbb{C}), \quad R := \kappa_1 \cdot r_1 + \cdots + \kappa_n \cdot r_n,$$

that maps a tuple $(m_1, \ldots, m_n)$ of matrices to the block diagonal $(R \times R)$-matrix in which the block m_1 is first repeated κ_1 times, then the block m_2 is repeated κ_2 times, and so on. More generally, one may let $\underline{\kappa}$ be a tuple of positive real numbers. The resulting notion of $(\underline{\kappa}, \underline{\chi})$-slope (semi)stability is reproduced in Section 2.1. First experiments with these notions of semistability for quivers with few vertices show that they are very hard to handle but also that the usual finiteness results [4, 54] might still be true.

In Section 2.2, we will show by means of an example that, admitting arbitrary values of $\underline{\kappa}$ yields new semistable objects and, thus, completes our understanding of the moduli problem. Then, we will explain how, using a suitable normalization of the stability parameters, the space of all normalized stability parameters can be subdivided by linear walls into a locally finite collection of locally closed chambers, such that the notion of semistability remains constant within each chamber. We will prove that one may pass to a finite collection of such chambers if and only if the family of all semistable quiver sheaves with fixed topological invariants is bounded. To get an example where we can apply the techniques developed in the following sections, we fix a natural number $n \geq 2$ and look at the quiver A_n with vertex set $\{1, \ldots, n\}$, arrow set $\{a_1, \ldots, a_{n-1}\}$, $t(a_i) = i$, $h(a_i) = i + 1$, $i = 1, \ldots, n-1$, i.e.,

$$A_n: \quad 1 \xrightarrow{a_1} 2 \xrightarrow{a_2} \cdots \xrightarrow{a_{n-2}} n-1 \xrightarrow{a_{n-1}} n.$$

Furthermore, we let $\underline{r} = (r_i, i = 1, \ldots, n)$ be a tuple of integers with

$$r_1 > r_2 > \cdots > r_{n-1} > r_n > 0. \tag{23}$$

We also require

$$\mathrm{rk}(M_i) = 1, \quad M_i := M_{a_i}, \quad i = 1, \ldots, n. \tag{24}$$

Finally, we let $\underline{P} = (P_i, i = 1, \ldots, n)$ be a tuple of Hilbert polynomials, such that the coefficient of $x^{\dim(X)-1}$ in P_i is $r_i/(\dim(X) - 1)!$, $i = 1, \ldots, n$.

Theorem 4. *In the above situation, the set of isomorphism classes of torsion-free coherent $\mathscr{O}_X$-modules $\mathscr{F}$, such that there exist stability parameters $\underline{\kappa} \in (\mathbb{R}_{>0})^{\times n}$, $\underline{\chi} \in \mathbb{R}^n$, a $(\underline{\kappa}, \underline{\chi})$-slope semistable A_n-sheaf $(\mathscr{E}_i, i = 1, \ldots, n, \varphi_{a_i}, i = 1, \ldots, n-1)$ with $P(\mathscr{E}_i) = P_i$, $i = 0, \ldots, n$, and an index $i_0 \in \{0, \ldots, n\}$ with $\mathscr{F} \cong \mathscr{E}_{i_0}$ is bounded.*

In the case $M_a = \mathscr{O}_X$, $a \in A$, the above quiver sheaves play an important role in the investigation of the moduli stack of Higgs bundles on a curve (see [31, Section 4.3]). We refer the reader to [55] for another example and to [56] for the complete solution.

Conventions

We will use freely the notation and the results from the paper [54]. In addition, we will adopt the following terminology: Given a tuple $\underline{k} = (k_v, v \in V)$ of integers, we set $|\underline{k}| := \sum_{v \in V} k_v$.

A *quiver* will be a quadruple $Q = (V, A, t, h)$ in which V and A are finite sets and $t, h\colon A \longrightarrow V$ are maps. The elements of V are the *vertices* of the quiver Q and the elements of A its *arrows*. The vertices $t(a)$ and $h(a)$ are the *tail* and the *head*, respectively, of the arrow a, $a \in A$.

We will work on a polarized smooth projective variety $(X, \mathscr{O}_X(1))$, defined over the field $\mathbb{C}$ of complex numbers. Given a coherent $\mathscr{O}_X$-module $\mathscr{G}$, its *degree* will be the one computed with respect to the ample line bundle $\mathscr{O}_X(1)$, i.e., $\deg(\mathscr{G}) = (c_1(\mathscr{O}_X(1))^{\dim(X)-1}.c_1(\mathscr{G}))[X]$.

Given a coherent $\mathscr{O}_X$-module $\mathscr{E}$, a submodule $\mathscr{F} \subset \mathscr{E}$ is said to be *saturated*, if the quotient module $\mathscr{E}/\mathscr{F}$ is torsion-free. For an arbitrary submodule $\mathscr{F} \subset \mathscr{E}$, we define

$$\mathscr{F}^{\text{sat}} := \ker\big(\mathscr{E} \longrightarrow (\mathscr{E}/\mathscr{F})/\text{Tors}(\mathscr{E}/\mathscr{F})\big).$$

Then, $\mathscr{F}^{\text{sat}}$ is a saturated submodule and $\mathscr{F} \subset \mathscr{F}^{\text{sat}}$ is generically an isomorphism, so that $\deg(\mathscr{F}) \le \deg(\mathscr{F}^{\text{sat}})$.

We will say that a family $\mathfrak{S}$ of isomorphism classes of torsion-free coherent $\mathscr{O}_X$-modules of fixed rank r and degree d is *slope bounded*, if there is a constant C with $\mu_{\max}(\mathscr{E}) \le C$, for every torsion-free $\mathscr{O}_X$-module $\mathscr{E}$ with $[\mathscr{E}] \in \mathfrak{S}$. This is equivalent to the existence of a constant C' with $\mu_{\min}(\mathscr{E}) \ge C'$, for every torsion-free $\mathscr{O}_X$-module $\mathscr{E}$ with $[\mathscr{E}] \in \mathfrak{S}$.

We will often be in the situation that we have fixed some topological background data $\underline{r}$, $\underline{d}$. If this is the case and we speak of "constants", "bounds" or similar concepts, it is understood that they do only depend on $\underline{r}$ and $\underline{d}$.

2.1 Split sheaves, quiver sheaves, and semistability

Let V be a finite set. Recall from [51, Section 3.3], that a V-*split sheaf* is a tuple $(\mathscr{E}_v, v \in V)$ consisting of coherent $\mathscr{O}_X$-modules $\mathscr{E}_v$, $v \in V$. We say that a V-split sheaf $(\mathscr{E}_v, v \in V)$ is *non-trivial*, if there is an index $v_0 \in V$ with $\mathscr{E}_{v_0} \neq 0$. Split sheaves form in a natural way an abelian category. For a V-split sheaf $(\mathscr{E}_v, v \in V)$, we define

$$\mathscr{E}_{\text{total}} := \bigoplus_{v \in V} \mathscr{E}_v.$$

Remark 5.

(i) Let $\underline{r} = (r_v, v \in V)$ be a vector of positive natural numbers. The datum of a V-split vector bundle $(E_v, v \in V)$ with $\text{rk}(E_v) = r_v$, $v \in V$, is equivalent to the datum of a principal $\text{GL}_{\underline{r}}(\mathbb{C})$-bundle, $\text{GL}_{\underline{r}}(\mathbb{C}) := \bigtimes_{v \in V} \text{GL}_{r_v}(\mathbb{C})$.

(ii) A V-split sheaf is a quiver sheaf for the quiver with vertex set V and no arrows.

There will be two sets of stability parameters occurring in our theory. The first set is a tuple

$$\underline{\kappa} = (\kappa_v, v \in V) \in (\mathbb{R}_{>0})^{\times \# V}.$$

The second set will be a tuple

$$\underline{\chi} = (\chi_v, v \in V) \in \mathbb{R}^{\# V}.$$

For a non-trivial V-split sheaf $(\mathcal{E}_v, v \in V)$, the $\underline{\kappa}$-*rank* is

$$\mathrm{rk}_{\underline{\kappa}}\big(\mathcal{E}_v, v \in V\big) := \sum_{v \in V} \kappa_v \cdot \mathrm{rk}(\mathcal{E}_v),$$

and the $(\underline{\kappa}, \underline{\chi})$-*degree*

$$\deg_{\underline{\kappa},\underline{\chi}}\big(\mathcal{E}_v, v \in V\big) := \sum_{v \in V} \kappa_v \cdot \deg(\mathcal{E}_v) + \chi_v \cdot \mathrm{rk}(\mathcal{E}_v).$$

If $\mathrm{rk}_{\underline{\kappa}}(\mathcal{E}_v, v \in V) \neq 0$, then we also introduce the $(\underline{\kappa}, \underline{\chi})$-*slope*

$$\mu_{\underline{\kappa},\underline{\chi}}\big(\mathcal{E}_v, v \in V\big) := \frac{\deg_{\underline{\kappa},\underline{\chi}}\big(\mathcal{E}_v, v \in V\big)}{\mathrm{rk}_{\underline{\kappa}}\big(\mathcal{E}_v, v \in V\big)}.$$

Remark 6. Note that, for $\underline{\kappa} \in (\mathbb{R}_{>0})^{\# V}$, a V-split sheaf $(\mathcal{E}_v, v \in V)$ whose entries are torsion-free and not all trivial has positive $\underline{\kappa}$-rank.

Let $\mathcal{R} = (\mathcal{E}_v, v \in V, \varphi_a, a \in A)$ be an $\underline{M}$-twisted Q-sheaf. A *Q-subsheaf* of $\mathcal{R}$ is a V-split sheaf $(\mathcal{F}_v, v \in V)$ in which $\mathcal{F}_v$ is an $\mathcal{O}_X$-submodule of $\mathcal{E}_v$, such that $\varphi_a(M_a \otimes \mathcal{F}_{t(a)}) \subset \mathcal{F}_{h(a)}$, $a \in A$. It is *non-trivial* (*proper*), if there is an index $v_0 \in V$ with $\mathcal{F}_{v_0} \neq 0$ ($\mathcal{F}_{v_0} \neq \mathcal{E}_{v_0}$). Given stability parameters $\underline{\kappa} \in (\mathbb{R}_{>0})^{\# V}, \underline{\chi} \in \mathbb{R}^{\# V}$, we define $(\mathcal{E}_v, v \in V, \varphi_a, a \in A)$ to be $(\underline{\kappa}, \underline{\chi})$-*slope (semi)stable*, if (a) $\mathcal{E}_v$ is torsion-free, $v \in V$, and (b) the inequality

$$\mu_{\underline{\kappa},\underline{\chi}}\big(\mathcal{F}_v, v \in V\big) (\leq) \mu_{\underline{\kappa},\underline{\chi}}\big(\mathcal{E}_v, v \in V\big)$$

holds for all non-trivial, proper Q-subsheaves $(\mathcal{F}_v, v \in V)$ of $(\mathcal{E}_v, v \in V, \varphi_a, a \in A)$.

Remark 7.

(i) It is sufficient to test the above inequality for saturated Q-subsheaves, i.e., Q-subsheaves $(\mathcal{F}_v, v \in V)$, such that $\mathcal{E}_v / \mathcal{F}_v$ is torsion-free, $v \in V$.

(ii) A *quotient Q-sheaf* of $(\mathcal{E}_v, v \in V, \varphi_a, a \in A)$ is a V-split sheaf $(\mathcal{Q}_v, v \in V)$ for which there exist surjections $\pi_v \colon \mathcal{E}_v \longrightarrow \mathcal{Q}_v$, $v \in V$, such that $(\ker(\pi_v), v \in V)$ is a Q-subsheaf of $(\mathcal{E}_v, v \in V, \varphi_a, a \in A)$. It is *proper* (*non-trivial*), if $(\ker(\pi_v), v \in V)$

is non-trivial (proper). We will write a quotient Q-sheaf also in the form $(\mathcal{Q}_v, v \in V)$. Since the $\underline{\kappa}$-rank and the $(\underline{\kappa}, \underline{\chi})$-degree behave additively on short exact sequences, the definition and Part (i) of the remark imply that a Q-sheaf $(\mathcal{E}_v, v \in V, \varphi_a, a \in A)$ in which $\mathcal{E}_v$ is torsion-free, $v \in V$, is $(\underline{\kappa}, \underline{\chi})$-slope (semi)stable if and only if the inequality

$$\mu_{\underline{\kappa},\underline{\chi}}\big(\mathcal{E}_v, v \in V\big)(\leq)\mu_{\underline{\kappa},\underline{\chi}}\big(\mathcal{Q}_v, v \in V\big)$$

holds for all proper quotient Q-sheaves $(\mathcal{Q}_v, v \in V)$ of $(\mathcal{E}_v, v \in V, \varphi_a, a \in A)$ of positive $\underline{\kappa}$-rank. As before, it suffices to look at quotient Q-sheaves $(\mathcal{Q}_v, v \in V)$ in which $\mathcal{Q}_v$ is torsion-free, $v \in V$.

(iii) Let $(\mathcal{E}_v, v \in V, \varphi_a, a \in A)$ be an $\underline{M}$-twisted Q-sheaf. The *dual $\underline{M}$-twisted Q-sheaf* is $(\mathcal{E}_v^\vee, v \in V, \varphi_a^\vee, a \in A)$ with $\mathcal{E}_v^\vee := \underline{\mathrm{Hom}}(\mathcal{E}_v, \mathcal{O}_X)$, $v \in V$, and $\varphi_a^\vee\colon M_a \otimes \mathcal{E}_{t^\vee(a)}^\vee \longrightarrow \mathcal{E}_{h^\vee(a)}^\vee$ being the dual of φ_a tensorized by id_{M_a} followed by the evaluation $M_a \otimes M_a^\vee \longrightarrow \mathcal{O}_X$ tensorized by $\mathrm{id}_{\mathcal{E}_{h^\vee(a)}^\vee}$, $a \in A$. It is an $\underline{M}$-twisted $Q^\vee$-sheaf. Since the slope is insensitive to changes in codimension two, it follows from Part (ii) of the remark that a Q-sheaf $(\mathcal{E}_v, v \in V, \varphi_a, a \in A)$ is $(\underline{\kappa}, \underline{\chi})$-slope (semi)stable if and only if the dual Q-sheaf $(\mathcal{E}_v^\vee, v \in V, \varphi_a^\vee, a \in A)$ is $(\underline{\kappa}, -\underline{\chi})$-slope (semi)stable. Here, we set $-\underline{\chi} = (-\chi_v, v \in V)$, for $\underline{\chi} = (\chi_v, v \in V) \in \mathbb{R}^{\#V}$. The analogous property for Gieseker semistability (see [51, 53]) is not true.

(iv) Let $\underline{\kappa} \in (\mathbb{R}_{>0})^{\#V}$, $\underline{\chi} \in \mathbb{R}^{\#V}$, and $c \in \mathbb{R}$. Define

$$\underline{\chi}^c := (\chi_v^c, v \in V) := (\chi_v + \kappa_v \cdot c, v \in V).$$

Then, it is easy to see that the concepts of $(\underline{\kappa}, \underline{\chi})$-(semi)stability and $(\underline{\kappa}, \underline{\chi}^c)$-(semi)stability are equivalent. Suppose $\underline{r} = (r_v, v \in V)$ is fixed. Using

$$c_0 := -\frac{\sum_{v \in V} \chi_v \cdot r_v}{\sum_{v \in V} \kappa_v \cdot r_v},$$

we have

$$\sum_{v \in V} \chi_v^{c_0} \cdot r_v = 0. \tag{25}$$

If both $\underline{r} = (r_v, v \in V)$ and $\underline{d} = (d_v, v \in V)$ are fixed, we set

$$c_0' := -\frac{\sum_{v \in V} (\kappa_v \cdot d_v + \chi_v \cdot r_v)}{\sum_{v \in V} \kappa_v \cdot r_v}.$$

This gives

$$\sum_{v\in V}(\kappa_v \cdot d_v + \chi_v^{c_0'} \cdot r_v) = 0. \tag{26}$$

The latter normalization will turn out useful in Section 2.5.

2.2 About the stability parameter κ

Explicit investigations of moduli spaces of $\underline{M}$-twisted Q-sheaves have been carried out mainly for the quiver $\bullet \circlearrowleft$ and quivers of type A_n, $n \geq 1$. Quiver sheaves for the first quiver are the famous Higgs bundles. According to eq. (25), we may require $\chi_a = 0$. Then, κ_a cancels out in the slope. So, there is no stability parameter in the theory of slope semistable Higgs bundles. For the quiver A_n and the twisting bundles $M_i := \mathscr{O}_X$, $i = 1, \ldots, n$, one arrives at the theory of holomorphic chains. Concrete investigations of the variation of moduli spaces of holomorphic chains with the stability parameter were carried out, for example, in [4, 14, 30]. In those papers, the stability parameter $\underline{\kappa} = (\kappa_1, \ldots, \kappa_n)$ was the one with $\underline{\kappa}_i = 1$, $i = 1, \ldots, n$. Here, we will give a simple example which shows that allowing more general values for $\underline{\kappa}$ yields new slope semistable quiver sheaves. The quiver will be A_2.

Example 8. Let $\mathscr{L}_1, \mathscr{L}_2$ be invertible $\mathscr{O}_X$-modules of degrees a_1, a_2, respectively, and $\varphi: \mathscr{L}_1 \longrightarrow \mathscr{L}_2$ an injective homomorphism. Note that this implies $a_1 \leq a_2$. Let $\kappa_1, \kappa_2 \in \mathbb{R}$ be real numbers and $\chi \in \mathbb{R}_{>0}$ a positive real number. We set $\underline{\kappa} := (\kappa_1, \kappa_2)$ and $\underline{\chi} := (\chi, -\chi)$. We are interested in $(\underline{\kappa}, \underline{\chi})$-(semi)stability of the A_2-sheaf $(\mathscr{L}_1, \mathscr{L}_2, \varphi)$. The only A_2-subsheaf of $(\mathscr{L}_1, \mathscr{L}_2, \varphi)$ for which $(\underline{\kappa}, \underline{\chi})$-(semi)stability has to be checked is $(0, \mathscr{L}_2)$. So, $(\mathscr{L}_1, \mathscr{L}_2, \varphi)$ will be $(\underline{\kappa}, \underline{\chi})$-(semi)stable if and only if

$$a_2 - \frac{\chi}{\kappa_2} (\leq) \frac{\kappa_1 \cdot a_1 + \kappa_2 \cdot a_2}{\kappa_1 + \kappa_2}.$$

For fixed values of κ_1 and κ_2, this will be true for all $\chi \gg 0$.

In addition to the data in Example 8, we fix invertible $\mathscr{O}_X$-modules $\mathscr{M}_1, \mathscr{M}_2$ of degrees b_1, b_2 and an injective homomorphism $\psi: \mathscr{M}_1 \longrightarrow \mathscr{M}_2$. Suppose that $(\mathscr{L}_1, \mathscr{L}_2, \varphi)$ and $(\mathscr{M}_1, \mathscr{M}_2, \psi)$ are both $(\underline{\kappa}, \underline{\chi})$-semistable. Then, $(\mathscr{L}_1, \mathscr{L}_2, \varphi) \oplus (\mathscr{M}_1, \mathscr{M}_2, \psi)$ will be $(\underline{\kappa}, \underline{\chi})$-semistable if and only if

$$\mu_{\underline{\kappa},\underline{\chi}}(\mathscr{L}_1, \mathscr{L}_2, \varphi) = \mu_{\underline{\kappa},\underline{\chi}}(\mathscr{M}_1, \mathscr{M}_2, \psi),$$

i.e.,

$$\kappa_2 \cdot (a_2 - b_2) = \kappa_1 \cdot (b_1 - a_1). \tag{27}$$

We assume $a_2 > b_2$. Since κ_1 and κ_2 have to be positive, we must choose $b_1 > a_1$. Altogether, we need $a_2 > b_2 \geq b_1 > a_1$. We can solve eq. (27) by $\kappa_2 = 1$ and $\kappa_1 = (a_2 - b_2)/(b_1 - a_1)$, or $\kappa_1 = 1$ and $\kappa_2 = (b_1 - a_1)/(a_2 - b_2)$. According to Example 8, we can find $\chi \in \mathbb{R}_{>0}$, such that $(\mathcal{L}_1, \mathcal{L}_2, \varphi)$, $(\mathcal{M}_1, \mathcal{M}_2, \psi)$, and $(\mathcal{L}_1, \mathcal{L}_2, \varphi) \oplus (\mathcal{M}_1, \mathcal{M}_2, \psi)$ will be semistable with respect to the stability parameter

$$\left(\frac{a_2 - b_2}{b_1 - a_1}, 1, \chi, -\chi\right) \quad \text{or} \quad \left(1, \frac{b_1 - a_1}{a_2 - b_2}, \chi, -\chi\right).$$

If $a_2 - b_2 \neq b_1 - a_1$, then $(\mathcal{L}_1, \mathcal{L}_2, \varphi) \oplus (\mathcal{M}_1, \mathcal{M}_2, \psi)$ will be unstable for all stability parameters of the form $(1, 1, \chi, -\chi)$, $\chi \in \mathbb{R}$.

Example 9. To get a concrete numerical example, let us pick $a_1 := 0$, $a_2 := 4$, $b_1 := 1$, and $b_2 := 2$. Then, we obtain $\kappa_1 = 1$ and $\kappa_2 = 1/2$. Furthermore, using Example 8, we check that it suffices to take $\chi = 2$. In this way, we have constructed an example of a $(1, 1/2, 2, -2)$-semistable A_2-sheaf which isn't semistable for any stability parameter of the form $(1, 1, \chi, -\chi)$, $\chi \in \mathbb{R}_{>0}$.

2.3 The main problem

We say that $(\mathcal{E}_v, v \in V, \varphi_a, a \in A)$ is *slope semistable*, if there exist stability parameters $\underline{\kappa} \in (\mathbb{R}_{>0})^{\times \# V}$ and $\underline{\chi} \in \mathbb{R}^{\# V}$, such that $(\mathcal{E}_v, v \in V, \varphi_a, a \in A)$ is $(\underline{\kappa}, \underline{\chi})$-slope semistable. The central question we would like to investigate is the following.

Problem 10. Fix a tuple $\underline{r} = (r_v, v \in V)$ of positive integers and a tuple $\underline{d} = (d_v, v \in V)$ of integers. Does there exist a constant C, such that, for every slope semistable $\underline{M}$-twisted Q-sheaf $(\mathcal{E}_v, v \in V, \varphi_a, a \in A)$ with $\mathrm{rk}(\mathcal{E}_v) = r_v$, $\deg(\mathcal{E}_v) = d_v$, $v \in V$, one has

$$\mu_{\max}(\mathcal{E}_v) \leq C, \quad v \in V?$$

Let us briefly explain why we are interested in this result. For this, let us assume that X is a curve. By the results of [51], there is for every stability parameter $(\underline{\kappa}, \underline{\chi})$ a moduli space $\mathcal{M}^{(\underline{\kappa},\underline{\chi})\text{-ss}}_{\underline{M}/\underline{r}/\underline{d}}$ for $(\underline{\kappa}, \underline{\chi})$-semistable $\underline{M}$-twisted Q-bundles[3] $(\mathcal{E}_v, v \in V, \varphi_a, a \in A)$ with $\mathrm{rk}(\mathcal{E}_v) = r_v$, $\deg(\mathcal{E}_v) = d_v$, $v \in V$. So, a priori, we have infinitely many distinct notions of semistability and infinitely many distinct moduli spaces for $\underline{M}$-twisted Q-bundles with topological invariants $(\underline{r}, \underline{d})$. If the answer to the above question were yes, then, as we shall see in Section 2.6, there would be only finitely many distinct notions of semistability and, thus, only finitely many distinct moduli spaces. In addition, one could divide the space of all stability parameters into finitely many locally closed chambers,

3 Recall that, on a smooth curve, any torsion-free coherent $\mathcal{O}_X$-module is locally free, i.e., a bundle.

such that the notion of semistability and the associated moduli space remained constant within each chamber. The adjacency of these chambers contains information on how the variation of moduli spaces looks like (see Proposition 19).

2.4 The Harder–Narasimhan filtration

We will follow Nitsure's strategy from [47] to study boundedness properties of twisted quiver bundles. This strategy rests on the properties of Harder–Narasimhan filtrations which we will, therefore, discuss now.

Theorem 11 (Harder–Narasimhan filtration). *Let* $Q = (V, A, t, h)$ *be a quiver,* $\underline{\kappa} \in (\mathbb{R}_{>0})^{\#V}$, $\underline{\chi} \in \mathbb{R}^{\#V}$, *and* $(\mathscr{E}_v, v \in V, \varphi_a, a \in A)$ *a non-trivial* $\underline{M}$*-twisted Q-sheaf. Then* $(\mathscr{E}_v, v \in V, \varphi_a, a \in A)$ *possesses a unique filtration*

$$0 = \left(\mathscr{E}_v^0, v \in V\right) \subsetneq \left(\mathscr{E}_v^1, v \in V\right) \subsetneq \cdots \subsetneq \left(\mathscr{E}_v^{s-1}, v \in V\right) \subsetneq \left(\mathscr{E}_v^s, v \in V\right) = \left(\mathscr{E}_v, v \in V\right)$$

by Q-subsheaves, such that

$$\left(\mathscr{F}_v^j, v \in V\right) := \left(\mathscr{E}_v^j, v \in V\right) / \left(\mathscr{E}_v^{j-1}, v \in V\right)$$

together with the induced homomorphisms $\overline{\varphi}_a^j \colon M_a \otimes \mathscr{F}_{t(a)}^j \longrightarrow \mathscr{F}_{h(a)}^j$, $a \in A$, *is a non-trivial* $(\underline{\kappa}, \underline{\chi})$*-slope semistable* $\underline{M}$*-twisted Q-sheaf,* $j = 1, \ldots, s$, *and*

$$\mu_{\underline{\kappa},\underline{\chi}}\left(\mathscr{F}_v^1, v \in V\right) > \mu_{\underline{\kappa},\underline{\chi}}\left(\mathscr{F}_v^2, v \in V\right) > \cdots \mu_{\underline{\kappa},\underline{\chi}}\left(\mathscr{F}_v^{s-1}, v \in V\right) > \mu_{\underline{\kappa},\underline{\chi}}\left(\mathscr{F}_v^s, v \in V\right).$$

Proof. This is obtained by the usual arguments (compare [35, Theorem 1.3.4]). □

We will study semistability of V-split sheaves more closely.[4]

Proposition 12. *Let* $\underline{\kappa} \in (\mathbb{R}_{>0})^{\times \#V}$, $\underline{\chi} \in \mathbb{R}^{\#V}$, *and* $(\mathscr{E}_v, v \in V)$ *be a* $(\underline{\kappa}, \underline{\chi})$*-slope semistable* V*-split sheaf.*

(i) *For* $v_0 \in V$ *with* $\mathscr{E}_{v_0} \neq 0$, $\mathscr{E}_{v_0}$ *is a slope semistable sheaf with*

$$\mu(\mathscr{E}_{v_0}) = \mu_{\underline{\kappa},\underline{\chi}}\left(\mathscr{E}_v, v \in V\right) - \frac{\chi_{v_0}}{\kappa_{v_0}}.$$

4 Bear in mind that, by Remark 5, (ii), V-split sheaves are quiver sheaves, so that there is a notion of (semi)stability for them.

(ii) *The maximal and minimal slope of* $\mathscr{E}_{\mathrm{total}}$ *are*

$$\mu_{\max}(\mathscr{E}_{\mathrm{total}}) = \mu_{\underline{\kappa},\underline{\chi}}\big(\mathscr{E}_v, v \in V\big) - \min\left\{ \frac{\chi_v}{\kappa_v} \,\middle|\, v \in V \right\},$$

$$\mu_{\min}(\mathscr{E}_{\mathrm{total}}) = \mu_{\underline{\kappa},\underline{\chi}}\big(\mathscr{E}_v, v \in V\big) - \max\left\{ \frac{\chi_v}{\kappa_v} \,\middle|\, v \in V \right\}.$$

Proof.

(i) Let $(\mathscr{F}_v, v \in V)$ be the non-trivial V-split sheaf with $\mathscr{F}_{v_0} = \mathscr{E}_{v_0}$ and $\mathscr{F}_v = 0$, $v \in V \setminus \{v_0\}$. It is both a subobject and a quotient object of $(\mathscr{E}_v, v \in V)$. The condition of $(\underline{\kappa}, \underline{\chi})$-slope semistability, therefore, yields

$$\mu(\mathscr{E}_{v_0}) + \frac{\chi_{v_0}}{\kappa_{v_0}} = \mu_{\underline{\kappa},\underline{\chi}}\big(\mathscr{F}_v, v \in V\big) = \mu_{\underline{\kappa},\underline{\chi}}\big(\mathscr{E}_v, v \in V\big).$$

Obviously, this and the $(\underline{\kappa}, \underline{\chi})$-slope semistability of $(\mathscr{E}_v, v \in V)$ imply that $\mathscr{E}_{v_0}$ is a slope semistable sheaf.

(ii) We clearly have

$$\mu_{\max}(\mathscr{E}_{\mathrm{total}}) = \max\big\{ \mu_{\max}(\mathscr{E}_v) \,|\, v \in V \big\}, \quad \mu_{\min}(\mathscr{E}_{\mathrm{total}}) = \min\big\{ \mu_{\min}(\mathscr{E}_v) \,|\, v \in V \big\}.$$

Together with (i), this implies the claim. □

Remark 13. Assume more generally that Q is a disconnected quiver. A similar argument reduces the study of semistable twisted Q-sheaves to the study of semistable twisted quiver sheaves for the connected components of Q.

Proposition 14. *Let* $\underline{\kappa} \in (\mathbb{R}_{>0})^{\times \# V}$, $\underline{\chi} \in \mathbb{R}^{\# V}$, $(\mathscr{E}_v, v \in V)$ *be a* V*-split sheaf, and*

$$0 = \big(\mathscr{E}_v^0, v \in V\big) \subsetneq \big(\mathscr{E}_v^1, v \in V\big) \subsetneq \cdots \subsetneq \big(\mathscr{E}_v^{s-1}, v \in V\big) \subsetneq \big(\mathscr{E}_v^s, v \in V\big) = \big(\mathscr{E}_v, v \in V\big)$$

its Harder–Narasimhan filtration.[5] *We have*

$$\mu_{\max}(\mathscr{E}_{\mathrm{total}}) \le \mu_{\underline{\kappa},\underline{\chi}}\big(\mathscr{E}_v^1, v \in V\big) - \min\left\{ \frac{\chi_v}{\kappa_v} \,\middle|\, v \in V \right\},$$

$$\mu_{\min}(\mathscr{E}_{\mathrm{total}}) \ge \mu_{\underline{\kappa},\underline{\chi}}\big(\mathscr{F}_v^s, v \in V\big) - \max\left\{ \frac{\chi_v}{\kappa_v} \,\middle|\, v \in V \right\}$$

5 Recall Remark 5, (ii).

and

$$\mu_{\underline{\kappa},\underline{\chi}}\big(\mathscr{E}_\nu^1, \nu \in V\big) \le \mu_{\max}(\mathscr{E}_{\text{total}}) + \max\left\{ \frac{\chi_\nu}{\kappa_\nu} \,\middle|\, \nu \in V \right\},$$

$$\mu_{\underline{\kappa},\underline{\chi}}\big(\mathscr{F}_\nu^s, \nu \in V\big) \ge \mu_{\min}(\mathscr{E}_{\text{total}}) + \min\left\{ \frac{\chi_\nu}{\kappa_\nu} \,\middle|\, \nu \in V \right\}.$$

Proof. Let us verify the first inequality for the maximal slope. We have

$$\mu_{\max}(\mathscr{E}_{\text{total}}) = \max\big\{ \mu_{\max}(\mathscr{E}_\nu) \,|\, \nu \in V \big\}.$$

Let $\nu_0 \in V$ be an index at which the maximum is attained and $\mathscr{F}_{\nu_0} \subset \mathscr{E}_{\nu_0}$ a subsheaf with $\mu(\mathscr{F}_{\nu_0}) = \mu_{\max}(\mathscr{E}_{\nu_0})$. As in the proof of Proposition 12, we form a V-split subsheaf $(\mathscr{F}_\nu, \nu \in V)$ of $(\mathscr{E}_\nu, \nu \in V)$. We then find

$$\mu_{\max}(\mathscr{E}_{\text{total}}) + \frac{\chi_{\nu_0}}{\kappa_{\nu_0}} = \mu(\mathscr{F}_{\nu_0}) + \frac{\chi_{\nu_0}}{\kappa_{\nu_0}} = \mu_{\underline{\kappa},\underline{\chi}}\big(\mathscr{F}_\nu, \nu \in V\big) \le \mu_{\underline{\kappa},\underline{\chi}}\big(\mathscr{E}_\nu^1, \nu \in V\big).$$

For the second inequality concerning the maximal slope, let $\nu_0 \in V$ be an index at which $\max\{\mu(\mathscr{E}_\nu^1) \,|\, \nu \in V\}$ is taken. By Proposition 12, (i),

$$\mu_{\underline{\kappa},\underline{\chi}}\big(\mathscr{E}_\nu^1, \nu \in V\big) = \mu(\mathscr{E}_{\nu_0}^1) + \frac{\chi_{\nu_0}}{\kappa_{\nu_0}} \le \mu_{\max}(\mathscr{E}_{\nu_0}) + \frac{\chi_{\nu_0}}{\kappa_{\nu_0}} \le \mu_{\max}(\mathscr{E}_{\text{total}}) + \frac{\chi_{\nu_0}}{\kappa_{\nu_0}}.$$

The arguments for the minimal slope are similar. □

2.5 A locally finite chamber decomposition

We now turn to the setting outlined in the introduction. Throughout this section, we will fix a tuple $(\underline{r} = (r_\nu, \nu \in V), \underline{d} = (d_\nu, \nu \in V))$ and look only at $\underline{M}$-twisted Q-sheaves $(\mathscr{E}_\nu, \nu \in V, \varphi_a, a \in A)$ in which $\mathscr{E}_\nu$ is torsion-free, $\mathrm{rk}(\mathscr{E}_\nu) = r_\nu$, and $\deg(\mathscr{E}_\nu) = d_\nu$, $\nu \in V$.

Remark 15.

(i) Note that we may find integers $m, b \gg 0$ and surjections $\mathscr{O}_X(-m)^{\oplus b} \longrightarrow M_a$, $a \in A$. Using these surjections, we may associate with a Q-sheaf with twisting bundles $\underline{M} = (M_a, a \in A)$ a Q-sheaf in which all twisting bundles equal $\mathscr{O}_X(-m)^{\oplus b}$. This construction is compatible with the notions of semistability and stability that we have just discussed. In the following considerations, we will, therefore, assume that all twisting bundles are of the form $\mathscr{O}_X(-m)^{\oplus b}$ for fixed positive natural numbers m and b.

(ii) Define the quiver $Q^b := (V, A^b, t^b, h^b)$ with $A^b := A \times [b]$, $t^b(a,\beta) := t(a)$, $h^b(a,\beta) := h(a)$, $(a,\beta) \in A \times [b]$. This is the quiver that is obtained from Q by replacing the arrow a by b copies of it, $a \in A$. Let H be the dual space of $\operatorname{Rep}_{\underline{r}}(Q^b)$ and $\sigma\colon \mathrm{GL}_{\underline{r}}(\mathbb{C}) \longrightarrow \mathrm{GL}(H)$ the natural representation. Given the above forms of the twisting bundles, an $\underline{M}$-twisted Q-sheaf is the same as an $\mathcal{O}_X(m)$-twisted affine σ-bump.

Proposition 16. *Let $\underline{\kappa} \in (\mathbb{R}_{>0})^{\times \# V}$, $\underline{\chi} \in \mathbb{R}^{\# V}$ be stability parameters and $(\mathcal{E}_v, v \in V, \varphi_a, a \in A)$ a $(\underline{\kappa}, \underline{\chi})$-slope semistable quiver sheaf. For every index $v_0 \in V$, the maximal slope $\mu_{\max}(\mathcal{E}_{v_0})$ is bounded from above by*

$$\frac{\sum\limits_{v\in V}(\kappa_v \cdot d_v + \chi_v \cdot r_v)}{\sum\limits_{v\in V}\kappa_v \cdot r_v} + (|\underline{r}| - 1)\cdot\Big(\deg(\mathcal{O}_X(m)) + \max_{v\in V}\Big\{\frac{\chi_v}{\kappa_v}\Big\} - 2\cdot\min_{v\in V}\Big\{\frac{\chi_v}{\kappa_v}\Big\}\Big).$$

Proof. In principle, this is shown in [51, Page 44], following ideas of Nitsure [47]. Since we use more general values of $\underline{\kappa}$ here, we explain it. We look at the Harder–Narasimhan filtration

$$0 = \big(\mathcal{E}_v^0, v \in V\big) \subsetneq \big(\mathcal{E}_v^1, v \in V\big) \subsetneq \cdots \subsetneq \big(\mathcal{E}_v^{s-1}, v \in V\big) \subsetneq \big(\mathcal{E}_v^s, v \in V\big) = \big(\mathcal{E}_v, v \in V\big)$$

of the V-split sheaf $(\mathcal{E}_v, v \in V)$. Keeping in mind the convention of Remark 15, (i), we put the homomorphisms $\varphi_a \otimes \mathrm{id}_{\mathcal{O}_X(m)^{\oplus b}}\colon \mathcal{E}_{t(a)} \longrightarrow \mathcal{E}_{h(a)} \otimes \mathcal{O}_X(m)^{\oplus b}$ together to a homomorphism

$$\Phi\colon \bigoplus_{v\in V}\mathcal{E}_v \longrightarrow \Big(\bigoplus_{v\in V}\mathcal{E}_v\Big)(m)^{\oplus b}.$$

Claim: *In the Harder–Narasimhan filtration, we have*

$$\mu_{\underline{\kappa},\underline{\chi}}\big(\mathcal{E}_v^j, v \in V\big) > \mu_{\underline{\kappa},\underline{\chi}}\big(\mathcal{E}_v, v \in V\big), \quad j = 1, \ldots, s-1.$$

Proof. This can be seen by induction on s. In fact, the defining property of the Harder–Narasimhan filtration yields

$$\mu_{\underline{\kappa},\underline{\chi}}\big(\mathcal{E}_v^1, v \in V\big) > \mu_{\underline{\kappa},\underline{\chi}}\big(\mathcal{E}_v, v \in V\big), \tag{28}$$

and this settles the case $s = 2$. For the induction step, we abbreviate

$$R_j := \mathrm{rk}_{\underline{\kappa}}\big(\mathcal{E}_v^j, v \in V\big),\ D_j := \deg_{\underline{\kappa},\underline{\chi}}\big(\mathcal{E}_v^j, v \in V\big),\ j = 1, \ldots, s, \quad R := R_s,\ D := D_s.$$

Now,

$$0 \subsetneq \big(\mathcal{E}_v^2/\mathcal{E}_v^1, v \in V\big) \subsetneq \cdots \subsetneq \big(\mathcal{E}_v^{s-1}/\mathcal{E}_v^1, v \in V\big) \subsetneq \big(\mathcal{E}_v^s/\mathcal{E}_v^1, v \in V\big)$$

is the Harder–Narasimhan filtration of $(\mathscr{E}_v^s/\mathscr{E}_v^1, v \in V) = (\mathscr{E}_v/\mathscr{E}_v^1, v \in V)$. The induction hypothesis shows that, for $j = 2, \ldots, s-1$,

$$\frac{D_j - D_1}{R_j - R_1} > \frac{D - D_1}{R - R_1}.$$

We will be done, if we can prove that

$$\frac{D_1}{R_1} > \frac{D - D_j}{R - R_j}, \quad j = 2, \ldots, s-1.$$

For this, we observe

$$\mu_{\underline{\kappa},\underline{\chi}}\big(\mathscr{E}_v^1, v \in V\big) > \mu_{\underline{\kappa},\underline{\chi}}\big(\mathscr{E}_v^{j+1}/\mathscr{E}_v^j, v \in V\big) \geq \mu_{\underline{\kappa},\underline{\chi}}\big(\mathscr{E}/\mathscr{E}_v^j, v \in V\big), \quad j = 2, \ldots, s-1.$$

This is a consequence of the properties of the Harder–Narasimhan filtration and the induction hypothesis. □

By the claim, $\bigoplus_{v\in V} \mathscr{E}_v^j$ cannot be invariant under Φ, so that we have an induced non-trivial homomorphism

$$\overline{\Phi}^j \colon \bigoplus_{v\in V} \mathscr{E}_v^j \longrightarrow \Big(\bigoplus_{v\in V} \mathscr{E}_v/\mathscr{E}_v^j\Big)(m)^{\oplus b}, \quad j = 1, \ldots, s-1.$$

Let $1 \leq \iota \leq j-1$ be the maximal index, such that $\bigoplus_{v\in V} \mathscr{E}_v^\iota$ is contained in the kernel of $\overline{\Phi}^j$ and $j+1 \leq \kappa \leq s$ the minimal index, such that the image of $\overline{\Phi}^j$ is contained in $(\bigoplus_{v\in V} \mathscr{E}_v^\kappa/\mathscr{E}_v^j)(m)^{\oplus b}$, $j = 1, \ldots, s-1$. So, there is a non-trivial homomorphism

$$\widetilde{\Phi}^j \colon \bigoplus_{v\in V} \mathscr{E}_v^{\iota+1}/\mathscr{E}_v^\iota \longrightarrow \Big(\bigoplus_{v\in V} \mathscr{E}_v^\kappa/\mathscr{E}_v^{\kappa-1}\Big)(m)^{\oplus b}.$$

We set

$$M := \max\left\{ \frac{\chi_v}{\kappa_v} \,\middle|\, v \in V \right\} - \min\left\{ \frac{\chi_v}{\kappa_v} \,\middle|\, v \in V \right\}.$$

Now, $(\mathscr{E}_v^{\iota+1}/\mathscr{E}_v^\iota, v \in V)$ and $(\mathscr{E}_v^\kappa/\mathscr{E}_v^{\kappa-1}, v \in V)$ are $(\underline{\kappa}, \underline{\chi})$-slope semistable V-split sheaves. This and Proposition 12, (ii) imply

$$\begin{aligned}\mu_{\underline{\kappa},\underline{\chi}}\big(\mathscr{E}_v^{\iota+1}/\mathscr{E}_v^\iota, v \in V\big) &\leq \mu_{\underline{\kappa},\underline{\chi}}\big(\big(\mathscr{E}_v^\kappa/\mathscr{E}_v^{\kappa-1}\big)(m)^{\oplus b}, v \in V\big) + M \\ &= \mu_{\underline{\kappa},\underline{\chi}}\big(\mathscr{E}_v^\kappa/\mathscr{E}_v^{\kappa-1}, v \in V\big) + M + \deg\big(\mathscr{O}_X(m)\big)\end{aligned}$$

and

$$\mu_{\underline{\kappa},\underline{\chi}}\big(\mathscr{E}_v^j/\mathscr{E}_v^{j-1}, v \in V\big) \le \mu_{\underline{\kappa},\underline{\chi}}\big(\mathscr{E}_v^{j+1}/\mathscr{E}_v^j, v \in V\big) + M + \deg\big(\mathscr{O}_X(m)\big),$$

$j = 1, \ldots, s-1$. These inequalities sum up to

$$\begin{aligned}\mu_{\underline{\kappa},\underline{\chi}}\big(\mathscr{E}_v^1, v \in V\big) &\le \mu_{\underline{\kappa},\underline{\chi}}\big(\mathscr{E}_v/\mathscr{E}_v^{s-1}, v \in V\big) + (s-1)\cdot\Big(M + \deg\big(\mathscr{O}_X(m)\big)\Big)\\ &\le \mu_{\underline{\kappa},\underline{\chi}}\big(\mathscr{E}_v/\mathscr{E}_v^{s-1}, v \in V\big) + \big(|\underline{r}|-1\big)\cdot\Big(M + \deg\big(\mathscr{O}_X(m)\big)\Big).\end{aligned}$$

By eq. (28), $\mu_{\underline{\kappa},\underline{\chi}}(\mathscr{E}_v/\mathscr{E}_v^{s-1}, v \in V) < \mu_{\underline{\kappa},\underline{\chi}}(\mathscr{E}_v, v \in V)$, so that we finally obtain

$$\mu_{\underline{\kappa},\underline{\chi}}\big(\mathscr{E}_v^1, v \in V\big) \le \mu_{\underline{\kappa},\underline{\chi}}\big(\mathscr{E}_v, v \in V\big) + \big(|\underline{r}|-1\big)\cdot\Big(M + \deg\big(\mathscr{O}_X(m)\big)\Big). \tag{29}$$

This and Proposition 14 yield the assertion. □

Together with the basic observations from the last section, we may use this result to get some uniform boundedness results. To this end, we fix a norm $\|\cdot\|$ on $\mathbb{R}^{\#V}$. The condition of $(\underline{\kappa},\underline{\chi})$-slope semistability clearly does not change, if we replace the stability parameter $(\underline{\kappa},\underline{\chi})$ by $(\lambda\cdot\underline{\kappa},\lambda\cdot\underline{\chi})$ for some positive real number $\lambda \in \mathbb{R}_{>0}$. For this reason, we may assume without loss of generality that $\|\underline{\kappa}\| = 1$. We set

$$\Xi := \Big\{\underline{\nu} = (\nu_v, v \in V) \in (\mathbb{R}_{>0})^{\times \#V} \;\Big|\; \|\underline{\nu}\| = 1\Big\}.$$

We use the normalization in eq. (26). So, we define the parameter region

$$\Pi := \Big\{(\underline{\nu},\underline{\psi}) \in \Xi \times \mathbb{R}^{\#V} \;\Big|\; \sum_{v\in V}(\nu_v \cdot d_v + \psi_v \cdot r_v) = 0\Big\}. \tag{30}$$

In that region, we introduce the compact sets

$$S_{mn} := \Big\{(\underline{\nu},\underline{\psi}) \in \Pi \;\Big|\; \frac{1}{m} \le \nu_v,\ v \in V,\ \|\underline{\psi}\| \le n\Big\},\ 1 \le m, n. \tag{31}$$

The key result for our construction is the following.

Proposition 17. *Fix natural numbers $m, n \ge 1$. Then, there exists a constant C_1, such that, for every stability parameter $(\underline{\kappa},\underline{\chi}) \in S_{mn}$ and every $(\underline{\kappa},\underline{\chi})$-slope semistable $\underline{M}$-twisted Q-sheaf $(\mathscr{E}_v, v \in V, \varphi_a, a \in A)$, one has*

$$\mu_{\max}(\mathscr{E}_v) \le C_1, \quad v \in V.$$

Proof. Let A_v and a_v be the maximum and the minimum of the function $\underline{\psi} \longmapsto \psi_v$ on the compact set $\{\underline{\psi} \in \mathbb{R}^{\#V} \mid \|\underline{\psi}\| \leq n\}$, $v \in V$, $A := \max\{A_v \mid v \in V\}$ and $a := \min\{a_v \mid v \in V\}$. Proposition 16 shows that

$$\mu_{\max}(\mathscr{E}_v) \leq m \cdot \left(\frac{|\underline{d}|}{|\underline{r}|} + 2 \cdot A - 2 \cdot a\right) + \left(|\underline{r}| - 1\right) \cdot \deg\left(\mathscr{O}_X(m)\right).$$

This estimate settles the claim. □

Now, we will define a locally finite chamber decomposition of the set Π. In the following, we will refer to a pair $(\underline{s}, \underline{e})$ consisting of a tuple $\underline{s} = (s_v, v \in V)$ with $0 \leq s_v \leq r_v$, $v \in V$, and $1 \leq |\underline{s}| < |\underline{r}|$, and a tuple $\underline{e} = (e_v, v \in V)$ of integers as a *test type*. With a test type $(\underline{s}, \underline{e})$, we associate the *wall*

$$W(\underline{s}, \underline{e}) := \left\{ (\underline{\nu}, \underline{\psi}) \in \Pi \,\middle|\, \sum_{v \in V} (\nu_v \cdot e_v + \psi_v \cdot s_v) = 0 \right\}.$$

Remark 18. Here, one sees that the effect of the normalization (26) is that the walls are defined by linear equations (compare [54, concluding remark]). It becomes also clear that any $\underline{M}$-twisted Q-sheaf which is (semi)stable with respect to a real stability parameter is also so for a rational parameter.

Next, we include the artificial walls

$$K^i_m := \left\{ (\underline{\nu}, \underline{\psi}) \in \Pi \,\middle|\, \frac{1}{m} = \nu_v \right\}, \quad L_n := \left\{ (\underline{\nu}, \underline{\psi}) \in \Pi \,\middle|\, \|\underline{\psi}\| = n \right\},$$

$v \in V$, $m, n \geq 1$. The walls K^i_m and L_n, $v \in V$, $m, n \geq 1$, define a locally finite decomposition

$$\Pi = \bigsqcup_{j \in J} \mathscr{K}_j$$

into a disjoint union of locally closed subsets.

We fix an index $j \in J$. By Proposition 17, there is a constant B_1, such that, for every stability parameter $(\underline{\kappa}, \underline{\chi}) \in \mathscr{K}_j$, every $(\underline{\kappa}, \underline{\chi})$-slope semistable $\underline{M}$-twisted Q-sheaf $(\mathscr{E}_v, v \in V, \varphi_a, a \in A)$, every index $v_0 \in V$, and every subsheaf $\mathscr{F} \subset \mathscr{E}_{v_0}$,

$$\deg(\mathscr{F}) \leq B_1. \tag{32}$$

Since there are only finitely many possibilities for the tuple $\underline{s}$ and the closure of $\mathscr{K}_j$ is compact, we find constants B_2 and B_3 with

$$\sum_{v \in V} \chi_v \cdot s_v \leq B_2 \quad \text{and} \quad \sum_{v \in V \setminus \{v_0\}} \kappa_v \cdot e_v \leq B_3,$$

for every test type $(\underline{s}, \underline{e})$ with $e_v \le B_1$, $v \in V$, every element $v_0 \in V$, and every stability parameter $(\underline{\kappa}, \underline{\chi}) \in \mathscr{K}_j$. Using this observation and eq. (32), it is easy to see that there is a constant B_4, such that, for every test type $(\underline{s}, \underline{e})$ with $e_v \le B_1$, $v \in V$, and for which there exists an index $v_0 \in V$ with $e_{v_0} < B_4$, one has

$$\sum_{v \in V} (\kappa_v \cdot e_v + \chi_v \cdot s_v) < 0,$$

for all stability parameters $(\underline{\kappa}, \underline{\chi}) \in \mathscr{K}_j$. Finally, using the intersections of $\mathscr{K}_j$ with the walls $W(\underline{s}, \underline{e})$ assigned to test types $(\underline{s}, \underline{e})$ with $0 \le s_v \le r_v$, $v \in V$, $1 \le |\underline{s}| < |\underline{r}|$, and $B_4 \le e_v \le B_1$, $v \in V$, we define a decomposition of $\mathscr{K}_j$ into finitely many locally closed subsets, such that the concepts of $(\underline{\kappa}, \underline{\chi})$-stability and $(\underline{\kappa}, \underline{\chi})$-slope semistability are constant within each of these locally closed subsets. Performing this construction for every chamber $\mathscr{K}_j$, $j \in J$, we arrive at a decomposition

$$\Pi = \bigsqcup_{k \in K} \mathscr{C}_k$$

into locally closed subsets. This decomposition is locally finite, and satisfies (compare [4, Section 2.4]) the following properties.

Proposition 19.

(i) *For an index $k \in K$ and stability parameters $(\underline{\kappa}, \underline{\chi}), (\underline{\kappa}', \underline{\chi}') \in \mathscr{C}_k$, an $\underline{M}$-twisted Q-sheaf $(\mathscr{E}_v, v \in V, \varphi_a, a \in A)$ is $(\underline{\kappa}, \underline{\chi})$-(semi)stable if and only if it is $(\underline{\kappa}', \underline{\chi}')$-(semi)stable.*

(ii) *Let $k, k' \in K$ be indices, such that $\mathscr{C}_{k'}$ intersects the closure of $\mathscr{C}_k$. For stability parameters $(\underline{\kappa}, \underline{\chi}) \in \mathscr{C}_k$, $(\underline{\kappa}', \underline{\chi}') \in \mathscr{C}_{k'}$, and an $\underline{M}$-twisted Q-sheaf $(\mathscr{E}_v, v \in V, \varphi_a, a \in A)$, the following holds true:*
   - *If $(\mathscr{E}_v, v \in V, \varphi_a, a \in A)$ is $(\underline{\kappa}, \underline{\chi})$-slope semistable, it is also $(\underline{\kappa}', \underline{\chi}')$-slope semistable.*
   - *If $(\mathscr{E}_v, v \in V, \varphi_a, a \in A)$ is $(\underline{\kappa}', \underline{\chi}')$-slope stable, then it is also $(\underline{\kappa}, \underline{\chi})$-slope stable.*

Remark 20. Note that, for $m, n \ge 1$, the region S_{mn} from eq. (31) is a finite union of chambers.

2.6 Chamber decomposition and boundedness

Let $n \in \mathbb{N}$ be a natural number and

$$\Pi_{\le n} = \left\{ (\underline{v}, \underline{\psi}) \in \Pi \,\middle|\, \|\underline{\psi}\| \le n \right\}.$$

If there is a decomposition of $\Pi_{\le n}$ into finitely many locally closed subsets, such that the analog of Proposition 19 holds, then there exist only finitely many distinct semistability concepts as the stability parameter $(\underline{\kappa}, \underline{\chi})$ varies over $\Pi_{\leq n}$, and it is clear that there is a constant C such that, for every stability parameter $(\underline{\kappa}, \underline{\chi}) \in \Pi_{\leq n}$ and every $(\underline{\kappa}, \underline{\chi})$-slope semistable $\underline{M}$-twisted Q-sheaf $(\mathscr{E}_v, v \in V, \varphi_a, a \in A)$ with $\mathrm{rk}(\mathscr{E}_v) = r_v$ and $\deg(\mathscr{E}_v) = d_v$, $v \in V$, one has $\mu_{\max}(\mathscr{E}_v) \leq C$, $v \in V$. This implies that, if one fixes the tuple $\underline{P} = (P_v, v \in V)$ of Hilbert polynomials, the family of $\underline{M}$-twisted Q-sheaves with this tuple of Hilbert polynomials that are slope semistable with respect to some stability parameter $(\underline{\kappa}, \underline{\chi}) \in \Pi_{\leq n}$ is bounded [35, Theorem 3.3.7]. The converse is already non-obvious.

Proposition 21. *Let $\underline{P} = (P_v, v \in V)$ be a tuple of Hilbert polynomials. Assume that $n \in \mathbb{N}$ is a natural number and C_2 is a constant, such that, for every stability parameter $(\underline{\kappa}, \underline{\chi}) \in \Pi_{\leq n}$ and every $(\underline{\kappa}, \underline{\chi})$-slope semistable $\underline{M}$-twisted Q-sheaf $(\mathscr{F}_v, v \in V, \varphi_a, a \in A)$ with $\mathrm{rk}(\mathscr{F}_v) = r_v$, $\deg(\mathscr{F}_v) = d_v$, $v \in V$, the estimate*

$$\mu_{\max}(\mathscr{E}_v) \leq C_2, \quad v \in V,$$

holds true. Then, there are finitely many stability parameters $(\underline{\kappa}_1, \underline{\chi}_1), \ldots, (\underline{\kappa}_s, \underline{\chi}_s) \in \Pi_{\leq n}$ with the property that, for any stability parameter $(\underline{\kappa}, \underline{\chi}) \in \Pi_{\leq n}$, there is an index $j_0 \in \{1, \ldots, s\}$, such that an $\underline{M}$-twisted Q-sheaf $(\mathscr{E}_v, v \in V, \varphi_a, a \in A)$ with $P(\mathscr{E}_v) = P_v$, $v \in V$, is $(\underline{\kappa}, \underline{\chi})$-(semi)stable if and only if it is $(\underline{\kappa}_{j_0}, \underline{\chi}_{j_0})$-(semi)stable.

Proof. As we recalled before, given a Hilbert polynomial P and a constant C_3, the family of torsion-free sheaves $\mathscr{E}$ with $P(\mathscr{E}) = P$ and $\mu_{\max}(\mathscr{E}) \leq C_3$ is bounded. So, our assumption implies that there are a quasi-projective scheme $\mathfrak{S}$, flat families $\mathscr{E}_{\mathfrak{S},v}$ of torsion-free sheaves with Hilbert polynomial P_v on $\mathfrak{S} \times X$, $v \in V$, and homomorphisms

$$\varphi_{\mathfrak{S},a} \colon \mathscr{E}_{\mathfrak{S},t(a)} \otimes \pi_X^*\big(\mathscr{O}_X(-m)^{\oplus b}\big) \longrightarrow \mathscr{E}_{\mathfrak{S},h(a)}, \quad a \in A,$$

such that, for every stability parameter $(\underline{\kappa}, \underline{\chi}) \in \Pi_{\leq n}$ and every $(\underline{\kappa}, \underline{\chi})$-slope semistable $\underline{M}$-twisted Q-sheaf $(\mathscr{E}_v, v \in V, \varphi_a, a \in A)$ with $P(\mathscr{E}_v) = P_v$, $v \in V$, there is a point $s \in \mathfrak{S}$, such that the $\underline{M}$-twisted Q-sheaf $(\mathscr{E}_v, v \in V, \varphi_a, a \in A)$ is isomorphic to $(\mathscr{E}_{\mathfrak{S},v|\{s\}\times X}, v \in V, \varphi_{\mathfrak{S},a|\{s\}\times X}, a \in A)$.

For every $(\underline{\kappa}, \underline{\chi}) \in \Pi_{\leq n}$, there is an open subscheme $\mathfrak{S}(\underline{\kappa}, \underline{\chi}) \subset \mathfrak{S}$ whose closed points are exactly those corresponding to $(\underline{\kappa}, \underline{\chi})$-slope semistable $\underline{M}$-twisted Q-sheaves. Thus,

$$\mathfrak{S}^* := \bigcup_{(\underline{\nu}, \underline{\psi}) \in \Pi_{\leq n}} \mathfrak{S}(\underline{\nu}, \underline{\psi}) \quad \text{and} \quad \mathfrak{S}_m := \bigcup_{(\underline{\nu}, \underline{\psi}) \in S_{mn}} \mathfrak{S}(\underline{\nu}, \underline{\psi}), \quad m \in \mathbb{N},$$

are open subschemes of $\mathfrak{S}$.

Since $\mathfrak{S}^*$ is quasi-compact and $\mathfrak{S}_m \subset \mathfrak{S}_{m'}$, $m \leq m'$, there is a natural number $m_0 \in \mathbb{N}$ with

$$\mathfrak{S}^* = \mathfrak{S}_{m_0}. \tag{33}$$

Recall from our previous discussion, especially Remark 20, that there are finitely many test types $(\underline{s}_1, \underline{e}_1), \ldots, (\underline{s}_t, \underline{e}_t)$ which define the walls in the chamber decomposition of $S_{m_0 n}$ described in Section 2.5. These test types define hyperplanes which we can use to define a chamber decomposition

$$\bigsqcup_{j \in J_{m_0 n}} \mathscr{K}_j^{m_0 n}$$

of $\Pi_{\leq n}$.

Now, assume that there are an index $j_0 \in J_{m_0 n}$ and stability parameters $(\underline{\kappa}, \underline{\chi})$, $(\underline{\kappa}', \underline{\chi}') \in \mathscr{K}_j^{m_0 n}$, such that the condition of $(\underline{\kappa}, \underline{\chi})$-slope semistability is not equivalent to the one of $(\underline{\kappa}', \underline{\chi}')$-slope semistability or the condition of $(\underline{\kappa}, \underline{\chi})$-stability is not equivalent to the one of $(\underline{\kappa}', \underline{\chi}')$-stability. There is a natural number $m' > m_0$ with $(\underline{\kappa}, \underline{\chi})$, $(\underline{\kappa}', \underline{\chi}') \in S_{m'n}$. We need to add some test types $(\underline{s}_{t+1}, \underline{e}_{t+1}), \ldots, (\underline{s}_u, \underline{e}_u)$, such that the walls $W(\underline{s}_1, \underline{e}_1), \ldots, W(\underline{s}_u, \underline{e}_u)$ define the chamber decomposition of $S_{m'n}$ discussed before. Our assumption implies that $(\underline{\kappa}, \underline{\chi})$ and $(\underline{\kappa}', \underline{\chi}')$ are separated by one or more walls of the collection $W(\underline{s}_{t+1}, \underline{e}_{t+1}), \ldots, W(\underline{s}_u, \underline{e}_u)$. It is clearly sufficient to treat the case when there is only one wall. For simplicity, let it be $W(\underline{s}_u, \underline{e}_u)$. We study two cases

Case A. Suppose there are a $(\underline{\kappa}, \underline{\chi})$-slope semistable $\underline{M}$-twisted Q-sheaf $(\mathscr{E}_v, v \in V, \varphi_a, a \in A)$ which is not $(\underline{\kappa}', \underline{\chi}')$-slope semistable and a Q-subsheaf $(\mathscr{F}_v, v \in V)$ with $\operatorname{rk}(\mathscr{F}_v) = s_{u,v}$, $\deg(\mathscr{F}_v) = e_{u,v}$, $v \in V$, such that

$$\sum_{v \in V} \left(\kappa_v' \cdot \deg(\mathscr{F}_v) + \chi_v' \cdot \operatorname{rk}(\mathscr{F}_v)\right) > 0.$$

Recall that we assume that the test type $(\underline{s}_u, \underline{e}_u)$ does not belong to $\{ (\underline{s}_1, \underline{e}_1), \ldots, (\underline{s}_t, \underline{e}_t) \}$. This means

$$\forall (\underline{\nu}, \underline{\psi}) \in S_{m_0 n}: \quad \sum_{v \in V} \left(\nu_v \cdot \deg(\mathscr{F}_v) + \psi_v \cdot \operatorname{rk}(\mathscr{F}_v)\right) < 0. \tag{34}$$

Let (κ_0, χ_0) be the point of intersection of a suitable path joining $(\underline{\kappa}, \underline{\chi})$ and $(\underline{\kappa}', \underline{\chi}')$ and the wall $W(\underline{s}_u, \underline{e}_u)$. According to Proposition 19, $(\mathscr{E}_v, v \in V, \varphi_a, a \in A)$ is (κ_0, χ_0)-slope semistable. Note that

$$\sum_{v \in V} \left(\kappa_v^0 \cdot \deg(\mathscr{F}_v) + \chi_v^0 \cdot \operatorname{rk}(\mathscr{F}_v)\right) = 0. \tag{35}$$

General properties and eq. (35) imply that the $\underline{M}$-twisted Q-sheaf

$$(\mathscr{F}_v, v \in V, \varphi_{a|\mathscr{F}_{t(a)}}, a \in A) \oplus (\mathscr{E}_v/\mathscr{F}_v, v \in V, \overline{\varphi}_a, a \in A)$$

is also (κ_0, χ_0)-slope semistable. By eq. (34), this $\underline{M}$-twisted Q-sheaf cannot be semistable with respect to any stability parameter in $S_{m_0 n}$. This contradicts our choice of m_0.

Case B. There are a $(\underline{\kappa}, \underline{\chi})$-slope stable $\underline{M}$-twisted Q-sheaf $(\mathscr{E}_v, v \in V, \varphi_a, a \in A)$ which is not $(\underline{\kappa}', \underline{\chi}')$-slope stable and a Q-subsheaf $(\mathscr{F}_v, v \in V)$ with $\mathrm{rk}(\mathscr{F}_v) = s_v$, $\deg(\mathscr{F}_v) = e_v$, $v \in V$, such that

$$\sum_{v \in V} \left(\kappa_v \cdot \deg(\mathscr{F}_v) + \chi_v \cdot \mathrm{rk}(\mathscr{F}_v)\right) < 0 \quad \text{and} \quad \sum_{v \in V} \left(\kappa'_v \cdot \deg(\mathscr{F}_v) + \chi'_v \cdot \mathrm{rk}(\mathscr{F}_v)\right) \geq 0.$$

This case can be dealt with as Case A. □

2.7 An example

In this section, we will prove Theorem 4. To simplify notation, we will write φ_i instead of φ_{a_i}, $i = 1, \dots, n-1$. We will fix the following data

- a tuple $\underline{r} = (r_i, i = 1, \dots, n)$ of natural numbers, subject to Condition (23),
- a tuple $\underline{d} = (d_i, i = 1, \dots, n)$ of integers.

We define $\mathfrak{R}$ to be the family of isomorphism classes of slope semistable $\underline{M}$-twisted A_n-sheaves $\mathscr{R} = (\mathscr{E}_i, i = 1, \dots, n, \varphi_i, i = 1, \dots, n-1)$ with $\mathrm{rk}(\mathscr{E}_i) = r_i$, $\deg(\mathscr{E}_i) = d_i$, $i = 1, \dots, n$. Then, $\mathscr{R} \in \mathfrak{R}$ means that $\mathscr{R} = (\mathscr{E}_i, i = 1, \dots, n, \varphi_i, i = 1, \dots, n-1)$ is a slope semistable $\underline{M}$-twisted A_n-sheaf with $\mathrm{rk}(\mathscr{E}_i) = r_i$ and $\deg(\mathscr{E}_i) = d_i$, $i = 1, \dots, n$.

Remark 22. Suppose we are also given a tuple $\underline{\kappa} = (\kappa_i, i = 1, \dots, n)$ of positive real numbers. Then, we set

$$\delta := \min\left\{ \frac{d_i}{r_i} \,\middle|\, i = 1, \dots, n \right\}, \quad \Delta := \max\left\{ \frac{d_i}{r_i} \,\middle|\, i = 1, \dots, n \right\},$$

and note

$$\delta \leq \frac{\sum_{i=1}^{n} \kappa_i \cdot d_i}{\sum_{i=1}^{n} \kappa_i \cdot r_i} \leq \Delta.$$

In this section, we will use Normalization (25). So, we set

$$N := \left\{ (\psi_1, \dots, \psi_n) \in \mathbb{R}^n \mid r_1 \cdot \psi_1 + \cdots + r_n \cdot \psi_n = 0 \right\}.$$

The next set we need to consider consists of all stability parameters for which there do exist slope semistable $\underline{M}$-twisted A_n-sheaves of type $(\underline{r}, \underline{d})$, i.e.,

$$\mathfrak{S} := \left\{ (\underline{\kappa}, \underline{\chi}) \in (\mathbb{R}_{>0})^n \times N \,\middle|\, \exists \mathscr{R} \in \mathfrak{R} \text{ which is } (\underline{\kappa}, \underline{\chi})\text{-slope semistable} \right\}.$$

The questions we will address are the following:

- Given $j \in \{1, \dots, n\}$, is there a constant C_0^j with $\mu_{\max}(\mathscr{E}_j) \le C_0^j$, for every $\mathscr{R} = (\mathscr{E}_i, i = 1, \dots, n, \varphi_i, i = 1, \dots, n-1) \in \mathfrak{R}$?
- Given $j \in \{1, \dots, n\}$, is there a constant C_1^j with $\chi_j/\kappa_j \le C_1^j$, for every $(\underline{\kappa}, \underline{\chi}) = (\kappa_i, \chi_i, i = 1, \dots, n) \in \mathfrak{S}$?

Our first result explains how these two questions are related and how to obtain answers to them.

Proposition 23.

(i) *Suppose that $(\underline{\kappa}, \underline{\chi}) \in \mathfrak{S}$, that $\mathscr{R} = (\mathscr{E}_i, i = 1, \dots, n, \varphi_i, i = 1, \dots, n-1)$ is a $(\underline{\kappa}, \underline{\chi})$-slope semistable $\underline{M}$-twisted A_n-sheaf, satisfying $\mathrm{rk}(\mathscr{E}_i) = r_i$, $\deg(\mathscr{E}_i) = d_i$, $i = 1, \dots, n$, and that $j \in \{1, \dots, n-1\}$ is an index with*

$$\mu_{\max}(\mathscr{E}_j) > \mu_{\max}(\mathscr{E}_{j+1}) - \deg(M_j).$$

Then, χ_j/κ_j is bounded from above by $-\mu_{\max}(\mathscr{E}_j) + \Delta$.

(ii) *Assume that $j \in \{1, \dots, n-1\}$ is an index, such that $\mathscr{E}_{j+1}$ does not move in a slope bounded family as $\mathscr{R}$ varies over $\mathfrak{R}$, and that C_1^{j+1} is a constant with $\chi_{j+1}/\kappa_{j+1} \le C_1^{j+1}$, for all $(\underline{\kappa}, \underline{\chi}) \in \mathfrak{S}$. Then, $\mathscr{E}_j$ does not belong to a slope bounded family, either, and there is also a constant C_1^j with $\chi_j/\kappa_j \le C_1^j$, for all $(\underline{\kappa}, \underline{\chi}) \in \mathfrak{S}$.*

Proof.

(i) Let $\mathscr{F}_j$ be the maximal destabilizing subsheaf of $\mathscr{E}_j$. Because of the assumption, there is no homomorphism from $M_j \otimes \mathscr{F}_j$ to $\mathscr{E}_{j+1}$, i.e., $M_j \otimes \mathscr{F}_j \subset \ker(\varphi_j)$. We complete $\mathscr{F}_j$ by zeroes to an A_n-subsheaf $(\mathscr{F}_i, i = 1, \dots, n)$. Slope semistability of $(\mathscr{E}_i, i = 1, \dots, n, \varphi_i, i = 1, \dots, n-1)$ yields

$$\mu_{\max}(\mathscr{E}_j) + \frac{\chi_j}{\kappa_j} = \mu(\mathscr{F}_j) + \frac{\chi_j}{\kappa_j} = \mu_{\underline{\kappa},\underline{\chi}}\big(\mathscr{F}_i, i = 1, \dots, n\big) \le \mu_{\underline{\kappa},\underline{\chi}}\big(\mathscr{E}_i, i = 1, \dots, n\big) \le \Delta,$$

i.e.,

$$\frac{\chi_j}{\kappa_j} \le -\mu_{\max}(\mathscr{E}_j) + \Delta.$$

(ii) Suppose that $\mathscr{E}_j$ moves in a slope bounded family. Let $\mathscr{Q}_{j+1}$ be a torsion-free quotient sheaf of $\mathscr{E}_{j+1}$ and $\mathscr{F}_{j+1} := \ker(\mathscr{E}_{j+1} \longrightarrow \mathscr{Q}_{j+1})$. We have a commutative diagram

$$\begin{array}{ccccccccc} 0 & \longrightarrow & \mathrm{im}(\varphi_j) \cap \mathscr{F}_{j+1} & \longrightarrow & \mathscr{F}_{j+1} & \longrightarrow & \mathscr{C}_{j+1} & \longrightarrow & 0 \\ & & \downarrow & & \downarrow & & \downarrow & & \\ 0 & \longrightarrow & \mathrm{im}(\varphi_j) & \longrightarrow & \mathscr{E}_{j+1} & \longrightarrow & \mathrm{coker}(\varphi_j) & \longrightarrow & 0 \\ & & \downarrow & & \downarrow & & \downarrow & & \\ 0 & \longrightarrow & \mathscr{R}_{j+1} & \longrightarrow & \mathscr{Q}_{j+1} & \longrightarrow & \mathscr{D}_{j+1} & \longrightarrow & 0 \end{array}$$

with exact rows and not necessarily exact left and right column. Now, $\mathscr{R}_{j+1}$ is a torsion-free quotient sheaf of $\mathscr{E}_j$. By assumption, its slope is bounded from below by some constant C_2^{j+1}. If $\mathscr{D}_{j+1}$ is a torsion sheaf, then $\mu(\mathscr{Q}_{j+1}) \geq \mu(\mathscr{R}_{j+1}) \geq C_2^{j+1}$. Otherwise, note that $(\mathscr{D}_i, i = 1, \ldots, n)$ with $\mathscr{D}_i = 0$, $i \in \{1, \ldots, n\} \setminus \{j+1\}$, is a quotient A_n-sheaf of positive $\underline{\kappa}$-rank. Slope semistability gives

$$\delta \leq \mu(\mathscr{D}_{j+1}) + \frac{\chi_{j+1}}{\kappa_{j+1}} \leq \mu(\mathscr{D}_{j+1}) + C_1^{j+1}. \tag{36}$$

This gives a lower bound on $\mu(\mathscr{D}_{j+1})$ and, thus, a lower bound on $\mu(\mathscr{Q}_{j+1})$ and on $\mu_{\min}(\mathscr{E}_{j+1})$. This contradicts our assumption on $\mathscr{E}_{j+1}$.

Finally, we have to explain how to obtain the upper bound on χ_j/κ_j. Note that φ_j is not injective. If φ_j is not generically surjective, the previous computations give a lower bound on the slope of $\mathrm{coker}(\varphi_j)$. This turns into an upper bound on the slope of $\mathrm{im}(\varphi_j)$ and a lower bound on the slope of $\ker(\varphi_j)$. As we have seen before, this yields an upper bound on χ_j/κ_j. If φ_j is generically surjective, we have

$$\mu\big(\ker(\varphi_j)\big) \geq \frac{d_j + r_j \cdot \deg(M_j) - d_{j+1}}{r_j - r_{j+1}}.$$

Plugging the A_n-subsheaf $(\mathscr{F}_i, i = 1, \ldots, n)$ with $\mathscr{F}_j := M_j^\vee \otimes \ker(\varphi_j)$ and $\mathscr{F}_i := 0$, $i \in \{1, \ldots, n\} \setminus \{j\}$ into the semistability condition gives an upper bound for χ_j/κ_j. □

Proof of Theorem 4. We have to verify the existence of a constant C with $\mu_{\max}(\mathscr{E}_j) < C$, for every $\underline{M}$-twisted A_n-sheaf $(\mathscr{E}_i, i = 1, \ldots, n, \varphi_i, i = 1, \ldots, n-1) \in \mathfrak{R}$ and every index $j \in \{1, \ldots, n\}$.

First, we point out that χ_n/κ_n is bounded from above. To see this, fix $\underline{\kappa}$, $\underline{\chi}$, a $(\underline{\kappa}, \underline{\chi})$-slope semistable $\underline{M}$-twisted A_n-sheaf $\mathscr{R} = (\mathscr{E}_i, i = 1, \ldots, n, \varphi_i, i = 1, \ldots, n-1)$ with $\mathrm{rk}(\mathscr{E}_i) = r_i$ and $\deg(\mathscr{E}_i) = d_i$, $i = 1, \ldots, n$, and look at the A_n-subsheaf $(\mathscr{F}_i, i = 1, \ldots, n)$ with $\mathscr{F}_i = 0$, $i = 1, \ldots, n-1$, and $\mathscr{F}_n = \mathscr{E}_n$. Slope semistability implies

$$\frac{d_n}{r_n} + \frac{\chi_n}{\kappa_n} = \mu_{\underline{\kappa},\underline{\chi}}(\mathscr{F}_i, i = 1, \ldots, n) \leq \Delta,$$

i.e.,

$$\frac{\chi_n}{\kappa_n} \leq \Delta - \frac{d_n}{r_n}. \tag{37}$$

Case A. Here, we assume that, for every constant C, there exist stability parameters $\underline{\kappa}$, $\underline{\chi}$, and a $(\underline{\kappa}, \underline{\chi})$-slope semistable $\underline{M}$-twisted A_n-sheaf $\mathscr{R} = (\mathscr{E}_i, i = 1, \ldots, n, \varphi_i, i = 1, \ldots, n-1)$ with $\mathrm{rk}(\mathscr{E}_i) = r_i$ and $\deg(\mathscr{E}_i) = d_i$, $i = 1, \ldots, n$, such that

$$\mu_{\max}(\mathscr{E}_n) > C.$$

Using this fact and eq. (37), we may repeatedly apply Proposition 23, (ii), to infer the following:

- There is a constant C' with $\chi_1/\kappa_1 \leq C'$, for every stability parameter $(\underline{\kappa}, \underline{\chi}) \in \mathfrak{S}$.
- For every constant C, there exists an $\underline{M}$-twisted A_n-sheaf $\mathscr{R} = (\mathscr{E}_i, i = 1, \ldots, n, \varphi_i, i = 1, \ldots, n-1) \in \mathfrak{R}$ with $\mu_{\max}(\mathscr{E}_1) > C$.

Fix C, choose $\mathscr{R}$ as in the second item, and let $\mathscr{F}$ be the maximal destabilizing subsheaf of $\mathscr{E}_1$. Now, $(\mathscr{Q}_i, i = 1, \ldots, n)$ with $\mathscr{Q}_1 = \mathscr{E}_1/\mathscr{F}$ and $\mathscr{Q}_i = 0$, $i = 2, \ldots, n$, is a quotient A_n-sheaf of $\mathscr{R}$. We apply slope semistability and find

$$\delta \leq \mu_{\underline{\kappa},\underline{\chi}}(\mathscr{Q}_i, i = 1, \ldots, n) = \mu(\mathscr{E}_1/\mathscr{F}) + \frac{\chi_1}{\kappa_1} \leq \frac{d_1 - \mathrm{rk}(\mathscr{F}) \cdot C}{r_1 - \mathrm{rk}(\mathscr{F})} + C'.$$

We see that this is impossible, if we choose C sufficiently large.

Case B. We assume that we are not in Case A. Then, it makes sense to define $i_0 \in \{1, \ldots, n\}$ to be minimal among those indices $i \in \{1, \ldots, n\}$ for which there exists a constant C with the property that

$$\mu_{\max}(\mathscr{E}_j) \leq C$$

holds for every $\underline{M}$-twisted A_n-sheaf $\mathscr{R} = (\mathscr{E}_i, i = 1, \ldots, n, \varphi_i, i = 1, \ldots, n-1) \in \mathfrak{R}$ and every index $j \in \{i, \ldots, n\}$ and. If $i_0 = 1$, we are done. Otherwise, we apply Proposition 23, first Part (i) to $j = i_0$, and then Part (ii), starting with $j = i_0 - 2$, and conclude as in Case A. □

Bibliography

[1] Adams A. Conformal field theory and the Reid conjecture, 2007 arXiv:hep-th/0703048, 4pp.

[2] Álvarez-Cónsul L, García-Prada O. Dimensional reduction, SL(2;C)-equivariant bundles and stable holomorphic chains. Int J Math 2001;12:159–201.

[3] Álvarez-Cónsul L, García-Prada O. Hitchin–Kobayashi correspondence, quivers, and vortices. Comm Math Phys 2003;238:1–33.

[4] Álvarez-Cónsul L, García-Prada O, Schmitt A. On the geometry of moduli spaces of holomorphic chains over compact Riemann surfaces. IMRP Int Math Res Pap 2006, Art. ID 73597:82 pp.

[5] Andreas B. On vector bundles and chiral matter in N = 1 heterotic compactifications. J High Energ Phys 1998;9812:0–24.
[6] Andreas B, Curio G. Heterotic models without fivebranes. J Geom Phys 2007;57 :2136–45.
[7] Andreas B, Curio G. Stable bundle extensions on elliptic Calabi–Yau threefolds. J Geom Phys 2007;57:2249–62.
[8] Andreas B, García-Fernández M. Solutions of the Strominger system via stable bundles on Calabi– Yau threefolds. Comm Math Phys 2012;315:153–68.
[9] Andreas B, García-Fernández M. Heterotic non-Kähler geometries via polystable bundles on Calabi–Yau threefolds. J Geom Phys 2012;62:183–8.
[10] Andreas B, García-Fernández M. Note on solutions of the Strominger system from unitary representations of cocompact lattices of SL(2; C). Comm Math Phys 2014;332:1381–3.
[11] Biswas I, Mukherjee A. Solutions of Strominger system from unitary representations of cocompact lattices of SL(2;C). Comm Math Phys 2013;322:373–84.
[12] Beck N. A generalization of principal bundles with a parabolic or level structure. J Transformation Groups 2018;1:1–40.
[13] Bradlow SB. Special metrics and stability for holomorphic bundles with global sections. J Diff Geom 1991;33:169–213.
[14] Bradlow SB, García-Prada O, Gothen PB, Moduli spaces of holomorphic triples over compact Riemann surfaces. Math Ann 2004;328:299–351.
[15] Brambilla MC. Semistability of certain bundles on a quintic Calabi–Yau threefold. Rev Mat Complut 2009;22:53–61.
[16] Candelas P, Horowitz G, Strominger A, Witten E. Vacuum configurations for superstrings. Nucl Phys B 1985;258:46–74.
[17] Casnati G, Ekedahl T. Covers of algebraic varieties. I. A general structure theorem, covers of degree 3, 4 and Enriques surfaces. J Algebraic Geom 1996;5:439–60.
[18] Diaconescu D-E. Moduli of ADHM sheaves and the local Donaldson-Thomas theory. J Geom Phys 2012;62:763–99.
[19] Distler J, Greene B. Aspects of (2; 0) compactifications. Nucl Phys B 1988;304:1–62.
[20] Donaldson SK. Anti-self-dual Yang–Mills connections on a complex algebraic surface and stable vector bundles. Proc London Math Soc 1985;50:1–26.
[21] Douglas MR, Zhou C. Chirality change in string theory. J High Energ Phys 2004;6:14–42.
[22] Fei T, Yau S-T, Invariant solutions to the Strominger system on complex Lie groups and their quotients. Comm Math Phys 2015;338:1183–95.
[23] Fernández M, Ivanov S, Ugarte L, Villacampa R. Non-Kaehler heterotic string compactifications with non-zero fluxes and constant dilaton. Comm Math Phys 2009;288:677–97.
[24] Friedman R, Morgan JW, Witten E. Vector bundles and F-theory. Comm Math Phys 1997;187:679–743.
[25] Friedman R, Morgan JW, Witten E. Principal G-bundles over elliptic curves. Math Res Lett 1998;5:97–118.
[26] Friedman R, Morgan JW, Witten E. Vector bundles over elliptic fibrations. J Algebra Geom 1999;8:279–401.
[27] Fu J-X, Li J, Yau S-T. Balanced metrics on some families of non-Kähler Calabi–Yau threefolds. J Diff Geom 2012;90:81–129.
[28] Fu J-X, Tseng L-S, Yau S-T. Local heterotic torsional models. Comm Math Phys 2009;289:1151–69.
[29] Fu J-X, Yau S-T. The theory of superstring with flux on non-Kähler manifolds and the complex Monge–Ampère equation. J Diff Geom 2008;78:369–428.
[30] García-Prada O, Heinloth J. The y-genus of the moduli space of PGLn-Higgs bundles on a curve (for degree coprime to n). Duke Math J 2013;162:2731–49.

[31] García-Prada O, Heinloth J, Schmitt A. On the motives of moduli of chains and Higgs bundles. J Eur Math Soc (JEMS) 2014;16:2617–68.

[32] Greb D, Ross J, Toma M. Variation of Gieseker–Maruyama moduli spaces via quiver GIT. Geom Topol 2016;20(3):1539–610.

[33] Hitchin NJ. The self-duality equations on a Riemann surface. Proc London Math Soc (3) 1987;55:59–126.

[34] Huybrechts D. The tangent bundle of a Calabi–Yau manifold – deformations and restriction to rational curves. Comm Math Phys 1995;171:139–58.

[35] Huybrechts D, Lehn M. The geometry of moduli spaces of sheaves, 2nd ed. Cambridge Mathematical Library. Cambridge: Cambridge University Press, 2010:xviii+325 pp.

[36] Ivanov S. Heterotic supersymmetry, anomaly cancellation and equations of motion. Phys Lett B 2010;685:190–6.

[37] Kim B. Stable quasimaps. Comm Korean Math Soc 2012;27:571–81.

[38] Kim B, Lee H. Wall-crossings for twisted quiver bundles. Int J Math 2013;24:16 pp, Article ID 1350038.

[39] Lafforgue L. Une compactification des champs classifiant les chtoucas de Drinfeld. J Am Math Soc 1998;11:1001–36.

[40] Lechtenfeld O, Popov AD, Szabo RJ. Rank two quiver gauge theory, graded connections and noncommutative vortices. J High Energ Phys 2006;054:46 pp.

[41] Li J, Yau S-T. Hermitian–Yang–Mills connections on non-Kähler manifolds. In: Mathematical aspects of string theory (San Diego, CA, 1986). Adv Ser Math Phys, vol. 1. Singapore: World Scientific Publishing, 1987:560–73.

[42] Li J, Yau S-T. The existence of supersymmetric string theory with torsion. J Diff Geom 2005;70:143–81.

[43] Lübke M, Teleman A. The Kobayashi–Hitchin correspondence. River Edge, NJ: World Scientific Publishing Co., Inc., 1995.

[44] Lübke M, Teleman A. The universal Kobayashi–Hitchin correspondence on Hermitian manifolds. Mem Am Math Soc 2006;183:vi+97 pp.

[45] Maruyama M. Moduli of stable sheaves II. J Math Kyoto 1978;18:557–614.

[46] Ngo Dac T. Compactification des champs de chtoucas et théorie géométrique des invariants. Astérisque 2007;313:124 pp.

[47] Nitsure N. Moduli space of semistable pairs on a curve. Proc London Math Soc (3) 1991;62:275–300.

[48] Qin Zh. Equivalence classes of polarizations and moduli spaces of sheaves. J Diff Geom 1993;37:397–415.

[49] Reid M. The moduli space of 3-folds with K = 0 may nevertheless be irreducible. Math Ann 1987;278:329–34.

[50] Schmitt AHW. Walls for Gieseker semistability and the Mumford–Thaddeus principle for moduli spaces of sheaves over higher dimensional bases. Comment Math Helv 2000;75:216–31.

[51] Schmitt AHW. Moduli for decorated tuples of sheaves and representation spaces for quivers. Proc Indian Acad Sci, Math Sci 2005;115:15–49.

[52] Schmitt AHW. Geometric invariant theory and decorated principal bundles. Zurich Lectures in Advanced Mathematics Zürich: European Mathematical Society, 2008 vii+389 pp.

[53] Schmitt AHW. A remark on semistability of quiver bundles. Eurasian Math J 2012;3:110–38.

[54] Schmitt AHW. Global boundedness for semistable decorated principal bundles with special regard to quiver sheaves. J Ramanujan Math Soc 2013;28A:443–90.

[55] Schmitt AHW. Semistability and instability in products and applications String-Math 2014, 201–14, Proc Sympos Pure Math., 93, Amer Math Soc. Providence, RI, 2016.

[56] Schmitt AHW. Stability parameters for quiver sheaves. 2017, 23 pp, hal-01699111.

[57] Schottenloher M. Geometrie und Symmetrie in der Physik. Leitmotiv der Mathematischen Physik. (German), Vieweg Lehrbuch Mathematische Physik, Braunschweig, 1995, xxii+408 pp.

[58] Strominger A. Superstrings with torsion. Nucl Phys B 1986;274:253–84.

[59] Szendroi B. Sheaves on fibered threefolds and quiver sheaves. Comm Math Phys 2008;278:627–41.

[60] Uhlenbeck K, Yau S-T. On the existence of Hermitian–Yang–Mills connections in stable vector bundles, in Frontiers of the mathematical sciences: 1985 (New York, 1985). Comm Pure Appl Math 1986;39(S):S257–93.

[61] Witten E. New issues in manifolds of SU(3) holonomy. Nucl Phys B 1986;268:79–112.

[62] Witten E. Strong coupling expansion of Calabi–Yau compactification. Nucl Phys B 1996;471:135–58.

[63] Witten E, Witten L. Large radius expansion of superstring compactification. Nucl Phys B 1987;281:109–26.

[64] Woodward Ch. Quantum Kirwan morphism and Gromov–Witten invariants of quotients I. Trans Group 2015;20:507–56.

[65] Woodward Ch. Quantum Kirwan morphism and Gromov–Witten invariants of quotients II. Trans Group 2015;20:881–920.

[66] Woodward Ch. Quantum Kirwan morphism and Gromov–Witten invariants of quotients III. Trans Group 2015;20:1155–93.

[67] Wu X, Witten L. Space-time supersymmetry in large radius expansion of superstring compactification. Nucl Phys B 1987;289:385–96.

Dirk Kreimer

Dyson–Schwinger equations: Fix-point equations for quantum fields

Abstract: We consider the structure of renormalizable quantum field theories from the viewpoint of their corresponding equations of motions for Wightman functions – the Dyson–Schwinger equations. These equations originate from fix-point equations in Hochschild cohomology. This overview gives an update on recent results.

Keywords: Hopf Algebras, renormalization, Hochschild Cohomology, fixed point equations, Dyson Schwinger equations, non-perturbative physics, gauge symmetry, Hopf ideals

Mathematics Subject Classification 2010: MSC2010: 81T15, 81T16, 81T17, 81T18, 16T05, 57T05

1 Introduction

The notion of locality, the structure of Dyson–Schwinger equations and number-theoretic properties of Green functions in quantum field theory (QFT) are intimately related. Here, we emphasize properties of Dyson–Schwinger equations which emerged since, for example, the review [22] was written. We do not include a discussion of algebra-geometric and number-theoretic aspects of field theory which emerged recently thanks to efforts of Francis Brown [6], Erik Panzer [30], Oliver Schetz [33] and others.

We start from a perturbative approach provided through the usual expansion in Feynman diagrams.

Physics beyond perturbation theory is governed by the equations of motion which the full Green functions have to fulfill. These systems of Dyson–Schwinger equations (DSEs) reflect the self-similarity of correlators or amplitudes in quantum field theory.

Such systems are driven by a skeleton expansion: the computation of propagation or interaction amplitudes proceeds by taking into account that the same propagation or interaction can happen in internal processes. Hence the notion of internal process demands objects which possess internal structure: the skeleton graphs of a theory. They are crucial in understanding at the same time the algebraic as well as the number-theoretic properties of a field theory.

The resulting fix-point equations for physics amplitudes illuminate how the algebraic structure of a perturbative expansion leads to non-perturbative solutions. Deriving such equations from algebraic structures is crucial in understanding local renormalization as well as the above self-similar structure of Green functions.

DOI 10.1515/9783110452150-004_s_001

The Collaborative Research Center 647 Space–Time–Matter in Berlin and Potsdam, which involved many scientists from Freie Universität and Humboldt Universität, provided a fruitful research environment to the author, through its seminar program and through partial support of own group seminar.

2 Algebraic structure in local QFT

In recent years, we have collected a considerable amount of algebraic structure underlying the computational practice of quantum field theory, clarifying the mathematical foundations of these computations and coming to new insights in QFT, see [1, 4, 5, 11, 19–21, 24–26, 28, 34–36, 38] and references there. In this section, we review the set-up which already was underlying [22, 25] which we follow.

2.1 Hopf and Lie algebras of graphs

There is a Hopf and Lie algebra coming with 1PI Feynman graphs Γ:

$$\Delta_{2n}(\Gamma) = \Gamma \otimes \mathbb{I} + \mathbb{I} \otimes \Gamma + \sum_{[\gamma]_{2n}} \gamma \otimes \Gamma/\gamma. \tag{1}$$

The sum is over all proper subsets $[\gamma]_{2n} \subset \Gamma$, which constitute disjoint unions $\gamma = \prod \gamma_i$ of 1PI graphs such that each γ_i fulfills $\omega_{2n}(\gamma_i) \geq 0$. Here,

$$\omega_{2n}(\Gamma) = 2n|\Gamma| - \sum_{\text{weights } w} w(\Gamma),$$

is the powercounting weight and we sum over edge- and vertex-weights w.

These Hopf algebras H include the Hopf algebra of renormalization for a theory renormalizable at $2n = D$ dimensions which then gives the forest formula of renormalization theory. With an antipode

$$S(\Gamma) = -\Gamma - \sum S(\gamma)\Gamma/\gamma,$$

we obtain renormalized Feynman rules Φ^R and counterterms S_R as

$$\Phi^R = m(S_R \otimes \Phi)\Delta, \; S_R = -R(S_R \otimes \Phi P)\Delta, \tag{2}$$

from unrenormalized Feynman rules Φ and a projection P into the augmentation ideal $P : H \to \text{Aug of } H$.

Such Hopf algebras H are dual to universal enveloping algebras $U(L)$ of suitable Lie algebras L, which arise from pre-Lie algebras of graph insertions.

Here is an example of the coproduct $\Delta = \Delta_4$ ion, the Hopf algebra of quantum electrodynamics for the sum of the two-loop vertex graphs $c_2^{\bar{\psi}A\psi}$ in four-dimensional quantum electrodynamics:

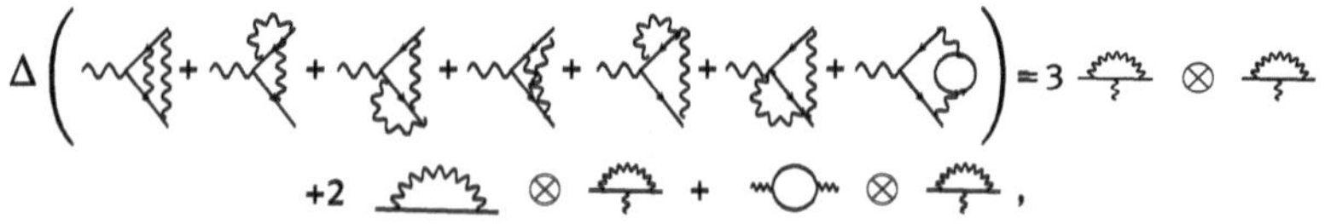

which reveals a sub-Hopf algebra structure.

2.2 Sub-Hopf algebras

Along with such Hopf algebras comes the corresponding Hochschild cohomology and the sub-Hopf algebras H_{grad}, generated by the grading and the quantum equations of motion. This was outlined for example in [25], from which we now quote the set-up.

Let c_k^r be the sum of all graphs contributing to an amplitude r at k loops

$$c_k^r = \sum_{\mathbf{res}(\Gamma)=r, |\Gamma|=k} \frac{1}{\mathrm{Aut}(\Gamma)} \Gamma. \tag{3}$$

The co-product Δ gives to these generators a sub-Hopf algebra structure: Let $\mathcal{R}$ be the set of amplitudes which need renormalization, and let H_{grad} be the Hopf algebra spanned by generators $\{\mathbb{I}, c_k^r\}$, $k \in \mathbb{N}$, $r \in \mathcal{R}$. The c_k^r form indeed a sub-Hopf algebra:

$$\Delta(c_k^r) = \sum_{j=0}^{k} P_{j,k}^s \otimes c_{k-j}^r, \tag{4}$$

with $P_{j,k}^s$ a polynomial in generators c_m^t of degree j.

Consider the core Hopf algebra which contains the renormalization Hopf algebra as a quotient Hopf algebra [23]. We are particularly interested in the structure of Green functions with respect to this Hopf algebra.

We introduce the set-up of [25]: Write $G^r \equiv G^r(\{Q\}, \{M\}, \{g\}; R)$ for a generic Green function, where

- r indicates the amplitude under consideration and we write $E \equiv |r|$ for its number of external legs. Amongst all possible amplitudes, there is a set of amplitudes provided by the free propagators and vertices of the theory. We write $\mathcal{R}$ for this set. It is in one-to-one correspondence with field monomials in a Lagrangian approach to field theory. The set of all amplitudes is denoted by $\mathcal{A} = \mathcal{F} \cup \mathcal{R}$, which defines $\mathcal{F}$ as those amplitudes only present through quantum corrections.
- $\{Q\}$ is the set of E external momenta q_j subject to the condition $\sum_{j=1}^{E} q_j = 0$.

- $\{M\}$ is the set of masses in the theory.
- $\{g\}$ is the set of coupling constants specifying the theory. Below, we proceed for the case of a single coupling constant g, the general case posing no principal new problems.
- R indicates the chosen renormalization scheme [23].

Each G^r can be obtained by the evaluation of a series of 1PI graphs

$$X^r(g) = \mathbb{I} - \sum_{E(\Gamma)\sim r} g^{|\Gamma|}\frac{\Gamma}{\mathrm{Sym}(\Gamma)}, \forall r \in \mathcal{R}, |r| = 2, \tag{5}$$

$$X^r(g) = \mathbb{I} + \sum_{E(\Gamma)\sim r} g^{|\Gamma|}\frac{\Gamma}{\mathrm{Sym}(\Gamma)}, \forall r \in \mathcal{R}, |r| > 2, \tag{6}$$

$$X^r(g) = \sum_{E(\Gamma)\sim r} g^{|\Gamma|}\frac{\Gamma}{\mathrm{Sym}(\Gamma)}, \forall r \notin \mathcal{R}, \tag{7}$$

where we take the minus sign for two-point functions, $|r| = 2$, and the plus sign for vertex functions, $|r| > 2$.

Furthermore, the notation $E(\Gamma) \sim r$ indicates a sum over graphs with external leg structure in accordance with r.

We write Φ, Φ_R for the unrenormalized and renormalized Feynman rules regarded as a map: $H \to \mathbb{C}$ from the Hopf algebra to $\mathbb{C}$.

We have

$$G^r_{t(r)} = \Phi_R(X^r(g))(\{Q\}, \{M\}, \{g\}; R), \tag{8}$$

where each non-empty graph is evaluated by the renormalized Feynman rules

$$\Phi_R(\Gamma) := (1 - R)m(S_R^\Phi \otimes \Phi P)\Delta(\Gamma) \tag{9}$$

and $\Phi_R(\mathbb{I}) = 1$, and P the projection into the augmentation ideal of H, and R the renormalization map.

It is in the evaluation (9) that the coproduct of the renormalization Hopf algebra appears.

The above sum over all graphs simplifies when one takes the Hochschild cohomology of the (renormalization) Hopf algebra into account:

$$X^r(g) = \delta_{r,\mathcal{R}}\mathbb{I} \pm \sum_{E(\gamma)\sim r;\Delta(\gamma)=\gamma\otimes\mathbb{I}+\mathbb{I}\otimes\gamma} \frac{1}{\mathrm{Sym}(\Gamma)} g^{|\gamma|} B_+^\gamma(X^r(g)Q(g)), \tag{10}$$

(– sign for $|r| = 2$, + sign for $|r| > 2$, $\delta_{r,\mathscr{R}} = 1$ for $r \in \mathscr{R}$, 0 else) with $Q(g)$ being the formal series of graphs assigned to an invariant charge of the coupling g:

$$Q^r(g) = \left[\frac{\overline{X}^{r,|r|>2}}{\prod_{e \in E(r)} \sqrt{X^e}} \right]^{\frac{1}{|r|-2}}, \tag{11}$$

where $\overline{X}^r = X^r$ for $r \in \mathscr{R}$ and $\overline{X}^r = X^r + \mathbb{I}$ else. Also, B_+^γ are *grafting operators* which are Hochschild cocycles.

The existence of a unique such invariant charge depends on the existence of suitable coideals in the renormalization Hopf algebra [25].

There is a tower of quotient Hopf algebras

$$H_4 \subset H_6 \cdots \subset H_{2n} \cdots H_{\text{core}} = H, \tag{12}$$

obtained by restricting the coproduct to sums over graphs which are superficially divergent in

$$D = 4, 6, \ldots, 2n, \ldots, \infty$$

dimensions. They are defined via a coproduct which restricts to superficially divergent graphs $\omega_D(\Gamma) \leq 0$ in an even number of dimension D greater than the critical dimension $D = 4$.

$$\Delta(\Gamma) = \Gamma \otimes 1 + 1 \otimes \Gamma + \sum_{\emptyset \subsetneq \gamma \subsetneq \Gamma; \omega_D} \gamma \otimes \Gamma/\gamma, \tag{13}$$

where ω_D restricts to disjoint unions $\gamma = \cup_i \gamma_i$, such that $\omega_D(\gamma_i) \leq 0$ for all γ_i.

The above algebraic structures given by this tower of Hopf algebras underly many familiar aspects of field theory. For example, in effective field theories, one consider couplings for any interaction in accordance with the symmetries of the theory, often suppressed by a scale which is large compared to the scales which are experimentally observable. The set $\mathscr{R}$ can then exhaust the full set $\mathscr{A}$. Still, the Hopf algebra renormalizing the corresponding Green functions is a quotient Hopf algebra of the core Hopf algebra, as some amplitudes might only demand counterterms from a suitably high loop number onwards.

Also, in operator product expansions we effectively enlarge the set $\mathscr{R}$ to contain any local amplitude which appears in the high-momentum Taylor expansion of a given amplitude in terms of local operator insertions, and the corresponding renormalizations in this expansion form again a quotient Hopf algebra of the core Hopf algebra.

Finally, for theories which obey a gravity power-counting, the core Hopf algebra becomes the renormalization Hopf algebra with a corresponding co-ideal structure [19].

Of special interest is the case when the system of DSEs is linear. Henry Kißler gave a comprehensive treatment in his master thesis [16]. Typically, the sub-Hopf algebra becomes co-commutative, in the dual the underlying Lie algebra becomes abelian, and Feynman rules lead to systems which can be solved by scaling solutions, reflecting the fact that the invariant charge becomes a constant, $Q(g) = cg\mathbb{I}$, $c \in \mathbb{C}$, and exhibits no renormalization group flow. This flow is indeed captured naturally from the underlying co-radical filtration of the Hopf algebra.

Linear or not, the co-radical filtration and Dynkin operators govern the renormalization group and leading log expansion: Indeed, in a leading-log expansion, terms $\sim \ln^j s$ (for s a suitable kinematical parameter) are obtained by the evaluation of

$$\sigma_k = \frac{1}{k!}\Phi^R m^{k-1}[S \otimes Y]^k \Delta^{k-1}. \tag{14}$$

This allows to turn DSEs to ordinary (non-)linear differential equations, see [10, 14, 28, 34, 35, 38].

Finally, we observe that there is a semi-direct product structure between the cocommutative Hopf algebra H_{ab} of superficially convergent amplitudes and the Hopf algebra H of amplitudes in $\mathcal{R}$ needing renormalization, which is very handy in the organization of those Dyson–Schwinger equations:

$$H_{\mathrm{full}} = H_{\mathrm{ab}} \times H, \tag{15}$$

reflecting the decomposition of amplitudes $\mathcal{A} = \mathcal{F} \cup \mathcal{R}$. This goes back to early work in the subject [8, 9], and is particularly useful when one tries to understand how physics amplitudes behave differently with regard to the dependence on scales and on scattering angles.

3 Kinematics as cohomology

Let us consider a chosen Green function G^r and let us identify a suitable variable s such that

$$G_R^r(g, s, \Theta) = 1 \pm s^{\omega_r} \gamma_j^r(\{g\}, \Theta, \Theta_0) \ln^j \frac{s}{s_0}, \tag{16}$$

where Θ, Θ_0 stand for sets of generalized angles $p_i \cdot p_j/s$, $p_i' \cdot p_j'/s$, with p_i' momenta which determine the renormalization point. We have

$$G_R^r = \Phi^R_{\ln s/s_0, \Theta, \Theta_0}(X^r) \tag{17}$$

for renormalized Feynman rules $\Phi^R_{\ln s,\Theta}$ and X^r a series which is a formal fix-point of the equation

$$X^r = \mathbb{I} \pm \sum_k g^k B_+^{r;k}(X^r Q^k). \tag{18}$$

Let us answer how the Hochshild cohomology of the perturbation expansion relates to the scale variations. We assume R is a kinematical subtraction scheme which renormalizes by subtraction at a fixed kinematical point. Such kinematical renormalization schemes are much preferred in our approach. We have

$$\Phi^R_{\ln s_1/s_0+\ln s_2/s_0,\Theta;\Theta_0} = \phi^R_{\ln s_1/s_0,\Theta;\Theta_0} \star \phi^R_{\ln s_2/s_0,\Theta;\Theta_0}. \tag{19}$$

As will be discussed elsewhere, this reveals an intimate connection to Tannaka categories and the representation theory of the additive group $\mathbb{G}_A$ [2].

More detailed work also reveals the behaviour under a variation of scattering angles [7]. We find

$$\Phi^R = \Phi^{-1}_{\text{fin}}(\Theta_0) \star \Phi^R_{\text{1-s}}(s/s_0) \star \Phi_{\text{fin}}(\Theta)$$

the decomposition of renormalized Feynman rules Φ^R into angle- and scale-variations in the very thorough analysis of [7].

As an interesting application, the six-loop β-function in Φ^4_4 theory was recently obtained by Kompaniets and Panzer [18] using this decomposition, pairing it with parametric integration via hyperlogarithms [30].

This approach to Green functions and Feynman rules also underlies recent work on Cutkosky rules [3]. >From there, unexpected connections to the work of Karen Vogtmann (see, e.g. [13]) emerge which are currently under investigation.

4 Symmetry

By now, the emergence of sub-Hopf algebras from fix-point equations in Hochschild cohomology is a well-studied problem. Typically, internal symmetries of a quantum field theory lead to the existence of co-ideals in the Hopf algebra, and the existence of Ward- or Slavnov–Taylor identities then implies that Feynman rules vanish on these ideals, so that we can work in a suitable quotient.

We refer in particular to [11, 21, 24, 36] for more details including detailed studies of gauge theories.

A crucial observation is that co-ideals in the Hopf algebra of perturbation theory correspond to symmetries in the Lagrangian: Internal symmetries are in accordance with the Hochschild cohomology structure of these Hopf algebras. For example, the

Ward identity in quantum electrodynamivs is equivalent to the statement that renormalized Feynman rules can be defined on the quotient H/I, where I is the ideal and co-ideal given by $c_k^{\bar{\psi}A\psi} + c_k^{\bar{\psi}\psi} = 0$.

For non-abelian gauge theories, this leads to the remarkable fact that amplitudes in scalar field theory for a theory with 3-regular vertices generate the full gauge theory amplitudes, including ghost sectors and 4-valent interactions. In [24], see also [32], the authors review quantization of gauge fields using algebraic properties of 3-regular graphs.

The Feynman integrand at n loops for a non-abelian gauge theory quantized in a covariant gauge is generated from scalar integrands for connected 3-regular graphs, obtained from the two Symanzik polynomials.

The transition to the full gauge theory amplitude is obtained by the use of a third, new, graph polynomial, the corolla polynomial [27]. All the relevant signs of the ghost sector are incorporated in a double complex furnished by the corolla polynomial – we call it cycle homology – and by graph homology. These homologies illuminate the interplay between the colour sector and the kinematics of Green functions and hence shed light on topics such as colour-kinematics dualities [12].

For an abelian gauge theory, Henry Kißler exhibited all the details of a Hopf algebraic approach in [15]. He worked in the context of massless quantum electrodynamics with a linear covariant gauge fixing. The coproduct formula of Green's functions reveals two invariant charges, which give rise to different renormalization group functions for the transversal and longitudinal couplings in a very conceptual manner. In [17] this was used to obtain a DSE for the gauge dependence of the fermion propagator in quantum electrodynamics quantized in a linear covariant gauge.

5 Non-perturbative aspects

As we saw, Dyson-Schwinger equations determine the Green functions $G^r(\alpha, L)$ in quantum field theory. Their solutions are known to be triangular series in a coupling constant α and an external scale parameter L for a chosen amplitude r, with the order in L bounded by the order in the coupling.

Here, we mention recent progress with non-perturbative physics.

The leading-log expansion gives a series in the combined variable $(\alpha \ln s/s_0)$, whilst the much harder to get non-leading log expansions were traced back to differential equations in [28], making full use of the Hopf as well as Lie algebraic properties of a perturbative expansion in Feynman graphs. The results are in accordance with recent results of Courtiel and Yeats [10]. The latter are themselves based on the Mellin transform and rooted chord diagram approach of Yeats and collaborators [14, 29, 38].

In [4], the structure of overlapping subdivergences is analysed using algebraic lattice theory for partial ordered sets. Typically the sets of subdivergences of Feynman diagrams form algebraic lattice. In kinematic renormalization schemes, in which tadpole diagrams vanish, these lattices are semimodular. Hence the Hopf algebra of Feynman diagrams is graded by the coradical degree. As an application, a formula for the counterterms in zero-dimensional QFT is given leading to the enumeration of primitive or skeleton diagrams long sought after.

In [5], the algebraic properties of certain formal power series with factorial growth were studied. They form a subring of $R[[x]]$ which is closed under composition. An "asymptotic derivation" is defined which maps a power series to its asymptotic expansion. Leibniz and chain rules for this derivation are deduced. As an application the full asymptotic expansions of the number of connected chord diagrams and the number of simple permutations are given, with more results upcoming. These type of results are Taylor-made to be combined with estimates on the growth of individual Feynman diagrams as suggested by Panzer recently [31].

Finally, let us mention [26]. Starting from free field theories, a non-linear field diffeomorphisms was applied. For the first finite orders of perturbation theory, it was found that tree-level amplitudes for the transformed fields must satisfy recursion relations for the S-matrix of Britto–Cachazo–Feng–Witten type to remain trivial. A necessary condition for the Feynman rules to respect the maximal ideal and co-ideal defined by the core Hopf algebra of the transformed theory is that upon renormalization all massive tadpole integrals (defined as all integrals independent of the kinematics of external momenta) are mapped to zero in accordance with a kinematic renormalization scheme. This result can be established to all orders rigorously and also in the context of interacting QFT [37], allowing to establish diffeomorphism invariance (a property taken for granted somewhat carelessly in manipulations of the path-integral) of QFT at least in such renormalization schemes as preferred from an algebraic geometry viewpoint anyhow.

In this context, let us note that putting matter fields in the adjoint representation, it is easy to see that even for non-abelian gauge theories one can construct a linear covariant gauge in which Slavnov-Taylor identities for counterterms take the form

$$1 = \frac{Z^{\bar{\psi}A\psi}}{Z^{\bar{\psi}\psi}} = \frac{Z^{AAA}}{Z^{AA}} = \frac{Z^{AAAA}}{Z^{AAA}} = \frac{Z^{\bar{g}Ag}}{Z^{\bar{g}g}}.$$

In integrable field theories in such a gauge, one has a vanishing anomalous dimension for the gauge field A, as a consequence of a vanishing beta function. This implies a vanishing forward decay of the gauge particle into anything. It is a tempting thought that this indicates the existence of a field superdiffeomorphism which renders an integrable field theory unitarily equivalent to a free theory.

Bibliography

[1] Bergbauer C, Kreimer D. Hopf algebras in renormalization theory: Locality and Dyson-Schwinger equations from Hochschild cohomology. IRMA Lect Math Theor Phys 2006;10:133. DOI:10.4171/028-1/4, hep-th/0506190.

[2] Bloch S, Kreimer D, Yeats K. in preparation.

[3] Bloch S, Kreimer D. Cutkosky rules and outer space, arXiv:1512.01705, hep-th.

[4] Borinsky M. Algebraic lattices in QFT renormalization. Lett Math Phys 2016;106(7):879. DOI:10.1007/s11005-016-0843-9, arXiv:1509.01862, hep-th.

[5] Borinsky M. Generating asymptotics for factorially divergent sequences. 2016, Available at: http://arxiv.org/abs/1603.01236.

[6] Brown F. Periods and Feynman amplitudes. 2015, arXiv:1512.09265, math-ph.

[7] Brown F, Kreimer D. Angles, scales and parametric renormalization. Lett Math Phys 2013;103:933. DOI:10.1007/s11005-013-0625-6, arXiv:1112.1180, hep-th.

[8] Connes A, Kreimer D. Renormalization in quantum field theory and the Riemann-Hilbert problem. I: The Hopf algebra structure of graphs and the main theorem. Comm Math Phys 2000;210:249, arXiv:hep-th/9912092.

[9] Connes A, Kreimer D. Renormalization in quantum field theory and the Riemann-Hilbert problem. II: The beta-function, diffeomorphisms and the renormalization group. Comm Math Phys 2001;216:215. hep-th/0003188.

[10] Courtiel J, Yeats K. Terminal chord in connected chord diagrams. Available at: http://arxiv.org/abs/1603.08596. 10.

[11] Foissy L. Mulitgraded Dyson-Schwinger systems. 2015, arXiv:1511.06859, math.RA.

[12] Fu CH, Krasnov K. 2016, arXiv:1603.02033, hep-th.

[13] Hatcher A, Vogtmann K. Rational Homology of Aut(Fn). Math Res Lett 1998;5:759–80.

[14] Hihn M, Yeats K. Generalized chord diagram expansions of Dyson-Schwinger equations. 2016, arXiv:1602.02550, math-ph.

[15] Kißler H. Hopf-algebraic Renormalization of QED in the linear covariant Gauge. Ann Phys 2016;372:159. DOI:10.1016/j.aop.2016.05.008, arXiv:1602.07003, hep-th.

[16] Kißler H. On linear systems of Dyson Schwinger equations, Master Thesis. 2012, MaPhy-AvH/2012-15.

[17] Kißler H, Kreimer D. Diagrammatic cancellations and the gauge dependence of QED. Phys Lett 2017;B764:318–321. DOI:10.1016/j.physletb.2016.11.052.

[18] Kompaniets M, Panzer E. Renormalization group functions of $\varphi 4$ theory in the MS-scheme to six loops. 2016, arXiv:1606.09210, hep-th.

[19] Kreimer D. A remark on quantum gravity. Ann Phys 2008;323:49. DOI:10.1016/j.aop.2007.06.005, arXiv:0705.3897, hep-th.

[20] Kreimer D. Algebraic structures in local QFT. Nucl Phys Proc Suppl 2010;205–6 :122. DOI:10.1016/j.nuclphysbps.2010.08.030, arXiv:1007.0341, hep-th.

[21] Kreimer D. Anatomy of a gauge theory. Ann Phys 2006;321:2757–81. DOI:10.1016/j.aop.2006.01.004, hep-th/0509135.

[22] Kreimer D. Dyson Schwinger equations: from Hopf algebras to number theory. Fields Inst Comm 2007;50:225, hep-th/0609004.

[23] Kreimer D. The core Hopf algebra. Clay Math Proc 2010;11:313, arXiv:0902.1223, hep-th.

[24] Kreimer D, Sars M, van Suijlekom WD. Quantization of gauge fields, graph polynomials and graph homology. Ann Phys 2013;336:180. DOI:10.1016/j.aop.2013.04.019, arXiv:1208.6477, hep-th.

[25] Kreimer Dvan Suijlekom WD. Recursive relations in the core Hopf algebra. Nucl Phys B 2009;820:682. DOI:10.1016/j.nuclphysb.2009.04.025, arXiv:0903.2849, hep-th.

[26] Kreimer D, Velenich A. Field diffeomorphisms and the algebraic structure of perturbative expansions. Lett Math Phys 2013;103:171. DOI:10.1007/s11005-012-0589-y, arXiv:1204.3790, hep-th.
[27] Kreimer D, Yeats K. Properties of the corolla polynomial of a 3-regular graph, arXiv:1207.5460, math.CO. Electron J Combin 2013;20(1):41.
[28] Krüger O, Kreimer D. Filtrations in DysonSchwinger equations: Next-to^j-leading log expansions systematically. Ann Phys 2015;360:293. DOI:10.1016/j.aop.2015.05.013, arXiv:1412.1657, hep-th.
[29] Marie N, Yeats K. A chord diagram expansion coming from some Dyson-Schwinger equations. Comm Num Theory Phys 2013;07:251. DOI:10.4310/CNTP.2013.v7.n2.a2, arXiv:1210.5457, math.CO.
[30] Panzer E. Feynman integrals and hyperlogarithms. 2015, arXiv:1506.07243, math-ph.
[31] Panzer E. The Hepp bound a rational period, talk given at Humboldt University, 13 June 2016.
[32] Sars M. Parametric representation of Feynman amplitudes in gauge theories, Thesis. Humboldt University. 2015, MaPhy-AvH/2015-01.
[33] Schnetz O. Numbers and functions in quantum field theory. 2016, arXiv:1606.08598, hep-th.
[34] van Baalen G, Kreimer D, Uminsky D, Yeats K. The QED beta-function from global solutions to Dyson-Schwinger equations. Ann Phys 2009;324:205. DOI:10.1016/j.aop.2008.05.007, arXiv:0805.0826, hep-th.
[35] van Baalen G, Kreimer D, Uminsky D, Yeats K. The QCD beta-function from global solutions to Dyson-Schwinger equations. Ann Phys 2010;325:300. DOI:10.1016/j.aop.2009.10.011, arXiv:0906.1754, hep-th.
[36] van Suijlekom W. The Hopf algebra of Feynman graphs in QED. Lett Math Phys 2006;77:265, arXiv:hep-th/0602126.
[37] van Suijlekom WD. The structure of renormalization Hopf algebras for gauge theories. I: representing Feynman graphs on BV-algebras. Commun Math Phys 2009;290:291. DOI:10.1007/s00220-009-0829-x, arXiv:0807.0999, math-ph.
[38] Kreimer D, Yeats K. Diffeomorphisms of quantum fields. Math Phys Ann Geom 2017;20(2):16. DOI:10.1007/s11040-017-9246-0, arXiv:1610.01837, math-ph.
[39] Yeats KA. Growth estimates for Dyson-Schwinger equations, arXiv:0810.2249, math-ph.

Dhritiman Nandan and Gang Yang

Hidden structure in the form factors of $\mathcal{N} = 4$ SYM

Abstract: On-shell techniques have been the cornerstone of tremendous progress in our understanding of quantum field theory over the last couple of decades. Initially such ideas were only applied to computing scattering amplitudes but in recent years they have also been extended in computing and finding novel structure in off-shell quantities like form factors. As a close cousin of quantum chromodynamics (QCD), $\mathcal{N} = 4$ super Yang-Mills (SYM) theory has been at the center of much of the activity in the above-mentioned on-shell-based approach. On the other hand, in recent years various other aspects of this theory has also been subject to intense investigation, particularly in the framework of the gauge/string duality. Notably, the one-loop dilatation operator of this theory was identified with the Hamiltonian of an integrable spin chain which has lead to non-perturbative solutions of anomalous dimensions based on the conjectured full integrability in the planar limit. In this note, we will focus on the form factors of general gauge invariant operators in $\mathcal{N} = 4$ SYM. Using the on-shell techniques we present novel methods to compute the dilatation operator of this theory. We also present new on-shell diagrams and Grassmannian formulation of form factors, as well as study finite physical observables from form factors.

Keywords: Form factors, maximally supersymmetric Yang-Mills theory, scattering amplitudes, on-shell, Grassmannian, IR safe observables, energy–energy correlation, integrability, dilatation operator, BPS operator

Mathematics Subject Classification 2010: 81T18, 81T13, 81Q30, 81T60, 81T40, 81Q60, 81Q80, 70S10, 53C28

1 Introduction

In recent years, the resurgence in the study of scattering amplitudes in gauge theories, driven by the new knowledge of on-shell- and unitarity-based techniques, has led to a lot of progress in understanding the structure of scattering amplitudes. The very interesting facet of this approach is that this both has been a way to probe formal aspects of quantum field theories and also been a great boon in finding new computational techniques that would assist in decoding the signals in the collider experiments. The Large Hadron Collider (LHC) is one of the most ambitious endeavors to unravel the mysteries of fundamental interactions. It is expected to fill in the missing pieces in the Standard Model as well as push the frontiers of our knowledge and throw some light on the new

DOI 10.1515/9783110452150-004_s_002

physics beyond the Standard Model. In order to achieve this goal, one needs a solid understanding of the quantum chromodynamics (QCD) processes forming the background to new physics, While scattering amplitudes in QCD are notoriously difficult to compute, $\mathcal{N} = 4$ super Yang-Mills (SYM) theory with gauge group $SU(N_c)$ in four dimensions has played an important role in the aforementioned developments. This theory shares many properties with QCD but the maximal supersymmetry imbues it with many simplifying structures.

On the other hand, according to the AdS/CFT correspondence discovered by Maldacena [45], $\mathcal{N} = 4$ SYM has a dual description in terms of a string theory, allowing its study also at strong coupling. Moreover, in the planar limit first studied by 't Hooft [37], it shows signs of integrability at weak as well as at strong coupling, which is believed to be present even at any coupling. Based on the conjectured integrability, new predictions for the spectrum, i.e. for the anomalous scaling dimensions of gauge-invariant composite operators, were made; see e.g. Beisert et al. [8] for a review. This rises the hope that the theory is exactly solvable, and it is hence sometimes even referred to as the "harmonic oscillator of the 21st century".

Given the success of the aforementioned on-shell techniques for amplitudes, it is an intriguing question whether they can be applied for determining off-shell quantities such as anomalous dimensions or more generally correlation functions as well. A bridge between the purely on-shell amplitudes and the purely off-shell correlation functions is provided by *form factors*. In particular, they also contain the information necessary to determine the anomalous dimensions. An n-point form factor describes the overlap of an off-shell initial state, described by a composite operator, into an on-shell final state consisting of n elementary fields. It is given by

$$\mathcal{F}_{n,\mathcal{O}} = \int d^D x e^{-iq\cdot x} \langle 1 \cdots n|\mathcal{O}(x)|0\rangle = (2\pi)^D \delta^{(D)}\Big(q - \sum_{i=1}^{n} p_i\Big)\langle 1 \cdots n|\mathcal{O}(0)|0\rangle, \tag{1}$$

where the particles labeled by $i = 1, \ldots, n$ carry individual on-shell momenta p_i and the operator $\mathcal{O}$ carries off-shell momentum q.

In $\mathcal{N} = 4$ SYM theory, the most intensively studied form factors are the ones of the half-BPS operator which belongs to the stress-tensor supermultiplet. Its Sudakov form factor was first studied by van Neerven [53]. The form factors of the stress-tensor multiplet with general n external legs can be analyzed in analogy to scattering amplitudes with modern on-shell techniques. The n-point form factor was first studied in Brandhuber et al. [18], Bork et al. [15] and later generalized to the full stress-tensor multiplet in Brandhuber et al. [17] and Bork et al. [16]. There have been extended study of form factors at loop level, see e.g. Boels et al. [14] and references therein. Form factors have also been studied at strong coupling via the AdS/CFT correspondence by Alday and Maldacena [1], and a Y-system formulation was given in Maldacena and Zhiboedov [46] for AdS_3 and in Gao and Yang [33] for AdS_5.

The aforementioned studies have shown that form factors share very similar recursive and analytic properties with scattering amplitudes, at least for the protected operators. Moreover, the robust set of on-shell techniques for computing on-shell objects is also applicable in this case. This rises the hope that fully off-shell quantities may be studied using on-shell methods, and that such an enhancement of the toolkit allows us to detect new features of the theory.

The study of form factors with nonprotected operators as well as their connection between form factors and the spectral problem of $\mathcal{N} = 4$ SYM was recently pushed forward in Wilhelm [54], Nandan et al. [51] and Loebbert et al. [43]. In Wilhelm [54], form factors for generic operators were investigated. It was shown that the complete one-loop dilatation operator first derived by Beisert [6] can be derived using one-loop minimal form factors, which explains the relation between the one-loop dilatation operator and the four-point scattering amplitude derived from symmetry in Zwiebel [55]. In Nandan et al. [51] and Loebbert et al. [43], it was demonstrated that form factors can also be used to calculate anomalous dimensions at two-loop order by investigating the Konishi primary operator as well as the operators in the $SU(2)$ sector. In these studies, on-shell amplitude techniques have played a major role, in particular the (generalized) unitarity method developed in Bern et al. [11, 12] and Britto et al. [23].

Interesting on-shell approaches towards the computation of correlation functions and the dilatation operator were also applied in the following works: see Engelund and Roiban [31] and Brandhuber et al. [21] for the application of generalized unitarity, Engelund [30] for a spacetime version thereof, Koster et al. [39–41], Chicherin et al. [25], Chicherin and Sokatchev [26, 27] for twistor techniques and Brandhuber et al. [20] for the application of maximally helicity violating (MHV) diagrams.

This report is structured as follows. In Section 2, we give a brief review of spinor helicity formalism and some facts of form factors. Section 3 discusses the formulation of any local operators in terms of minimal form factors and how to obtain dilatation operator from form factors via unitarity method. In Section 4, we discuss the construction of IR finite observables from form factors. Section 5 introduces on-shell diagrams for form factor as well as its Grassmannian formulation.

2 Review

Here we review some of the key aspects of the modern on-shell amplitude techniques that would be relevant for our discussion in this note. A more comprehensive review of this subject can be found in Elvang and Huang [29], Henn and Plefka [35].

Let us first introduce the spinor helicity formalism. Massless four-momentum can be written in a bi-spinor form as

$$p_i^{\alpha\dot{\alpha}} = \lambda_i^{\alpha}\tilde{\lambda}_i^{\dot{\alpha}}, \tag{2}$$

where $\lambda, \tilde{\lambda}$ are two Weyl spinors. The Lorentz product of four-momenta is given in terms of spinor contraction as

$$2p_i \cdot p_j = \langle ij\rangle[ji], \qquad \langle ij\rangle = \epsilon_{\alpha\beta}\lambda_i^\alpha\lambda_j^\beta, \quad [ji] = \epsilon_{\dot\alpha\dot\beta}\tilde\lambda_i^{\dot\alpha}\tilde\lambda_j^{\dot\beta}\,. \tag{3}$$

Furthermore, polarization vectors of gluons can be expressed in the bi-spinor representation as

$$\varepsilon_{+,i}^{\alpha\dot\alpha} = \frac{\xi_i^\alpha\tilde\lambda_i^{\dot\alpha}}{\langle\xi_i\lambda_i\rangle}, \qquad \varepsilon_{-,i}^{\alpha\dot\alpha} = \frac{\lambda_i^\alpha\tilde\xi_i^{\dot\alpha}}{[\lambda_i\xi_i]}, \tag{4}$$

where $\xi, \tilde\xi$ are *arbitrary* reference spinors.

$\mathcal{N} = 4$ SYM has PSU(2,2|4) super-spacetime symmetry. Similar to on-shell momentum, we can write on-shell supermomentum $q_i^{\alpha A}$ of the ith external particle as

$$q_i^{\alpha A} = \lambda_i^\alpha\eta_i^A, \tag{5}$$

where η_i^A's are Grassmann variables and $A = 1,\ldots,4$ is the $SU(4)$ R-symmetry index. Together with the conjugate supermomentum $\bar q_A^{\dot\alpha} = \tilde\lambda^{\dot\alpha}\frac{\partial}{\partial\eta^A}$, they give the supersymmetry anti-commutation relation $\{q^{\alpha A}, \bar q_B^{\dot\alpha}\} = \delta_B^A p^{\alpha\dot\alpha}$.

The external on-shell states can be described using the $\mathcal{N} = 4$ on-shell superfield introduced by Nair [50]:

$$\Phi(p,\eta) = g_+(p) + \eta^A\,\bar\psi_A(p) + \frac{\eta^A\eta^B}{2!}\,\phi_{AB}(p) + \frac{\varepsilon_{ABCD}\eta^A\eta^B\eta^C}{3!}\,\psi^D(p) + \eta^1\eta^2\eta^3\eta^4\,g_-(p). \tag{6}$$

Note that η encodes the flavor and helicity of the component particles. The power of this formalism is that one can combine amplitudes with different external fields into one super amplitudes. For example, the super MHV amplitudes can be given as

$$\mathcal{A}_n^{\mathrm{MHV}} = \frac{\delta^{(4)}(\sum_i\lambda_i\tilde\lambda_i)\delta^{(8)}(\sum_i\lambda_i\eta_i^A)}{\langle12\rangle\langle23\rangle\ldots\langle n-1n\rangle\langle n1\rangle}\,. \tag{7}$$

Form factors, despite being partially off-shell, can take the advantage of the powerful on-shell techniques. In particular, it was found that the MHV form factor with chiral supermultiplet takes the surprisingly simple form as MHV amplitudes, given in Brandhuber et al. [17],

$$\mathcal{F}_{\mathcal{T},n}^{\mathrm{MHV}} = \frac{\delta^{(4)}(q-\sum_i\lambda_i\tilde\lambda_i)\delta^{(8)}(\gamma-\sum_i\lambda_i\eta_i^A)}{\langle12\rangle\langle23\rangle\ldots\langle n-1n\rangle\langle n1\rangle}\,. \tag{8}$$

Because the tree-level results are building blocks of loop corrections, this further indicates the simplicity of form factors at loop level.

As amplitudes, the full form factor including color factors can be given as

$$\hat{\mathcal{F}}^{(\ell)}_{\mathcal{O},n}(\{a_i, p_i, \eta_i\}) = g_{\rm YM}^{n-2} \sum_{\sigma \in S_n/Z_n} \operatorname{tr}(\mathrm{T}^{a_{\sigma(1)}} \cdots \mathrm{T}^{a_{\sigma(n)}}) \mathcal{F}^{(\ell)}_{\mathcal{O},n}(\{p_{\sigma(i)}, \eta_{\sigma(i)}\}) + \text{multi-trace terms}, \quad (9)$$

where T^a, $a = 1, \ldots, N_c^2 - 1$ are the gauge-group generators of $SU(N_c)$ and $\mathcal{F}_{\mathcal{O},n}$ is the color-stripped form factor.

Since form factors contain “off-shell” information, i.e. the associated operators, it serves as a bridge to bring powerful on-shell techniques to compute off-shell quantities, such as the dilatation operator. This is the topic of next section.

3 Form factors and dilatation operator

3.1 Minimal tree-level form factors and local operators

Local gauge invariant operators can be constructed as trace of products of covariant fields:

$$\mathcal{O}(x) = \operatorname{Tr}\left(\mathcal{W}_1^{(m_1)} \mathcal{W}_2^{(m_2)} \ldots \mathcal{W}_n^{(m_n)}\right)(x), \quad (10)$$

where the covariant fields $\mathcal{W}_i$ can be any of the following field[1]

$$\mathcal{W}_i \in \left\{\phi^{AB}, F^{\alpha\beta}, \bar{F}^{\dot{\alpha}\dot{\beta}}, \bar{\psi}^{\dot{\alpha}A}, \psi^{\alpha ABC}\right\}, \quad (11)$$

which under gauge transformation behave as $\mathcal{W} \to U\mathcal{W}U^\dagger$. Besides, each covariant field can be dressed with covariant derivative

$$\mathcal{W}^{(m)} := D^m \mathcal{W}, \qquad D_{\alpha\dot{\alpha}}\mathcal{W} = \partial_{\alpha\dot{\alpha}}\mathcal{W} - i g_{\rm YM}[A_{\alpha\dot{\alpha}}, \mathcal{W}]. \quad (12)$$

Note that different D's can carry different Lorentz indices.

The above construction of operators have redundancy. Due to $[D_\mu, D_\nu] = F_{\mu\nu}$, it is enough to consider the symmetric product of D_μ acting on any field. Furthermore, using equations of motions and Bianchi identities such as

$$D^\mu F_{\mu\nu} = 0, \quad \text{or} \qquad D^\mu D_\mu \phi = \text{non-derivative terms}, \quad (13)$$

one can get rid of these covariant derivative terms.

A basis of independent operator can be chosen such that spinor indices $\alpha, \dot{\alpha}$ are totally symmetric while R-charge indices are totally antisymmetric. Such a basis can

---

1 We use the same notation for scalars and fermions as in eq. (6) for on-shell states, but one should understand the fields in the operator are the same as fields in the Lagrangian which are off-shell.

be represented by states of a supersymmetric Harmonic oscillator as in Gunaydin and Marcus [34]. For the fields in the operators, there is the following correspondence:

$$\begin{array}{ccc}\bar{F}^{\dot{\alpha}\dot{\beta}} & \longrightarrow & \mathrm{b}^{\dagger\dot{\alpha}}\mathrm{b}^{\dagger\dot{\beta}}|0\rangle \\ \bar{\psi}^{\dot{\alpha}A} & \longrightarrow & \mathrm{b}^{\dagger\dot{\alpha}}\mathrm{d}^{\dagger A}|0\rangle \\ \phi^{AB} & \longrightarrow & \mathrm{d}^{\dagger A}\mathrm{d}^{\dagger B}|0\rangle \\ \psi^{\alpha ABC} & \longrightarrow & \mathrm{a}^{\dagger\alpha}\mathrm{d}^{\dagger A}\mathrm{d}^{\dagger B}\mathrm{d}^{\dagger C}|0\rangle \\ F^{\alpha\beta} & \longrightarrow & \mathrm{a}^{\dagger\alpha}\mathrm{a}^{\dagger\beta}\mathrm{d}^{\dagger 1}\mathrm{d}^{\dagger 2}\mathrm{d}^{\dagger 3}\mathrm{d}^{\dagger 4}|0\rangle \\ D^{\alpha\dot{\alpha}} & \longrightarrow & \mathrm{a}^{\dagger\alpha}\mathrm{b}^{\dagger\dot{\alpha}}|0\rangle \end{array} .$$

An analogy between the oscillator and (super) spinor helicity variables was observed in Beisert [5] as

$$\mathrm{a}^{\dagger} \sim \lambda\,, \mathrm{b}^{\dagger} \sim \tilde{\lambda}\,, \mathrm{d}^{\dagger} \sim \eta\,. \tag{14}$$

The correspondence was put into practical use by Zwiebel [55]. The connection of the picture to form factors was pointed out later by Wilhelm [54]. The new correspondence can be formulated as

$$\begin{array}{ccc}\bar{F}^{\dot{\alpha}\dot{\beta}} & \xrightarrow{g_+} & \tilde{\lambda}^{\dot{\alpha}}\tilde{\lambda}^{\dot{\beta}} \\ \bar{\psi}^{\dot{\alpha}A} & \xrightarrow{\bar{\psi}_{\dot{\alpha}A}} & \tilde{\lambda}^{\dot{\alpha}}\eta^{A} \\ \phi^{AB} & \xrightarrow{\phi_{AB}} & \eta^{A}\eta^{B} \\ \psi^{\alpha ABC} & \xrightarrow{\psi_{\alpha ABC}} & \lambda^{\alpha}\eta^{A}\eta^{B}\eta^{C} \\ F^{\alpha\beta} & \xrightarrow{g_-} & \lambda^{\alpha}\lambda^{\beta}\eta^{1}\eta^{2}\eta^{3}\eta^{4} \\ D^{\alpha\dot{\alpha}} & \longrightarrow & \lambda^{\alpha}\tilde{\lambda}^{\dot{\alpha}} \end{array} .$$

Note that $F_{\alpha\beta}$ corresponds to out-going negative helicity gluon state, and $F_{\dot{\alpha}\dot{\beta}}$ to out-going positive helicity gluon state. This can be seen as follows. First, the field strength $F_{\mu\nu}$ can be decomposed as self-dual and anti-self-dual parts:

$$F_{\mu\nu} \to \epsilon_{\dot{\alpha}\dot{\beta}}F_{\alpha\beta} + \epsilon_{\alpha\beta}\bar{F}_{\dot{\alpha}\dot{\beta}}\,, \tag{15}$$

or more explicitly

$$F_{\alpha\beta} = \epsilon^{\dot{\alpha}\dot{\beta}}\left(\partial_{\alpha\dot{\alpha}}A_{\beta\dot{\beta}} - \partial_{\beta\dot{\beta}}A_{\alpha\dot{\alpha}}\right), \qquad F_{\dot{\alpha}\dot{\beta}} = \epsilon^{\alpha\beta}\left(\partial_{\alpha\dot{\alpha}}A_{\beta\dot{\beta}} - \partial_{\beta\dot{\beta}}A_{\alpha\dot{\alpha}}\right). \tag{16}$$

In the form factor picture, after Wick contraction and LSZ reduction, $A_{\alpha\dot{\alpha}}$ in eq. (16) is effectively replaced by polarization vectors. Recall the polarization vectors given in eq. (4) and replacing $\partial_{\alpha\dot{\alpha}}$ by $\lambda_\alpha\tilde{\lambda}_{\dot{\alpha}}$, we have

$$F_{\alpha\beta} \to \epsilon^{\dot{\alpha}\dot{\beta}}\left(\lambda_\alpha\tilde{\lambda}_{\dot{\alpha}}\varepsilon^{(-)}_{\beta\dot{\beta}} - \lambda_\beta\tilde{\lambda}_{\dot{\beta}}\varepsilon^{(-)}_{\alpha\dot{\alpha}}\right) \propto \lambda_\alpha\lambda_\beta\,, \tag{17}$$

which reproduce the relation given in eq. (14). Similar argument applies to scalar and fermion fields. See also [54].

Apply the above correspondence to local operators, one gets for example

$$\mathrm{tr}(F_{\alpha\beta}F^{\alpha\beta}) \to \lambda_1^\alpha \lambda_1^\beta \lambda_{2\alpha}\lambda_{2\beta}(\eta_1)^4(\eta_2)^4 = \langle 1\,2\rangle^2(\eta_1)^4(\eta_2)^4\,, \tag{18}$$

$$\mathrm{tr}(\bar{F}_{\dot\alpha}^{\ \dot\beta}\bar{F}_{\dot\beta}^{\ \dot\gamma}\bar{F}_{\dot\gamma}^{\ \dot\alpha}) \to \tilde\lambda_1^{\dot\alpha}\tilde\lambda_{1\dot\beta}\tilde\lambda_2^{\dot\beta}\tilde\lambda_{2\dot\gamma}\tilde\lambda_3^{\dot\gamma}\tilde\lambda_{3\dot\alpha} = [1\,2][2\,3][3\,1]\,. \tag{19}$$

The right-hand side are exactly the result of corresponding *minimal* form factors with the given operator.[2, 3]

This correspondence provides a useful "translator" to change any off-shell local operator to on-shell data, which is a minimal form factor. Although the form factors are the minimal cases, on-shell recursion relations may be applied to non-minimal form factors. Furthermore, they provide building blocks for computing loop level form factors via unitarity cut method, which will be studied in the next section.

3.2 Dilatation operator from form factors

In this subsection, we consider how to use form factors to obtain dilation operator. The basic idea is straightforward. Given the renormalization matrix $\mathcal{Z}$ defined in the renormalized local operator

$$\mathcal{O}^I_{\text{ren}} = \mathcal{Z}^I_J\, \mathcal{O}^J_{\text{bare}}\,, \tag{20}$$

the dilatation operator can be obtained as

$$\delta\mathfrak{D} = \frac{\mathrm{d}}{\mathrm{d}\mu}\log\mathcal{Z} = 2\varepsilon g^2\frac{\partial}{\partial g^2}\log\mathcal{Z} = \sum_{\ell=1}^{\infty} g^{2\ell}\mathfrak{D}^{(\ell)}\,. \tag{21}$$

We use a modified minimal subtraction scheme with coupling constant $g^2 = (4\pi e^{-\gamma_E})^\varepsilon \frac{g^2_{\mathrm{YM}}N_c}{(4\pi)^2}$.

The problem is then reduced to computing the renormalization matrix. This can be extracted from the ultraviolet (UV) divergence of two-point functions as in the usual strategy. Here we will consider form factors. The advantage is that powerful on-shell methods can be applied efficiently.

Let us mention that the crucial observation introducing integrability to planar $\mathcal{N} = 4$ SYM theory was that the one-loop anomalous dilatation operator takes the

---

2 If the number n of the external fields exactly match the number of fields contained in $\mathcal{O}$, the form factor is called *minimal*.

3 Note that for operators the cyclicity is automatic, while for form factor a cyclic sum is also required.

form of the integrable Heisenberg spin-chain Hamiltonian within the SO(6) sector, discovered by Minahan and Zarembo [48]:

$$\mathfrak{D}^{(1)} = \mathbb{H}_{SO(6)} = \sum_i 2(\mathbb{1} - \mathbb{P})_{i\,i+1} + \mathbb{T}_{i\,i+1}. \tag{22}$$

We will reproduce this using form factor results.

One-loop SO(6) sector

Form factors in the loop expansion can be written in the following form:

$$\mathcal{F}_{\mathcal{O}} = \left(1 + g^2 \mathcal{I}^{(1)} + g^4 \mathcal{I}^{(2)} + \cdots\right) \mathcal{F}_{\mathcal{O}}^{(0)} . \tag{23}$$

For operators that are eigenstates under renormalization, such as BPS operators or the Konishi primary, $\mathcal{I}^{(\ell)}$ is simply the ratio of the ℓ-loop and tree-level form factor. However, for form factors of operators that renormalize non-diagonally, this is no longer the case, for example, the loop corrections to vanishing tree-level form factors can be non-vanishing. As we will see in eq. (26), to overcome this problem, it is necessary to promote $\mathcal{I}^{(\ell)}$ to an operator that acts on the tree-level form factor $\mathcal{F}_{\mathcal{O}}^{(0)}$ and creates a different tree-level form factor from it. This also makes it convenient to study the symmetry properties of form factors, as we will show in the end.

In the planar limit, ℓ-loop interactions can maximally involve $\ell + 1$ neighboring fields in the color-ordered form factor at a time. Hence, $\mathcal{I}^{(\ell)}$ can be written as an interaction density that is summed over all insertion points. At one-loop order, the maximal interaction range is two, and we can write

$$\mathcal{I}^{(1)} = \sum_{i=1}^{L} I^{(1)}_{i\,i+1}. \tag{24}$$

Here, L denotes the length of the operator $\mathcal{O}$, $I^{(1)}_{i\,i+1}$ acts on the external fields i and $i+1$ and cyclic identification $i + L \sim i$ is understood.

We will focus on the $SO(6)$ sector. The general form of the operator is given by trace of the product of all possible scalar fields, for which we introduce the notation:

$$X = \phi_{14}, \quad Y = \phi_{24}, \quad Z = \phi_{34}, \tag{25}$$

and the barred fields are their conjugates such as $\bar{X} = \phi_{23}$.

In SO(6) sector, it is sufficient to consider the following range-two interactions between different fields that are allowed by R-charge conservation: $XX \to XX$, $XY \to XY$, $XY \to YX$, $X\bar{X} \to X\bar{X}$, $X\bar{X} \to \bar{X}X$, $X\bar{X} \to Y\bar{Y}$, the others can be obtained from them by replacing $X \leftrightarrow Y, Z$, or with the similar ones of their conjugates. We denote

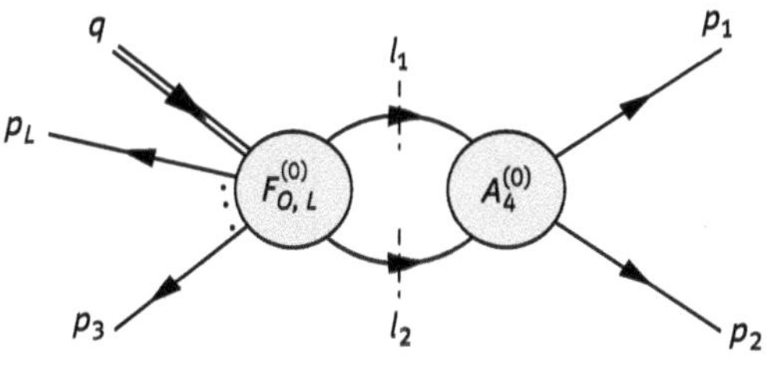

Figure 1: The $(p_1 + p_2)^2$ double cut of one-loop form factors.

the contribution to a given combination of external fields $\phi_{AB}\phi_{CD} \rightarrow \phi_{A'B'}\phi_{C'D'}$ by $(I_i^{(1)})_{\phi_{AB}\phi_{CD}}^{\phi_{A'B'}\phi_{C'D'}}$.

In terms of these matrix elements, the operator $I_{i\,i+1}^{(1)}$ is explicitly given by

$$I_{i\,i+1}^{(1)} = \sum_{\substack{\alpha_i,\beta_i,\gamma_i,\delta_i=1 \\ \alpha\equiv\{\alpha_1,\alpha_2\}}}^{4} (I_i^{(1)})_{\phi_\alpha\phi_\beta}^{\phi_\gamma\phi_\delta} \eta_i^{\gamma_1}\eta_i^{\gamma_2} \frac{\partial}{\partial\eta_i^{\alpha_1}} \frac{\partial}{\partial\eta_i^{\alpha_2}} \eta_{i+1}^{\delta_1}\eta_{i+1}^{\delta_2} \frac{\partial}{\partial\eta_{i+1}^{\beta_1}} \frac{\partial}{\partial\eta_{i+1}^{\beta_2}} . \tag{26}$$

The matrix elements $(I_i^{(1)})_{\phi_\alpha\phi_\beta}^{\phi_\gamma\phi_\delta}$ can be efficiently computed by unitarity method. Since only the interactions between two adjacent field are relevant, we only need to consider the double cut shown in Figure 1. Let us briefly consider the $(I_1^{(1)})_{XY}^{YX}$ case. The cut integrand is given by

$$\int \mathrm{dPS}(l_1, l_2) \mathrm{d}^4\eta_{l_1} \mathrm{d}^4\eta_{l_2} \mathcal{F}_{\mathcal{O}}^{(0)}(l_1^X, l_2^Y, p_3, \ldots, p_L; q) \mathcal{A}_4^{(0)}(-l_2, -l_1, p_1^Y, p_2^X) , \tag{27}$$

where the minimal tree-level form factor can be obtained as in Section 3.1 and the four-point amplitude is given by the standard MHV expression. Also dPS is the two-particle phase space measure whose explicit form is not important here. The labeling of the external legs with X, Y in the tree-level amplitude and form factor means to take the corresponding η components; for example, $\mathcal{A}_4(-l_2, -l_1, p_1^Y, p_2^X)$ is the component of $\mathcal{A}_4(-l_2, -l_1, p_1, p_2)$ containing the $(\eta_1^2\eta_1^4)(\eta_2^1\eta_2^4)$ factor. Integrating out the η_{l_i} variables, the cut integrand is given by

$$(\eta_1^2\eta_1^4)(\eta_2^1\eta_2^4) \mathcal{F}_{\mathcal{O}}^{(0)}(p_1^X, p_2^Y, p_3, \ldots, p_L; q)\big|_{\eta_1^A=\eta_2^A=1} \int \mathrm{dPS}(l_1, l_2) . \tag{28}$$

So, the cut integrand for this case is $(I_1^{(1)})_{XY}^{YX}|_{q^2} = -1$. The variables η_1 and η_2 indicate that the result is not necessarily proportional to the tree-level form factor of the original operator but to the operator in which the corresponding X and Y fields are permuted. The occurring phase space integral is simply the cut of a scalar bubble integral: More details of such computations can be found, for example, in Wilhelm [54] and Nandan et al. [51].

The results are summarized in Figure 2. The one-mass triangle integral is infrared (IR) divergent and UV finite. The bubble integral, on the other hand, is

$(I_i^{(1)})$	$\frac{XX}{XX}$	$\frac{XY}{XY}$	$\frac{YX}{XY}$	$\frac{X\bar{X}}{X\bar{X}}$	$\frac{\bar{X}X}{X\bar{X}}$	$\frac{Y\bar{Y}}{X\bar{X}}$
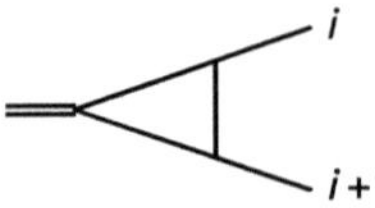 s_{ii+1}	−1	−1	0	−1	0	0
	0	−1	+1	−3/2	+1/2	−1/2

Figure 2: Linear combinations of diagrams contributing to the minimal one-loop form factors in the SO(6) sector.

IR finite but UV divergent. Hence, the IR and UV divergences can be separated naturally.

The UV divergences require the renormalization of the operators. The renormalized operators are defined in terms of the bare operators and the renormalization constant $\mathscr{Z}$ as shown in (20). The renormalized form factor is nothing but the form factor of the renormalized operator.

Since the form factor is linear in the operator, we can write in the case of the minimal form factor:

$$\mathscr{F}^{(0)}_{\mathscr{Z}\mathscr{O}}(1,\ldots,L;q) = \mathscr{Z}\mathscr{F}^{(0)}_{\mathscr{O}}(1,\ldots,L;q)\,, \tag{29}$$

where, on the right-hand side, $\mathscr{Z}$ acts as an *operator* on the tree-level form factor, similar to $\mathscr{I}^{(\ell)}$ discussed before, cf. eq. (23).

At one-loop level, $\mathscr{Z}^{(1)}$ has to render the renormalized one-loop interaction

$$\underline{\mathscr{I}}^{(1)} = \mathscr{I}^{(1)} + \mathscr{Z}^{(1)} \tag{30}$$

UV finite, which means that $\mathscr{Z}^{(1)}_{ii+1}$ has to cancel the UV divergence of the bubble integrals occurring in $I^{(1)}_{ii+1}$. Using the results in Figure 2, the one-loop renormalization constant density is given by the matrix elements

$$(\mathscr{Z}_i^{(1)})^{XX}_{XX} = 0\,, \qquad (\mathscr{Z}_i^{(1)})^{XY}_{XY} = \frac{1}{\epsilon}\,, \qquad (\mathscr{Z}_i^{(1)})^{YX}_{XY} = -\frac{1}{\epsilon}\,, \qquad (\mathscr{Z}_i^{(1)})^{Y\bar{Y}}_{X\bar{X}} = -\frac{1}{2\epsilon}\,. \tag{31}$$

It can be written in the compact operatorial form

$$\mathscr{Z}_{ii+1} = \frac{1}{\epsilon}\left(\mathbb{1} - \mathbb{P} + \frac{1}{2}\mathbb{T}\right)_{ii+1}\,, \tag{32}$$

where $\mathbb{1}$ is the identity operator and

$$\mathbb{P}_{ii+1} = \sum_{A,B,C,D=1}^{4} \eta_i^A \eta_i^B \frac{\partial}{\partial \eta_i^C} \frac{\partial}{\partial \eta_i^D} \eta_{i+1}^C \eta_{i+1}^D \frac{\partial}{\partial \eta_{i+1}^A} \frac{\partial}{\partial \eta_{i+1}^B} \tag{33}$$

denotes the permutation operator and,

$$\mathbb{T}_{ii+1} = \sum_{A,B,C,D=1}^{4} \frac{\epsilon_{ABCD}\epsilon_{A'B'C'D'}}{16} \eta_i^{A'} \eta_i^{B'} \frac{\partial}{\partial \eta_{i+1}^A} \frac{\partial}{\partial \eta_{i+1}^B} \eta_{i+1}^{C'} \eta_{i+1}^{D'} \frac{\partial}{\partial \eta_i^C} \frac{\partial}{\partial \eta_i^D} \tag{34}$$

is the trace operator.

Applying eqs. (21) to (31), we find the one-loop dilatation operator density

$$(\mathfrak{D}_i^{(1)})_{XX}^{XX} = 0\,, \qquad (\mathfrak{D}_i^{(1)})_{XY}^{XY} = 2\,, \qquad (\mathfrak{D}_i^{(1)})_{XY}^{YX} = -2 \qquad (\mathfrak{D}_i^{(1)})_{X\bar{X}}^{Y\bar{Y}} = -1. \tag{35}$$

These expressions can be combined into the well-known form (22).

Higher-loop cases

Unlike for the one-loop case where the UV divergences stem only from the bubble integrals, for two- and higher-loop form factors, the integrals in general contain a mixing of IR and UV divergences. Luckily, IR divergences have a well-understood universal structure obtained from the study of Sudakov form factors, see Mueller [49], Collins [28], Sen [52] and Magnea and Sterman [44]. This allows us to subtract the IR divergences systematically using the Bern, Dixon and Smirnov (BDS) ansatz Bern et al. [13] and the logarithm of form factor (normalized by the tree factor) can be given as an expansion in the coupling constant g and at an arbitrary mass scale μ by

$$\log \mathscr{I} = \sum_{\ell=1}^{\infty} g^{2\ell} \left(-\frac{\gamma_{\text{cusp}}^{(\ell)}}{(2\ell\epsilon)^2} - \frac{\mathscr{G}_0^{(\ell)}}{2\ell\epsilon} \right) \sum_{i=1}^{n} \left(\frac{\mu^2}{-s_{ii+1}} \right)^{\ell\epsilon} - \log \mathscr{Z} + \text{Fin} + \mathscr{O}(\epsilon), \tag{36}$$

where $\gamma_{\text{cusp}}^{(\ell)}$ and $\mathscr{G}_0^{(\ell)}$ are the cusp and collinear anomalous dimensions respectively along with a finite part defined as Fin. The UV divergence takes the form of the renormalization matrix term which is related to the dilatation operator as $\log \mathscr{Z} = \sum_{\ell=1}^{\infty} g^{2\ell} \frac{\mathscr{D}^{(\ell)}}{2\epsilon\ell}$, see also eq. (21).

We note that at the two-loop order, connected interactions involve at most three fields of the composite operator, which have to be adjacent at the planar level i.e we have both interactions of range 2 and 3. This is reflected in the fact that for the unitarity cuts we need both the double and triple cuts as shown in Figure 3. As mentioned before, the unitarity cuts result in a more involved set of master integrals at the two-loop level where the IR and UV divergences are mixed up which can be sorted by a systematic application of the BDS ansatz Anastasiou et al. [2] and Bern et al. [13] for IR-subtraction.

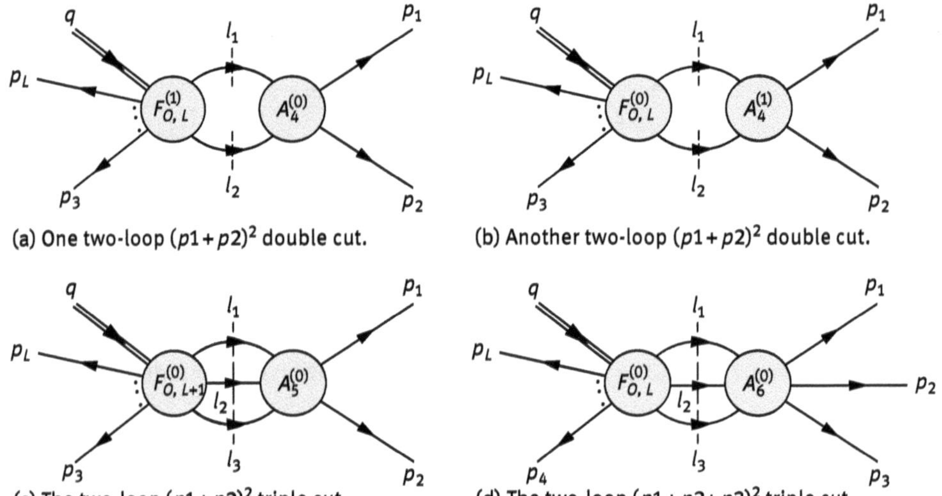

(a) One two-loop $(p1+p2)^2$ double cut.

(b) Another two-loop $(p1+p2)^2$ double cut.

(c) The two-loop $(p1+p2)^2$ triple cut.

(d) The two-loop $(p1+p2+p3)^2$ triple cut.

Figure 3: Unitarity cuts of the minimal two-loop form factor.

As a result, similar to the one-loop case (30), the two-loop renormalized form factor is given by

$$\underline{\mathcal{I}}^{(2)} = \mathcal{I}^{(2)} + \mathcal{I}^{(1)}\mathcal{Z}^{(1)} + \mathcal{Z}^{(2)}, \tag{37}$$

where

$$\mathcal{Z}^{(2)} = \sum_{i=1}^{L}\Big(\mathcal{Z}^{(2)}_{i\,i+1\,i+2} + \frac{1}{2}\sum_{j=i+2}^{L+i-2}\mathcal{Z}^{(1)}_{i\,i+1}\mathcal{Z}^{(1)}_{j\,j+1}\Big), \tag{38}$$

where the first term inside the summation above is due to the range 3 interactions. Now using eq. (21), we have for the two-loop dilatation operator

$$\mathfrak{D}^{(2)} = 4\epsilon\Big(\mathcal{Z}^{(2)} - \frac{1}{2}\big(\mathcal{Z}^{(1)}\big)^2\Big). \tag{39}$$

We find form factor results reproduce exactly the known result in Beisert et al. [7] for operators in the $SU(2)$ sector, where details can be found in Loebbert et al. [43].

In addition to the dilatation operator, we can also extract the kinematic-dependent finite part of the two loop form factor which has very interesting properties and is analogous to the remainder function studied for scattering amplitudes for two (and higher) loops. The two-loop remainder is given as

$$\mathcal{R}^{(2)} = \underline{\mathcal{I}}^{(2)}(\epsilon) - \frac{1}{2}\big(\underline{\mathcal{I}}^{(1)}(\epsilon)\big)^2 - f^{(2)}(\epsilon)\underline{\mathcal{I}}^{(1)}(2\epsilon) + \mathcal{O}(\epsilon), \tag{40}$$

where

$$f^{(2)}(\epsilon) = -2\zeta_2 - 2\zeta_3\epsilon - 2\zeta_4\epsilon^2 . \tag{41}$$

The two-loop remainder function is a transcendental function of the kinematic invariants of maximal degree 4. Due to the existence of UV divergences, the remainder function is not of uniform transcendentality, except for the BPS operators which have only the highest transcendentality [19]. Interestingly, in SU(2) sector the maximally transcendental part of the remainder function (for the non-shuffling case, see below) turns out to be the same as the BPS case. These results and further evidence in other sectors lead us to conjecture that the two-loop remainder of every minimal form factor has the same degree-four part as the BPS one. Another interesting observation is that the highest degree of transcendentality $t = 4 - s$ is related to the shuffling number s of the respective remainder density, i.e. to the number indicating by how many legs the field flavors are shuffled. For instance, $(R_i^{(2)})_{XXY}^{XYX}$ has shuffling number $s = 1$ and maximal transcendentality degree $t = 3$.

As a closing remark, we comment on the symmetry properties of form factors that are manifest in these computations. The PSU(2,2|4) symmetry of $\mathcal{N} = 4$ SYM theory leads to the following Ward identity of form factors:

$$\sum_{i=1}^{n} \mathfrak{J}_i^A \mathscr{F}_{\mathcal{O}}(1, \ldots, n; q) = \mathscr{F}_{\mathfrak{J}^A\mathcal{O}}(1, \ldots, n; q) , \tag{42}$$

which holds for any generator $\mathfrak{J}_i^A$ of PSU(2,2|4); see e.g. Brandhuber et al. [17] for a derivation. Let us consider the generators of the SU(2) subgroup

$$\mathfrak{J}_i^1 = \eta_i^1 \frac{\partial}{\partial \eta_i^2} + \eta_i^2 \frac{\partial}{\partial \eta_i^1} , \quad \mathfrak{J}_i^2 = -i\eta_i^1 \frac{\partial}{\partial \eta_i^2} + i\eta_i^2 \frac{\partial}{\partial \eta_i^1} , \quad \mathfrak{J}_i^3 = \eta_i^1 \frac{\partial}{\partial \eta_i^1} - \eta_i^2 \frac{\partial}{\partial \eta_i^2} . \tag{43}$$

Applying eq. (42) to eq. (23) for the minimal tree-level and one-loop form factor, we find

$$\left[\mathfrak{J}^A, \mathscr{I}^{(1)}\right] = 0 , \tag{44}$$

where $\mathfrak{J}^A = \sum_{i=1}^{L} \mathfrak{J}_i^A$. Inserting eq. (26) into eq. (44) yields

$$(I_i^{(1)})_{XY}^{XY} + (I_i^{(1)})_{XY}^{YX} = (I_i^{(1)})_{XX}^{XX} , \tag{45}$$

as well as similar identities, which are all consistent with explicit form factor results. At two-loop level we can derive similar relations from the Ward identity and notably such relations also hold for the remainder function density.

4 IR safe observables from form factors

Scattering amplitudes and form factors are themselves not physical observables, as they contain IR divergences from the integration of loop momenta. The IR divergences coming from virtual loop corrections can be canceled by IR divergences from the real emissions of soft and collinear particles in the so-called bremsstrahlung contributions by Kinoshita [38] and Lee and Nauenberg [42]. Cross sections are constructed in this way and are free of IR divergences, and hence physical observables.

Alternatively, we can consider the decay of an initial off-shell state created by an operator $\mathcal{O}(q)$ with timelike momentum ($q^2 > 0$) into any final on-shell multi-particle state. The probability of this inclusive decay is the total decay rate of $\mathcal{O}(q)$, which is defined by

$$\sigma_{\text{tot},\mathcal{O}}(q) = \sum_X \delta^{(D)}(q - p_X)\,|\langle X|\mathcal{O}(0)|0\rangle|^2\,, \tag{46}$$

where X is a final state with total momentum $q = p_X$ and the sum ensures that the quantity is inclusive. The matrix element $\langle X|\mathcal{O}(0)|0\rangle$ is the form factor of $\mathcal{O}$ with final state X.

A more general class of observables closely related to total cross section (46) are the so-called "event shapes" functions:

$$\sigma_{w,\mathcal{O}}(q) = \sum_X \delta^{(D)}(q - p_X)\,w(X)|\langle X|\mathcal{O}(0)|0\rangle|^2\,, \tag{47}$$

where the weight factor $w(X)$ depends on the quantum numbers of the final states that one selects in the detector apparatus. Different choices of $w(X)$ correspond to various different event shape functions, such as thrust, heavy mass, energy–energy correlations. In $\mathcal{N} = 4$ SYM theory, the energy–energy correlations have been considered in Hofman and Maldacena [36], Engelund and Roiban [31] and Belitsky et al. [9, 10]. In the following subsections, we present one-loop examples for the total cross section and the energy correlation function. For simplicity, we will consider the case with half-BPS operator $\mathcal{O} = \text{tr}(\phi_{12}^2)$.

Let us write eq. (47) in a form convenient for explicit perturbative calculation. Expanded in powers of g, eq. (47) can be given as follows:

$$\sigma_w = \sum_{\ell=0}^{\infty} g^{2\ell}\sigma_w^{(\ell)}\,, \qquad \sigma_w^{(\ell)} = \sum_{n=2}^{\ell+2} g^{2(2-n)} \int d\text{PS}_n w_n \mathcal{M}_n^{(\ell+2-n)}. \tag{48}$$

The squared matrix elements are given by

$$\mathcal{M}_n^{(\ell)} = \frac{1}{n!}\int \prod_{i=1}^{n} d^4\eta_i \sum_{a_i,k,l} \hat{\mathcal{F}}_{\mathcal{O},n}^{\text{N}^k\text{MHV},(l)}(\{a_i,p_i,\eta_i\})\hat{\mathcal{F}}_{\bar{\mathcal{O}},n}^{*,\text{N}^{m-k}\text{MHV},(\ell-l)}(\{a_i,p_i,\eta_i\})\,, \tag{49}$$

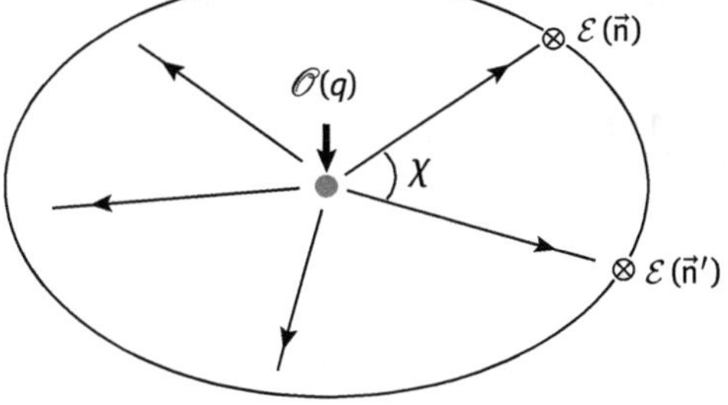

Figure 4: Graphical representation of energy–energy correlations: particles produced by source $\mathcal{O}(q)$ out of the vacuum, and two detectors located at infinity in the direction of $\vec{n}$ and $\vec{n}'$.

in which $\hat{\mathcal{F}}_n^{(\ell)}(\{a_i, p_i, \eta_i\})$ is the ℓ-loop n-point non-color-ordered super form factor defined in eq. (9), and the summation involves the MHV degrees, color indices and the sum over the types of particles given in terms of integrations over ηs.

The respective measure for the integration over the phase space of the n particles in the final state is given by

$$\mathrm{dPS}_n = \prod_{\ell=1}^{n} \frac{d^D p_\ell}{(2\pi)^D} 2\pi\delta_+(p_\ell^2) \cdot (2\pi)^D \delta^{(D)}\Big(q - \sum_{\ell=1}^{n} p_\ell\Big). \tag{50}$$

For energy–energy correlation, the weight factor w_n is given as

$$w_n = \sum_{i \neq j} p_i^0 p_j^0 \delta^2(\Omega_{\vec{p}_i} - \Omega_{\vec{n}}) \delta^2(\Omega_{\vec{p}_j} - \Omega_{\vec{n}'}), \tag{51}$$

where there are two detectors oriented along the direction $\vec{n}$ and $\vec{n}'$, as shown in Figure 4. The rotation symmetry implies that the observable depends only on the relative angle θ of the vectors, $\vec{n} \cdot \vec{n}' = \cos\chi$. Total cross section corresponds to the simplest case that $w_n = 1$.

4.1 Total cross section

The leading order squared matrix element is

$$\mathcal{M}^{(0)}_{\mathrm{BPS},2} = \frac{1}{2!} \sum_{a_1, a_2} \int d^4\eta_1 d^4\eta_2 \, \hat{\mathcal{F}}^{(0)}_{\mathrm{BPS}}(1,2) \, \hat{\mathcal{F}}^{*(0)}_{\mathrm{BPS}}(1,2) = \frac{N_c^2 - 1}{2}. \tag{52}$$

The tree-level cross section is given by the integral of $\mathcal{M}^{(0)}_{\mathrm{BPS},2}$ over the two-particle phase space. This yields

$$\sigma^{(0)}_{\mathrm{tot}} = \int \mathrm{dPS}_2 \, \mathcal{M}^{(0)}_{\mathrm{BPS},2} = \left(\frac{\mu^2}{q^2}\right)^{\epsilon} \frac{1}{4(16\pi)^{\frac{1}{2}-\epsilon} \Gamma\left(\frac{3}{2} - \epsilon\right)} \frac{N_c^2 - 1}{2}. \tag{53}$$

The one-loop cross section is given by the sum of a two-particle and a three-particle channel

$$\sigma^{(1)}_{\text{tot}} = \int d\text{PS}_2\, \mathcal{M}^{(1)}_{\text{BPS},2} + \frac{1}{g^2}\int d\text{PS}_3\, \mathcal{M}^{(0)}_{\text{BPS},3}. \tag{54}$$

Here we consider explicitly the three-particle channel. We parametrize the momenta and mandelstam variables as

$$\begin{aligned} &q^\mu = (q_0, 0, 0, 0)\,, \quad k_1^\mu = \frac{\tau_1 q_0}{2}\left(1, 1, \vec{\mathbf{0}}_{D-2}\right), \quad k_2^\mu = \frac{\tau_2 q_0}{2}\left(1, \cos\theta, \sin\theta, \vec{\mathbf{0}}_{D-3}\right), \\ &s_{12} = q_0^2(\tau_1 + \tau_2 - 1)\,, \qquad s_{23} = q_0^2(1-\tau_1)\,, \qquad s_{31} = q_0^2(1-\tau_2)\,. \end{aligned} \tag{55}$$

The positive energy condition requires that

$$\tau_1 < 1\,, \qquad \tau_2 < 1\,, \qquad \tau_1 + \tau_2 > 1\,. \tag{56}$$

The three-particle phase space measure is then given as

$$\begin{aligned} \int d\text{PS}_3 &= \frac{(q_0/2)^{2D-6}}{16(2\pi)^{2D-3}} \int \prod_{a=1}^{2} d\tau_a\, \tau_a^{D-3}\, d\Omega_a^{D-2} \cdot \delta_+\!\left(1 - \tau_1 - \tau_2 + \tau_1\tau_2 z\right) \\ &= \frac{q_0^{2-4\epsilon}/2}{(4\pi)^{3-2\epsilon}\Gamma(2-2\epsilon)} \int_0^1 d\tau_1 d\tau_2 [(1-\tau_1)(1-\tau_2)(\tau_1+\tau_2-1)]^{-\epsilon}\,, \end{aligned}$$

where $z = \frac{1-\cos\theta}{2}$. The phase space integrand can be computed from eq. (49) which gives

$$\mathcal{M}_3^{(0)} = \frac{2}{3} g_{\text{YM}}^2 N_c\, (N_c^2 - 1)\, \frac{(q^2)^2}{s_{12}s_{23}s_{31}}\,. \tag{57}$$

After the phase space integration, three-particle channel contribution is:

$$\frac{1}{g^2}\int d\text{PS}_3\, \mathcal{M}_3^{(0)} = \sigma^{(0)} \left(\frac{\mu^2}{q^2}\right)^{\epsilon} \left(\frac{4}{\epsilon^2} - \frac{7\pi^2}{3}\right) + \mathcal{O}(\epsilon)\,. \tag{58}$$

This exactly cancels the two-particle channel as required by the BPS condition, for which one can find details in Nandan et al. [51].

4.2 Energy–energy correlation

Next, we consider the energy–energy correlation. We consider the relative angle χ of detectors in the range $0 < \chi < \pi$. At one-loop, there is only contribution from the three particle channel. Since the integral is finite, we can work in four-dimensional phase space. Compared to the total cross section case, the new weight factors constrain the angle variables as:

$$\int dPS_3 \delta^2(\Omega_{\vec{p}_1} - \Omega_{\vec{n}})\delta^2(\Omega_{\vec{p}_2} - \Omega_{\vec{n}'}) = \frac{q_0^2}{64(2\pi)^5}\int_0^1 d\tau_1 d\tau_2\, \tau_1\tau_2\delta_+(1-\tau_1-\tau_2+\tau_1\tau_2\tilde{z}),$$

where $\tilde{z} = \frac{1-\cos\chi}{2}$. Plugging in the phase space integrand (57) and p_i^0, p_j^0 factors of eq. (51), and also taking into account the symmetry factor of exchanging particles, one obtains the one-loop energy–energy correlations as (see also [9])

$$\begin{aligned}\sigma_{\mathrm{EEC}}^{(1)} &= \frac{\sigma_{\mathrm{tot}}}{8\pi^2}\int_0^1 d\tau_1 d\tau_2 \frac{\tau_1^2\tau_2^2\delta_+(1-\tau_1-\tau_2+\tau_1\tau_2\tilde{z})}{(1-\tau_1)(1-\tau_2)(\tau_1+\tau_2-1)}\\ &= \frac{\sigma_{\mathrm{tot}}}{8\pi^2}\frac{1}{\tilde{z}(1-\tilde{z})}\int_0^1 d\tau_1 \frac{1}{1-\tilde{z}\tau_1} = -\frac{\sigma_{\mathrm{tot}}}{8\pi^2}\frac{\log(1-\tilde{z})}{\tilde{z}^2(1-\tilde{z})}\,.\end{aligned} \tag{59}$$

Two-loop results are also obtained in Belitsky et al. [9, 10] using correlation function method. It will be interesting to have an explicit cross section type computation along the procedure outlined above. This is under investigation and we hope to report this in some other place.

5 On-shell diagrams and Grassmannian for form factors

In the quest for reformulating quantum field theory, several remarkable advances have been made in understanding scattering amplitudes using novel ideas in geometry, combinatorics and other fields of mathematics, the overarching goal being the search for a mathematical formulation where locality and unitarity are not manifest. These new ideas have been highly dependent on the on-shell states used to study scattering problems and in this section we will review two such approaches, the *Grassmannian formulation* and the *on-shell diagrams*, and extend their applicability to form factors and their interplay.

5.1 On-shell diagrams

The on-shell diagrams can be used to reconstruct scattering processes using only on-shell amplitudes as building blocks. Hence, unlike in usual Feynman diagram approach there are no off-shell particles even for loop-level processes thus making gauge invariance manifest at all stages of a computation. They are built from two different elements, namely the three-point MHV amplitude $\mathcal{A}_{3,2}$ and the three-point $\overline{\mathrm{MHV}}$ amplitude $\mathcal{A}_{3,1}$:

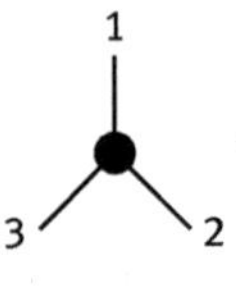

$$= \mathscr{A}_{3,2} = \frac{\delta^4(\lambda_1\tilde{\lambda}_1 + \lambda_2\tilde{\lambda}_2 + \lambda_3\tilde{\lambda}_3)\delta^8(\lambda_1\tilde{\eta}_1 + \lambda_2\tilde{\eta}_2 + \lambda_3\tilde{\eta}_3)}{\langle 12\rangle\langle 23\rangle\langle 31\rangle},$$

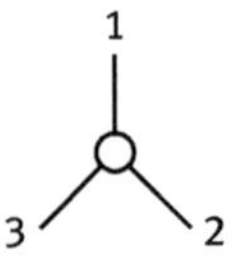

$$= \mathscr{A}_{3,1} = \frac{\delta^4(\lambda_1\tilde{\lambda}_1 + \lambda_2\tilde{\lambda}_2 + \lambda_3\tilde{\lambda}_3)\delta^4([12]\,\tilde{\eta}_3 + [23]\,\tilde{\eta}_1 + [31]\,\tilde{\eta}_2)}{[12]\,[23]\,[31]}.$$

As shown in the paper of Britto et al. [22, 24], all tree amplitudes can be built from BCFW recursion relations. This recursion relation can be naturally depicted using the on-shell diagram picture as given by Arkani-Hamed et al. [3], and tree amplitudes can be depicted

$$\mathscr{A}_{n,k} = \sum_{\substack{n',n'',k',k'' \\ n'+n''=n+2 \\ k'+k''=k+1}}$$

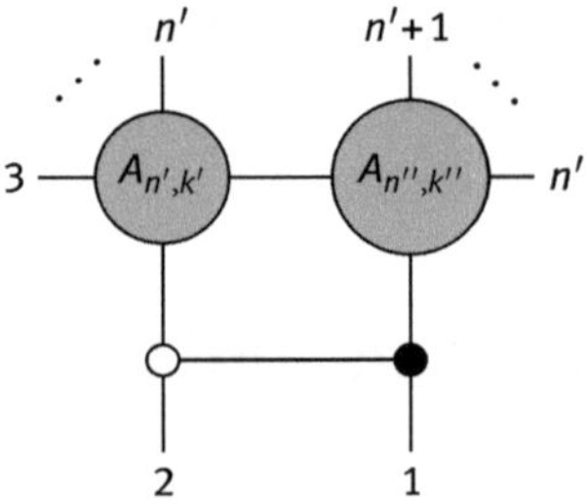

, (60)

where the BCFW bridge attached at positions 1 and 2 implements the BCFW shift, see Arkani-Hamed et al. [3]. The left and right sub-amplitudes in eq. (60) can be recursively built in terms of on-shell diagrams starting from the three point on-shell diagrams and acting with successive BCFW bridges.

In order to extend this picture and to depict tree-level form factors via on-shell diagrams, we need to add another building block to our dictionary: the minimal form factor introduced earlier. We will focus on the form factor with stress tensor supermultiplet $\mathscr{T}$, and its minimal form factor is the two-point case of eq. (8):

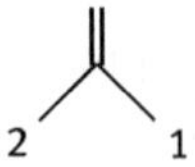

$$= \mathscr{F}_{2,2} = \frac{\delta^4(\lambda_1\tilde{\lambda}_1 + \lambda_2\tilde{\lambda}_2 - q)\delta^{(8)}(\gamma - \sum_{i=1}^{2}\lambda_i\eta_i^A)}{\langle 12\rangle\langle 21\rangle}. \tag{61}$$

We can then use the construction of the form factors via BCFW recursion relations as in Brandhuber et al. [17, 18], which we depict as

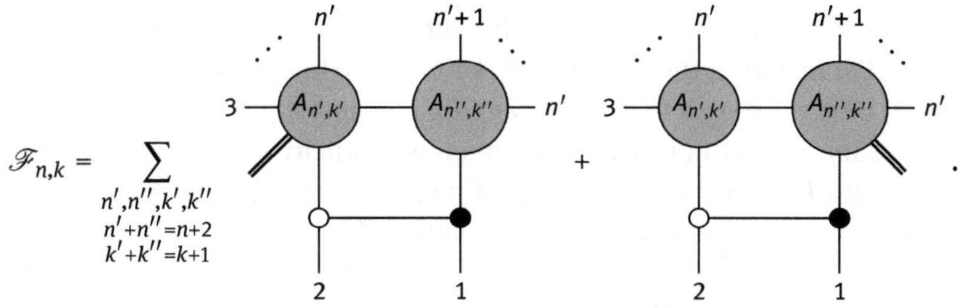

(62)

Let us now turn to the Grassmannian integral representation for form factors (and amplitudes) which provides a systematic way to express the diagram as a function of the kinematics.

5.2 General considerations on the Grassmannian

The fundamental idea behind Grassmannian integral representations of scattering amplitudes by Arkani-Hamed et al. [3, 4], and Mason and Skinner [47] is to express momentum conservation in a geometric way. We can combine the external kinematic data for n particles and combine all λ_i into a 2 by n matrix λ, as a two-plane in n-dimensional space. Similarly for $\tilde{\lambda}$. Momentum conservation is expressed as the orthogonality of these planes:

$$\lambda \cdot \tilde{\lambda} \equiv \sum_{i=1}^{n} \lambda_i \tilde{\lambda}_i = 0 \,. \tag{63}$$

The Grassmannian representation linearizes this constraint by introducing an auxiliary hyperplane $C \in G(k, n)$ such that

$$(C \cdot \tilde{\lambda})_I^{\dot{\alpha}} = \sum_{i=1}^{n} C_{Ii} \tilde{\lambda}_i^{\dot{\alpha}} = 0 \quad \text{and} \quad (C^{\perp} \cdot \lambda)_J^{\alpha} = \sum_{i=1}^{n} C_{Ji}^{\perp} \lambda_i^{\alpha} = 0 \quad \Longrightarrow \quad \lambda \cdot \tilde{\lambda} = 0 \,, \tag{64}$$

where $C^{\perp}$ is the orthogonal complement of C fulfilling $C(C^{\perp})^T = 0$ and $I = 1, \dots, k$, $J = 1, \dots, n-k$. In fact the two three-point amplitudes can be written as a Grassmannian namely $G(2, 3)$ and $G(1, 3)$ for the MHV and the $\overline{\text{MHV}}$ case respectively. Hence we can associate each on-shell diagram to a Grassmannian integral, which for scattering amplitudes integrates a holomorphic form on $G(k, n)$ on the support of the constraints (64).

We can similarly geometrize momentum conservation for form factors, but the presence of the off-shell momenta carried by the operator makes the situation different than the case of the amplitude. For the MHV case, a naive way of introducing an auxiliary Grassmannian works. One can simply use the same Grassmannian and the same form as one would use for the MHV amplitude with the same number of legs. It is clear, however, that this way of linearizing the geometrical constraint cannot work beyond MHV. For instance, the MHV degree k ranges up to n for form factors, while for amplitudes it only ranges up to $n-2$. Since $G(n,n)$ is just a point, a larger Grassmannian is necessary: in fact not just for the maximal helicity degree case of N^{k-2}MHV form factors with k counting the number of negative helicity gluons, but also starting from NMHV form factors. The different range of MHV degrees already suggests that the correct Grassmannian is $G(k,n+2)$; this also fits nicely with the observation with the general fact that an off-shell momentum can be parametrized by two on-shell ones. Indeed, we can define new kinematic variables as a pair of two-planes in an $(n+2)$-dimensional space as

$$\begin{aligned} \underline{\lambda}_i &= \lambda_i\,, \quad i=1,\dots,n\,, \quad \underline{\lambda}_{n+1} = \xi_A\,, \qquad \underline{\lambda}_{n+2} = \xi_B\,, \\ \underline{\tilde{\lambda}}_i &= \tilde{\lambda}_i\,, \quad i=1,\dots,n\,, \quad \underline{\tilde{\lambda}}_{n+1} = -\frac{\langle \xi_B | q}{\langle \xi_B \xi_A \rangle}\,, \quad \underline{\tilde{\lambda}}_{n+2} = -\frac{\langle \xi_A | q}{\langle \xi_A \xi_B \rangle}\,, \end{aligned} \tag{65}$$

where ξ_A and ξ_B are arbitrary non-collinear reference spinors. Momentum conservation is then expressed as $\underline{\lambda} \cdot \underline{\tilde{\lambda}} = 0$ such that the additional on-shell momenta indeed encode the off-shell momentum: $\underline{\lambda}_{n+1}\underline{\tilde{\lambda}}_{n+1} + \underline{\lambda}_{n+2}\underline{\tilde{\lambda}}_{n+2} = -q$. Analogously, we can also define fermionic variables which encode the off-shell supermomentum conservation. In order to determine the form which is integrated over the Grassmannian $G(k,n+2)$ for a particular form-factor on-shell diagram, we break the corresponding diagram into two pieces: the minimal form factor (61), and a purely on-shell piece with $n+2$ legs for which a Grassmannian integral representation is given by Arkani-Hamed et al. [3] as

$$I = \int \frac{\mathrm{d}\alpha_1}{\alpha_1} \cdots \frac{\mathrm{d}\alpha_m}{\alpha_m}\, \delta^{k\times 2}(C \cdot \tilde{\lambda})\, \delta^{k\times 4}(C \cdot \tilde{\eta})\, \delta^{(n+2-k)\times 2}(C^{\perp} \cdot \lambda)\,, \tag{66}$$

where the matrix C depends on the α_i's, $C = C(\alpha_i) \in G(k,n+2)$ and m is the dimension of the corresponding cell in the Grassmannian. We then glue these two pieces together, i.e. we perform the on-shell phase space integration.

Let us use the above ideas and explicitly write down the Grassmannian integral corresponding to an on-shell diagram for a MHV amplitude. Gluing the minimal form factor to the legs $n+1$ and $n+2$ corresponds to calculating

$$\int \prod_{i=n+1}^{n+2} \left(\frac{\mathrm{d}^2\lambda_i\, \mathrm{d}^2\tilde{\lambda}_i}{\mathrm{Vol}[GL(1)]}\, \mathrm{d}^4\tilde{\eta}_i \right)\, \mathscr{F}_{2,2}(-(n+1), -(n+2))\;\, I(1,\dots,n+1,n+2)\,, \tag{67}$$

where $\mathscr{F}_{2,2}$ is defined in eq. (61). To remove the $GL(1)^2$ redundancy in the remaining λ integrations, we parametrize

$$\lambda_{n+1} = \xi_A - \beta_1 \xi_B\,, \qquad \lambda_{n+2} = \xi_B - \beta_2 \xi_A\,, \tag{68}$$

where ξ_A and ξ_B will be identified with the ones in eq. (65). With this, $\langle n+1\ n+2\rangle = (\beta_1\beta_2 - 1)\langle \xi_B \xi_A \rangle$.

Finally, the Grassmannian formula for all n-particle MHV form factors is given by

$$\langle \xi_A \xi_B \rangle^2 \int \frac{\mathrm{d}^{2(n+2)} C'}{\mathrm{Vol}[GL(2)]} \frac{Y(1-Y)^{-1}}{(12)(23)\cdots(n+1\ n+2)(n+2\ 1)} \delta^4(C' \cdot \underline{\underline{\tilde{\lambda}}})\, \delta^8(C' \cdot \underline{\underline{\tilde{\eta}}})\, \delta^{2n}(C'^{\perp} \cdot \underline{\underline{\lambda}}), \tag{69}$$

where

$$Y = \frac{(n\ n+1)(n+2\ 1)}{(n\ n+2)(n+1\ 1)} . \tag{70}$$

We note that this is a contour integral on the support of the delta functions and the residues compute the form factors as we have checked. We have also been able to extend the above procedure and present a similar formula for all n-particle form factors even for all the higher MHV degree, more details regarding this can be found in Frassek et al. [32]. We note that unlike in the MHV case, for the higher MHV degree form factors we need to combine several Grassmannian forms connected by shifts of the external on-shell legs in order to make the cyclic symmetry manifest.

Acknowledgments: This report is based on the works in collaboration with Rouven Frassek (Durham University), and members of SFB 647 from Humboldt University at Berlin namely, Florian Loebbert, David Meidinger, Christoph Sieg, and Matthias Wilhelm. We greatly acknowledge their important contribution in discovering the results which we present here in the review.

It is a pleasure to thank Andreas Brandhuber, Simon Caron-Huot, Lance Dixon, Burkhard Eden, Jan Fokken, Gregory Korchemsky, Brenda Penante, Jan Plefka, Matthias Staudacher, Gabriele Travaglini, Peter Uwer, Vitaly Velizhanin, Christian Vergu and Congkao Wen for useful discussions.

We would like to whole-heartedly thank the DFG grant in the framework of the SFB 647 "Raum-Zeit-Materie. Analytische und Geometrische Strukturen" which provided financial support to explore the research topics reviewed here. The first author would especially like to thank the SFB 647 grant for providing full financial support for his post-doctoral research at HU Berlin. The generous travel support from the grant for both authors enabled fruitful collaborations, exchange of knowledge as well as dissemination of the new results at various international conferences. Moreover, we thank the SFB for providing an active and productive research environment in Berlin for novel and interdisciplinary research in Mathematical Physics.

We also thank the Marie Curie network GATIS (gatis.desy.eu) of the European Union's Seventh Framework Programme FP7/2007–2013/ under REA Grant Agreement No 317089 for support.

Bibliography

[1] Alday LF, Maldacena J. Comments on gluon scattering amplitudes via AdS/CFT. JHEP 2007;0711:068. DOI:10.1088/1126-6708/2007/11/068.

[2] Anastasiou C, Bern Z, Dixon LJ, Kosower DA. Planar amplitudes in maximally supersymmetric Yang-Mills theory. Phys Rev Lett 2003;91:251602. DOI:10.1103/PhysRevLett.91.251602.

[3] Arkani-Hamed N, Bourjaily JL, Cachazo F, Goncharov AB, Postnikov A, Trnka J. Grassmannian geometry of scattering amplitudes. Cambridge University Press, 2016, http://inspirehep.net/record/1208741/files/arXiv:1212.5605.pdf.

[4] Arkani-Hamed N, Cachazo F, Cheung C, Kaplan J. A Duality For The S Matrix. JHEP 2010;03:020. DOI:10.1007/JHEP03(2010)020.

[5] Beisert N. On Yangian symmetry in planar N = 4 SYM. Quantum chromodynamics and beyond: Gribov-80 memorial volume. Proceedings, Memorial Workshop devoted to the 80th birthday of V.N. Gribov, Trieste, Italy, 26–28 May 2010. 2011:175–203, http://inspirehep.net/record/853644/files/arXiv:1004.5423.pdf. DOI:10.1142/9789814350198_0039.

[6] Beisert N. The complete one-loop dilatation operator of N = 4 super Yang-Mills theory. Nucl Phys 2004;B676:3–42. DOI:10.1016/j.nuclphysb.2003.10.019.

[7] Beisert N, Kristjansen C, Staudacher M. The Dilatation operator of conformal N = 4 super Yang-Mills theory. Nucl Phys 2003;B664:131–84. DOI:10.1016/S0550-3213(03)00406-1.

[8] Beisert N, et al. Review of AdS/CFT Integrability: an overview. Lett Math Phys 2012;99:3–32. DOI:10.1007/s11005-011-0529-2.

[9] Belitsky AV, Hohenegger S, Korchemsky GP, Sokatchev E, Zhiboedov A. Event shapes in N = 4 super-Yang-Mills theory. Nucl Phys 2014a;B884:206–56. DOI:10.1016/j.nuclphysb.2014.04.019.

[10] Belitsky AV, Hohenegger S, Korchemsky GP, Sokatchev E, Zhiboedov A. From correlation functions to event shapes. Nucl Phys 2014b;B884:305–43. DOI:10.1016/j.nuclphysb.2014.04.020.

[11] Bern Z, Dixon LJ, Dunbar DC, Kosower DA. One-loop n-point gauge theory amplitudes, unitarity and collinear limits. Nucl Phys 1994;B425:217–60. DOI:10.1016/0550-3213(94)90179-1.

[12] Bern Z, Dixon LJ, Dunbar DC, Kosower DA. Fusing gauge theory tree amplitudes into loop amplitudes. Nucl Phys 1995;B435:59–101. DOI:10.1016/0550-3213(94)00488-Z.

[13] Bern Z, Dixon LJ, Smirnov VA. Iteration of planar amplitudes in maximally supersymmetric Yang-Mills theory at three loops and beyond. Phys Rev 2005;D72:085001. DOI:10.1103/PhysRevD.72.085001.

[14] Boels R, Kniehl BA, Yang G. Master integrals for the four-loop Sudakov form factor. Nucl Phys 2016;B902:387–414. DOI:10.1016/j.nuclphysb.2015.11.016.

[15] Bork LV, Kazakov DI, and Vartanov GS. On form factors in N = 4 SYM. JHEP 2011a;1102:063. DOI:10.1007/JHEP02(2011)063.

[16] Bork LV, Kazakov DI, Vartanov GS. On MHV form factors in superspace for N = 4 SYM theory. JHEP 2011b;1110:133. DOI:10.1007/JHEP10(2011)133.

[17] Brandhuber A, Gurdogan O, Mooney R, Travaglini G, Yang G. Harmony of super form factors. JHEP 2011a;1110:046. DOI:10.1007/JHEP10(2011)046.

[18] Brandhuber A, Spence B, Travaglini G, Yang G. Form factors in N = 4 super Yang-Mills and periodic Wilson loops. JHEP 2001b;1101:134. DOI:10.1007/JHEP01(2011)134.

[19] Brandhuber A, Penante B, Travaglini G, Wen C. The last of the simple remainders. JHEP 2014;1408:100. DOI:10.1007/JHEP08(2014)100.

[20] Brandhuber A, Penante B, Travaglini G, Young D. Integrability and MHV diagrams in N = 4 supersymmetric Yang-Mills theory. Phys Rev Lett 2015a;114:071602. DOI:10.1103/PhysRevLett.114.071602.

[21] Brandhuber A, Penante B, Travaglini G, Young D. Integrability and unitarity. JHEP 2015;5:5. DOI:10.1007/JHEP05(2015)005.

[22] Britto R, Cachazo F, Feng B. New recursion relations for tree amplitudes of gluons. Nucl Phys 2005a;B715:499–522. DOI:10.1016/j.nuclphysb.2005.02.030.

[23] Britto R, Cachazo F, Feng B. Generalized unitarity and one-loop amplitudes in N = 4 super-Yang-Mills. Nucl Phys 2005b;B725:275–305. DOI:10.1016/j.nuclphysb.2005.07.014.

[24] Britto R, Cachazo F, Feng B, Witten E. Direct proof of tree-level recursion relation in Yang-Mills theory. Phys Rev Lett 2005c;94:181602. DOI:10.1103/PhysRevLett.94.181602.

[25] Chicherin D, Doobary R, Eden B, Heslop P, Korchemsky GP, Mason L, Sokatchev E. Correlation functions of the chiral stress-tensor multiplet in $\mathcal{N} = 4$ SYM. JHEP 2015;6:198. DOI:10.1007/JHEP06(2015)198.

[26] Chicherin D, Sokatchev E. Composite operators and form factors in $\mathcal{N} = 4$ SYM. J Phys 2017;A50(27):275402.DOI:10.1088/1751-8121/aa72fe.

[27] Chicherin D, Sokatchev E. Demystifying the twistor construction of composite operators in $\mathcal{N} = 4$ super-Yang–Mills theory. J Phys 2017;A50(20):205402. DOI:10.1088/1751-8121/aa6b95.

[28] Collins JC. Algorithm to compute corrections to the Sudakov form-factor. Phys Rev 1980;D22:1478. DOI:10.1103/PhysRevD.22.1478.

[29] Elvang H, Huang Yu-tin. Scattering amplitudes in gauge theory and gravity. Cambridge University Press, 2015. http://www.cambridge.org/mw/academic/subjects/physics/theoreticalphysics-and-mathematical-physics/scattering-amplitudes-gauge-theory-andgravity?format=AR.

[30] Engelund, OT. Lagrangian insertion in the light-like limit and the super-correlators/super-amplitudes duality. JHEP 2016;02:30. DOI:10.1007/JHEP02(2016)030.

[31] Engelund OT, Roiban R. Correlation functions of local composite operators from generalized unitarity. JHEP 2013;1303:172. DOI:10.1007/JHEP03(2013)172.

[32] Frassek R, Meidinger D, Nandan D, Wilhelm M. On-shell diagrams, Grassmannians and integrability for form factors. JHEP 2016;01:182. DOI:10.1007/JHEP01(2016)182.

[33] Gao Z, Yang G. Y-system for form factors at strong coupling in AdS5 and with multi-operator insertions in AdS3. JHEP 2013;1306:105. DOI:10.1007/JHEP06(2013)105.

[34] Gunaydin M, Marcus N. The Spectrum of the s**5 Compactification of the Chiral N=2, D=10 supergravity and the unitary supermultiplets of U(2, 2/4). Class. Quant Grav 1985;2(2):L11–17.

[35] Henn JM, Plefka JC. Scattering amplitudes in gauge theories. Lect Notes Phys 2014;883:1–195. DOI:978-3-642-54021-9, 10.1007/978-3-642-54022-6.

[36] Hofman DM, Maldacena J. Conformal collider physics: energy and charge correlations. JHEP 2008;0805:012. DOI:10.1088/1126-6708/2008/05/012.

[37] ’t Hooft G. A planar diagram theory for strong interactions. Nucl Phys 1974;B72:461. DOI:10.1016/0550-3213(74)90154-0.

[38] Kinoshita T. Mass singularities of Feynman amplitudes. J Math Phys 1962;3:650–77. DOI:10.1063/1.1724268.

[39] Koster L, Mitev V, Staudacher M. A twistorial approach to Integrability in N = 4 SYM. Fortsch Phys 2015;63(2):142–7. DOI:10.1002/prop.201400085.

[40] Koster, L, Mitev V, Staudacher, M, Wilhelm M. Composite operators in the twistor formulation of N = 4 supersymmetric Yang-Mills theory. Phys Rev Lett 2016;117(1):011601. DOI:10.1103/PhysRevLett.117.011601.

[41] Koster L, Mitev V, Staudacher M, Wilhelm M. All tree-level MHV form factors in $\mathcal{N} = 4$. SYM from twistor space. JHEP 2016;6:162. DOI:10.1007/JHEP06(2016)162.

[42] Lee TD, Nauenberg M. Degenerate systems and mass singularities. Phys Rev 1964;133:B1549–62. DOI:10.1103/PhysRev.133.B1549.

[43] Loebbert F, Nandan D, Sieg C, Wilhelm M, Yang G. On-shell methods for the two-loop dilatation operator and finite remainders. JHEP 2015;10:012. DOI:10.1007/JHEP10(2015)012.
[44] Magnea L, Sterman GF. Analytic continuation of the Sudakov form-factor in QCD. Phys Rev 1990;D42:4222–7. DOI:10.1103/PhysRevD.42.4222.
[45] Maldacena JM. The large n limit of superconformal field theories and supergravity. Adv Theor Math Phys 1998;2:231–52.
[46] Maldacena J, Zhiboedov A. Form factors at strong coupling via a Y-system. JHEP 2010;1011:104. DOI:10.1007/JHEP11(2010)104.
[47] Mason LJ, Skinner D. Dual superconformal invariance, momentum twistors and Grassmannians. JHEP 2009;11:045. DOI:10.1088/1126-6708/2009/11/045.
[48] Minahan JA, Zarembo K. The Bethe ansatz for N = 4 super Yang-Mills. JHEP 2003;0303:013.
[49] Mueller AH. On the asymptotic behavior of the Sudakov form-factor. Phys Rev 1979;D20:2037. DOI:10.1103/PhysRevD.20.2037.
[50] Nair VP. A current algebra for some gauge theory amplitudes. Phys Lett 1988;B214:215. DOI:10.1016/0370-2693(88)91471-2.
[51] Nandan D, Sieg C, Wilhelm M, Yang G. Cutting through form factors and cross sections of non-protected operators in N = 4 SYM. JHEP 2015;06:156. DOI:10.1007/JHEP06(2015)156.
[52] Sen A. Asymptotic behavior of the Sudakov form-factor in QCD. Phys Rev 1981;D24:3281. DOI:10.1103/PhysRevD.24.3281.
[53] van Neerven WL. Infrared behavior of on-shell form-factors in a N = 4 supersymmetric Yang-Mills field theory. Z Phys 1986;C30:595. DOI:10.1007/BF01571808.
[54] Wilhelm M. Amplitudes, form factors and the dilatation operator in N = 4 SYM theory. JHEP 2015;02:149. DOI:10.1007/JHEP02(2015)149.
[55] Zwiebel BI. From scattering amplitudes to the dilatation generator in N = 4 SYM. J Phys 2012;A45:115401. DOI:10.1088/1751-8113/45/11/115401.

Valentina Forini

On regulating the AdS superstring

Abstract: We discuss perturbative and non-perturbative approaches to the quantization of the Green-Schwarz string in anti de Sitter backgrounds with Ramond-Ramond fluxes, where the guiding thread is the use of genuine field theory methods, the search for a good regularization scheme associated with them and the generality of the analysis carried out. We touch upon various computational setups, both analytical and numerical, and on the role of their outcomes in understanding the detailed structure of the AdS/CFT correspondence.

Keywords: string theory, sigma models, AdS/CFT correspondence, perturbation theory, unitarity methods, ordinary differential equations, lattice field theory methods, numerical analysis.

Mathematics Subject Classification 2010: 34, 65, 81, 83

Introduction

Over the previous decade there has been beautiful progress in obtaining exact results in the framework of the duality between superconformal gauge theories and string theory in anti de Sitter (AdS) backgrounds with Ramond-Ramond (RR) fluxes, or AdS/CFT correspondence. Several examples of physical observables exist by now, whose functional behavior with the coupling is known – explicitly or implicitly – not only in the regimes which are naturally under control perturbatively (both from a gauge theory and sigma-model perspective) but also at finite coupling. Essentially two methods are decisive here, the first relying on the integrability of the underlying system [4] and the second on supersymmetric localization [78, 79].

However, not only integrability is in the finite-coupling region an assumption, and supersymmetric localization is only accessible in a limited set of cases (for those observables protected by supersymmetry[1]). Importantly, from the point of view of the string worldsheet theory – which is ours in this note – integrability is a solid fact only classically, and supersymmetric localization is not even formulated. The Green-Schwarz superstring on AdS backgrounds with RR fluxes remains, beyond its supergravity approximation, a complicated interacting two-dimensional field theory which presents subtleties also at the perturbative level. Its action, when explicitly

1 See, however, [24] for a relevant extension of this set.

DOI 10.1515/9783110452150-005_s_001

expanded in terms of independent fermionic degrees of freedom, is highly non-linear and usually quantized in a semiclassical approach [47, 54], expanding around a classical solution in powers of the (effective) string tension [73]. Here difficulties may arise due to the fact that fermionic string coordinates, which are space-time spinors, appear in the Lagrangian always through their two-dimensional projection involving derivatives of the classical background (which, in order to define fermion propagators and perform perturbation theory, must be non-trivial). Such derivatives introduce a dimensional scale and appear nonlinearly in the quartic fermionic terms, leading to non-renormalizable interactions and higher-power divergences beyond one loop.[2] Verifying the cancellation of the ultraviolet (UV) divergences with suitable regularization schemes – crucial for a well-defined expansion – may be then non-trivial. The search of regularization which is "good," i.e. equivalent to the one (implicitly) assumed by the calculations performed via integrability or localization, characterizes the work reviewed in the first part of this note. Quantizing the theory in a semiclassical approximation implies, beyond the leading order which defines minimal string surfaces (to be suitably regularized at the *AdS* boundary), solving the spectral problem of highly non-trivial differential operators of Laplace and Dirac type, namely evaluating the zeta-function determinant in the context of elliptic boundary value problems, a procedure that we illustrate on two relevant examples below. We also sketch the evaluation of next-to-leading or two-loop order corrections, and comment on an efficient alternative to Feynman diagrammatics, based on unitarity cuts, which may be used in the case of on-shell objects such as worldsheet scattering amplitudes.

At a non-perturbative level, a natural way to regularize a theory and perform ab initio calculations within it is to define it on a discretized spacetime or lattice. Lattice field theory methods have recently become a subject of study also in the framework of worldsheet string models [9, 37, 72]. This approach bypasses the subtleties of realizing supersymmetry on the lattice – which characterize the lattice approach to the duality from the gauge theory side [22] – in that the Green-Schwarz superstring formulation that we use displays supersymmetry only in the target space. In the two-dimensional string worldsheet model under analysis supersymmetry appears as a flavor symmetry. Importantly, local symmetries (diffeomorphism and fermionic kappa-symmetry) are all fixed, and only scalar fields (some of which anticommuting) appear, assigned to sites. This rather simplified setting – useful to have at most quartic fermionic interactions – still retains the sophisticated dynamics of relevant observables in this framework.

Below we will be mostly dealing with the $AdS_5 \times S^5$ superstring; with few exceptions, a majority of the observations generalizes to other AdS/CFT-relevant backgrounds.

2 This is already true for the flat space case [80]. See also discussion in [83].

1 Sigma model and perturbation theory

When evaluating the $AdS_5 \times S^5$ string partition function in a semiclassical quantization, it is possible and useful to remain extremely general at least in writing down the fluctuation spectrum about such solutions, applying elementary concepts of intrinsic and extrinsic geometry to the properties of string worldsheet embedded in a D-dimensional curved space-time. Taking full advantage of the equations of Gauss, Codazzi, and Ricci for surfaces embedded in a general background one obtains simple and general expressions for perturbations over them.[3] For example, writing down the complete mass matrix $\mathscr{M}$ in the bosonic fluctuation sector only requires as an input generic properties of the classical configuration, basic information about the space-time background and the inclusion of a suitable choice of orthonormal vectors which are orthogonal to the surface spanned by the string solution

$$\begin{aligned} \mathscr{M}_{ij} &= -\, m^2_{AdS_5}(\hat{N}_i \cdot \hat{N}_j) - m^2_{S^5}\,(\bar{N}_i \cdot \bar{N}_j) + K_{i\alpha\beta}K_j^{\alpha\beta}\,, \\ m^2_{AdS_5} &\equiv \gamma^{\rho\sigma}(\hat{t}_\rho \cdot \hat{t}_\sigma) \quad \text{and} \quad m^2_{S_5} \equiv -\gamma^{\rho\sigma}(\bar{t}_\rho \cdot \bar{t}_\sigma). \end{aligned} \tag{1}$$

Above, $\gamma_{\alpha\beta}$, $\alpha, \beta = 1, 2$ is the induced metric (pullback of the $AdS_5 \times S^5$ target space metric); hats and bars refer to the projections onto AdS_5 and S^5 of vectors tangent (t) and orthogonal (N) to the worldsheet; $K^i_{\alpha\beta} \equiv K^A_{\alpha\beta}N^i_A$ is the extrinsic curvature of the embedding; $i, j, \ldots, 8$ are transverse space-time indices.

To proceed in the one-loop analysis, one is to explicitly compute the functional determinants associated to the fluctuations operators. This is in general difficult, except in the case of rational rigid string solutions, so-called "homogeneous" [5, 48–50, 76], for which the Lagrangian has coefficients constant in the worldsheet coordinates and the one-loop partition function results in a sum over characteristic frequencies which are relatively simple to calculate. For the non-homogeneous case, one is to restrict to problems which are effectively one-dimensional. This step, which may involve regularization subtleties and other issues – as the appropriate definition of integration measure, kappa-symmetry ghosts, Jacobians due to change of fluctuation basis – is often feasible with standard techniques, such as the Gelfand-Yaglom method for the evaluation of functional determinants (stated originally in [51] and later improved in [44, 45, 66–68, 71]).[4] This algorithm has the advantage of computing ratios of determinants bypassing the computation of the full set of eigenvalues and is based on the solution of an auxiliary initial value problem. Considering the pair of n-order ordinary differential operators in one variable

$$\mathcal{O} = P_0(\sigma)\frac{d^n}{d\sigma^n} + \sum_{k=0}^{n-1} P_{n-k}(\sigma)\frac{d^k}{d\sigma^k}\,, \quad \hat{\mathcal{O}} = P_0(\sigma)\frac{d^n}{d\sigma^n} + \sum_{k=0}^{n-1} \hat{P}_{n-k}(\sigma)\frac{d^k}{d\sigma^k} \tag{2}$$

3 This follows and enlarges earlier investigations, see [16, 31].

4 See for example [32, 68], or the concise review in Appendix B of [40].

with coefficients being $r \times r$ complex matrices and continuous functions of σ on the finite interval $I = [a, b]$. The principal symbol (proportional to the coefficient $P_0(\sigma)$ of the highest-order derivative) is assumed to be equal and invertible ($\det P_0(\sigma) \neq 0$) on the whole interval[5]. The operators act on the space of square-integrable r-component functions $\vec{f} \equiv (f_1, f_2, ..., f_r)^T \in \mathscr{L}^2(I)$, and $nr \times nr$ constant matrices M, N implement the linear boundary conditions at the extrema of I

$$M \begin{pmatrix} \vec{f}(a) \\ \frac{d}{d\sigma}\vec{f}(a) \\ \vdots \\ \frac{d^{n-1}}{d\sigma^{n-1}}\vec{f}(a) \end{pmatrix} + N \begin{pmatrix} \vec{f}(b) \\ \frac{d}{d\sigma}\vec{f}(b) \\ \vdots \\ \frac{d^{n-1}}{d\sigma^{n-1}}\vec{f}(b) \end{pmatrix} = \begin{pmatrix} 0 \\ 0 \\ \vdots \\ 0 \end{pmatrix}. \tag{3}$$

The significance of the Gel'fand-Yaglom theorem is that it drastically reduces the complexity of finding the spectrum of the operators of interest

$$\mathcal{O}\vec{f}_\lambda(\sigma) = \lambda \vec{f}_\lambda(\sigma), \qquad \hat{\mathcal{O}}\hat{\vec{f}}_{\hat{\lambda}}(\sigma) = \hat{\lambda}\hat{\vec{f}}_{\hat{\lambda}}(\sigma), \tag{4}$$

encoding it into the elegant formula for (e.g. even-order) differential operators

$$\frac{\mathrm{Det}_\omega \mathcal{O}}{\mathrm{Det}_\omega \hat{\mathcal{O}}} = \frac{\exp\left\{\frac{1}{2}\int_a^b \mathrm{tr}\left[P_1(\sigma)P_0^{-1}(\sigma)\right] d\sigma\right\} \det\left[M + NY_{\mathcal{O}}(b)\right]}{\exp\left\{\frac{1}{2}\int_a^b \mathrm{tr}\left[\hat{P}_1(\sigma)P_0^{-1}(\sigma)\right] d\sigma\right\} \det\left[M + NY_{\hat{\mathcal{O}}}(b)\right]}. \tag{5}$$

This result agrees with the one obtained via ζ-function regularization for elliptic differential operators. Above, the $nr \times nr$ matrix

$$Y_{\mathcal{O}}(\sigma) = \begin{pmatrix} \vec{f}_{(I)}(\sigma) & \vec{f}_{(II)}(\sigma) & \cdots & \vec{f}_{(nr)}(\sigma) \\ \frac{d}{d\sigma}\vec{f}_{(I)}(\sigma) & \frac{d}{d\sigma}\vec{f}_{(II)}(\sigma) & \cdots & \frac{d}{d\sigma}\vec{f}_{(nr)}(\sigma) \\ \vdots & \vdots & \ddots & \vdots \\ \frac{d^{n-1}}{d^{n-1}\sigma}\vec{f}_{(I)}(\sigma) & \frac{d^{n-1}}{d^{n-1}\sigma}\vec{f}_{(II)}(\sigma) & \cdots & \frac{d^{n-1}}{d^{n-1}\sigma}\vec{f}_{(nr)}(\sigma) \end{pmatrix} \tag{6}$$

uses all the independent homogeneous solutions of

$$\mathcal{O}\vec{f}_{(i)}(\sigma) = 0 \qquad i = I, II, ..., 2r \tag{7}$$

chosen such that $Y_{\mathcal{O}}(a) = \mathbb{I}_{nr}$. In a number of relevant cases this method has been strikingly efficient in combination with the underlying classical integrable structure of

5 This assumption ensures that the leading behavior of the eigenvalues is comparable, thus the ratio is well defined despite the fact each determinant is formally the product of infinitely many eigenvalues of increasing magnitude.

the Green-Schwarz superstring on $AdS_5 \times S^5$, revealed by the presence of a class of integrable differential operators – typically of Lamé type [2, 41] – for which solutions are known in the literature. In some other problems highly non-trivial second-order *matrix* 2d differential operators appear, whose coefficients have a complicated coordinate-dependence, for example in the (effectively bosonic) mixed-modes case of a folded string spinning in S^5 with two large angular momenta (J_1, J_2), solution of the Landau-Lifshitz effective action of [69]. In this case one has to build the ingredients of the Gelf'and Yaglom method, studying ex novo fourth-order differential equations with doubly periodic coefficients, see [41]. Among other findings, this study allows the *analytic* proof of equivalence between the full exact one-loop string partition function (for the one-spin folded string) in conformal and static gauge – a non-trivial statement which finds its counterpart only in flat space [46].

1.1 Sigma-model perturbation theory and localization

The computation of the disk partition function for the $AdS_5 \times S^5$ superstring appears to be subtle, beyond the supergravity approximation, in the cases of classical solutions corresponding to supersymmetric Wilson loops. For euclidean minimal surfaces ending at the boundary on circular loops – the maximal 1/2 Bogomol'nyi-prasad-Sommerfield (BPS) [15, 31, 70], the 1/4 BPS family of "latitudes" [26, 28, 29, 36, 39, 40], the k-wound case in the fundamental representation [6], as well as loops in k-symmetric and k-antisymmetric representations [15] – the first correction to the partition function gives a result which disagrees with the gauge theory result, conjectured in [27, 30, 35, 85] and proven in [77, 78] via supersymmetric localization. To eliminate ambiguities due to the absolute normalization of the string partition function, and under the assumption that the latter is independent of the geometry of the classical worldsheet, one should consider the ratio between the partition functions for two supersymmetric Wilson loops with the same topology. This was done in [40],[6] where the one-loop determinants for fluctuations about the classical solutions corresponding to a generic "latitude" – the 1/4 BPS Wilson loops of [26, 28, 29] – and the maximal 1/2-BPS circle were evaluated with the Gel'fand-Yaglom method, confirming the disagreement with the exact gauge theory result. Recent developments suggest that to reconcile sigma-model perturbation theory and localization one should consider such ratios and use heat kernel techniques in a perturbative approach about the case of the maximal circle [43]. The relevant string worldsheet for the latter is AdS_2, where explicit heat kernel expressions for the spectra of Laplace and Dirac operators are available in [17–20], and in this case one explicitly evaluates their corrections due to the near-AdS_2

6 See also [36].

geometry induced by the generic latitude in $S^2 \subset S^5$ parametrized by a small angle θ_0. Then one considers the perturbative expansion

$$
\begin{aligned}
g_{ij} &= \bar{g}_{ij} + \theta_0^2\, \tilde{g}_{ij} + O\left(\theta_0^4\right) \\
\mathcal{O} &= \bar{\mathcal{O}} + \theta_0^2\, \tilde{\mathcal{O}} + O\left(\theta_0^4\right) , \\
K_{\mathcal{O}}(x, x^{'} ; t) &= \bar{K}_{\mathcal{O}}(x, x^{'} ; t) + \theta_0^2\, \tilde{K}_{\mathcal{O}}(x, x^{'} ; t) + O\left(\theta_0^4\right)
\end{aligned}
\tag{8}
$$

in the heat equation

$$
(\partial_t + \mathcal{O}_x)\, K_{\mathcal{O}}(x, x^{'} ; t) = 0 \qquad K_{\mathcal{O}}(x, x^{'} ; 0) = \frac{1}{\sqrt{g}} \delta^{(d)} \left(x - x^{'}\right) \mathbb{I} , \tag{9}
$$

and finds for the correction $\tilde{K}_{\mathcal{O}}$ to the functional trace $K_{\mathcal{O}}\, (t) = \bar{K}_{\mathcal{O}}\, (t) + \theta_0^2\, \tilde{K}_{\mathcal{O}}\, (t) + O\left(\theta_0^4\right)$

$$
\tilde{K}_{\mathcal{O}}\, (t) = -t \int_x \sqrt{\bar{g}}\ \mathrm{tr} \left[\tilde{\mathcal{O}}_x\, \bar{K}_{\mathcal{O}} \left(x, x^{'} ; t\right)\right]_{x = x^{'}} . \tag{10}
$$

This translates in the perturbative evaluation of each determinant in the partition function as [43],

$$
\log(\det \mathcal{O}) = -\bar{\zeta}^{'}_{\mathcal{O}}\, (0) - \theta_0^2\, \tilde{\zeta}^{'}_{\mathcal{O}}\, (0) + O(\theta_0^4) , \tag{11}
$$

$$
\bar{\zeta}_{\mathcal{O}}\, (s) = \frac{1}{\Gamma\, (s)} \int_0^\infty dt\, t^{s-1}\, \bar{K}_{\mathcal{O}}\, (t) \ , \quad \tilde{\zeta}_{\mathcal{O}}\, (s) = \frac{1}{\Gamma\, (s)} \int_0^\infty dt\, t^{s-1} \tilde{K}_{\mathcal{O}}\, (t) .
$$

This approach turns out to be successful: the gauge theory exact result is indeed reproduced, at one loop and at order $\mathcal{O}(\theta_0^2)$, by the analysis in sigma-model perturbation theory. Despite being both based on zeta-function regularization, the two procedures illustrated here for the evaluation of functional determinants differ substantially on few aspects. In this context, where the spectral problem is effectively (after Fourier-transforming in, say, τ) one-dimensional, the Gelfand-Yaglom uses a zeta-function-like regularization in σ – whose outcome is equivalent to the solution (5) above – and a cutoff regularization in the sum over the Fourier τ-modes, the latter being a priori arbitrary. As mentioned above, it also requires considering ratios of determinants for differential operators with the same principal symbol, which in turns implies a functional rescaling by the conformal factor. One may certainly quantify[7] how such conformal rescaling of the operators affects the finite part of the

7 See for example Appendix A of [31].

regularized determinants. However, a simple check for (the ratio of) two bosonic operators in [36, 40] reveals that adding this contribution does not "solve" the discrepancy with the corresponding result obtained here. Together with the arbitrariness of the sum over modes mentioned above, what may account for this is the fictitious boundary – a cut at the origin of the disk – introduced in [36, 40, 70] to allow the calculation of determinants on the finite interval (see also [25, 50]). Since the cancellation of such unphysical cutoff in the partition function is a subtle effect of the regularization scheme [40], it would be interesting to perform an explicit comparison between the calculations eliminating the need of this regulator, which in fact does not appear in the heat kernel approach.[8]

1.2 Higher orders

Beyond one loop, one has to further restrict the class of feasible problems to homogeneous configurations, and trade the standard conformal gauge with the so-called *AdS* light-cone gauge, where the light cone is entirely in *AdS* [74]. This setup – where propagators are in general simple, and (in the bosonic case) diagonal, a fact that drastically reduces the number of Feynman diagrams to be evaluated – was efficiently used in [52] to evaluate the strong coupling corrections to the $\mathcal{N} = 4$ super Yang Mills (SYM) cusp anomaly up to two-loop order. In [10] a very similar calculation was done in the considerably more involved case of the *AdS* light-cone gauge-fixed action derived via double dimensional reduction from a $D = 11$ membrane action based on the supercoset $OSp(8|4)/(SO(7) \times SO(1,3))$. As the relevant classical solution is homogeneous, the one-loop partition function is a sum of simple frequencies. At two loops, the possible topologies of connected vacuum diagrams (sunset, double bubble, double tadpole) occurring when studying the effective string action at two loops in this setup are given in Figure 1.

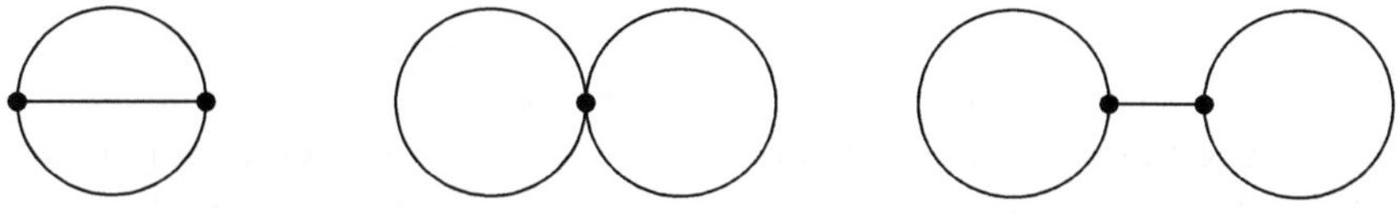

Figure 1: Sunset, double bubble and double tadpole are the diagrams appearing in the two-loop contribution to the partition function.

8 A more general application of the Gelf'and Yaglom method [33] suggests that in the case of a non-compact interval one may try to proceed selecting suitably, "well-behaved" eigenfunctions of the auxiliary initial value problem at the basis of the procedure.

When combining vertices and propagators in the sunset diagrams various non-covariant integrals are originated, but standard reduction techniques allow to rewrite every integral as a linear combination of the two following scalar ones

$$I\left(m^2\right) \equiv \int \frac{d^2p}{(2\pi)^2}\,\frac{1}{p^2+m^2} \tag{12}$$

$$I\left(m_1^2,m_2^2,m_3^2\right) \equiv \int \frac{d^2p\,d^2q\,d^2r}{(2\pi)^4}\,\frac{\delta^{(2)}(p+q+r)}{(p^2+m_1^2)(q^2+m_2^2)(r^2+m_3^2)}\,. \tag{13}$$

In this process it is standard [10, 52, 82, 83] to set to zero power UV divergent massless tadpoles, as in dimensional regularization

$$\int \frac{d^2p}{(2\pi)^2}\left(p^2\right)^n = 0\,, \qquad\qquad n \geq 0, \tag{14}$$

so that all manipulations in the numerators are performed in $d = 2$, which has the advantage of simpler tensor integral reductions. While UV finiteness is not obvious, as each diagram in eq.(12) is separately divergent (the last one in the IR, the former in both UV and IR), all logarithmically divergent integrals – remaining after the power-like are set to zero via (14) – happen to cancel out in the computation and there is no need to pick up an explicit regularization scheme to compute them. Such a non-trivial result, together with establishing the quantum consistency of the string action proposed in [86, 87], has been the first non-trivial check at strong coupling of the conjectured [53] all-order expression of the interpolating function $h(\lambda)$ appearing in terms of which all calculations based on the integrability of the AdS_4/CFT_3 system are based.

1.3 Unitarity methods in $d = 2$ dimensions

As extremely efficient alternative to Feynman diagrammatics – however only well-established for on-shell objects – unitarity-cut techniques are a powerful tool in non-abelian gauge theories for the evaluation of space-time scattering amplitudes (see e.g. [81]). In [12, 34] their use was initiated for the one-loop, perturbative study of the S-matrix for massive two-dimensional field theories, describing the scattering of the Lagrangian excitations. Here the method boils down to a reverse application of Cutkowsky rules, allowing the extraction of the discontinuity of a Feynman diagram across its branch cut. In applying the standard unitarity rules (derived from the optical theorem) [8] to the example of a one-loop four-point amplitude, one considers *two-particle cuts*, obtained by putting two intermediate lines on-shell.

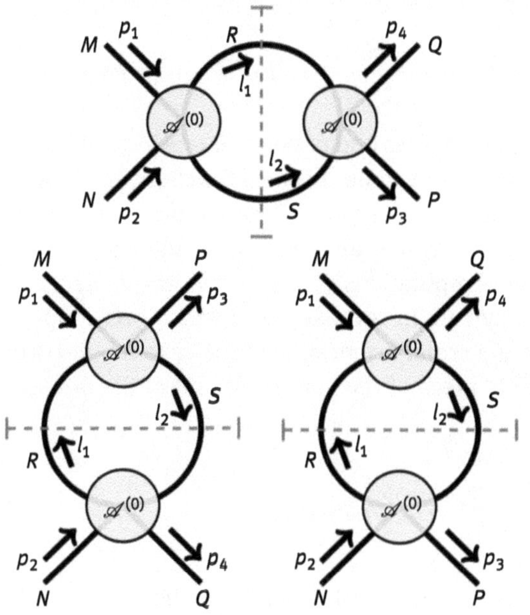

Figure 2: Diagrams representing s-, t- and u-channel cuts contributing to the four-point one-loop amplitude.

The contributions to the imaginary part of the amplitude are therefore given by the sum of s- t- and u- channel cuts illustrated in Figure 2, explicitly

$$\mathscr{A}^{(1)PQ}_{MN}(p_1, p_2, p_3, p_4)|_{s\text{-cut}} = \int \frac{d^2l_1}{(2\pi)^2} \int \frac{d^2l_2}{(2\pi)^2}\, i\pi\delta^+(l_1{}^2 - 1)\, i\pi\delta^+(l_2^2 - 1)$$
$$\times \mathscr{A}^{(0)RS}_{MN}(p_1, p_2, l_1, l_2)\mathscr{A}^{(0)PQ}_{SR}(l_2, l_1, p_3, p_4),$$

$$\mathscr{A}^{(1)PQ}_{MN}(p_1, p_2, p_3, p_4)|_{t\text{-cut}} = \int \frac{d^2l_1}{(2\pi)^2} \int \frac{d^2l_2}{(2\pi)^2}\, i\pi\delta^+(l_1{}^2 - 1)\, i\pi\delta^+(l_2{}^2 - 1)$$
$$\times \mathscr{A}^{(0)SP}_{MR}(p_1, l_1, l_2, p_3)\mathscr{A}^{(0)RQ}_{SN}(l_2, p_2, l_1, p_4),$$

$$\mathscr{A}^{(1)PQ}_{MN}(p_1, p_2, p_3, p_4)|_{u\text{-cut}} = \int \frac{d^2l_1}{(2\pi)^2} \int \frac{d^2l_2}{(2\pi)^2}\, i\pi\delta^+(l_1{}^2 - 1)\, i\pi\delta^+(l_2{}^2 - 1)$$
$$\times \mathscr{A}^{(0)SQ}_{MR}(p_1, l_1, l_2, p_4)\mathscr{A}^{(0)RP}_{SN}(l_2, p_2, l_1, p_3),$$

where $\mathscr{A}^{(0)}$ are tree-level amplitudes and a sum over the complete set of intermediate states R, S (all allowed particles for the cut lines) is understood.

Notice that tadpole graphs, having no physical two-particle cuts, are by definition ignored in this procedure. To proceed, in each case one uses the momentum conservation at the vertex involving the momentum p_1 to integrate over l_2, for instance for the s-channel

$$\widetilde{\mathscr{A}}^{(1)PQ}_{MN}(p_1, p_2, p_3, p_4)|_{s-cut} = \int \frac{d^2 l_1}{(2\pi)^2}\, i\pi\delta^+(l_1{}^2 - 1)\, i\pi\delta^+((l_1 - p_1 - p_2)^2 - 1)$$
$$\times\, \widetilde{\mathscr{A}}^{(0)RS}_{MN}(p_1, p_2, l_1, -l_1 + p_1 + p_2)\, \widetilde{\mathscr{A}}^{(0)PQ}_{SR}(-l_1 + p_1 + p_2, l_1, p_3, p_4)\,.$$

The simplicity of the two-dimensional kinematics and of being at one loop plays now its role, since in each of the integrals the set of zeroes of the δ-functions is a discrete set, and the cut loop momenta are frozen to specific values.[9] This allows us to pull out the tree-level amplitudes with the loop momenta evaluated at those zeroes.[10] In what remains, following standard unitarity computations [8], we apply the replacement $i\pi\delta^+(l^2 - 1) \longrightarrow \frac{1}{l^2-1}$ (*i.e.* the Cutkowsky rule in reverse order) which sets loop momenta back off-shell, thus reconstructing scalar bubbles. This allows to rebuild, from its imaginary part, the cut-constructible piece of the amplitude and of the S-matrix, via [12]:

$$S^{PQ}_{MN}(p_1, p_2) \equiv \frac{J(p_1, p_2)}{4\epsilon_1\epsilon_2}\, \widetilde{\mathscr{A}}^{PQ}_{MN}(p_1, p_2, p_1, p_2)\,. \tag{15}$$

where the Jacobian $J(p_1, p_2) = 1/(\partial\epsilon_{\mathrm{p}_1}/\partial \mathrm{p}_1 - \partial\epsilon_{\mathrm{p}_2}/\partial \mathrm{p}_2)$ depends on the dispersion relation ϵ_{p}, on-shell energy associated to p (the *spatial* momentum) for the theory at hand. The expression for the one-loop S-matrix elements is given by the following simple sum of products of two tree-level amplitudes[11]:

$$S^{(1)PQ}_{MN}(p_1, p_2) = \frac{1}{4(\epsilon_2\, \mathrm{p}_1 - \epsilon_1\, \mathrm{p}_2)} \Big[\tilde{S}^{(0)RS}_{MN}(p_1, p_2)\tilde{S}^{(0)PQ}_{RS}(p_1, p_2)\, I_{p_1+p_2} \tag{16}$$
$$+\, \tilde{S}^{(0)SP}_{MR}(p_1, p_1)\tilde{S}^{(0)RQ}_{SN}(p_1, p_2)\, I_0 + \tilde{S}^{(0)SQ}_{MR}(p_1, p_2)\tilde{S}^{(0)PR}_{SN}(p_1, p_2)\, I_{p_1-p_2}\Big],$$

where the coefficients are given in terms of the bubble integral

$$I_p = \int \frac{d^2 q}{(2\pi)^2}\, \frac{1}{(q^2 - 1 + i\epsilon)((q - p)^2 - 1 + i\epsilon)} \tag{17}$$

9 At two loops, to constrain completely the four components of the two momenta circulating in the loops one needs four cuts, each one giving an on-shell δ-function. Two-particle cuts at two loops would result in a manifold of conditions for the loop momenta.

10 This is like using $f(x)\delta(x - x_0) = f(x_0)\delta(x - x_0)$, where $f(x)$ are the tree-level amplitudes in the integrals.

11 In (16), $\tilde{S}^{(0)}(p_1, p_2) = 4(\epsilon_2\, p_1 - \epsilon_1\, p_2)S^{(0)}(p_1, p_2)$ and the denominator on the right-hand side comes from the Jacobian $J(p_1, p_2)$ assuming a standard relativistic dispersion relation (for the theories we consider, at one loop this is indeed the case).

and read explicitly[12]

$$I_{p_1+p_2} = \frac{i\pi - \operatorname{arsinh}(\epsilon_2 p_1 - \epsilon_1 p_2)}{4\pi i\,(\epsilon_2 p_1 - \epsilon_1 p_2)}, \quad I_0 = \frac{1}{4\pi i}, \quad I_{p_1-p_2} = \frac{\operatorname{arsinh}(\epsilon_2 p_1 - \epsilon_1 p_2)}{4\pi i\,(\epsilon_2 p_1 - \epsilon_1 p_2)}\,.$$

As it only involves the scalar bubble integral in two dimensions, the result (16) following from our procedure is inherently *finite*. No additional regularization is required and the result can be compared directly with the $2 \to 2$ particle S-matrix (following from the finite or renormalized four-point amplitude) found using standard perturbation theory. Of course, this need not be the case for the original bubble integrals before cutting – due to factors of loop momentum in the numerators. These divergences, along with those coming from tadpole graphs, which are not considered in this procedure, should be taken into account for the renormalization of the theory. We have not investigated this issue, since all the theories considered in [12][13] are either UV-finite or renormalizable. The method, applied to various models, has shown enough evidence to postulate that supersymmetric, integrable two-dimensional theories should be cut-constructible via standard unitarity methods. For bosonic theories with integrability, agreement was found with perturbation theory up to a finite shift in the coupling. In the case of the superstring worldsheet models in, $AdS_5 \times S^5$ [12] and $AdS_3 \times S^3 \times M^4$ [14] the method allowed non-trivial confirmations of the integrability prediction together with conjectures (then confirmed) on the one-loop phases.

2 $AdS_5 \times S^5$ superstring on a lattice

The natural, genuinely field-theoretical way to investigate the finite-coupling region and in general the non-perturbative realm of a quantum field theory is to discretize the spacetime where the model lives, and proceed with numerical methods for the lattice field theory so defined. A rich and interesting program of putting $\mathcal{N} = 4$ SYM a spacetime lattice is being carried out for some years [7, 23, 63, 65, 84].[14] Alternatively, one could discretize the worldsheet spanned by the Green-Schwarz string embedded in $AdS_5 \times S^5$. If the aim is a test of the AdS/CFT correspondence and/or the integrability of the string sigma model, it is is obviously computationally cheaper to use a two-dimensional grid, rather than a four-dimensional one, where no gauge degrees of freedom are present and all fields are assigned to sites – so that only scalar fields (some of which anticommuting) appear in the relevant action. Also, although we are dealing with superstrings, there is here no subtlety involved with putting supersymmetry on

12 The t-channel cut requires a prescription [12].

13 They include, among relativistic theories, a class of generalized sine-Gordon models, defined by a gauged WZW model for a coset G/H plus a potential. Notable non-relativistic cases are the superstring worldsheet models in $AdS_5 \times S^5$ and $AdS_3 \times S^3 \times M^4$, see [14].

14 See also the numerical, non-lattice formulation of $\mathcal{N} = 4$ SYM on $R \times S^3$ as plane-wave (BMN) matrix model given in [56–62].

the lattice (see e.g. [22]), both because of the Green-Schwarz formulation of the action (with supersymmetry only manifest in the target space) and because κ-symmetry is gauge-fixed. In general, one merit of this analysis is to explore another route via which lattice simulations[15] could become a potentially efficient tool in numerical holography. Following the earlier proposal of [72], such a route has been taken in [9, 37] to investigate relevant observables in AdS/CFT, discretizing the dual two-dimensional string worldsheet. There, the focus is on particularly important observables completely "solved" via integrability [3]: the cusp anomalous dimension of $\mathcal{N} = 4$ SYM – measured by the path integral of an open string bounded by a null-cusped Wilson loop at the AdS boundary – and the spectrum of excitations around the corresponding string minimal surface. The relevant string worldsheet theory, an AdS-light-cone-gauge-fixed action [74, 75], is a highly non-trivial 2d non-linear sigma model with rich non-perturbative dynamics. On the lattice, several subtleties appear (fermion doublers, complex phases) which require special treatment, as we sketch below.

2.1 The observable in the continuum

The cusp anomaly of $\mathcal{N} = 4$ SYM governs the renormalization of a cusped Wilson loop, and according to AdS/CFT should be represented by the path integral of an open string ending on the loop at the AdS boundary

$$\begin{aligned}\langle W[C_{\rm cusp}]\rangle \equiv Z_{\rm cusp} &= \int [D\delta X][D\delta\Psi]\, e^{-S_{\rm cusp}[X_{\rm cl}+\delta X,\delta\Psi]} \\ &= e^{-\Gamma_{\rm eff}} \equiv e^{-\frac{1}{8}f(g)\,V_2}\,.\end{aligned} \tag{18}$$

Above, $X_{\rm cl} = X_{\rm cl}(t,s)$ – with t, s the temporal and spatial coordinates spanning the string worldsheet – is the relevant classical solution [52], $S_{\rm cusp}[X + \delta X, \delta\Psi]$ is the action for field fluctuations over it – the fields being both bosonic and fermionic string coordinates $X(t,s)$, $\Psi(t,s)$ – and is reported below in equation (20) in terms of the effective bosonic and fermionic degrees of freedom remaining after gauge-fixing. Being a homogeneous solution, the worldsheet volume simply factorizes out[16] in front of the function of the coupling $f(g)$, as in the last equivalence above. Rather than partition functions, in a lattice approach it is natural to study vacuum expectation values. In simulating the vacuum expectation value of the "cusp" action

$$\langle S_{\rm cusp}\rangle = \frac{\int [D\delta X][D\delta\Psi]\, S_{\rm cusp}\, e^{-S_{\rm cusp}}}{\int [D\delta X][D\delta\Psi]\, e^{-S_{\rm cusp}}} = -g\,\frac{d\ln Z_{\rm cusp}}{dg} \equiv g\,\frac{V_2}{8}\,f'(g), \tag{19}$$

15 See for example [55] and reference therein on possible further uses of lattice techniques in AdS/CFT.
16 The normalization of V_2 with a 1/4 factor follows the convention of [52].

one therefore obtains information on the *derivative* of the scaling function. In the continuum, the $AdS_5 \times S^5$ superstring action S_{cusp} describing quantum fluctuations around the null-cusp background is [52] (after Wick-rotation)

$$\begin{aligned} S_{\text{cusp}} = g \int dtds \Big\{ & |\partial_t x + \tfrac{1}{2}x|^2 + \tfrac{1}{z^4}|\partial_s x - \tfrac{1}{2}x|^2 - \tfrac{1}{z^2}\left(\eta^i\eta_i\right)^2 + \tfrac{1}{z^4}\left(\partial_s z^M - \tfrac{1}{2}z^M\right)^2 \\ & + \left(\partial_t z^M + \tfrac{1}{2}z^M + \tfrac{i}{z^2}z_N\eta_i\left(\rho^{MN}\right)^i{}_j\eta^j\right)^2 + i\left(\theta^i\partial_t\theta_i + \eta^i\partial_t\eta_i + \theta_i\partial_t\theta^i + \eta_i\partial_t\eta^i\right) \\ & + 2i\Big[\tfrac{1}{z^3}z^M\eta^i\left(\rho^M\right)_{ij}\left(\partial_s\theta^j - \tfrac{1}{2}\theta^j - \tfrac{i}{z}\eta^j\left(\partial_s x - \tfrac{1}{2}x\right)\right) \\ & + \tfrac{1}{z^3}z^M\eta_i(\rho_M^\dagger)^{ij}\left(\partial_s\theta_j - \tfrac{1}{2}\theta_j + \tfrac{i}{z}\eta_j\left(\partial_s x - \tfrac{1}{2}x\right)^*\right)\Big]\Big\}. \end{aligned} \tag{20}$$

Above, x, x^* are the two bosonic AdS_5 (coordinate) fields transverse to the AdS_3 subspace of the classical solution, and z^M ($M = 1, \ldots, 6$) are the bosonic coordinates of the $AdS_5 \times S^5$ background in Poincaré parametrization, with $z = \sqrt{z_M z^M}$, remaining after fixing the AdS light-cone gauge. The fields θ_i, η_i, $i = 1, 2, 3, 4$ are 4+4 complex anticommuting variables for which $\theta^i = (\theta_i)^\dagger$, $\eta^i = (\eta_i)^\dagger$. They transform in the fundamental representation of the $SU(4)$ R-symmetry and do not carry (Lorentz) spinor indices. The matrices ρ^M_{ij} are the off-diagonal blocks of $SO(6)$ Dirac matrices γ^M in the chiral representation and $(\rho^{MN})_i{}^j = (\rho^{[M}\rho^{\dagger N]})_i{}^j$ are the $SO(6)$ generators. In eq. (20) – where a massive parameter $m \sim P^+$, usually set to one, is restored – local bosonic (diffeomorphism) and fermionic (κ-) symmetries originally present have been fixed. With this action one can directly proceed to the perturbative evaluation of the cusp anomaly (K is the Catalan constant) and of the dispersion relation for the field excitations

$$\begin{aligned} f(g) &= 4g\left(1 - \frac{3\log 2}{4\pi g} - \frac{K}{16\pi^2 g^2} + \mathcal{O}(g^{-3})\right) \\ m_x^2(g) &= \frac{m^2}{2}\left(1 - \frac{1}{8g} + \mathcal{O}(g^{-2})\right). \end{aligned} \tag{21}$$

While the bosonic part of eq. (20) can be easily discretized and simulated, Graßmann-odd fields are formally integrated out, letting their determinant to become part – via exponentiation in terms of pseudo-fermions, see (26) below – of the Boltzmann weight of each configuration in the statistical ensemble. In the case of higher-order fermionic interactions – as in eq. (20), where they are at most quartic – this is possible via the introduction of auxiliary fields realizing a linearization. The most natural linearization [72] introduces seven real auxiliary fields, one scalar ϕ and a $SO(6)$ vector field ϕ_M, with a Hubbard-Stratonovich transformation

$$\begin{aligned} & e^{-g\int dtds\left[-\frac{1}{z^2}\left(\eta^i\eta_i\right)^2 + \left(\frac{i}{z^2}z_N\eta_i\rho^{MN\,i}{}_j\eta^j\right)^2\right]} \\ & \sim \int D\phi D\phi^M\, e^{-g\int dtds\left[\frac{1}{2}\phi^2 + \frac{\sqrt{2}}{z}\phi\,\eta^2 + \frac{1}{2}(\phi_M)^2 - i\frac{\sqrt{2}}{z^2}\phi^M z_N\left(i\eta_i\rho^{MN\,i}{}_j\eta^j\right)\right]}. \end{aligned} \tag{22}$$

Above, in the second line we have written the Lagrangian for ϕ^M so to emphasize that it has an imaginary part, due to the fact that the bilinear form in round brackets is hermitian

$$\left(i\,\eta_i\rho^{MN^i}{}_j\eta^j\right)^\dagger = -i(\eta^j)^\dagger(\rho^{MN^i}{}_j)^*(\eta_i)^\dagger = -i\eta_j\,\rho^{MN}{}_i^j\,\eta^i = i\eta_j\,\rho^{MN^j}{}_i\,\eta^i\,, \tag{23}$$

as follows from the properties of the $SO(6)$ generators. Since the auxiliary vector field ϕ^M has real support, the Yukawa-term for it sets *a priori* a phase problem, the only question being whether the latter is treatable via standard reweighing. After the transformation (22), the corresponding Lagrangian reads

$$\begin{aligned}\mathscr{L} &= |\partial_t x + \tfrac{m}{2}x|^2 + \tfrac{1}{z^4}|\partial_s x - \tfrac{m}{2}x|^2 + (\partial_t z^M + \tfrac{1}{2}z^M)^2 + \tfrac{1}{z^4}(\partial_s z^M - \tfrac{m}{2}z^M)^2 \\ &\quad + \frac{1}{2}\phi^2 + \frac{1}{2}(\phi_M)^2 + \psi^T O_F\psi \end{aligned} \tag{24}$$

with $\psi \equiv (\theta^i, \theta_i, \eta^i, \eta_i)$ and

$$O_F = \begin{pmatrix} 0 & i\partial_t & -i\rho^M\left(\partial_s + \frac{m}{2}\right)\frac{z^M}{z^3} & 0 \\ i\partial_t & 0 & 0 & -i\rho^\dagger_M\left(\partial_s + \frac{m}{2}\right)\frac{z^M}{z^3} \\ i\frac{z^M}{z^3}\rho^M\left(\partial_s - \frac{m}{2}\right) & 0 & 2\frac{z^M}{z^4}\rho^M\left(\partial_s x - m\frac{x}{2}\right) & i\partial_t - A^T \\ 0 & i\frac{z^M}{z^3}\rho^\dagger_M\left(\partial_s - \frac{m}{2}\right) & i\partial_t + A & -2\frac{z^M}{z^4}\rho^\dagger_M\left(\partial_s x^* - m\frac{x^*}{2}\right) \end{pmatrix},$$

$$A = \frac{1}{\sqrt{2}z^2}\phi_M\rho^{MN}z_N - \frac{1}{\sqrt{2}z}\phi + i\,\frac{z_N}{z^2}\rho^{MN}\,\partial_t z^M\,. \tag{25}$$

The quadratic fermionic contribution resulting from linearization gives then formally a Pfaffian $\mathrm{Pf}\,O_F$, which – in order to enter the Boltzmann weight and thus be interpreted as a probability – should be positive definite. For this reason, one proceeds as follows

$$\int D\Psi\, e^{-\int dt ds\, \Psi^T O_F \Psi} = \mathrm{Pf}\,O_F \equiv (\det O_F\, O_F^\dagger)^{\frac{1}{4}} = \int D\xi D\bar{\xi}\, e^{-\int dt ds\, \bar{\xi}(O_F O_F^\dagger)^{-\frac{1}{4}}\xi}\,, \tag{26}$$

where the second equivalence obviously ignores potential phases or anomalies. The values of the discretized (scalar) fields are assigned to each lattice site, with periodic boundary conditions for all the fields except for antiperiodic temporal boundary conditions in the case of fermions. The discrete approximation of continuum derivatives are finite difference operators defined on the lattice. A Wilson-like lattice operator must be introduced, such that fermion doublers are suppressed and the one-loop constant $-3\ln 2/\pi$ in eq. (21) is recovered in lattice perturbation theory.

2.2 Simulations

The Monte Carlo evolution of the action (24) is generated by the standard Rational Hybrid Monte Carlo algorithm, with a rational approximation (Remez algorithm) for the inverse fractional power in the last equation of (26), as in [72]. In the continuum model there are two parameters, the dimensionless coupling $g = \frac{\sqrt{\lambda}}{4\pi}$ and the mass scale m. In taking the continuum limit, the dimensionless physical quantities that it is natural to keep constant are the physical masses of the field excitations rescaled by L, the spatial lattice extent. This is our line of constant physics. For the example in eq. (21), this means

$$L^2\, m_x^2 = \text{const}\,, \qquad \text{which leads to} \qquad L^2\, m^2 \equiv (NM)^2 = \text{const}\,, \tag{27}$$

where we defined the dimensionless $M = ma$ with the lattice spacing a. The second equation in (27) relies first on the assumption that g is *not* renormalized, which is suggested by lattice perturbation theory. Second, one should investigate whether the second relation in eq. (21), and the analog ones for the other fields of the model, are still true in the discretized model – i.e. the physical masses undergo only a *finite* renormalization. In this case, at each fixed g fixing $L^2\, m^2$ constant would be enough to keep the rescaled physical masses constant, namely no tuning of the "bare" parameter m would be necessary. In [9], we considered the example of bosonic x, x^* correlators, whose asymptotic exponential decay is governed by the physical mass $m_{x\text{LAT}}$, as from the partially Fourier transformed

$$C_x(t;\,0) \overset{t\gg 1}{\sim} e^{-t\, m_{x\text{LAT}}}\,, \qquad m_{x\text{LAT}} = \lim_{T,\,t\to\infty} m_x^{\text{eff}} \equiv \lim_{T,\,t\to\infty,} \frac{1}{a} \log \frac{C_x(t;\,0)}{C_x(t+a;\,0)}\,. \tag{28}$$

On the lattice $m_{x\text{LAT}}$ is usefully obtained as a limit of an effective mass, the discretized logarithmic derivative above, that in Figure 3 is measured as a function of the time t (in units of $m_{x\text{LAT}}$) for different lattice sizes. No $(1/a)$ divergence is found, and in the large g region that we investigate the ratio considered approaches the expected continuum value 1/2. Having this as hint corroborating the choice of the line of constant physics, and because with the proposed discretization we recover in perturbation theory the one-loop cusp anomaly, we assume that in the discretized model no further scale but the lattice spacing a is present. Any observable F_{LAT} is therefore a function $F_{\text{LAT}} = F_{\text{LAT}}(g, N, M)$ of the input (dimensionless) parameters $g = \frac{\sqrt{\lambda}}{4\pi}$, $N = \frac{L}{a}$ and $M = a\,m$. At fixed coupling g and fixed $m\,L \equiv M\,N$ (large enough so to keep finite volume effects $\sim e^{-mL}$ small), F_{LAT} is evaluated for different values of N. The continuum limit – which we do not attempt here – is then obtained extrapolating to infinite N.

In measuring the action (19) on the lattice, we are supposed to recover the following general behavior

$$\frac{\langle S_{\text{LAT}}\rangle}{N^2} = \frac{c}{2} + \frac{1}{8} M^2\, g f'(g)\,, \tag{29}$$

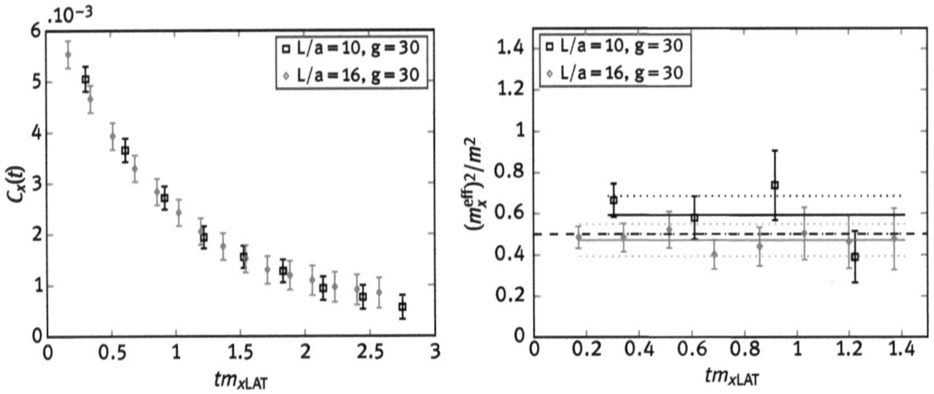

Figure 3: Correlator $C_x(t) = \sum_{s_1,s_2} \langle x(t, s_1)x^*(0, s_2)\rangle$ of bosonic fields x, x^* (**left panel**) and corresponding effective mass $m_x^{\mathrm{eff}} = \frac{1}{a} \ln \frac{C_x(t)}{C_x(t+a)}$ normalized by m^2 (**right panel**), plotted as functions of the time t in units of $m_{x\mathrm{LAT}}$ for different g and lattice sizes. The flatness of the effective mass indicates that the ground state saturates the correlation function, and allows for a reliable extraction of the mass of the x-excitation. Data points are masked by large error bars for time scales greater than unity because the signal of the correlator degrades exponentially compared with the statistical noise.

where we have reinserted the parameter m, used that $V_2 = a^2 N^2$ and added a constant contribution in g which takes into account possible coupling-dependent Jacobians relating the (derivative of the) partition function on the lattice to the one in the continuum. Measurements for the ratio $\frac{\langle S_{LAT}\rangle - c\,N^2/2}{M^2 N^2 g/2} = \frac{f'(g)}{4}$ are, at large g, in good agreement with $\frac{c}{2} = 7.5(1)$, consistently with the counting of those degrees of freedom which appear quadratically, and multiplying g, in the action – the number of bosons.[17] Having determined with good precision the coefficient of the divergence, one proceeds first fixing it to be exactly $c = 15$ and subtracting it from the action. At large g, a good agreement is found with the leading-order prediction in eq. (21) for which $f'(g) = 4$. For lower values of g one observes deviations that obstruct the continuum limit and signal the presence of further quadratic ($\sim N^2$) divergences. It seems natural to relate these power divergences to those arising in continuum perturbation theory and mentioned in the previous Section, where they are usually set to zero using dimensional regularization. From the perspective of a hard cutoff regularization like the lattice one, this is related to the emergence in the continuum limit of power divergences – quadratic, in the present two-dimensional case – induced by mixing of the (scalar) Lagrangian with the identity operator under UV renormalization. One may proceed with a non-perturbative subtraction of these divergences.

---

17 In lattice codes, it is conventional to omit the coupling from the (pseudo)fermionic part of the action, since this is quadratic in the fields and hence its contribution in g can be evaluated by a simple scaling argument.

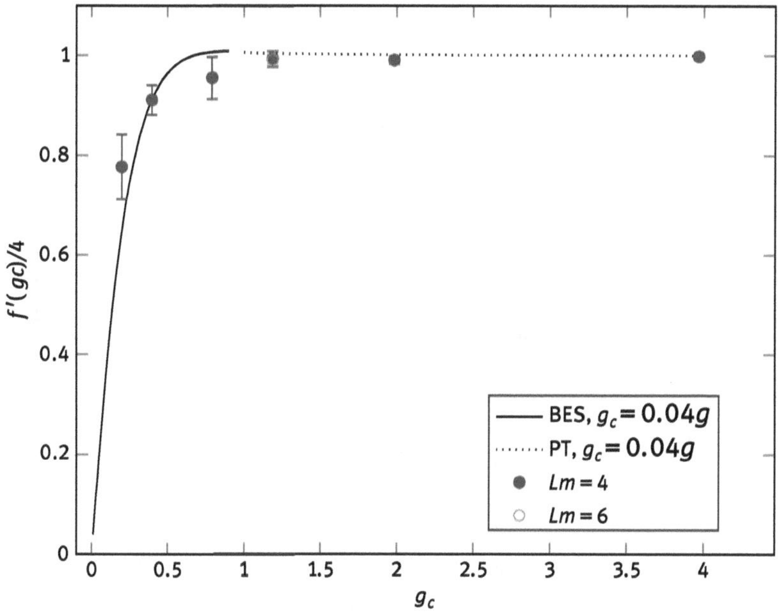

Figure 4: Plot for $f'(g)/4$ as a function of the (bare) continuum coupling g_c under the hypothesis that the latter is just a finite rescaling of the lattice bare coupling g ($g_c = 0.04\,g$). The dashed line represents the first few terms in the perturbative series in eq. (21), the continuous line is obtained from a numerical solution of the BES equation and represents therefore the prediction from the integrability of the model. The simulations at $g = 30$, $mL = 6$ (orange point) are used for a check of the finite volume effects, that appear to be within statistical errors.

A simple look at Figure 4 shows that, in the perturbative region, our analysis – and the related assumption for the finite rescaling of the coupling – is in good qualitative agreement with the integrability prediction. About direct comparison with the perturbative series, the plot in Figure 4 does not catch the minimal upward trend of the first correction to the expected large g behavior $f'(g)/4 \sim 1$ – we are considering the derivative of eq. (21) and the first correction is (positive and) too small, about 2 percent, if compared to the statistical error. Notice that, again under the assumption that such simple relation between the couplings exists – something that within our error bars cannot be excluded – the non-perturbative regime beginning with $g_c = 1$ would start at $g = 25$, implying that our simulations at $g = 10, 5$ would already test a fully non-perturbative regime of the string sigma model under investigation. In proximity to $g \sim 1$, severe fluctuations appear in the averaged *complex* phase of the Pfaffian – see Figure 5 – signaling the sign problem mentioned above. Interestingly, at least some steps in the direction of solving this problem can be done analytically [11, 13]. A new auxiliary field representation of the four-fermi term may be realized, following an algebraic manipulation from which a hermitian Lagrangian linearized in fermions results, leading to a Pfaffian of the quadratic fermionic operator O_F which is *real*,

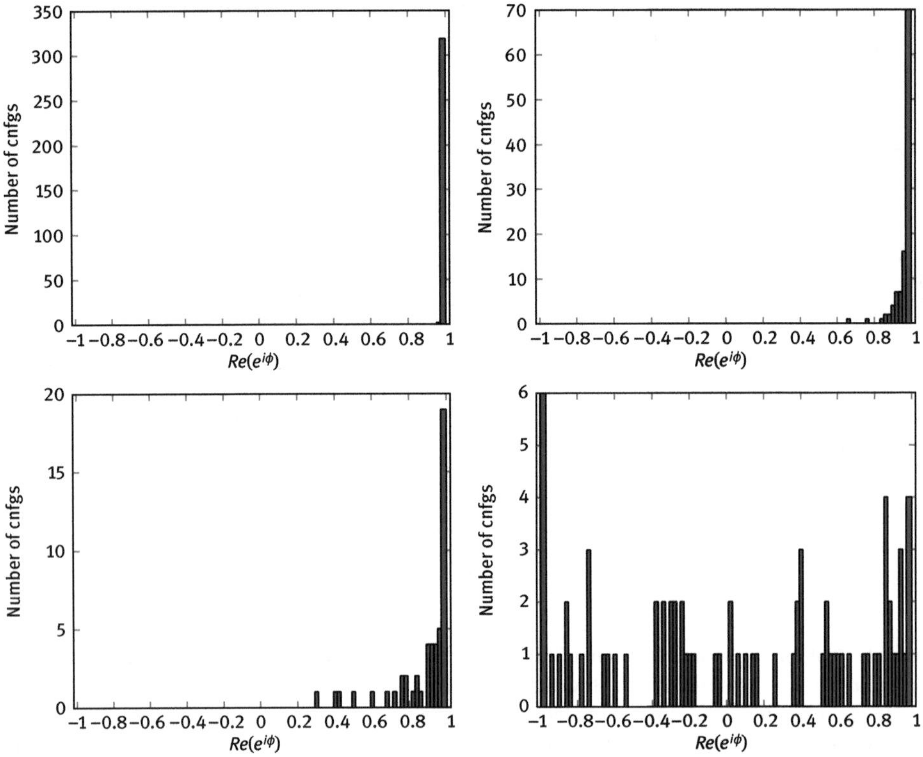

Figure 5: Histograms for the frequency of the real part of the reweighting phase factor $e^{i\theta}$ of the Pfaffian $\mathrm{Pf}\,O_F = |(\det O_F)^{\frac{1}{2}}|\,e^{i\theta}$, based on the ensembles generated at $g = 30, 10, 5, 1$ (from left to right, top to down) for $L/a = 8$.

$(\mathrm{Pf}\,O_F)^2 = \det O_F \geq 0$. Although a sign ambiguity remains, as the Pfaffian is still not positive definite, $\mathrm{Pf}\,O_F = \pm \det O_F$, this is an important advancement in the efficiency of the simulations, as it allows eliminating systematic errors and identifying with precision the region of parameter space where information on non-perturbative physics may be captured.

3 Discussion and outlook

We have reviewed and discussed perturbative and non-perturbative approaches to the quantization of the Green-Schwarz string in AdS backgrounds with RR fluxes, with an emphasis on the use of direct quantum field theory methods and on the cross-fertilization of theoretical tools well established in gauge field theories to the string worldsheet context.

In dealing with sigma-model perturbation theory, crucial subtleties appear in evaluating regularized functional determinants for string fluctuations and the computational technology for them has to be carefully adjusted to the problem at hand. It would be important to develop a diffeomorphism-preserving regularization scheme which retains the efficiency of the Gelf'and-Yaglom method for one-dimensional cases, and extend such techniques to regularized super-traces and super-determinants so to address a uniform way of treating BPS and non-BPS observables. It would be extremely interesting to elucidate the role of the measure (structure and normalization) in the string path integral for supersymmetric configurations, for example investigating the string dual of the anomaly of [30].

Going beyond perturbation theory, a new research line has been addressed, which employs Monte Carlo simulations to investigate observables defined on suitably discretized euclidean string worldsheets. At a fundamental level, this is the natural setup for verifying with unequaled definiteness the holographic conjecture *and* the exact methods that "solve" various sectors of the AdS/CFT system. Lattice methods are also the most suitable candidates for the study of several observables and backgrounds for which alternative techniques to go beyond perturbation theory are not existing (string backgrounds which are not classically integrable) or yet at a preliminary stage (correlators of string vertex operators and dual gauge theory correlation functions).

It is important to emphasize that the analysis here carried out is far from being a non-perturbative definition, à la Wilson lattice Quantum Chromodynamics (QCD), of the Green-Schwarz worldsheet string model. For this purpose one should work with a Lagrangian which is invariant under the local symmetries – bosonic diffeomorphisms and κ-symmetry – of the model, while as mentioned we make use of an action which fixes them all. There is however a number of reasons which make this model interesting for lattice investigations, within and hopefully beyond the community interested in holographic models.

As computational playground this is an interesting one on its own, allowing in principle for explicit investigations/improvements of algorithms: a highly non-trivial two-dimensional model with four-fermion interactions, for which relevant observables have not only, through AdS/CFT, an explicit analytic strong coupling expansion – the perturbative series in the dual gauge theory – but also, through AdS/CFT *and* the assumption of integrability, an explicit numerical prediction at all couplings.

The results discussed here open the way to a variety of further explorations and developements. A natural evolution consists in treating strings propagating in those backgrounds (the ten-dimensional $AdS_4{\times}CP^3$, $AdS^3{\times}S^3{\times}T^4$, $AdS_3{\times}S^3{\times}S^3{\times}S^1$ supported by RR fluxes) relevant for lower-dimensional formulations of the correspondence, for which several predictions exist from integrability, and for which an independent AdS-light-cone gauge-fixed Lagrangians are expected to be considerably more involved than in the prototypical case, but still with vertices at most quartic in fermions. In all the novel cases of study the presence of massless fermionic modes is expected to require an ad hoc treatment, one possibility being to work in a finite volume setting like the Schrödinger functional scheme. A thorough study of the possible sign problem – related to the absence of positive definiteness of the fermion Pfaffian – that is likely to appear at large values of the string tension as in the prototypical case would be crucial. This may consist in carving out the region of parameter space where the sign ambiguity is not severe and clarifying whether non-perturbative physics is obtainable. One may then verify the possibility of tracking down this ambiguity to the behavior of a smaller set of degrees of freedom – such analysis may profit, at least pedagogically, from recent progress on the analysis of the sign problem in (considerably simpler) models with quartic fermionic interactions [1, 21]. Also, it would be very interesting to explore the discretization of the gauge-fixed string action of [64], whose relevance from the point of view of the string/gauge gravity correspondence is far less clear, but that being only quadratic in fermions may lead to considerable simplifications in the general analysis.

Acknowledgments: It is a pleasure to thank Lorenzo Bianchi, Marco S. Bianchi, Alexis Brés, Ben Hoare, Valentina Giangreco M. Puletti, Luca Griguolo, Bjoern Leder, Michael Pawellek, Domenico Seminara, Arkady A. Tseytlin and Edoardo Vescovi for the very nice collaboration on [9, 10, 12, 37–42], on which this review is based. In particular, I am grateful to the long-period members of the Emmy Noether Group "Gauge Fields from Strings" – Ben Hoare, Lorenzo Bianchi and Edoardo Vescovi – for joining me and my research program, and making possible to achieve several relevant goals in our given time frame. This research was largely supported by the German Research Foundation (DFG) through the Emmy Noether Group 31408816. The novel, interdisciplinary program of discretization and simulation of the Green-Schwarz superstring would have not been possible without the support of the Collaborative Research Centre "Space-time-matter" SFB 647 – subproject C5 "AdS/CFT Correspondence: Integrable Structures and new Observables" – involving several scientists at Berlin and Postdam Universities.

Bibliography

[1] Alexandru A, Basar G, Bedaque PF, Ridgway GW, Warrington NC. Monte Carlo calculations of the finite density Thirring model. Phys Rev 2017;D95:014502, [1609.01730].

[2] Beccaria M, Dunne GV, Forini V, Pawellek M, Tseytlin AA. Exact computation of one-loop correction to energy of spinning folded string in ads5xs5. J Phys 2010;A43:165402, [1001.4018].

[3] Beisert N, Eden B, Staudacher M, Transcendentality and crossing. J Stat Mech 2007;0701:P01021, [hep-th/0610251].

[4] Beisert N, et al. Review of AdS/CFT integrability: an overview. Lett Math Phys 2016;99:3–32, [1012.3982].

[5] Beisert N, Tseytlin AA, Zarembo K. Matching quantum strings to quantum spins: one-loop versus finite-size corrections. Nucl Phys 2005;B715:190–210, [hep-th/0502173].

[6] Bergamin R, Tseytlin AA. Heat kernels on cone of AdS2 and k-wound circular Wilson loop in $AdS_5 \times S^5$ superstring. J Phys 2016;A49:14LT01, [1510.06894].

[7] Bergner G, Catterall S. Supersymmetry on the lattice, [1603.04478].

[8] Bern Z, Dixon LJ, Dunbar DC, Kosower DA. One loop n point gauge theory amplitudes, unitarity and collinear limits. Nucl Phys 1994;B425:217–60, [hep-ph/9403226].

[9] Bianchi L, Bianchi MS, Forini V, Leder B, Vescovi E. Green-Schwarz superstring on the lattice. JHEP 2016;07:014, [1605.01726].

[10] Bianchi L, Bianchi MS, Bres A, Forini V, Vescovi E. Two-loop cusp anomaly in ABJM at strong coupling. JHEP 2014;10:13, [1407.4788].

[11] Bianchi L, Forini V, Leder B, Toepfer P, Vescovi E. Strings on the lattice and AdS/CFT. PoS 2016;LATTICE2016:206.

[12] Bianchi L, Forini V, Hoare B. Two-dimensional S-matrices from unitarity cuts. JHEP 2013;1307:088, [1304.1798].

[13] Bianchi L, Forini V, Leder B, Toepfer P, Vescovi E, Superstring worldsheet on a lattice: taming the complex phase (to appear).

[14] Bianchi L, Hoare B. $AdS_3 \times S^3 \times M^4$ string S-matrices from unitarity cuts. JHEP 2014;08:097, [1405.7947].

[15] Buchbinder EI, Tseytlin AA, 1=N correction in the D3-brane description of a circular Wilson loop at strong coupling. Phys Rev 2014;D89:126008, [1404.4952].

[16] Callan CG, Jr., Thorlacius L. Sigma models and string theory, TASI Lectures 1988.

[17] Camporesi R. Harmonic analysis and propagators on homogeneous spaces. Phys Rept 1990;196:1–134.

[18] Camporesi R. The spinor heat kernel in maximally symmetric spaces. Commun Math Phys 1992;148:283–308.

[19] Camporesi R, Higuchi A. Spectral functions and zeta functions in hyperbolic spaces. J Math Phys 1994;35:4217–46.

[20] Camporesi R, Higuchi A. On the eigenfunctions of the Dirac operator on spheres and real hyperbolic spaces. J Geom Phys 1996;20:1–18, [gr-qc/9505009].

[21] Catterall S. Fermion mass without symmetry breaking. JHEP 2016;01:121, [1510.04153].

[22] Catterall S, Giedt J, Schaich D, Damgaard PH DeGrand T. Results from lattice simulations of N = 4 supersymmetric Yang-Mills. PoS 2014;LATTICE2014:267, [1411.0166].

[23] Catterall S, Schaich D, Damgaard PH, DeGrand T, Giedt J. N = 4 supersymmetry on a space-time lattice. Phys Rev 2014;D90:065013, [1405.0644].

[24] Correa D, Henn J, Maldacena J, Sever A. An exact formula for the radiation of a moving quark in N = 4 super Yang Mills. JHEP 2012;1206:048, [1202.4455].

[25] Dekel A, Klose T. Correlation function of circular Wilson loops at strong coupling. JHEP 2013;1311:117, [1309.3203].

[26] Drukker N. 1/4 BPS circular loops, unstable world-sheet instantons and the matrix model. JHEP 2006;09:004, [hep-th/0605151].

[27] Drukker N, Fiol B, All-genus calculation of Wilson loops using D-branes. JHEP 2005;02:010, [hep-th/0501109].

[28] Drukker N, Fiol B. On the integrability of Wilson loops in ads5×s5: some periodic ansatze. JHEP 2006;0601:056, [hep-th/0506058].

[29] Drukker N, Giombi S, Ricci R, Trancanelli D. Supersymmetric wilson loops on S3. JHEP 2008;0805:017, [0711.3226].

[30] Drukker N, Gross DJ. An exact prediction of N = 4 SUSYM theory for string theory. J Math Phys 2001;42:2896–914, [hep-th/0010274].

[31] Drukker N, Gross DJ, Tseytlin AA. Green-schwarz string in ads5×s5: Semiclassical partition function. JHEP 2000;0004:021, [hep-th/0001204].

[32] Dunne GV. Functional determinants in quantum field theory. J Phys 2008;A41:304006, [0711.1178].

[33] Dunne GV, Kirsten K. Functional determinants for radial operators. J Phys 2006;A39:11915–28, [hep-th/0607066].

[34] Engelund OT, McKeown RW, Roiban R. Generalized unitarity and the worldsheet S matrix in $AdS_n \times S^n \times M^{10-2n}$. JHEP 2013;1308:023, [1304.4281].

[35] Erickson J, Semenoff G, Zarembo K. Wilson loops in N = 4 supersymmetric Yang-Mills theory. Nucl Phys 2000;B582:155–75, [hep-th/0003055].

[36] Faraggi A, Pando Zayas LA, Silva GA, Trancanelli D. Toward precision holography with supersymmetric Wilson loops. JHEP 2016;04:053, [1601.04708].

[37] Forini V, Bianchi L, Bianchi MS, Leder B, Vescovi E. Lattice and string worldsheet in AdS/CFT: a numerical study. PoS 2015;LATTICE2015:244, [1601.04670].

[38] Forini V, Bianchi L, Hoare B. Scattering and unitarity methods in two dimensions. Springer Proc Phys 2016;170:169–77, [1401.0448].

[39] Forini V, Puletti VGM, Griguolo L, Seminara D, Vescovi E. Remarks on the geometrical properties of semiclassically quantized strings. J Phys 2015;A48:475401, [1507.01883].

[40] Forini V, Puletti VGM, Griguolo L, Seminara L, Vescovi E. Precision calculation of 1/4-BPS Wilson loops in $AdS_5 \times S^5$. JHEP 2016;02:105, [1512.00841].

[41] Forini V, Puletti VGM, Pawellek M, Vescovi E. One-loop spectroscopy of semiclassically quantized strings: bosonic sector. J Phys 2015;A48:085401, [1409.8674].

[42] Forini V, Puletti VGM, Sax OO, The generalized cusp in $AdS_4 \times CP^3$ and more one-loop results from semiclassical strings. J Phys 2013;A46:115402, [1204.3302].

[43] Forini V, Tseytlin AA, Vescovi E. Perturbative computation of string one-loop corrections to Wilson loop minimal surfaces in $AdS_5 \times S^5$. JHEP 2017;1703:003, [1702.02164].

[44] Forman R. Functional determinants and geometry. Invent Math 1987;88:447–93.

[45] Forman R. Functional determinants and geometry (erratum). Invent Math 1992;108:453–54.

[46] Fradkin ES, Tseytlin AA. On quantized string models. Ann Phys 1982;143:413.

[47] Frolov S, Tseytlin AA. Semiclassical quantization of rotating superstring in $AdS_5 \times S^5$. JHEP 2002;0206:007, [hep-th/0204226].

[48] Frolov S, Tseytlin AA. Multispin string solutions in $AdS_5 \times S^5$. Nucl Phys 2003;B668:77–110, [hep-th/0304255].

[49] Frolov S, Tseytlin AA. Quantizing three spin string solution in $AdS_5 \times S^5$. JHEP 2003;0307:016, [hep-th/0306130].

[50] Frolov S, Park I, Tseytlin AA, On one-loop correction to energy of spinning strings in S5. Phys Rev 2005;D71:026006, [hep-th/0408187].

[51] Gelfand IM, Yaglom AM. Integration in functional spaces and it applications in quantum physics. J Math Phys 1960;1:48.

[52] Giombi S, Ricci R, Roiban R, Tseytlin A, Vergu C. Quantum $AdS_5 \times S^5$ superstring in the AdS light-cone gauge. JHEP 2010;1003:003, [0912.5105].

[53] Gromov N, Sizov G. Exact slope and interpolating functions in N = 6 supersymmetric Chern-Simons theory. Phys Rev Lett 2014;113:121601, [1403.1894].

[54] Gubser SS, Klebanov IR, Polyakov AM. A semi-classical limit of the gauge/string correspondence. Nucl Phys 2002;B636:99–114, [hep-th/0204051].

[55] Hanada M. What lattice theorists can do for quantum gravity, [1604.05421].

[56] Hanada M, Hyakutake Y, Ishiki G, Nishimura J. Holographic description of quantum black hole on a computer. Science 2014;344:882–5, [1311.5607].

[57] Hanada M, Matsuura S, Sugino F. Two-dimensional lattice for four-dimensional N = 4 supersymmetric Yang-Mills. Prog Theor Phys 2011;126:597–611, [1004.5513].

[58] Honda M, Ishiki G, Nishimura J, Tsuchiya A. Testing the AdS/CFT correspondence by Monte Carlo calculation of BPS and non-BPS Wilson loops in 4d N = 4 super-Yang-Mills theory. PoS 2011;LATTICE2011:244, [1112.4274].

[59] Honda M, Ishiki G, Kim S-W, Nishimura J, Tsuchiya A. Direct test of the AdS/CFT correspondence by Monte Carlo studies of N = 4 super Yang-Mills theory. JHEP 2013;11:200, [1308.3525].
[60] Ishii T, Ishiki G, Shimasaki S, Tsuchiya A. N = 4 Super Yang-Mills from the plane wave matrix model. Phys Rev 2008;D78:106001, [0807.2352].
[61] Ishiki G, Kim S-W, Nishimura J, Tsuchiya A. Deconfinement phase transition in N = 4 super Yang-Mills theory on $R \times S^3$ from supersymmetric matrix quantum mechanics. Phys Rev Lett 2009;102:111601, [0810.2884].
[62] Ishiki G, Kim S-W, Nishimura J, Tsuchiya A. Testing a novel large-N reduction for N = 4 super Yang-Mills theory on $R \times S^3$. JHEP 2009;09:029, [0907.1488].
[63] Joseph A. Review of lattice supersymmetry and gauge-gravity duality. Int J Mod Phys 2015;A30:1530054, [1509.01440].
[64] Kallosh R, Tseytlin AA. Simplifying superstring action on AdS(5) × S**5. JHEP 1998;10:016, [hep-th/9808088].
[65] Kaplan DB, Unsal M. A Euclidean lattice construction of supersymmetric Yang-Mills theories with sixteen supercharges. JHEP 2005;09:042, [hep-lat/0503039].
[66] Kirsten K, Loya P. Computation of determinants using contour integrals. Am J Phys 2008;76:60–4, [0707.3755].
[67] Kirsten K, McKane AJ. Functional determinants by contour integration methods. Ann Phys 2003;308:502–27, [math-ph/0305010].
[68] Kirsten K, McKane AJ. Functional determinants for general Sturm-Liouville problems. J Phys 2004;A37:4649–70, [math-ph/0403050].
[69] Kruczenski M. Spin chains and string theory. Phys Rev Lett 2004;93:161602, [hep-th/0311203].
[70] Kruczenski M, Tirziu A. Matching the circular Wilson loop with dual open string solution at 1-loop in strong coupling. JHEP 2008;0805:064, [0803.0315].
[71] McKane AJ, Tarlie MB. Regularization of functional determinants using boundary perturbations. J Phys 1995;A28:6931–42, [cond-mat/9509126].
[72] McKeown RW, Roiban R. The quantum $AdS_5 \times S^5$ superstring at finite coupling, [1308.4875].
[73] McLoughlin T. Review of AdS/CFT integrability, Chapter II.2: quantum strings in $AdS_5 \times S^5$ Lett Math Phys 2012;99:127–48, [1012.3987].
[74] Metsaev R, Thorn CB, Tseytlin AA. Light cone superstring in AdS space-time. Nucl Phys 2001;B596:151–84, [hep-th/0009171].
[75] Metsaev R, Tseytlin AA. Superstring action in $AdS_5 \times S^5$. Kappa symmetry light cone gauge. Phys Rev 2001;D63:046002, [hep-th/0007036].
[76] Park I, Tirziu A, Tseytlin AA. Spinning strings in $AdS_5 \times S^5$: one-loop correction to energy in SL(2) sector. JHEP 2005;0503:013, [hep-th/0501203].
[77] Pestun V, Localization of the four-dimensional N = 4 SYM to a two-sphere and 1/8 BPS Wilson loops. JHEP 2012;12:067, [0906.0638].
[78] Pestun V. Localization of gauge theory on a four-sphere and supersymmetric Wilson loops. Commun Math Phys 2012;313:71–129, [0712.2824].
[79] Pestun V, et al. Localization techniques in quantum field theories, [1608.02952].
[80] Polyakov AM. Conformal fixed points of unidentified gauge theories. Mod Phys Lett 2004;A19:1649–60, [hep-th/0405106].
[81] Roiban R. Review of AdS/CFT integrability, Chapter V.1: scattering amplitudes – a brief introduction. Lett Math Phys 2012;99:455–79, [1012.4001].
[82] Roiban R, Tirziu A, Tseytlin AA. Two-loop world-sheet corrections in $AdS_5 \times S^5$ superstring. JHEP 2007;0707:056, [0704.3638].

[83] Roiban R, Tseytlin AA. Strong-coupling expansion of cusp anomaly from quantum superstring. JHEP 2007;0711:016, [0709.0681].
[84] Schaich D. Aspects of lattice N = 4 supersymmetric Yang-Mills. PoS 2015;LATTICE2015:242, [1512.01137].
[85] Semenoff GW, Zarembo K. Wilson loops in SYM theory: from weak to strong coupling. Nucl Phys Proc Suppl 2002;108:106–12, [hep-th/0202156].
[86] Uvarov D. $AdS_4 \times CP^3$ superstring in the light-cone gauge. Nucl Phys 2010;B826:294–312, [0906.4699].
[87] Uvarov D. Light-cone gauge Hamiltonian for $AdS_4 \times CP^3$ superstring. Mod Phys Lett 2010;A25:1251–65, [0912.1044].

Pedro Liendo

Constraints on CFT observables from the bootstrap program

Abstract: In this chapter, we report on recent developments on the conformal bootstrap approach to conformal theories. The text is divided into two main sections. In the first part we discuss $\mathcal{N} = 2$ superconformal field theories. We review the existence of a protected subsector of operators whose correlators are captured by a $2d$ chiral algebra and then apply numerical techniques in an attempt to solve a specific $\mathcal{N} = 2$ superconformal fixed point. In the second part of this report we change gears and apply bootstrap techniques to the interesting case of conformal field theories in the presence of a boundary. In particular, we estimate surface critical exponents for the $O(N)$ universality classes.

Keywords: CFT, supersymmetry

Mathematics Subject Classification 2010: 81T40, 83C47, 81Q60, 82B27

1 The bootstrap philosophy

The "bootstrap" has been a recurring dream in theoretical physics. It is the ambitious aspiration that, starting from a few basic spectral assumptions, symmetries and general consistency requirements (such as unitarity and crossing) will be powerful enough to fix the form of the theory, with no reference to a Lagrangian model. The dual models of the strong interactions emerged as an incarnation of the S-matrix bootstrap attempts of the 1960s and eventually led to the discovery of string theory. The bootstrap program for conformal field theories (CFTs) in d dimensions was formulated in the early 1970s [25, 26, 41]. Despite important formal developments such as the operator product expansion (OPE) and the conformal block decomposition (see e.g. the early books [24, 47]), attempts to solve CFTs in arbitrary dimensions were not successful. For two-dimensional CFTs, the revolution came in the 1980s with the discovery of many exactly solvable "rational" models. While this is a beautiful incarnation of the bootstrap idea, the methods that work in $2d$ rational CFTs[1] are too specialized to be imitated in higher dimensions or even in two dimensions for the generic nonrational model.

The interest in CFT in various dimensions is nowadays stronger than ever, sustained by phenomenological questions in condensed matter physics ($d = 3$) and

1 Or in closely related models such as Liouville theory

DOI 10.1515/9783110452150-005_s_002

particle physics ($d = 4$), as well as by more formal motivations such as the Anti-de Sitter (AdS)/CFT correspondence and the rich integrability structures of superconformal field theories (SCFTs) ($d \leq 6$).

In this report we apply the bootstrap philosophy to two classes of systems, and the questions we attempt to answer are of a very different nature depending on the system. In the first part we will study CFTs that also contain $\mathcal{N} = 2$ supersymmetry and therefore belong to the realm of SCFTs. Recently, there has been an explosion of results in $\mathcal{N} = 2$ superconformal dynamics (mainly motivated by the seminal paper by Gaiotto [28]), where many of the newly discovered models have no known Lagrangian description. Having such an ample catalogue of theories, one of the main goals of the bootstrap approach would be to organize the landscape of $\mathcal{N} = 2$ theories. The bootstrap approach is based only on the operator algebra, with no reference to a Lagrangian formulation, and is therefore well suited for such a task. As we will see in this report, a crucial role is played by the existence of a solvable subsector of crossing symmetry, which is only present in $\mathcal{N} = 2$ supersymmetric theories. Once the protected subsector is understood, we will proceed with a numerical analysis of the bootstrap equations.

The second part is more phenomenologically oriented and attempts to describe the critical behavior of condensed matter systems with a flat boundary. The correct framework to describe this type of system is that of CFT with defects, we call it "boundary CFT" or BCFT for short. Unlike the supersymmetric case, there is no solvable subsector of the bootstrap equations, and the approach will be numerical from the start. The highlight of the second part will be a numerical estimate of surface critical exponents for the $O(N)$ universality classes.

2 The $\mathcal{N} = 2$ superconformal bootstrap

In this section we describe the conformal bootstrap program for four-dimensional CFTs with $\mathcal{N} = 2$ supersymmetry. These theories are extraordinarily rich, both physically and mathematically, and have been studied intensively from many viewpoints. Nevertheless, it seems that a coherent picture is still missing, and we hope that the generality of the conformal bootstrap framework will allow such a picture to be developed. The recent explosion of results for $\mathcal{N} = 2$ SCFTs calls out for a more systematic approach, while the methods first introduced in [44] have reinvigorated the conformal bootstrap with a powerful and flexible toolkit for studying CFTs.

The first examples of $\mathcal{N} = 2$ SCFTs were relatively simple gauge theories with matter representations chosen, so that the beta functions for all gauge couplings would vanish. Since then, the library of known theories has grown in size, with the new additions including many Lagrangian models [46], but remarkably also many theories that appear to admit no such description. In particular, the class $\mathcal{S}$ construction of [28, 30] gives rise to an enormous landscape of theories, most of which resist description by conventional Lagrangian field theoretic techniques. Despite this abundance, the current catalogue seems fairly structured, and one may reasonably suspect that a complete classification of $\mathcal{N} = 2$ SCFTs will ultimately be possible. The development of the $\mathcal{N} = 2$ superconformal bootstrap seems an indispensable step towards this ambitious goal.

Our first task is to introduce an abstract operator-algebraic language for $\mathcal{N} = 2$ SCFTs. In this reformulation, we retain only the vector space of local operators (organized into representations of the superconformal algebra), and the algebraic structure on this vector space defined by the operator product expansion. From this viewpoint, we can see that a theory is free (or contains a free factor) if its operator spectrum includes higher spin currents; we can see that a theory has a Higgs branch of vacua if its operator algebra includes an appropriate chiral ring that is the coordinate ring of an affine algebraic variety and so on. Representation theory of the $\mathcal{N} = 2$ superconformal algebra proves an invaluable tool, as its shortened representations neatly encode different facets of physics. Once equipped with the proper language, we can make an informed decision on where and how to employ bootstrap methods.

There are two broad types of questions that we can hope to address by bootstrap methods. First of all, we can *constrain the space of consistent $\mathcal{N} = 2$ SCFTs*. There are a number of universal structures that appear throughout the $\mathcal{N} = 2$ catalogue that cannot be satisfactorily explained in the abstract bootstrap language. Are Coulomb branch chiral rings always freely generated? Are central charges bounded from below by those of free theories, or are there exotic theories with even lower central charges? Is every $\mathcal{N} = 2$ conformal manifold parametrized by gauge couplings? As we will see, these questions can sometimes be connected with the constraints of crossing symmetry.

Our second motivation is to *learn more about specific $\mathcal{N} = 2$ SCFTs*. There are many cases where supersymmetry can tell us a lot about an $\mathcal{N} = 2$ SCFT even when we have no Lagrangian description. In many examples we know, e.g., the central charges (including flavor central charges), the spectrum of protected operators, and some of its OPE coefficients. Optimistically, we may hope that protected data and the constraints of crossing symmetry are enough to determine the theory uniquely. The bootstrap may then allow us to effectively solve the theory along the lines of what has been done for the three-dimensional Ising CFT [22, 23, 35]. Because the bootstrap is completely nonperturbative in nature, it is a natural tool for studying intrinsically strongly coupled (non-Lagrangian) theories. In fact, when it comes to studying unprotected operators in a non-Lagrangian theory, the bootstrap is really the only game in town.

A key role in our analysis will be played by the existence of a protected subsector of operators present in any $\mathcal{N} = 2$ SCFT, whose correlators are captured by the holomorphic sector of a two-dimensional CFT, also known as a 2*d* chiral algebra. The bootstrap program can then be thought of as a two-step process: First we solve for the protected sector analytically using the power of 2*d* CFTs and then we tackle the harder task of studying nonprotected quantities using modern numerical bootstrap techniques. In the following sections we will see an example of this two-step process approach, in which using the $2d$ chiral algebra description we carve the landscape of $\mathcal{N} = 2$ theories analytically and then we proceed to study numerically a specific model that sits in a very special corner of the parameter space of $\mathcal{N} = 2$ theories.

2.1 Chiral algebras in $\mathcal{N} = 2$ theories

It is possible to define a map that associates with any $\mathcal{N} = 2$ SCFTs a two-dimensional chiral algebra:

$$4d \text{ SCFT} \quad \rightarrow \quad 2d \text{ chiral algebra}$$

whose correlation functions describe a protected subsector of the original four-dimensional theory. The construction of the two-dimensional chiral algebra is obtained using the cohomology of a certain nilpotent supercharge:

$$\mathbb{Q} = \mathcal{Q}^1_- + \tilde{\mathcal{S}}^{2\dot{-}}, \tag{1}$$

where $\mathcal{Q}^i_\alpha$ and $\tilde{\mathcal{S}}^i$ are the standard supercharges of the $\mathcal{N} = 2$ superconformal algebra. Fixing a plane $\mathbb{R}^2 \in \mathbb{R}^4$ and defining complex coordinates $(z, \bar{z})$ on it, the conformal symmetry restricted to the plane acts as $\mathrm{SL}(2) \times \overline{\mathrm{SL}(2)}$. The supercharge $\mathbb{Q}$ can be used to define holomorphic translations that are $\mathbb{Q}$-closed and anti-holomorphic translations that are $\mathbb{Q}$-exact:

$$[\mathbb{Q}, \mathrm{SL}(2)] = 0, \qquad \{\mathbb{Q}, \text{something}\} = \widehat{\mathrm{SL}(2)}, \tag{2}$$

where $\widehat{\mathrm{SL}(2)} = \mathrm{diag}\left(\overline{\mathrm{SL}(2)} \times \mathrm{SL}(2)_R\right)$ and $\mathrm{SL}(2)_R$ is the complexification of the compact $\mathrm{SU}(2)_R$ R-symmetry. Operators that belong to the cohomology of $\mathbb{Q}$ transform in chiral representations of the $\mathrm{SL}(2) \times \widehat{\mathrm{SL}(2)}$ subalgebra. This implies that they have meromorphic OPEs (module $\mathbb{Q}$-exact terms) and their correlation functions are meromorphic functions of their positions when restricted to the plane.

In order to identify the cohomology of $\mathbb{Q}$ we will consider operators at the origin and then we will translate them across the plane using the $\mathrm{SL}(2) \times \widehat{\mathrm{SL}(2)}$ generators. As shown in [7], a necessary and sufficient condition for an operator to be in the cohomology of $\mathbb{Q}$ is given by:

$$\frac{1}{2}(\Delta - (j + \bar{\jmath})) - R = 0, \qquad r + (j - \bar{\jmath}) = 0. \tag{3}$$

We call these operators as *Schur operators* because they contribute to the Schur limit of the superconformal index [27]. It can be shown that Schur operators occupy the highest weight of their respective $\mathrm{SU}(2)_R$ and Lorentz representations,

$$\mathcal{O}^{1\cdots 1}_{+\cdots+\dot{+}\cdots\dot{+}}(0). \tag{4}$$

Having identified the operator at the origin, we proceed to translate it using the $\mathrm{SL}(2) \times \widehat{\mathrm{SL}(2)}$ generators. Equation (2) implies that the anti-holomorphic dependence gets entangled with the $\mathrm{SU}(2)_R$ structure due to the twisted nature of the $\widehat{\mathrm{SL}(2)}$ generators. The coordinate dependence after translation is given by

$$\mathcal{O}(z,\bar{z}) = u_{i_1}(\bar{z}) \dots u_{i_k}(\bar{z}) \mathcal{O}^{(i_1 \dots i_k)}(z,\bar{z}) \qquad \text{where} \qquad u_i(\bar{z}) = (1,\bar{z})\,. \tag{5}$$

By construction, these operators define cohomology classes with meromorphic correlators. For each cohomology class we define,

$$\mathcal{O}(z) = [\mathcal{O}(z,\bar{z})]_{\mathbb{Q}}\,. \tag{6}$$

That is, to any $4d$ Schur operator there is an associated $2d$ dimensional holomorphic operator. Schur operators have protected conformal dimension and therefore sit in shortened multiplets of the superconformal algebra. In Table 1 we present the list of multiplets (in the notation of [19]) that contain a Schur operator and the holomorphic dimension h of the corresponding two-dimensional operator.

2.1.1 Enhancement to Virasoro and Affine Kac-Moody (AKM) symmetry

Among the list of multiplets in Table 1 is the stress-tensor multiplet $\hat{\mathcal{C}}_{0(0,0)}$, and its Schur operator is the $\mathrm{SU}(2)_R$ conserved current J^{11}_{++}. Its corresponding holomorphic operator is defined as $T(z) = [J_{++}(z,\bar{z})]_{\mathbb{Q}}$, and the four-dimensional $J_{++}(x)J_{++}(0)$ OPE implies,

$$T(z)T(0) \sim -\frac{6\,c_{4d}}{z^4} + 2\frac{T(0)}{z^2} + \frac{\partial T(0)}{z} + \cdots\,. \tag{7}$$

We can therefore identify $T(z)$ as the $2d$ stress tensor. The $2d$ central charge is given by:

$$c_{2d} = -12\,c_{4d}\,. \tag{8}$$

Unitarity of the four-dimensional theory implies that the two-dimensional theory is nonunitary.

Table 1: Four-dimensional superconformal multiplets that contain Schur operators. We denote the superconformal primary by Ψ. The second column indicates where in the multiplet the Schur operator sits. The third and fourth columns give the two-dimensional quantum numbers in terms of $(R, j, \bar{j})$.

Multiplet	$\mathcal{O}_{\text{Schur}}$	h	r
$\hat{\mathcal{B}}_R$	$\Psi^{11\dots1}$	R	0
$\mathcal{D}_{R(0,\bar{j})}$	$\bar{\mathcal{Q}}^1_{\dot{+}}\Psi^{11\dots1}_{\dot{+}\dots\dot{+}}$	$R+\bar{j}+1$	$\bar{j}+\frac{1}{2}$
$\bar{\mathcal{D}}_{R(j,0)}$	$\mathcal{Q}^1_{+}\Psi^{11\dots1}_{+\dots+}$	$R+j+1$	$-j-\frac{1}{2}$
$\hat{\mathcal{C}}_{R(j,\bar{j})}$	$\mathcal{Q}^1_{+}\bar{\mathcal{Q}}^1_{\dot{+}}\Psi^{11\dots1}_{+\dots+\dot{+}\dots\dot{+}}$	$R+j+\bar{j}+2$	$\bar{j}-j$

A similar analysis can be done for the flavor current multiplet $\hat{\mathscr{B}}_1$. The associated Schur operator is the moment map $J^a(z) = [M^{11\,a}(z,\bar{z})]_Q$, the $4d$ OPEs implies,

$$J^a(z)J^b(0) \sim -\frac{k_{4d}}{2}\frac{\delta^{ab}}{z^2} + if^{abc}\frac{J^c(0)}{z} + \cdots . \tag{9}$$

The global flavor symmetry gives rise to the holomorphic current $J^a(z)$ of an affine Kac-Moody algebra at level k_{2d}. The $2d$ central charges is

$$k_{2d} = -\frac{1}{2}k_{4d} . \tag{10}$$

In some cases the chiral algebras of selected $\mathcal{N} = 2$ SCFTs have been identified with well-known models. Thanks to these analytic results, it is possible to solve for certain OPE coefficients which in turn can be related to the central charge c and the flavor central charge k. Unitarity of the $4d$ theory then implies analytic bounds for c and k. Let us now obtain the simplest example of such a bound.

The holomorphic correlator of the stress tensor can be completely fixed in terms of the central charge, and its relation to the parent theory in four dimensions gives a bound on c. The holomorphic correlator of the stress tensor is

$$g(z) = 1 + z^4 + \frac{z^4}{(1-z)^4} + \frac{8}{c_{2d}}\left(z^2 + z^3 + \frac{z^4}{(1-z)^2} + \frac{z^4}{1-z}\right), \tag{11}$$

and admits the following expansion in SL(2) blocks,

$$g(z) = \sum_{\ell=0}^{\infty} a_\ell z^\ell \, {}_2F_1(\ell,\ell,2\ell,z) \qquad \ell \ \text{ even}, \tag{12}$$

where ${}_2F_1$ is the standard hypergeometric function. Thanks to the $4d/2d$ correspondence, we can interpret the SL(2) blocks as contributions from four-dimensional multiplets containing Schur operators. The super OPE selection rules imply only two possible choices:

$$\hat{\mathscr{C}}_{0(\frac{\ell}{2},\frac{\ell}{2})} \qquad \text{and} \qquad \hat{\mathscr{C}}_{1(\frac{\ell}{2},\frac{\ell}{2})} . \tag{13}$$

The $\hat{\mathscr{C}}_{0(\frac{\ell}{2},\frac{\ell}{2})}$ multiplets contain higher spin currents and we do not expect them in an interacting theory [3, 38]. The only candidate then is $\hat{\mathscr{C}}_{1(\frac{\ell}{2},\frac{\ell}{2})}$, the exact proportionality constant α between the OPE coefficients $\lambda^2_{\hat{\mathscr{C}}_{1(\frac{\ell}{2},\frac{\ell}{2})}}$ and the SL(2) coefficients a_ℓ can be carefully worked out, but we will not need it. The explicit expansion of eq. (11) in terms of SL(2) blocks was worked out in [9], in particular,

$$\lambda^2_{\hat{\mathscr{C}}_{1(\frac{1}{2},\frac{1}{2})}} = \alpha\left(2 - \frac{11}{15c_{4d}}\right) . \tag{14}$$

Unitarity of the four-dimensional theory implies $\lambda^2_{\hat{\mathcal{C}}_{1(\frac{1}{2},\frac{1}{2})}} \geq 0$, then,[2]

$$c_{4d} \geq \frac{11}{30}\,. \tag{15}$$

Let us note that in order to obtain this bound we only assumed $\mathcal{N} = 2$ superconformal symmetry, existence of a stress tensor, and the absence of higher spin currents. Bounds of this type were obtained in [7] using the $\hat{\mathcal{B}}_1$ four-point function; in that case, however, it is necessary to assume the existence of flavor symmetries whose conserved currents sit in $\hat{\mathcal{B}}_1$ multiplets. In the present case, our assumptions are weaker. A similar bound was also obtained for $\mathcal{N} = 4$ theories in [10], where absence of higher spin currents imply $c \geq \frac{3}{4}$.

Going through the $\mathcal{N} = 2$ literature one can check that the simplest rank one Argyres-Douglas fixed point (sometimes denoted as H_0 due to its construction in F-theory) has central charge $c = \frac{11}{30}$ [1, 2, 5, 6], which precisely saturates our bound. The analytic bounds of [7] turned out to have interesting consequences for four-dimensional physics: The saturation of a bound was identified as a relation in the Higgs branch chiral ring due to the decoupling of the associated multiplet. It would be interesting to explore whether the absence of the $\hat{\mathcal{C}}_{1(\frac{1}{2},\frac{1}{2})}$ multiplet is associated with some intrinsic structure that characterizes the H_0 theory.

From the two-dimensional point of view, the $2d$ chiral algebra that describes the H_0 theory has been conjectured to be the Yang-Lee minimal model [43]. Indeed, the $2d$ value of the central charge is $c_{2d} = -\frac{22}{5}$. Saturation of the bound implies the absence of the $\hat{\mathcal{C}}_{1(\frac{1}{2},\frac{1}{2})}$ multiplet. In Table 1 the associated $2d$ operator has holomorphic dimension 4. Hence, the absence of $\hat{\mathcal{C}}_{1(\frac{1}{2},\frac{1}{2})}$ translate to the existence of a null state of dimension 4. Remarkably, one of the hallmarks of the Yang-Lee minimal a model is a level 4 null descendant of the identity, $\left(L^2_{-2} - \frac{3}{5}L_{-4}\right)|0\rangle$. Our results are then consistent with the conjectured correspondence. The Schur index of Argyres-Douglas fixed points and its relation to $2d$ chiral algebras were recently studied in [12, 16].

The vanishing of certain OPE coefficients has also been instrumental in characterizing the $3d$ critical Ising model using numerical bootstrap techniques [22, 23, 35]. One can then label the rank one H_0 theory as the "Ising model" of $\mathcal{N} = 2$ superconformal theories, in the sense that it shares two of its most prominent features: minimum value of the central charge and vanishing of certain OPE coefficients. Both features indicate that this superconformal fixed point sits in a very special place in the parameter space of $\mathcal{N} = 2$ theories. In the next section we will study this fixed point numerically [36].

---

2 Because we have not calculated the exact proportionality constant, one could complain that an overall minus sign will invalidate our bound. However, common sense dictates that the sign should be positive, otherwise we will rule out every known interacting $\mathcal{N} = 2$ SCFT.

2.2 Nonperturbative numerical constraints

Until this point we have avoided making assumptions that are not shared by all $\mathcal{N} = 2$ SCFTs because our goal so far has been to obtain generic constraints on the CFT data. In this subsection we will focus on a specific theory instead. A natural candidate for this analysis is the rank one H_0 theory mentioned above. It corresponds to the $N = 1$ case in the (A_1, A_{2N}) family of Argyres-Douglas theories, and it also appears as the low-energy description of a single D3-brane probing an F-theory singularity of type H_0 [2].

2.2.1 The rank one H_0 theory

As already discussed, it has been conjectured that the H_0 theory has a closed subsector of operators described by the two-dimensional Yang-Lee minimal model. The central charge of this $2d$ chiral algebra has the right value, and the Schur index of the $4d$ theory seems to match the $2d$ vacuum character of the minimal model [16].

The H_0 theory then sits in a special place in the landscape of $\mathcal{N} = 2$ SCFTs. It has the lowest possible value of the central charge c and does not have flavor symmetries, which simplifies its operator content. It is also suggestive that it has a subsector described by the Yang-Lee model, which is one of the paradigmatic examples of a solvable model. A further piece of information is that it does not have a mixed branch, and therefore the $\mathcal{B}_{1,2r_1-1(0,0)}$ multiplet is absent from the $\phi_{r_1} \times \phi_{r_1}$ OPE, where ϕ_{r_0} is the highest weight scalar of the $\mathcal{N} = 2$ chiral multiplet with r-charge r_0. Also, crucial CFT data such as the external dimension $r = \frac{6}{5}$ and the already mentioned central charge $c = \frac{11}{30}$ are known. Altogether, this makes the bootstrap program more likely to succeed for this theory, and we will try to leverage this information in order to corner H_0 and make specific statements about its operator spectrum.

This subsection is divided into two parts. In the first part we obtain upper bounds for operator dimensions and speculate where inside the bound H_0 sits. To make definite statements, we will work under the hypothesis that the numerical minimum of the central charge will converge to $\frac{11}{30}$. We will skip the details of the numerical implementation in this report and we refer the reader to the original paper [36]. In particular, the parameter Λ measures how strong the bound is, the bigger the Λ, the stronger the bound. Let us point out though that in all our plots we show "exclusion regions," lower Λ does not imply that the bound is wrong, but only that the bound is "nonoptimal" but nevertheless true. In the second part we constrain the OPE coefficients of the $\mathcal{E}_{2r_1}$ and $\mathcal{C}_{0,2r_1-1(0,1)}$ multiplets for the rank one H_0 theory, now setting $c = \frac{11}{30}$ by hand. The allowed ranges for the last of these coefficients will be quite narrow.

Let us start by asking what characterizes the solution with the minimum central charge (for $r_1 = \frac{6}{5}$), and whether its features are consistent with the H_0 theory. Because H_0 does not have a mixed branch, we actually know where to look. In Fig. 1 we present

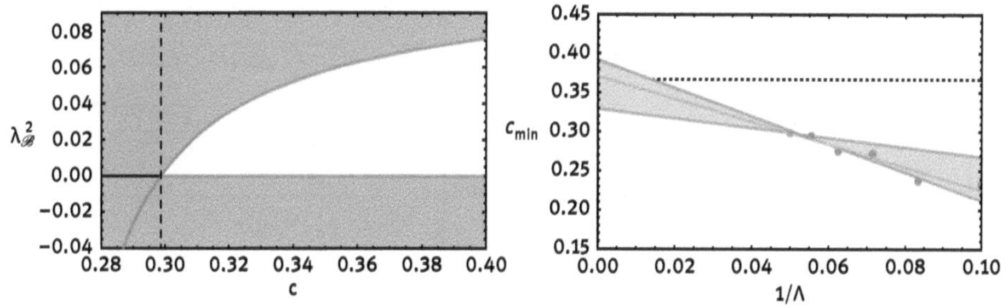

Figure 1: Left: Upper bound on the OPE coefficient of the $\mathcal{B}_{R=1,r=2r_1-1(0,0)}$ multiplet, for external dimension $r_1 = \frac{6}{5}$, as a function of the central charge at $\Lambda = 12$. The vertical dashed line corresponds to the minimum central charge allowed numerically with $\Lambda = 12$. Right: Minimum allowed central charge for varying Λ, the dashed horizontal line marks the central charge of the rank one H_0 theory. The middle orange line shows a linear fit to all the data points, while the top and bottom blue lines show fits to different subsets of the points.

an upper bound on the OPE coefficient of the $\mathcal{B}_{1,2r_1-1(0,0)}$ multiplet as a function of the central charge, keeping the external dimension fixed at $r_1 = \frac{6}{5}$. Unitarity, which requires $\lambda^2_{\mathcal{B}} \geqslant 0$, combined with the numerical upper bound restricts the coefficient to lie in the unshaded region. The upper bound, plotted for $\Lambda = 12$, crosses zero precisely at the numerical central charge lower bound, c_{min}, for the same Λ. The bound becomes negative to the left of this point, implying that there is no unitary solution to crossing symmetry for $c < c_{\text{min}}$. The simplest interpretation is that the vanishing of this OPE coefficient is responsible for the central charge bound. This is reminiscent of what happens with the six-dimensional A_1 theory studied in [9]. The hypothesis then is that when $\Lambda \to \infty$, $c_{\text{min}} \to \frac{11}{30}$, and at that point there will be a unique solution to crossing [21]. This solution will have $r_1 = \frac{6}{5}$, $c = \frac{11}{30}$, and $\lambda^2_{\mathcal{B}} = 0$, and should therefore correspond to the H_0 theory.

2.2.2 Scalar bound for H_0

In [36] it was shown that for $r_1 = \frac{6}{5}$ the dimension of the first long scalar operator in the $\phi_{r_1} \times \phi_{r_1}$ channel must obey $4.4 \leq \Delta_0 \lesssim 4.92$, where the lower end follows from the unitarity bound for this multiplet. As in the previous subsection we present the dimension of the superconformal descendant that appears in the OPE. We now want to understand which solution to crossing symmetry saturates the bound, and if it corresponds to the H_0 theory. For that we assume the *first* long scalar Δ_0 has a given dimension, lying in the allowed range, and ask what is the bound on the *second* long scalar Δ_0'. This is shown in Fig. 2 in black and blue.

The figure has a characteristic kink which signals the transition between two regimes. To the right of the kink (black curve), the $\mathcal{B}_{1,2r_1-1(0,0)}$ multiplet is always

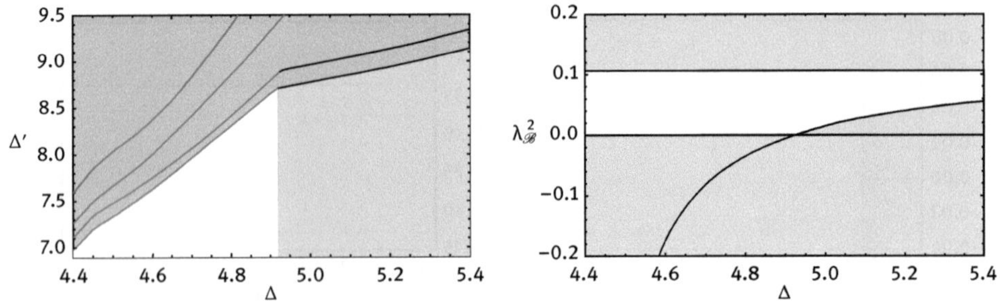

Figure 2: Left: Bound on the dimension of the second long scalar Δ'_0 in the $\phi\phi$ channel as a function of the dimension of the first long scalar Δ_0, for external dimension $r_1 = \frac{6}{5}$. The central charge is left arbitrary and $\Lambda = 14, 16, 18$, and 20. The black curve corresponds to the region where the short multiplet is required to have a positive OPE coefficient. The shaded black region is always excluded, and the shaded blue region is excluded only if one demands the absence of the $\mathscr{B}_{1,2r_1-1(0,0)}$ short multiplet. Right: Upper and lower bounds on the OPE coefficient of the $\mathscr{B}_{1,2r_1-1(0,0)}$ multiplet, as a function of the dimension of the first long scalar Δ_0.

present, because for $\Delta_0 \gtrsim 4.92$, its OPE coefficient is required to be positive, as it must lie in the unshaded region of the plot on the right side of Fig. 2.[3] To the left of the kink (blue curve), the lower bound for $\mathscr{B}_{1,2r_1-1(0,0)}$ disappears and the multiplet can be safely removed. This is a common phenomenon in bootstrap studies: A kink in a scalar bound is a consequence of the vanishing of an OPE coefficient.

The results so far do not clarify where inside the bound the H_0 theory sits. Let us repeat the analysis, but now fixing the central charge to several values. In Fig. 3 we show the same arbitrary central charge bound (at $\Lambda = 16$) overlapped with curves in red, yellow, and green obtained by fixing c to $\frac{3}{10}$, $\frac{11}{30}$, and $\frac{17}{12}$. The first of these values is close the numerical central charge bound at $\Lambda = 16$ (~ 0.275), while the last two correspond to the rank one and rank two H_0 theories, respectively. Because the long multiplet $\mathscr{A}^{\Delta}_{0,0(0,0)}$ mimics the contribution of the stress-tensor multiplet when $\Delta \sim 2$, the lines in Fig. 3 should be interpreted as representing a range of central charges: $c \leqslant \frac{3}{10}$, $c \leqslant \frac{11}{30}$, and $c \leqslant \frac{17}{12}$.

One conclusion that can be immediately drawn from these results is that the bound for arbitrary central charge to the left of the kink (in blue) is being controlled by the large central charges. At the minimum allowed numerical central charge, a big portion to the left of the kink is ruled out. Reducing the central charge has the effect of carving the allowed region away from the kink, while keeping the region near the kink untouched. For the minimum numerical central charge there is a unique solution to the truncated crossing equations [21], and from Fig. 1 we see that the $\mathscr{B}_{1,2r_1-1(0,0)}$ multiplet should be absent in said solution, this suggests the position of the kink

3 With a gap of Δ_0 being imposed we can obtain both lower and upper bounds on the $\mathscr{B}_{1,2r_1-1(0,0)}$ OPE coefficient.

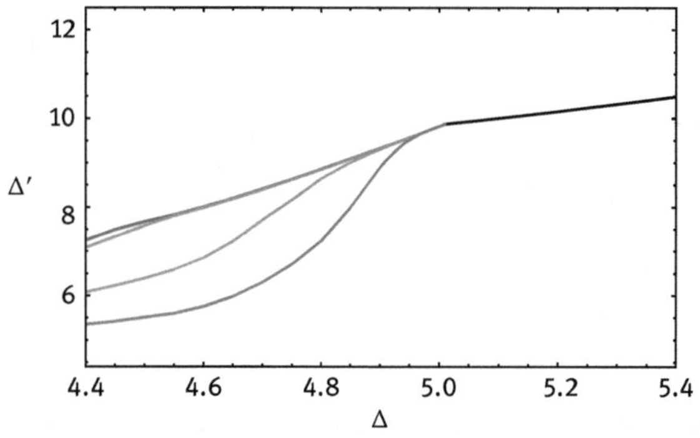

Figure 3: Bound on the dimension of the second long scalar Δ'_0 in the $\phi_{r_1} \times \phi_{r_1}$ channel as a function of the dimension of the first long scalar Δ_0, for external dimension $r_1 = \frac{6}{5}$ with $\Lambda = 16$. The different colours correspond to different central charges.

corresponds to the minimum central charge theory. If the minimum central charge reaches $\frac{11}{30}$ as $\Lambda \to \infty$, then this last statement applies to the rank one H_0 theory, implying that the theory is the *unique* solution to the crossing equations for $r_1 = \frac{6}{5}$ and $c = \frac{11}{30}$, with the position of the kink giving an estimate on the dimension of the two lowest operators in the chiral channel. Altogether, the possibility that the H_0 theory saturates the numerical bound of $\Delta \lesssim 4.92$ seems plausible, and at the least it warrants further investigation.

2.2.3 OPE bounds for H_0

We now turn to bounding the OPE coefficients of $\mathscr{C}$-type short multiplets, where lower and upper bounds can be obtained. These OPE coefficients are bounded to very narrow ranges, confirming once again the usefulness of the numerical bootstrap program. In this section we set the central charge to $c = \frac{11}{30}$, and the results do not rely on any assumption regarding the $\Lambda \to \infty$ limit. The bounds obtained here are rigorous and can only improve for higher Λ.

The $\mathscr{E}_{2r_1}$ multiplet was bounded in Fig. 24 of [8] for various central charges and external dimensions. We now provide in Fig. 4 a slice of that plot with the central charge constrained to $c \leq \frac{11}{30}$, and with the external dimension close to $r_1 = \frac{6}{5}$. This implies the following bound:

$$2.13 < \lambda^2_{\mathscr{E}_{2\times 6/5}} < 2.20 \qquad \text{for the rank one } H_0 \text{ theory.} \tag{16}$$

The situation gets better if one looks at the OPE coefficient of the $\mathscr{C}_{0\,2r_1-1(0,1)}$ multiplet:

$$0.467 < \lambda^2 < 0.470 \qquad \text{for the rank one } H_0 \text{ theory.} \tag{17}$$

All other $\mathscr{C}_{0,2r_1-1(j-1,j)}$ multiplets can be bootstrapped in a similar manner, and it seems plausible that they will lead to strong bounds as well. Finally, we note that if the speculations of the previous subsection prove to be true, the ranges provided here will

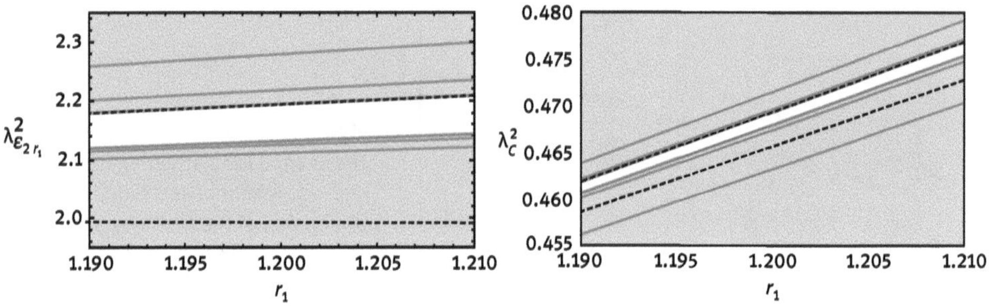

Figure 4: Upper and lower bounds on the $\mathcal{E}_{2r_1}$ (left) and $\mathcal{C}_{R=0\,r=2r_1-1(0,1)}$ (right) OPE coefficients for $c \leqslant \frac{11}{30}$, with various derivatives $\Lambda = 12, 16, 20$. The dashed black lines indicate the position of the bounds at $\Lambda = 20$ with arbitrary central charge.

shrink to a point for $\Lambda \to \infty$, as there will be a unique solution to crossing symmetry at $c = \frac{11}{30}$ and $r_1 = \frac{6}{5}$.

2.3 A quick look at the landscape

The analysis presented above for the H_0 theory can be generalized in several directions. If we consider $\mathcal{N} = 2$ SCFTs with flavor more general analytic bounds can be obtained. In particular, studying several combinations of correlators between the stress-tensor multiplet and the flavor-current multiplet implies,

$$\frac{1}{k} \leqslant \frac{12c + \dim_G}{24ch^\vee} . \tag{18}$$

$$k\left(-180c^2 + 66c + 3\dim_G\right) + 60c^2h^\vee - 22ch^\vee \leqslant 0 . \tag{19}$$

As before, after an analytic understanding of the landscape, we can proceed with a numerical analysis of some selected models. A landscape plot for $\mathcal{N} = 2$ theories with $SU(2)_F$ flavor group is shown in Fig. 5. The natural candidate to be studied numerically in this case is the H_1 theory, which sits at the intersection of both bounds. This has not been done yet, but it would be very interesting to explore in the future.

3 Applications to condensed matter

This section is mostly based on [32], and its aim is to apply the conformal bootstrap program to some examples of defect CFTs. These are theories in which the conformal group is broken down to the stabilizer of some hypersurface. We shall be concerned only with the case of a codimension one hyperplane, alias a flat interface, but one

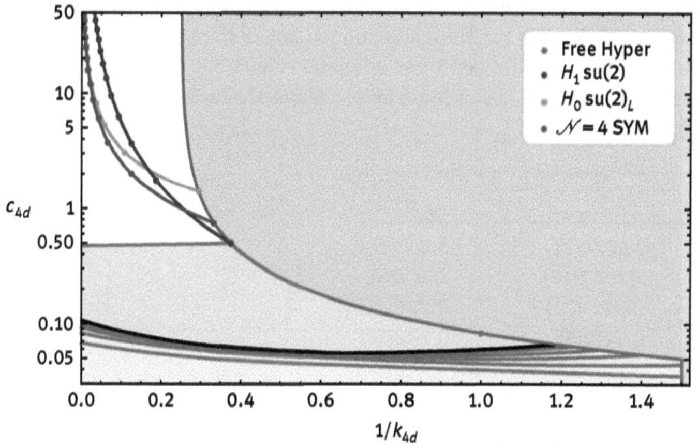

Figure 5: Constraints on the (c_{4d}, k_{4d}) plane for (2) flavor symmetries arising as a combination of the analytic bound (19) (blue), the analytic bounds (18) (red), and the numerical bound of [8] (gray). We have marked the position of some known = 2 theories which contain a (2) flavor symmetry, namely the free hyper multiplet, $\mathcal{N}$ = 4 SU(N) SYM, and the theories obtained by N $D3$-branes probing an F-theory singularity of type H_1.

can apply a similar logic to generic flat conformal defects. Motivations for studying conformal defects are both phenomenological and abstract. For instance, conformal defects describe modifications of a d dimensional Quantum Field Theory (QFT) localized near a p dimensional plane, with $p < d$, in the infrared limit, provided these modifications are not swept away by coarse graining, and scale invariance is enhanced to invariance under the conformal group SO(p+1, 1). The simplest example is of course a conformal boundary – that is, an interface between a nontrivial and the trivial CFT. Lower dimensional defects may correspond to magnetic-like impurities in a spin system, see for instance [11], or to dispersionless fermions, acting as a source for the order parameter of some bosonic system [4], or to vortices in holographic superfluids and superconductors [17], etc. On the more abstract side, extended defects are probes of a system, and may be used to constrain properties of the bulk CFT. We shall in fact see this happening in the present study. Moreover, interfaces are a natural way to "compare" two theories and may provide information on the geometric structure of the space of CFTs [20].

The conformal bootstrap was first applied to the boundary setup in [37], while the twist line defect defined in [11] was tackled in [29]. Both papers are concerned with the 3d Ising model, and both used the linear functional method of [44]. In the latter, four-point functions of defect operators were considered, while the former focused on two-point functions of bulk operators. Correlators of defect operators are blind to bulk-to-defect couplings, but correlators of bulk primaries do not satisfy in general the positivity constraints required by the linear functional method, and *ad hoc* assumptions were made in [37], motivated by computations in 2d and in ϵ-expansion. In this

Table 2: The first table collects the input parameters. The second one is a comparison between two-loop calculations [18] and our bootstrap results for the scaling dimension of the surface operator $\partial_z\hat{\sigma}$ in the ordinary transition of 3d $O(N)$ models. The last three columns collect our results for the OPE coefficients. The critical indices η and ν for $N = 0, 1, 2, 3$ are taken, respectively, from [13, 14, 34, 42]. Those for ω are taken from [33].

N	η	ν	ω
0	0.0314(32)	0.5874(2)	0.812(16)
1	0.03627(10)	0.63002(10)	0.832(6)
2	0.0380(4)	0.67155(27)	0.789(11)
3	0.0364(6)	0.7112(5)	0.782(13)

	$\Delta_{\partial_z\hat{\sigma}}$	
N	**Two-loop**	**Bootstrap**
0	1.33	1.332(6)
1	1.26	1.276(2)
2	1.211	1.2342(9)
3	1.169	1.198(1)

N	$a_\varepsilon\lambda_{\sigma\sigma\varepsilon}$	$a_{\varepsilon'}\lambda_{\sigma\sigma\varepsilon'}$	$\mu^2_{\hat{\Delta}}$
0	0.8447(34)	0.0366(17)	0.692(1)
1	0.789(3)	0.042(1)	0.755(13)
2	0.747(1)	0.0488(4)	0.80022(5)
3	0.710(1)	0.0509(6)	0.8395(6)

report we concentrate on the two-point function of bulk scalar primaries, using the method of determinants developed in [31], which can be safely applied to this case. We will be mostly concerned with solutions to the crossing equations corresponding to the ordinary transition, which cannot be studied with the linear functional method. In the latter case we extended the analysis to the $O(N)$ models with $N = 0, 2, 3$, where a comparison can be made with two-loop calculations. The main results are summarized in Table 2.

3.1 Defect CFTs and the method of determinants

The constraints imposed by conformal symmetry on correlation functions near a boundary were analyzed in [39] (see also [40]). Here we review the necessary material and then introduce the method of determinants. A general p-dimensional defect differs from the codimension one case for the residual $SO(d-p)$ symmetry generated by rotations around the defect. This is just a flavor symmetry for the defect operators but induces some differences when it comes to bulk-to-defect couplings. Although most of what we shall say applies to a generic flat defect, in this chapter we shall be concerned with the codimension one case. Therefore, further reference to the general case is limited to some side comments.

Correlation functions of excitations living at the defect are the same as in an ordinary $(d-1)$-dimensional CFT and are completely characterized by the spectrum of scale dimensions $(\widehat{\Delta}_l)$ and the coefficients of three-point functions $(\widehat{\lambda}_{lmn})$. We shall later need one more piece of information. While no conserved stress tensor is expected to exist on the defect, a protected scalar operator of dimension $d-1$ or $p+1$ in the general case – is always present: The displacement operator, which we call $D(x^a)$, measures the breaking of translational invariance and is defined by the Ward identity for the stress tensor:

$$\partial_\mu T^{\mu d}(x) = -D(x^a)\,\delta(x^d). \tag{20}$$

Here we denoted by Latin indices the directions along the defect, which is placed at $x^d = 0$, while Greek letters run from 1 to d. Similarly, for every bulk current whose conservation is violated by the defect, a protected defect operator exists.

In the bulk, there is of course the usual OPE. For scalar primaries,

$$O_1(x)O_2(y) = \frac{\delta_{12}}{(x-y)^{2\Delta_1}} + \sum_k \lambda_{12k} C[x-y, \partial_y] O_k(y)\,, \tag{21}$$

where $C[x-y, \partial_y]$ are determined by conformal invariance, and we isolated the contribution of the identity. One can also fuse a local operator with the defect. The bulk operator is thus turned into a sum over defect primaries. The bulk-to-defect OPE for a scalar primary can be written as

$$O_1(x) = \frac{a_1}{2x^{d\Delta_1}} + \sum_l \mu_{1l} D[x^d, \partial_a] \widehat{O}_l(x^b)\,, \tag{22}$$

where we denoted defect operators with a hat. Again, the differential operators $D[x^d, \partial_a]$ are fixed by conformal invariance. Similar OPEs can be written for bulk tensors. The λ_{12k}'s in eq. (21) are the coefficients of three-point functions without the defect, while μ_l is the coefficient of the correlator $O(x)\widehat{O}_l(y^a)$, otherwise fixed by conformal symmetry. Even if, for the sake of simplicity, some abuse of notation is present, all OPE coefficients refer to canonically normalized operators, with one exception: The normalization of the displacement operator is fixed by eq. (20). Taking the expectation value of both sides in eq. (22) one sees that a scalar acquires a one-point function proportional to a_O, the coefficient of the identity in the bulk-to-defect OPE. It is not difficult to prove that tensors do not acquire an expectation value in the presence of a codimension one defect. They do, instead, if they are even spin representations and the defect is lower dimensional.

Let us now derive the easiest crossing equation involving the OPEs (21) and (22). Consider the two-point function $O_1(x)O_2(x')$. One can decompose it into the bulk channel by plugging in eq. (21): a sum over one-point functions is obtained, that is, a sum over the coefficients $\lambda_{12k}a_k$ multiplying some known functions of the kinematic variables. Or, one can substitute both operators with their Defect OPE, and in this case the sum involves the quantities $\mu_{1l}\mu_{2l}$. In order to write explicitly the equality of the two conformal block decompositions, let us introduce the conformal invariant combination

$$\xi = \frac{(x-x')^2}{4x^d x'^d}. \tag{23}$$

This cross-ratio is conveniently positive when both points are chosen in the half-plane $x^d > 0$. Conformal symmetry justifies the following parametrization:

$$O_1(x)O_2(x') = \frac{1}{(2x^d)^{\Delta_1}(2x'^d)^{\Delta_2}}\,\xi^{-(\Delta_1+\Delta_2)/2}G_{12}(\xi). \tag{24}$$

Then the crossing equation can be written as a double decomposition of the function $G_{12}(\xi)$:

$$\begin{aligned} G_{12}(\xi) &= \delta_{12} + \sum_k \lambda_{12k}\, a_k\, f_{\text{bulk}}(\Delta_{12}, \Delta_k; \xi) \\ &= \xi^{(\Delta_1+\Delta_2)/2}\left(a_1 a_2 + \sum_l \mu_{1l}\,\mu_{2l}\, f_{\text{bdy}}(\widehat{\Delta}_l; \xi)\right), \end{aligned} \tag{25}$$

where [39]

$$f_{\text{bulk}}(\Delta_{12}, \Delta, \xi) = \xi^{\Delta/2}\, {}_2F_1\left(\frac{1}{2}(\Delta_1 - \Delta_2 + \Delta), \frac{1}{2}(\Delta_2 - \Delta_1 + \Delta); \Delta + 1 - \frac{d}{2}, -\xi\right), \tag{26a}$$

$$f_{\text{bdy}}(\Delta, \xi) = \xi^{-\Delta}\, {}_2F_1\left(\Delta, \Delta + 1 - \frac{d}{2}; 2\Delta + 2 - d; -\frac{1}{\xi}\right). \tag{26b}$$

Before describing how to extract information from eq. (25), we make some side remarks. The set $\{\widehat{\Delta}_l, \widehat{\lambda}_{lmn}, \Delta_i, \lambda_{ijk}, a_i, \mu_l\}$ is in fact redundant: By repeatedly applying the bulk-to-defect OPE, one can reduce all correlators to correlators of defect operators, therefore the λ_{ijk} are in principle unnecessary to solve the theory. However, it is easy to realize that all crossing equations constraining the bulk-to-defect couplings μ_l also involve the bulk three-point function coefficients. One is naturally led to the following question: What is the minimal set of correlators encoding all the crossing symmetry constraints of a Defect CFT? All the four-point functions of defect operators are surely in the number, the proof being the usual one (see for instance [45]). A similar argument shows that all the other crossing equations of a generic correlator of bulk and defect primaries are automatically satisfied once the three-point functions $O_1 O_2 \widehat{O}$ are crossing symmetric. In the rest of this chapter we explore the case $\widehat{O} = \mathscr{I}$, leaving for future work the general case.

In the next section we will present approximate solutions to the crossing eq. (25) using the method of determinants developed in [31]. The results are taken from [32] and we refer the reader to the original paper for more details.

3.2 Surface critical exponents

In this section we shall consider the boundary conformal field theories (BCFTs) associated with the Ising model and other magnetic systems. Specifically, the infrared (IR) properties of the surface transitions in these systems are controlled by Renormalization Group (RG) fixed points, which are of course described by just as many defect CFTs. We denote with $\sigma(x)$ the scalar field (i.e. the order parameter of the theory) and with $\widehat{\sigma}$ the corresponding surface operator. The surface Hamiltonian associated with a flat $d-1$ dimensional boundary of a semi-infinite system can be written in terms of the three relevant surface operators

$$H = \int d^{d-1}x \left(c\widehat{\sigma}^2 + h_1\widehat{\sigma} + h_2\partial_z\widehat{\sigma}\right). \tag{27}$$

Here $z \equiv x^d$ is the coordinate orthogonal to the boundary. This Hamiltonian has three fixed points

$$O: \quad h_1 = h_2 = 0,\ c = +\infty\,; \tag{28}$$
$$E: \quad h_1 = h_2 = 0,\ c = -\infty\,; \tag{29}$$
$$S: \quad h_1 = h_2 = c = 0\,. \tag{30}$$

Near the first fixed point the configurations with $\widehat{\sigma} \neq 0$ are exponentially suppressed, then $\widehat{\sigma} = 0$ (i.e. Dirichlet boundary condition). This fixed point controls the ordinary transition. The only relevant surface operator in this phase is $\partial_z\widehat{\sigma}$. The fixed point with $c = -\infty$ favors the configurations with $\widehat{\sigma} \neq 0$: It is associated with the extraordinary transition, where the $\mathbb{Z}_2$ symmetry is broken and no relevant surface operator can couple with it; the lowest dimensional surface operator, besides the identity, is the displacement, whose scaling dimension is d. The fixed point with $c = 0$ controls the special transition, a multicritical phase with two relevant primaries. The even operator $\widehat{\sigma}^2$ is responsible for the flow of c to ∞ or $-\infty$ according to the initial sign, while the odd one, $\widehat{\sigma}$, is the symmetry breaking operator of this phase, characterized by the Neumann boundary condition $\partial_z\widehat{\sigma} = 0$. We omitted a classically marginal coupling, $\partial_z\widehat{\sigma}^2$, because it vanishes with both Neumann and Dirichlet boundary conditions, and it cannot be turned on in the extraordinary transition, where there is no local odd relevant excitation. We shall come back to this operator when considering the RG domain wall.

One important question to address within a BCFT is how to find the scale dimensions of the surface operators and their OPE coefficients in terms of the bulk data. This problem has been completely solved in 2d [15], thanks to the modular invariance. In $d > 2$ useful information can be extracted by the epsilon expansion and other perturbative methods. Recently, the conformal bootstrap approach has been shown to be very promising [37]. Here we face this problem with the method of determinants.

We study the two-point function $\sigma(x)\sigma(y)$. In order to proceed we will consider truncations of the crossing symmetry eq. (25) labeled by (n_{bulk}, n_{bdy}, s). Here n_{bulk} and n_{bdy} are the number of bulk and boundary blocks we keep, and $s = 1, 0$ indicates whether the boundary identity is present or not. The general criterion we use to classify the surface transition associated with a specific truncation (n_{bulk}, n_{bdy}, s) is based on three steps. First, we verify that the solution is compatible with a unitary theory by requiring the positivity of all the nonvanishing couplings μ_a^2 $(a = 1, 2, \ldots, n_{bdy})$. Then we look at the sign of the couplings to the bulk blocks $a_k\lambda_{\sigma\sigma k}$ $(k = 1, \ldots, n_{bulk})$. As in [37], we will assume that the ordinary transition is signaled by the presence of at least one negative coupling in the bulk channel. On the other hand, positivity of the couplings indicates the extraordinary or the special transition, depending on the presence or absence of the surface identity. We should point out that these assumptions have not been proven. However, the results of this work seem to confirm them, serving as a consistency check on the whole setup. For the purpose of this report, however, we will only present the analysis for the ordinary transition, where our results are quite precise and seem to be consistent with other techniques.

3.3 The ordinary transition

We start by considering what is perhaps the simplest successful truncation of eq. (25), corresponding to the fusion rules

$$\begin{aligned} &\sigma \times \sigma \sim 1 + \varepsilon + \varepsilon', && \text{bulk channel,} \\ &\sigma \sim \widehat{O}, && \text{boundary channel.} \end{aligned} \tag{31}$$

This truncation is denoted by the triple (2,1,0). System (25) admits a solution if and only if the 3×3 determinants made with the derivatives of the conformal blocks associated with $\varepsilon, \varepsilon', \widehat{O}$ vanish. We assume that the scale dimensions of σ, ε, and ε' are known ($\Delta_\sigma = \frac{1}{2} + \frac{\eta}{2}$; $\Delta_\varepsilon = 3 - 1/\nu$; $\Delta_{\varepsilon'} = 3 + \omega$, see Table 2) and in this particular case the only unknown scale dimension is $\Delta_{\widehat{O}}$. Fig. 6 shows the values of few determinants of this kind. Clearly they all apparently vanish at the same point. In fact there is a microscopic spread of the solutions and we find $\Delta_{\widehat{O}} = 1.276(2)$. The solution of the complete linear system yields a negative $a_\varepsilon\lambda_{\sigma\sigma\varepsilon}$, thus, according to the above criterion, we are faced with the ordinary transition of the 3d Ising model. Hence, $\widehat{O}$ has to be identified with $\partial_z\widehat{\sigma}$. A two-loop calculation in the 3d ϕ^4 model yields [18] $\Delta_{\partial_z\widehat{\sigma}} \simeq 1.26$ in good agreement with our result.

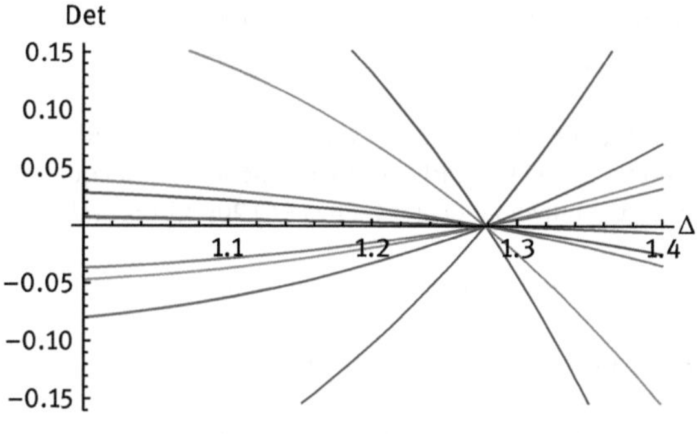

Figure 6: Plot of the ten 3×3 minors made with the first five derivatives of the conformal blocks associated with ε, ε', and $\widehat{O}$ as functions of $\Delta_{\widehat{O}}$. They all vanish approximately at the same point, selecting the allowed value of $\Delta_{\widehat{O}}$.

This solution admits a straightforward generalization to any 3d $O(N)$ model by simply replacing the critical indices with the appropriate values. Table 2 shows our results for $N = 0$ (the nonunitary self-avoiding walk model), $N = 1$ (Ising), $N = 2$ (XY model), and $N = 3$ (Heisenberg model), where we can compare our results with the two-loop calculation of [18].

Acknowledgments: P. L. research at Humboldt University was financed by the SFB 647 "Raum-Zeit-Materie. Analytische und Geometrische Strukturen" for a period of three years starting September 2013.

Bibliography

[1] Aharony O, Tachikawa Y. A Holographic computation of the central charges of d = 4, N = 2 SCFTs. JHEP 2008;01:037, arXiv:0711.4532.

[2] Aharony O, Fayyazuddin A, Maldacena JM. The large N limit of N = 2, N = 1 field theories from three-branes in F theory. JHEP 1998;07:013, hep-th/9806159.

[3] Alba V, Diab K. Constraining conformal field theories with a higher spin symmetry in d = 4, 10.1007/JHEP03(2016)044, arXiv:1307.8092.

[4] Allais A, Sachdev S. Spectral function of a localized fermion coupled to the Wilson-Fisher conformal field theory. Phys Rev 2014;B90:035131, arXiv:1406.3022.

[5] Argyres PC, Douglas MR. New phenomena in SU(3) supersymmetric gauge theory. Nucl Phys 1995;B448:93–126, hep-th/9505062.

[6] Argyres PC, Plesser MR, Seiberg N, Witten E. New N = 2 superconformal field theories in four-dimensions. Nucl Phys 1996;B461:71–84, hep-th/9511154.

[7] Beem C, Lemos M, Liendo P, Peelaers W, Rastelli L, van Rees BC. Infinite chiral symmetry in four dimensions. Comm Math Phys 2015;336(3):1359–1433, arXiv:1312.5344.

[8] Beem C, Lemos M, Liendo P, Rastelli L, van Rees BC. The N = 2 superconformal bootstrap. JHEP 2016;03:183, arXiv:1412.7541.

[9] Beem C, Lemos M, Rastelli L, van Rees BC. The (2; 0) superconformal bootstrap, arXiv:1507.05637. 10.1103/PhysRevD.93.025016.

[10] Beem C, Rastelli L, van Rees BC. The N = 4 superconformal bootstrap. Phys Rev Lett 2013;111:071601, arXiv:1304.1803.

[11] Billó M, Caselle M, Gaiotto D, Gliozzi F, Meineri M, et al. Line defects in the 3d Ising model. JHEP 2013;1307:055, arXiv:1304.4110.

[12] Buican M, Nishinaka T. On the superconformal index of Argyres-Douglas theories, arXiv:1505.05884. 10.1088/1751-8113/49/1/015401.
[13] Campostrini M, Hasenbusch M, Pelissetto A, Rossi P, Vicari E. Critical behavior of the three-dimensional xy universality class. Phys Rev 2001;B63:214503, cond-mat/0010360.
[14] Campostrini M, Hasenbusch M, Pelissetto A, Rossi P, Vicari E. Critical exponents and equation of state of the three-dimensional Heisenberg universality class. Phys Rev 2002;B65:144520, cond-mat/0110336.
[15] Cardy JL, Lewellen DC. Bulk and boundary operators in conformal field theory. Phys Lett 1991;B259:274–78.
[16] Cordova C, Shao SH. Schur indices, BPS particles, and Argyres-Douglas theories, arXiv:1506.00265. 10.1007/JHEP01(2016)040.
[17] Dias, ÒJC, Horowitz GT, Iqbal N, Santos JE. Vortices in holographic superfluids and superconductors as conformal defects. JHEP 2014;1404:096, arXiv:1311.3673.
[18] Diehl H, Shpot M. Massive field theory approach to surface critical behavior in three-dimensional systems. Nucl Phys 1998;B528:595–647, cond-mat/9804083.
[19] Dolan FA, Osborn H. On short and semi-short representations for four-dimensional superconformal symmetry. Annals Phys 2003;307:41–89, hep-th/0209056.
[20] Douglas MR. Spaces of quantum field theories. J Phys Conf Ser 2013;462(1):012011, arXiv:1005.2779.
[21] El-Showk S, Paulos MF. Bootstrapping conformal field theories with the extremal functional method. Phys Rev Lett 2013;111(24):241601, arXiv:1211.2810.
[22] El-Showk S, Paulos MF, Poland D, Rychkov S, Simmons-Duffin D, Vichi A. Solving the 3D ising model with the conformal bootstrap. Phys Rev 2012;D86:025022, arXiv:1203.6064.
[23] El-Showk S, Paulos MF, Poland D, Rychkov S, Simmons-Duffin D, Vichi A. Solving the 3d ising model with the conformal bootstrap II. c-minimization and precise critical exponents. J Stat Phys 2014;157:869, arXiv:1403.4545.
[24] Ferrara S, Gatto R, Grillo A. Conformal algebra in space-time and operator product expansion. Springer, Berlin Heidelberg, 1973.
[25] Ferrara S, Grillo A, Gatto R. Tensor representations of conformal algebra and conformally covariant operator product expansion. Annals Phys 1973;76:161–88.
[26] Ferrara S, Grillo A, Parisi G, Gatto R. Covariant expansion of the conformal four-point function. Nucl Phys 1972;B49:77–98.
[27] Gadde A, Rastelli L, Razamat SS, Yan W. Gauge theories and Macdonald polynomials. Comm Math Phys 2013;319:147–93, arXiv:1110.3740.
[28] Gaiotto D. N = 2 dualities. JHEP 2012;08:034, arXiv:0904.2715.
[29] Gaiotto D, Mazac D, Paulos MF. Bootstrapping the 3d Ising twist defect. JHEP 2014;1403:100, arXiv:1310.5078.
[30] Gaiotto D, Moore GW, Neitzke A. Wall-crossing, Hitchin systems, and the WKB approximation, arXiv:0907.3987. 10.1007/s00023-013-0239-7.
[31] Gliozzi F. More constraining conformal bootstrap. Phys Rev Lett 2013;111:161602, arXiv:1307.3111.
[32] Gliozzi F, Liendo P, Meineri M, Rago A. Boundary and interface CFTs from the conformal bootstrap. JHEP 2015;05:036, arXiv:1502.07217.
[33] Guida R, Zinn-Justin J. Critical exponents of the N vector model. J Phys 1998;A31:8103–21, cond-mat/9803240.
[34] Hasenbusch M. Finite size scaling study of lattice models in the three-dimensional Ising universality class. Phys Rev 2010;B82:174433, arXiv:1004.4486.
[35] Kos F, Poland D, Simmons-Duffin D. Bootstrapping mixed correlators in the 3D Ising model. JHEP 2014;11:109, arXiv:1406.4858.

[36] Lemos M, Liendo P. Bootstrapping N = 2 chiral correlators. JHEP 2016;01:025, arXiv:1510.03866.
[37] Liendo P, Rastelli L, van Rees BC. The bootstrap program for boundary CFTd. JHEP 2013;1307:113, arXiv:1210.4258.
[38] Maldacena J, Zhiboedov A. Constraining conformal field theories with a higher spin symmetry, arXiv:1112.1016. 10.1007/JHEP03(2016)044.
[39] McAvity D, Osborn H. Conformal field theories near a boundary in general dimensions. Nucl Phys 1995;B455:522–76, cond-mat/9505127.
[40] McAvity D, Osborn H. Energy momentum tensor in conformal field theories near a boundary. Nucl Phys 1993;B406:655–80, hep-th/9302068.
[41] Polyakov A. Nonhamiltonian approach to conformal quantum field theory. Zh Eksp Teor Fiz 1974;66:23–42.
[42] Prellberg T. Scaling of self-avoiding walks and self-avoiding trails in tree dimensions. J Phys A: Math Gen 2001;34:L599, cond-mat/0108538.
[43] Rastelli L. Private communication.
[44] Rattazzi R, Rychkov VS, Tonni E, Vichi A. Bounding scalar operator dimensions in 4D CFT. JHEP 2008;0812:031, arXiv:0807.0004.
[45] Rychkov S. EPFL lectures on conformal field theory in D _ 3 dimensions, arXiv:1601.05000. 10.1007/978-3-319-43626-5.
[46] Tachikawa Y. N = 2 supersymmetric dynamics for pedestrians. Lect Notes Phys 2014;890:2014, 2013. arXiv:1312.2684.
[47] Todorov IT, Mintchev MC, Petkova VB. Conformal invariance in quantum field theory. Edizioni della Normale, Pisa, 1978.

Marco Chiodaroli

Simplifying amplitudes in Maxwell-Einstein and Yang-Mills-Einstein supergravities

Abstract: This chapter reviews recent progress in formulating and extending double-copy constructions for scattering amplitudes in supergravities with eight supercharges ($\mathcal{N} = 2$ supersymmetry) in five and four spacetime dimensions. In particular, we consider amplitudes in Maxwell-Einstein theories with symmetric and homogeneous target spaces and in Yang-Mills-Einstein theories with compact gauge groups, identifying the gauge-theory factors entering the construction. Extension of the construction to theories with spontaneously broken gauge symmetry is also discussed.

Keywords: Scattering amplitudes, Supergravity, Supersymmetry

Physics Subject Classification 2010: 04.65.+e, 11.30.Pb, 11.55.Bq, 11.10.Kk, 11.15.Ex

1 Introduction and background

Over the previous decade, scattering amplitudes in quantum field theories involving gravity have been the object of renewed interest and intense investigation. Calculations in the maximal and half-maximal supergravities have brought into focus simpler-than-expected structures and revealed improved ultraviolet (UV) behaviors. While most of the explicit computations thus far have been within the purview of theories with a large number of supersymmetries, a growing number of researches aim to extend this progress to theories with reduced supersymmetry. In this contribution, I will discuss the extension of modern computational techniques, in particular the double-copy construction, to infinite families of supergravities with eight supercharges.

The canonical work of Kawai, Lewellen, and Tye (KLT) [66] established that tree-level gauge-theory amplitudes are sufficient for constructing tree-level amplitudes in the gravity theories which can be obtained from toroidal compactifications of string theory. The structure underlying the KLT relations has achieved a more modern formulation through the work of Bern, Carrasco, and Johansson (BCJ), who expressed loop-level gravity amplitudes at the integrand level as "double copies" of amplitudes in suitably chosen gauge theories [12, 13]. Their construction relies on the availability of gauge-theory amplitudes in which color and kinematic factors obey a duality known as color/kinematics (C/K) duality. This double-copy procedure constitutes a breath-taking computational advance, as it directly relates loop-level gravity amplitudes with gauge-theory amplitudes, which are significantly easier to obtain. At the

DOI 10.1515/9783110452150-006_s_001

same time, the double copy has proven itself to be particularly well suited for studying amplitudes in broader classes of theories with respect to its KLT precursor.

A combination of the KLT property with unitarity-based methods [23] and the BCJ double-copy construction has both been instrumental in performing impressive multi-loop computations in the gravitational theories whose results are most amenable to perturbative calculations, that is theories which possess a large number of supersymmetries. The theory with a maximal number of supersymmetries, $\mathcal{N} = 8$ supergravity, plays a key role in this context. This theory was first constructed by Cremmer and Julia and by de Wit and Nicolai [48, 49, 52]. Explicit computations have shown that its UV behavior matches the one of the $\mathcal{N} = 4$ super-Yang-Mills (sYM) theory at least through four loops [8–11]. These results lend support to the conjecture that the theory might be perturbatively UV-finite in four dimensions [24] and thus constitute the first known example of a mathematically consistent quantum field theory of gravity.[1]

Similar calculations have also been carried out for the half-maximal theory – $\mathcal{N} = 4$ supergravity [16–19, 21] – and more recently for $\mathcal{N} = 5$ supergravity [15]. In the absence of additional matter, the former theory appears to be finite at three loops in four dimensions and to diverge at four loops. The latter theory is UV-finite through at least four loops. At the moment, there are no widely accepted symmetry arguments explaining the finiteness of pure $\mathcal{N} = 4$ supergravity at three loops and of $\mathcal{N} = 5$ supergravity at four loops. The improved UV behavior of these theories is linked to the presence of enhanced cancellations between different terms of a diagrammatic presentation of the amplitude. Interestingly, the four-loop divergence of pure $\mathcal{N} = 4$ supergravity appears to be related to a $U(1)$ quantum anomaly [21, 40]. Additionally, recent calculations in gravity same-helicity amplitudes have shown that adding evanescent operators to the gravity action can alter the UV divergence of the theory while keeping the amplitude's dependence on the renormalization scale unchanged [14]. This result suggests that the latter quantity should be the one regarded as the physically relevant.

A better understanding of enhanced cancellations, the link between anomalies and divergences, and the role of evanescent contributions will be critical in exploring the UV properties of $\mathcal{N} = 8$ supergravity and in determining its fate as a potential theory of quantum gravity, at least in the perturbative context. In turn, achieving this understanding requires the ability for performing perturbative calculations in more general gravity theories, i.e., in theories that are not as special as maximal or half-maximal supergravity. Unsurprisingly, one of the major research directions established since the advent of the double-copy construction has been its extension and application to broader arrays of theories. In particular, it is natural to ask: Is the double-copy structure a general property of gravitational interactions?

1 Reviewing arguments for and against finiteness is beyond the scope of this note. Ultimately, calculations at even higher loop order are necessary to establish conclusively the UV properties of the theory.

Earlier studies on double copies with reduced ($\mathcal{N} < 4$) supersymmetry have investigated very special theories, such as the ones that can be obtained as truncations of $\mathcal{N} = 8$ supergravity [39, 46, 50], pure supergravities [65], and Einstein gravity coupled to a dilaton and an antisymmetric tensor [20]. This contribution focuses on extending the construction to infinite classes of supergravities with eight supercharges – $\mathcal{N} = 2$ theories in four and five spacetime dimensions. Unlike their more supersymmetric relatives, these theories are no longer completely specified by their matter content alone, in the sense that two theories with the same spectra can have different interaction terms while still preserving $\mathcal{N} = 2$ supersymmetry. Moreover, many of these theories cannot be obtained by taking field-theory limits of toroidal compactifications of string theory. The aim of this contribution is to provide a summary and a pedagogical introduction; the reader should consult [43–45] for a complete treatment. Particular attention will be given to discussing how some physical features that can arise in supergravities with reduced supersymmetry, such as the introduction of nonabelian gauge interactions and the supergravity Higgs mechanism, are translated in the double-copy language.

2 Maxwell-Einstein and Yang-Mills-Einstein supergravities

Maxwell-Einstein theories with $\mathcal{N} = 2$ supersymmetry in five dimensions have been explicitly known since the early 1980s due to the work of Günaydin, Sierra, and Townsend [57–60]. These theories give the coupling of the gravity multiplet to n matter vector multiplets. To fix the notation, we write their bosonic fields as $\left(h_{\mu\nu}, A^0_\mu\right) \oplus (A^x_\mu, \phi^x)$, where $x = 1, 2, \ldots, n$, and A^0_μ is the vector in the gravity multiplet. Their bosonic Lagrangian is

$$e^{-1}\mathcal{L} = -\frac{1}{2}R - \frac{1}{4}\mathring{a}_{IJ}F^I_{\mu\nu}F^{J\mu\nu} - \frac{1}{2}g_{xy}\partial_\mu\phi^x\partial^\mu\phi^y + \frac{e^{-1}}{6\sqrt{6}}C_{IJK}\varepsilon^{\mu\nu\rho\sigma\lambda}F^I_{\mu\nu}F^J_{\rho\sigma}A^K_\lambda , \quad (1)$$

where the $F^I_{\mu\nu}$ ($I = 0, 1, \ldots, n$) are abelian field strengths. We note that the symmetric tensor C_{IJK} that appears in the $F \wedge F \wedge A$ term needs to be constant to preserve gauge invariance. The matrices $\mathring{a}_{IJ}$ and g_{xy} are functions of the physical scalars. The authors of [58] employ an ansatz for supergravity Lagrangian and supersymmetry transformations which depend on generic functions of the scalar fields and use invariance of the Lagrangian under supersymmetry and closure of the supersymmetry algebra to derive a set of algebraic and differential constraints. The most general solution to these constraints is found introducing an auxiliary ambient space with coordinates ξ^I and defining a cubic polynomial $\mathcal{V}(\xi)$ in terms of the C-tensor,

$$\mathcal{V}(\xi) \equiv C_{IJK}\xi^I\xi^J\xi^K . \quad (2)$$

The cubic polynomial is used to introduce the ambient-space metric

$$a_{IJ}(\xi) \equiv -\frac{1}{3}\frac{\partial}{\partial \xi^I}\frac{\partial}{\partial \xi^J}\ln \mathcal{V}(\xi)\,. \tag{3}$$

Then, the matrices $\overset{\circ}{a}_{IJ}$ and g_{xy}, as well as the other quantities in the Lagrangian, have a geometrical interpretation:

- The n-dimensional target space $\mathcal{M}_5$ with coordinates ϕ^x is defined as the hypersurface with

$$\mathcal{V}(h) = C_{IJK}h^I h^J h^K = 1, \qquad h^I = \sqrt{\frac{2}{3}}\xi^I; \tag{4}$$

- The matrix $\overset{\circ}{a}_{IJ}(\phi)$ which appears in the kinetic-energy term for the vector fields is the restriction of the ambient-space metric to $\mathcal{M}_5$,

$$\overset{\circ}{a}_{IJ}(\phi) = a_{IJ}\big|_{\mathcal{V}(h)=1}\,; \tag{5}$$

- The metric $g_{xy}(\phi)$ in the kinetic-energy term for the scalars is the induced metric on $\mathcal{M}_5$,

$$g_{xy}(\phi) = \frac{3}{2}\frac{\partial \xi^I}{\partial \phi^x}\frac{\partial \xi^J}{\partial \phi^y}a_{IJ}\bigg|_{\mathcal{V}(h)=1}\,. \tag{6}$$

The key result is that all quantities in the Maxwell-Einstein Lagrangian can be expressed in terms of the C-tensor. Since the C-tensor can be obtained by inspecting three-point amplitudes, $\mathcal{N} = 2$ Maxwell-Einstein theories in five dimensions are uniquely specified by their three-point interactions. This is in contrast to Maxwell-Einstein theories that only exist in four dimensions, for which supersymmetry is not as constraining.

Requiring positive-definiteness of the scalar and vector kinetic terms at a base-point imposes a constraint relating the base-point $\xi^I = c^I$ and the C-tensor,

$$c^I = \sqrt{\frac{2}{3}}C_{IJK}c^J c^K\,. \tag{7}$$

Choosing the pull-back of the C-tensor to be covariantly constant results in a locally symmetric target space. In turn, C-tensors with this property can be obtained using the theory of Jordan algebras and identifying $\mathcal{V}(\xi)$ as the norm of a euclidean Jordan algebra of degree three. This construction has permitted to obtain two classes of supergravities based on symmetric target spaces. The first, named generic Jordan family, is based on an infinite family of reducible Jordan algebras. The second is based on 3×3 hermitian matrices with entries in the four division algebras, and gives the so-called magical supergravities.

We give here only the cubic polynomials relevant to the cases of interest. Theories belonging to the generic Jordan family have [58]

$$\mathcal{V}(\xi) = \sqrt{2}\big(\xi^0(\xi^1)^2 - \xi^0(\xi^i)^2\big)\,, \qquad i = 2, 3, \ldots, n\,. \tag{8}$$

Their target spaces in five and four dimensions are the symmetric spaces

$$\mathcal{M}_5 = \frac{SO(n-1,1)}{SO(n-1)} \times SO(1,1)\,, \qquad \mathcal{M}_4 = \frac{SO(n,2)}{SO(n)\times SO(2)} \times \frac{SU(1,1)}{U(1)}\,. \tag{9}$$

A more general class of theories is associated with homogeneous target spaces. Van Proeyen and de Wit [53] have shown that the requirement of a transitive group of target-space isometries permits to write the cubic polynomial as

$$\mathcal{V}(\xi) = \sqrt{2}\big(\xi^0(\xi^1)^2 - \xi^0(\xi^i)^2\big) + \xi^1(\xi^\alpha)^2 + \tilde{\Gamma}^i_{\alpha\beta}\xi^i\xi^\alpha\xi^\beta\,, \tag{10}$$

where $i, j = 2, 3, \ldots, q+2$, α, β are indices with range r, and the total number of vector multiplets in 5D is $n = 2 + q + r$. The matrices $\tilde{\Gamma}^i_{\alpha\beta}$ are symmetric gamma matrices and form a representation of the euclidean Clifford algebra $\mathcal{C}(q+1, 0)$. The parameter r is a multiple of the dimension of the irreducible representation of $\mathcal{C}(q+1, 0)$, which is denoted as $\mathcal{D}_q$,

$$r(P, \dot{P}, q) = \mathcal{D}_q(P + \dot{P}) \text{ if } q = 0, 4 \ (\mathrm{mod}\ 8)\,, \qquad r(P, q) = \mathcal{D}_q P \text{ otherwise}, \tag{11}$$

where P and $\dot{P}$ are nonnegative integers. The parameter $\dot{P}$ is introduced for the values of q such that there exist two inequivalent irreducible representations of $\mathcal{C}(q+1, 0)$. The generic Jordan family corresponds to either $q = 1$ and P arbitrary or to $P = 0$ and q arbitrary. The magical theories correspond to $P = 1$, $\dot{P} = 0$, and $q = 1, 2, 4, 8$.

So far we have considered only supergravities of the Maxwell-Einstein type. However, it is possible to promote a subgroup of the isometry group and/or of the R-symmetry group to a nonabelian gauge group as discussed in [57, 59]. Isometry transformations act linearly on the ambient-space coordinates as

$$\delta_\alpha \xi^I = (M_r)^I{}_J \xi^J \alpha^r\,, \qquad [M_r, M_s] = f_{rs}{}^t M_t\,, \tag{12}$$

where the matrices M_r leave the C-tensor invariant. The gauging procedure is particularly simple when we consider compact isometry gaugings (i.e., we do not gauge part of the R-symmetry group) and we further restrict to the case in which the vector fields furnish the adjoint representation of the gauge group plus additional singlets (i.e., nontrivial representations other than the adjoint are not present). This class of gaugings is obtained by introducing covariant derivatives and field strengths in the Lagrangian in eq. (1),

$$\mathcal{D}_\mu \phi^x = \partial_\mu \phi^x + g_s A^r_\mu K^x_r\,, \qquad K^x_r = -\sqrt{\frac{3}{2}} f^{rIJ} h_I h^{Jx}\,. \tag{13}$$

$$\mathscr{F}^I_{\mu\nu} = 2\partial_{[\mu}A^I_{\nu]} + g_s f^I{}_{JK}A^J_\mu A^K_\nu\,, \tag{14}$$

where g_s is the gauge coupling constant. The antisymmetric tensors f^{IJK} are equal to the structure constants of the gauge group when the three indices I, J, K assume values corresponding to the vectors promoted to gluons and to zero otherwise. Additionally, we need to covariantize the $F \wedge F \wedge A$ term in eq. (1) and add a Yukawa-like term to the fermionic part of the Lagrangian. However, supersymmetry does not require a nontrivial scalar potential, and hence the vacua of the theory are still of the Minkowski class.

This procedure yields large classes of $\mathcal{N} = 2$ Yang-Mills-Einstein supergravities with compact gauge groups. According to the choice of base-point, the nonabelian gauge symmetry can be unbroken or spontaneously broken. This contribution focuses on Yang-Mills-Einstein theories of the generic Jordan family, for which the gauge group is a subgroup of $SO(n)$. All gauge groups can be accommodated in this construction, provided that n is taken to be large enough. The canonical choice of base-point is

$$c^I_{V_s} = \left(\frac{1}{\sqrt{2}}, 1, V_s, 0, 0\right). \tag{15}$$

With $V_s = 0$, the Yang-Mills-Einstein theory is in the unbroken gauge phase. When $V_s \neq 0$, the theory is on the Coulomb branch and contains massive vector multiplets transforming in matter (nonadjoint) representations of the unbroken gauge group. A $SO(n)$ transformation can be employed to bring all base-points to the form (15). Since, in general, the theory will be invariant only under the subgroup of $SO(n)$ which is gauged, this transformation will change the form of the structure constants f^{IJK}.

It is convenient to label the massive vectors with an index α running over all (not necessarily irreducible) matter representations, so that the bosonic spectrum of the theory becomes

$$\left(A^a_\mu, \phi^a\right) \oplus W_{\alpha\mu} \oplus \overline{W}^\alpha_\mu\,. \tag{16}$$

The fields $W_{\alpha\mu}$ and $\overline{W}^\alpha_\mu$ transform in conjugate representations. This choice corresponds to redefining the gauge-group generators that do not commute with T^2 so that $(T^\alpha)^\dagger = T_\alpha$ in a similar fashion to the treatment of root generators in the Chevalley basis of a simple Lie group. The mass matrix for the massive vector multiplets is proportional to the preferred $U(1)$ generator corresponding to the direction singled out by the nonzero V_s,

$$m_\beta{}^\alpha = ig_s\frac{V_s}{\sqrt{1-V_s^2}}(f^2{}_\beta{}^\alpha)\,. \tag{17}$$

The following sections will discuss how this plethora of theories can be obtained from the double copy of suitably chosen gauge theories. It is important to point out that several theories of interest have not (for now) been considered. These include theories with some matter hypermultiplets, supergravities with R-symmetry gaugings, theories with noncompact gauge groups, five-dimensional supergravities with massive tensors in matter representations, and theories with spontaneously broken supersymmetry.

3 Double-copy construction

We start by considering L-loop, n-point amplitudes in a gauge theory with fields in the adjoint representation and in a set of matter representations. For very large classes of theories, such amplitudes can be expressed in a form based on a set of cubic graphs in which all external or internal lines are labeled by a gauge-group representation:

$$\mathscr{A}_n^{(L)} = i^{L-1} g^{n-2+2L} \sum_{i \in \text{cubic}} \int \frac{d^{LD}\ell}{(2\pi)^{LD}} \frac{1}{S_i} \frac{c_i n_i}{D_i} . \tag{18}$$

S_i are symmetry factors relevant for loop amplitudes. The color factors c_i are obtained by contracting the invariant tensors with indices in the representations carried by the lines joining at each vertex. The denominators D_i are given by the products of propagators associated to the internal lines of the graph. Finally, the numerator factors n_i are functions of loop and external momenta, as well as polarization vectors (or spinor-helicity brackets) and, when applicable, invariant tensors of the global-symmetry group.

Color/kinematics duality [12, 13] is the requirement that the color factors in eq. (18) have the same algebraic properties as the numerator factors. In particular, due to the gauge-group Jacobi relations and the representation-matrices commutation relations, there will be triplets of graphs whose color factors add to zero. In this case, an amplitude presentation is said to obey C/K duality manifestly if the corresponding numerator factors obey the same relations:

$$n_i - n_j = n_k \quad \Leftrightarrow \quad c_i - c_j = c_k .$$

This formulation requires the gauge group to be kept general since avoiding a specific choice for the gauge group prevents the color factors from obeying extra identities aside from the one stemming from the Jacobi and commutation relations.

The existence of C/K-satisfying amplitude presentations has been proven at tree level for $\mathcal{N} = 4$ sYM [25, 72, 79] and, by extension, for some other theories that can be obtained directly from string theory, such as pure sYM theories with varying amounts of supersymmetry. While at loop level this property has a conjectural status, there is a strong and growing body of evidence in its favor, at least in certain theories. For $\mathcal{N} = 4$ sYM, amplitude presentations that obey the duality manifestly have been constructed up to four loops at four points [10] and up to two loops at five points [37]. General methods for constructing BCJ numerators at one loop have been investigated in [26, 61].

This property has also been established for a variety of other theories, including self-dual Yang-Mills (YM) [7, 28, 73], theories with higher-dimension operators [28], Quantum Chromodynamics (QCD) (including supersymmetric versions) [51, 64], non-supersymmetric YM theories with various matter [77], the nonlinear sigma model [42], the Bagger-Lambert-Gustavsson theory in three dimensions [62, 63], and several non-supersymmetric theories that we will later review.

Duality-satisfying structures arise naturally in string theory [25, 79], particularly employing the pure-spinor formalism [31, 69–72].

Note that gauge-theory amplitudes will obey the Ward identities dictated by gauge invariance, that is, when a gluon polarization vector for the ith external particle is changed as $\epsilon_i \to \epsilon_i + f(k,\epsilon)k_i$, the corresponding amplitude will be unchanged. At tree level, this requires the variations of the numerator factors to obey,

$$n_i \to n_i + \Delta_i\,, \qquad \sum_{i\in\text{cubic}} \frac{c_i\Delta_i}{D_i} = 0\,. \tag{19}$$

This identity relies only on the algebraic relations between the color factors.

We now turn to the double-copy prescription for amplitudes in theories involving gravity and consider pairs or gauge theories with identical gauge groups and the same set of representations. A gravity asymptotic state is associated with each gauge-invariant bilinear build with a pair of states from the two gauge theories. Some states in the gravity theory, most notably the physical polarizations of the graviton and the other states in the same multiplet, will be associated with bilinears with adjoint gauge-theory states. For theories involving nonadjoint matter representations, there will be additional sectors corresponding to tensor products involving conjugate matter representations. Due to the requirement of gauge invariance, bilinears made of one adjoint and one nonadjoint state will be disregarded.

Once the double copy has been established at the level of the free theory, amplitudes in the interacting theory can be obtained replacing the color factors of the first theory with the numerator factors of the second:

$$\mathcal{M}_n^{(L)} = i^{L-1}\left(\frac{\kappa}{2}\right)^{n-2+2L} \sum_{i\in\text{cubic}} \int \frac{d^{LD}\ell}{(2\pi)^{LD}} \frac{1}{S_i} \frac{n_i\tilde{n}_i}{D_i}\,. \tag{20}$$

Gauge invariance under linearized diffeomorphisms follows from the gauge invariance of the gauge theories entering the construction. The corresponding Ward identity dictates that replacing the graviton polarization tensor for the ith external particle as $\epsilon^i_{\mu\nu} \to \epsilon^i_{\mu\nu} + f(k,\epsilon)k^i_\mu q_\nu + f(k,\epsilon)k^i_\nu q_\mu$, with $q\cdot k^i = 0$, does not change the amplitude. However, this follows directly from eq. (19) since color factors are replaced with numerator factors that have the same algebraic properties, as required by C/K duality. While both gauge theories need to obey C/K duality, the construction leads to a sensible theory as long as the numerators of one of the two theories obey the duality manifestly.

When a gauge theory possesses a global symmetry, that symmetry will be inherited by the gravity theory from the double copy. However, the symmetry that is manifest in the construction will be only a subgroup of the symmetry of the resulting gravitational theory. In particular, amplitudes are covariant only under a compact subgroup of the U-duality group, while the noncompact symmetries of the Lagrangian manifest themselves through the amplitudes' vanishing soft limits [6, 30]. An alternative approach to understanding the gauge-theory origin of gravitational symmetries requires a formulation of the double copy at the level of off-shell linearized supermultiplets [3–5].

At tree level, the double-copy construction is equivalent to the KLT relations for theories that can be seen as the field-theory limit of some toroidal compactification of string theory. However, the KLT relations do not apply to theories not belonging to this class or, alternatively, when the gauge theories entering the construction have nonadjoint fields.

The double-copy property is a feature of other modern approaches to scattering amplitudes, including most notably the scattering equation formalism [33, 35, 36]. Beyond the realm of scattering amplitudes, a growing body of literature has identified double-copy structures in the context of classical solutions. The reader should consult [67, 68, 74, 78] for exciting developments in this direction. Finally, a comprehensive review of gauge-gravity relations can be found in [38].

4 Amplitudes from the double copy

4.1 Generic Jordan family of Maxwell-Einstein supergravities

This section applies the double-copy prescription to the specific theories we have introduced in Section 2. At tree level and for a modest number of external particles, it is still possible, although increasingly cumbersome, to obtain amplitudes with a conventional Feynman-rule computation. For technical reasons, it is easier to carry out computations in four spacetime dimensions. Hence, we start by dimensionally reducing the Lagrangian (1). Denoting as $e^{2\sigma}$, A^{-1}_μ, and A^I the g_{55} component of the metric, the graviphoton field from dimensional reduction, and the components of the vector fields along the fifth dimension, the bosonic Lagrangian reduced to four dimensions is expressed as follows [56]:

$$
\begin{aligned}
e^{-1}\mathscr{L}^{4D} = {} & -\frac{1}{2}R - \frac{1}{16}e^{3\sigma}F^{-1}_{\mu\nu}F^{-1\mu\nu} - \frac{3}{4}\partial_\mu\sigma\partial^\mu\sigma - \frac{3}{4}\mathring{a}_{IJ}\partial_\mu h^I\partial^\mu h^J \\
& -\frac{1}{2}e^{-2\sigma}\mathring{a}_{IJ}\,\partial_\mu A^I\,\partial^\mu A^J - \frac{1}{4}e^{\sigma}\mathring{a}_{IJ}\Big(F^I_{\mu\nu} - \frac{1}{\sqrt{2}}F^{-1}_{\mu\nu}A^I\Big)\Big(F^{J\mu\nu} - \frac{1}{\sqrt{2}}F^{-1\mu\nu}A^J\Big) \\
& + \frac{e^{-1}}{2\sqrt{6}}C_{IJK}\,\epsilon^{\mu\nu\rho\sigma}\Big\{F^I_{\mu\nu}F^J_{\rho\sigma}A^K - \frac{1}{\sqrt{2}}F^I_{\mu\nu}F^{-1}_{\rho\sigma}A^JA^K + \frac{1}{6}F^{-1}_{\mu\nu}F^{-1}_{\rho\sigma}A^IA^JA^K\Big\}.
\end{aligned}
\tag{21}
$$

The real scalars from the five-dimensional Lagrangian pair with the components of the vectors along the fifth dimension to give complex scalars in four dimensions,

$$z^I = \frac{1}{\sqrt{2}}A^I + \frac{\sqrt{3}}{2}ie^{\sigma}h^I \,. \tag{22}$$

An alternative route to write the four-dimensional Lagrangian involves a symplectic formulation based on a prepotential which, in turn, depends on the C-tensor, as explained in [55].

Focusing on the case of the generic Jordan family with cubic polynomial (4), we expand around the base-point (15) and, following [44], carry out the following steps:

1. Dualize the graviphoton field A_μ^{-1}.
2. Redefine the vector fields as

$$\begin{aligned} A_\mu^{-1} &\to \frac{1}{4}\left(A_\mu^{-1} - A_\mu^0 - \sqrt{2}A_\mu^1\right), \\ A_\mu^0 &\to \frac{1}{2}\left(-A_\mu^{-1} + A_\mu^0 - \sqrt{2}A_\mu^1\right), \\ A_\mu^1 &\to -\frac{1}{\sqrt{2}}\left(A_\mu^{-1} + A_\mu^0\right). \end{aligned} \tag{23}$$

3. Dualize the new A_μ^1 field and redefine $z^1 \to -iz^1$.

Amplitudes at three and four points can be straightforwardly obtained with a standard Feynman-rule computation. One of the advantages of working in four dimensions is the possibility of employing the spinor-helicity formalism [54]. As an example, the vector-vector-scalar amplitude is

$$\mathcal{M}_3^{(0)}\left(1A_-^0, 2A_-^A, 3\bar{z}^B\right) = -\frac{\kappa}{2}\langle 12\rangle^2\delta^{AB} \,. \tag{24}$$

The large symmetry group of the supergravities belonging to the generic Jordan family facilitates the identification of the correct gauge theories entering their double-copy construction. First, one of the gauge theories needs to provide the desired amount of supersymmetry. Hence, a natural candidate is a pure $\mathcal{N} = 2$ sYM theory. As before, we write only the bosonic part of the Lagrangian in four dimensions:

$$\mathcal{L}_{bos}^{\mathcal{N}=2} = -\frac{1}{4}F_{\mu\nu}^{\hat{A}}F^{\hat{A}\mu\nu} + (\overline{D_\mu\phi})^{\hat{A}}(D^\mu\phi)^{\hat{A}} + \frac{g^2}{2}f^{\hat{A}\hat{B}\hat{C}}\phi^{\hat{B}}\bar{\phi}^{\hat{C}}f^{\hat{A}\hat{D}\hat{E}}\phi^{\hat{D}}\bar{\phi}^{\hat{E}} \,, \tag{25}$$

where $\hat{A}, \hat{B}, \hat{C}$ are adjoint gauge indices and ϕ is a complex scalar. The second gauge theory is the dimensional reduction of a $(4+n)$-dimensional pure YM theory:

$$\mathcal{L}^{\mathcal{N}=0} = -\frac{1}{4}F_{\mu\nu}^{\hat{A}}F^{\hat{A}\mu\nu} + \frac{1}{2}(D_\mu\phi^A)^{\hat{A}}(D^\mu\phi^B)^{\hat{A}} - \frac{g^2}{4}f^{\hat{A}\hat{B}\hat{C}}\phi^{\hat{B}B}\phi^{\hat{C}C}f^{\hat{A}\hat{D}\hat{E}}\phi^{\hat{D}B}\phi^{\hat{E}C}, \tag{26}$$

where ϕ^A are real scalars labeled by the global indices $A, B, C = 1, \ldots, n$. This theory has a manifest $SO(n)$ global symmetry. In this case, both gauge theories are known to obey C/K duality at least at tree level. However, in the more involved examples discussed in subsequent sections, C/K duality will pose nontrivial constraints.

At this point, it is straightforward to check that the three-point amplitudes from the double-copy formula (20) reproduce the ones from the Lagrangian, provided that we identify the supergravity states with the $(\mathcal{N}=2)\otimes(\mathcal{N}=0)$ bilinears as [44]:

$$
\begin{aligned}
A_-^{-1} &= \bar{\phi} \otimes A_- \,, & h_- &= A_- \otimes A_- \,, \\
A_-^0 &= \phi \otimes A_- \,, & i\bar{z}^0 &= A_+ \otimes A_- \,, \\
A_-^A &= A_- \otimes \phi^A \,, & i\bar{z}^A &= \bar{\phi} \otimes \phi^A \,,
\end{aligned}
\tag{27}
$$

with similar expressions for the states obtained by parity, charge conjugation, or time reversal. With this map, the amplitude (24) is obtained as

$$
\mathcal{M}_3^{(0)}(1A_-^0, 2A_-^A, 3\bar{z}^B) = \mathcal{A}_3^{(0)}(1\phi, 2A_-, 3\bar{\phi})\Big|_{\mathcal{N}=2} \otimes \mathcal{A}_3^{(0)}(1A_-, 2\phi^A, 3\phi^B)\Big|_{\mathcal{N}=0} . \tag{28}
$$

The purpose of the dualization and field redefinitions enacted after reducing the theory to four dimensions has been precisely to "align" the physical states from the supergravity Lagrangian with the ones from the double copy. Since the three-point amplitudes from the double copy match the ones from the Lagrangian, the construction will give the amplitudes of the correct theory at a higher point also, at least at tree level (provided that we are able to find amplitude presentations obeying C/K duality manifestly). As explained before, this is a consequence of supersymmetry and of the existence of a five-dimensional uplift for the theories.

From a gauge-theory perspective, there are several interaction terms we can envisage adding to the Lagrangians (25) and (26). Indeed, the following sections will discuss simple deformations of the gauge theories which have interesting interpretations from the vantage point of the resulting supergravity.

4.2 Isometry gaugings

A very natural deformation of the non-supersymmetric Lagrangian (26) is the introduction of cubic scalar couplings of the form,

$$
\delta\mathcal{L}^{\mathcal{N}=0} = \frac{\lambda}{3!} g\, f^{\hat{A}\hat{B}\hat{C}} F^{ABC} \phi^{\hat{A}A} \phi^{\hat{B}B} \phi^{\hat{C}C} . \tag{29}
$$

Here F^{ABC} are antisymmetric tensors with three global indices and λ is a real parameter. It is important to verify that the theory still obeys C/K duality after adding the term (29). At four points, the order-λ^2 part of the four-scalar amplitude has the expression

$$\mathscr{A}_4^{(0)}(1\phi^{A_1}, 2\phi^{A_2}, 3\phi^{A_3}, 4\phi^{A_4})\Big|_{\lambda^2} = g^2\lambda^2 \Big(\frac{1}{s} F^{A_1A_2B}F^{A_3A_4B} f^{\hat{A}_1\hat{A}_2\hat{B}} f^{\hat{A}_3\hat{A}_4\hat{B}} + \frac{1}{u} F^{A_3A_1B}F^{A_2A_4B} f^{\hat{A}_3\hat{A}_1\hat{B}} f^{\hat{A}_2\hat{A}_4\hat{B}} + \frac{1}{t} F^{A_2A_3B}F^{A_1A_4B} f^{\hat{A}_2\hat{A}_3\hat{B}} f^{\hat{A}_1\hat{A}_4\hat{B}} \Big). \tag{30}$$

Requiring that this amplitude obeys the duality produces a nontrivial constraint on the F-tensors, which need to obey Jacobi relations. We have verified for up to six points that C/K duality does not pose additional constraints in the YM-scalar theory modified by the term (29) [44].

The interaction term (29) produces a nonvanishing amplitude between three scalars. Taking the double copy of this amplitude with a three-gluon amplitude in the supersymmetric gauge theory produces a nonvanishing gravity amplitude between three vector fields,

$$\mathscr{M}_3^{(0)}(1A_-^A, 2A_-^B, 3A_+^C) = -\left(\frac{\kappa}{2}\right)\frac{\lambda}{\sqrt{2}}\frac{\langle 12\rangle^3}{\langle 23\rangle\langle 31\rangle}F^{ABC}. \tag{31}$$

Hence, the interaction term (29) produces nonabelian gauge interactions in the supergravity Lagrangian [44]. Inspection of the amplitudes from the double copy leads to the identification of F^{ABC} and λ with the supergravity gauge-group structure constants and coupling constant,

$$\lambda\left(\frac{\kappa}{2}\right)F^{ABC} = 2g_s f_{sg}^{ABC}, \qquad A, B, C = 1, \dots, n. \tag{32}$$

It is interesting to note that, in this case, the double copy has promoted a global symmetry in one of the gauge theories to a local symmetry in the resulting supergravity.

This construction extends previous single-trace amplitude results from [22]. Amplitudes in Yang-Mills-Einstein theories have also been recently investigated from the point of view of scattering equations [32, 34], ambitwistor string constructions [1, 41], and string amplitudes [80–82].

4.3 Spontaneously broken theories

Having formulated a double-copy construction for Yang-Mills-Einstein theories, it is paramount to probe its validity away from the unbroken gauge phase, i.e., for $V_s \neq 0$. In this case, a sensible candidate for the first gauge theory is a spontaneously broken version of the Lagrangian (25). This is achieved by considering a scalar vacuum expectation value of the form

$$\langle\phi^a\rangle = Vt^0\,\delta^{a0}, \tag{33}$$

with V real. Reality of the expectation value is a consequence of the existence of a five-dimensional uplift for the spontaneously broken theory. We then write the gauge-group generators as $\{t^{\hat{A}}\} = \{t^{\hat{a}}, t^{\hat{\alpha}}, t_{\hat{\alpha}}\}$, where $(t^{\hat{a}})^\dagger = t^{\hat{a}}$, $(t_{\hat{\alpha}})^\dagger = t^{\hat{\alpha}}$, and the index $\hat{a}$ runs over all unbroken generators. The scalar and vector fields of the resulting spontaneously broken gauge theory split accordingly as $A^{\hat{A}}_\mu = \left(A^{\hat{a}}_\mu, W_{\mu\hat{\alpha}}, \overline{W}^{\hat{\alpha}}_\mu\right)$ and $\phi^{\hat{A}} = \left(\phi^{\hat{a}}, \varphi_{\hat{\alpha}}, \overline{\varphi}^{\hat{\alpha}}\right)$. With this notation, the structure constants $f^{\hat{A}\hat{B}\hat{C}}$ of the gauge group before symmetry breaking yield the structure constants of the unbroken gauge group, representation matrices for the massive vector multiplets, and Clebsh-Gordan coefficients entering couplings between three matter representations,

$$f^{\hat{a}\hat{b}\hat{c}} = -i\,\mathrm{Tr}([t^{\hat{a}}, t^{\hat{b}}]t^{\hat{c}}), \quad f^{\hat{a}\ \hat{\alpha}}_{\ \hat{\beta}} = -i\,\mathrm{Tr}([t^{\hat{a}}, (t^{\hat{\beta}})^\dagger]t^{\hat{\alpha}}), \quad f^{\hat{\alpha}\ \hat{\gamma}}_{\ \hat{\beta}} = -i\,\mathrm{Tr}([t^{\hat{\alpha}}, (t^{\hat{\beta}})^\dagger]t^{\hat{\gamma}}). \tag{34}$$

The tensors (34) obey color relations inherited from the Jacobi identities of the gauge group before symmetry breaking. Aside from the Jacobi relations for the structure constants of the unbroken gauge group and the commutation relations for the representation matrices, the Clebsh-Gordan coefficients obey two extra relations,

$$\begin{aligned} f^{\hat{\alpha}\ \hat{\gamma}}_{\ \hat{\epsilon}} f^{\hat{\epsilon}\ \hat{\beta}}_{\ \hat{\delta}} - f^{\hat{\alpha}\ \hat{\beta}}_{\ \hat{\epsilon}} f^{\hat{\epsilon}\ \hat{\gamma}}_{\ \hat{\delta}} &= f^{\hat{\alpha}\ \hat{\epsilon}}_{\ \hat{\delta}} f^{\hat{\gamma}\ \hat{\beta}}_{\ \hat{\epsilon}}\,, \\ \left(f^{\hat{\beta}\ \hat{\epsilon}}_{\ \hat{\gamma}} f^{\ \hat{\alpha}}_{\hat{\epsilon}\ \hat{\delta}} + f^{\hat{\alpha}\ \hat{\epsilon}}_{\ \hat{\delta}} f^{\ \hat{\beta}}_{\hat{\epsilon}\ \hat{\gamma}} + f^{\hat{a}\ \hat{\beta}}_{\ \hat{\gamma}} f^{\hat{a}\ \hat{\alpha}}_{\ \hat{\delta}}\right) - (\hat{\alpha} \leftrightarrow \hat{\beta}) &= f^{\hat{\alpha}\ \hat{\beta}}_{\ \hat{\epsilon}} f^{\ \hat{\epsilon}}_{\hat{\delta}\ \hat{\gamma}}\,. \end{aligned} \tag{35}$$

These identities arise only for representations obtained from the symmetry breaking of a larger gauge group. The seven-term identity can be thought of as a set of three-term identities since, for any assignment of external masses, at most three terms can be nonzero.

The mass spectrum of the theory is given by $m_{\hat{\alpha}}^{\ \hat{\beta}} = igVf^{0\ \hat{\beta}}_{\ \hat{\alpha}}$, which can be taken in a block-diagonal form with blocks corresponding to different irreducible representations. Massive fields can be further organized into representations labeled by the $U(1)$ charge associated with their mass. The number of such representations needs to be kept general in order to cover all possible symmetry-breaking patterns. Amplitudes in the simple supersymmetric theory discussed here can be obtained considering higher-dimensional amplitudes with massless fields and assigning compact momenta proportional to the masses to the external particles, as done in [2, 27, 47, 75, 76].

The non-supersymmetric gauge theory entering the construction is an extension of the YM-scalar theory (26). In particular, the theory has a set of complex scalars φ_α which have the same masses as the ones in the spontaneously broken theory and transform in conjugate representations. We then write the most general cubic couplings involving three scalars, obtaining the Lagrangian:

$$\begin{aligned} \mathscr{L}' = &-\frac{1}{4}F^{\hat{a}}_{\mu\nu}F^{\mu\nu\hat{a}} + \frac{1}{2}(D_\mu\phi^a)^{\hat{a}}(D^\mu\phi^a)^{\hat{a}} + (\overline{D_\mu\varphi^\alpha})_{\hat{\alpha}}(D^\mu\varphi_\alpha)^{\hat{\alpha}} - (m^2)_\alpha^{\ \beta}\,\overline{\varphi}^\alpha_{\hat{\alpha}}\,\varphi_\beta^{\hat{\alpha}} \\ &+ \frac{1}{3!}g\lambda F^{abc} f^{\hat{a}\hat{b}\hat{c}}\,\phi^{a\hat{a}}\,\phi^{b\hat{b}}\,\phi^{c\hat{c}} + g\lambda\Delta^{ab}\,F_\alpha^{\ a\ \beta} f^{\hat{a}\ \hat{\beta}}_{\ \hat{\gamma}}\,\phi^{b\hat{a}}\,\overline{\varphi}^\alpha_{\hat{\beta}}\,\varphi_\beta^{\hat{\gamma}} \\ &+ \frac{1}{2}g\lambda\,F^{\alpha\ \gamma}_{\ \beta} f^{\ \hat{\beta}}_{\hat{\alpha}\ \hat{\gamma}}\,\varphi_\alpha^{\hat{\alpha}}\,\overline{\varphi}^\beta_{\hat{\beta}}\,\varphi_\gamma^{\hat{\gamma}} + \frac{1}{2}g\lambda\,F_{\alpha\ \gamma}^{\ \beta} f^{\hat{\alpha}\ \hat{\gamma}}_{\ \hat{\beta}}\,\overline{\varphi}^\alpha_{\hat{\alpha}}\,\varphi_\beta^{\hat{\beta}}\,\overline{\varphi}^\gamma_{\hat{\gamma}} + \mathscr{L}_{\text{contact}}\,, \end{aligned} \tag{36}$$

where Δ^{ab} is a diagonal matrix and F^{abc}, $F^{a}{}_{\alpha}{}^{\beta}$, $F^{\alpha}{}_{\beta}{}^{\gamma}$ are tensors in the global indices. Imposing relations (35) on the numerators of amplitudes with four massive scalars fixes the contact terms in the Lagrangian (36) and produces the following constraints:

- The F-tensors need to collectively obey the same algebraic relations as the structure constants, representation matrices, and Clebsh-Gordan symbols of the gauge group. They can be interpreted as pieces of the structure constants of a larger group, which is spontaneously broken down to a subgroup with structure constants F^{abc}. The symmetry-breaking pattern of the global group needs to be the same as the one of the gauge group.
- The massive scalars need to assume a block-diagonal form such that

$$2Vf^{0}{}_{\hat{\beta}}{}^{\hat{\alpha}}\varphi^{\hat{\beta}}_{\alpha} = \lambda\rho F^{0}{}_{\alpha}{}^{\beta}\varphi^{\hat{\alpha}}_{\beta}\,, \tag{37}$$

  where ρ is a free parameter, i.e., for each block, the charge of the preferred $U(1)$ gauge generator is proportional to the charge of a preferred global $U(1)$ generator.
- The diagonal matrices Δ are fixed to $\Delta^{ab} = \delta^{ab} + (\sqrt{1+\rho^2} - 1)\delta^{a0}\delta^{0b}$.

The reader should consult [45] for a complete treatment. As before, identifying supergravity three-point amplitudes from the double copy using (20) with the ones from a Feynman-rule computation leads to the $(\mathcal{N} = 2) \times (\mathcal{N} = 0)$ field map:

$$\begin{aligned}
A^{-1}_{-} &= \bar{\phi} \otimes A_{-}\,, & h_{-} &= A_{-} \otimes A_{-}\,,\\
A^{0}_{-} &= \phi \otimes A_{-}\,, & i\bar{z}^{0} &= A_{+} \otimes A_{-}\,,\\
A^{a}_{-} &= A_{-} \otimes \phi^{a}\,, & i\bar{z}^{a} &= \bar{\phi} \otimes \phi^{a}\,,\\
\varphi_{\alpha} &= \varphi \otimes \varphi_{\alpha}\,, & W_{\alpha} &= W \otimes \varphi_{\alpha}\,.
\end{aligned} \tag{38}$$

Similarly, the two free parameters λ and ρ in the non-supersymmetric gauge theory are related to the supergravity parameters as

$$\left(\frac{\kappa}{2}\right)\lambda F^{ABC} = 2g_{s}f^{ABC}_{\mathrm{sg}}\,, \qquad \rho = \frac{V_s}{\sqrt{1 - V_s^2}}\,. \tag{39}$$

Note that the massless supergravity states are obtained as double copies of adjoint gauge-theory states, while the massive sector is obtained from the double copy of gauge-theory states in matter representations. While this contribution considers supergravities with eight supercharges, analogous constructions can be set forth for spontaneously broken theories with $\mathcal{N} = 4$ or no supersymmetry by adjusting the spontaneously broken gauge-theory factor.

4.4 Homogeneous supergravities

The analysis of the supergravity Higgs mechanism has underlined the key role played by matter (nonadjoint) representations in extending the double copy to larger classes of supergravities. Along these lines, another option for modifying the construction of Section 4.1 is to include in the non-supersymmetric gauge-theory fermions transforming in a matter representation. Adding adjoint fermions without at the same time introducing extra supersymmetries is forbidden by C/K duality, as discussed in [46]. To have a nontrivial result, we also need to add to the supersymmetric theory some fields transforming in the conjugate representation. In this regard, the minimal set of fields that can be added is two fermions and two scalars, corresponding to a half-hypermultiplet. To fix notation, we denote the physical states in the supersymmetric gauge theory as

$$\left(A_+^{\hat{a}}, \psi_+^{\hat{a}}, \phi^{\hat{a}}\right)_G \oplus \left(A_-^{\hat{a}}, \psi_-^{\hat{a}}, \bar{\phi}^{\hat{a}}\right)_G \oplus \left(\chi_+, \varphi_1, \varphi_2, \chi_-\right)_R ,$$

where R and G label the matter representation and the adjoint representation. For generic R, we also need to add the CPT-conjugate states. However, here it is convenient to consider the case in which the representation R is pseudo-real, i.e., there exists a unitary matrix V such that $VT^{\hat{a}}V^\dagger = -(T^{\hat{a}})^*$, $VV^* = -1$. In this case, the half-hypermultiplet alone is CPT-self-conjugate.

The non-supersymmetric gauge theory entering the construction is a YM-scalar theory with extra fermions in the representation R,

$$\begin{aligned}\mathcal{L} = & -\frac{1}{4}F^{\hat{a}}_{\mu\nu}F^{\hat{a}\mu\nu} + \frac{1}{2}(D_\mu\phi^a)^{\hat{a}}(D^\mu\phi^a)^{\hat{a}} + \frac{i}{2}\bar{\lambda}^\alpha D_\mu\gamma^\mu\lambda_\alpha \\ & + \frac{g}{2}\phi^{a\hat{a}}\Gamma^{a\ \beta}_{\alpha}\bar{\lambda}^\alpha\gamma_5 T^{\hat{a}}\lambda_\beta - \frac{g^2}{4}f^{\hat{a}\hat{b}\hat{e}}f^{\hat{c}\hat{d}\hat{a}}\phi^{a\hat{a}}\phi^{b\hat{b}}\phi^{a\hat{c}}\phi^{b\hat{d}} .\end{aligned} \tag{40}$$

$\hat{a}, \hat{b}$ are adjoint indices of the gauge group while $\alpha, \beta = 1, \ldots, r$ and $a, b = 1, \ldots, q+2$ are global indices. Spacetime spinor indices and gauge-group indices for the representation R are not explicitly displayed. As before, we introduce unconstrained matrices $\Gamma^{a\ \beta}_{\alpha}$ in the global indices and let C/K duality establish their algebraic properties.

Imposing C/K duality on the four-point amplitudes with two adjoint scalars and two matter fermions gives the constraint [43]:

$$n_u - n_t = n_s \quad \rightarrow \quad \{\Gamma^a, \Gamma^b\} = 2\delta^{ab} , \tag{41}$$

i.e., that the matrices Γ^a form a $(q+2)$-dimensional Clifford algebra. In turn, thanks to this relation, the non-supersymmetric theory can be regarded as the dimensional reduction of a YM + fermions theory in $D = (q+6)$ dimensions. The problem of charting all possible supergravities obtained with this construction is thus equivalent to listing irreducible spinor representations in $D = (q+6)$ dimensions. Parameter P is equal to the number of irreducible spinors introduced in the gauge theory. When P is greater than one, an additional flavor symmetry will be manifest in the non-supersymmetric gauge theory. The only difference with the standard treatment of spinors in D dimensions is the presence of the matrix V acting on the gauge indices, which enters reality (R) and pseudo-reality (PR) conditions of the form [43]

$$\bar{\lambda} = \lambda^t \mathscr{C}_4 C V\,, \qquad \text{R: } C = \mathscr{C}_q\,, \quad \text{PR: } C = \mathscr{C}_q \Omega\,, \tag{42}$$

where $\mathscr{C}_q$ and $\mathscr{C}_4$ are the global and spacetime charge-conjugation matrices which obey the relations $\mathscr{C}_q \Gamma^a \mathscr{C}_q^{-1} = -\zeta(\Gamma^a)^t$, $\mathscr{C}_4 \gamma^\mu \mathscr{C}_4^{-1} = -\zeta(\gamma^\mu)^t$, $\zeta = \pm 1$. Ω is an antisymmetric real matrix acting on the flavor indices. These conditions can be employed to obtain irreducible spinor representations with a q-by-q analysis. Note that the reality (R) condition is the combination of a pseudo-reality condition on the gauge-group indices with a pseudo-Majorana condition on the spinor indices. In the particular case of $q = 0, 4 \pmod 8$, there are two inequivalent irreducible spinors, and the parameters $P, \dot{P}$ count the number of each. The analysis is summarized in Table 1. The total range of the global indices α, β is fixed to $r = \mathscr{D}_q P$ or $r = \mathscr{D}_q(P + \dot{P})$, where $4\mathscr{D}_q$ is the dimension of the irreducible $SO(q + 5, 1)$ spinor. The number of vector multiplets in the four-dimensional supergravity theory obtained with the double-copy construction is equal to $(3 + q + r)$. This construction reproduces the classification of homogeneous supergravities by de Wit and van Proeyen [53]. As for the previous cases, three-point amplitudes from the double copy (20) are compared with the ones from the supergravity Lagrangian. The two sets of amplitudes agree provided that the field map

$$\begin{aligned} A_-^{-1} &= \bar{\phi} \otimes A_-\,, & h_- &= A_- \otimes A_-\,, \\ A_-^0 &= \phi \otimes A_-\,, & i\bar{z}^0 &= A_+ \otimes A_-\,, \\ A_-^a &= A_- \otimes \phi^a\,, & i\bar{z}^a &= \bar{\phi} \otimes \phi^a\,, \\ A_{\alpha-} &= \chi_- \otimes (U\lambda_-)_\alpha\,, & i\bar{z}_\alpha &= \chi_+ \otimes (U\lambda_-)_\alpha\,, \end{aligned} \tag{43}$$

Table 1: Parameters in the double-copy construction for homogeneous supergravities [43].

q	$\mathscr{D}_q$	4D fermions $r(q, P, \dot{P})$	Conditions	Flavor group
−1	1	P	R	$SO(P)$
0	1	$P + \dot{P}$	RW	$SO(P) \times SO(\dot{P})$
1	2	$2P$	R	$SO(P)$
2	4	$4P$	R or W	$U(P)$
3	8	$8P$	PR	$USp(2P)$
4	8	$8P + 8\dot{P}$	PRW	$USp(2P) \times USp(2\dot{P})$
5	16	$16P$	PR	$USp(2P)$
6	16	$16P$	R or W	$U(P)$
k+8	$16\,\mathscr{D}_k$	$16\,r(k, P, \dot{P})$	as for k	as for k

The third column gives the number of four-dimensional irreducible spinors in the non-supersymmetric gauge theory, which can obey a reality (R), pseudo-reality (PR), or Weyl (W) conditions.

is employed. U is a unitary matrix whose exact form depends on the choice of Γ^1. With an appropriate choice for U, some of the entries of the Γ^a matrices in the Yukawa couplings of the non-supersymmetric gauge theory reproduce the real $\tilde{\Gamma}^i$ matrices in the cubic polynomial (10),

$$(U^t \Gamma^a C^{-1} U) = \left(-\mathbf{1}, i\,\tilde{\Gamma}^i \right) . \tag{44}$$

It should be noted that the four magical supergravities can be recovered as particular cases of this construction. Finally, this framework can be modified to include supergravities with hypermultiplets by introducing in the non-supersymmetric gauge-theory scalars transforming in the representation R.

5 Discussion and outlook

This contribution has discussed how the double-copy construction can be formulated in theories with $\mathcal{N} = 2$ supersymmetry in four and five dimensions. Our starting point has been an infinite family of supergravities with symmetric target spaces, the generic Jordan family of Maxwell-Einstein theories. In this case, two very simple gauge theories enter the construction: one is a pure sYM theory, the other a YM-scalar theory obtained by dimensional reduction. Simple modifications of the above theories produce intriguing effects in the resulting supergravity. In particular, some physical features of supergravities with reduced supersymmetry, such as the possibility of gauging part of their isometry groups, are straightforwardly incorporated in the double-copy framework. Results are summarized in Table 2.

Extension of the double copy has relied on the possibility of incorporating non-adjoint representations in the gauge theories entering the construction. In all cases, it

Table 2: Summary of double-copy constructions for supergravities with eight supercharges.

Gauge theories	**Supergravity**
GT1: Pure $\mathcal{N} = 2$ sYM theory **GT2:** YM + scalar theory from dim. red.	Generic Jordan family of Maxwell-Einstein supergravities
GT1: As before **GT2:** Add trilinear scalar couplings	Yang-Mills-Einstein theories (compact gaugings)
GT1: Spont. broken $\mathcal{N} = 2$ sYM theory **GT2:** Add explicit masses	Higgsed supergravities
GT1: Add hypers in representation R **GT2:** Add fermions in representation R with Yukawa couplings	Homogeneous supergravities
GT1: Add hypers in representation R **GT2:** Add scalars in representation R	Supergravities with hypermultiplets

The first row describes the basic construction, while rows 2–5 list some variants.

has been possible to identify the supergravity given by the double-copy construction from its amplitudes at three points. This is a consequence of supersymmetry combined with the existence of a five-dimensional uplift for both the supergravity theory and the gauge theories entering the construction. Nevertheless, amplitudes at higher points have been considered as a consistency check in [43–45], where sample amplitudes at one loop have also been displayed.

These developments open the door to loop-level computations in large families of supergravities with eight supercharges. As UV divergences are expected already at one loop for generic theories with matter, results will likely give some insight into the abundance and role of enhanced cancellations and into the connection between higher-loop divergences and lower-loop quantum anomalies. A systematic study of loop-level amplitudes is ongoing.

The success in extending the double copy strongly suggests that the construction has a significant role to play in computations for generic gravity theories with reduced (or no) supersymmetry. In this respect, the extension of the construction to the supergravity Higgs mechanism is particularly relevant, as spontaneously-broken gauge symmetry is a general feature of Yang-Mills-Einstein theories, which generically have noncompact gauge groups that are broken down to a compact subgroup. Finally, the last extension discussed in this note is critically important as homogeneous supergravities now constitute the largest family of theories previously studied in the supergravity literature for which a double-copy construction is explicitly known. Such family includes theories that cannot be obtained as a toroidal compactification of string theory, marking a substantial departure from the setting in which the KLT relations and the double-copy construction were first introduced.

Acknowledgments: I would like to thank the Collaborative Research Center (CRC) 647 Space-Time-Matter in Berlin and Potsdam (Teilgruppe C6) for funding my postdoctoral position, including a particularly generous travel and collaboration budget, and providing a very fruitful research environment which enabled substantial progress in the areas surveyed here. I am also very grateful to Murat Günaydin, Henrik Johansson, and Radu Roiban for collaboration on [43–45], on which this contribution is based.

Bibliography

[1] Adamo T, Casali E, Roehrig KA, Skinner D. On tree amplitudes of supersymmetric Einstein-Yang-Mills theory, arXiv:1507.02207 [hep-th].

[2] Alday LF, Henn JM, Plefka J, Schuster T. Scattering into the fifth dimension of N=4 super Yang-Mills. JHEP 2010;1001:077, arXiv:0908.0684 [hep-th].

[3] Anastasiou A, Borsten L, Duff MJ, Hughes LJ, Nagy S. A magic pyramid of supergravities. JHEP 2014;1404:178, arXiv:1312.6523 [hep-th].

[4] Anastasiou A, Borsten L, Duff MJ, Hughes LJ, Nagy S. Yang-Mills origin of gravitational symmetries. Phys Rev Lett 2014;113(23):231606, arXiv:1408.4434 [hep-th].

[5] Anastasiou A, Borsten L, Hughes MJ, Nagy S. Global symmetries of Yang-Mills squared in various dimensions. JHEP 2016;1601:148, arXiv:1502.05359 [hep-th].

[6] Arkani-Hamed N, Cachazo F, Kaplan J. What is the simplest quantum field theory? JHEP 2010;1009:016, arXiv:0808.1446 [hep-th].
[7] Badger S, Mogull G, Ochirov A, O'Connell D. A complete two-loop, five-gluon helicity amplitude in Yang-Mills theory. JHEP 2015;1510:064, arXiv:1507.08797 [hep-ph].
[8] Bern Z, Carrasco JJ, Dixon LJ, Johansson H, Kosower DA, Roiban R. Three-loop superfiniteness of N=8 supergravity. Phys Rev Lett 2007;98:161303 [hep-th/0702112].
[9] Bern Z, Carrasco JJM, Dixon LJ, Johansson H, Roiban R. Manifest ultraviolet behavior for the three-loop four-point amplitude of N=8 supergravity. Phys Rev 2008;D78:105019, arXiv:0808.4112 [hep-th].
[10] Bern Z, Carrasco JJ, Dixon LJ, Johansson H, Roiban R. The ultraviolet behavior of N=8 supergravity at four loops. Phys Rev Lett 2009;103:081301, arXiv:0905.2326 [hep-th].
[11] Bern Z, Carrasco JJM, Dixon LJ, Johansson H, Roiban R. Simplifying multiloop integrands and ultraviolet divergences of gauge theory and gravity amplitudes. Phys Rev 2012;D85:105014, arXiv:1201.5366 [hep-th].
[12] Bern Z, Carrasco JJM, Johansson H. New relations for gauge-theory amplitudes. Phys Rev 2008;D78:085011, arXiv:0805.3993 [hep-ph].
[13] Bern Z, Carrasco JJM, Johansson H. Perturbative quantum gravity as a double copy of gauge theory. Phys Rev Lett 2010;105:061602, arXiv:1004.0476 [hep-th].
[14] Bern Z, Cheung C, Chi HH, Davies S, Dixon L, Nohle J. Evanescent effects can alter ultraviolet divergences in quantum gravity without physical consequences, arXiv:1507.06118 [hep-th].
[15] Bern Z, Davies S, Dennen T. Enhanced ultraviolet cancellations in N = 5 supergravity at four loops. Phys Rev D 2014;90(10):105011, arXiv:1409.3089 [hep-th].
[16] Bern Z, Davies S, Dennen T. The ultraviolet critical dimension of half-maximal supergravity at three loops, arXiv:1412.2441 [hep-th].
[17] Bern Z, Davies S, Dennen T. The ultraviolet structure of half-maximal supergravity with matter multiplets at two and three loops. Phys Rev 2013;D88:065007, arXiv:1305.4876 [hep-th].
[18] Bern Z, Davies S, Dennen T, Huang YT. Absence of three-loop four-point divergences in N=4 supergravity. Phys Rev Lett 2012;108:201301, arXiv:1202.3423 [hep-th].
[19] Bern Z, Davies S, Dennen T, Huang Yt. Ultraviolet cancellations in half-maximal supergravity as a consequence of the double-copy structure. Phys Rev 2012;D86:105014, arXiv:1209.2472 [hep-th].
[20] Bern Z, Davies S, Dennen T, Huang Yt, Nohle J. Color-kinematics duality for pure Yang-Mills and gravity at one and two loops, arXiv:1303.6605 [hep-th].
[21] Bern Z, Davies S, Dennen T, Smirnov AV, Smirnov VA. Ultraviolet properties of N=4 supergravity at four loops. Phys Rev Lett 2013;111(23):231302, arXiv:1309.2498 [hep-th].
[22] Bern Z, De Freitas A, Wong HL. On the coupling of gravitons to matter. Phys Rev Lett 2000;84:3531 [hep-th/9912033].
[23] Bern Z, Dixon LJ, Dunbar DC, Kosower DA. One loop n point gauge theory amplitudes, unitarity and collinear limits. Nucl Phys 1994;B425:217 [hep-ph/9403226].
[24] Bern Z, Dixon LJ, Roiban R. Is N = 8 supergravity ultraviolet finite? Phys Lett 2007;B644:265 [hep-th/0611086].
[25] Bjerrum-Bohr NEJ, Damgaard PH, Vanhove P. Minimal basis for gauge theory amplitudes. Phys Rev Lett 2009;103:161602, arXiv:0907.1425 [hep-th].
[26] Bjerrum-Bohr NEJ, Dennen T, Monteiro R, O'Connell D. Integrand oxidation and one-loop colour-dual numerators in N=4 gauge theory. JHEP 2013;1307:092, arXiv:1303.2913 [hep-th].
[27] Boels RH. No triangles on the moduli space of maximally supersymmetric gauge theory. JHEP 2010;1005:046, arXiv:1003.2989 [hep-th].
[28] Boels RH, Isermann RS, Monteiro R, O'Connell D. Colour-kinematics duality for one-loop rational amplitudes. JHEP 2013;1304:107, arXiv:1301.4165 [hep-th].

[29] Broedel J, Dixon LJ. Color-kinematics duality and double-copy construction for amplitudes from higher-dimension operators. JHEP 2012;1210:091, arXiv:1208.0876 [hep-th].
[30] Broedel J, Dixon LJ. R**4 counterterm and E(7)(7) symmetry in maximal supergravity. JHEP 2010;1005:003, arXiv:0911.5704 [hep-th].
[31] Broedel J, Schlotterer O, Stieberger S. Polylogarithms, multiple zeta values and superstring amplitudes. Fortsch Phys 2013;61:812, arXiv:1304.7267 [hep-th].
[32] Cachazo F, He S, Yuan EY. Einstein-Yang-Mills scattering amplitudes from scattering equations. JHEP 2015;1501:121, arXiv:1409.8256 [hep-th].
[33] Cachazo F, He S, Yuan EY. Scattering equations and KLT orthogonality. arXiv:1306.6575 [hep-th].
[34] Cachazo F, He S, Yuan EY. Scattering equations and matrices: From Einstein To Yang-Mills, DBI and NLSM. JHEP 2015;1507:149, arXiv:1412.3479 [hep-th].
[35] Cachazo F, He S, Yuan EY. Scattering of massless particles in arbitrary dimension, arXiv:1307.2199 [hep-th].
[36] Cachazo F, He S, Yuan EY. Scattering of massless particles: Scalars, gluons and gravitons, arXiv:1309.0885 [hep-th].
[37] Carrasco JJ, Johansson H. Five-point amplitudes in N=4 super-Yang-Mills theory and N=8 supergravity. Phys Rev D 2012;85:025006, arXiv:1106.4711 [hep-th].
[38] Carrasco JJM. Gauge and gravity amplitude relations, arXiv:1506.00974 [hep-th].
[39] Carrasco JJM, Chiodaroli M, Günaydin M, Roiban R. One-loop four-point amplitudes in pure and matter-coupled N _ 4 supergravity. JHEP 2013;1303:056, arXiv:1212.1146 [hep-th].
[40] Carrasco JJM, Kallosh R, Roiban R, Tseytlin AA. On the U(1) duality anomaly and the S-matrix of N=4 supergravity. JHEP 2013;1307:029, arXiv:1303.6219 [hep-th].
[41] Casali E, Geyer Y, Mason L, Monteiro R, Roehrig KA. New ambitwistor string theories, arXiv:1506.08771 [hep-th].
[42] Chen G, Du YJ. Amplitude relations in non-linear sigma model. JHEP 2014;1401:061, arXiv:1311.1133 [hep-th].
[43] Chiodaroli M, Gunaydin M, Johansson H, Roiban R. Complete construction of magical, symmetric and homogeneous N=2 supergravities as double copies of gauge theories, arXiv:1512.09130 [hep-th].
[44] Chiodaroli M, Günaydin M, Johansson H, Roiban R. Scattering amplitudes in N = 2 Maxwell-Einstein and Yang-Mills-Einstein supergravity. JHEP 2015;1501:081, arXiv:1408.0764 [hep-th].
[45] Chiodaroli M, Gunaydin M, Johansson H, Roiban R. Spontaneously broken Yang-Mills-Einstein supergravities as double copies, arXiv:1511.01740 [hep-th].
[46] Chiodaroli M, Jin Q, Roiban R. C/K duality for general abelian orbifolds of N=4 super Yang-Mills theory. JHEP 2014;1401:152, arXiv:1311.3600 [hep-th].
[47] Craig N, Elvang H, Kiermaier M, Slatyer T. Massive amplitudes on the coulomb branch of N=4 SYM. JHEP 2011;1112:097, arXiv:1104.2050 [hep-th].
[48] Cremmer E, Julia B. The N=8 supergravity theory. 1. The Lagrangian. Phys Lett 1978;B80:48.
[49] Cremmer E, Julia B. The SO(8) supergravity. Nucl Phys 1979;B159:141.
[50] Damgaard PH, Huang R, Sondergaard T, Zhang Y. The complete KLT-map between gravity and gauge theories. JHEP 2012;1208:101, arXiv:1206.1577 [hep-th].
[51] de la Cruz L, Kniss A, Weinzierl S. Proof of the fundamental BCJ relations for QCD amplitudes. JHEP 2015;1509:197, arXiv:1508.01432 [hep-th].
[52] de Wit B, Nicolai H. N=8 supergravity. Nucl Phys 1982;B208:323.
[53] de Wit B, Van Proeyen A. Special geometry, cubic polynomials and homogeneous quaternionic spaces. Commun Math Phys 1992;149:307 [hep-th/ 9112027].
[54] Elvang H, Huang Yt. Scattering amplitudes, arXiv:1308.1697 [hep-th].

[55] Freedman DZ, Van Proeyen A. Supergravity, Cambridge University Press, 2012. ISBN: 9780521194013

[56] Gunaydin M, McReynolds S, Zagermann M. The R-map and the coupling of N=2 tensor multiplets in 5 and 4 dimensions. JHEP 2006;0601:168 [hep-th/0511025].

[57] Günaydin M, Sierra G, Townsend PK. Quantization of the gauge coupling constant in a five-dimensional Yang-Mills / Einstein supergravity theory. Phys Rev Lett 1984;53:322.

[58] Günaydin M, Sierra G, Townsend PK. The geometry of N=2 Maxwell-Einstein supergravity and Jordan Algebras. Nucl Phys 1984;B242:244.

[59] Günaydin M, Sierra G, Townsend PK. Gauging the d = 5 Maxwell-Einstein supergravity theories: More on Jordan Algebras. Nucl Phys 1985;B253:573.

[60] Gunaydin M, Sierra G, Townsend PK. More on d = 5 Maxwell-Einstein supergravity: Symmetric spaces and kinks. Class Quant Grav 1986;3:763.

[61] He S, Monteiro R, Schlotterer O. String-inspired BCJ numerators for one-loop MHV amplitudes, arXiv:1507.06288 [hep-th].

[62] Huang Yt, Johansson H. Equivalent D=3 supergravity amplitudes from double copies of three-algebra and two-algebra gauge theories. Phys Rev Lett 2013;110:171601, arXiv:1210.2255 [hep-th].

[63] Huang Yt, Johansson H, Lee S. On three-algebra and bi-fundamental matter amplitudes and integrability of supergravity. JHEP 2013;1311:050, arXiv:1307.2222 [hep-th].

[64] Johansson H, Ochirov A. Color-Kinematics duality for QCD amplitudes, arXiv:1507.00332 [hep-ph].

[65] Johansson H, Ochirov A. Pure gravities via color-kinematics duality for fundamental matter, arXiv:1407.4772 [hep-th].

[66] Kawai H, Lewellen DC, Tye SHH. A relation between tree amplitudes of closed and open strings. Nucl Phys 1986;B269:1.

[67] Luna A, Monteiro R, Nicholson I, O'Connell D, White CD. The double copy: Bremsstrahlung and accelerating black holes, arXiv:1603.05737 [hep-th].

[68] Luna A, Monteiro R, O'Connell D, White CD. The classical double copy for Taub-NUT spacetime, arXiv:1507.01869 [hep-th].

[69] Mafra CR, Schlotterer O. Multiparticle SYM equations of motion and pure spinor BRST blocks. JHEP 2014;1407:153, arXiv:1404.4986 [hep-th].

[70] Mafra CR, Schlotterer O. The structure of n-point one-loop open superstring amplitudes. JHEP 2014;1408:099, arXiv:1203.6215 [hep-th].

[71] Mafra CR, Schlotterer O. Towards one-loop SYM amplitudes from the pure spinor BRST cohomology. Fortsch Phys 2015;63(2):105, arXiv:1410.0668 [hep-th].

[72] Mafra CR, Schlotterer O, Stieberger S. Explicit BCJ numerators from pure spinors. JHEP 2011;1107:092, arXiv:1104.5224 [hep-th].

[73] Mogull G, O'Connell D. Overcoming obstacles to colour-kinematics duality at two loops. JHEP 2015;1512:135, arXiv:1511.06652 [hep-th].

[74] Monteiro R, O'Connell D, White CD. Black holes and the double copy. JHEP 2014;1412:056, arXiv:1410.0239 [hep-th].

[75] Naculich SG. CHY representations for gauge theory and gravity amplitudes with up to three massive particles. JHEP 2015;1505:050, arXiv:1501.03500 [hep-th].

[76] Naculich SG. Scattering equations and BCJ relations for gauge and gravitational amplitudes with massive scalar particles. JHEP 2014;1409:029, arXiv:1407.7836 [hep-th].

[77] Nohle J. Color-Kinematics duality in one-loop four-gluon amplitudes with matter. Phys Rev D 2014;90:025020, arXiv:1309.7416 [hep-th].

[78] Ridgway AK, Wise MB. Static spherically symmetric Kerr-Schild metrics and implications for the classical double copy, arXiv:1512.02243 [hep-th].

[79] Stieberger S. Open & closed vs. pure open string disk amplitudes, arXiv:0907.2211 [hep-th].
[80] Stieberger S, Taylor TR. Graviton as a pair of collinear gauge bosons. Phys Lett 2014;B739:457, arXiv:1409.4771 [hep-th].
[81] Stieberger S, Taylor TR. Graviton amplitudes from collinear limits of gauge amplitudes. Phys Lett 2015;B744:160, arXiv:1502.00655 [hep-th].
[82] Stieberger S, Taylor TR. Subleading terms in the collinear limit of Yang-Mills amplitudes. Phys Lett 2015;B750:587, arXiv:1508.01116 [hep-th].

Livia Ferro, Jan Plefka, and Matthias Staudacher

Yangian symmetry in maximally supersymmetric Yang-Mills theory

Abstract: In physics, symmetries are of invaluable help in studying the structure and form of observables of a theory. Much progress has been made in the last decade concerning symmetries in maximally supersymmetric planar Yang-Mills theory. It has been realized that the ordinary superconformal symmetry is not all there is. A second, dual superconformal symmetry may be exposed. Together they form a Yangian structure, the hallmark of the underlying integrability of the theory. Here we give an overview of the Yangian algebra and of its action on scattering amplitudes, at tree and loop level. We also discuss the emergence of the Yangian for supersymmetric Wilson loops.

Keywords: Yang-Mills and other gauge theories, supersymmetric field theories, quantum groups and related algebraic methods, relations with integrable systems

Mathematics Subject Classification 2010: 81T13, 81T60, 81R50, 81R12

Introduction

In mathematical physics, much attention has been devoted in recent years to $\mathcal{N} = 4$ supersymmetric Yang-Mills theory (SYM) on a worldwide scale. This model gauge quantum field theory reveals a host of interesting features, which turn out to be very telling and useful even for less supersymmetric theories. In the course of the running period of the SFB 647 scientists funded by this research network have been making important and even seminal contributions to the field, in particular with respect to demonstrating the quantum integrability of the theory. In the present article, we describe some more recent contributions obtained in the third and final funding period.

Apart from its SU(N) non-abelian gauge symmetry, this model is a conformal field theory in four space-time dimensions with $\mathcal{N} = 4$ copies of supersymmetry. In consequence, it is invariant under the superconformal group PSU(2, 2|4). At finite N it furthermore possesses a weak-strong coupling SL(2, $\mathbb{Z}$) symmetry acting on the complexified gauge coupling. This symmetry is destroyed in the planar large-N limit. However, in compensation, an even larger and quite hidden symmetry not visible at the Lagrangian level appears: It is a second, distinct copy of PSU(2, 2|4) termed dual superconformal symmetry. First discovered at weak coupling by considering scattering amplitudes in a kinematically "dual" space, it was soon interpreted as a duality between maximally helicity violating (MHV) amplitudes and light-like Wilson loops.

DOI 10.1515/9783110452150-006_s_002

The combination of the two superconformal symmetry algebras forms a *Yangian* structure. It may be interpreted as the algebraic manifestation of the quantum integrability of $\mathcal{N} = 4$ SYM.

The exactness of the Yangian symmetry at the level of the scattering amplitudes is obscured by a plethora of infrared divergences. Indeed, even at tree level the amplitudes are not strictly invariant due to singular kinematic configurations. At loop level the divergences naively break it even for generic external kinematic data since now the loop momenta may align to singular configurations. Much effort has therefore been put into the understanding of the conjectured Yangian invariance at higher loop orders, and it is fair to state that no full understanding has yet been reached. Nevertheless, the veracity of this conjecture is generally expected, and has certainly been used to strongly constrain the detailed structure of scattering amplitudes. To give one non-trivial example, the anomalous Ward identity stemming from the dual conformal symmetry indicated that the Bern-Dixon-Smirnov ansatz for the finite part of the n-point MHV scattering amplitude could fail starting from six points, which turned out to be indeed the case.

This chapter aims at presenting an overview of some of these topics, paying particular attention to the essentials of Yangian symmetry, as well as to the specific research work of the authors.

In Section 1 we succinctly review basic notions of Yangian algebras and their representations, with special attention to certain oscillator realizations. Then, in Section 2, we give some details on the structure of superamplitudes and on the superconformal and dual superconformal symmetry of $\mathcal{N} = 4$ SYM. We show how they combine into a Yangian algebra structure and how the latter acts on amplitudes. We also discuss the duality between Wilson loops and MHV amplitudes. In Section 3 we deal with Yangian invariants. In particular, we begin in Section 3.1 by reviewing certain multi-dimensional contour integrals on Graßmannian manifolds that compute tree-level amplitudes, and furthermore contain information on leading singularities at loop level. In Section 3.2 we come to the main joint collaborative effort of the authors of this review: The study of certain mathematically and, as we hope, even physically useful deformations of Yangian symmetry. Introducing inhomogeneities into the Yangian generators, we show how this deforms scattering amplitudes by certain parameters, allowing for their interpretation as concatenations of R-matrices satisfying the Yang-Baxter equation. This leads to natural and, hopefully, useful generalizations of the said contour integrals. We next discuss in Section 3.3 a certain finite function constructed from a combination of Wilson loops. It is Yangian invariant at one-loop order, at least in special kinematical configurations. Finally, in Section 4, we turn our attention to the sector of smooth Wilson loops in $\mathcal{N} = 4$ SYM. In particular, we discuss how the established Wilson loop operator of Yang-Mills theory may be suitably supersymmetrized in an on-shell superspace formulation. This operator represents the natural Wilson loop from the anti-de Sitter (AdS)-dual string construction and, as we review, is superconformal invariant and enjoys a hidden Yangian symmetry. We close in Section 5 with a short outlook.

1 Yangian algebras and representations

We begin our overview by recalling some major notions about Yangian algebras, which will be useful later on. We focus on the non-graded case, related to ordinary Lie algebras, but everything can be extended straightforwardly to superalgebras. Detailed introductions to this subject can be found for instance in [27, 34, 75].

1.1 Generalities

The definition of Yangian algebra was first introduced by Drinfeld in [38, 39]. Let us call $\mathfrak{g}$ the simple Lie algebra spanned by the generators J_a:

$$[J_a, J_b] = f_{ab}{}^c J_c\,, \tag{1}$$

where $f_{ab}{}^c$ are the structure constants of $\mathfrak{g}$ and[1] $a = 1, \ldots, \dim \mathfrak{g}$. The Lie algebra is a Hopf subalgebra of the Yangian with trivial coproduct $\delta : \mathfrak{g} \to \mathfrak{g} \otimes \mathfrak{g}$, which acts on the generators in the following way:

$$\delta(J_a) = J_a \otimes 1 + 1 \otimes J_a\,. \tag{2}$$

The Yangian $Y(\mathfrak{g})$ of a Lie algebra $\mathfrak{g}$ is then the Hopf algebra generated by the set of J_as (which form the so-called level zero) together with another set $J_a^{(1)}$, the level one, which obeys

$$[J_a, J_b^{(1)}] = f_{ab}{}^c J_c^{(1)}\,, \tag{3}$$

therefore transforming in the adjoint representation of $\mathfrak{g}$. The coproduct $\delta : Y(\mathfrak{g}) \to Y(\mathfrak{g}) \otimes Y(\mathfrak{g})$ acts non-trivially on the level-one generators

$$\delta(J_a^{(1)}) = J_a^{(1)} \otimes 1 + 1 \otimes J_a^{(1)} + \gamma f_a{}^{cb} J_b \otimes J_c\,, \tag{4}$$

where γ is any number different from zero[2]. By requiring $\delta(J_a^{(1)})$ to be a homomorphism, the commutator between level-one generators is constrained by the Serre relations:

$$\begin{aligned}&[J_a^{(1)}, [J_b^{(1)}, J_c]] + [J_b^{(1)}, [J_c^{(1)}, J_a]] + [J_c^{(1)}, [J_a^{(1)}, J_b]]\\ &= \{J_l, J_m, J_n\} f_{ar}{}^l f_{bs}{}^m f_{ct}{}^n f^{rst}\,,\end{aligned} \tag{5}$$

where $\{A, B, C\}$ is the symmetrized product of the three generators A, B and C. Given these relations, the infinite number of levels of the Yangian algebra is completely

1 In the following we will use lowercase letters a for the adjoint representation and uppercase A for the fundamental one.

2 Since it does not change the relations below, we will choose it to be 1.

generated. For instance the second level will be constructed from the commutation relations of level-one generators:

$$[J_a^{(1)}, J_b^{(1)}] = f_{ab}{}^{c} J_c^{(2)} + h(J, J^{(1)}) \tag{6}$$

with $h(J, J^{(1)})$ a function of the level zero and one being constrained by eq. (5). The coproduct helps to build general representations from simple ones. For instance, if the space where we act is a multi-site space, given by the tensor product representation of n vector spaces, then the generators will be defined as follows:

$$J_a = \sum_{i=1}^{n} J_{a,i} , \tag{7}$$

$$J_a^{(1)} = \sum_{1 \leqslant i < j \leqslant n} f_a{}^{cb} J_{b,i} J_{c,j} + \sum_{j=1}^{n} v_j J_{a,j} , \tag{8}$$

where definition (8) of level-one generators in terms of the level-zero set is the so-called bilocal formula and it is given by the coproduct (4). We will show explicitly how the coproduct actually works in the next section. The v_j which enter in the bilocal formula are called inhomogeneities and are free parameters which will prominently appear again in Section 3.2 when considering deformed amplitudes.

For the reader familiar with loop algebras, we note that the Yangian is in the following sense a "half-loop" algebra, being defined by

$$[J_a^{(n)}, J_b^{(m)}] = f_{ab}{}^{c} J_c^{(n+m)} , \tag{9}$$

where $n, m \in \mathbb{Z}^+$ define the levels of the generators.

Another presentation of $Y(\mathfrak{g})$ is given by the R-matrix. Here we just mention that the generators of the Yangian algebra can be arranged in a generating function, the *transfer matrix*

$$t^{AB}(u) = \sum_{n=0}^{\infty} u^{-n} t_n^{AB} , \tag{10}$$

where t_n^{AB} are the relabelling of the generators $J_a^{(n)}$, replacing adjoint indices with fundamental ones, $u \in \mathbb{C}$ is the spectral parameter, n is the level and $t_0^{AB} = \delta^{AB}$. If we consider $Y(\mathfrak{gl}(N))$ we can define a matrix $T(u)$, the *monodromy matrix*

$$T(u) = \sum_{A,B=1}^{N} t^{AB}(u) \otimes E_{AB} , \tag{11}$$

where E_{AB} is a basis for an auxiliary space $\mathfrak{gl}(N)$ with $(E_{AB})_{CD} = \delta_{AC}\delta_{BD}$, and a matrix $R_{12}(u - v)$

$$R_{12}(u-v) = 1_N \otimes 1_N - \frac{1}{u-v} P_{12} \tag{12}$$

where P_{12} is the permutation operator $P_{12} = \sum_{A,B=1}^{N} E_{AB} \otimes E_{BA}$. Then the so-called RTT relation

$$R_{12}(u-v)T_1(u)T_2(v) = T_2(v)T_1(u)R_{12}(u-v) \tag{13}$$

is equivalent to all the Yangian relations when expanded in series of $\frac{1}{u}$ and $\frac{1}{v}$. The same R-matrix given in equation (12) appears in spin chains and it is the building block of the monodromy matrix $T(u)$ obeying eq. (13). When the Hamiltonian of a spin chain can be obtained with a monodromy matrix, the system is integrable, in the sense that it is completely solvable. See in particular Section 3.2. This is one of the reasons why the Yangian symmetry in $\mathcal{N} = 4$ SYM has received a lot of attention in the last years, as we will discuss in the following section.

1.2 Representations of Yangian algebras

Let us consider as an example $\mathfrak{g} = \mathfrak{sl}(m)$, which will be useful in discussing scattering amplitudes. An interesting representation π satisfying all the properties listed above is the oscillator representation, see [38, 40]. The one-particle (single-site) level-zero generators are written in terms of the W^A oscillator variables

$$\pi(J^A{}_B) = W^A \frac{\partial}{\partial W^B} - \frac{\delta^A_B}{m} h\,, \tag{14}$$

where $A, B = 1, \ldots, m$ and h is the generator of $U(1)$ in $\mathfrak{gl}(m)$:

$$h = \pi(J^C{}_C) = W^C \frac{\partial}{\partial W^C}\,, \tag{15}$$

where summation over contracted indices is implicit. The operator h being central, we can decompose the space of functions of the W^A into those of fixed degrees of homogeneity h. Thus we can think of W as homogeneous coordinates on $\mathbb{CP}^{m-1}$. Our representation acts on functions with fixed degrees of homogeneity on this space (which we will denote as $\mathcal{F}(\mathbb{CP}^{m-1})$). The representation of the level-one generators can be obtained through the evaluation map, see [34], which constructs the level-one operator $J^{(1)}$ in terms of the algebra generators J:

$$\pi_\mu\left(J^{(1)a}\right) = \mu\, \pi(J^a) + \frac{1}{4} \sum_{bc} d^a{}_{bc} \pi(J^b J^c) \tag{16}$$

where μ is a free parameter and

$$d^a_{bc} = Tr\left[J^a\left(J_b J_c + J_c J_b\right)\right]\,. \tag{17}$$

We could compute this explicitly but there is a shortcut to the answer. In order to represent the level-one generator $J^{(1)A}{}_B$ we need to write down an operator in the adjoint representation. Since the operator h is central (and so can be assigned some fixed numerical value) our only choice is

$$\pi_v\left(J_v^{(1)A}{}_B\right) = v\left(W^A \frac{\partial}{\partial W^B} - \frac{1}{m}\delta_B^A W^C \frac{\partial}{\partial W^C}\right). \tag{18}$$

As we have previously mentioned, we construct further representations acting on the tensor product $\mathscr{F}(\mathbb{CP}^{m-1}) \otimes \cdots \otimes \mathscr{F}(\mathbb{CP}^{m-1})$ by using the coproduct. To construct now a two-parameter representation for two sites, we use (2) for the level-zero

$$\left(\pi_{v_1} \otimes \pi_{v_2}\right)(J^A{}_B) = \pi_{v_1}(J^A{}_B) \otimes 1 + 1 \otimes \pi_{v_2}(J^A{}_B) \tag{19}$$

$$= W_1^A \frac{\partial}{\partial W_1^B} + W_2^A \frac{\partial}{\partial W_2^B} - \frac{\delta_B^A}{m} \sum_{i=1}^{2} W_i^C \frac{\partial}{\partial W_i^C} \tag{20}$$

and (4) for the level-one operators

$$\begin{aligned}
&\left(\pi_{v_1} \otimes \pi_{v_2}\right)\left(J^{(1)A}{}_B\right) = \\
&= \pi_{v_1}\left(J^{(1)A}{}_B\right) \otimes 1 + 1 \otimes \pi_{v_2}\left(J^{(1)A}{}_B\right) + f_{BE}^{A}{}_{C}^{F}{}^{D}\,\pi_{v_1}(J^C{}_D) \otimes \pi_{v_2}(J^E{}_F) && (21)\\
&= v_1 W_1^A \frac{\partial}{\partial W_1^B} + v_2 W_2^A \frac{\partial}{\partial W_2^B} - \frac{\delta_B^A}{m} \sum_{i=1}^{2} v_i h_i^C \frac{\partial}{\partial W_i^C} \\
&+ \left(W_1^A \frac{\partial}{\partial W_1^C} W_2^C \frac{\partial}{\partial W_2^B} - (1,2)\right), && (22)
\end{aligned}$$

where the f denotes the structure constants of the Lie algebra and by (i,j) we mean the exchange of i and j indices. By repeated application of the coproduct and the projection with π_{v_i} on the ith site, the representation for n sites reads

$$\pi_{\vec{v}}(J^A{}_B) = \sum_{i=1}^{n}\left(W_i^A \frac{\partial}{\partial W_i^B} - \frac{1}{m}\delta_B^A h_i\right), \tag{23}$$

$$\begin{aligned}
\pi_{\vec{v}}(J^{(1)A}{}_B) &= \sum_{i<j}\left(W_i^A \frac{\partial}{\partial W_i^C} W_j^C \frac{\partial}{\partial W_j^B} - (i,j)\right) \\
&+ \sum_{i=1}^{n} v_i\left(W_i^A \frac{\partial}{\partial W_i^B} - \frac{1}{m}\delta_B^A h_i\right), && (24)
\end{aligned}$$

where $\vec{v} = (v_1, \ldots, v_n)$ is a list of free parameters, the inhomogeneities. Equation (24) is the bilocal formula (8). For the sake of simplicity, in the following we will use the symbols J and $J_{\vec{v}}^{(1)}$ to denote this representation of the level-zero and level-one generators. Together the operators $J^{A}{}_{B}$ and $J^{(1)A}_{\vec{v}}{}_{B}$ generate the Yangian $Y(\mathfrak{sl}(m))$. The operators

$$h_i = W_i^C \frac{\partial}{\partial W_i^C} \tag{25}$$

are central and so we can decompose the space of functions of the W_i into spaces of fixed homogeneity in each of the W_i separately. In the following we will explicitly use these results in the context of scattering amplitudes.

2 Scattering amplitudes, symmetries and Wilson loops in $\mathcal{N} = 4$ SYM

In this section we describe some properties of (colour-ordered) scattering amplitudes in $\mathcal{N} = 4$ SYM. In particular we will focus on the symmetries of the theory and on the duality between Wilson loops and MHV amplitudes.

2.1 Superamplitudes and symmetries

The on-shell supermultiplet of $\mathcal{N} = 4$ super Yang-Mills theory is conveniently described by a superfield Φ, dependent on Graßmann parameters η^A which transform in the fundamental representation of $\mathfrak{su}(4)$. The on-shell superfield can be decomposed as follows:

$$\begin{aligned}\Phi(p,\eta) &= G^+(p) + \eta^A\Gamma_A(p) + \frac{1}{2!}\eta^A\eta^B S_{AB}(p) + \frac{1}{3!}\eta^A\eta^B\eta^C\epsilon_{ABCD}\overline{\Gamma}^D(p)\\ &+\frac{1}{4!}\eta^A\eta^B\eta^C\eta^D\epsilon_{ABCD}G^-(p).\end{aligned} \tag{26}$$

Here $G^+, \Gamma_A, S_{AB} = \frac{1}{2}\epsilon_{ABCD}\overline{S}^{CD}, \overline{\Gamma}^A, G^-$ are the positive-helicity gluon, gluino, scalar, anti-gluino and negative-helicity gluon states respectively. Each state $\phi \in \{G^+, \Gamma_A, S_{AB}, \overline{\Gamma}^A, G^-\}$ carries a definite on-shell momentum. In the following we will use the spinor-helicity formalism, which is a very useful tool for describing scattering amplitudes of massless particles. The condition $p^2 = 0$ translates into writing the momentum as a bi-spinor

$$p^{\alpha\dot{\alpha}} = \lambda^\alpha\tilde{\lambda}^{\dot{\alpha}}, \tag{27}$$

where λ^α and $\tilde{\lambda}^{\dot{\alpha}}$ transform in the (2, 0) and (0, 2) representations of the Lorentz group. The space defined by $(\lambda^\alpha, \tilde{\lambda}^{\dot{\alpha}}, \eta^A)$ is called on-shell superspace. Each state ϕ has a definite weight h (called helicity) under the rescaling

$$\lambda \longrightarrow \alpha\lambda, \qquad \tilde{\lambda} \longrightarrow \alpha^{-1}\tilde{\lambda}, \qquad \phi(\lambda,\tilde{\lambda}) \longrightarrow \alpha^{-2h}\phi(\lambda,\tilde{\lambda}). \tag{28}$$

The helicities of the states appearing in eq. (26) are $\{+1, +\frac{1}{2}, 0, -\frac{1}{2}, -1\}$ respectively. If, in addition, we now assign η to formally transform in the same way as $\tilde{\lambda}$, thereby replacing the notion of helicity (a representation label of only the conformal algebra $\mathfrak{su}(2,2)$) by the notion of superhelicity (a label of the full superconformal algebra $\mathfrak{psu}(2,2|4)$), i.e.

$$\eta^A \longrightarrow \alpha^{-1}\eta^A \,, \tag{29}$$

then the whole superfield Φ has superhelicity 1. In other words, the superhelicity generator

$$h = -\frac{1}{2}\lambda^{\alpha}\frac{\partial}{\partial\lambda^{\alpha}} + \frac{1}{2}\tilde{\lambda}^{\dot{\alpha}}\frac{\partial}{\partial\tilde{\lambda}^{\dot{\alpha}}} + \frac{1}{2}\eta^{A}\frac{\partial}{\partial\eta^{A}} \tag{30}$$

acts on Φ in the following way:

$$h\Phi = \Phi. \tag{31}$$

When we consider scattering amplitudes of the on-shell superfields, then the helicity condition (or "homogeneity condition") is satisfied for each particle, i.e.

$$h_i\mathscr{A}_n(\Phi_1,\dots,\Phi_n) = \mathscr{A}_n(\Phi_1,\dots,\Phi_n), \qquad i = 1,\dots,n. \tag{32}$$

The superamplitude $\mathscr{A}_n$ can be expanded in terms of the Graßmann parameters η_i^A, with the terms of order η^{4k} corresponding to super-helicity k amplitudes. The case $k = 2$ is the MHV amplitude, $k = 3$ is the next-to-MHV (NMHV) one and so on. We can then write

$$\mathscr{A}_n = \mathscr{A}_n^{\text{MHV}} + \mathscr{A}_n^{\text{NMHV}} + \ldots + \mathscr{A}_n^{\overline{\text{MHV}}} = \frac{\delta^4(p)\delta^8(q)}{\langle 12\rangle\langle 23\rangle\ldots\langle n1\rangle}\mathscr{P}_n(\lambda_i,\tilde{\lambda}_i,\eta_i)\,, \tag{33}$$

where $\delta^4(p)$ is a four-dimensional δ-function of the, now, total momentum $p = \sum_{i=1}^n \lambda_i^{\alpha}\tilde{\lambda}_i^{\dot{\alpha}}$ and $\delta^8(q) = \Pi_{A=1}^4\Pi_{\alpha=1,2}\left(\sum_{i=1}^n \lambda_i^{\alpha}\eta_i^A\right)$ is a Graßmann δ-function, consequence of supersymmetry, and $\langle ij\rangle = \epsilon_{\alpha\beta}\lambda_i^{\alpha}\lambda_j^{\beta}$. The function $\mathscr{P}_n$ is a polynomial in the η_is and carries no helicity

$$h_i\,\mathscr{P}_n = 0, \qquad i = 1,\dots,n. \tag{34}$$

At tree level, the function $\mathscr{P}_n$ is finite and its explicit form was given in [45]. At loop level, it contains infrared divergences which need to be regularized. The regularization procedure has the effect of breaking the symmetries of the theory, as we are going to discuss now.

Since $\mathscr{N} = 4$ SYM is a superconformal field theory, one expects this to be reflected in the structure of the scattering amplitudes. This turns out to be true for tree-level amplitudes but not at loop level, where the presence of infrared divergences spoils the symmetry. Therefore one should consider the invariance of the full S-matrix under $\mathfrak{psu}(2,2|4)$ rather than of the amplitudes individually. If we denote with j_a any generator of the superconformal algebra $\mathfrak{psu}(2,2|4)$

$$j_a \in \{p^{\alpha\dot\alpha}, q^{\alpha A}, \bar q^{\dot\alpha}_A, m_{\alpha\beta}, \bar m_{\dot\alpha\dot\beta}, r^A{}_B, d, s^\alpha_A, \bar s^A_{\dot\alpha}, k_{\alpha\dot\alpha}\} \tag{35}$$

then we can write at tree level

$$j_a \mathscr{A}_n^{\text{tree}} = 0. \tag{36}$$

The invariance was shown directly by applying the generators to the explicit form of the amplitudes in [85] for MHV amplitudes and [73] for NMHV amplitudes. In fact, eq. (36) is not completely exact, because of the so-called holomorphic anomaly [10, 74, 82]. This is given by special configurations of the external momenta, and specifically when two adjacent momenta become collinear. Even if at tree level this effect can be thought as negligible, since it is given just by some particular configuration, at loop level it must be taken into account due to the integration over the loop momentum running over the whole phase space. Moreover, at loop level there are effects due to the infrared regularization procedure. These effects can be taken into account by deforming the superconformal generators and in [10, 13, 82] it was shown how to redefine them in such a way to restore symmetry at one loop.

The ordinary superconformal symmetry is not the full story. It has been revealed [50] that there exists also a hidden symmetry, which is not manifest in the on-shell superspace coordinates, the so-called dual superconformal symmetry with generators J_a. This second $\mathfrak{psu}(2,2|4)$ algebra is the T-dual version of the standard one, and this property corresponds very naturally to the T-self-duality of the AdS sigma model, see [17, 20, 23]. The existence of such a symmetry was shown by defining a dual superspace $(x^{\alpha\dot\alpha}, \theta^{\alpha A})$ through the relations

$$x_i^{\alpha\dot\alpha} - x_{i+1}^{\alpha\dot\alpha} = \lambda_i^\alpha \tilde\lambda_i^{\dot\alpha}, \quad \theta_i^{\alpha A} - \theta_{i+1}^{\alpha A} = \lambda_i^\alpha \eta_i^A \tag{37}$$

and re-expressing the amplitudes in terms of the dual coordinates

$$\mathscr{A}_n = \frac{\delta^4(x_1 - x_{n+1})\delta^8(\theta_1 - \theta_{n+1})}{\langle 12\rangle\langle 23\rangle \dots \langle n1\rangle} \mathscr{P}_n(x_i, \theta_i). \tag{38}$$

In this space, dual superconformal symmetry acts canonically and it also acts on the on-shell superspace variables in order to be compatible with the constraints (37). The dual superconformal symmetry is generated by the set of J_as being the dual copy of $\mathfrak{psu}(2,2|4)$

$$J_a \in \{P_{\alpha\dot\alpha}, Q_{\alpha A}, \bar Q^A_{\dot\alpha}, M_{\alpha\beta}, \overline{M}_{\dot\alpha\dot\beta}, R^A{}_B, D, S^A_\alpha, \overline{S}^{\dot\alpha}_A, K^{\alpha\dot\alpha}\}. \tag{39}$$

In fact, some of the generators J_a do not leave the amplitudes invariant. Under the dual superconformal $S^{\alpha A}$, special conformal $K^{\alpha\dot{\alpha}}$ and dilatation D operators the amplitudes are covariant, see [28, 50]. By redefining these generators as in [51] through suitable additions of terms leaving the amplitudes invariant, the covariance can be rephrased as an invariance of $\mathscr{A}_n$ in such a way that at tree level

$$J'_a \mathscr{A}_n^{\text{tree}} = 0\,, \tag{40}$$

where J'_a differs from J_a only for $K^{\alpha\dot{\alpha}}$, $S^{\alpha A}$ and D, these being shifted by $\sum_i x_i^{\alpha\dot{\alpha}}$, $\sum_i \theta_i^{\alpha a}$ and n, respectively. In order to have both symmetries acting on the same space, it is useful to restrict the dual superconformal generators to act only on the on-shell superspace variables $(\lambda_i, \tilde{\lambda}_i, \eta_i)$. The generators $P_{\alpha\dot{\alpha}}, Q_{\alpha A}$ become trivial, while the generators $\{\bar{Q}, M, \bar{M}, R, D', \bar{S}\}$ coincide (up to signs) with generators of the standard superconformal symmetry. The non-trivial generators which are not part of the j_a are K' and S'.

In [51] it was shown that the generators j_a (35) and S' (or K') together generate the Yangian of the superconformal algebra, $Y(\mathfrak{psu}(2,2|4))$. As we have already commented, the presence of a Yangian structure is the hallmark of the integrability of $\mathcal{N} = 4$ SYM. The existence of integrability had already been pointed out from the evaluation of the spectrum of anomalous dimensions [14, 19, 79] and, through the anti-de Sitter/conformal field theory (AdS/CFT) correspondence, in the classical integrability of Type IIB superstring on $AdS_5 \times S^5$ [22].

In particular, the generators j_a form the level-zero $\mathfrak{psu}(2,2|4)$ subalgebra as in eq. (1), while the level one is generated by considering S' or K' in eq. (3). Let us stress that the choice of S' or K' does not change the Yangian, since $[K', \bar{Q}] = S'$. The level-zero generators are represented by the sum over single particle generators (7),

$$j_a = \sum_{k=1}^{n} j_{ka}\,, \tag{41}$$

while the level-one generators are obtained via the bilocal formula (8)

$$j_a^{(1)} = f_a{}^{cb} \sum_{k<k'} j_{kb} j_{k'c}\,, \tag{42}$$

where for the time being the parameters v_j are set to zero. In [37] it was demonstrated that under certain conditions, which turn out to be satisfied by the oscillator representation in exam, formula (42) is sufficient to define the Yangian, so that the Serre relations are automatically satisfied. The bilocal formula also respects the cyclicity of the amplitudes for the case of $\mathfrak{psl}(n|n)$. Thus, finally the full symmetry of the tree-level amplitudes can be rephrased as

$$y \mathscr{A}_n^{\text{tree}} = 0, \tag{43}$$

for any $y \in Y(\mathfrak{psu}(2,2|4))$. As we have already mentioned, at tree level there are correction terms of the superconformal symmetry given by the holomorphic anomaly. With these deformations, the algebra (1) does close only modulo gauge transformations [10]. By using the bilocal formula, the deformations are directly inserted in a new definition of level-one generators. At loop level, one can again define deformed level-zero and level-one generators [13, 82]. In both tree- and loop-level cases, it has not been checked so far that the deformed generators close into a Yangian algebra.

In [42] it was demonstrated that there exists an alternative representation of the symmetry, which can be viewed as the Yangian version of the T-self-duality property of the full $AdS_5 \times S^5$ background of the string sigma model, see [17, 20, 23]. In this case, it is the unmodified dual superconformal symmetry J_a which plays the role of a level-zero subalgebra. Now it is the function $\mathscr{P}_n$, i.e. the amplitude with the MHV part factored out, to be left invariant at tree level

$$J_a \mathscr{P}_n^{\text{tree}} = 0 \,. \tag{44}$$

The level one is then provided by some of the (modified) standard superconformal generators. Similarly to the previous case, the generators of translations $p^{\alpha\dot{\alpha}}$ and supertranslations $q^{\alpha A}$ are trivialized while $k_{\alpha\dot{\alpha}}$ and s_A^α are redefined in such a way to annihilate $\mathscr{P}_n^{\text{tree}}$. Then J_a and k' (or s') generate the Yangian of the dual superconformal algebra. Therefore there are two equivalent ways of looking at the full symmetry algebra of the scattering amplitudes. We summarize them in Table 1, where we introduce the super-twistors $\mathscr{Z}^{\mathscr{A}}$ and momentum super-twistors $\mathscr{W}^{\mathscr{A}}$. The formulation in these spaces has a particularly simple form, see [42, 51], and in particular the level-zero generators are linearized in terms of these variables in both spaces. Moreover, the identical formulations show explicitly the T-self-duality property. The twistor coordinates are defined as $\mathscr{Z}^{\mathscr{A}} = (\tilde{\mu}^\alpha, \tilde{\lambda}^{\dot{\alpha}}, \eta^A)$, where $\tilde{\mu}$ is the Fourier conjugate of λ. The momentum twistors are instead $\mathscr{W}^{\mathscr{A}} = (\lambda^\alpha, \mu^{\dot{\alpha}}, \chi^A)$, where $\mu^{\dot{\alpha}}$ and χ^A are defined through the incidence relations

$$\mu_i^{\dot{\alpha}} = x_i^{\alpha\dot{\alpha}} \lambda_{i\alpha}, \qquad \chi_i^A = \theta_i^{\alpha A} \lambda_{i\alpha} \,. \tag{45}$$

This is the standard twistor space associated with Wilson loops and was introduced in [65]. One of the most useful properties of the momentum twistors is that the

Table 1: Yangian generators in twistor and momentum twistor coordinates.

Twistor space	**Momentum twistor space**
$j^{\mathscr{A}}{}_{\mathscr{B}} = \sum_i \mathscr{Z}_i^{\mathscr{A}} \frac{\partial}{\partial \mathscr{Z}_i^{\mathscr{B}}}$	$J^{\mathscr{A}}{}_{\mathscr{B}} = \sum_i \mathscr{W}_i^{\mathscr{A}} \frac{\partial}{\partial \mathscr{W}_i^{\mathscr{B}}}$
$j^{(1)\mathscr{A}}{}_{\mathscr{B}} = \sum_{i<j} (-1)^{\mathscr{C}} \left[\mathscr{Z}_i^{\mathscr{A}} \frac{\partial}{\partial \mathscr{Z}_i^{\mathscr{C}}} \mathscr{Z}_j^{\mathscr{C}} \frac{\partial}{\partial \mathscr{Z}_j^{\mathscr{B}}} - (i,j) \right]$	$J^{(1)\mathscr{A}}{}_{\mathscr{B}} = \sum_{i<j} (-1)^{\mathscr{C}} \left[\mathscr{W}_i^{\mathscr{A}} \frac{\partial}{\partial \mathscr{W}_i^{\mathscr{C}}} \mathscr{W}_j^{\mathscr{C}} \frac{\partial}{\partial \mathscr{W}_j^{\mathscr{B}}} - (i,j) \right]$
$j \mathscr{A}_n = j^{(1)} \mathscr{A}_n = 0$	$J \mathscr{P}_n = J^{(1)} \mathscr{P}_n = 0$

light-likeness constraint and momentum conservation are already implemented via the incidence relations (45).

The T-dual relationship between twistor and momentum twistor space is exactly the property which appears in the Graßmannian formulas of [8, 77], as we will explain in section 3.

Before moving to the Wilson loop/amplitude duality, we want to point out some references about various results related to the topics discussed above. In particular, we want to mention that there is another alternative regularization which preserves the dual conformal symmetry at loop level [2, 62, 63] by going to the Coulomb branch of the theory. See [24, 33, 36] for higher dimensional related theories. Another very interesting result is the enhancement of the $\mathfrak{psu}(2,2|4)$ algebra through the hypercharge generator $\mathfrak{B}$, which measures the helicity of scattering amplitudes. While $\mathfrak{B}$ is broken, the level-one helicity generator associated to it is a symmetry of the amplitudes in $\mathcal{N} = 4$ SYM. See [18] for more details.

2.2 MHV amplitude/Wilson loop duality

In $\mathcal{N} = 4$ SYM, MHV scattering amplitudes are dual to light-like Wilson loops. For extension of the duality to non-MHV amplitudes at weak coupling see [21, 31, 78].

The MHV amplitude is one term of the superamplitude $\mathcal{A}_n$ when expanded in terms of the Graßmann variables η_is given in eq. (33) and can be written as follows:

$$\mathcal{A}_n^{\text{MHV}} = \mathcal{A}_{n;tree}^{\text{MHV}} M_n \,. \tag{46}$$

Like $\mathcal{P}_n$, the function M_n is the loop-correction factor and does not carry any helicity, which is completely encoded in the prefactor. The exponentiation of infrared divergences (IR) leads to

$$\log M_n = [\text{IR divergences}] + F_n^{\text{MHV}}(p_1, \ldots, p_n; a) + O(\epsilon) \tag{47}$$

where $a = \frac{g^2 N}{8\pi^2}$ is the 't Hooft coupling. The first term on the right-hand side, encoding the infrared divergences, is well understood in $\mathcal{N} = 4$ SYM. In particular, the leading IR singularity is a double pole, proportional to the cusp anomalous dimension. This quantity is known to be related to ultraviolet (UV) divergences of Wilson loops with cusps in quantum chromodynamics (QCD), see [66, 67, 71]. The finite part has been subject to a lot of analysis in the last years, and many developments in its computation have been achieved. Bern, Dixon and Smirnov (BDS) [26] proposed an ansatz for F_n^{MHV} based on calculations of the four-point amplitude and verified it to hold for five-point scatterings. In the meanwhile, using the AdS/CFT correspondence, in [3] it was shown that in the strong-coupling regime there is a relation between planar n-gluon scattering amplitudes (the dependence on the helicity configuration of the amplitude appears as a subleading effect at strong coupling) and the expectation value of Wilson loops defined on closed light-like contours C_n:

$$W_n = \frac{1}{N}\langle 0|\operatorname{Tr}\mathcal{P}\exp\left(ig\oint_{C_n} dx^\mu A_\mu(x)\right)|0\rangle. \tag{48}$$

This means that not only the IR divergences of the scattering amplitude match the UV divergences of the light-like Wilson loop given by the cusps, but also implies that the finite parts have to coincide. It was then conjectured and verified that such a duality exists also in the perturbative regime for MHV amplitudes [29, 46–49, 52]. It formally arises from the identification of particle momenta with the dual coordinates (37)

$$x_i - x_{i+1} = p_i, \qquad (x_i - x_{i+1})^2 = 0\,, \tag{49}$$

and a formula similar to eq. (47) can be written for the Wilson loops in dual space:

$$\log W_n = [\text{UV divergences}] + F_n^{\text{WL}}(x_1,\dots,x_n;a) + O(\epsilon)\,. \tag{50}$$

Therefore the Wilson loop W_n can be seen as the MHV scattering amplitude with the tree-level part factored out, upon going from on-shell to dual coordinates. The duality then is really a statement on the finite parts (up to an additive constant)

$$F_n^{\text{MHV}} = F_n^{\text{WL}} + \text{const}\,, \tag{51}$$

upon the identification (49). The dual conformal symmetry for the scattering amplitudes, whose supersymmetric formulation we have discussed in the previous section, is the ordinary conformal symmetry of Wilson loops. The dual space is the configuration space for the Wilson loops and for this reason, the breaking of the symmetry is under control through anomalous Ward identities. In particular it has been found [46, 49] that

$$K^\mu F_n^{\text{WL}}(x_1,\dots,x_n) = \frac{1}{2}\Gamma_{\text{cusp}}(a)\sum_{i=1}^{n} x^\mu_{i,i+1}\log\frac{x^2_{i,i+2}}{x^2_{i-1,i+1}}\,. \tag{52}$$

Using these Ward identities, it was shown that the BDS ansatz can differ from the Wilson loop result by a non-trivial function of $(3n-15)$ conformal invariants u_{ijkl}

$$u_{ijkl} = \frac{x^2_{ij}x^2_{kl}}{x^2_{ik}x^2_{jl}}\,. \tag{53}$$

Since the first case where the conformal cross-ratios (53) can appear is at six points, the BDS ansatz, based on four- and five-gluon amplitudes, was not able to detect this term. The discrepancy, named as the remainder function $\mathcal{R}$, has been verified to be non-zero at two loops and beyond, and for six or more points [32, 47, 48]. Thus we can write the finite part as the sum of two contributions

$$F_n^{\mathrm{WL}} = F_n^{\mathrm{anom}}(x_1, \ldots, x_n; a) + I_n(u_1, \ldots, u_m; a) . \tag{54}$$

If we choose F_n^{anom} to coincide with the BDS ansatz for the MHV amplitude, then I_n coincide with the standard definition of $\mathscr{R}$. But there are alternative definitions of the anomalous part which modify it by adding some function of invariants and subtracting the same function from the remainder function. We will come back to this later on in this paper.

3 Yangian invariants

We now want to address the topic of Yangian invariants. We will begin by reviewing recent results on the connection between Graßmannians, Yangian invariants and scattering amplitudes. Deformed amplitudes, introduced in [55] and further studied in [56, 58, 59, 68, 69] depending on inhomogeneities will be presented. Then we will focus on a particular function constructed through Wilson loops, which turns out to be Yangian invariant at one loop.

3.1 Superamplitudes and Graßmannians

In [8] a remarkable formula was proposed which expresses the general tree-level amplitude $\mathscr{A}_{n,k}$ and furthermore contains the information on all leading singularities of N^{k-2}MHV scattering amplitudes in $\mathscr{N} = 4$ super Yang-Mills theory. It was formulated in super-twistor space, where the external kinematical data are given by the $n \times 4|4$ supermatrix $\mathscr{Z}$, and take the form of an integral over the Graßmannian $\mathrm{Gr}(k, n)$, the space of complex k-planes through the origin in $\mathbb{C}^n$, of certain superconformally invariant delta functions

$$\mathscr{A}_{n,k} = \int \frac{d^{k \cdot n} C}{\mathrm{vol}(\mathrm{GL}(k))} \frac{\delta^{4k|4k}(C \cdot \mathscr{Z})}{(1, \ldots, k)(2, \ldots, k{+}1) \ldots (n, \ldots, n{+}k{-}1)} . \tag{55}$$

Here the elements of the $k \times n$ matrix C are complex parameters parametrizing the elements of the Graßmannian, in generalization of projective coordinates. The $\mathrm{GL}(k)$ symmetry of the integration measure expresses the freedom of choosing a basis for the k-planes. The denominator is the cyclic product of consecutive $(k \times k)$ minors made from the columns $p, \ldots, p + k - 1$ of the $(k \times n)$ matrix of the C. The integrand is a specific $k(n - k)$-form to be integrated over cycles, with the integral being treated as a multi-dimensional integral along suitable "contours". The result obtained depends on the choice of the latter and is non-vanishing for closed contours because the form has poles located on certain hyperplanes in the Graßmannian. Formula (55) has a T-dual version [77], expressed in terms of momentum twistors, where the external kinematical data are given by the $n\times 4|4$ supermatrix $\mathscr{W}$. The momentum twistor Graßmannian formula for general tree-level amplitudes $\mathscr{A}_{n,k}$ is very similar to the one in twistor space (55), but now with the MHV tree-level amplitude factored out, cf. eq. (33),

$$\mathscr{A}_{n,k} = \frac{\delta^4(p^{\alpha\dot{\alpha}})\delta^8(q^{\alpha A})}{\langle 12\rangle\langle 23\rangle\ldots\langle n1\rangle} \int \frac{d^{\hat{k}\cdot n}\hat{C}}{\mathrm{vol}(\mathrm{GL}(\hat{k}))} \frac{\delta^{4\hat{k}|4\hat{k}}(\hat{C}\cdot\mathscr{W})}{(1,\,\ldots,\hat{k})\ldots(n,\,\ldots,n+\hat{k}-1)}. \tag{56}$$

This time around it is the dual superconformal symmetry that is manifest. The integration variables $\hat{C}$ are now coordinates on a smaller Graßmannian space $\mathrm{Gr}(\hat{k}, n)$ of $\hat{k}$-planes in $\mathbb{C}^n$ where $\hat{k} = k-2$, they are $(\hat{k}\times n)$ matrices with complex entries. The equivalence of the two formulations (55) and (56) was shown in [7], see also [53], through a change of variables. Therefore, since each of the formulas has a different superconformal symmetry manifest, they should both possess an invariance under the Yangian $Y(\mathfrak{psl}(4|4))$! The Yangian symmetry of these formulas was explicitly demonstrated in [42] by directly applying the undeformed (i.e. $v_j = 0$) Yangian level-one "bilocal" generators (8) to the Graßmannian integrals. In particular it was shown that $J^{(1)\mathscr{A}}{}_{\mathscr{B}}$ on the integrand of the expression in (56) yields a total derivative, which guarantees that $\mathscr{A}_{n,k}$ is invariant when the contour is closed. It was later shown [43, 72] that the form of the Graßmannian integral is uniquely fixed by requiring Yangian invariance. Specifically, in [43], by using the methods developed in [42], it was demonstrated that one cannot modify the integrand by a non-constant multiplicative function without breaking Yangian invariance (in the case of zero-homogeneity condition). In [6] it was shown that the Yangian invariance of scattering amplitudes is related to the diffeomorphisms of the positive Graßmannian.

At loop level it has recently been realized that the above statements hold at the level of the (unregularized) all-loop planar integrand [5]. The integrand at a given loop order has been constructed from its singularities via a generalization of the Britto-Cachazo-Feng-Witten (BCFW) recursion relations, whose each term is individually invariant under the full Yangian symmetry up to a total derivative. But at the level of the actual amplitudes the situation with the full symmetry is less clear. The Yangian invariance is broken by the action of integrating over specific contours, which leads to infrared divergences.

In [44] it was shown that Yangian invariance survives at one loop for Wilson loops (or equivalently MHV amplitudes), at least in a particular kinematic regime.

3.2 Deformed Yangian invariants and amplitudes

As stated in the preceding section, the contour integrals (55) and (56) are invariant under the undeformed ($v_j = 0$) "bilocal" Yangian generators (8). Interestingly, the Yangian algebra permits a "local" deformation, where one switches on complex-valued constants v_j in eq. (8). This possibility was first noticed and pointed out in the context of scattering amplitudes in [55, 56]. It implies that there should be a mathematically natural deformation of Yangian invariants in general, and thereby of scattering amplitudes in particular. The physical interpretation is also interesting: It leads to the consideration of deformed helicities of the scattering particles [55, 56], with the exciting logical possibility of gaining a novel set of parameters potentially able to regulate infrared divergences while preserving Yangian invariance.

In fact, one might want to turn things around, assume Yangian invariance as a fundamental symmetry, and attempt to mathematically derive the Graßmannian contour integrals (55) and (56) employing integrability techniques. That this is indeed possible was demonstrated in [59], see also [35]. If we consider a scattering amplitude of n particles as a kind of "spin chain state" $|\Psi\rangle$, then invariance under the Yangian generators of all levels reads $J_a|\Psi\rangle = J_a^{(1)}|\Psi\rangle = \cdots = 0$, or, in the transfer matrix notation in eq. (10) simply $t_n^{AB}|\Psi\rangle = \delta_{n,0}^{AB}|\Psi\rangle$. This then leads to a very elegant equation for the monodromy matrix (11) of the spin chain:

$$T(u)|\Psi\rangle = |\Psi\rangle\,. \tag{57}$$

One may now, in principle, apply the complete machinery of the quantum inverse scattering method, see e.g. [54], to the Yangian-invariant scattering amplitude problem. This includes the Bethe ansatz [59] or a similar method based on intertwining R-matrices [30, 35, 69]. Very importantly, one finds that these constructions *require* the introduction of degeneracy lifting, regulating deformation parameters v_j. The latter may only be removed on the level of the final result for the invariants *after* the quantum inverse scattering method is applied.

What then is the generalized, deformed version of the Graßmannian contour integrals (55) and (56) one obtains? The answer is quite elegant, after one introduces shifted deformation parameters [12]

$$v_j^\pm = v_j \pm \tfrac{c_j}{2} \qquad \text{with} \qquad v_{j+k}^+ = v_j^- \qquad \text{for} \qquad j = 1, \dots, n\,, \tag{58}$$

where the "local" central charge quantum numbers c_j are related to the superhelicities h_j in (30) through the relation $c_j = 2 - 2h_j$. The contour integral with twistor space data (55) becomes [11, 58]

$$\mathscr{A}_{n,k}\left(\{v_j^\pm\}\right) = \int \frac{d^{k\cdot n}C}{\mathrm{vol}(\mathrm{GL}(k))} \frac{\delta^{4k|4k}(C\cdot\mathscr{Z})}{(1,\ \dots,k)^{1+v_k^+-v_1^-}\dots(n,\ \dots,k-1)^{1+v_{k-1}^+-v_n^-}}\,, \tag{59}$$

while the integral with momentum twistor space data (56) deforms to [11, 58]

$$\mathscr{A}_{n,k}\left(\{v_j^\pm\}\right) = \frac{\delta^4(p^{\alpha\dot\alpha})\delta^8(q^{\alpha A})}{\langle 12\rangle^{1+v_2^+-v_1^-}\dots\langle n1\rangle^{1+v_1^+-v_n^-}} \times A_{n,k}\left(\{v_j^\pm\}\right), \tag{60}$$

with

$$A_{n,k}\left(\{v_j^\pm\}\right) = \int \frac{d^{\hat{k}\cdot n}\hat{C}}{\mathrm{vol}(\mathrm{GL}(\hat{k}))} \frac{\delta^{4\hat{k}|4\hat{k}}(\hat{C}\cdot\mathcal{W})}{(1,\ldots,\hat{k})^{1+v^+_{\hat{k}+1}-v^-_n}\ldots(n,\ldots,\hat{k}-1)^{1+v^+_{\hat{k}}-v^-_{n-1}}}. \tag{61}$$

Note that in the case of MHV amplitudes, where $k = 2$, $\hat{k} = 0$, the prefactor of eq. (60) is the exact deformed amplitude. In this case no further contour integral has to be performed.

However, in the non-MHV case the question remains, how to rigorously and explicitly define the contour of the integrals (59) and (61). In the undeformed case, where the minors lead to a set of multiple poles on the Graßmannian manifold, it has been suggested in [8, 77], see also [7, 53], that a multi-residue theorem compatible with the BCFW recursion relations is to be employed. This construction is however quite implicit. Furthermore, and quite interestingly, it becomes questionable in the deformed case, where the multi-poles turn into multi-branchpoints, and where the BCFW relations, themselves based on the residue theorem, no longer hold. Let us illustrate the problem in the simplest case of the NMHV 6-point amplitude. We then have $n = 6$ and $\hat{k} = 1$, and the Graßmannian manifold Gr(1, 6) simplifies to the projective space $\mathbb{CP}^5$. Focussing for example on the integral (61), it becomes

$$A_{6,3}\left(\{v_j^\pm\}\right) = \int \prod_{i=2}^{6} \frac{dc_{1i}}{c_{1i}^{1-v^-_{i-1}+v^+_{i+1}}}\, \delta^{4|4}(\mathcal{W}_1 + c_{12}\mathcal{W}_2 + \cdots + c_{16}\mathcal{W}_6), \tag{62}$$

where we have fixed the $\mathrm{GL}(1) = \mathbb{C}$ invariance by setting $c_{11} = 1$. In the undeformed case there are six poles (one of them at infinity, due to the gauge fixing). The single, ordinary contour that remains after performing four of the five integrations in eq. (62) using the bosonic δ-functions has to circle the correct set of three poles out of the total set of six, in line with the BCFW relations. The integral is then found using the (ordinary, in this case) residue theorem for the one remaining integration. However, switching on the deformation parameters, the contour always seems to cross branchcuts in complicated ways, and the integral (62) becomes incomplete and, potentially, even ill-defined due to the non-polynomial nature of the integrand.

It has been suggested in [68] that the way out of this problem is to consider the proper *reality conditions* of Minkowski space $\mathbb{R}^{1,3}$ for the kinematic variables and, in consequence, for the multi-contour on the Graßmannian. This is indeed the case [70]! The multi-contour can be found, and interestingly only weakly depends on the data: The contour is completely fixed in a finite number of certain kinematical sectors. In the special "MHV case" $n = 2k$ it is particularly simple, and turns out to be the group manifold of the unitary group $\mathrm{U}(k) \subset \mathrm{Gr}(k, 2k)$. The contour integration measure is just the Haar measure on the group. Excitingly, the branchcut problem is thereby solved: The integrand becomes *single valued* for arbitrary deformation parameters satisfying (58), and the contour avoids all branchcuts. The interesting exception are collinear kinematical configurations, where the tree-level amplitudes (58) are infrared divergent. For more details see [70].

The question whether the deformation parameters, which was the original motivation [51], may serve as a powerful regulator for infrared-divergent loop integrals still remains an interesting open research topic.

3.3 Polygonal Wilson loops and the ratio function

In Section 2.2 we have reviewed the duality between MHV scattering amplitudes and Wilson loops. As we have pointed out, much effort has been put into evaluating the finite part F_n^{WL}, as the divergent part of Wilson loops is already well understood. However the definition of F_n^{anom} in (54) is ambiguous because it can be modified by any function of the available conformal invariants. An example is the definition of the "BDS-like" piece of [4] where the anomalous part depends only on the shortest distances $x_{i,i+2}^2$. A particularly interesting definition for the decomposition was made in [1]. In this case one picks two of the light-like edges and forms a light-like square by picking two more light-like lines intersecting them both. Then one can consider four different Wilson loops. The original Wilson loop W_n, the Wilson loop on the square W_{sq} and the Wilson loops formed by replacing the top or bottom set of intermediate edges by the corresponding part of the square, W_{top} and W_{bottom}, respectively. This is best illustrated by Figure 1. A conformally-invariant quantity, function of the cross-ratios $u_1, \ldots, u_k$, can be defined by the following ratio of the Wilson loops

$$r_n = \log\left(\frac{W_n W_{\text{sq}}}{W_{\text{top}} W_{\text{bottom}}}\right). \tag{63}$$

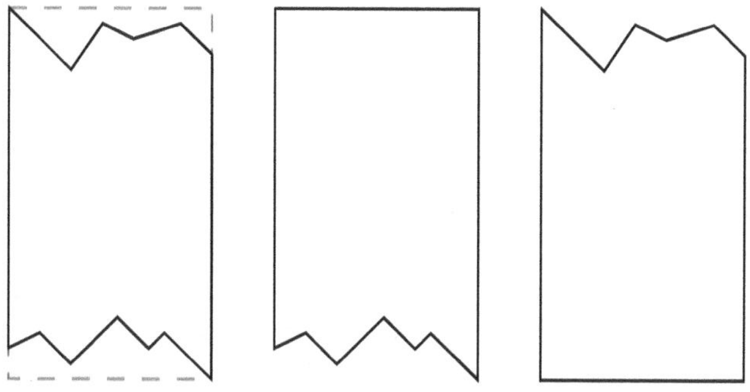

Figure 1: The four different Wilson loops entering the definition of the ratio (63). The Wilson loop on the square is shown by the dashed line. The bottom and top Wilson loops are obtained by replacing a sequence of edges by the corresponding part of the square.

Unlike the usual definition of the remainder function $\mathcal{R}$, r_n is non-zero already at one loop. It is also not cyclic invariant since its definition requires a choice of two special lines from which to form the square. This choice essentially corresponds to the choice of operator product expansion (OPE) "channel" in which one expands the Wilson loop over exchanged intermediate excited flux tube states [1]. The quantity r_n is particularly simple at one loop. It corresponds to the connected part of the correlation between the two Wilson loops. A further simplification is obtained when considering restricted two-dimensional kinematics as in [4]. In this case one needs an even number of sides to the Wilson loops, alternating in orientation between the x^+ direction and the x^- direction as one travels round the loop. The number of independent cross-ratios is reduced in the two-dimensional kinematics. In fact there are $(n-6)$ independent ratios left from the original $(3n-15)$. Since n is always even, the first non-trivial ratio is therefore at eight points.

As we are delving with Wilson loops, thus with the function M_n defined in eq. (46), we consider the $\mathfrak{sl}(4)$ bosonic subspace of momentum twistors $W_i^A = (\lambda^\alpha, \mu^{\dot\alpha})$ and functions with no helicity

$$h_i = W_i^C \frac{\partial}{\partial W_i^C} = 0\,. \tag{64}$$

Actually, when restricting to the special two-dimensional kinematics, the twistor variables are also restricted and preserve two commuting copies of $\mathfrak{sl}(2)$ inside $\mathfrak{sl}(4)$. Specifically we can decompose the W_i^A into upper and lower components, each transforming under its own $\mathfrak{sl}(2)$. The alternating orientations of the lines correspond to an alternating between twistors transforming under the two copies of $\mathfrak{sl}(2)$. We take the odd-numbered twistors to transform under the first copy and the even-numbered ones to transform under the second copy,

$$W_{2i+1} = \begin{pmatrix} w_{2i+1} \\ 0 \end{pmatrix}, \qquad W_{2i} = \begin{pmatrix} 0 \\ \bar{w}_{2i} \end{pmatrix}. \tag{65}$$

In this two-dimensional case the cross-ratios are simply

$$u = \frac{(15)(37)}{(17)(35)}, \qquad v = \frac{(26)(48)}{(28)(46)}, \tag{66}$$

with $(ij) = \epsilon_{A_1A_2} W_i^{A_1} W_j^{A_2}$. The ratio r_n is simple to compute since the relevant Wilson loops are known at one loop [29] to coincide with the one-loop MHV amplitudes [25] through the duality. In the special two-dimensional kinematics all quantities can be expressed in terms of logarithms [4, 64]. Remarkably the function r_8 at one loop is none other than [60]

$$r_8 = g^2 \log u \log v + \text{const}\,. \tag{67}$$

Also the Yangian symmetry is restricted to the $Y(\mathfrak{sl}(2))\oplus Y(\mathfrak{sl}(2))$ subalgebra. Following section 1, in particular formulas (7) and (8), we can write the generators of the two copies of $Y(\mathfrak{sl}(2))$ as (we recall that $h_i = 0$)

$$\begin{aligned}
J^a{}_b &= \sum_{i\,\text{odd}} w_i^a \frac{\partial}{\partial w_i^b} \\
J^{(1)a}{}_b &= \sum_{i<j\,\text{odd}} \left(w_i^a \frac{\partial}{\partial w_i^c} w_j^c \frac{\partial}{\partial w_j^b} - (i,j) \right) + \sum_{i\,\text{odd}} v_i w_i^a \frac{\partial}{\partial w_i^b} \\
\bar{J}^{\bar{a}}{}_{\bar{b}} &= \sum_{i\,\text{even}} \bar{w}_i^{\bar{a}} \frac{\partial}{\partial \bar{w}_i^{\bar{b}}} \\
\bar{J}^{(1)\bar{a}}{}_{\bar{b}} &= \sum_{i<j\,\text{even}} \left(\bar{w}_i^{\bar{a}} \frac{\partial}{\partial \bar{w}_i^{\bar{c}}} \bar{w}_j^{\bar{c}} \frac{\partial}{\partial \bar{w}_j^{\bar{b}}} - (i,j) \right) + \sum_{i\,\text{even}} \bar{v}_i \bar{w}_i^{\bar{a}} \frac{\partial}{\partial \bar{w}_i^{\bar{b}}} \,.
\end{aligned} \tag{68}$$

Here a, b and $\bar{a}, \bar{b}$ run from 1 to 2 and the $v_i, \bar{v}_i$s are the free parameters we had indicated by v_j in eq. (8). Now let us consider functions of the w_i which are invariant under the action of the Yangian generators. Let us start with the level-zero conformal algebra. As we are just interested in $\mathfrak{sl}(2)$-invariant functions of the w_i, we can have any function of the invariant quantities,

$$(i_1 i_2) = \epsilon_{a_1 a_2} w_{i_1}^{a_1} w_{i_2}^{a_2} \,. \tag{69}$$

By requiring homogeneous functions with degree zero in all of the w_i, we must consider functions of homogeneous ratios of the invariants in equation (69). The first possibility to form a homogeneous ratio is at four sites (1,2,3,4) where we can write

$$u = \frac{(13)(24)}{(14)(23)} \,, \tag{70}$$

which is the only independent invariant. The only other possibility is related to u using the cyclic identity $(ij)w_k^a + (jk)w_i^a + (ki)w_j^a = 0$,

$$\frac{(12)(34)}{(41)(23)} = 1 - u \,. \tag{71}$$

Thus the $\mathfrak{sl}(2)$-invariant functions on four copies of $(\mathbb{CP}^1)$ are functions of u. Requiring that they are also Yangian-invariant functions means that we have to solve the equations

$$J_{\vec{v}}^{(1)a}{}_b f(u) = 0 \,. \tag{72}$$

Obviously a constant function is always a solution of the equations. There exists also a non-trivial one-parameter solution given by hypergeometric functions,

$$f_\mu(u) = \frac{(1-u)^{1+\mu}}{1+\mu}\ {}_2F_1(1, 1+\mu, 2+\mu; 1-u)\,, \qquad \mu = \frac{1}{2}(\nu_2 - \nu_1)\,, \tag{73}$$

with the constraint $\nu_1 - \nu_3 = \nu_2 - \nu_4 = 2$. These functions represent the only homogeneous functions at four sites which are also Yangian invariants. If we consider the special case $\mu = 0$, whose related values of the weights ν_is correspond to demanding cyclicity of the four-site level-one generators, we get

$$f_0(u) = (1-u)\ {}_2F_1(1, 1, 2; 1-u) = \log u\,, \tag{74}$$

which is exactly one of the terms in eq. (67) after renumbering of the sides. Thus finally we can write[3]

$$J^a{}_b \log u = 0\,, \qquad J^{(1)a}_{\vec{\nu}}{}_b \log u = 0\,, \qquad \vec{\nu} = (1, 1, -1, -1)\,. \tag{75}$$

Therefore r_8 is Yangian invariant under two copies of the Yangian $Y(\mathfrak{sl}(2))$ for the choices $\vec{\nu} = (1, 1, -1, -1)$ and $\vec{\bar{\nu}} = (1, 1, -1, -1)$ where the entries range over the odd and even values of the light-like segments respectively. Moreover, as pointed out in [60], at one loop there is a very simple relation between r_n and r_{n-2} for a given choice of reference square. This amounts to the fact that at one loop the Wilson loops are additive in nature. We can use this reduction argument to write

$$r_n = f(r_{n-2}) = g(r_{n-4}) = \cdots = h(r_8)\,. \tag{76}$$

Thus we conclude that r_n is always invariant under the two commuting Yangians with a natural representation corresponding to the choice of OPE channel (i.e. choice of reference square).

This analysis is the first indication that the Yangian symmetry seen at the level of the tree amplitudes (or the integrand for loop amplitudes) exhibits itself in a simple and natural way on the functions at one loop. Of course in the case presented, there are only two copies of the bosonic Yangian $Y(\mathfrak{sl}(2))$, not the full $Y(\mathfrak{psl}(4|4))$. The fate of the symmetry beyond one loop and the special momentum configuration is still unknown. However the above results are a strong indication that it should be still present in the general case.

3 For the details of the computation see [44].

4 Yangian symmetry in smooth Wilson loops

4.1 Maldacena-Wilson loop

Wilson loops constitute the prime non-local observables in gauge field theories, they serve as an order parameter to detect confinement through an area law for the vacuum expectation value of rectangular shaped loops. In the superconformal $\mathcal{N} = 4$ super Yang-Mills model the standard Wilson loop operator [84] defined as the holonomy of the gauge connection $A_\mu(x)$ along the loop path may be augmented by a coupling to the adjoint scalar fields resulting in a so-called Maldacena-Wilson loop [76]:

$$W(C) = \frac{1}{N}\,\mathrm{Tr}\,\mathcal{P}\exp\left(i\oint_C ds\,(A_\mu(x)\,\dot{x}^\mu + \Phi_i(x)\,\eta^i\,|\dot{x}|)\right). \tag{77}$$

Here $x^\mu(s) : [0, 1] \to \mathbb{R}^{1,3}$ is a closed path in space-time and the functions $\eta^i(s)$ with $i = 1, \ldots, 6$ represent a path on the internal S^5, i.e. $(\eta^i)^2 = 1$. This operator is a locally 1/2 BPS quantity in the following sense: Using a 10D description of the $\mathcal{N} = 4$ super Yang-Mills model with the gauge field via $A_{\hat{\mu}} = (A_\mu, \Phi_i)$ and the sixteen-component Majorana-Weyl gluino Ψ one has the supersymmetry transformation

$$\delta_\epsilon A_{\hat{\mu}} = -\epsilon\Gamma_{\hat{\mu}}\Psi\,, \qquad \delta_\epsilon\Psi = \frac{1}{2}\Gamma^{\hat{\mu}\hat{\nu}}\epsilon\,F_{\hat{\mu}\hat{\nu}}\,. \tag{78}$$

For the Maldacena-Wilson loop operator this implies

$$\delta_\epsilon W \sim \oint ds(\epsilon\dot{x}^{\hat{\mu}}\,\Gamma_{\hat{\mu}}\Psi)\;W(C), \qquad \dot{x}^{\hat{\mu}} := (\dot{x}^\mu, \eta^i\,\sqrt{\dot{x}\cdot\dot{x}})\,. \tag{79}$$

If we now choose a supersymmetry parameter ϵ of the form $\epsilon = \kappa\,\dot{x}^{\hat{\mu}}\,\tilde{\Gamma}_{\hat{\mu}}$ locally, we see that $\delta_\epsilon W = 0$ by virtue of the effective 10D light-likeness of $\dot{x}^{\hat{\mu}}$

$$x^{\hat{\mu}}\,x_{\hat{\mu}} = \dot{x}\cdot\dot{x} - (\eta^i)^2\,\dot{x}\cdot\dot{x} = 0\,. \tag{80}$$

As κ has $16/2 = 8$ independent degrees of freedom, we speak of a locally 1/2 BPS quantity. The Maldacena-Wilson loop operator defined in eq. (77) has remarkable properties. Its vacuum expectation value is finite and it is invariant under conformal transformations. At leading order in the perturbative expansion it takes the form

$$\langle W(C)\rangle = 1 - \frac{\lambda}{16\pi^2}\int ds_1\,ds_2\,\frac{\dot{x}_1\cdot\dot{x}_2 - |\dot{x}_1|\,|\dot{x}_2|}{(x_1 - x_2)^2} + \ldots\,, \qquad x_i = x(s_i)\,. \tag{81}$$

Here we see that the potential ultraviolet divergence arising from coinciding point $x_1 \to x_2$ cancels between the gluon ($\sim\ \dot{x}_1\cdot\dot{x}_2$) and adjoint scalar ($\sim\ |\dot{x}_1|\,|\dot{x}_2|$) contributions.

How are the conformal symmetries realized for this observable? The relevant representation of the conformal generators acting as variational derivative operator on the space of paths $x^\mu(s)$ take the form

$$\begin{aligned} P_\mu &:= \oint ds\, p_\mu(s) = \oint ds \frac{\delta}{\delta x^\mu(s)}, \\ M_{\mu\nu} &:= \oint ds\, m_{\mu\nu}(s) = \oint ds \left(x_\mu(s) \frac{\delta}{\delta x^\nu(s)} - \mu \leftrightarrow \nu \right) \\ D &:= \oint ds\, d(s) = \oint ds\, x^\rho(s) \frac{\delta}{\delta x^\rho(s)}, \\ K_\mu &:= \oint ds\, k_\mu(s) = \oint ds \left(x^2(s) \frac{\delta}{\delta x^\mu(s)} - 2x_\mu(s)\, x^\nu(s) \frac{\delta}{\delta x^\nu(s)} \right). \end{aligned} \tag{82}$$

They are first-order variational derivative operators and represent symmetries of the expectation values of the Wilson loop operator

$$\begin{aligned} P_\mu \langle W(C) \rangle &= 0 \qquad M_{\mu\nu} \langle W(C) \rangle = 0 \\ K_\mu \langle W(C) \rangle &= 0 \qquad D \langle W(C) \rangle = 0\,. \end{aligned} \tag{83}$$

The proof of the last two relations is non-trivial, see [41]. Following our discussions in the previous chapters the immediate question now arises whether there exists a Yangian symmetry for the Maldacena-Wilson loops as well. In fact, given the form of the conformal generators in eq. (82) we are led to an immediate guess for a level-one Yangian generator to take the form

$$J_a = \oint ds\, j_a(s), \qquad \hat{J}_a = \oint ds\, \hat{j}_a(s) + f_a{}^{bc} \int_{s_1<s_2} ds_1\, ds_2\, j_b(s_1)\, j_c(s_2),$$

where the $j_a(s)$ denote the level-zero densities appearing in eq. (82). This is the natural continuum version of eq. (8). At this point the local level-one density $\hat{j}_a(s)$ is left undetermined. Explicitly for $\hat{P}^\mu \in Y[\mathfrak{psu}(2,2|4)]$ one then derives the expression for the non-local part

$$\begin{aligned} \hat{P}^\mu = \int ds_1\, ds_2 \Big\{ & (m^{\mu\nu}(s_1) - d(s_1)\, \eta^{\mu\nu})\, p_\nu(s_2) + \bar{q}^{A\dot{\alpha}}(s_1)\, \sigma^\mu_{\alpha\dot{\alpha}}\, q^\alpha_A(s_2) \\ & - (s_1 \leftrightarrow s_2) \Big\}\, \theta(s_2 - s_1) + \oint ds\, \hat{p}^\mu(s)\,. \end{aligned} \tag{84}$$

However, as long as we consider the Maldacena-Wilson loop operator in (77) the supercharge densities $\bar{q}(s)$ and $q(s)$ are unclear, as there are no fermionic coordinates appearing in the definition (77). Acting with only the bosonic non-local terms in the above expression one finds as the result of a lengthy computation [80]

$$\hat{P}^{\mu}_{\mathrm{bos},\varepsilon}\langle W(C)\rangle_{\text{1-loop}} = -\frac{\lambda}{16\pi^2}\left\{\frac{1}{6}\int \mathrm{d}s\, \dot{x}^{\mu}\left(\frac{\ddot{x}^2}{\dot{x}^4} - \frac{(\dot{x}\cdot\ddot{x})^2}{\dot{x}^6}\right) + \right.$$
$$\left. + 16\int \mathrm{d}s_1 \mathrm{d}s_2\, \frac{\dot{x}_1\dot{x}_2 - |\dot{x}_1||\dot{x}_2|}{(x_1 - x_2)^4}\,(x_1^{\mu} - x_2^{\mu})\,\theta(\tau_2 - \tau_1 - \varepsilon)\right\} \tag{85}$$

where a point-splitting regulator $s_2 > s_1 + \epsilon$ was employed in order to regularize the expression.[4] We see that the first integral in the above is a welcome local term, which could serve as the definition of $\hat{j}_a(s)$ when put on the left-hand side. Its precise form is, however, regulator dependent. Looking at the second double integral in eq. (85), we see that it destroys the possibility of hidden Yangian symmetry for Maldacena-Wilson loops as it is still a bilocal expression and cannot be absorbed into $\hat{j}_a(s)$. Of course, the source of this mismatch is to be searched in the missing fermionic contributions to the Maldacena-Wilson loop operator of eq. (77). In line with this the term inside the double integral takes the form of a fermion propagator, just as it would arise due to a gluino-gluino contraction along the loop. We should therefore seek for a supersymmetrized version of the Maldacena-Wilson loop operator (77).

4.2 $\mathcal{N} = 4$ SYM in 4d on-shell superspace

In order to define the super Maldacena-Wilson loop operator we turn to a formulation of the $\mathcal{N} = 4$ SYM in an on-shell superspace. This space is parametrized by 4 bosonic coordinates x^{μ} and 16 fermionic coordinates $\theta^{a\alpha}, \bar{\theta}^{\dot{\alpha}}_a$ which we combine as

$$X^{\mathscr{A}} = \left(x^{\dot{\alpha}\alpha} = x^{\mu}\bar{\sigma}_{\mu}^{\dot{\alpha}\alpha}, \theta^{a\alpha}, \bar{\theta}^{\dot{\alpha}}_a\right) \qquad \begin{matrix} \mu = 0,\dots,3 & \alpha = 1,2 \\ a = 1,,\dots,4 & \dot{\alpha} = 1,2. \end{matrix} \tag{86}$$

We have 16 supercharges and 16 SUSY-covariant derivatives

$$\begin{aligned} \mathrm{Q}_{a\alpha} &= \bar{\theta}^{\dot{\alpha}}_a \partial_{\alpha\dot{\alpha}} - \partial_{a\alpha} & \qquad \mathrm{D}_{a\alpha} &= \bar{\theta}^{\dot{\alpha}}_a \partial_{\alpha\dot{\alpha}} + \partial_{a\alpha} \\ \bar{\mathrm{Q}}^a_{\dot{\alpha}} &= \theta^{a\alpha}\partial_{\alpha\dot{\alpha}} - \bar{\partial}^a_{\dot{\alpha}} & \qquad \bar{\mathrm{D}}^a_{\dot{\alpha}} &= \theta^{a\alpha}\partial_{\alpha\dot{\alpha}} + \bar{\partial}^a_{\dot{\alpha}}. \end{aligned} \tag{87}$$

obeying the anti-commutation relations

$$\left\{\mathrm{Q}_{a\alpha}, \bar{\mathrm{Q}}^b_{\dot{\alpha}}\right\} = 2\delta^b_a \mathrm{P}_{\alpha\dot{\alpha}} \quad \text{and} \quad \left\{\mathrm{D}_{a\alpha}, \bar{\mathrm{D}}^b_{\dot{\alpha}}\right\} = -2\delta^b_a \mathrm{P}_{\alpha\dot{\alpha}}. \tag{88}$$

4 In eq. (85) $\mathcal{O}(\epsilon)$ and $\delta(\epsilon)$ were dropped.

We furthermore introduce the supersymmetry-invariant one-forms

$$e^{\dot{\alpha}\alpha} = dx^{\dot{\alpha}\alpha} - 2d\bar{\theta}^{\dot{\alpha}}_{a}\theta^{a\alpha} + 2\bar{\theta}^{\dot{\alpha}}_{a}d\theta^{a\alpha}\,, \quad d\theta^{a\alpha}\,, \quad d\bar{\theta}^{\dot{\alpha}}_{a}\,. \tag{89}$$

Next, a gauge connection one-form is introduced via

$$\mathscr{A} = \mathscr{E}^{\mathscr{A}}\,\mathscr{A}_{\mathscr{A}} = \tfrac{1}{2}e^{\dot{\alpha}\alpha}\,\mathscr{A}_{\alpha\dot{\alpha}}(x,\theta,\bar{\theta}) + d\theta^{a\alpha}\,\mathscr{A}_{a\alpha}(x,\theta,\bar{\theta}) + d\bar{\theta}^{\dot{\alpha}}_{a}\,\bar{\mathscr{A}}^{a}_{\dot{\alpha}}(x,\theta,\bar{\theta}) \tag{90}$$

which allows us to write down the gauge-covariant derivatives

$$\mathscr{D}_{\alpha\dot{\alpha}} = \partial_{\alpha\dot{\alpha}} + \mathscr{A}_{\alpha\dot{\alpha}}\,, \qquad \mathscr{D}_{a\alpha} = \mathrm{D}_{a\alpha} + \mathscr{A}_{a\alpha}\,, \qquad \bar{\mathscr{D}}^{a}_{\dot{\alpha}} = \bar{\mathrm{D}}^{a}_{\dot{\alpha}} + \bar{\mathscr{A}}^{a}_{\dot{\alpha}}\,. \tag{91}$$

With these the field strength two-form follows as $\mathscr{F} = d\mathscr{A} + \mathscr{A}\wedge\mathscr{A}$. In fact, the field strength superfield has many more component fields than the $\mathscr{N}=4$ SYM theory and one needs to impose constraints on them in order to match the degrees of freedom [61, 86]. The constraints we put are

$$\mathscr{F}_{a\alpha,\,b\beta} = -4\,\varepsilon_{\alpha\beta}\,\Phi_{ab}\,, \quad \mathscr{F}_{a\alpha,}{}^{b}{}_{\dot{\beta}} = 0\,, \quad \mathscr{F}^{a}{}_{\dot{\alpha},}{}^{b}{}_{\dot{\beta}} = -2\,\varepsilon_{\dot{\alpha}\dot{\beta}}\,\varepsilon^{abcd}\,\Phi_{cd}\,, \tag{92}$$

where a super-scalar field appears being natural from a 10d reduction perspective. An intriguing fact is that these constraints imply the equations of motion of $\mathscr{N}=4$ SYM, see [61, 86]. This is why the superspace introduced is an on-shell $\mathscr{N}=4$ superspace. Using the constraints and the Bianchi identities allows one to express the superfield strength via the super-scalar field [16] via

$$\begin{aligned}\mathscr{F} = {} & \Big[2(d\theta\varepsilon d\theta^{\mathsf{T}})^{ab} + \tfrac{1}{3}(e\varepsilon d\theta^{\mathsf{T}})^{\dot{\alpha}a}\,\bar{\mathscr{D}}^{b}{}_{\dot{\alpha}} + \tfrac{1}{96}(e\varepsilon e^{\mathsf{T}})^{\dot{\alpha}\dot{\beta}}\,\bar{\mathscr{D}}^{a}{}_{\dot{\alpha}}\bar{\mathscr{D}}^{b}{}_{\dot{\beta}}\Big]\Phi_{ab} \\ & + \Big[2(d\bar{\theta}^{\mathsf{T}}\varepsilon d\bar{\theta})_{ab} + \tfrac{1}{3}(e^{\mathsf{T}}\varepsilon d\bar{\theta})^{\alpha}{}_{a}\mathscr{D}_{\alpha b} + \tfrac{1}{96}(e^{\mathsf{T}}\varepsilon e)^{\alpha\beta}\,\mathscr{D}_{a\alpha}\mathscr{D}_{\beta b}\Big]\bar{\Phi}^{ab}.\end{aligned} \tag{93}$$

Note that in the above we are using a index-free condensed matrix notation with ε denoting the Levi-Civita symbol, for details see [16].

An important question now is how do the component fields of $\mathscr{N}=4$ SYM enter in the θ and $\bar{\theta}$ expansion of the superfield strength. The answer depends on a gauge choice, as the gauge connection is defined only up to transformations of the type

$$\delta\mathscr{A}_{\alpha\dot{\alpha}} = \mathscr{D}_{\alpha\dot{\alpha}}\Lambda\,, \quad \delta\mathscr{A}_{a\alpha} = \mathscr{D}_{a\alpha}\Lambda\,, \quad \delta\bar{\mathscr{A}}^{a}_{\dot{\alpha}} = \bar{\mathscr{D}}^{a}_{\dot{\alpha}}\Lambda\,. \tag{94}$$

One may greatly simplify the component field expansion by fixing the fermionic gauge freedom via the Harnad-Shnider gauge constraint [61]

$$\theta^{a\alpha}\,\mathscr{A}_{a\alpha} + \bar{\theta}^{\dot{\alpha}}_{a}\bar{\mathscr{A}}^{a}_{\dot{\alpha}} = 0\,. \tag{95}$$

Using this gauge and the Bianchi identities one establishes a recursive algorithm to construct the component field expansion of all relevant superfields involved. The explicit results read to the first few orders [16],

$$\begin{aligned}
\mathcal{A}^{\dot{\alpha}\alpha}(x,\theta,\bar{\theta}) &= A^{\dot{\alpha}\alpha} + 2\bar{\psi}^{\dot{\alpha}}_{a}\theta^{a\alpha} - 2\bar{\theta}^{\dot{\alpha}}_{a}\psi^{a\alpha} + \mathcal{O}(\theta^2),\\
\Phi_{ab}(x,\theta,\bar{\theta}) &= \phi_{ab} - \varepsilon_{abcd}\,\theta^{c\alpha}\varepsilon_{\alpha\beta}\psi^{d\beta} - 2\bar{\theta}^{\dot{\alpha}}_{[a|}\varepsilon_{\dot{\alpha}\dot{\beta}}\bar{\psi}^{\dot{\beta}}_{|b]} + \mathcal{O}(\theta^2),\\
\mathcal{A}_{a\alpha}(x,\theta,\bar{\theta}) &= A_{\alpha\dot{\alpha}}\bar{\theta}^{\dot{\alpha}}_{a} - \phi_{ab}\theta^{b\beta}\varepsilon_{\beta\alpha} + \mathcal{O}(\theta^2),\\
\bar{\mathcal{A}}^{a}_{\dot{\alpha}}(x,\theta,\bar{\theta}) &= \theta^{a\alpha}A_{\alpha\dot{\alpha}} - \bar{\phi}^{ab}\bar{\theta}^{\dot{\beta}}_{b}\varepsilon_{\dot{\beta}\dot{\alpha}} + \mathcal{O}(\theta^2).
\end{aligned} \tag{96}$$

4.3 The supersymmetric Maldacena-Wilson loop operator

Equipped with this on-shell superspace formalism it is straightforward to write down a Wilson loop operator through a path in $X^{\mathcal{A}}$ as

$$\mathrm{tr}\mathcal{W} = \frac{1}{N}\,\mathrm{Tr}\,\mathcal{P}\,\exp\left[\int d\tau\left(\tfrac{1}{2}p^{\dot{\alpha}\alpha}\mathcal{A}_{\alpha\dot{\alpha}} + \dot{\theta}^{a\alpha}\mathcal{A}_{a\alpha} + \dot{\bar{\theta}}^{\dot{\alpha}}_{a}\bar{\mathcal{A}}^{a}_{\dot{\alpha}}\right)\right] \tag{97}$$

where $p^{\dot{\alpha}\alpha}$ is the super-momentum being defined as

$$\begin{aligned}
e^{\dot{\alpha}\alpha} &= d\tau\, p^{\dot{\alpha}\alpha} = d\tau\left(\dot{x}^{\dot{\alpha}\alpha} - 2\dot{\bar{\theta}}^{\dot{\alpha}}_{a}\theta^{a\alpha} + 2\bar{\theta}^{\dot{\alpha}}_{a}\dot{\theta}^{a\alpha}\right),\\
d\theta^{a\alpha} &= d\tau\,\dot{\theta}^{a\alpha}, \qquad d\bar{\theta}^{\dot{\alpha}}_{a} = d\tau\,\dot{\bar{\theta}}^{\dot{\alpha}}_{a}.
\end{aligned} \tag{98}$$

The operator $\mathrm{tr}\mathcal{W}$ is manifestly supersymmetric. However, it clearly does not limit to the Maldacena-Wilson loop (77) as one sees by truncation to the bosonic sector. Hence, it is not the operator we are seeking. In order to supersymmetrize the Maldacena-Wilson loop operator (77) we need to extend the on-shell superspace by a bosonic coordinate η^i with $i = 1,\dots,6$ with $(\eta^i)^2 = 1$, as was done in [16]:

$$\left(x^{\dot{\alpha}\alpha}(\tau), \theta^{a\alpha}(\tau), \bar{\theta}^{\dot{\alpha}}_{a}(\tau)\right) \quad\longrightarrow\quad \left(x^{\dot{\alpha}\alpha}(\tau), \theta^{a\alpha}(\tau), \bar{\theta}^{\dot{\alpha}}_{a}(\tau), n^{i}(\tau)\right). \tag{99}$$

We then define the supersymmetric Maldacena-Wilson loop through

$$\mathrm{tr}\tilde{\mathcal{W}} = \frac{1}{N}\,\mathrm{tr}\,\mathcal{P}\,\exp\left[\int d\tau\left(\tfrac{1}{2}p^{\dot{\alpha}\alpha}\mathcal{A}_{\alpha\dot{\alpha}} + \dot{\theta}^{a\alpha}\mathcal{A}_{a\alpha} + \dot{\bar{\theta}}^{\dot{\alpha}}_{a}\bar{\mathcal{A}}^{a}_{\dot{\alpha}} + q^{i}\Phi_{i}\right)\right], \tag{100}$$

an expression that also appeared in [81]. Here the Φ_i is the scalar superfield and $q^2 - p^2 = 0$, i.e. $q^i = n^i\sqrt{p^2}$. The so defined Wilson loop operator enjoys a number of important properties [15]. For one, if we consider a smooth path C for our loop the vacuum expectation value of this operator is free of divergences (at least at the one-loop order)

$$\left\langle \mathrm{tr}\tilde{\mathcal{W}} \right\rangle_{(1)} \sim \iint \left\langle (\mathcal{A} + \Phi)(1), (\mathcal{A} + \Phi)(2) \right\rangle \tag{101}$$

$$\sim \iint \mathrm{d}\tau\, \mathrm{d}\varepsilon \left[\frac{1}{\varepsilon^2} \frac{p^2 - q^2}{p^2} - \frac{1}{\varepsilon} \frac{q^2}{p^2} \left(\frac{\dot{q} \cdot q}{q^2} - \frac{\dot{p} \cdot p}{p^2} \right) + \mathcal{O}(1) \right]. \tag{102}$$

The potential singularities above cancel due to $q^2 = p^2$ – reflecting the effective 10d light-like property of $p^{\hat{\mu}} = (p^{\mu}, q^{i})$. Here, as before, the one-loop computation is regularized via point-splitting.

Moreover the 1/2 BPS property of the bosonic Maldacena-Wilson loop operator, discussed in section 4.1, generalizes to the invariance of the supersymmetric Maldacena-Wilson loop under a local fermionic symmetry, known as the κ-symmetry of the 10d superparticle. To see this we define the κ-symmetry transformations

$$\delta_{\kappa} \theta^{\hat{\beta}} = \kappa_{\hat{\alpha}}\, p^{\hat{\mu}} \Gamma_{\hat{\mu}}^{\hat{\alpha}\hat{\beta}}, \qquad \delta_{\kappa} p^{\hat{\mu}} = -2\theta \Gamma^{\hat{\mu}} \delta_{\kappa} \theta . \tag{103}$$

Note that there is also a non-trivial transformation on the S^5 coordinates with η^i acting. One then derives the transformation of the supersymmetric Maldacena-Wilson loop operator to take the form

$$\delta_{\kappa} \mathrm{tr}\tilde{\mathcal{W}} = \frac{1}{N} \mathrm{tr}\mathcal{P} \left\{ \exp\left(\oint_{\tilde{Z}} \mathcal{A} \right) \oint_{\tilde{Z}} \mathrm{d}\tau \left(p^{\hat{\mu}}\, \delta_{\kappa} \theta^{\hat{\alpha}}\, \mathcal{F}_{\hat{\alpha}\hat{\mu}} + \dot{\theta}^{\hat{\beta}}\, \delta_{\kappa} \theta^{\hat{\alpha}}\, \mathcal{F}_{\hat{\alpha}\hat{\beta}} \right) \right\} \tag{104}$$

The constraints (92) translated to 10d language with coordinates $\hat{\mu} = 0, \ldots, 9$ and $\hat{\alpha} = 1, \ldots, 16$ result in $\mathcal{F}_{\hat{\alpha}\hat{\beta}} = 0$ killing the second term. Moreover we have $\mathcal{F}_{\hat{\alpha}\hat{\mu}} = (\tilde{\Gamma}_{\hat{\mu}})_{\hat{\alpha}\hat{\beta}}\, \Psi^{\hat{\beta}}$, for a superfield with fermionic body $\Psi^{\hat{\beta}}$ which follows from the Bianchi identities. This leads to a vanishing also of the first term in the above due to $p^{\hat{\mu}} p_{\hat{\mu}} = 0$. In summary, we see that the famous 1/2 BPS property of the Maldacena-Wilson loop operator is embedded into the κ symmetry invariance of the full supersymmetric loop operator. A formulation of these super Maldaena-Wilson loops from a twistorial point of view has recently been obtained in [83].

4.4 Superconformal invariance

In order to properly define the Yangian symmetry generators for the smooth Maldacena-Wilson loops let us reconsider the conformal transformations for the purely bosonic case first. We had defined in eq. (82) a path action of the generators via

$$\mathbf{J}^{\kappa} = \int_0^1 \mathrm{d}\tau\, \xi_{\mu}^{\kappa}(\tau) \frac{\delta}{\delta x_{\mu}(\tau)} \tag{105}$$

with the conformal Killing vectors $\xi^\kappa_\mu(\tau)$. To begin with we apply this to an ordinary Wilson loop to get

$$\begin{aligned}\mathrm{J}^\kappa \mathrm{Tr}\,(W) &= N\mathrm{tr}\int_0^1 \mathrm{d}\tau\, W_{[0,\tau]}\Big(\dot{\xi}^\kappa_{\mathcal{A}}\mathcal{A}^{\mathcal{A}} + \dot{x}_\nu \xi^\kappa_\mu \partial^\mu A^\nu\Big)W_{[\tau,1]} \\ &= N\mathrm{tr}\int_0^1 \mathrm{d}\tau\, W_{[0,\tau]}\Big(\xi^\kappa_\mu \dot{x}_\nu F^{\mu\nu} + \mathrm{D}_\tau\big(\xi^\kappa_\mu A^\mu\big)\Big)W_{[\tau,1]}\end{aligned} \tag{106}$$

where $\mathscr{D}_\tau\mathscr{O} = \partial_\tau + \dot{x}^\sigma\big[A_\sigma, \mathscr{O}\big]$ is the covariant derivative acting along the loop. Here $W_{[\tau_1,\tau_2]}$ denotes a Wilson line operator integrated between τ_1 and τ_2. Equivalently, we can work with action by insertion rather than path variation defined as

$$\mathbb{J}^\kappa \mathrm{tr} W := N\mathrm{tr}\int_0^1 W_{[0,\tau]}\big(\mathbb{J}^\kappa A\big)W_{[\tau,1]}\,, \tag{107}$$

where from above $\mathbb{J}^\kappa A = \xi^\kappa_\mu \mathrm{d}x_\nu F^{\mu\nu} + \mathscr{D}\big(\xi^\kappa_\mu A^\mu\big)$. Note that the last term is a gauge transformation.

We now define an alternative action of the conformal symmetry generators on the Wilson loop which has the virtue of being gauge covariant. To do so we simply subtract the derivative term and define

$$\mathbb{J}^\kappa_* A := \mathbb{J}^\kappa A - \mathscr{D}\big(\xi^\kappa_\mu A^\mu\big) = \xi^\kappa_\mu \mathrm{d}x_\nu F^{\mu\nu}. \tag{108}$$

Note that since tr W is gauge invariant due to the closed trace structure we have

$$\mathbb{J}^\kappa \mathrm{tr}\, W = N\mathrm{tr}\int_0^1 W_{[0,\tau]}\big(\xi^\kappa_\mu \mathrm{d}x_\nu F^{\mu\nu} + \mathscr{D}\big(\xi^\kappa_\mu A^\mu\big)\big)W_{[\tau,1]} = \mathbb{J}^\kappa_* \mathrm{tr}\, W\,, \tag{109}$$

i.e. both representations are equivalent upon action on the Wilson loop operator. These considerations may be directly lifted to the supersymmetric Maldacena-Wilson loop. The gauge-covariant action of the superconformal generators $\mathbb{J}^\kappa_*$ on the loop reads

$$\mathbb{J}^\kappa_* \mathrm{tr}\tilde{\mathscr{W}} = N\mathrm{tr}\int_0^1 \tilde{\mathscr{W}}_{[0,\tau]}\mathbb{J}^\kappa_*\big(\mathscr{A} + \Phi\big)\tilde{\mathscr{W}}_{[\tau,1]}\,, \tag{110}$$

where now we have

$$\mathbb{J}^\kappa_* \mathscr{A} = \mathrm{J}^\kappa X^{\mathscr{B}} \mathrm{d}X^{\mathscr{A}}\, \mathscr{F}_{\mathscr{A}\mathscr{B}}\, \mathbb{J}^\kappa_* \Phi = -\mathrm{J}^\kappa X^{\mathscr{A}}\, \mathscr{D}_{\mathscr{A}}\Phi - \mathrm{J}^\kappa q^i\, \Phi_i\,. \tag{111}$$

Superconformal symmetry of the Maldacena-Wilson loop is then the statement that $\langle \mathbb{J}^\kappa_* \mathrm{tr}\mathcal{W} \rangle = 0$. The covariantized generators satisfy the algebra relation

$$\left[\mathbb{J}^\kappa_*, \mathbb{J}^\rho_*\right] = f^{\kappa\rho}{}_\delta \mathbb{J}^\delta_* - \mathbb{G}[\mathcal{G}^{\kappa\rho}] \tag{112}$$

with $\mathcal{G}^{\kappa\rho}$ being defined as

$$\mathcal{G}^{\kappa\rho} = -(-1)^{|\kappa||\mathcal{A}|} \mathrm{J}^\kappa X^{\mathcal{A}}\, \mathrm{J}^\rho X^{\mathcal{B}}\, \mathcal{F}_{\mathcal{B}\mathcal{A}}\,. \tag{113}$$

Here $\mathbb{G}[\Lambda]$ is the generator of gauge transformations which acts as

$$\mathbb{G}[\Lambda]\mathcal{A} = \mathcal{D}\Lambda\,, \qquad \mathbb{G}[\Lambda]\Phi = [\Phi, \Lambda]\,. \tag{114}$$

For the fields of $\mathcal{N} = 4$ super Yang-Mills theory $\mathcal{G}^{\kappa\rho}$ satisfies a curious identity

$$f^\kappa{}_{\rho\sigma}\mathcal{G}^{\sigma\rho} = 0\,, \tag{115}$$

which was termed the "$\mathcal{G}$-identity" in [15].

4.5 Yangian invariance

The necessity of using the gauge-covariant action of the symmetry generators becomes evident when one constructs the bilocal piece of the level-one Yangian generators. The naive form based on J^δ reads

$$\widehat{\mathbb{J}}^\kappa_{\mathrm{bi}}\mathrm{tr}\tilde{\mathcal{W}} = N f^\kappa{}_{\rho\delta}\mathrm{tr}\iint_{1<2} \tilde{\mathcal{W}}_{[0,\tau_1]}\Big(\mathrm{J}^\delta\big(\mathcal{A} + \Phi\big)(1)\Big)\tilde{\mathcal{W}}_{[\tau_1,\tau_2]}\Big(\mathrm{J}^\rho\big(\mathcal{A} + \Phi\big)(2)\Big)\tilde{\mathcal{W}}_{[\tau_2,1]}\,. \tag{116}$$

However, this expression has two problems: For one it is not gauge invariant, due to the presence of the non-covariant derivative term $\mathcal{D}\big(\xi^\kappa_\mu A^\mu\big)$ in the action of J^κ, c.f. eq. (106). This is very bad, as it implies that J^κ maps a gauge-invariant expression ($\mathrm{tr}\tilde{\mathcal{W}}$) to a gauge-variant one! Secondly, the resulting expression destroys cyclicity.

Luckily, both problems may be cured by using the gauge-covariant generators $\mathbb{J}^\kappa_*$ introduced in the previous section for the level-one generators. We thus define the gauge-covariant bilocal contribution to the Yangian level-one generators as

$$\widehat{\mathbb{J}}^\kappa_{*,\mathrm{bi}}\mathrm{tr}\tilde{\mathcal{W}} = \\ N f^\kappa{}_{\rho\delta}\mathrm{tr}\iint_{1<2} \tilde{\mathcal{W}}_{[0,\tau_1]}\Big(\mathbb{J}^\delta_*\big(\mathcal{A} + \Phi\big)(1)\Big)\tilde{\mathcal{W}}_{[\tau_1,\tau_2]}\Big(\mathbb{J}^\rho_*\big(\mathcal{A} + \Phi\big)(2)\Big)\tilde{\mathcal{W}}_{[\tau_2,1]}\,. \tag{117}$$

One also establishes the algebra relations of this bilocal generator

$$\left[\mathbb{J}^{\kappa}_{*}, \widehat{\mathbb{J}}^{\rho}_{*,\mathrm{bi}}\right]\mathrm{tr}\mathscr{W} = f^{\kappa\rho}{}_{\delta}\widehat{\mathbb{J}}^{\delta}_{*,\mathrm{bi}}\mathrm{tr}\mathscr{W} + f^{\rho}{}_{\sigma\omega}\mathbb{J}^{\sigma}_{*}\mathrm{tr}\left(\mathscr{W}\mathscr{G}^{\kappa\omega}(1) + \mathscr{G}^{\kappa\omega}(0)\mathscr{W}\right). \tag{118}$$

On the level of expectation values $\langle\cdots\rangle$ this closes.

The central question to be addressed now is whether $\widehat{\mathbb{J}}^{\delta}_{*,\mathrm{bi}}$ is a symmetry of the expectation value of the Wilson loop operator. For this one has at leading perturbative order

$$\begin{aligned}\left\langle\widehat{\mathbb{J}}^{\kappa}_{*,\mathrm{bi}}\left(\mathscr{A}(1)\mathscr{A}(2)\right)\right\rangle &= f^{\kappa}{}_{\rho\delta}\left\langle \mathrm{J}^{\delta}X_1^{\mathscr{A}}\,\mathscr{E}_1^{\mathscr{B}}\,\mathscr{F}_{1\mathscr{B}\mathscr{A}}\,\mathrm{J}^{\rho}X_2^{\mathscr{C}}\,\mathscr{E}_2^{\mathscr{D}}\,\mathscr{F}_{2\mathscr{D}\mathscr{C}}\right\rangle \\ &=: f^{\kappa}{}_{\rho\delta}\, i[\mathrm{J}_1^{\delta}]\, i[\mathrm{J}_2^{\rho}]\left\langle\mathscr{F}(1)\mathscr{F}(2)\right\rangle .\end{aligned} \tag{119}$$

Here $\mathscr{E}^{\mathscr{A}} = (e, d\theta, d\bar{\theta})$ and $i[J]\mathscr{F}$ denotes the contraction of $\mathrm{J}^{\delta}X^{\mathscr{A}}\,D_{\mathscr{A}}$ with the field strength two-form. In order to evaluate this, one may make use of the result (93) which expresses the field strength in terms of the scalar superfield. Hence the right-hand side of eq. (119) follows from the two-point function

$$\left\langle\bar{\Phi}^{ab}(1)\,\Phi_{cd}(2)\right\rangle = -\frac{g^2}{\pi^2}\frac{C^a{}_{[c}C^b{}_{d]}}{(x_{12}^{-+})^2}, \qquad C^a{}_c = (1 - 4\theta_{12}x_{12}^{-+,-1}\bar{\theta}_{12})^a{}_c, \tag{120}$$

which in fact is determined up to normalization by conformal symmetry. Here $x_{12}^{-+,-1}$ is the inverse of

$$x_{12}^{-+} = x_{12} + 2\bar{\theta}_{12}\theta_{12}. \tag{121}$$

One finds for the bilocal part of the level-one translation operator

$$\widehat{\mathbb{P}}_{*,\mathrm{bi},\alpha\dot{\alpha}} = \mathbb{P}_{*,\beta\dot{\alpha}} \wedge \mathbb{L}^{\beta}_{*\alpha} + \mathbb{P}_{*,\alpha\dot{\beta}} \wedge \bar{\mathbb{L}}^{\dot{\beta}}_{*\dot{\alpha}} + 2\mathbb{P}_{*,\alpha\dot{\alpha}} \wedge \mathbb{D}_{*} + \mathbb{Q}_{*,a\alpha} \wedge \bar{\mathbb{Q}}^{a}_{*\dot{\alpha}}. \tag{122}$$

the result

$$\left\langle\widehat{\mathbb{P}}_{*,\mathrm{bi},\alpha\dot{\alpha}}\left(\mathscr{A}(1)\,\mathscr{A}(2)\right)\right\rangle = \frac{2g^2}{\pi^2}\mathrm{d}_1\mathrm{d}_2R^{\mathrm{P}}_{12}, \qquad R^{\mathrm{P}}_{12} = x^{-+,-1}_{12,\alpha\dot{\alpha}} - x^{-+,-1}_{21,\alpha\dot{\alpha}}. \tag{123}$$

Where R^{P}_{12} is the remainder function. The other remainder functions corresponding to the further generators follow from the superconformal algebra. This then results in

$$\left\langle\widehat{\mathbb{J}}^{\kappa}_{*,\mathrm{bi}}\left(\mathscr{A}(1)\mathscr{A}(2)\right)\right\rangle = \frac{2g^2}{\pi^2}\,\mathrm{d}_1\mathrm{d}_2R^{\kappa}_{12}. \tag{124}$$

Using this result and the fact that Φ_{ab} may be related to the superfield strength via eq. (93) one deduces that $\langle \widehat{\mathbb{J}}^{\kappa}_{*,\mathrm{bi}}(\Phi(1)\,\Phi(2)) \rangle = 0$. Putting everything together we have shown that

$$\widehat{\mathbb{J}}^{\kappa}_{*,\mathrm{bi}}\langle \mathrm{tr}\tilde{\mathscr{W}} \rangle = \frac{N}{2} f^{\kappa}{}_{\rho\delta} \iint_{1<2} \langle \mathbb{J}^{\delta}_{*}(\mathscr{A}+\Phi)(1)\, \mathbb{J}^{\rho}_{*}(\mathscr{A}+\Phi)(2) \rangle$$
$$= \frac{\lambda}{\pi^2} \iint_{1<2} \mathrm{d}_1 \mathrm{d}_2 R^{\kappa}_{12} \,. \tag{125}$$

In order to evaluate this path-ordered integral we need to regularize the remainder function. Doing so with a point-splitting procedure yields an ε-expansion of the form

$$R^{\kappa}(\tau, \tau+\varepsilon) = \frac{1}{\varepsilon} R^{\kappa}_{-1}(\tau) + \tfrac{1}{2}\dot{R}^{\kappa}_{-1}(\tau) + R^{\kappa}_{1}(\tau)\varepsilon + \mathscr{O}(\varepsilon^2) \,. \tag{126}$$

We then find for eq. (125) the local terms

$$\iint_{\mathrm{reg}(\tau_1<\tau_2)} \mathrm{d}_1 \mathrm{d}_2 R^{\kappa}_{12} = \int_0^1 \mathrm{d}\tau \left(-\frac{1+\dot{\varepsilon}}{\varepsilon^2} R^{\kappa}_{-1}(\tau) + R^{\kappa}_{1}(\tau) \right) + \frac{4+2\dot{\varepsilon}}{\varepsilon} R^{\kappa}_{-1}(0) + \mathscr{O}(\varepsilon^2). \tag{127}$$

The resulting local term is then chosen to cancel the remainder function contribution:

$$(\widehat{\mathbb{J}}^{\kappa}_{*}\mathscr{A})_{\varepsilon} := -\frac{N}{2}\mathrm{d}\tau(\partial_2 R^{\kappa})(\tau - \bar{\varepsilon}, \tau) \tag{128}$$

Summarizing, the complete level-one action takes the form

$$(\widehat{\mathbb{J}}^{\kappa}_{*}\mathrm{tr}\tilde{\mathscr{W}})_{\varepsilon} = \mathrm{tr}\tilde{\mathscr{W}}\left[(\widehat{\mathbb{J}}^{\kappa}_{*}\mathscr{A})_{\varepsilon}\right] + f^{\kappa}{}_{\rho\delta}\mathrm{tr}\tilde{\mathscr{W}}\left[\mathbb{J}^{\delta}_{*}\mathscr{A}; \mathbb{J}^{\rho}_{*}\mathscr{A}\right]_{\varepsilon}$$
$$- N\big(R^{\kappa}(0,\varepsilon) - R^{\kappa}(0,-\bar{\varepsilon})\big)\mathrm{tr}\tilde{\mathscr{W}} \,, \tag{129}$$

and annihilates the vacuum expectation value of super Maldacena-Wilson loop operator. We note in closing that the form of the local terms in the level-one Yangian generators depends on the regulator.

5 Outlook

In this review we have exposed multiple aspects of Yangian symmetry in $\mathscr{N} = 4$ SYM that emerge both in scattering amplitudes and their deformations as well as in smooth and light-like Wilson loop expectation values.

This infinite-dimensional symmetry is a hallmark of integrability and gives hope to completely solve the theory. Importantly, it also allows us to employ techniques proper to integrable theories. In particular, as discussed in Section 3.2, a deformation of scattering amplitudes in terms of a spectral parameter was proposed in [55, 56]. Yangian invariants relevant for amplitudes have been systematically constructed in [59, 69] from an underlying spin chain description. Furthermore, very recently, it has been shown [57], that Yangian invariance emerges also in the amplituhedron formulation [9] of tree-level scattering amplitudes.

One of the main open questions concerns the implications and constraints from Yangian symmetry and integrability for loop-level scattering amplitudes. It is crucial to establish whether one can indeed use the spectral parameter to regulate infrared divergences. It would furthermore be very interesting to extend previous results found at tree level and explore whether the amplituhedron integrand is Yangian invariant at loop level.

As to the smooth super Maldacena-Wilson loop operators that we have reviewed in Section 4, the next pressing question is whether we may use the exposed Yangian symmetry to establish new (exact) results for their vacuum expectation values $\langle \mathrm{tr}\tilde{\mathcal{W}} \rangle$? What roles do the geometries of the super-loops play? What are the invariants under these transformations? And in consequence which objects are related by Yangian symmetry transformations?

We hope that the methodologies reported in this article will lead in the near future to a new tool for the practical computation of scattering amplitudes and Wilson loops, where one departs directly from Yangian invariance and integrability, as opposed to merely "observing" it. The gold standard here is the spectral problem, where integrability by now is the method of choice for high-order-perturbative, strong-coupling, and even non-perturbative results, vastly outdoing all "traditional" field theoretical methods. We would certainly like to repeat this success story for the quantities studied in this review!

Acknowledgments: The SFB 647 Space-Time-Matter was the key intellectual enterprise that enabled the work reported in this chapter. The SFB in fact was the crucial structural entity for first preserving and subsequently fostering research in Mathematical Physics at Humboldt University Berlin. All the papers of the present authors cited in this article were influenced or made possible via this SFB. We are very grateful to the DFG for providing these opportunities. In addition, we would like to thank all our collaborators in the various projects funded by the SFB 647 constituting the basis of this review: Niklas Beisert, James Drummond, Johannes Henn, Rouven Frassek, Nils Kanning, Yumi Ko, Tomasz Łukowski, Carlo Meneghelli, Dennis Müller, Hagen Münkler, Eric Ragoucy, Cristian Vergu. Incidentally, nearly all of them also benefitted directly or indirectly from the funding and supporting structures of the SFB 647.

Bibliography

[1] Alday LF, Gaiotto D, Maldacena J, Sever A, Vieira P. An operator product expansion for polygonal null Wilson Loops. JHEP 2011;1104:088 arxiv:1006.2788

[2] Alday LF, Henn JM, Plefka J, Schuster T. Scattering into the fifth dimension of N = 4 super Yang-Mills. JHEP 2010;1001:077 arxiv:0908.0684

[3] Alday LF, Maldacena JM. Gluon scattering amplitudes at strong coupling. JHEP 2007;0706:064 arxiv:0705.0303

[4] Alday LF, Maldacena J. Null polygonal Wilson loops and minimal surfaces in Anti-de-Sitter space. JHEP 2009;0911:082 arxiv:0904.0663

[5] Arkani-Hamed N, Bourjaily JL, Cachazo F, Caron-Huot S, Trnka J. The all-loop integrand for scattering amplitudes in planar N=4 SYM. JHEP 2011;1101:041 arxiv:1008.2958

[6] Arkani-Hamed N, Bourjaily JL, Cachazo F, Goncharov AB, Postnikov A, Trnka J. Scattering amplitudes and the positive Grassmannian. arxiv:1212.5605

[7] Arkani-Hamed N, Cachazo F, Cheung C. The Grassmannian origin of dual superconformal invariance. JHEP 2010;1003:036 arxiv:0909.0483

[8] Arkani-Hamed N, Cachazo F, Cheung C, Kaplan J. A duality for the S Matrix. JHEP 2010;1003:020 arxiv:0907.5418

[9] Arkani-Hamed N, Trnka J. The amplituhedron. JHEP 2014;1410:030 arxiv:1312.2007

[10] Bargheer T, Beisert N, Galleas W, Loebbert F, McLoughlin T. Exacting N = 4 Superconformal Symmetry. JHEP 2009;0911:056 arxiv:0905.3738

[11] Bargheer T, Huang Yt, Loebbert F, Yamazaki M. Integrable amplitude deformations for N=4 Super Yang-Mills and ABJM Theory. Phys Rev 2015;D91:026004 arxiv:1407.4449

[12] Beisert N, Broedel J, Rosso M. On Yangian-invariant regularisation of deformed on-shell diagrams in N = 4 super-Yang-Mills theory. arxiv:1401.7274

[13] Beisert N, Henn J, McLoughlin T, Plefka J. One-loop superconformal and Yangian symmetries of scattering amplitudes in N=4 Super Yang-Mills. JHEP 2010;1004:085 arxiv:1002.1733

[14] Beisert N, Kristjansen C, Staudacher M. The dilatation operator of N = 4 conformal super Yang-Mills theory. Nucl Phys 2003;B664:131 hep-th/0303060

[15] Beisert N, Muller D, Plefka J, Vergu C. Integrability of smooth Wilson loops in N = 4 superspace. JHEP 2015;1512:141 arxiv:1509.05403

[16] Beisert N, Muller D, Plefka J, Vergu C. Smooth Wilson loops in N = 4 non-chiral superspace. JHEP 2015;1512:140 arxiv:1506.07047

[17] Beisert N, Ricci R, Tseytlin AA, Wolf M. Dual superconformal symmetry from AdS5 _ S5 Superstring Integrability. Phys Rev 2008;D78:126004 arxiv:0807.3228

[18] Beisert N, Schwab BUW. Bonus Yangian symmetry for the planar S-Matrix of N=4 Super Yang-Mills. Phys Rev Lett 2011;106:231602 arxiv:1103.0646

[19] Beisert N, Staudacher M. Long-range PSU(2,2=4) Bethe Ansaetze for gauge theory and strings. Nucl Phys 2005;B727:1 hep-th/0504190

[20] Beisert N. T-duality, dual conformal symmetry and integrability for strings on AdS5 _ S5. Fortschr Phys 2009;57:329 arxiv:0903.0609

[21] Belitsky AV, Korchemsky GP, Sokatchev E. Are scattering amplitudes dual to super Wilson loops? Nucl Phys 2012;B855:333 arxiv:1103.3008

[22] Bena I, Polchinski J, Roiban R. Hidden symmetries of the AdS5 _ S5 superstring. Phys Rev 2004;D69:046002 hep-th/0305116

[23] Berkovits N, Maldacena J. Fermionic T-duality, dual superconformal symmetry, and the Amplitude/Wilson Loop Connection. JHEP 2008;0809:062 arxiv:0807.3196

[24] Bern Z, Carrasco JJ, Dennen T, Huang Yt, Ita H. Generalized unitarity and six-dimensional helicity. Phys Rev 2011;D83:085022 arxiv:1010.0494

[25] Bern Z, Dixon LJ, Dunbar DC, Kosower DA. One-loop n-point gauge theory amplitudes, Unitarity and collinear limits. Nucl Phys 1994;B425:217, hep-ph/9403226
[26] Bern Z, Dixon LJ, Smirnov VA. Iteration of planar amplitudes in maximally supersymmetric Yang-Mills theory at three loops and beyond. Phys Rev 2005;D72:085001 hep-th/0505205
[27] Bernard D. An Introduction to Yangian Symmetries. Int J Mod Phys 1993;B7:3517 hep-th/9211133, In: Advanced Research Workshop on Integrable Quantum Field Theories: Conformal Field Theories and Current Algebra, Integrable Models, 2D Quantum Gravity, Matrix Models and String Theory Como, Italy, September 14-19, 1992:3517-3530
[28] Brandhuber A, Heslop P, Travaglini G. A note on dual superconformal symmetry of the N = 4 super Yang-Mills S-matrix. Phys Rev 2008;D78:125005 arxiv:0807.4097
[29] Brandhuber A, Heslop P, Travaglini G. MHV amplitudes in N = 4 super Yang–Mills and Wilson loops. Nucl Phys 2008;B794:231 arxiv:0707.1153
[30] Broedel J, de Leeuw M, Rosso M. A dictionary between R-operators, on-shell graphs and Yangian algebras. JHEP 2014;1406:170 arxiv:1403.3670
[31] Caron-Huot S. Notes on the scattering amplitude / Wilson loop duality. JHEP 2011;1107:058 arxiv:1010.1167
[32] Caron-Huot S. Superconformal symmetry and two-loop amplitudes in planar N=4 super Yang-Mills. JHEP 2011;1112:066 arxiv:1105.5606
[33] Caron-Huot S, O'Connell D. Spinor helicity and dual conformal symmetry in ten dimensions. JHEP 2011;1108:014 arxiv:1010.5487
[34] Chari V, Pressley A. Fundamental representations of Yangians and singularities of R-matrices. J Ang Math 1991;417:87. Available at: http://eudml.org/doc/153325
[35] Chicherin D, Derkachov S, Kirschner R. Yang-Baxter operators and scattering amplitudes in N=4 super-Yang-Mills theory. Nucl Phys 2014;B881:467 arxiv:1309.5748
[36] Dennen T, Huang Yt. Dual conformal properties of six-dimensional maximal super Yang-Mills amplitudes. JHEP 2011;1101:140 arxiv:1010.5874
[37] Dolan L, Nappi CR, Witten E. Yangian symmetry in D =4 superconformal Yang-Mills theory. hep-th/0401243, In: Argyres PC et al, editor. Quantum theory and symmetries. Singapore: World Scientific, 2004
[38] Drinfel'd VG. Hopf algebras and the quantum Yang-Baxter equation. Sov Math Dokl 1985;32:254
[39] Drinfel'd VG. Quantum groups. J Math Sci 1988;41:898
[40] Drinfeld V. A new realization of Yangians and quantized affine algebras. Sov Math Dokl 1988;36:212
[41] Drukker N, Gross DJ. An exact prediction of N = 4 SUSYM theory for string theory. J Math Phys 2001;42:2896 hep-th/0010274
[42] Drummond J, Ferro L. Yangians, Grassmannians and T-duality. JHEP 2010;1007:027 arxiv:1001.3348
[43] Drummond JM, Ferro L. The Yangian origin of the Grassmannian integral. JHEP 2010;1012:010 arxiv:1002.4622
[44] Drummond JM, Ferro L, Ragoucy E. Yangian symmetry of light-like Wilson loops. JHEP 2011;1111:049 arxiv:1011.4264
[45] Drummond JM, Henn JM. All tree-level amplitudes in N = 4 SYM. JHEP 2009;0904:018 arxiv:0808.2475
[46] Drummond JM, Henn J, Korchemsky GP, Sokatchev E. On planar gluon amplitudes/Wilson loops duality. Nucl Phys 2008;B795:52 arxiv:0709.2368
[47] Drummond JM, Henn J, Korchemsky GP, Sokatchev E. The hexagon Wilson loop and the BDS ansatz for the six-gluon amplitude. Phys Lett 2008;B662:456 arxiv:0712.4138

[48] Drummond JM, Henn J, Korchemsky GP, Sokatchev E. Hexagon Wilson loop = six-gluon MHV amplitude. Nucl Phys 2009;B815:142 arxiv:0803.1466
[49] Drummond JM, Henn J, Korchemsky GP, Sokatchev E. Conformal ward identities for Wilson loops and a test of the duality with gluon amplitudes. Nucl Phys 2010;B826:337 arxiv:0712.1223
[50] Drummond JM, Henn J, Korchemsky GP, Sokatchev E. Dual superconformal symmetry of scattering amplitudes in N = 4 super-Yang–Mills theory. Nucl Phys 2010;B828:317 arxiv:0807.1095
[51] Drummond JM, Henn JM, Plefka J. Yangian symmetry of scattering amplitudes in N=4 super Yang-Mills theory. JHEP 2009;0905:046 arxiv:0902.2987
[52] Drummond JM, Korchemsky GP, Sokatchev E. Conformal properties of four-gluon planar amplitudes and Wilson loops. Nucl Phys 2008;B795:385 arxiv:0707.0243
[53] Elvang H, Huang Yt, Keeler C, Lam T, Olson TM, Roland SB, Speyer DE. Grassmannians for scattering amplitudes in 4d N = 4 SYM and 3d ABJM. JHEP 2014;1412:181 arxiv:1410.0621
[54] Faddeev LD. How algebraic Bethe Ansatz works for integrable model. hep-th/9605187, In: Marck JA, Lasota JP, editor. Relativistic gravitation and gravitational radiation. Cambridge: Cambridge University Press, 1997
[55] Ferro L, Lukowski T, Meneghelli C, Plefka J, Staudacher M. Harmonic R-matrices for scattering amplitudes and spectral regularization. Phys Rev Lett 2013;110:121602 arxiv:1212.0850
[56] Ferro L, Lukowski T, Meneghelli C, Plefka J, Staudacher M. Spectral parameters for scattering amplitudes in N=4 Super Yang-Mills theory. JHEP 2014;1401:094 arxiv:1308.3494
[57] Ferro L, Lukowski T, Orta A, Parisi M. Yangian symmetry for the tree amplituhedron. J Phys A 50, no. 29, 294005 (2017) arxiv:1612.04378
[58] Ferro L, Lukowski T, Staudacher M. N = 4 scattering amplitudes and the deformed Grassmannian. Nucl Phys 2014;B889:192 arxiv:1407.6736
[59] Frassek R, Kanning N, Ko Y, Staudacher M. Bethe Ansatz for Yangian Invariants: Towards super Yang-Mills scattering amplitudes. Nucl Phys 2014;B883:373 arxiv:1312.1693
[60] Gaiotto D, Maldacena J, Sever A, Vieira P. Bootstrapping null polygon Wilson Loops. JHEP 2011;1103:092 arxiv:1010.5009
[61] Harnad JP, Shnider S. Constraints and field equations for ten-dimensional super Yang-Mills theory. Commun Math Phys 1986;106:183
[62] Henn JM, Naculich SG, Schnitzer HJ, Spradlin M. Higgs-regularized three-loop four-gluon amplitude in N=4 SYM: Exponentiation and Regge limits. JHEP 2010;1004:038 arxiv:1001.1358
[63] Henn JM, Naculich SG, Schnitzer HJ, Spradlin M. More loops and legs in Higgs-regulated N=4 SYM amplitudes. JHEP 2010;1008:002 arxiv:1004.5381
[64] Heslop P, Khoze VV. Analytic results for MHV Wilson Loops. JHEP 2010;1011:035 arxiv:1007.1805
[65] Hodges A. Eliminating spurious poles from gauge-theoretic amplitudes. JHEP 2013;1305:135 arxiv:0905.1473
[66] Ivanov SV, Korchemsky GP. Some supplements of nonperturbative gauges. Phys Lett 1985;B154:197
[67] Ivanov SV, Korchemsky GP, Radyushkin AV. Infrared asymptotics of perturbative QCD: Contour gauges. Yad Fiz 1986;44:230 [Sov. J. Nucl. Phys.44,145(1986)]
[68] Kanning N, Ko Y, Staudacher M. Graßmannian integrals as matrix models for non-compact Yangian invariants. Nucl Phys 2015;B894:407 arxiv:1412.8476
[69] Kanning N, Lukowski T, Staudacher M. A shortcut to general tree-level scattering amplitudes in N = 4 SYM via integrability. Fortsch Phys 2014;62:556 arxiv:1403.3382
[70] Bourjaily J, Kanning N, Richter G, Staudacher M. Work in progress.

[71] Korchemsky GP, Radyushkin AV. Loop space formalism and renormalization group for the infrared asymptotics of QCD. Phys Lett 1986;B171:459
[72] Korchemsky G, Sokatchev E. Superconformal invariants for scattering amplitudes in N=4 SYM theory. Nucl Phys 2010;B839:377 arxiv:1002.4625
[73] Korchemsky GP, Sokatchev E. Twistor transform of all tree amplitudes in N=4 SYM theory. Nucl Phys 2010;B829:478 arxiv:0907.4107
[74] Korchemsky GP, Sokatchev E. Symmetries and analytic properties of scattering amplitudes in N = 4 SYM theory. arxiv:0906.1737
[75] MacKay N. Introduction to Yangian symmetry in integrable field theory. Int J Mod Phys 2005;A20:7189 hep-th/0409183
[76] Maldacena JM. Wilson loops in large N field theories. Phys Rev Lett 1998;80:4859 hep-th/9803002
[77] Mason L, Skinner D. Dual superconformal invariance, momentum twistors and Grassmannians. JHEP 2009;0911:045 arxiv:0909.0250
[78] Mason L, Skinner D. The complete planar S-matrix of N=4 SYM as a Wilson Loop in Twistor Space. JHEP 2010;1012:018 arxiv:1009.2225
[79] Minahan JA, Zarembo K. The Bethe-ansatz for N = 4 super Yang-Mills. JHEP 2003;0303:013 hep-th/0212208
[80] Muller D, Munkler H, Plefka J, Pollok J, Zarembo K. Yangian symmetry of smooth Wilson loops in N = 4 super Yang-Mills Theory. JHEP 2013;1311:081 arxiv:1309.1676
[81] Ooguri H, Rahmfeld J, Robins H, Tannenhauser J. Holography in superspace. JHEP 2000;0007:045 hep-th/0007104
[82] Sever A, Vieira P. Symmetries of the N = 4 SYM S-matrix. arxiv:0908.2437
[83] Vergu C. Twistor parametrization of locally BPS super-Wilson loops. arxiv:1605.03165
[84] Wilson KG, Confinement of quarks. Phys Rev 1974;D10:2445, [45(1974)]
[85] Witten E. Perturbative gauge theory as a string theory in twistor space. Commun Math Phys 2004;252:189 hep-th/0312171
[86] Witten E. Twistor - like transform in ten-dimensions. Nucl Phys 1986;B266:245

Lars Andersson and Christian Bär

Wave and Dirac equations on manifolds

Abstract: We review some recent results on geometric equations on Lorentzian manifolds such as the wave and Dirac equations. This includes well-posedness and stability for various initial value problems, as well as results on the structure of these equations on black hole spacetimes (in particular, on the Kerr solution), the index theorem for hyperbolic Dirac operators and properties of the class of Green-hyperbolic operators.

Keywords: wave equation, Dirac equation, globally hyperbolic Lorentzian manifold, Cauchy problem, Goursat problem, black hole spacetime, Kerr solution, Killing spinor, index theorem, chiral anomaly, Green-hyperbolic operator

Mathematics Subject Classification 2010: 53C27, 53C50, 83C57, 83C60, 58J45.

Introduction

Spacetime in general relativity is modeled by a Lorentzian manifold. The physically relevant field equations are geometric partial differential equations on these manifolds. The most prominent examples are wave and Dirac equations. This chapter surveys some progress in the study of these geometric equations obtained in the past years within the research network SFB 647.

Wave and Dirac equations are hyperbolic partial differential equations (PDEs) and the other equations under consideration share some aspects of hyperbolicity as well. In order to obtain a reasonable analytic solution theory, we have to exclude causal pathologies like closed timelike loops in the underlying manifold. The right class of Lorentzian manifolds turns out to be that of globally hyperbolic ones. In the first section we quickly review their most important properties.

Considering equations of hyperbolic type on globally hyperbolic spacetimes leads to the study of initial value problems. The second section provides a discussion of the Cauchy problem for wave equations and for Dirac equations, the Goursat problem (i.e. the characteristic initial value problem) for wave equations and the Cauchy problem for the vacuum Einstein equation. From a physical point of view, well-posedness of these initial value problems is essential for predictability.

The singularity theorems of Hawking and Penrose indicate that vacuum spacetimes with strong gravitational fields will typically be singular in the sense that they have a nontrivial Cauchy horizon. According to the weak cosmic censorship conjecture such singularities in isolated systems must be hidden behind event horizons. Such models are known as black hole solutions. The most prominent examples are the Schwarzschild solution describing a static black hole and the Kerr solutions

DOI 10.1515/9783110452150-007_s_001

describing rotating black holes (in the subextreme case). The form of the spacetime metric of these models is known explicitly.

The Kerr black hole solution is conjectured to be the unique stationary, asymptotically flat, vacuum spacetime containing a non-degenerate black hole. In addition to being unique, the Kerr spacetime is conjectured to be dynamically stable, in the sense that the maximal Cauchy development of any sufficiently small perturbation of subextreme Kerr Cauchy data is asymptotic to the future, to a subextreme member of the Kerr family.

We may refer to the above conjectures as the *black hole uniqueness* and *black hole stability* conjectures. Both of these conjectures, which are of central importance in general relativity and astrophysics are open in full generality, in spite of much work on the topic during the last 50 years.

The importance of the Kerr solution, and an analysis of its properties, is explained by the fact that according to our current understanding, black holes are ubiquitous in the universe, in particular most galaxies have a supermassive black hole at their center, and these play an important role in the life of the galaxy. Also our galaxy has at its center a very compact object, Sagittarius A*, with a diameter of less than one astronomical unit, and a mass estimated to be $10^6\ m_{\odot}$. Evidence for this includes observations of the orbits of stars in its vicinity.

In Section 3, we provide background on massless field equations on black hole spacetimes and on the geometry of the Kerr solution. The subsequent section contains new results on integrated local energy decay for solutions of massless field equations on Kerr spacetimes. Moreover, higher-order conservation laws for Maxwell equations are derived and integrability conditions on the linearized Einstein equations on Kerr spacetime are established. All of this relies on the existence of Killing-Yano forms on Kerr spacetimes.

In Section 5, we describe a recently found analogue to the Atiyah-Patodi-Singer (APS) index theorem on globally hyperbolic Lorentzian spacetimes. In contrast to the original Riemann version of this index theorem, the spectral boundary conditions admit a natural physical interpretation in the Lorentzian case in terms of a particle-antiparticle splitting. We indicate how the Lorentzian index theorem can be applied to compute the chiral anomaly in quantum field theory. The chiral anomaly is concerned with the fact that particles may be created by gravity or an external field without also creating their corresponding antiparticles.

Well-posedness of the Cauchy problem implies the existence of Green's operators which are solution operators satisfying certain causality conditions. In the last section, we turn things around and deal with those differential operators which possess Green's operators even if they do not have a well-posed Cauchy problem. This class is very large; it is closed under many natural operations producing new operators out of given ones. Yet much of the analytic theory is shown to carry over to these equations. This is important for quantizing these fields in the framework of algebraic quantum field theory on curved backgrounds.

1 Globally hyperbolic Lorentzian manifolds

We describe the class of Lorentzian manifolds on which wave equations have reasonable analytic properties. For a general introduction to Lorentzian geometry see one of the standard textbooks such as [16, 22, 24].

Throughout this section let M denote a connected time-oriented Lorentzian manifold. For $A \subset M$ let $J^+(A)$ be the *causal future* of A, i.e. the set of all points of M which can be reached by future-directed causal curves starting in A. Similarly, one defines the *causal past* $J^-(A)$. We also write $J(A) = J^+(A) \cup J^-(A)$. A subset $A \subset M$ is called *past compact* if $A \cap J^-(p)$ is compact for all $p \in M$. Similarly, one defines *future compact* subsets.

A subset $S \subset M$ is called a *Cauchy hypersurface* if every inextendible timelike curve meets S exactly once. There is no a priori regularity assumption on Cauchy hypersurfaces but it turns out that they are Lipschitz hypersurfaces, i.e. can locally be written as graphs of a Lipschitz function. Moreover, any Cauchy hypersurface is a closed subset of M, every inextendible causal curve meets S (possibly more often than once) and any two Cauchy hypersurfaces of M are homeomorphic.

We say that M satisfies the *causality condition* if M does not contain a closed causal loop.

Theorem 1 (Bernal, Sánchez). *The following are equivalent:*
(1) *There exists a Cauchy hypersurface in M;*
(2) *M satisfies the causality condition and the intersection $J^+(p) \cap J^-(p)$ is compact for all $p \in M$;*
(3) *The manifold M is isometric to $\mathbb{R} \times S$ with metric $-N^2dt + g_t$, where N is a smooth positive function (the* lapse function*), g_t is a Riemannian metric on S depending smoothly on the parameter t and all sets $\{t_0\} \times S$ are Cauchy hypersurfaces in M.*

The implication (3)⇒(1) is trivial and the implication (1)⇒(2) has been known for a long time, see e.g. [24, Cor. 39, p. 422]. The implication (2)⇒(3) is a remarkable structural result due to A. Bernal and M. Sánchez (combine [17, Thm. 1.1] with [18, Thm. 3.2]).

Manifolds satisfying the conditions in Theorem 1 are called *globally hyperbolic*. Many important relativistic models are globally hyperbolic, such as Minkowski spacetime, Robertson-Walker spacetimes (in particular, Friedman cosmological models), the interior and the exterior Schwarschild models, deSitter spacetime and many more.

2 Initial value problems

We now describe some analytical facts on initial value problems for linear wave equations.

2.1 Cauchy problem for linear wave equations

Throughout this section let M be a globally hyperbolic Lorentzian manifold and let $E \to M$ be a vector bundle. A linear differential operator P of second order with smooth coefficients acting on sections of E is called *normally hyperbolic* if its principal symbol is given by the Lorentzian metric, i.e. in local coordinates P takes the form

$$P = -\sum_{i,j=1}^{n} g^{ij}(x)\frac{\partial^2}{\partial x^i \partial x^j} + \sum_{j=1}^{n} A_j(x)\frac{\partial}{\partial x^j} + B(x).$$

A linear wave equation is an equation of the form $Pu = f$ with given f and an unknown section u. By the *Cauchy problem* we mean the problem of solving such a wave equation while imposing initial value conditions of zeroth and first order. More precisely, let $S \subset M$ be a smooth spacelike Cauchy hypersurface and let $\mathfrak{n}$ be the future-directed unit normal vector field along S. We fix a connection ∇ on E. Now the Cauchy problem consists of finding a section u such that

$$\begin{cases} Pu = f & \text{on } M, \\ u = u_0 & \text{along } S, \\ \nabla_{\mathfrak{n}} u = u_1 & \text{along } S, \end{cases} \tag{1}$$

for given f, u_0, and u_1.

Theorem 2. *Under the assumptions of this section the following holds:*

(i) (Existence and uniqueness) *For each $u_0, u_1 \in C_c^\infty(S, E)$ and for each $f \in C_c^\infty(M, E)$ there exists a unique $u \in C^\infty(M, E)$ solving eq.* (1).

(ii) (Stability) *The solution u depends continuously on the data f, u_0, and u_1.*

(iii) (Finite propagation speed) *The solution u satisfies* $\operatorname{supp}(u) \subset J(K)$ *where* $K = \operatorname{supp}(u_0) \cup \operatorname{supp}(u_1) \cup \operatorname{supp}(f)$.

A proof can be found in [11, Thms. 3.2.11 and 3.2.12]. Assertions (i) and (ii) are often combined by stating that the Cauchy problem is *well posed*.

This theorem also holds for data of lower regularity; we can, for instance, impose Sobolev regularity. More precisely, the theorem holds for $u_0 \in H_c^k(S, E)$, $u_1 \in H_c^{k-1}(S, E)$, and f being locally L^2 in time, H^{k-1} in space and having spatially compact support, $k \in \mathbb{R}$. The solution u will then be in a suitable space of finite-energy sections, see [14, Thm. 8] for details.

2.2 Goursat problem for linear wave equations

There is an interesting borderline case between initial value problems and boundary value problems, namely the *characteristic initial value problem* also known as the *Goursat problem*. Here one replaces the smooth spacelike Cauchy hypersurface along which the initial conditions have been required in the Cauchy problem by a partial Cauchy hypersurface[1] S which is characteristic, meaning that the induced metric on S is degenerate. Typical examples are of the form $S = \partial J^+(A)$ for any nonempty subset $A \subset M$. It turns out that one can prescribe the initial values along S but one cannot impose any condition on the derivative.

Theorem 3 (Bär, Tagne Wafo [14, Thm. 23]). *Let $S \subset M$ be a characteristic partial Cauchy hypersurface. Assume that $J^+(S)$ is past compact.*

Then for any locally L^2-section f on M with spatially compact support and any $u_0 \in H^1_c(S, E)$ there exists a section u on M in a suitable finite-energy space such that $Pu = f$ on $J^+(S)$ and $u|_S = u_0$. On $J^+(S)$, u is unique.

Examples show easily that the assumption on $J^+(S)$ being past compact cannot be dropped.

2.3 Cauchy problem for Dirac equations

Theorem 2 allows us to derive well-posedness also for the Cauchy problem for Dirac equations. A linear differential operator D of first order acting on sections of E is called a *Dirac-type operator* if $P = D^2$ is normally hyperbolic. Examples are the classical Dirac operators acting on sections of the spinor bundle $E = \Sigma M$ (see [15] for details) or, more generally, on sections of a twisted spinor bundle $E = \Sigma M \otimes F$ where F is any "coefficient bundle" equipped with a connection.

Particular examples are the *Euler operator* $P = i(d - \delta)$ on $E = \bigoplus_k \Lambda^k T^*M$ and, in dimension $\dim(M) = 4$, the *Buchdahl operators* on $\Sigma M \otimes \Sigma_+^{\odot k} M$, see [10, Sec. 2.5] for details. The following is the basic well-posedness property for the Cauchy problem for Dirac operators.

Theorem 4. *Let D be a Dirac-type operator acting on sections of a vector bundle E over a globally hyperbolic manifold M. Let $S \subset M$ be a smooth spacelike Cauchy hypersurface. Then*

(i) (Existence and uniqueness) *For each $u_0 \in C_c^\infty(S, E)$ and for each $f \in C_c^\infty(M, E)$ there exists a unique $u \in C^\infty(M, E)$ solving $Du = f$ on M and $u = u_0$ along S.*

(ii) (Stability) *The solution u depends continuously on the data f and u_0.*

(iii) (Finite propagation speed) *The solution u satisfies* $\mathrm{supp}(u) \subset J(K)$ *where* $K = \mathrm{supp}(u_0) \cup \mathrm{supp}(f)$.

Proof. Along S we can write $D = -\sigma\nabla_{\mathfrak{n}} + Q$ where the coefficient field σ is invertible because $\mathfrak{n}$ is timelike and hence non-characteristic (σ is Clifford multiplication with $\mathfrak{n}$)

1 A *partial Cauchy hypersurface* is a closed Lipschitz hypersurface which is met by each timelike curve at most once.

and Q is a differential operator involving only derivatives tangential to S. Note that if a section v vanishes along S then Qv vanishes along S as well.

To show existence of solutions let $u_0 \in C_c^\infty(S, E)$ and $f \in C_c^\infty(M, E)$. We apply Theorem 2 with $P = D^2$ and solve $D^2 v = f$ on M with $v|_S = 0$ and $\nabla_n v = -\sigma^{-1} u_0$ along S. Now we put $u := Dv$. Then $Du = D^2 v = f$ and along S we have

$$u = Dv = -\sigma \nabla_n v + Qv = -\sigma \nabla_n v = u_0$$

as required.

As to uniqueness, let $Du = 0$ on M with $u = 0$ along S. Then, clearly, $D^2 u = 0$ on M and along S we have

$$\nabla_n u = -\sigma^{-1}(Du - Qu) = -\sigma^{-1}(0 - 0) = 0.$$

Thus, by Theorem 2, $u = 0$. This proves assertion (i).

The other two assertions follow for $u = Dv$ directly from the corresponding statements for v in Theorem 2. □

Again, it is possible to treat data with less regularity. If $u_0 \in H_c^k(S, E)$ and f is locally L^2 in time and H^k in space, then the solution u will be continuous in time and H^k in space.

2.4 Cauchy problem for the Einstein equation

The vacuum Einstein equation

$$\mathrm{Ric} = 0 \tag{2}$$

is the field equation of general relativity. Let S be a smooth spacelike Cauchy hypersurface in M. Then the induced Riemannian metric h on S and the second fundamental form k must satisfy the *constraint equations*

$$\mathrm{scal}_h + (\mathrm{tr}_h k)^2 - |k|^2 = 0, \tag{3a}$$

$$\mathrm{div}(k) - d(\mathrm{tr}_h k) = 0. \tag{3b}$$

A 3-manifold S with fields (h, k) solving the vacuum constraint equations (3) is called a *vacuum Cauchy data set*. A vacuum spacetime (M, g) containing a Cauchy surface S with induced data (h, k) is called a *Cauchy development* of (S, h, k). The vacuum Cauchy developments of (S, h, k) can be partially ordered by inclusion and, by Zorn's Lemma, there is thus a *maximal* element of the set of Cauchy developments.

In a suitable gauge eq. (2) becomes a system of nonlinear wave equations. Constraint and gauge conditions can be shown to propagate. This allows one to solve the Cauchy problem for eq. (2) locally. The fact that the field equations are covariant, together with the local uniqueness, implies global uniqueness up to isometry.

Theorem 5 (Choquet-Bruhat, Geroch [20, Thm. 3]). *Let (S, h, k) be a vacuum Cauchy data set. Then there is a unique, up to isometry, maximal vacuum Cauchy development (M, g) of (S, h, k).*

3 Waves on black hole spacetimes

3.1 Spin geometry in 3 + 1 dimensions

An orientable, globally hyperbolic 3+1-dimensional spacetime is spin; so for our purposes, we may assume without loss of generality that M is spin and fix a spin structure $P_{\mathrm{Spin}} \to M$. The spin group is $\mathrm{Spin}(1,3) = \mathrm{SL}(2,\mathbb{C})$, the universal covering of the Lorentz group. It has the inequivalent spinor representations $\mathbb{C}^2$ and $\bar{\mathbb{C}}^2$. The identification $\mathbb{C} \otimes \mathbb{R}^4 \equiv \mathbb{C}^2 \otimes \bar{\mathbb{C}}^2$ provides a correspondence between tensors and spinors[2]. The irreducible representations of $\mathrm{Spin}(1,3)$ are the spaces of symmetric spinors of valence (k, l), where $k, l \in \mathbb{N}_0$. We write $S_{k,l} := (\mathbb{C}^2)^{\odot k} \otimes (\bar{\mathbb{C}}^2)^{\odot l}$ to denote the space of symmetric spinors of valence (k, l). Moreover, we use the convention $S_{k,l} = 0$ if $k < 0$ or $l < 0$.

The associated vector bundles will be denoted by

$$S_{k,l}M := P_{\mathrm{Spin}} \times_{\mathrm{Spin}(1,3)} S_{k,l}$$

and the space of sections of $S_{k,l}M$ by $\mathscr{S}_{k,l}$.

If $k+l$ is even, then sections of $S_{k,l}M$ can be identified with tensor fields. Examples are given by vector fields (or 1-forms), traceless symmetric tensor fields of rank 2, self-dual differential 2-forms and tensor fields with the symmetries and trace properties of the Weyl tensor. They correspond to elements of $\mathscr{S}_{k,l}$ with $(k, l) = (1, 1), (2, 2), (2, 0)$, and $(4, 0)$, respectively.

Decomposing spinor and tensor expressions into irreducible pieces yields a powerful tool for simplification and canonicalization. The tensor product decomposition

$$S_{1,1} \otimes S_{k,l} = S_{k-1,l-1} \oplus S_{k+1,l-1} \oplus S_{k-1,l+1} \oplus S_{k+1,l+1} \tag{4}$$

yields four *fundamental operators* [4]. Composing the covariant derivative $\nabla : C^\infty(M, S_{k,l}M) \to C^\infty(M, T^*M \otimes S_{k,l}M) = C^\infty(M, S_{1,1}M \otimes S_{k,l}M)$ with the corresponding projections gives

2 Here we use the term spinor to refer to spin-tensors of general valence.

$$\mathscr{D}_{k,l} : \mathscr{S}_{k,l} \to \mathscr{S}_{k-1,l-1}, \tag{5a}$$

$$\mathscr{C}_{k,l} : \mathscr{S}_{k,l} \to \mathscr{S}_{k+1,l-1}, \tag{5b}$$

$$\mathscr{C}^\dagger_{k,l} : \mathscr{S}_{k,l} \to \mathscr{S}_{k-1,l+1}, \tag{5c}$$

$$\mathscr{T}_{k,l} : \mathscr{S}_{k,l} \to \mathscr{S}_{k+1,l+1}\,. \tag{5d}$$

The operators are called the *divergence*, *curl*, *curl-dagger*, and *twistor operators*, respectively. Unless otherwise stated, we shall only consider symmetric spinors in the following.

The equations for a massless field of spin $\mathfrak{s} \in \frac{1}{2}\mathbb{N}$ are

$$\mathscr{C}^\dagger_{2\mathfrak{s},0}\phi = 0, \quad \phi \in \mathscr{S}_{2\mathfrak{s},0} \quad \text{(left handed fields)}, \tag{6a}$$

$$\mathscr{C}_{0,2\mathfrak{s}}\varphi = 0, \quad \varphi \in \mathscr{S}_{0,2\mathfrak{s}} \quad \text{(right handed fields)}. \tag{6b}$$

For $\mathfrak{s} = 1/2$ equation (6a) is the Dirac equation for massless left-handed spinors, also known as Dirac-Weyl equation, and in case $\mathfrak{s} = 1$ (6) are the left and right Maxwell equations. The Maxwell equation for a real field strength F_{ab} is equivalent to either of the equations (6).

The irreducible components of the Riemann curvature tensor are the scalar curvature, the traceless Ricci tensor, and the Weyl tensor. The Weyl tensor corresponds to the Weyl spinor $\Psi \in \mathscr{S}_{4,0}$. The vacuum Einstein equations imply that the Weyl spinor satisfies the spin-2, or Bianchi, equation[3]

$$\mathscr{C}^\dagger_{4,0}\Psi = 0. \tag{7}$$

For $\mathfrak{s} \geq 3/2$, the existence of a nontrivial solution to the spin-$\mathfrak{s}$ equation implies curvature conditions, a fact known as the Buchdahl constraint [19]. For example, the spin-2 field on a general background is proportional to the Weyl spinor on the background.

A spinor $\varkappa$ of valence (k, l) satisfying

$$\mathscr{T}_{k,l}\varkappa = 0 \tag{8}$$

is called a *Killing spinor*. Killing spinors of valence $(1, 1)$, $(2, 0)$, and $(2, 2)$ correspond to conformal Killing fields, conformal Killing-Yano 2-forms, and traceless conformal Killing tensors of rank 2, respectively. Recall that a conformal Killing-Yano 2-form is a tensor $Y_{ab} = Y_{[ab]}$, satisfying

$$\nabla_{(a}Y_{b)c} = -\tfrac{1}{3}g_{ab}\nabla_d Y_c{}^d + \tfrac{1}{3}g_{(a|c|}\nabla^d Y_{b)d}. \tag{9}$$

Similarly, a Killing tensor is a tensor field $K_{ab\cdots d} = K_{(ab\cdots d)}$ satisfying $\nabla_{(a}K_{bc\cdots d)} = 0$.

3 One conventionally chooses the left-handed Weyl spinor $\Psi \in \mathscr{S}_{4,0}$ to represent the real Weyl tensor.

The Petrov classification gives a classification of spacetimes according to the algebraic type of the Weyl spinor. The following result is a consequence of the Bianchi identity.

Theorem 6 (Walker, Penrose [26, Lemma 1]). *Assume (M, g) is a vacuum spacetime of Petrov type D. Then (M, g) admits a one-dimensional space of Killing spinors of valence* $(2, 0)$.

As mentioned above, a Killing spinor of valence $(2, 0)$ is equivalent to a real conformal Killing-Yano 2-form. The existence of a Killing spinor implies symmetry operators for the massless field equations, as well as conservation laws. By a *symmetry operator*, we mean an operator which takes solutions to solutions.

Given a Killing spinor $\kappa \in \mathscr{S}_{2,0}$ we may define a complex scalar field κ_1 by choosing κ_1^2 to be proportional to the $\mathscr{S}_{0,0}$ term in the expansion of $\kappa \otimes \kappa \in \mathscr{S}_{2,0} \otimes \mathscr{S}_{2,0}$ into irreducible components, analogous to (4), and choosing the principal root to define κ_1. If (M, g) is of Petrov type D, then κ_1 defined in this manner is nonzero.

Let

$$U_a = -\nabla_a \log(\kappa_1), \tag{10}$$

and define, using the same procedure as discussed above, the extended fundamental operators $\mathscr{D}_{k,l,m,n}$, $\mathscr{C}_{k,l,m,n}$, $\mathscr{C}^{\dagger}_{k,l,m,n}$, $\mathscr{T}_{k,l,m,n}$ by projections of $\nabla_a + mU_a + n\bar{U}_a$; see [1, §2.2] for details. Here and below, $\bar{\ }$ denotes complex conjugation.

Finally, given a Killing spinor $\kappa \in \mathscr{S}_{2,0}$ in a Petrov D spacetime, we define for $\varphi \in \mathscr{S}_{k,l}$, the operators $\varphi \mapsto \mathscr{K}^{j}_{k,l}\varphi$, $j = 0, 1, 2$ as the $\mathscr{S}_{k+2-2j,l}$ components in the irreducible decomposition of $\kappa_1^{-1}\kappa \otimes \varphi$. Here we restrict to parameters such that $k + 2 - 2j \geq 0$. The operators $\bar{\mathscr{K}}^{j}_{k,l}$ are defined analogously in terms of projections of $\bar{\kappa}_1^{-1}\bar{\kappa} \otimes \varphi$.

We emphasize that symbolic computations have played an essential role in the results on fields with non-zero spin presented here, in particular in Sections 4.2, 4.3. The packages *SymManipulator* [8] and *SpinFrames* [2], developed for the Mathematica-based symbolic differential geometry suite *xAct* [23] makes it possible to systematically exploit decompositions in terms of irreducible representations of the spin group Spin(1, 3), and allows one to carry out investigations that are not feasible by hand.

3.2 Geometry of black hole spacetimes

A spacetime (M, g) is *asymptotically flat at null and spatial infinity* if there is a spacetime $(\tilde{M}, \tilde{g})$ which is C^∞ except possibly at a point i^0 called *spatial infinity*, and a conformal diffeomorphism $\psi : M \to \psi(M) \subset \tilde{M}$,

$$\tilde{g} = \Omega^2 \psi^* g, \quad \text{in } \psi(M)$$

with conformal factor Ω, such that

$$\overline{J^+(i^0)} \cup \overline{J^-(i^0)} = \tilde{M} - M$$

Here we have, for simplicity, identified M with $\psi(M)$. The boundaries of the causal future and past of i^0 are called null infinity and denoted by $\mathscr{I}^\pm$,

$$\mathscr{I}^\pm = \partial J^\pm(i^0) \setminus i^0 .$$

We further require that $\Omega = 0$ and $\tilde{\nabla}\Omega \neq 0$ on $\mathscr{I}^\pm$; see [25, §11.1] for details. If in addition, there is a neighborhood $\tilde{V} \subset \tilde{M}$ of $\overline{M \cap J^-(\mathscr{I}^+)}$ which is globally hyperbolic, then M is said to be *strongly asymptotically predictable.*

Given the notions introduced above, we say that (M, g) is a *black hole spacetime* if it is asymptotically flat at null and spatial infinity, strongly asymptotically predicable, and such that $M \setminus J^-(\mathscr{I}^+) \neq \emptyset$. In this case, $H = \partial J^-(\mathscr{I}^+) \cap M$ is called the *event horizon,* and $O = I^+(\mathscr{I}^-) \cap I^-(\mathscr{I}^+)$ is called the *domain of outer communication* (DOC).

An example of these notions is provided by the Schwarzschild spacetime, i.e. the unique static, spherically symmetric black hole spacetime satisfying the Einstein vacuum equations

$$\mathrm{Ric} = 0.$$

In Schwarzschild coordinates (t, r, θ, ϕ), the Schwarzschild metric takes the form

$$g = -f dt^2 + f^{-1} dr^2 + r^2 d\Omega^2_{S^2} \tag{11}$$

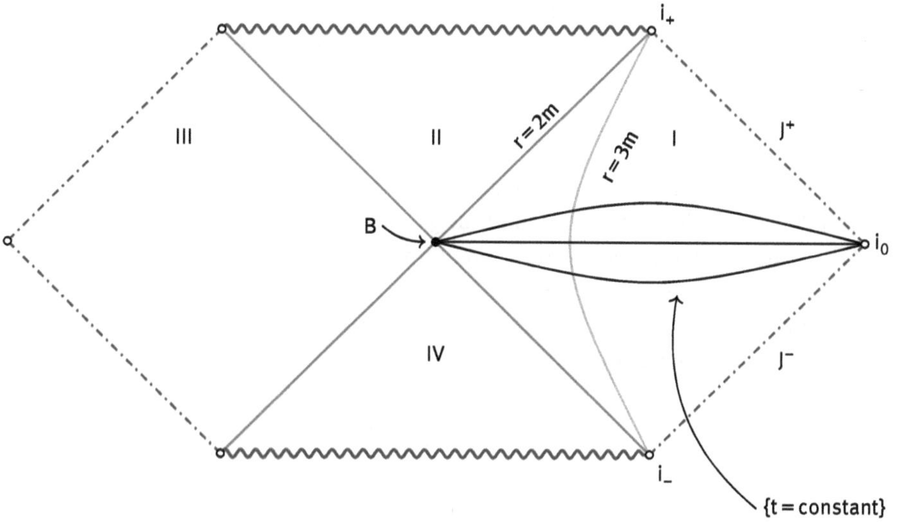

Figure 1: The maximal extension of the Schwarzschild spacetime. Region I is the domain of outer communication.

with $f = 1 - 2m/r$. Here $d\Omega^2_{S^2} = d\theta^2 + \sin^2\theta d\phi^2$ is the standard metric on the unit 2-sphere. An explicit construction of the conformal diffeomorphism ψ and conformal factor Ω satisfying the above conditions for the Schwarzschild spacetime is provided by the Kruskal-Szekeres extension. Referring to Figure 1, the DOC is represented by region I. The parameter m is the ADM mass of the Schwarzschild black hole.

3.3 The Kerr solution

The Kerr family of vacuum, rotating black hole spacetimes, provides the most important example for our purposes. In Boyer-Lindquist coordinates (t, r, θ, ϕ), the Kerr metric takes the form

$$\begin{aligned} g = &\left(-1 + \frac{2mr}{\Sigma}\right) dt^2 - \frac{4mra\sin^2\theta}{\Sigma} dt d\phi \\ &+ \frac{\Pi \sin^2\theta}{\Sigma} d\phi^2 + \frac{\Sigma}{\Delta} dr^2 + \Sigma d\theta^2, \end{aligned} \tag{12}$$

where

$$\begin{aligned} \Sigma &= r^2 + a^2\cos^2\theta, \\ \Delta &= r^2 - 2mr + a^2, \\ \Pi &= (r^2 + a^2)^2 - \Delta a^2 \sin^2\theta. \end{aligned}$$

The Kerr spacetime admits two Killing vector fields, the stationary Killing field ∂_t which is timelike near infinity, and the axial field ∂_ϕ. The Kerr family of metrics is parametrized by the mass m and the angular momentum per unit mass a. The mass and angular momentum am are the ADM momenta corresponding to ∂_t and ∂_ϕ respectively. In the subextreme case $|a| < m$, the Kerr spacetime contains a nondegenerate black hole[4].

The Kerr metric is algebraically special, of Petrov type D. In particular, a principal null tetrad $l^a, n^a, m^a, \bar{m}^a$ normalized so that $g_{ab} = 2(l_{(a}n_{b)} - m_{(a}\bar{m}_{b)})$ can be chosen, where l^a, n^a are repeated principal null directions.

By Theorem 6, the Kerr spacetime admits a Killing spinor κ of valence $(2, 0)$, or equivalently a conformal Killing-Yano 2-form. Associated with Y_{ab} is the complex 1-form $\xi = \mathscr{C}^\dagger\kappa$, or in terms of Y_{ab},

$$\xi_a = \tfrac{1}{3} i \nabla_b Y_a{}^b - \tfrac{1}{3}\nabla_b {*Y_a}^b. \tag{13}$$

4 A stationary black hole is said to be *nondegenerate* if the surface gravity κ of the event horizon is nonzero, where κ is defined by $\kappa^2 = -\frac{1}{2}(\nabla^a\chi^b)(\nabla_a\chi_b)$. Here χ^a is a null generator of the event horizon.

In the case of Kerr, or more generally in the *generalized Kerr-NUT class of type D spacetimes*, Y_{ab} can be chosen so that $\nabla^a Y_{ab} = 0$, in which case Y_{ab} is a Killing-Yano 2-form satisfying

$$\nabla_{(a} Y_{b)c} = 0. \tag{14}$$

In particular, in this case ξ_a is real. It is convenient to normalize Y_{ab} (or equivalently κ) so that in Kerr, in terms of Boyer-Lindquist coordinates, $\xi^a = (\partial_t)^a$, and $\kappa_1 = -\frac{1}{3}(r - ia\cos\theta)$.

If eq. (14) holds, the square of Y_{ab}, $K_{ab} = Y_{ac}Y^c{}_b$ is a symmetric Killing tensor,

$$\nabla_{(a} K_{bc)} = 0.$$

The Killing tensor K_{ab} is the Carter tensor in the Kerr spacetime.

In a general spacetime with only two Killing vector fields, the geodesic motion is chaotic. The conserved quantities for geodesics in Kerr associated to ∂_t and ∂_ϕ are energy $\boldsymbol{e} = \dot\gamma^a(\partial_t)_a$ and azimuthal angular momentum $\boldsymbol{\ell}_z = \dot\gamma^a(\partial_\phi)_a$. However, the presence of the Carter Killing tensor provides an additional conserved quantity $\boldsymbol{k} = K_{ab}\dot\gamma^a\dot\gamma^b$ for geodesics in the Kerr spacetime, the Carter constant. Hence, by the Liouville theorem the geodesic equations in Kerr can be integrated by quadratures. Further, as will be discussed below, it provides separability and decoupling properties which allow one to analyze fields on the Kerr spacetime.

4 Fields on Kerr spacetime

A necessary prerequisite for solving the black hole stability problem is to be able to prove decay estimates for linear test fields on the Kerr background. The fields of interest for the black hole stability problem are massless test fields of spins up to 2.

The field equations for massless test fields of integer spin $\mathfrak{s} = 0, 1, 2$ are

($\mathfrak{s} = 0$) Scalar wave equation

$$\nabla^a \nabla_a \phi = 0,$$

($\mathfrak{s} = 1$) Maxwell equation

$$F_{ab} = F_{[ab]}, \quad \nabla^a F_{ab} = 0, \quad \nabla_{[a} F_{bc]} = 0,$$

($\mathfrak{s} = 2$) Linearized gravity

$$D\mathrm{Ric}(\delta g) = 0.$$

Here $D\mathrm{Ric}$ is the linearization of the Ricci tensor at the Kerr background. Let $g(\lambda)$ be a 1-parameter family of metrics with $g(0)$ the Kerr metric and $\frac{d}{d\lambda}\big|_{\lambda=0} g(\lambda) = \delta g$. Then $D\mathrm{Ric}$ is defined as

$$D\mathrm{Ric}(\delta g) = \frac{d}{d\lambda}\bigg|_{\lambda=0} \mathrm{Ric}[g(\lambda)].$$

In contrast to the scalar field, the field equations of nonzero spin admit bound states, i.e. nontrivial time-independent solutions, on the Kerr spacetime. We refer to these as nonradiating modes. For the Maxwell field, the nonradiating modes are the time-independent Coloumb solutions, and for linearized gravity, the nonradiating modes correspond to deformations within the Kerr family. Thus, for a field of nonzero integer spin, a local integrated energy estimate must be independent of the nonradiating modes, which constitutes a significant additional difficulty compared to the spin-0 case.

4.1 Integrated energy decay

The essential step needed for a proof of decay estimates for massless fields is to prove dispersion. By dispersion we mean that the energy density of the field in a local region decays in time. Dispersion is expressed by an integrated local energy estimate, or Morawetz estimate, of the form[5]

$$\int_{t_0}^{t_1} \int_R W \lesssim I$$

where W dominates the energy density of some order k, up to a weight depending on r (a degeneration of W near trapping is acceptable) and I depends only on an energy of order k evaluated on initial data. For a nonlinear stability result, pointwise decay estimates are essential. Once a Morawetz estimate is available, such estimates can be proved using standard techniques. We will here only discuss Morawetz estimates.

Two features of a black hole spacetime constitute important obstacles to proving such a result, namely trapping and superradiance. In a black hole spacetime, there are future-directed null geodesics emanating from every point which either cross the horizon or null infinity. By continuity, there must therefore be geodesics which do neither, and hence there are trapped null geodesics which orbit the black hole for all time. In the Schwarzschild spacetime these are located at $r = 3m$, and in the Kerr spacetime the location of the trapped null geodesic depends on the conserved quantities $\boldsymbol{e}, \boldsymbol{\ell_z}, \boldsymbol{k}$.

In view of the geometric optics approximation, there are high-frequency wave packets near these trapped orbits that tend to disperse slowly, and hence trapping is an obstacle to dispersion.

5 We say that $a \lesssim b$ if there is a suitably universal constant $C > 0$ so that $a \leq Cb$.

In the Kerr spacetime, the stationary Killing vector field ∂_t which is timelike near infinity becomes spacelike near the horizon. The region where this happens is called the ergoregion. Due to the existence of the ergoregion, there is no positive-definite conserved energy for fields in the Kerr spacetime. One may show that there are wave packets whose energy measured near infinity increases, when scattered off the black hole.

Finally, the group of isometries of Kerr spacetimes is only two-dimensional, which is an obstacle to constructing a sufficient number of conserved currents based on Lie symmetries of the spacetime, which can be used to control fields on Kerr spacetimes. The existence of the Carter Killing tensor and related conservation laws and symmetry operators (which may be termed *hidden symmetries*) can be used to circumvent the problems caused by the lack of Lie symmetries.

As mentioned above, the existence of the Carter Killing tensor is deeply related to separability properties of the field equations. The equations for massless fields on the Kerr spacetime admit a second-order symmetry operator related to the Carter Killing tensor, which are not reducible to the first-order Lie symmetries $\mathcal{L}_{\partial_t}, \mathcal{L}_{\partial_\phi}$.

Define Q by

$$Q = \frac{1}{\sin\theta}\partial_\theta \sin\theta\partial_\theta + \frac{\cos^2\theta}{\sin^2\theta}\partial_\phi^2 + a^2\sin^2\theta\partial_t^2. \tag{15}$$

The Carter symmetry operator which commutes with $\nabla^a\nabla_a$ involves derivatives with respect to both r, θ. The operator Q is a symmetry operator for $\nabla^a\nabla_a$ but does not commute; instead Q commutes with $\Sigma\nabla^a\nabla_a$. Here Σ may be viewed as a separating factor.

We denote the set of order-n generators of the symmetry algebra generated by ∂_t, ∂_ϕ, and Q by

$$\mathbb{S}_n = \{\partial_t^{n_t}\partial_\phi^{n_\phi}Q^{n_Q} \mid n_t + n_\phi + 2n_Q = n; n_t, n_\phi, n_Q \in \mathbb{N}\}. \tag{16}$$

In particular,

$$\mathbb{S}_0 = \{\mathrm{Id}\}, \quad \mathbb{S}_1 = \{\partial_t, \partial_\phi\}.$$

Of particular importance in our analysis will be the set of second-order symmetry operators,

$$\mathbb{S}_2 = \{\partial_t^2, \partial_t\partial_\phi, \partial_\phi^2, Q\} = \{S_{\underline{a}}\},$$

and underlined indices always refer to the index in this set.

A key observation is that one may use generalized vector fields

$$A^a = \mathcal{F}^{a\underline{ab}}\partial_a S_{\underline{a}} S_{\underline{b}}, \quad S_{\underline{a}} \in \mathbb{S}_2,$$

to define currents using the formula

$$J_a = T_{ab\underline{a}\underline{b}} A^b$$

where for a field ϕ, $T_{ab\underline{a}\underline{b}}$ is defined via a polarization of the standard stress-energy tensor T_{ab},

$$T_{ab\underline{a}\underline{b}}[S_{\underline{a}}\phi, S_{\underline{b}}\phi] = \tfrac{1}{4}(T_{ab}[S_{\underline{a}}\phi + S_{\underline{b}}\phi] - T_{ab}[S_{\underline{a}}\phi - S_{\underline{b}}\phi] .$$

This approach yields a *local spacetime proof* of a Morawetz estimate for the wave equation on slowly rotating Kerr black holes with $|a| \ll m$.

Higher-order pointwise norms are defined in terms of $\mathbb{S}_n$ by

$$|\psi|_n^2 = \sum_{j=0}^{n} \sum_{S \in \mathbb{S}_j} |S\psi|^2. \tag{17}$$

We now state our main results and briefly compare them with previous results. In formulating our estimates, we shall make use of the following model energy,

$$\begin{aligned} &E_{\text{model},3}[\psi](\Sigma_t) \\ &\quad = \int_{\Sigma_t} \left(\frac{(r^2+a^2)^2}{\Delta} |\partial_t \psi|_2^2 + \Delta |\partial_r \psi|_2^2 + |\partial_\theta \psi|_2^2 + \frac{1}{\sin^2\theta} |\partial_\phi \psi|_2^2 \right) \mathrm{d}^3\mu, \end{aligned}$$

where $|\cdot|_2$ is the second-order pointwise norm introduced in (17) above, and $\mathrm{d}^3\mu = \sin\theta \mathrm{d}r\mathrm{d}\theta\mathrm{d}\phi$ is a reference volume element on the Cauchy slice Σ_t.

To deal with the necessary degeneracy in the Morawetz estimate near trapping, we introduce a cutoff function $\mathbf{1}_{r \neq 3M}$ which is supported away from trapping.

Theorem 7 (Andersson, Blue [6, Thm. 1.2]). *Let ψ be a solution of the wave equation on the Kerr exterior spacetime, $\nabla^a \nabla_a \psi = 0$. Then*

$$\begin{aligned} &\int_{-\infty}^{\infty} \int_{r_+}^{\infty} \int_{S^2} \left(\left(\frac{\Delta^2}{r^4} \right) |\partial_r \psi|_2^2 + r^{-2} |\psi|_2^2 + \mathbf{1}_{r\neq 3M} \frac{1}{r} \left(|\partial_t \psi|_2^2 + |\nabla\!\!\!/ \psi|_2^2 \right) \right) \mathrm{d}^4\mu \\ &\quad \lesssim E_{model,3}(\Sigma_0). \end{aligned}$$

Recently, a Morawetz and pointwise decay estimate proved for the scalar field equation on Kerr spacetime for the whole subextreme range $|a| < m$ has been given, using Fourier techniques and separation of variables [21]. The problem of giving a local, spacetime-based proof of such a result is open.

For massless fields with non-zero spin, any Morawetz estimate must eliminate the nonradiating modes, in particular the spacetime energy density W must cancel the nonradiating mode. For the spin-1 or Maxwell case, the best result so far is the following decay estimate for a test Maxwell field on a slowly rotating Kerr background.

To state this result, we define, for a charge-free Maxwell field with components ϕ_i and $\Upsilon = \Upsilon[\mathrm{F}] = (r - ia\cos\theta)\phi_1$, the following bulk spacetime integrals

$$B_\pm = \int_0^\infty \int_{\Sigma_t} \frac{m\Delta}{(r^2+a^2)^2}(|\phi_0|^2 + |\phi_2|^2)\sqrt{|g|}\mathrm{d}^4x, \tag{18a}$$

$$B_0 = \int_0^\infty \int_{\Sigma_t} \frac{m|\phi_1|^2}{r^2}\sqrt{|g|}\mathrm{d}^4x = \int_0^\infty \int_{\Sigma_t} \frac{m}{r^4}|\Upsilon|^2 r^2 \mathrm{d}^4\mu_{\mathrm{FI}}, \tag{18b}$$

$$B_{2,0} = \int_0^\infty \int_{\Sigma_t} \left(\frac{m\Delta^2}{(r^2+a^2)r^2}|\partial_r\Upsilon|^2 + \mathbf{1}_{r\neq 3M}\frac{m^2|\partial_t\Upsilon|^2 + |\not\nabla\Upsilon|^2}{r}\right)\mathrm{d}^4\mu_{\mathrm{FI}}, \tag{18c}$$

$$B_1 = \int_{r_+}^\infty (1 - \mathbf{1}_{r\neq 3M})\left| \int_0^T \int_{\mathbb{S}^2} \Im(\bar{\Upsilon}\partial_t\Upsilon)\sin\theta\mathrm{d}\theta\mathrm{d}\phi\mathrm{d}t \right| \mathrm{d}r, \tag{18d}$$

where $\sqrt{|g|}\mathrm{d}^4x$ is the geometrically defined volume form $\Sigma\sin\theta\mathrm{d}\theta\mathrm{d}\phi\mathrm{d}r\mathrm{d}t$, and $\mathrm{d}^4\mu_{\mathrm{FI}}$ is the coordinate volume form $\sin\theta\mathrm{d}\theta\mathrm{d}\phi\mathrm{d}r\mathrm{d}t$. For $T < 0$, we reverse the sign in these bulk terms, so that they remain nonnegative. The indexing is chosen so that $B_\pm$ involves ϕ_0, ϕ_2, and B_0 involves ϕ_1 or, equivalently, Υ with no derivatives, B_1 involves Υ with one derivative in the integral, and $B_{2,0}$ involves Υ with two derivatives but with a degeneracy near the orbiting null geodesics.

Theorem 8 (Andersson, Blue [7, Thm. 1.3]). *There are positive constants $\bar{\epsilon}_a$, C, such that if $\frac{|a|}{M} \le \bar{\epsilon}_a$, F_{total} is a regular solution of the Maxwell equation for which $E[\mathrm{F}_{total}](0)$ and $E_{FI}[\Upsilon_{total}](0)$ are finite, and the quantities $B_\pm$, B_0, B_1, and $B_{2,0}$ are defined in terms of the uncharged part* F, *then* $\forall T \in \mathbb{R}$:

$$B_\pm + B_0 + B_1 + B_{2,0} \le C\left(E[\mathrm{F}_{total}](0) + E_{FI}[\Upsilon_{total}](0)\right). \tag{19}$$

Although this type of decay estimate is relatively weak, it is sufficiently robust to have formed the foundation for many further decay results in the study of fields outside black holes.

4.2 Structure of massless field equations on Kerr

Let (M, g) be a vacuum spacetime with a valence $(2,0)$ Killing spinor κ. Let $\xi = \mathscr{C}^\dagger_{2,0}\kappa$, and assume further that ξ is real. This holds in particular in the Kerr spacetime for a suitable normalization of κ.

By a *symmetry operator* for a field equation, we mean an operator which takes solutions to solutions. As shown by Carter, a rank 2 Killing tensor K_{ab} in a vacuum spacetime provides a *commuting* symmetry operator for the wave equation

$$[\nabla_a K^{ab}\nabla_b, \nabla^a\nabla_a] = 0.$$

As we shall now discuss, the Maxwell equation on the Kerr spacetime (and more generally on vacuum spacetimes of type D) admits symmetry operators.

Recall that we defined the extended fundamental spinor operators $\mathscr{D}_{k,l,m,n}$, $\mathscr{C}_{k,l,m,n}$, $\mathscr{C}^{\dagger}_{k,l,m,n}$, $\mathscr{T}_{k,l,m,n}$ as well as the operators $\mathscr{K}^{j}_{k,l}$, $\bar{\mathscr{K}}^{j}_{k,l}$ in section 3.1.

Definition 9. Define the first-order 1-form linear concomitants A and B by

$$A[\kappa,\phi] = \overline{\mathscr{K}}^{1}_{1,1}\mathscr{C}^{\dagger}_{2,0,0,2}\mathscr{K}^{1}_{2,0}(\kappa_1\bar{\kappa}_{1'}\phi), \tag{20a}$$

$$B[\kappa,\phi] = \mathscr{C}^{\dagger}_{2,0,2,0}\mathscr{K}^{1}_{2,0}\mathscr{K}^{1}_{2,0}(\kappa_1^2\phi). \tag{20b}$$

When there is no room for confusion, we suppress the arguments, and write simply A and B. The following result shows that A, B solves the adjoint Maxwell equations, provided ϕ solves the Maxwell equation.

Theorem 10 (Andersson, Bäckdahl, Blue [5, Prop. 5.3]). *Assume that κ is a Killing spinor of valence* (2, 0) *and that ϕ is a Maxwell field. Then, with A, B given by* (20) *the spinors χ and ω, defined by*

$$\chi = Q\phi + \mathscr{C}_{1,1}A, \tag{21}$$

$$\omega = \mathscr{C}^{\dagger}_{1,1}B, \tag{22}$$

are solutions of the left and right Maxwell equations (6), *respectively.*

The energy-momentum tensor for the Maxwell field, $T = \phi\bar{\phi}$, or in terms of the field strength tensor F_{ab},

$$T_{ab} = F_{ac}F_{b}{}^{c} - \frac{1}{4}F_{cd}F^{cd}g_{ab}$$

is conserved and traceless. Hence, if v^a is a conformal Killing vector field, $J_a = T_{ab}v^a$ is conserved. However, in a spacetime with hidden symmetry in the form of a Killing spinor, there are higher-order conserved tensors for the Maxwell field.

If v^a is a conformal Killing field, then one may define a first-order operator $\varphi \to \hat{\mathscr{L}}_v\varphi$ providing a lift of the Lie-derivative. The operator $\mathscr{L}_v$ defines a symmetry operator of first order. For the equations of spins 0 and 1, the only first-order symmetry operators are given by conformal Killing fields.

Consider the operator $F_{ab} \to Z_{ab}$ defined by

$$Z_{ab} = -\tfrac{4}{3}(*F)_{[a}{}^{c}Y_{b]c}, \tag{23}$$

This is proportional to $\kappa_1\mathscr{K}^{1}_{2,0}$.

Theorem 11 (Andersson, Bäckdahl, Blue [3, Thm. 1.1]). *Let (M, g) be a vacuum spacetime of dimension 4, and let Y_{ab} and F_{ab} be real 2-forms. Define the complex 1-form η_a by*

$$\eta_a = \nabla_b Z_a{}^b + i\nabla_b {*Z_a}^b, \tag{24}$$

where Z_{ab} is given by eq. (23), and the real symmetric 2-tensor V_{ab} by

$$\begin{aligned} V_{ab} = {} & \eta_{(a}\bar{\eta}_{b)} - \tfrac{1}{2} g_{ab}\eta^c\bar{\eta}_c - \tfrac{1}{3}(\mathscr{L}_{Re\xi}F)_{(a}{}^c Z_{b)c} + \tfrac{1}{12} g_{ab}(\mathscr{L}_{Re\xi}F)^{cd} Z_{cd} \\ & + \tfrac{1}{3}(\mathscr{L}_{Im\xi} {*F})_{(a}{}^c Z_{b)c} - \tfrac{1}{12} g_{ab}(\mathscr{L}_{Im\xi} {*F})^{cd} Z_{cd}, \end{aligned} \tag{25}$$

where ξ_a is given by eq. (13).

If Y_{ab} is a conformal Killing-Yano tensor and F_{ab} satisfies the Maxwell equations, then V_{ab} is conserved,

$$\nabla^a V_{ab} = 0.$$

The leading order part of the conserved tensor V_{ab} satisfies the dominant energy condition, and hence one may use V_{ab} to construct energy currents which are positive definite to leading order. This can be expected to yield an approach to decay estimates for the Maxwell field on the Kerr spacetime which is more systematic than the approach used in the proof of Theorem 8, which relied on a Fourier-based Morawetz estimate for the wave equation for the middle component of the Maxwell field.

An examination of the proof of Theorem 11 shows that the fact that V_{ab} is conserved follows from the Teukolsky Master Equation (TME) and the Teukolsky-Starobinsky Identities (TSI), which are integrability conditions implied by the spin-$\mathfrak{s}$ field equations. In view of this fact, a deeper understanding of the TME and TSI systems for the spin-2 or linearized gravity case is fundamental in order to generalize conservation laws of the type exhibited in Theorem 11 to the spin-2 case. We shall now briefly mention recent work which provides the initial step in this direction.

4.3 TME and TSI for linearized gravity

Let δg be a solution to the linearized vacuum Einstein equation on the Kerr background and let $\kappa \in \mathscr{S}_{2,0}$ be the Killing spinor of valence (2, 0). Let $\dot{\Psi}$ be the linearized Weyl spinor defined with respect to δg, and let

$$\widehat{\phi} = \mathscr{K}^1_{4,0}\dot{\Psi},$$

where $\mathscr{K}^1_{4,0}$ was introduced in section 3.1. Defining $\mathscr{M} \in \mathscr{S}_{2,2}$ by

$$\mathscr{M} = \mathscr{C}^\dagger_{3,1}\mathscr{C}^\dagger_{4,0,4,0}\widehat{\phi},$$

one finds that the traceless symmetric rank 2 tensor $\mathcal{M}$ is a *complex* solution of the linearized Einstein equation. The linearized metric $\mathcal{M}$ is generated using $\widehat{\phi}$ as a Hertz potential.

Lemma 12 (Aksteiner, Andersson, Bäckdahl [1, Theorem 1.1]). *There is a complex vector field A so that*

$$\mathcal{M} = \tfrac{m}{27}\mathcal{L}_\xi \delta g + \mathcal{L}_A g. \tag{26}$$

The fact that $\mathcal{M}$ is pure gauge apart from the term involving $\mathcal{L}_\xi \delta g$ yields, after applying $(\mathscr{C}^\dagger)^2$ to both sides of eq. (26), and recalling that $(\mathscr{C}^\dagger)^2$ is the complex conjugate of the map to linearized curvature, the following result.

Theorem 13 (Aksteiner, Andersson, Bäckdahl [1, Cor. 1.4]).

$$\mathscr{C}^\dagger_{1,3}\mathscr{C}^\dagger_{2,2}\mathscr{C}^\dagger_{3,1}\mathscr{C}^\dagger_{4,0,4,0}\widehat{\phi} = \tfrac{m}{27}\mathcal{L}_\xi \overline{\phi} + \mathcal{L}_A \overline{\Psi}.$$

This is the full, covariant form of the TSI for linearized gravity.

5 Index theorem for Dirac operators on Lorentzian manifolds

Index theory is a huge and well-developed field in the Riemannian setting where Dirac operators are elliptic. That the hyperbolic Dirac operator on a Lorentzian spin manifold can also be Fredholm under suitable assumptions and hence possess an index is a rather recent insight. In fact, applications in physics demand such an index formula; in the past physicists have often resorted to a so-called Wick rotation (in most cases, a heuristic argument at best) in order to apply Riemannian index theorems in a Lorentzian setting. This can now be avoided; we will come back to an application in quantum field theory at the end of this section.

To describe the setup let M be a globally hyperbolic manifold whose Cauchy hypersurfaces are compact. In other words, M is spatially compact. Let M carry a spin structure so that the spinor bundle $\Sigma M \to M$ and the classical Dirac operator acting on sections of ΣM are defined. Furthermore, we assume that $n = \dim(M)$ is even. In this case the spinor bundle splits into subbundles of left-handed and right-handed spinors, $\Sigma M = \Sigma^L M \oplus \Sigma^R M$. In 4 dimensions, $\Sigma^L M = S_{1,0}M$ and $\Sigma^R M = S_{0,1}M$. The Dirac operator interchanges these two subbundles, i.e. with respect to the splitting $\Sigma M = \Sigma^L M \oplus \Sigma^R M$ it takes the form

$$D = \begin{pmatrix} 0 & D^- \\ D^+ & 0 \end{pmatrix}.$$

If $S \subset M$ is a smooth spacelike hypersurface then along S we can write

$$D = -\sigma \cdot \left(\nabla_{\mathfrak{n}} + iA - \frac{n-1}{2}H\right)$$

where σ is an invertible endomorphism field, $\mathfrak{n}$ is the past-directed unit normal field along S, A is the Riemannian Dirac operator on S, and H is the mean curvature of S.

Since A is a self-adjoint elliptic differential operator on S it has discrete real spectrum with all eigensections being smooth. In particular, for any interval $I \subset \mathbb{R}$ we have the L^2-orthogonal projections $P_I(A)$ onto the sum of all eigenspaces of A to the eigenvalues in I.

Now we assume that M is a manifold with boundary, where the boundary is the disjoint union of two smooth spacelike Cauchy hypersurfaces S_1 and S_2. Let S_1 lie in the past of S_2. Denote the corresponding Riemannian Dirac operators by A_1 and A_2, respectively. Now we can formulate the APS boundary conditions. A sufficiently regular left-handed spinor field u on M is said to satisfy the APS boundary conditions if $P_{[0,\infty)}(A_1)(u|_{S_1}) = 0$ and $P_{(-\infty,0]}(A_2)(u|_{S_2}) = 0$.

By $F\!E_{\mathrm{APS}}(M, \Sigma^L M)$ we denote the space of all left-handed spinor fields u on M which are continuous in time, L^2 in space, satisfy the APS boundary conditions, and are such that Du is L^2. This space of "finite-energy sections" naturally forms a Banach space.

In the same manner, we can define the space $F\!E_{\mathrm{APS}}(M, \Sigma^L M \otimes F)$ of twisted finite-energy spinors satisfying the APS conditions where F is a Hermitian vector bundle equipped with a metric connection.

Theorem 14 (Bär, Strohmaier [12, Main thm.]). *Let (M, g) be a compact globally hyperbolic Lorentzian manifold with boundary $\partial M = S_1 \sqcup S_2$. Here S_1 and S_2 are smooth spacelike Cauchy hypersurfaces, with S_2 lying in the future of S_1. Assume that M is even dimensional and comes equipped with a spin structure. Let F be a Hermitian vector bundle over M equipped with a metric connection.*

Then the twisted Dirac operator

$$D_{\mathrm{APS}} : F\!E_{\mathrm{APS}}(M; \Sigma^L M \otimes F) \to L^2(M; \Sigma^R M \otimes F)$$

under Atiyah-Patodi-Singer boundary conditions is Fredholm and its index is given by

$$\operatorname{ind}[D_{\mathrm{APS}}] = \int_M \widehat{\mathrm{A}}(g) \wedge \mathrm{ch}(F) + \int_{\partial M} T(\widehat{\mathrm{A}}(g) \wedge \mathrm{ch}(F)) - \frac{h(A_1) + h(A_2) + \eta(A_1) - \eta(A_2)}{2}. \tag{27}$$

The right-hand side in the index formula is formally exactly the same as in the original Riemannian APS index theorem. Here $\widehat{\mathrm{A}}(g)$ is the $\widehat{\mathrm{A}}$-form manufactured from the curvature of the Levi-Civita connection of the Lorentzian manifold, $\mathrm{ch}(F)$ is the Chern character form of the curvature of F, $T(\widehat{\mathrm{A}}(g)\wedge \mathrm{ch}(F))$ is the corresponding transgression form which also depends on the second fundamental form of the boundary. Moreover, h denotes the dimension of the kernel, and η the η-invariant of the corresponding operator.

The Dirac operator on a Lorentzian manifold is far from hypoelliptic; solutions of the Dirac equation $Du = 0$ can in general have very low regularity. Theorem 3.5 in [12] tells us that under APS boundary conditions this is no longer so. Solutions of $Du = 0$ which satisfy the APS boundary conditions are always smooth as if we had elliptic regularity at our disposal. Moreover,

$$\mathrm{ind}[D_{\mathrm{APS}}] = \dim\ker\left[D_{\mathrm{APS}}|_{C^\infty(M;\Sigma^L M\otimes F)}\right] - \dim\ker\left[D_{\mathrm{aAPS}}|_{C^\infty(M;\Sigma^L M\otimes F)}\right].$$

Here D_{aAPS} stands for the Dirac operator subject to *anti-Atiyah-Patodi-Singer* boundary conditions, the conditions complementary to the APS conditions. The occurrence of the aAPS boundary conditions and the fact that D_{aAPS} again maps sections of $\Sigma^L M \otimes F$ to those of $\Sigma^R M \otimes F$, and not in the reverse direction, are different from the corresponding formula in the Riemannian setting. In the elliptic case, the Dirac operator with anti-APS boundary conditions will in general have infinite-dimensional kernel and thus not be Fredholm. In contrast, in the Lorentzian situation, anti-APS boundary conditions work equally well as the APS conditions.

In the remainder of this section we sketch an application of Theorem 14 in the context of algebraic quantum field theory (QFT) on curved spacetimes. Details can be found in [13]. The so-called *2-point functions* are objects of central importance in such a QFT. They are distributional bi-solutions of the Dirac equation on $M \times M$. A particularly important class of 2-point functions is determined by the *Hadamard condition* which specifies the singular structure of these distributions. Hence the difference $\omega_1 - \omega_2$ of two Hadamard 2-point functions ω_1 and ω_2 is a smooth bi-solution of the Dirac equation. In this case, we can associate a smooth 1-form J^{ω_1,ω_2}, the *relative current*. The point here is that the definition of an absolute current J^{ω_i} would require a regularization procedure due to the singular nature of ω_i but the relative version is unambiguously defined and smooth.

A computation shows that J^{ω_1,ω_2} is coclosed, $\delta J^{\omega_1,\omega_2} = 0$. Now if $S \subset M$ is a smooth spacelike Cauchy hypersurface with future-directed unit normal field $\mathfrak{n}$ then we can define the *relative charge* by $Q^{\omega_1,\omega_2} = \int_S J^{\omega_1,\omega_2}(\mathfrak{n})\, dS$. Since J^{ω_1,ω_2} is coclosed the divergence theorem implies that the definition of Q^{ω_1,ω_2} is independent of the choice of S.

The Cauchy hypersurface S in the above definition of the relative charge was just an auxiliary tool. But using a Fock space construction we can also associate a 2-point function ω_S to any smooth spacelike Cauchy hypersurface S. This ω_S is to be thought of as the vacuum expectation value for an observer with spatial universe S. In case the metric of M and the connection of F are of product form near S, it is known that ω_S is of Hadamard form. Thus we can define the relative charge $Q^{\omega_{S_1},\omega_{S_2}}$ for the two boundary parts if the metric of M and the connection of F are of product form near ∂M which we now assume.

Working with left-handed spinors the main result in [13] relates $Q^{\omega_{S_1},\omega_{S_2}}$ to the index of the Dirac operator in Theorem 14 and we obtain

$$Q_L^{\omega_{S_1},\omega_{S_2}} := Q^{\omega_{S_1},\omega_{S_2}} = \int_M \widehat{\mathrm{A}}(g) \wedge \mathrm{ch}(F) - \frac{h(A_1) - h(A_2) + \eta(A_1) - \eta(A_2)}{2}. \tag{28}$$

The product assumption on the metric of M and connection of F near the boundary implies that the transgression form vanishes near the boundary so that the boundary integral in eq. (27) vanishes. The opposite sign in front of $h(A_2)$ in eqs. (27) and (28) is not a misprint but due to the convention on how the eigenvalue 0 is treated in the APS conditions.

Similarly, interchanging the roles of left-handed and right-handed spinors we obtain the "right-handed" relative charge

$$Q_R^{\omega_{S_1},\omega_{S_2}} = -\int_M \widehat{\mathrm{A}}(g) \wedge \mathrm{ch}(F) + \frac{h(A_1) - h(A_2) + \eta(A_1) - \eta(A_2)}{2}. \tag{29}$$

Hence the *total relative charge* vanishes, $Q^{\omega_{S_1},\omega_{S_2}} := Q_R^{\omega_{S_1},\omega_{S_2}} + Q_L^{\omega_{S_1},\omega_{S_2}} = 0$, while the *chiral relative charge* does not in general,

$$\begin{aligned} Q_{\mathrm{chir}}^{\omega_{S_1},\omega_{S_2}} &:= Q_R^{\omega_{S_1},\omega_{S_2}} - Q_L^{\omega_{S_1},\omega_{S_2}} \\ &= -2\int_M \widehat{\mathrm{A}}(g) \wedge \mathrm{ch}(F) + h(A_1) - h(A_2) + \eta(A_1) - \eta(A_2). \end{aligned}$$

Thus observers in S_1 and later in S_2 will disagree on the difference of numbers of left-handed and right-handed fermions. Such quantities which are classically preserved but whose quantum counterparts are not are called *anomalies*. The *chiral anomaly* treated here is a prominent example in the physics literature. It explains the rate of decay of the neutral pion into two photons. It is influenced by the gravitational field via $\widehat{\mathrm{A}}(g)$ and by an external field (e.g. electromagnetic) via $\mathrm{ch}(F)$.

6 Green-hyperbolic operators

In this section, we describe a rather large class of linear differential operators of various orders introduced in [9] which generalize normally hyperbolic and Dirac operators and are characterized by the existence of so-called Green's operators. It turns out that these operators share many of the good solution properties of normally hyperbolic and Dirac operators.

Let $E_1, E_2 \to M$ be vector bundles over a globally hyperbolic manifold M. Let $P : C^\infty(M, E_1) \to C^\infty(M, E_2)$ be a linear differential operator. There is a unique linear differential operator acting on the dual bundles, $P^* : C^\infty(M, E_2^*) \to C^\infty(M, E_1^*)$, characterized by

$$\int_M \phi(Pf)\, dV = \int_M (P^*\phi)(f)\, dV \tag{30}$$

for all $f \in C^\infty(M, E_1)$ and $\phi \in C^\infty(M, E_2^*)$ such that $\operatorname{supp} f \cap \operatorname{supp}(\phi)$ is compact. The operator P^* is called the *formally dual operator* of P.

Definition 15. An *advanced Green's operator* of P is a linear map $G_+ : C_c^\infty(M, E_2) \to C^\infty(M, E_1)$ such that

(i) $G_+Pf = f$ for all $f \in C_c^\infty(M, E_1)$;
(ii) $PG_+f = f$ for all $f \in C_c^\infty(M, E_2)$;
(iii) $\operatorname{supp}(G_+f) \subset J^+(\operatorname{supp} f)$ for all $f \in C_c^\infty(M, E_2)$.

A linear map $G_- : C_c^\infty(M, E_2) \to C^\infty(M, E_1)$ is called a *retarded Green's operator* of P if (i), (ii), and

(iii)' $\operatorname{supp}(G_-f) \subset J^-(\operatorname{supp} f)$ holds for every $f \in C_c^\infty(M, E_2)$.

Definition 16. The operator P is be called *Green hyperbolic* if P and P^* have advanced and retarded Green's operators.

It turns out that the advanced and retarded Green's operators of a Green-hyperbolic operator are automatically unique. Prime examples for Green-hyperbolic operators are normally hyperbolic operators [11, Cor. 3.4.3], Dirac-type operators [9, Cor. 3.15], and symmetric hyperbolic systems [9, Thm. 5.9].

Another example is provided by the *Proca operator* describing massive vector bosons. Here $E_1 = E_2 = T^*M$ and m is a positive constant. Then the Proca operator is given by $P = \delta d + m^2$ where d is the exterior differential and δ the codifferential. It is of second order but not normally hyperbolic. Yet it is Green hyperbolic.

The class of Green-hyperbolic operators is closed under a number of natural operations: composition, taking direct sums, dualizing, and restriction to suitable subregions.

The advanced Green's operator can be extended to sections with past-compact support in such way that the conditions in Definition 15 remain valid. By condition (iii) the resulting section will again have past-compact support, $G_+ : C_{pc}^\infty(M, E_2) \to C_{pc}^\infty(M, E_1)$. Hence on $C_{pc}^\infty(M, E_2)$ the operator G_+ is the inverse of P itself restricted to sections with past-compact support. In particular, P is invertible as an operator $C_{pc}^\infty(M, E_1) \to C_{pc}^\infty(M, E_2)$. Similarly, G_- extends to an operator on sections with future-compact support, $G_- : C_{fc}^\infty(M, E_2) \to C_{fc}^\infty(M, E_1)$.

The difference $G := G_+ - G_-$, sometimes called the *causal propagator*, maps $C_c^\infty(M, E_2)$ to $C_{sc}^\infty(M, E_1)$, the space of smooth sections with spatially compact support.

Important information about the solution theory of Green-hyperbolic operators is encoded in the following.

Theorem 17. *The sequence*

$$\{0\} \to C_c^\infty(M,E_1) \xrightarrow{P} C_c^\infty(M,E_2) \xrightarrow{G} C_{sc}^\infty(M,E_1) \xrightarrow{P} C_{sc}^\infty(M,E_2) \to \{0\}$$

is exact.

The operator P itself and its advanced and retarded Green's operators extend to distributional sections. These extensions have essentially the same properties; in particular, the analogue of Theorem 17 holds [9, Thm. 4.3]. For applications in algebraic quantum field (QFT) theory on curved spacetimes see, for example, [10, 11].

Bibliography

[1] Aksteiner S, Andersson L, Bäckdahl T. On the structure of linearized gravity on vacuum spacetimes of Petrov type D. arXiv:1601.06084, 2016.

[2] Aksteiner S, Bäckdahl T. SpinFrames, 2015. Available at: http://www.xact.es/SpinFrames.

[3] Andersson L, Bäckdahl T, Blue P. A new tensorial conservation law for Maxwell fields on the Kerr background. J Diff Geom 2017;105(2):163–176. arXiv:1412.2960.

[4] Andersson L, Bäckdahl T, Blue P. Second order symmetry operators. Classical Quant Grav 2014;31(13):135015.

[5] Andersson L, Bäckdahl T, Blue P. Spin geometry and conservation laws in the Kerr spacetime. In: Bieri L, Yau ST, editors. One hundred years of general relativity. Boston: International Press, 2015:183–226. arXiv.org:1504.02069.

[6] Andersson L, Blue P. Hidden symmetries and decay for the wave equation on the Kerr spacetime. Ann Math 2015;182(3):787–853.

[7] Andersson L, Blue P. Uniform energy bound and asymptotics for the Maxwell field on a slowly rotating Kerr black hole exterior. J Hyperb Differ Eq 2015;12(04):689–743.

[8] Bäckdahl T. SymManipulator, 2011–2014. Available at: http://www.xact.es/SymManipulator.

[9] Bär C. Green-hyperbolic operators on globally hyperbolic spacetimes. Comm Math Phys 2015;333(3):1585–1615.

[10] Bär C, Ginoux N. Classical and quantum fields on Lorentzian manifolds. In: Global differential geometry, volume 17 of Springer Proc. Math. Heidelberg: Springer, 2012:359–400.

[11] Bär C, Ginoux N, Pfäffle F. Wave equations on Lorentzian manifolds and quantization. ESI Lectures in Mathematics and Physics. Zürich: European Mathematical Society (EMS), 2007.

[12] Bär C, Strohmaier A. An index theorem for Lorentzian manifolds with compact spacelike Cauchy boundary, 2015. arXiv:1506.00959. To appear in Amer J Math.

[13] Bär C, Strohmaier A. A rigorous geometric derivation of the chiral anomaly in curved backgrounds. Comm Math Phys 2016;347(3):703–721.

[14] Bär C, Tagne Wafo R. Initial value problems for wave equations on manifolds. Math Phys, Anal Geom 2015;18(1):Art. 7, 29.

[15] Baum H. Spin-Strukturen und Dirac-Operatoren über pseudoriemannschen Mannigfaltigkeiten, volume 41 of Teubner-Texte zur Mathematik. Leipzig: Teubner Verlagsgesellschaft, 1981.

[16] Beem JK, Ehrlich PE, Easley KL. Global lorentzian geometry, volume 202 of monographs and textbooks in pure and applied mathematics, 2nd ed. New York: Marcel Dekker, 1996.
[17] Bernal AN, Sánchez M. Smoothness of time functions and the metric splitting of globally hyperbolic spacetimes. Comm Math Phys 2005;257(1):43–50.
[18] Bernal AN, Sánchez M. Globally hyperbolic spacetimes can be defined as "causal" instead of "strongly causal". Classical Quant Grav 2007;24(3):745–49.
[19] Buchdahl HA. On the compatibility of relativistic wave equations for particles of higher spin in the presence of a gravitational field. Il Nuovo Cimento 1958;10(1):96–103.
[20] Choquet-Bruhat Y, Geroch R. Global aspects of the cauchy problem in general relativity. Comm Math Phys 1969;14:329–35.
[21] Dafermos M, Rodnianski I, Shlapentokh-Rothman Y. Decay for solutions of the wave equation on Kerr exterior spacetimes III: The full subextremal case jaj < M. Ann Math 2016;183:787–913.
[22] Hawking SW, Ellis GFR. The large scale structure of space-time. London-New York: Cambridge University Press, 1973. Cambridge Monographs on Mathematical Physics, No. 1.
[23] Martín-García JM. xAct: Efficient tensor computer algebra for Mathematica, 2002- 2014. Available at: http://www.xact.es.
[24] O'Neill B. Semi-Riemannian geometry. With applications to relativity, volume 103 of Pure and Applied Mathematics. New York: Academic Press, 1983.
[25] Wald RM. General relativity. Chicago: University of Chicago Press, 1984.
[26] Walker M, Penrose R. On quadratic first integrals of the geodesic equations for type f2,2g spacetimes. Comm Math Phys 1970;18:265–74.

Francesco Bei, Jochen Brüning, Batu Güneysu,
and Matthias Ludewig

Geometric analysis on singular spaces

Abstract: We are interested in the analysis of Dirac and Schrödinger-type operators associated to certain stratified spaces known as *Smooth Thom–Mather Spaces*. These are topological spaces that consist of a smooth manifold as dense open subset to which manifolds of lower dimension are attached in a suitable way; they are briefly described in Section 2.2. Prominent examples are polyhedra, projective varieties, and connected orbit spaces of proper Lie group actions. They all come equipped with canonical metrics that are of a rather different character. We therefore discuss in Section 3 the geometric analysis of Dirac and Schrödinger-type operators on *arbitrary* Riemannian manifolds and the construction of resovents and heat semigroups. In Section 4, we describe the construction of certain geometric invariants which are expressed in terms of the spectral data of suitable self-adjoint extensions of these operators.

Keywords: Thom–Mather spaces, noncompact Riemannian manifolds, spectral theory, index theory, path integrals, covariant Schrödinger semigroups

MSC Classification 2010: 57P99, 60J65, 58J35, 58J05, 58J65

1 Introduction (Jochen Brüning)

The work of the projects B1 and its continuation C7 was concerned with geometric analysis on singular spaces during the whole period of twelve years. In this chapter, we give a perspectival survey of the work done with the support of the DFG, under the auspices of the Collaborative Research Center 647 *Space-Time-Matter*.

In the first few years, quantum graphs and hybrid spaces, built from smooth manifolds of different dimensions, played an important role, but a permanent central point of investigation was geometric analysis on Riemannian manifolds (M, g) *without any restrictions* on either M nor g. The ultimate goal of the work was to develop the spectral theory of elliptic differential operators – like Dirac operators, Bochner Laplacians, or covariant Schrödinger operators – to such an extent that spectral invariants with significant geometric meaning could be determined.

The class of spaces we concentrated on eventually is named after René Thom and John Mather who developed their theory motivated by the stability problem for smooth maps. These spaces can be thought of as smooth manifolds to which smooth manifolds of lower dimension are attached. The theory of these spaces is somewhat intricate, hence the first section gives a short introduction to it.

The next section summarizes analytic techniques, including various new aspects, which build the basis for the spectral theory of the differential operators to be studied.

DOI 10.1515/9783110452150-007_s_002

In the third section, we turn to our main constructive tools, the heat semigroup and the resolvent. Even though these operator functions are equivalent in the sense that they can be transformed one into the other, they have quite different interpretations and applications. We build a basis of analytic techniques on which most of our geometric results rest.

They appear in the last section and can reasonably be summarized as spectral geometry, which surpasses, in our view, the theory path integrals of Brownian motion, index theory, path integrals, and the analysis of spectral invariants defined by the geometric differential operators we use. The core of our results has been established during the years 2005–2016.

2 Thom–Mather spaces (Jochen Brüning)

The spaces we are interested in emerged over a period of some thirty years, starting with the work of Hassler Whitney [123], vigorously continued by Réne Thom [117] and John Mather in connection with the stability question of smooth maps, and culminating in John Mather's famous *Notes on topological stability* of 1970 [93]; for considerably more detailed information see [53]. Mather's set of abstract axioms (in [93]) defined the class of *Thom–Mather (TM) spaces* which, in the smooth compact case, can be thought of as a partition of a "nice" topological space into "nice" pieces which are smooth manifolds, glued together by "tubular neighborhoods" built after the model of tubular neighborhoods of submanifolds. Prominent examples are polyhedra, algebraic varieties, and orbit spaces of proper Lie group actions on manifolds. We leave aside in this chapter the rich theory that was built in the embedded case, where Mather's axioms are replaced by Whitney's conditions A and B. A careful treatment can be found e.g. in [52].

The purpose of this section is to present (and occasionally also to modify) Mather's axioms in a way that motivates their formulation. In comparison to other treatments like [50, Ch.3] or [118, Ch.1], we will develop the topological case carefully before we treat smooth TM spaces. We will emphasize the importance of the *condition of frontier*, and define the category of TM spaces which is behind several of the original axioms. We will quote Mather's axioms as they appear in [93], as (A1) to (A11) (where (A10) and (A11) are consequences of the other axioms).

2.1 Topology

A *TM* space, W, is a topological space whose structure is defined by a set of axioms, AI.1,2,3 (Stratified spaces), AII.1,2 (Tubular neighborhoods), AIII.1,2 (Smoothness).

Axiom I.1. (= (A1)) W is a locally compact, second-countable Hausdorff space.

Spaces with the properties listed in Axiom I.1 will be called *nice*. For each $x \in W$, we can find an open and relatively compact neighborhood, U_x. The open covering $(U_x)_{x \in W}$

has a countable subcover, $(U_i)_{i\in\mathbb{N}}$, since W is second countable (by Lindelöf's Theorem [43, Theorem VIII.6.3]) which implies that any nice space admits a useful filtration by relatively compact open sets, $(G_i)_{i\in\mathbb{Z}}$, where $G_i = \emptyset$ for $i \le 0$, for convenience. Then we have $\overline{G_i} \subset G_{i+1}$,

$$W = \bigcup_{i\in\mathcal{N}} G_i, \tag{1}$$

and we put $K_i := \overline{G_i}$, $i \in \mathbb{Z}$, such that $K_0 = \emptyset$. Therefore, nice spaces are paracompact, regular [43, Theorem VIII.2.2], and metrizable (by the Nagata-Smirnov Theorem).

Axiom I.2. (= (A4)+(A5)) A partition $\mathscr{S}(W)$ of W is called *admissible* if it is locally finite and if

$$X, Y \in \mathscr{S}(W),\ X \cap \bar{Y} \neq \emptyset \text{ implies that } X \subset \bar{Y}. \tag{2}$$

We will write $X \le Y$ if this condition is satisfied, and $X < Y$ if, in addition, $X \neq Y$; we also write $X \sim Y$ if $X \le Y$ or $Y > X$. $(\mathscr{S}(W), \le)$ will be called a *stratification* of W, its elements the *strata*. It is called *trivial* if $\mathscr{S}(W) = \{W\}$.

By eq. (1), $\mathscr{S}(W)$ is countable and, by [43, Proposition III.9.3]

$$(\bar{X})_{X\in\mathscr{S}(W)} \text{ is locally finite.} \tag{3}$$

This implies in particular that

$$\mathscr{S}_X(W) := \{Y \in \mathscr{S}(W) : Y \ge X\}$$

is a finite set.

Property (2) is commonly referred to as the *condition of frontier*. It will turn out to be the most powerful property of a stratification but, on the other hand, also the most difficult to verify. Here, for any subset X of W, its *frontier* is defined as

$$\mathrm{FR}\,X := \bar{X} \cap \overline{W - X},$$

while its *boundary* is

$$\partial X := \bar{X} - X,$$

such that $\partial X \subset \mathrm{FR}\,X \cap X'$, where X' denotes the set of cluster points of X.

Proposition 1. *If $\mathscr{S}(W)$ is admissible and $Y \in \mathscr{S}(W)$, then*

$$\partial Y = \coprod_{X<Y} X \tag{4}$$

is closed in W.

Proof. Condition (2) means that the boundary of any stratum is a union of strata, hence we get eq. (4). We obtain for $Z \in \mathscr{S}(W)$

$$\partial Z = \coprod_{Y<Z} Y,$$

hence, by eq. (3),

$$\overline{\partial Z} = \coprod_{Y<Z} \bar{Y} = \coprod_{X \le Y < Z} X = \partial Z.$$

□

The condition that ∂X is closed in W is equivalent to X being open and dense in $\bar{X}$, i.e. there is an open set, O_X, in W such that

$$X = \bar{X} \cap O_X. \tag{5}$$

It follows that for any $x \in X$ and a neighborhood $U_x \subset O_X$, we get

$$X \cap U_x = \bar{X} \cap U_x, \tag{6}$$

i.e. *the neighborhood $X \cap U_x$ of x is closed in U_x*. Subsets with property (5) are called *locally closed*; it is easy to see that eqs. (5) and (6) are equivalent. Thus we have a proof of (A2):

Proposition 2. *If $(W, \mathscr{S}(W))$ is a stratified space, then all strata are locally closed.*

The importance of locally closed sets is illustrated by the fact that a submanifold of a manifold is a locally closed subset. Another important aspect is revealed by the following fact which is proved in [43, Theorem XI.6.5.3].

Theorem 3. *The nice subspaces of a nice space are precisely the locally closed subsets.*

If $(W, \mathscr{S}(W))$ is a stratified space and V a locally closed subspace, then V inherits a locally finite partition

$$\mathscr{S}(V, W) := \{X \cap V : X \in \mathscr{S}(W), X \cap V \neq \emptyset\} \tag{7}$$

consisting of locally closed sets. It is readily seen from examples that this partition will not be admissible in general, hence will not turn V into a stratified subspace, but this may be true for a refinement of eq. (7). This motivates the following definition.

Definition 4. A locally closed subspace V of the stratified space $\big(W, \mathscr{S}(W)\big)$ will be called a *(stratified) subspace of* $\big(W, \mathscr{S}(W)\big)$ if there is a stratification $\mathscr{S}(V)$ of V and a map $\underline{j} : \mathscr{S}(V) \to \mathscr{S}(W)$ with the property that $X \subset \underline{j}(X)$ for all $X \in \mathscr{S}(V)$. The stratification will be called *regular* if $\mathscr{S}(V) = \mathscr{S}(V, W)$.

The following statements are easily verified.

Proposition 5. *If V is open in W, or if V is a locally closed union of strata in $\mathscr{S}(W)$, then $(V, \mathscr{S}(V, W))$ is a stratified subspace of $(W, \mathscr{S}(W))$.*

We turn to the relation $\leq$.

Proposition 6. *$(\mathscr{S}(W), \leq)$ defines a partially ordered set.*

Proof. It is obvious that the relation $\leq$ is reflexive and transitive, it remains to show anti-reflexivity. Assume then that $X, Y \in \mathscr{S}(W)$ satisfy $X < Y$ and $Y < X$. Then we have $X \subset \partial Y$ and $Y \subset \partial X$, hence

$$X \subset \partial Y \subset \bar{Y} \subset \overline{\partial X} = \partial X.$$

This contradiction gives the result. □

We may consider $\mathscr{S}(W)$ as a directed graph. Then the *vertex degree* of any $X \in \mathscr{S}(W)$ is given by

$$d(X, \mathscr{S}(W)) := \sharp \{Y \in \mathscr{S}(W) : X \leq Y\}. \tag{8}$$

which is a finite number. If, however, $\sharp\ \mathscr{S}(W)$ is not finite then

$$\kappa(W) := \sup \{d(X, \mathscr{S}(W)) : X \in \mathscr{S}(W)\} \tag{9}$$

may be infinite. To obtain an integer which will allow us to use inductive arguments, we introduce

Axiom I.3. $\kappa(W) < \infty$.

$\kappa(W)$ will be called the *complexity of W*. If $(W, S(W))$ fulfils Axioms I.1, to I.3, then $(W, S(W))$ is called a *stratified space*. Now we can construct a useful filtration of $\mathscr{S}(W)$ as follows. A stratum in $\mathscr{S}(W)$ is called *minimal* if it is closed such that there is no smaller stratum. Define

$$\mathscr{S}_1(W) := \{X \in \mathscr{S}(W) : X \text{ minimal}\}, \tag{10}$$

$$\mathscr{S}_{(1)}(W) := \mathscr{S}_1(W). \tag{11}$$

If $\mathscr{S}_i(W)$ and $\mathscr{S}_{(i)}(W)$ are defined for all $i \le j$ and some $j \in \mathbb{N}$, then we define

$$\mathscr{S}_{j+1}(W) := \{X \in \mathscr{S}(W) : X \text{ minimal in } \mathscr{S}(W) - \mathscr{S}_{(j)}(W)\}, \tag{12}$$

$$\mathscr{S}_{(j+1)}(W) := \mathscr{S}_{(j)}(W) \cup \mathscr{S}_{j+1}(W). \tag{13}$$

We also use the notation

$$W_{(i)} := \coprod_{X \in \mathscr{S}_{(i)}(W)} X, \tag{14}$$

$$\mathscr{S}^{(i)}(W) := \mathscr{S}(W) - \mathscr{S}_{(i)}(W), \tag{15}$$

$$W^{(i)} := W - W_{(i)}. \tag{16}$$

Proposition 7.

(a) $W_{(i)}$ *is closed and* $W^{(i)}$ *is open in* W.

(b) $W_{(i)}$ *and* $W^{(i)}$ *are regular stratified subspaces of* $(W, \mathscr{S}(W))$, *with*

$$\kappa\big(W_{(i)}, \mathscr{S}_{(i)}(W)\big) = i, \tag{17}$$

$$\kappa\big(W^{(i)}, \mathscr{S}^{(i)}(W)\big) = \kappa(W) - i. \tag{18}$$

Proof.

(a) We need only to show that $W_{(i)}$ is closed for all i. This follows from eq. (4) since, by definition, $X \in \mathscr{S}_{i+1}(W)$ implies $\partial X \in \mathscr{S}_{(i)}(W)$.

(b) The first two statements are obvious, in view of eq. (4). To prove eq. (17), consider any $X_{i+1} \in \mathscr{S}_{i+1}(W)$. Since X_{i+1} is not minimal in $W_{(i)}$ by definition, we can find $X_i \in \mathscr{S}_i(W)$ such that $X_i < X_{i+1}$. Proceeding inductively, we obtain a chain $X_1 < \cdots < X_{i+1}$ with $X_j < X_{j+1}$ for $j = 1, \cdots, i$, which proves the assertion. Eq. (18) is proved similarly. □

We write $W_{\text{reg}} := W(\kappa(W))$ and $W_{\text{sing}} := W - W_{\text{reg}}$.

We can use the decomposition of $\mathscr{S}(W)$ just described for finding good separations of the nonrelated strata.

Lemma 8 (= (A10) + (A11)). *We can find open neighborhoods,* $(T_X)_{X \in \mathscr{S}(W)}$, *of the strata with the following properties.*

(a) *The families* $(T_X)_{X \in \mathscr{S}(W)}$ *and* $(\overline{T_X})_{X \in \mathscr{S}(W)}$ *are locally finite.*

(b) *If* $T_X \cap T_Y \neq \emptyset$ *then* $X \le Y$ *or* $Y < X$.

(c) *If* $T_X \cap Y \neq \emptyset$ *then* $X \le Y$.

Proof.
(a) We number the strata of W using the filtration (1): The strata with $Y \cap K_1 \neq \emptyset$ are finite in number; we denote them by $(Y_{1j})_{1\le j\le n_1}$. We continue in this way by requiring that the strata $(Y_{ij})_{1\le j\le n_i}$ do intersect K_i *but not* K_{i-1}. Setting

$$T^1_{Y_{ij}} := W - (K_{i-1} \cup \partial Y_{ij}) \tag{19}$$

defines a family of neighborhoods of Y_{ij} with the asserted properties.
(b) If X and Y are not related in $\mathscr{S}$, then

$$X \cap \bar{Y} = \emptyset. \tag{20}$$

Now we use induction on $\kappa(W)$. If $\kappa(W) = 1$, then all strata are open and closed, hence the assertion is satisfied with $T_X := X$ for all X.

If the assertion is proved for $\kappa(W) = i$, we consider a stratified space, W, with $\kappa(W) = i + 1$. Then we can apply the induction hypothesis to the stratified space $W^{(1)}$, which has $\kappa(W) = i$, and we find open neighborhoods, $T^2_Y \subset T^1_Y$, of all strata $Y \in \mathscr{S}^{(1)}(W) = \mathscr{S}(W^{(1)})$ which satisfy the assertion.
(c) The strata in $\mathscr{S}_{(1)}(W)$ are closed and unrelated in $\mathscr{S}(W)$. We write them as $(X_i)_{1\le i\le \mathscr{N}}$, $\mathscr{N} \le \infty$, and separate the closed sets X_1 and $A_1 := (X_j)_{j\ge 2}$ by disjoint open neighborhoods $T^2_{X_1} \subset T^1_{X_1}$ and U_1. Inductively, we separate X_{i+1} and $A_{i+1} := (X_j)_{j\ge i+2}$ by disjoint open subsets $T^2_{X_{i+1}} \subset T^1_{X_{i+1}}$ of U_{i+1}, such that the family $(T^2_{X_i})_{X_i \in \mathscr{S}_1}$ separates the minimal strata.

It remains to separate any $X \in \mathscr{S}_{(1)}(W)$ from the unrelated strata in $\mathscr{S}^{(1)}(W)$. Then X is disjoint from the closed set

$$C_X := \bigcup_{Y \in \mathscr{S}^{(1)}(W),\, Y \nsim X} \overline{Y},$$

and we can separate X and C_X by disjoint open sets $T^3_X \subset T^2_X$ and U_X; this fixes the open neighborhood of X. We define a suitable open neighborhood of any unrelated Y, by putting $\mathscr{X}_i := \{X \in \mathscr{S}_{(1)}(W) : X \cap G_i \neq \emptyset\}$ and

$$T^2_Y \supset T^3_Y := \bigcup_{i\in\mathbb{N}} \bigcap_{X\in\mathscr{X}_{i+1},\, \bar{Y}\subset C_X} U_X \cap G_i.$$

Since all intersections are finite, we see that T^3_Y is an open neighborhood of Y, but we have to verify that it does not intersect T^3_X. It suffices to show this for all the sets $U^i_X := \bigcap_{X\in\mathscr{X}_{i+1},\bar{Y}\subset C_X} U_X \cap G_i$. Given $i \in \mathbb{N}$ and $X_0 \in \mathscr{S}_{(1)}(W)$, assume that $X_0 \in \mathscr{X}_{i+1}, U_{X_0}$. Then $U^i_{X_0}$ will contain U_{X_0}, hence has empty intersection with $T^3_{X_0}$. If, however, $X_0 \notin \mathscr{X}_{i+1}$ then $K_{X_0} \supset K_i \supset G_i$ and $T^1_{X_0} \cap G_i = \emptyset$. The proof is complete. □

Definition 9. An open cover of W by neighborhoods of the strata, $(T_X)_{X\in\mathscr{S}(W)}$, will be called *convenient* if it has the properties described in Lemma 8.

We turn to *morphisms* of stratified spaces.

Definition 10. A morphism of stratified spaces, $(W_i, \mathscr{S}(W_i))$, $i = 1, 2$, is a map

$$\phi := (\overline{\phi}, \underline{\phi}) : (W_1, \mathscr{S}(W_1)) \to (W_2, \mathscr{S}(W_2)),$$

where $\underline{\phi} : \mathscr{S}(W_1) \to \mathscr{S}(W_2)$ is order preserving, $\overline{\phi} : W_1 \to W_2$ is continuous and $\overline{\phi}(X) \subset \underline{\phi}(X)$, for all $X \in \mathscr{S}(W_1)$.

The order preservation of $\underline{\phi}$ is in fact a consequence of the other assumptions.

Proposition 11. *If the conditions on $\overline{\phi}$ are satisfied, then $\underline{\phi}$ is order preserving.*

Proof. Consider a pair $X_1 < X_2$ of strata in W_1, with images $\underline{\phi}(X_i) =: \underline{X}_i$, $i = 1, 2$. Since every $x \in X_1$ is a cluster point of X_2, then every neighborhood of x will contain points of X_2, and by continuity of $\overline{\phi}$, every neighborhood of $\overline{\phi}(x)$ will contain points of $\underline{X}_2$. □

If $\mathscr{S}(W_1)$ is nontrivial, then nontrivial morphisms may be difficult to construct, but it is easily verified that stratified spaces form a category with the above definition of morphisms.

Examples An important general class of stratified spaces is provided by the *topologically stratified spaces*, as defined in [52, Sec.1.1]; this class comprises polyhedra, algebraic varieties, and orbit spaces of proper Lie group actions on manifolds, among many others. In particular, its strata are all topological manifolds.

2.2 Control data

In Lemma 8, we have constructed a *convenient* family of open neighborhoods of strata with good global properties. We will now introduce additional assumptions which are modelled after the properties of submanifolds of a manifold.

Definition 12. Let $(W, \mathscr{S}(W))$ be a stratified space. Then a *tubular neighborhood* or *control data* for a stratum X is a triple $(T_X, \pi_X, \rho_X) =: \tau_X$ with the following properties.
(1) T_X is an open neighborhood of X in W;
(2) $\pi_X : T_X \to X$ is a continuous retraction;
(3) $\rho_X : T_X \to [0, \infty)$ is a continuous map with $\rho^{-1}(0) = X$.

Property (3) implies that $T_X \cap \partial X = \emptyset$. The choice of ρ_X is to some extent arbitrary, while the choice of π_X is more rigid in practice. In fact, given π_X, we can choose a metric, d_W, defining the topology of W and put

$$\rho_X^d(w) := d_W(w, \pi_X(w)), \;\; w \in T_X.$$

Moreover, we can replace ρ_X by many other functions with the same properties, e.g. by

$$\rho_X^{bd}(w) := \min(\rho_X(w), 2). \tag{21}$$

We will also abbreviate

$$\sigma_X := (\pi_X, \rho_X) : T_X \to X \times \mathbb{R}_{\geq}.$$

Next we collect some useful notation regarding a subset, U, of X, and a continuous function $\varepsilon : X \to \mathbb{R}_{>}$, where we will write $\varepsilon_a, a > 0$, for the constant function with value a.

$$X^\varepsilon := \{(x,t) \in X \times \mathbb{R}_{\geq} : 0 \leq t < \varepsilon(x)\}, \tag{22}$$

$$\dot{X}^\varepsilon := \{(x,t) \in X \times \mathbb{R}_{\geq} : 0 < t < \varepsilon(x)\}, \tag{23}$$

$$X^a := X^{\varepsilon_a}, \tag{24}$$

$$T_U := \pi_X^{-1}(U), \;\; U \subset X, \tag{25}$$

$$T_U^\varepsilon := \sigma_X^{-1}(X^\varepsilon), \tag{26}$$

$$\dot{T}_U^\varepsilon := \sigma_X^{-1}(\dot{X}^\varepsilon), \tag{27}$$

$$\dot{T}_X := T_X - X, \tag{28}$$

$$T_U^a := T_U^{\varepsilon_a}, \;\; a \geq 0. \tag{29}$$

Given control data τ_X and any open neighborhood, O, of X then the triple $(T_X' := T_X \cap O, \sigma_X' := \sigma_X | T_X') =: \tau_X'$ also has the properties stated in Definition 32, which suggests the following

Definition 13. Control data τ_X' and τ_X for a stratum X are called *equivalent* if there is an open neighborhood T_X'' of X such that

$$T_X'' \subset T_X' \cap T_X,$$

and

$$\sigma_X' | T_X'' = \sigma_X | T_X''.$$

It is easily seen that this defines an equivalence relation. We denote by $[\tau_X]$ the equivalence class of τ_X. If all $X \in \mathcal{S}(W)$ admit control data, then we denote by $\mathcal{T}(W)$ the set of all equivalence classes which we will refer to as *control data* for $\big(W, \mathcal{S}(W)\big)$.

Axiom II.1. A *Thom–Mather space* or briefly, a *TM space* is a stratified space $(W, \mathscr{S}(W))$ with control data $\mathscr{T}(W)$. It will be denoted by the triple $\mathscr{W} := (W, \mathscr{S}(W), \mathscr{T}(W))$.

Given a TM space with explicit control data, $(\tau_X)_{X\in\mathscr{S}(W)}$, we can refine this data by intersecting $(T_X)_{X\in\mathscr{T}(W)}$ with the corresponding members of a convenient family of open neighborhoods of the strata (cf. Lemma 8). Hence we will assume from now on that *$\mathscr{T}(W)$ is represented by a convenient family $(T_X)_{X\in\mathscr{T}(W)}$*.

We proceed to even more special representatives.

Lemma 14. *Let $\mathscr{W} := (W, \mathscr{S}(W), \mathscr{T}(W))$ be a Thom–Mather space. Then we can find convenient control data $(\tau_X)_{X\in\mathscr{S}(W)}$ with the following additional properties.*

(a) *If $K \subset X$ is compact then $\overline{T_K}$ is compact in W.*

(b) *We can find a function $\varepsilon \in C(X, \mathbb{R}_>)$ such that $\overline{T_X^\varepsilon} \subset T_X$ and*

$$\sigma_X^\varepsilon := \sigma_X | T_X^\varepsilon : T_X^\varepsilon \to X^\varepsilon$$

is proper.

(c) *If X is locally connected and if $\sigma_X | T_X \cap Y : T_X \cap Y \to \mathbb{R}_>$ is an open map for some $Y > X$, then ε can be chosen in such a way that $\sigma_X(T_X^\varepsilon \cap Y) = \dot{X}^\varepsilon$, in other words, $\sigma_X^\varepsilon : T_X^\varepsilon \cap Y \to \dot{X}^\varepsilon$ is surjective.*

Proof. We start with convenient control data $(\tau_X^0)_{X\in\mathscr{S}(W)}$ and fix $X \in \mathscr{S}(W)$. For technical reasons, we will construct open neighborhoods $T_X^0 \supset T_X^1 \supset T_X^2$ of X which generate equivalent control data; eventually, we will set $T_X := T_X^2$.

(a) For the construction of T_X^1, we consider the family of all open sets with compact closure in X. Since X is nice, we can select a locally finite, countable subfamily, $(U_i)_{i\in\mathbb{N}}$, which covers X. By [43, Theorem XI.6.2], we can find an open, relatively compact neighborhood, V_i^0, of $\overline{U_i}$ in T_X^0 for all i, and we put

$$V_i := V_i^0 \cap (\pi_X^0)^{-1}(U_i), \tag{30}$$

$$T_X^1 := \bigcup_{i\in\mathbb{N}} V_i. \tag{31}$$

Since the family $(V_i)_{i\in\mathbb{N}}$ is obviously locally finite in T_X^0, we obtain for $K \subset X$ compact

$$\overline{T_K} = \bigcup_{i\in\mathbb{N}} \overline{V_i \cap (\pi_X')^{-1}(K)} \tag{32}$$

$$= \bigcup_{i\in\mathbb{N}} \overline{V_i' \cap (\pi_X')^{-1}(U_i \cap K)}, \tag{33}$$

which is a finite union of compact sets, as claimed.

We can iterate the construction to get $T_X = T_X^2$.

(b) We need several steps.

Step 1. We show that for any U_i as above and ε_i sufficiently small but positive, we have

$$T^{1,\varepsilon_i}_{\overline{U_i}} = T^{2,\varepsilon_i}_{\overline{U_i}} \subset T^2_X. \tag{34}$$

If eq. (34) were wrong, we could find a sequence $(v_n) \subset (T^{1,1/n}_{\overline{U_i}} - T^2_X)$. This would entail that $\sigma^1_X(v_n) = (x_n, t_n)$, where $x_n \in \overline{U_i}$, $t_n \in [0, 1/n)$. Since $\overline{U_i}$ is compact, we can apply (a) to find that (v_n) converges in T^1_X, possibly after passing to a subsequence. Then $v_0 := \lim_{n\to\infty} v_n$ satisfies $\sigma^1_X(v_0) =: (x_0, 0)$, $x_0 \in \overline{U_i}$ which implies the contradiction $v_0 = x_0 \in \overline{U_i}$.

Step 2. By Step 1, we can find for each $i \in \mathbb{N}$ a positive ε_i such that eq. (34) holds. We then define a function $\varepsilon^0 : T^2_X \to X^{\varepsilon^0}$ by

$$\varepsilon^0(x) := \max\{\varepsilon_i | x \in U_i\}, \ \ x \in X.$$

It is easy to see that ε^0 is lower semicontinuous. Thus we can invoke Dowker's Theorem [43, Theorem VIII.4.3] to obtain a function $\varepsilon \in C(X, \mathbb{R}_>)$ satisfying the inequality $0 < \varepsilon < \varepsilon^0$.

We can now give the proof of assertion (b). With ε from Step 2 and eq. (34) we find

$$T^{1,\varepsilon}_X = T^{2,\varepsilon}_X,$$

and we have to show that $\sigma^2_X : T^{2,\varepsilon}_X \to X^\varepsilon$ is proper. Consider then a sequence $(v_n) \in T^{2,\varepsilon}_X$ with the property that $\lim_{n\to\infty} v_n = (x_0, t_0) \in X^\varepsilon$ exists, such that $t_0 < \varepsilon(x_0)$. In particular, $(\pi^1_X(v_n))_{n\in\mathbb{N}}$ is contained in a compact subset of X. Using (a), we deduce that (v_n) admits a convergent subsequence in T^1_X, and we may assume that $v_n \to v_0$. But then $\sigma^1_X(v_0) = (x_0, t_0) = \sigma^2(v_0)$ shows that $v_0 \in T^{2,\varepsilon}_X$, as desired.

(c) We choose ε as in (b). For any point $(x_0, t_0) \in \dot{X}^\varepsilon$, we choose a connected neighborhood U_{x_0} of x_0 in X. The assertion follows if we show that $\sigma_X(T^\varepsilon_{U_{x_0}} \cap Y) = \dot{U}_{x_0}$. By the assumed openness of $\sigma_X | T^\varepsilon_X \cap Y$, $\sigma_X(T^\varepsilon_{U_{x_0}})$ is open in $\dot{U}_{x_0}$, and it is enough to show that it is also closed. But this follows from the properness of σ^ε_X. □

The following example shows that surjectivity will not hold without some additional property of σ_X.

Example 15. Consider the stratified subspace of $\mathbb{R}_\geq$ given by

$$W := \{0\} \sqcup \bigcup_{i\in\mathbb{N}} \left[\frac{1}{2n}, \frac{1}{2n-1}\right],$$

where $W_{\text{sing}} := \{0\}$ is closed, and $W_{\text{reg}} := W - \{0\}$ is open with W_{sing} in its closure. We can also introduce control data for W_{sing} by defining $\pi_{\text{sing}}(x) := 0$, $\rho_{\text{sing}}(x) := x$, but the restriction of σ_{sing} to any $T^\varepsilon_{W_{\text{sing}}}$ will not map onto X^ε.

We remark that we can normalize ρ_X on T_X^{ε} by

$$\rho_X^{\mathrm{st}}(v) := 2\rho_X(v)\big(\varepsilon(\pi_X(v))\big)^{-1}, \tag{35}$$

such that

$$\rho_X^{\mathrm{st}} : T_X^2 \to X^2. \tag{36}$$

Next we discuss morphisms of TM spaces, $\phi = (\underline{\phi}, \overline{\phi})$. These are stratified maps $\big(W, \mathscr{S}(W)\big) \to \big(W', \mathscr{S}(W')\big)$ which have to respect the control data, $\mathscr{S}(W), \mathscr{S}(W')$ in the following sense.

Definition 16. A morphism of stratified spaces,

$$\phi = (\underline{\phi}, \overline{\phi}) : \big(W, \mathscr{S}(W)\big) \to \big(W', \mathscr{S}(W')\big),$$

will be called *controlled*, if for any $X \in \mathscr{S}(W)$ and $X' := \underline{\phi}(X) \in \mathscr{S}(W')$, we can find control data $\tau_X, \tau_{X'}$ such that the following diagram commutes

$$\begin{array}{ccc} T_X \cap (\overline{\phi})^{-1}(T_{X'}) & \xrightarrow{\ \overline{\phi}\ } & T_{X'} \\ \big\downarrow{\scriptstyle \pi_X} & & \big\downarrow{\scriptstyle \pi_{X'}} \\ X & \xrightarrow{\ \overline{\phi}\ } & X' \end{array} \tag{37}$$

and, if $\dot{T}_{X'} := T_{X'} - X'$, also

$$\begin{array}{ccc} \dot{T}_X \cap (\overline{\phi})^{-1}(\dot{T}_{X'}) & \xrightarrow{\ \overline{\phi}\ } & \dot{T}_{X'} \\ \big\downarrow{\scriptstyle \rho_X} & \swarrow{\scriptstyle \rho_{X'}} & \\ (0,\infty) & & \end{array} \tag{38}$$

A stratified and controlled morphism between TM spaces will be called a *TM morphism*.

The special case that $\mathscr{W}'$ is trivial, i.e. $\mathscr{S}(W') = \{W'\}$, $\mathscr{T}(W') = (W', id_{W'}, 0)$ is of some interest. Then a morphism $\phi = (\underline{\phi}, \overline{\phi})$ has $\underline{\phi}(X) = W'$ for all $X \in \mathscr{S}(W)$ such that $\overline{\phi}$ can be an arbitrary continuous map $\overline{W} \to W'$. Then diagram (37) reduces to

$$\overline{\phi}(w) = \overline{\phi}(\pi_X(w)), \quad w \in T_X. \tag{39}$$

In the discussion so far, only a single concrete map between TM spaces has appeared, namely

$$\sigma_X : T_X \to X \times \mathbb{R}_{\geq},$$

which we want to be a *TM map*. So we need to identify σ_X as a stratified map in the first place. Since T_X is open in W by assumption, for suitably chosen control data T_X is stratified by

$$\mathscr{S}(T_X) := \bigsqcup_{Y \geq X} T_X \cap Y,$$

while a natural stratification for $X \times \mathbb{R}_{\geq}$ is given by

$$\mathscr{S}(X \times \mathbb{R}_{\geq}) := \{X \times \mathbb{R}_{>}, X \times 0\}.$$

Then we see that we must set

$$\underline{\sigma_X}(T_X \cap Y) := X \times \mathbb{R}_{>}, \;\; Y > X; \underline{\sigma_X}(X) := X \times \{0\}. \tag{40}$$

To turn σ_X into a TM map, we define control data for T_X as follows, noting that $(\pi_Y)^{-1}(Y \cap T_X)$ is open in T_Y:

$$T_X^{T_X} := T_X, \tag{41}$$

$$T_{Y\cap T_X}^{T_X} := T_X \cap (\pi_Y)^{-1}(Y \cap T_X), \tag{42}$$

$$\sigma_X^{T_X} := \sigma_X, \tag{43}$$

$$\sigma_{Y\cap T_X}^{T_X} := \sigma_X | T_{Y\cap T_X}^{T_X}. \tag{44}$$

Finally, $\sigma_{Y\cap T_X}^{T_X}$ has to satisfy condition (39) which appears as Axiom (A9) in [93]:

$$\sigma_{Y\cap T_X}^{T_X} \circ \pi_Y = \sigma_{Y\cap T_X}^{T_X}. \tag{45}$$

Combining this with part (c) of we formulate the following *regularity axiom*.

Axiom II.2. A TM space $\mathscr{W} := \big(W, \mathscr{S}(W), \mathscr{T}(W)\big)$ will be called *regular* if, for suitable control data and all strata $X < Y$, the maps

$$\sigma_{Y\cap T_X}^{T_X} : T_{Y\cap T_X}^{T_X} \to X \times (0, 2) \tag{46}$$

are *surjective* TM morphisms.

For regular TM spaces, we now obtain another important class of TM subspaces.

Lemma 17. *Let $\mathscr{W}_0, \mathscr{W}_1$ be TM spaces with $\mathscr{W}_1$ regular and $\mathscr{W}_0$ trivial. Let $\phi : \mathscr{W}_1 \to \mathscr{W}_0$ be a TM map with $\overline{\phi}$ open and proper. Then for any $x_0 \in \overline{\phi}(W_1)$, the fiber $V := V_{x_0} := (\overline{\phi})^{-1}(x_0)$ is a compact TM space with $\kappa(V) < \kappa(W_1)$.*

Proof. We may assume that $\overline{\phi}$ is surjective. We have the partitions

$$V = \bigsqcup_{X \in \mathscr{S}(W_1);\ X \cap V \neq \emptyset} X \cap V.$$

and

$$T_X \cap V = \bigsqcup_{x_1 \in X} T_{x_1} \cap V = \bigsqcup_{Y \geq X} T_X \cap Y \cap V. \tag{47}$$

Now, since $\overline{\phi}$ satisfies eq. (39), we have $\pi_X(w_1) = x_1$ for $w_1 \in T_{x_1}$, and so $T_{x_1} \cap V = T_{x_1}$ if $x_1 \in V$, and $T_{x_1} \cap V = \emptyset$ otherwise. This implies that for $X < Y$ and $X \cap V,\ Y \cap V \neq \emptyset$, we get

$$T_X \cap Y \cap V = T_{X \cap V} \cap Y.$$

Since $\mathscr{W}_1$ is regular, it now follows from Axiom II.2 that $\mathscr{S}(T_X)$ is indeed a (regular) stratification. □

One may ask under what conditions $\overline{\phi}$ may be *locally trivial.* In the (most important) smooth case, to be treated below, *properness* will be enough, by a suitable extension of the Ehresmann Theorem, cf. [93, Cor. 10.2]. In the topological case treated so far, a convenient condition seems not to be available yet; instead, topologists often include local triviality in the definition, as in [52, Sec. 1.1].

2.3 Flows

Flows or *local one-parameter subgroups* are very important construction tools in differential topology. They can be defined, however, also on the TM spaces we have introduced, i.e. on the topological level. The following definition is taken from [93, Section 10].

Definition 18. Let W be a nice space. A flow on W is a pair (α, J) where J is an open subset of $\mathbb{R} \times W$ and

$$\alpha : J \to W$$

is a continuous map having the following properties.

(a) $\{0\} \times W \subset J$ and $\alpha(0, w) = w$, for all $w \in W$.
(b) $J_w := J \cap (\mathbb{R} \times \{w\}) = \big(a_-(w), a_+(w)\big)$, for all $w \in W$ and real numbers $-\infty \le a_-(w) < 0 < a_+(w) \le \infty$.
(c) For $w \in W$ and $t, s, t + s \in \big(a_-(w), a_+(w)\big)$ we have

$$\alpha(t, \alpha(s, w)) = \alpha(t + s, w).$$

(d) For any $w \in W$ and any compact set $K \subset W$, there is $\varepsilon > 0$ such that for $t \in \big(a_-(w), a_-(w) + \varepsilon\big) \cup \big(a_+(w) - \varepsilon, a_+(w)\big)$ we have $\alpha(t, w) \notin K$.

We also define

$$\mathscr{D}_t := \{w \in W : \ t \in J_w\}, \tag{48}$$

$$\alpha_t : \mathscr{D}_t \ni w \to \alpha(t, w) \in \mathscr{D}_{-t}. \tag{49}$$

Then $\mathscr{D}_t$ is open for all t, and α_t is a homeomorphism, with inverse α_{-t}. We now define *stratified* or *TM flows* in terms of the $(\alpha_t)_{t \in \mathbb{R}}$.

Definition 19.
(a) If $\big(W, \mathscr{S}(W)\big)$ is stratified, then a flow α on W is stratified if all homeomorphisms α_t are stratified, i.e. if for $X \in \mathscr{S}(W)$

$$\alpha_t(X \cap \mathscr{D}_t) \subset X \cap \mathscr{D}_{-t}.$$

(b) If $\mathscr{W} = \big(W, \mathscr{S}(W), \mathscr{T}(W)\big)$ is a TM space, then a flow α on W is a TM flow if α is stratified and if there exist good control data on $\mathscr{W}$ such that α_t is a TM map with respect to the induced TM structure on $\mathscr{D}_t$, for all t. Such control data are called *adapted* to α.

If $\mathscr{D}_t = \emptyset$, then the above conditions are satisfied by definition. Due to the properness of all σ_X, we can characterize a TM flow by a global property as follows.

Lemma 20. *Let $\mathscr{W}$ be a TM space and α a TM flow on $\mathscr{W}$. Then for $X \in \mathscr{S}(W)$ and $w \in T_X$, we have the identities*

$$\mathscr{I}_w = \mathscr{I}_{\pi_X(w)}, \tag{50}$$

$$\sigma_X \alpha(t, w) = \big(\rho_X(w), \alpha(t, \pi_X(w))\big), \ \ t \in \mathscr{I}_{\pi_X(w)}. \tag{51}$$

We will see below that TM flows are the counterpart to *controlled vectorfields*, as defined in [93, Sec. 9] in the smooth case.

2.4 Smoothness

The main examples for TM spaces that appear in applications have the following additional smoothness property, which we formulate as an axiom.

Axiom III.1. (= (A3) + (A8)). A *smooth TM space* is a TM space $\mathscr{W} = \big(W, \mathscr{S}(W), \mathscr{T}(W)\big)$, with the additional properties

(a) all strata are topological manifolds (as subspaces) and carry a smooth structure of class C^μ, $\mu \in \mathbb{N}$;

(b) the maps $\sigma^{T_X}_{Y \cap T_X}$ are C^μ submersions.

We then find,

Lemma 21. *Any smooth TM space is regular.*

Proof. This follows from Lemma 14,(c), since (topological) manifolds are locally connected and smooth submersions are open maps. □

We will have to require that W_{reg} is a manifold, i.e. that all open strata have the same dimension, since otherwise the analysis becomes very complicated. Fortunately, this is true for most of the interesting examples.

Axiom III.2. W_{reg} is a smooth manifold. Its dimension is called the dimension of W, dim W.

Strata of codimension one will usually introduce boundary conditions for the geometric differential operators we want to deal with. Therefore, the following definition is reasonable.

Definition 22. A smooth TM space of dimension m is called a *pseudomanifold* if there are no strata of dimension $m - 1$.

It is fairly obvious how to define smooth TM flows:

Definition 23. A TM flow, α, on the smooth TM space $\mathscr{W}$ is called a smooth TM flow if α restricts to a smooth flow on each stratum.

Mather has defined *controlled vector fields* on smooth TM spaces as follows (cf. [93, p. 493]); for symmetry, we will call them *TM vector fields.*

Definition 24. Let ξ_X be a (smooth) vector field on X, for all $X \in \mathscr{S}(W)$; then the family $\xi := (\xi_X)_{X \in \mathscr{S}(W)}$ is called a *stratified vector field* on W. ξ is called a *TM vector field* on W if there are control conditions $(\tau_X)_{X\in\mathscr{S}(W)}$ such that for $X, Y \in \mathscr{S}(W)$ with $X < Y$, and for any $w \in T_X \cap Y$ we have

$$\xi_Y(w)(\rho_X) = 0, \tag{52}$$

$$\pi_{X,*}(\xi_Y(w)) = \xi_X(\pi_X(w)). \tag{53}$$

The following Lemma can be expected but its proof is somewhat involved and will not be given here.

Lemma 25. *TM vector fields and smooth TM flows are equivalent in the sense that*

(a) *a TM flow generates a TM vector field by differentiation, and*
(b) *the flows on the strata generated by a TM vector field combine to a TM flow.*

Emphasizing the TM flows, we can give the following flow version of the Ehresmann Theorem which, however will require smoothness. Again, the proof will not be given for lack of space. The local triviality of ϕ is then an easy consequence, cf. [93, Cor.10.2].

Theorem 26. *Let* $\mathscr{W} = \big(W, \mathscr{S}(W), \mathscr{T}(W)\big)$ *be a smooth TM space and let M be a smooth manifold. Let* $\phi = \overline{\phi} : W \to M$ *be a smooth submersion, and let* α_M *be a smooth flow on M. Then there is a smooth TM flow,* α_W, *on W such that*

$$\phi\alpha_W(t, w) = \alpha_M(t, \phi(w)), \tag{54}$$

for all $w \in W$ *and all* $t \in \mathscr{I}_{\alpha_M,w}$; *in particular,*

$$\mathscr{I}_{\alpha_W,w} \subset \mathscr{I}_{\alpha,w}. \tag{55}$$

If ϕ *is proper, then we obtain equality in eq.* (55).

2.5 Metrics

We now consider *only smooth TM spaces* $\mathscr{W}$, with $\mu = \infty$ for simplicity. Then $W = W_{\text{reg}} \sqcup W_{\text{sing}}$, with W_{reg} open and dense in W. Since W is metrizable, we always can find a metric generating the topology, but in many cases a TM space comes equipped with a *canonical Riemannian* metric on W_{reg}. For polyhedra imbedded in some Euclidean space, for projective varieties, and for G-spaces equipped with an invariant metric, the canonical metrics are obvious. Unfortunately, these metrics may be very complicated and not easy to handle. For many purposes, however, one can use a “Cheeger metric” (the notion “iterated edge metric” is also in use) as defined in [36, 37] that depends on $m - 1$ parameters $(a_2, \ldots, a_m)$, $a_i \geq 1$; the natural metric of a Euclidean polyhedron will be of this type. In the papers mentioned, Cheeger has shown that a compact smooth pseudomanifold, W, admits a conical metric of type (1,...,1) on its regular part, such that the corresponding L^2-cohomology coincides with the middle perversity intersection homology of W. Other choices of the parameters will lead to intersection homology with other perversities. In the simplest case of isolated conic singularities, the conical metric takes the form

$$g = dt^2 + t^2 g_L(t), \tag{56}$$

where the compact Riemannian manifold L is the link of the singularity. A careful construction of Cheeger metrics can be found in [20, Sec.5].

Definition 27. Let $\mathscr{W} = \big(W, \mathscr{S}(W), \mathscr{T}(W)\big)$ be a smooth TM space with a Cheeger metric, g, on W_{reg}. Then $(\mathscr{W}, g)$ is called a Riemannian TM space (RTM).

In our further discussion we start laying the necessary foundations in atmost generality, i.e. we begin with the geometric-analytic properties of *arbitrary Riemannian spaces*. Later we will specialize the choice of metric as the problems treated require.

3 Geometric analysis on Riemannian manifolds

In this section, we describe the class of geometric differential operators we want to study, as very important tools in geometric analysis as well as mathematical physics. (M, g) will always be an arbitrary Riemannian manifold of dimension m. The symbol $d(x, y)$ denotes the geodesic distance of (M, g), and $B(x, r)$ the corresponding open metric balls. We further denote by ∇^{TM} its Levi-Civita connection (while an obvious notation will be used for the connections that are induced by ∇^{TM}, like ∇^{T^*M} and so on), and by vol its volume measure. Whenever necessary, we will make the dependence of these and other data explicit in the notation, writing vol_g and so on. We are mainly concerned with differential operators of Dirac or Laplace type. For the second-order operators we construct the heat semigroup and the resolvent in general, but get finer results using more special metrics that are adapted to specific singular situations, notably RTM spaces.

3.1 Differential operators (B Güneysu)

3.1.1 First-order operators

The fundamental differential operator of first order on any smooth manifold is of course the de Rham differential, d, acting on the space $\Omega(M)$ of all differential forms; we denote by d_k its restriction to the space of k-forms, $\Omega^k(M)$, so that

$$d_k : \Omega^k(M) \longrightarrow \Omega^{k+1}(M).$$

To obtain an *elliptic* operator, we need the Riemannian metric g on M to introduce the formal adjoint $d^\dagger$ with respect to the metric on $\wedge T^*M$ and vol, to obtain the *Hodge-de Rham operator*

$$D := d + d^\dagger .$$

It is the prototype for the more general class of *Dirac operators*:

Definition 28. A *Dirac bundle* over (M, g) is a quadruple (E, h, c, ∇), also written as $(E, h, c, \nabla) \to (M, g)$, such that

(a) $(E, h) \to M$ is a complex metric $\mathbb{C}$-vector bundle

(b) c is a *Clifford multiplication*, in the sense that c is a homomorphism of vector bundles

$$c : TM \longrightarrow \mathrm{End}(E),$$

such that for all smooth vector fields $X \in \mathscr{X}(M)$ one has

$$c(X) = -c(X)^*, \quad c(X)^* c(X) = |X|^2 ; \tag{57}$$

(c) ∇ is a Clifford connection, that is, ∇ is a smooth metric covariant derivative on $(E, h) \to M$ such that for all smooth vector fields $X, Y \in \mathscr{X}(M)$, and all sections $\psi \in C^\infty(M, E)$ one has

$$\nabla_X(c(Y)\psi) = c(\nabla_X^{TM} Y)\psi + c(Y)\nabla_X \psi .$$

Every such structure canonically induces its Dirac operator

$$D_\nabla^c \psi(x) := \sum_{j=1}^m c(e_j)(x) \nabla_{e_j} \psi(x), \quad \psi \in C^\infty(M, E),$$

where $(e_i)_{1 \le i \le m}$ will always denote a smooth orthonormal frame which is defined near $x \in M$, and $(e^i)_{1 \le i \le m}$ its dual frame. It is easily checked that Dirac operators are formally self-adjoint elliptic first-order differential operators with smooth coefficients.

Let us show that the Hodge-de Rham operator indeed arises as a Dirac operator in the above sense: Let $U \subset M$ be an open subset such that TU is trivial. Then, with

$$w : T^*M \to \mathrm{End}(\wedge T^*M), \; i : TM \to \mathrm{End}(\wedge T^*M)$$

the exterior respectively interior multiplication, we have

$$d = \sum_j w(e^j) \nabla_{e_j}^{\wedge T^*M}, \quad d^\dagger = -\sum_j i(e_j) \nabla_{e_j}^{\wedge T^*M} .$$

Therefore, over U we obtain

$$D = d + d^\dagger = \sum_j \left(w(e^j) - i(e_j)\right)\nabla_{e_j}^{\wedge T^*M},$$

and it is easily checked that the map

$$c : TM \to \operatorname{End}(\wedge T^*M),\ \ c(X) := w(X^\flat) - i(X),$$

is a Clifford multiplication and that $\nabla^{\wedge T^*M}$ is a Clifford connection, so that the Hodge-de Rham operator D is in fact a Dirac operator in the sense of Dirac bundles.

3.1.2 Second-order operators

The very large and important class of second-order elliptic differential operators of interest in Riemannian geometry and mathematical physics is provided by what we will call *covariant Schrödinger bundles*. These are data of the form

$$(E, h, \nabla, V) \longrightarrow (M, g),$$

where

(a) $(E, h) \to M$ is a smooth complex metric vector bundle,

(b) ∇ is a smooth metric covariant derivative on $(E, h) \to M$, with formal adjoint (with respect to h and g) denoted by $\nabla^\dagger$,

(c) V is a self-adjoint Borel section in $\operatorname{End}(E) \to M$, that is, $V : M \to \operatorname{End}(E)$ is a Borel measurable map with $V(x) \in \operatorname{End}(E_x)$ a linear self-adjoint map w.r.t. h, for every $x \in M$. We will also refer to the endomorphism V as *a potential on* $(E, h) \to M$.

For simplicity, we will assume in the sequel that the potential V is in $L^2_{\mathrm{loc}}(M, \operatorname{End}(E))$, that is, locally square integrable. Then we can consider the formally self-adjoint elliptic second-order differential operator $\nabla^\dagger\nabla + V$ (defined on smooth compactly supported sections) in the Hilbert space of square-integrable sections $L^2(M, E)$, which is defined using the metric on E and the measure $\mathrm{vol}(dx)$. While all applications in geometry lead to smooth potentials (which may be unbounded though), typical applications in quantum physics clearly lead to potentials having local singularities, such as, prototypically, the Coulomb potential. The above definition of covariant Schrödinger bundles has been chosen to treat these situations simultaneously. A systematic functional analytic study of this theory can be found in [68], where this concept is introduced.

In case $E = \wedge^k_{\mathbb{C}} TM$, we are going to use notations such as

$$L^2\Omega^k(M) = L^2(M, \wedge^k_{\mathbb{C}} TM).$$

The operator $\nabla^{\dagger}\nabla + V$ will be called the *formal covariant Schrödinger operator corresponding to*

$$(E, h, \nabla, V) \longrightarrow (M, g).$$

The simplest situation of this type is provided by scalar Schrödinger operators $\Delta + V$: Here one takes $E = M \times \mathbb{C}$ with h the standard metric on $\mathbb{C}$ and $V : M \to \mathbb{R}$ a locally square-integrable Borel function, where $\nabla = d_0$ is given by the de Rham differential acting on functions, that is, 0-forms. If further $V = 0$, then we obtain the *Laplace-Beltrami operator* Δ.

An important example of covariant Schrödinger bundles is provided by the *k-Hodge Laplacian*

$$\Delta_k : \Omega^k(M) \longrightarrow \Omega^k(M)$$

on (M, g) acting on k-forms. This is the nonnegative, formally self-adjoint and second-order elliptic operator with smooth coefficients defined by

$$\Delta_k := d_k^{\dagger} \circ d_k + d_{k-1} \circ d_{k-1}^{\dagger}.$$

Indeed, according to the well-known Bochner-Weitzenböck formula, we have

$$\Delta_k = \left(\nabla^{\wedge^k T^* M}\right)^{\dagger} \nabla^{\wedge^k T^* M} + V_k,$$

where

$$V_k \in C^{\infty}(M, \mathrm{End}(\wedge^k T^* M))$$

depends only on the curvature of (M, g). In particular, when $k = 1$, we have $V_1 = \mathrm{Ric}$ where Ric stands for the Ricci curvature of (M, g). This particular covariant Schrödinger bundle is linked with Dirac operators: If we define the total Hodge Laplacian by

$$\underline{\Delta} := \oplus_{k=1}^{m} \Delta_k : \Omega(M) \longrightarrow \Omega(M),$$

then one has $\underline{\Delta} = D^2$, and clearly $\underline{\Delta}$ is a formal covariant Schrödinger operator having a smooth potential. This situation is not a coincidence: In fact, given an arbitrary Dirac bundle, the square of its Dirac operator always becomes a formal covariant Schrödinger operator with a smooth potential, so that Dirac bundles canonically induce covariant Schrödinger bundles with smooth potentials in this case.

3.2 Some Hilbert space theory (F Bei, B Güneysu)

3.2.1 Self-adjoint operators

As we have noted above, formal covariant Schrödinger operators as well as Dirac operators are *symmetric* operators in their respective Hilbert spaces. However, as we will explain in a moment, one is actually interested in *self-adjoint operators*. The aim of this section is to put the geometric operators of interest into this perspective.

Let $\mathscr{H}$ be a complex separable Hilbert space. Given a linear operator S in $\mathscr{H}$ we denote with $\mathrm{Dom}(S) \subset \mathscr{H}$ its domain, with $\mathrm{Ran}(S) \subset \mathscr{H}$ its range, and with $\mathrm{Ker}(S) \subset \mathscr{H}$ its kernel. Let us assume in the sequel that S is a densely defined operator in $\mathscr{H}$. As S is densely defined, its adjoint S^* is well defined, and S is called *symmetric*, if $S \subset S^*$ (that is, if S^* is an extension of S), and S is called *self-adjoint*, if one even has $S = S^*$.

Although it is true that the eigenvalues of a symmetric operator are real with eigenvectors corresponding to different eigenvalues orthogonal, it turns out that self-adjoint operators have many important additional properties. For example, a symmetric operator is self-adjoint if and only if its spectrum is real, a question which is obviously of considerable interest for quantum mechanical measurements. It is also the class of self-adjoint operators on $\mathscr{H}$ which stands in a bijective correspondence with orthogonal-projection-valued Borel probability measures

$$P : \{\text{Borel sets on } \mathbb{R}\} \longrightarrow \mathscr{L}(\mathscr{H}).$$

Namely, given such a P and a Borel function $\phi : \mathbb{R} \to \mathbb{R}$, one can canonically construct a densely defined operator $\phi(P)$ in $\mathscr{H}$ by means of a spectral integral

$$\phi(P) := \int \phi(\lambda)\, dP(\lambda).$$

Then $\phi(P)$ is bounded (semibounded, self-adjoint) if and only if ϕ is bounded (semibounded, real-valued) on the spectrum $\mathrm{spec}(H)$ of

$$H := \int \lambda\, dP(\lambda). \tag{58}$$

The *spectral theorem* of J. von Neumann now states that, conversely, for every self-adjoint operator H in $\mathscr{H}$ there exists precisely one projection-valued Borel probability measure P_H on $\mathbb{R}$ which satisfies eq. (58). For an arbitrary Borel function ϕ one then sets $\phi(H) := \phi(P_H)$. In particular, this construction allows to study the Schrödinger equation and the heat equation from an abstract functional analytic point of view. Indeed, if H is a self-adjoint operator in $\mathscr{H}$, then for every $\psi \in \mathrm{Dom}(H)$, the path

$$\mathbb{R} \ni t \longmapsto \psi(t) := \mathrm{e}^{-itH}\psi \in \mathscr{H}$$

is the unique (norm-)differentiable path with $\psi(0) = \psi$ which solves the *abstract Schrödinger equation*

$$(d/dt)\psi(t) = i\,H\psi(t), \quad t \in \mathbb{R}.$$

If in addition H is semibounded, then for every $\psi \in \mathscr{H}$ the path

$$[0,\infty) \ni t \longmapsto \psi(t) := \mathrm{e}^{-tH}\psi \in \mathscr{H}$$

is the uniquely determined continuous path with $\psi(0) = \psi$ which is differentiable in $(0,\infty)$ and satisfies there the *abstract heat equation*

$$(d/dt)\psi(t) = -H\psi(t).$$

The collection $(\mathrm{e}^{-tH})_{t\geq 0}$ forms a semigroup of operators and is called the *heat semigroup of H*.

Let H be an arbitrary self-adjoint operator in $\mathscr{H}$. We recall that the *resolvent set of H* is defined to be the set of all $z \in \mathbb{C}$ such that $H - z$ is continuously invertible as a linear map $\mathrm{Dom}(H) \to \mathscr{H}$. For such z, $(H-z)^{-1}$ is automatically bounded as a linear operator from $\mathscr{H}$ to $\mathscr{H}$ by the closed graph theorem, and the bounded operator

$$R_H(z) := (H-z)^{-1} : \mathscr{H} \longrightarrow \mathscr{H}$$

is called the *resolvent of H*.

The *spectrum* $\mathrm{spec}(H)$ of H is defined as the complement of the resolvent set of H in $\mathbb{C}$. Resolvent sets are open subsets of $\mathbb{C}$, and so spectra are always closed. A number $z \in \mathbb{C}$ is called an *eigenvalue of H*, if $\mathrm{Ker}(H-z) \neq \{0\}$; then $\dim \mathrm{Ker}(H-z)$ is called the *multiplicity* of z, and each $f \in \mathrm{Ker}(H-z)\setminus\{0\}$ is called an *eigenvector of H corresponding to z*.

Remark 29. For many applications, resolvents and heat semigroups are "equivalent concepts" in the following sense: Let H be a self-adjoint and semibounded from below operator in $\mathscr{H}$. Then one has the Post-Wedder formula

$$\exp(-tH) = \lim_{k\to\infty} (k/t)^k (H + (k/t))^{-k} \quad \text{for all } t > 0,$$

where the convergence is in the strong operator topology. Conversely, there is the Laplace transform, which states that for all z with real part $< \inf\ \mathrm{spec}(H)$ one has

$$R_H(z) = (H-z)^{-1} = \int_0^\infty \mathrm{e}^{zt}\mathrm{e}^{-tH}\,dt,$$

the integral being strongly convergent. These results are also reflected by the fact that the spectrum of H is purely discrete iff *either* $R_H(z)$ is compact for some z *or* $\exp(-tH)$ is compact for some $t > 0$.

Next, we recall a canonical decomposition of the spectrum of a self-adjoint operator: define the essential spectrum $\mathrm{spec}_{\mathrm{ess}}(H) \subset \mathrm{spec}(H)$ to be the set of all real numbers λ such that λ is either an eigenvalue of H with an infinite multiplicity, or a limit point of $\mathrm{spec}(H)$. Then the *discrete spectrum*

$$\mathrm{spec}_{\mathrm{dis}}(H) := \mathrm{spec}(H) \setminus \mathrm{spec}_{\mathrm{ess}}(H)$$

is precisely the set of isolated points of $\mathrm{spec}(H)$ that are eigenvalues of H having a finite multiplicity. A natural question that arises in "noncompact situations" is that of the stability of the essential spectrum under a perturbation. In this context, the following definition plays a central role.

Definition 30. A possibly unbounded operator V in $\mathscr{H}$ is called a *relatively compact perturbation of H* if $VR_H(z)$ is a compact operator in $\mathscr{H}$ for some (equivalently: all) z in the resolvent set of H.

The importance of this definition stems from the following basic result from perturbation theory.

Theorem 31. *Assume that H is self-adjoint (and semibounded), and that the symmetric operator V is a relatively compact perturbation of an operator H. Then $H + V$ is self-adjoint (and semibounded) on its natural domain, and*

$$\mathrm{spec}_{\mathrm{ess}}(H) = \mathrm{spec}_{\mathrm{ess}}(H + V).$$

As in typical applications both operators H and V are unbounded, the determination of applicable conditions that ensure the relative compactness of V with respect to H is a very subtle business.

Another important part of (the spectrum of) a self-adjoint operator H in $\mathscr{H}$ is its *absolutely continuous part*: we recall that the *absolutely continuous subspace of $\mathscr{H}$ with respect to H, $\mathscr{H}_{\mathrm{ac}}(H)$*, is defined to be the space of all $f \in \mathscr{H}$, such that the Borel measure $\|P_H(\lambda)f\|^2\, d\lambda$ on $\mathbb{R}$ is absolutely continuous with respect to the Lebesgue measure. It turns out that $\mathscr{H}_{\mathrm{ac}}(H)$ is a closed subspace of $\mathscr{H}$ which reduces H. In particular, the restriction H_{ac} of H to $\mathscr{H}_{\mathrm{ac}}(H)$ is again a self-adjoint operator, and one defines the *absolutely continuous spectrum of H* to be

$$\mathrm{spec}_{\mathrm{ac}}(H) := \mathrm{spec}(H_{\mathrm{ac}}).$$

The vectors from $\mathscr{H}_{\mathrm{ac}}(H)$ typically correspond to *dynamic states*, by the so called *RAGE Theorem* (named after Ruelle, Amrein, Georgescu, Enss; cf. Theorem 12.10 in [122]):

Theorem 32. *Assume that*

$$\overline{\bigcup_{n\in\mathbb{N}} \mathscr{H}_n} = \mathscr{H}$$

is an exhaustion of $\mathscr{H}$ with closed subspaces, and let P_n denote the orthogonal projection onto $\mathscr{H}_n$. If the "cut-off" spectral projector $P_n P_H(I)$ is compact for all $n \in \mathbb{N}$ and all bounded intervalls $I \subset \mathbb{R}$, then every $\psi \in \mathscr{H}_{\mathrm{ac}}(H)$ is a scattering state *with respect to $(\mathscr{H}_n)$, in the sense that for every $f \in \mathscr{H}_{\mathrm{ac}}(H)$ and every $n \in \mathbb{N}$ one has*

$$\lim_{|t|\to\infty} \left\| P_n \mathrm{e}^{-\mathrm{i}tH} f \right\| \to 0.$$

In particular, there is a natural interest to determine $\mathscr{H}_{\mathrm{ac}}(H)$ for a given H. We illustrate the above RAGE theorem in the setting of covariant Schrödinger operators.

Example 33. Given a covariant Schrödinger bundle $(E, h, \nabla, V) \to (M, g)$ with a locally bounded potential V, assume that there exists a semibounded self-adjoint extension H of $\nabla^\dagger\nabla + V$ in the Hilbert space $L^2(M, E)$. It follows from local elliptic regularity that the heat semigroup e^{-tH} has a jointly smooth integral kernel

$$(0, \infty) \times M \times n \ni (t, x, y) \longmapsto \mathrm{e}^{-tH}(x, y) \in \mathrm{Hom}(E_y, E_y).$$

This easily implies that for every compact $K \subset M$, the operator $1_K \mathrm{e}^{-tH}$ is Hilbert-Schmidt, thus the "cut-off" resolvent

$$1_K R_H(-1) = \int_0^\infty \mathrm{e}^{-t} 1_K \mathrm{e}^{-tH} dt$$

is compact (being a norm-limit of compact operators). It now follows easily from an abstract result (Satz 12.12 in [122]) that for every bounded interval $I \subset \mathbb{R}$ the operator $1_K P_H(I)$ is compact. In particular, the RAGE theorem implies that every H-absolutely continuous state is a scattering state for

$$L^2(M, E) = \overline{\bigcup_{n\in\mathbb{N}} L^2(K_n, E)},$$

where (K_n) is any exhaustion of M with compact subsets, and $L^2(K_n, E)$ denotes the subspace of $L^2(M, E)$ given by all f such that $f = 0$ a.e. on $M \setminus K_n$. Note that we have used the local boundedness assumption on V only to get that $\mathrm{e}^{-tH}(x, y)$ is locally bounded in (x, y), a property which should also hold for many V's having local singularities. We will come back to this below.

One perturbative way to determine the absolutely continuous states of a self-adjoint operator is provided by proving the completeness of appropriate wave operators.

We formulate such a result in a two-Hilbert space variant, as it is natural for Riemannian geometry; one then considers two self-adjoint operators H_j in $\mathscr{H}_j$, $j = 1,2$. Assuming the existence of an appropriate bounded operator $I_{1,2} : \mathscr{H}_1 \to \mathscr{H}_2$ one considers the corresponding *wave operators*

$$W_\pm(H_2, H_1, I_{1,2}) = \mathrm{s} - \lim_{t\to\pm\infty} \mathrm{e}^{\mathrm{i}tH_2} I_{1,2} \mathrm{e}^{-\mathrm{i}tH_1} P_{\mathrm{ac}}(H_1)$$

with $P_{\mathrm{ac}}(H_j)$ the projection onto $\mathscr{H}_{j,\mathrm{ac}}(H_j)$. Assuming the existence of the wave operators (that is, of the strong limits in their definition), one calls the scattering problem $(H_1, H_2, I_{1,2})$ *complete*, if the following relations hold true:

$$(\mathrm{Ker}\, W_\pm(H_2, H_1, I_{1,2}))^\perp = \mathscr{H}_{1,\mathrm{ac}}(H_1), \quad \overline{\mathrm{Ran}\,(W_\pm(H_2, H_1, I_{1,2}))} = \mathscr{H}_{2,\mathrm{ac}}(H_2).$$

Then a classical result of Kato states:

Theorem 34. *Assume in the above situation that the wave operators $W_\pm(H_2, H_1, I_{1,2})$ exist and are complete. Then the operators $H_{1,\mathrm{ac}}$ and $H_{2,\mathrm{ac}}$ are unitarily equivalent. In particular, one has*

$$\mathrm{spec}_{\mathrm{ac}}(H_1) = \mathrm{spec}_{\mathrm{ac}}(H_2).$$

3.2.2 Self-adjoint extensions of a symmetric operator

Typically, one starts with a densely defined symmetric operator $\tilde{H}$ in a Hilbert space and is then faced with the following problems:

(a) Do there exist self-adjoint extensions of $\tilde{H}$?

(b) If so, how can one construct canonical self-adjoint extensions of $\tilde{H}$?

The most satisfactory situation is of course when $\tilde{H}$ has precisely one self-adjoint extension, in other words, when $\tilde{H}$ is essentially self-adjoint. The essential self-adjointness of the symmetric operators in Riemannian geometry typically requires geodesic completeness assumptions.

Example 35. Assume (M,g) is geodesically complete. Then a result of Chernoff [38] shows that every power $\tilde{H} = (D^c_\nabla)^n$ in $L^2(M,E)$, $n \in \mathbb{N}$, of the Dirac operator corresponding to a Dirac bundle $(E, h, c, \nabla) \to (M,g)$ is essentially self-adjoint (when defined on smooth compactly supported sections). Given a covariant Schrödinger bundle $(E, h, \nabla, V) \to (M,g)$ with $V \geq$ const. in $L^2(M, \mathrm{End}(E))$, where (M,g) is geodesically complete, then a result by Braverman, Shubin, Milatovic [21] shows that $\tilde{H} = \nabla^\dagger\nabla + V$ is essentially self-adjoint in $L^2(M,E)$, too. Of course such an assumption on the potential is too restrictive for applications in quantum mechanics, where typically $V(x) \sim -1/|x|$. We are going to address essential self-adjointness results that can also cover these situations below.

However, often the underlying Riemannian manifold is not geodesically complete (for example, if M is an open subset of $\mathbb{R}^m$) and one has to proceed differently. To this end, the following classical result of K. Friedrichs provides a satisfactory answer for symmetric operators that are semibounded from below, a result which often applies to covariant Schrödinger operators. Before recalling it, we introduce the following notations: assume that $\tilde{H}$ is a symmetric operator in the separable complex Hilbert space $\mathscr{H}$ such that $\tilde{H}$ is semibounded from below, or in other words, that there exists a constant $C \geq 0$ such that for all $f \in \mathrm{Dom}(\tilde{H})$ one has $\langle \tilde{H}f, f \rangle \geq -C \|f\|^2$. Let us denote by $Q_{\tilde{H}}$ the inner product on $\mathrm{Dom}(\tilde{H})$ given by

$$Q_{\tilde{H}}(f_1, f_2) := C\langle f_1, f_2\rangle + \langle \tilde{H}f_1, f_2 \rangle$$

and let $\mathrm{Dom}(Q_{\tilde{H}})$ be the completion of $\mathrm{Dom}D(\tilde{H})$ with respect to $Q_{\tilde{H}}$. Clearly, the identity $\mathrm{Dom}(\tilde{H}) \to \mathrm{Dom}(\tilde{H})$ extends to a bounded injective map

$$i_{Q_{\tilde{H}}} : \mathrm{Dom}(Q_{\tilde{H}}) \longrightarrow \mathscr{H}.$$

Therefore we can identify $\mathrm{Dom}(Q_{\tilde{H}})$ with its image in $\mathscr{H}$ through $i_{Q_{\tilde{H}}}$, which is precisely the space of all $u \in \mathscr{H}$ which admit a sequence $\{u_n\}_{n\in\mathbb{N}} \subset \mathrm{Dom}(\tilde{H})$ such that

$$\langle u_n - u, u_n - u\rangle \to 0 \text{ and } Q_{\tilde{H}}(u_n - u_m, u_n - u_m) \to 0 \text{ as } m, n \to \infty.$$

Now we can formulate:

Theorem 36. *As above assume that $\tilde{H}$ is a symmetric operator in the separable complex Hilbert space $\mathscr{H}$ such that $\tilde{H}$ is semibounded from below, or in other words, that there exists a constant $C \geq 0$ such that for all $f \in \mathrm{Dom}(\tilde{H})$ one has*

$$\langle \tilde{H}f, f \rangle \geq -C \|f\|^2.$$

Then there exist self-adjoint semibounded extensions of $\tilde{H}$. Moreover, a canonically given self-adjoint semibounded extension H of $\tilde{H}$ is provided by

$$\mathrm{Dom}(H) = \mathrm{Dom}(Q_{\tilde{H}}) \cap \mathrm{Dom}((\tilde{H})^*) \text{ and } Hu := (\tilde{H})^* u \text{ for each } u \in \mathrm{Dom}(H).$$

The operator H is called the Friedrichs extension *of $\tilde{H}$. It satisfies the following alternative characterization:* $\mathrm{Dom}(H)$ *is precisely the space of all $u \in \mathrm{Dom}(Q_{\tilde{H}})$ which admits a $v \in H$ with*

$$Q_{\tilde{H}}(u, w) = \langle v, w\rangle \text{ for any } w \in \mathrm{Dom}(Q_{\tilde{H}}),$$

and then $Hu = v$.

Example 37. Clearly, this result applies to Bochner Laplacians of the form $\tilde{H} = \nabla^\dagger\nabla$, defined on smooth compactly supported sections. It is thus tempting to apply this result also to general formal covariant Schrödinger operators $\nabla^\dagger\nabla + V$ with potentials V in $L^2_{\mathrm{loc}}(M, \mathrm{End}(E))$. However, the following question turns out to be rather delicate: When is such a covariant Schrödinger operator actually semibounded from below? This is certainly the case if $V \geq$ const, but in general the semiboundedness of $\nabla^\dagger\nabla + V$ turns out to provide a subtle condition of some negative part of the potential V. We will come back to this question later on.

As Dirac operators are never semibounded from below, one has to proceed differently in order to obtain canonical self-adjoint realizations. A construction that nevertheless always delivers two canonical self-adjoint extensions of the Dirac operator $d + d^\dagger$ is provided by the concept of *Hilbert complexes*, introduced in [28].

Definition 38. A *Hilbert complex* $(\mathscr{H}_\bullet, \tilde{D}_\bullet)$ is a complex of the form:

$$0 \to \mathscr{H}_0 \overset{\tilde{D}_0}{\to} \mathscr{H}_1 \overset{\tilde{D}_1}{\to} \mathscr{H}_2 \overset{\tilde{D}_2}{\to} \dots \overset{\tilde{D}_{n-1}}{\to} \mathscr{H}_n \to 0, \tag{59}$$

where each $\mathscr{H}_i$ is a separable Hilbert space and each map $\tilde{D}_i$ is a closed operator called the differential such that:

(a) $\mathrm{Dom}(\tilde{D}_i)$ is dense in $\mathscr{H}_i$.
(b) $\mathrm{Ran}(\tilde{D}_i) \subset \mathrm{Dom}(\tilde{D}_{i+1})$.
(b) $\tilde{D}_{i+1} \circ \tilde{D}_i = 0$ for all i.

The cohomology groups of the complex are $H^i(\mathscr{H}_\bullet, \tilde{D}_\bullet) := \mathrm{Ker}(\tilde{D}_i)/\mathrm{Ran}(\tilde{D}_{i-1})$. Given a Hilbert complex there is a dual Hilbert complex

$$0 \leftarrow \mathscr{H}_0 \overset{\tilde{D}_0^*}{\leftarrow} \mathscr{H}_1 \overset{\tilde{D}_1^*}{\leftarrow} \mathscr{H}_2 \overset{\tilde{D}_2^*}{\leftarrow} \dots \overset{\tilde{D}_{n-1}^*}{\leftarrow} \mathscr{H}_n \leftarrow 0, \tag{60}$$

defined using $\tilde{D}_i^* : \mathscr{H}_{i+1} \to \mathscr{H}_i$, the Hilbert space adjoint of the differential $\tilde{D}_i : \mathscr{H}_i \to \mathscr{H}_{i+1}$. The cohomology groups of $(\mathscr{H}_\bullet, \tilde{D}_\bullet^*)$, the dual Hilbert complex, are

$$\mathrm{H}^i(\mathscr{H}_\bullet, \tilde{D}_\bullet^*) := \mathrm{Ker}(\tilde{D}_{n-i-1}^*)/\mathrm{Ran}(\tilde{D}_{n-i}^*).$$

An important self-adjoint operator associated to eq. (59) is the following one: let us put $\mathscr{H} := \bigoplus_{i=0}^n \mathscr{H}_i$ and let

$$\tilde{D} + \tilde{D}^* : \mathscr{H} \to \mathscr{H} \tag{61}$$

be the self-adjoint operator with domain

$$\mathrm{Dom}(\tilde{D} + \tilde{D}^*) = \bigoplus_{i=0}^{n} (\mathrm{Dom}(\tilde{D}_i) \cap \mathrm{Dom}(\tilde{D}_{i-1}^*))$$

and defined as

$$\tilde{D}+\tilde{D}^* := \bigoplus_{i=0}^{n}(\tilde{D}_i+\tilde{D}^*_{i-1}).$$

Moreover, for all i, there is also a Laplacian

$$\tilde{\Delta}_i = \tilde{D}^*_i\tilde{D}_i + \tilde{D}_{i-1}\tilde{D}^*_{i-1}$$

which is a self-adjoint operator on $\mathscr{H}_i$ with domain

$$\mathrm{Dom}(\tilde{\Delta}_i) = \{v \in \mathrm{Dom}(\tilde{D}_i)\cap \mathrm{Dom}(\tilde{D}^*_{i-1}) : \tilde{D}_i v \in \mathrm{Dom}(\tilde{D}^*_i), \tilde{D}^*_{i-1}v \in \mathrm{Dom}(\tilde{D}_{i-1})\} \tag{62}$$

and nullspace:

$$\mathscr{H}^i(\mathscr{H}_\bullet, \tilde{D}_\bullet) := \mathrm{Ker}(\tilde{\Delta}_i) = \mathrm{Ker}(\tilde{D}_i)\cap \mathrm{Ker}(\tilde{D}^*_{i-1}). \tag{63}$$

The following propositions are well known. The first result is the weak Kodaira decomposition (cf. Lemma 2.1 in [28]).

Proposition 39. *Let $(\mathscr{H}_\bullet, \tilde{D}_\bullet)$ be a Hilbert complex and $(\mathscr{H}_\bullet, \tilde{D}^*_\bullet)$ its dual complex, then:*

$$\mathscr{H}_i = \mathscr{H}^i \oplus \overline{\mathrm{Ran}(\tilde{D}_{i-1})} \oplus \overline{\mathrm{Ran}(\tilde{D}^*_i)}. \tag{64}$$

The reduced cohomology groups of the complex are:

$$\overline{\mathrm{H}}^i(\mathscr{H}_\bullet, \tilde{D}_\bullet) := \mathrm{Ker}(\tilde{D}_i)/(\overline{\mathrm{Ran}(\tilde{D}_{i-1})}).$$

By the above proposition, there is a pair of weak de Rham isomorphism theorems:

$$\begin{cases} \mathscr{H}^i(\mathscr{H}_\bullet, \tilde{D}_\bullet) \cong \overline{\mathrm{H}}^i(\mathscr{H}_\bullet, \tilde{D}_\bullet) \\ \mathscr{H}^i(\mathscr{H}_\bullet, \tilde{D}_\bullet) \cong \overline{\mathrm{H}}^{n-i}(\mathscr{H}_\bullet, \tilde{D}^*_\bullet) \end{cases} \tag{65}$$

where in the second case we mean the cohomology of the dual Hilbert complex.

A natural question that arises in this framework is whether the groups $\mathrm{H}^i(\mathscr{H}_\bullet, \tilde{D}_\bullet)$, $\mathscr{H}^i(\mathscr{H}_\bullet, \tilde{D}_\bullet)$ are finite dimensional. When the former condition is fulfilled, that is $\mathrm{H}^i(\mathscr{H}_\bullet, \tilde{D}_\bullet)$ is finite dimensional for all i, then the Hilbert complex $(\mathscr{H}_\bullet, \tilde{D}_\bullet)$ is called a *Fredholm complex*. When the latter condition is satisfied, that is $\mathscr{H}^i(\mathscr{H}_\bullet, \tilde{D}_\bullet)$ is finite dimensional for each i, then the Hilbert complex $(\mathscr{H}_\bullet, \tilde{D}_\bullet)$ is called *weakly Fredholm*. By the next proposition we get immediately that each Fredholm complex is a weak Fredholm complex (cf. Corollary 2.5 in [28]).

Proposition 40. *If the cohomology of a Hilbert complex $(\mathscr{H}_\bullet, \tilde{D}_\bullet)$ is finite dimensional then, for all i, $\mathrm{Ran}(\tilde{D}_{i-1})$ is closed and $\mathrm{H}^i(\mathscr{H}_\bullet, \tilde{D}_\bullet) \cong \mathscr{H}^i(\mathscr{H}_\bullet, \tilde{D}_\bullet)$.*

The next proposition provides a characterization of the Fredholm/weakly Fredholm property (cf. Theorem 2.4 from [28]).

Proposition 41. *The following properties are equivalent:*

(a) *Eq.* (59) *is a Fredholm complex.*
(b) *The operator defined in eq.* (61) *is a Fredholm operator on its domain endowed with the graph norm.*
(c) *For all* $i = 0, \dots, n$, *the operator* $\tilde{\Delta}_i : \mathrm{Dom}(\tilde{\Delta}_i) \to \mathscr{H}_i$ *is a Fredholm operator on its domain endowed with the graph norm.*

Analogously the following three properties are equivalent:

(a) (59) *is a weakly Fredholm complex*
(b) *The operator defined in* (61) *has finite-dimensional nullspace.*
(c) *For all* $i = 0, \dots, n$, *the space* $\mathrm{Ker}(\tilde{\Delta}_i)$ *is finite dimensional.*

Finally, we recall how Hilbert complexes appear naturally in the context of Riemannian and Hermitian geometry, providing canonical self-adjoint extensions of the canonical Dirac operators.

Let (M, g) be an open and possibly incomplete Riemannian manifold. Consider the de Rham complex $(\Omega_c^{\bullet}(M), d_{\bullet})$ where each form $\omega \in \Omega_c^i(M)$ is a i-form with compact support. To turn this complex into a Hilbert complex we must specify a closed extension of d_i. With the two following definitions we will recall the two canonical closed extensions of d_i:
The *maximal extension* is the operator with domain

$$\mathrm{Dom}(d_{i,\max}) = \{\omega \in L^2\Omega^i_{L^2}(M) : \exists\, \eta \in L^2\Omega^{i+1}(M) \tag{66}$$

$$\text{s.t. } \langle \omega, d_i^\dagger \zeta \rangle_{L^2\Omega^i(M)} = \langle \eta, \zeta \rangle_{L^2\Omega^{i+1}(M)}\ \forall\, \zeta \in \Omega_c^{i+1}(M)\}. \tag{67}$$

In this case, $d_{i,\max}\omega = \eta$. In other words, $\mathrm{Dom}(d_{i,\max})$ is the largest set of forms $\omega \in L^2\Omega^i(M)$ such that $d_i\omega$, computed distributionally, is also in $L^2\Omega^{i+1}(M)$.

The *minimal extension* $d_{i,\min}$ is given by the graph closure of d_i on $\Omega_c^i(M)$ with respect to the norm of $L^2\Omega^i(M)$, that is, $\mathrm{Dom}(d_{\min,i})$ is given by all $\omega \in L^2\Omega^i(M)$ which admit a sequence $\{\omega_j\}_{j\in\mathbb{N}} \subset \Omega_c^i(M)$ such that $\omega_j \to \omega$ and

$$d_i\omega_j \to \eta \in L^2\Omega^{i+1}(M),$$

and then $d_{i,\min}\omega := \eta$.

Obviously $\mathrm{Dom}(d_{i,\min}) \subseteq \mathrm{Dom}(d_{i,\max})$. Furthermore, from these definitions, we have immediately that

$$d_{i,\min}(\mathrm{Dom}(d_{i,\min})) \subset \mathrm{Dom}(d_{i+1,\min}),\ d_{i+1,\min} \circ d_{i,\min} = 0$$

and that

$$d_{i,\max}(\mathrm{Dom}(d_{i,\max})) \subset \mathrm{Dom}(d_{i+1,\max}), \ d_{i+1,\max} \circ d_{i,\max} = 0.$$

Therefore $(L^2\Omega^\bullet(M,g), d_{\bullet,\max/\min})$ are both Hilbert complexes and their cohomology groups are denoted by $\mathrm{H}^\bullet_{2,\max/\min}(M)$. Likewise, we have the associated minimal and maximal Laplacians

$$\Delta_{i,\max/\min} := d^*_{i,\max/\min} d_{i,\max/\min} + d_{i-1,\max/\min} d^*_{i-1,\max/\min}, \tag{68}$$

where

$$\Delta_{\max/\min} := \Delta_{0,\max/\min}$$

on functions. In the sequel, we will be mostly interested in minimal extensions, so in order to keep the notation simple, we will stick to the following small abuse of notation:

Remark 42. 1. As $\underline{\Delta} = D^2$ and thus each Δ_i is semibounded, these operators have Friedrichs realizations which will be denoted with the same symbol again. In fact, it is easily checked that with this abuse of notation one has

$$\Delta_i = \Delta_{i,\min}, \quad \Delta = \Delta_{0,\min}$$

2. In case (M,g) is geodesically complete, Chernoff's result (cf. Example 35) implies that $\underline{\Delta} = D^2$ is essentially self-adjoint, so that each Δ_i is essentially self-adjoint. In particular, we then have

$$\Delta_i := \Delta_{i,\min} = \Delta_{i,\max}.$$

3.3 Heat semigroups (B Güneysu)

In this section, we want to provide some functional analytic results for the heat semigroups associated with covariant Schrödinger operators on arbitrary (possibly incomplete) Riemannian manifolds, which will later on be used in order to obtain some spectral theoretic results, as well as results for spectral geometry. So let (M,g) be a possibly incomplete connected Riemannian manifold and let

$$(E, h, \nabla, V) \longrightarrow (M, g)$$

be a covariant Schrödinger bundle with $V \in L^2_{\mathrm{loc}}(M, \mathrm{End}(E))$ and set $m := \dim(M)$. As we do not assume that (M,g) is geodesically complete (so that it cannot be expected that $\nabla^\dagger\nabla + V$ is essentially self-adjoint), it is again tempting to use the Friedrichs construction in order to get a canonical self-adjoint realization of $\nabla^\dagger\nabla + V$. To this end, we have to determine whether $\nabla^\dagger\nabla + V$ is semibounded from below on smooth compactly supported sections. At this point, we simply make the following definition.

Definition 43. Assume $\nabla^\dagger\nabla + V$ is semibounded from below. Then the Friedrichs realization of this operator is denoted with H_∇^V and called the *covariant Schrödinger operator* associated with

$$(E, h, \nabla, V) \longrightarrow (M, g).$$

It is a self-adjoint operator in $L^2(M, E)$ which is semibounded from below. The heat semigroup

$$\left(\exp\left(- tH_\nabla^V\right)\right)_{t>0} \subset \mathscr{L}(L^2(M, E))$$

is called the induced *covariant Schrödinger semigroup.*

Remark 44. In the scalar situation, we will write Δ^V for the Friedrichs realization of $\Delta + V$ in $L^2(M)$, whenever $V : M \to \mathbb{R}$ is locally square integrable with $\Delta + V$ semibounded from below. Using our previous conventions, we have

$$\Delta^V|_{V=0} = \Delta = \Delta_0 = \Delta_{0,\min}$$

for the Friedrichs realization of the Laplace-Beltrami operator in $L^2(M)$.

It will turn out in a moment that a very general criterion for the negative part V_- of the potential that ensures the lower boundedness of $\nabla^\dagger\nabla + V$ can be formulated in terms of the Kato class $\mathscr{K}(M)$. To this end, we recall that using local parabolic regularity, the heat semigroup

$$(\exp(-t\Delta))_{t>0} \subset \mathscr{L}(L^2(M))$$

turns out to have a jointly smooth integral kernel

$$(0, \infty) \times M \times M \ni (t, x, y) \longmapsto p(t, x, y) \in [0, \infty),$$

and one always has

$$\int_M p(t, x, y) d\mathrm{vol}(y) \leq 1 \quad \text{for all } (t, x) \in (0, \infty) \times M. \tag{69}$$

Now we can make the following definition.

Definition 45. A Borel function $w : M \to \mathbb{R}$ is said to be in the *Kato class* $\mathcal{K}(M)$ of (M,g), if

$$\lim_{t\to 0+}\sup_{x\in M}\int_0^t\int_M p(s,x,y)|w(y)|d\mathrm{vol}(y)ds = 0. \tag{70}$$

The Kato class is a linear space and in view of eq. (69), one always has $L^\infty(M) \subset \mathcal{K}(M)$. The importance of this class for us stems from the following result, which appeared first in [60].

Proposition 46. *Assume that the fiberwise taken operator norm of V_- (which is a Borel function on M) satisfies $|V_-| \in \mathcal{K}(M)$. Then $\nabla^\dagger\nabla + V$ is semibounded from below, in particular, H_∇^V is well defined. Here and in the sequel, $V = V_+ - V_-$ denotes the decomposition of V into its positive part $V_+ \geq 0$ and its negative part $V_- \geq 0$ (both of which can be defined using the spectral calculus on the fibers of $(E,h) \to M$).*

The striking fact about this result is that it does not require any control on the geometry of (M,g). The question that remains at this point is: Under which conditions on the geometry of (M,g) can one actually find large subspaces of $\mathcal{K}(M)$? In the Euclidean space, inclusions of the form $L^q(\mathbb{R}^m) \subset \mathcal{K}(\mathbb{R}^m)$ have been established in [1]. If (M,g) has Ricci curvature bounded from below and a strictly positive injectivity radius, then this inclusion remains valid [77]. However, many interesting Riemannian manifolds do not have these properties (noting e.g. that a strictly positive injectivity radius implies geodesic completeness). In order to deal with these singular situations, the following definition has been proposed in [68] (see also [65]):

Definition 47. An ordered pair (F,L) of functions

$$F : M \longrightarrow (0,\infty],\ \ L : (0,\infty) \longrightarrow (0,\infty)$$

is called a *heat kernel control pair for the Riemannian manifold* (M,g), if the following assumptions are satisfied:

- F is continuous with $\inf F > 0$, L is Borel
- for all $x \in M$, $t > 0$, one has

$$\sup_{y\in M} p(t,x,y) \leq F(x)L(t)$$

- for all $q' \geq 1$ in case $m = 1$, and all $q' > m/2$ in case $m \geq 2$, one has

$$\int_0^\infty L^{1/q'}(t)e^{-At}dt < \infty \ \text{ for some } A > 0.$$

It is now straightforward to see that for every function F which fits into a heat kernel control pair for (M,g), one has the generally valid weighted L^q-subspace inclusion

$$L^q(M, F\mathrm{dvol}) \subset \mathscr{K}(M), \tag{71}$$

where $q > m/2$ if $m \geq 2$, and $q \geq 1$ if $m = 1$. What is not clear at all is whether an arbitrary Riemannian manifold admits a heat kernel control pair. This, however, has been established in [65, 68], using a localized parabolic L^1-mean value inequality for $p(t,x,y)$.

Theorem 48 (B Güneysu). *Every Riemannian manifold (M,g) admits a heat kernel control pair. In particular (in view of (71)), one has*

$$L^q_{\mathrm{loc}}(M) \subset \mathscr{K}_{\mathrm{loc}}(M),$$

for every $q > m/2$ if $m \geq 2$, and $q \geq 1$ if $m = 1$. Here, $\mathscr{K}_{\mathrm{loc}}(M)$ denotes the local Kato class[1] *of (M,g) formed by all w's such that $1_K w \in \mathscr{K}(M)$ for all compact $K \subset M$.*

The above inclusion

$$L^q_{\mathrm{loc}}(M) \subset \mathscr{K}_{\mathrm{loc}}(M),$$

is actually much more subtle than it may seem, as the heat kernel is a global object. At this point, however, it becomes a simple consequence of the existence of a heat kernel control pair. Another important feature of a heat kernel control pair is that whenever one has some control on the geometry, one can typically pick very explicit pairs.

Example 49. Assume that (M,g) is geodesically complete with $\mathrm{Ric} \geq -K$ for some constant $K \geq 0$. Then for every $\delta_1, \delta_2 > 0$ which satisfy

$$\delta_1\delta_2 > ((m-1)^2K)/8,$$

there exists a constant $C_{\delta_1,\delta_2,K,m} > 0$, which only depends on δ_j, K, and m, such that

$$F(x) := C_{\delta_1,\delta_2,K,m}\mathrm{vol}(B(x,1))^{-1},$$
$$L(t) := \left(\mathrm{e}^{\sqrt{(m-1)K}}t^{-m/2} + 1\right)\exp\left((\delta_2 - \min\mathrm{spec}(H))t\right)$$

is a heat kernel control pair for (M,g). This result can be found in [68], and it relies on the Li-Yau heat kernel bound [84, 114] and the Cheeger-Gromov volume estimate.

1 It is an open problem whether the local Kato class depends on the metric or not.

It is natural to consider various L^p-estimates for covariant Schrödinger semigroups with Kato potentials. In this context, the following *Kato-Simon inequality* is of a fundamental importance.

Theorem 50. *Assume* $|V_-| \in \mathcal{K}(M)$ *and let the scalar potential* $w : M \to \mathbb{R}$ *be in* $L^2_{\mathrm{loc}}(M)$ *with* $w_- \in \mathcal{K}(M)$. *Then the eigenvalue domination* $V \geq w$ *implies that for every* $t \geq 0$, $f \in L^2(M, E)$ *one has the following pointwise valid* Kato-Simon inequality

$$\left|\exp\left(-tH^V_\nabla\right)f\right| \leq \exp\left(-t\Delta^w\right)|f|. \tag{72}$$

Using Remark 29 one easily finds that inequality (72) is equivalent to its resolvent version

$$\left|R_{H^V_\nabla}(z)f\right| \leq R_{\Delta^w}(z)|f|. \tag{73}$$

Theorem 50 can be traced back for $M = \mathbb{R}^m$ to [109], for closed M's and smooth V's to [71], and the general case has been established in [64] using probabilistic methods (cf. [68] for a nonprobabilistic proof).

A central application of the Kato-Simon inequality is to deduce L^p bounds for covariant Schrödinger semigroups from those of appropriately chosen scalar Schrödinger operators. In order to state such a result, for an arbitrary $U \subset M$ and $t > 0$ set

$$C_U(t) := \sup_{x \in U, y \in M} p(t, x, y),$$

which is finite, if U is relatively compact [68]. Moreover, given a Borel function $v : M \to [0, \infty)$, a set $U \subset M$, and $t \geq 0$ we let

$$C_U(v, t) := \sup_{x \in U} \int_{\Omega(M,t)} e^{\int_0^t v(\gamma(s))ds} d\mathbb{P}^x(\gamma) \in [0, \infty]. \tag{74}$$

Above, the $\mathbb{P}^x$ denotes the *Brownian motion* (or Wiener) probability measure on the space Ω of possibly explosive continuous paths $\gamma : [0, \infty) \to \hat{M} = M \cup \{\infty\}$, and $\Omega(M, t) \subset \Omega$ denotes the paths in Ω that remain on M until the time $t \geq 0$. The Brownian motion measures satisfy the fundamental relation

$$\mathbb{P}^x\{\gamma : \gamma(t) \in U\} = \int_U p(t, x, y) d\mathrm{vol}(y) \quad \text{for all } t > 0, x \in M, U \subset M \text{ Borel}. \tag{75}$$

Here we follow the convention that Brownian motion is a Δ-diffusion process, rather than a $(1/2)\Delta$. As it turns out, if the geometry of (M, g) is bad at ∞, it may happen that

$$\mathbb{P}^x\{\Omega(M, t)\} = \mathbb{P}^x\{\gamma : \gamma(t) \in M\} < 1, \tag{76}$$

while in nice situations like on compact manifolds or in Euclidean space one has $\mathbb{P}^x\{\Omega(M, t)\} = 1$, which means that Brownian motions paths do not explode in a finite time. We will come back to this issue ("stochastic incompleteness") below. In any case, one can show that $C_U(v,t) < \infty$ for arbitrary subsets U, as long as $v \in \mathcal{K}(M)$ [68], without any further assumptions on the geometry of (M,g). Now we can formulate the following result from [68], which generalizes earlier results from [64]:

Theorem 51 (B Güneysu). *Let $|V_-| \in \mathcal{K}(M)$, let $w : M \to \mathbb{R}$ be in $L^2_{\mathrm{loc}}(M)$ with $w_- \in \mathcal{K}(M)$, and assume the eigenvalue domination $V \geq w$.*

a) *For any $t \geq 0$, $q \in [1,\infty]$, and any Borel $U \subset M$ (in particular for $U = M$) one has*

$$\left\| 1_U \exp\left(-tH^V_\nabla\right)|_{L^q \cap L^2(M,E)} \right\|_{L^q \to L^q} \leq C_U(w_-,t) < \infty. \tag{77}$$

b) *For every $t > 0$, every Borel $U \subset M$ with $C_U(t/2) < \infty$, and every $1 < q < \infty$ one has*

$$\left\| 1_U \exp\left(-tH^V_\nabla\right)|_{L^q \cap L^2(M,E)} \right\|_{L^q \to L^\infty} \leq C_U\left(q^* w_-,t\right) C_U(t)^{\frac{1}{q}}, \tag{78}$$

where $1 < q^* < \infty$ is determined by $1/q + 1/q^* = 1$.

c) *For every $t > 0$ with*[2] *$C_M(t/2) < \infty$ and every $1 \leq q_1 \leq q_2 \leq \infty$ one has*

$$\left\| \exp\left(-tH^V_\nabla\right)|_{L^{q_1} \cap L^2(M,E)} \right\|_{L^{q_1} \to L^{q_2}} \leq C_M\left(2w_-,t\right) C_M(t/2)^{\frac{1}{q_1}-\frac{1}{q_2}}. \tag{79}$$

The general proof strategy of Theorem 51 is as follows: In view of the Kato-Simon inequality it suffices to estimate the scalar semigroup $\exp(-t\Delta^w)$. This semigroup, on the other hand, is given explicitly by the Feynman-Kac formula

$$\exp\left(-t\Delta^w\right)\Psi(x) = \int_{\Omega(M,t)} \Psi(\gamma(t)) \exp\left(-\int_0^t w(\gamma(s))ds\right) d\mathbb{P}^x(\gamma),$$

so that one uses a combination of Hölder inequality and L^q-interpolation techniques to estimate the norms of interest. The (scalar) Feynman-Kac formula also explains how the quantities $C_U\left(qw_-,t\right)$ come into play. We remark here that a covariant version of the scalar Feynman-Kac formula also holds for general covariant Schrödinger semigroups $\exp(-tH^\nabla_V)$ [64].

Using the Lebesgue differentiation theorem, the results from Theorem 51 imply [68] that under the same assumptions, the Schrödinger semigroup has a well-defined integral kernel

$$(0,\infty) \times M \times M \ni (t,x,y) \longmapsto \exp(-tH^\nabla_V)(x,y) \in \mathrm{Hom}(E_y,E_x) \subset E \boxtimes E^* \tag{80}$$

which is locally bounded in (t,x,y), and in L^2 with respect to y for all fixed (t,x). Using the terminology and the remarks from Example 33 we thus get:

---

2 Note that such a $t > 0$ need not exist at all.

Corollary 52. *Let $|V_-| \in \mathcal{K}(M)$ and let (K_n) be any exhaustion of M with compact subsets. Then every H_V^∇-absolutely continuous state is a scattering state for the exhaustion*

$$L^2(M,E) = \overline{\bigcup_{n\in\mathbb{N}} L^2(K_n,E)}.$$

As another application of the existence of integral kernels with precise estimates thereof, the following result has been proved in [68].

Theorem 53 (B Güneysu). *Assume that $|V_-| \in \mathcal{K}(M)$ and that (M,g) has a finite volume and satisfies*

$$C_M(t) = \sup_{x,y\in M} p(t,x,y) < \infty \quad \textit{for all } t \in (0,1).$$

Then there exists constants $C_1, C_2 \geq 0$ such that for all $t \in (0,1)$ one has

$$\mathrm{tr}(\mathrm{e}^{-tH_\nabla^V}) \leq C_1 \mathrm{e}^{tC_2} C_M(t/4) < \infty.$$

In particular, H_∇^V has a purely discrete spectrum.

Concerning the continuity properties of covariant Schrödinger semigroups, we have established the following result which again holds on every Riemannian manifold [30]:

Theorem 54 (B Güneysu). *Assume $|V| \in \mathcal{K}(M)$. Then for all $f \in L^2(M,E)$, the map*

$$(0,\infty)\times M \ni (t,x) \longmapsto \exp\big(-tH_\nabla^V\big)f(x) \in E_x \subset E$$

is jointly continuous. In particular, all eigensections of H_∇^V are continuous.

We refer the reader also to [26] for earlier continuity results for magnetic Schrödinger operators on Riemannian manifolds having a bounded geometry, and also to [98].

3.4 Path integrals by finite-dimensional approximations (M Ludewig)

As we have previously indicated, the Feynman-Kac formula can serve as a central tool for the study of Schrödinger semigroups. The aim of this section is to explain that this formula can be rewritten in a more intuitive way, at least for compact Riemannian manifolds, explaining the previous functional analytic results from a more geometric point of view.

3.4.1 The path integral formula

Let (M, g) be a compact connected Riemannian n-manifold and let $V : M \to \mathbb{R}$ be a smooth scalar potential. As M is compact, the strict inequality (76) cannot occur, and the space of paths that do not explode in a finite time have a full Browian motion measure. In particular, the underlying Schrödinger semigroup can be written for an initial state $u_0 \in L^2(M)$ in terms of the Feynman-Kac formula according to

$$\exp(-t\Delta^V)u_0(x) = \int_{C([0,t],M)} \mathrm{e}^{-\int_0^t V(\gamma(s))} u_0(\gamma(t)) d\mathbb{P}^x(\gamma). \tag{81}$$

Heuristically, or from a physics point of view, it is expected that the RHS of the above formula can be formally rewritten, resulting in

$$\exp(-t\Delta^V)u_0(x) \overset{\text{formally}}{=} \fint_{C([0,t],M)} u_0(\gamma(t)) \exp\big(-S_V[t;\gamma]\big)\mathscr{D}\gamma, \tag{82}$$

where $\fint_{C([0,t],M)} \dots \mathscr{D}\gamma$ denotes integration with respect to some normalized infinite-dimensional "Riemannian volume" measure, and where the action functional $S_V[t;\gamma]$ is given by

$$S_V[t;\gamma] = \frac{1}{2}\int_0^t |\dot{\gamma}(s)|^2 + \int_0^t V\big(\gamma(s)\big)ds. \tag{83}$$

Note that we have omitted the underlying Riemannian metric g in the notation. As it stands, formula (82) obviously cannot make sense for several reasons: Firstly, the action functional is not well defined on continuous paths and secondly there exists no well-defined analog of the Riemannian volume measure in infinite dimensions. The aim of this section is to explain that, nevertheless, it is possible to find a well-defined version of eq. (82) in terms of *finite-dimensional approximations* of the "measure"

$$\fint_{C([0,t],M)} \exp\big(-S_0[t;\gamma]\big)\mathscr{D}\gamma,$$

by approximating the space of paths $C([0,t], M)$ by appropriately chosen spaces of piecewise geodesics.

To set the stage, given a partition

$$\tau = \{0 = \tau_0 < \tau_1 < \dots < \tau_N = t\}$$

of the time interval $[0,t]$, we write $\delta_j\tau := \tau_j - \tau_{j-1}$ for the corresponding increments. Then we define

$$H_{x;\tau}(M) := \big\{\gamma \in C([0,t], M) \mid \gamma(0) = x, \gamma|_{(\tau_{j-1},\tau_j)} \text{ is a geodesic for } j = 1,\dots,N\big\}.$$

This is a smooth manifold of dimension nN, where the tangent space at a path γ consists of the continuous vector fields X along γ with $X(0) = 0$ and such that $X|_{(\tau_{j-1},\tau_j)}$ is a Jacobi field. On $H_{x;\tau}(M)$, one can consider the H^1-metric

$$(X, Y)_{H^1} := \int_0^t \langle \nabla_s X(s), \nabla_s Y(s) \rangle ds, \tag{84}$$

but it turns out that its discretized version

$$(X, Y)_{\Sigma\text{-}H^1} := \sum_{j=1}^{N} \langle \nabla_s X(\tau_{j-1}+), \nabla_s Y(\tau_{j-1}+) \rangle \delta_j \tau \tag{85}$$

is a better choice for finite-dimensional approximations, for we have:

Theorem 55. *Let (M, g) be a compact connected Riemannian manifold and let $V : M \to \mathbb{R}$ be a smooth scalar potential. Then for every $t > 0$, $x \in M$, one has*

$$\exp(-t\Delta^V)u_0(x) = \lim_{|\tau|\to 0} \fint_{H_{x;\tau}(M)} \exp\big(-S_V(t;\gamma)\big)u\big(\gamma(t)\big)d^{\Sigma\text{-}H^1}\gamma,$$

where the limit goes over any sequence of partitions of the interval $[0, t]$, the mesh of which goes to zero, and the slash over the integral sign denotes divison by $(4\pi)^{\dim(H_{x;\tau}(M))/2}$. This result is true for u_0 in any of the spaces $C^0(M)$ or $L^p(M)$ with $1 \le p < \infty$, with the convergence in the respective space.

Much as the Feynman-Kac formula (81), this result also has a natural generalization to covariant Schrödinger semigroups. We refer the reader to [4] and, in more general settings, to [9, 90].

There are various ways to prove Thm. 55. The original proof by Andersson and Driver relies on techniques from stochastic analysis. The proof by Bär and Pfäffle (see [9]) relies on operator techniques and in particular uses Chernoff's theorem, which asserts that if P_t, $t \in [0, \infty)$, is a uniformly bounded family of linear operators on a Banach spaces with $P_0 = \mathrm{id}$ which satisfies $\|P_t\| = O(t)$ and possesses an infinitesimal generator $-H$ (where the infinitesimal generator is defined just as in the case of a semigroup of operators), then one has

$$\lim_{|\tau|\to 0} P_{\delta_1\tau} \cdots P_{\delta_N\tau} = \mathrm{e}^{-tH},$$

where the limit is in the strong operator topology and goes over any sequence of partitions of the interval $[0, t]$ with mesh going to zero (see [9, Thm. 2.8], Prop. 1 in [39] or [111]).

The advantage of the latter approach is that it readily generalizes to the case that M is a compact manifold with boundary. In this case, one replaces the spaces $H_{x;\tau}(M)$

of piecewise geodesics with the spaces $H^{\mathrm{refl}}_{x;\tau}(M)$ of *reflected* piecewise geodesics, and depending on the underlying boundary conditions, one has to include some additional factors in the path integral formula. One can treat a variety of boundary conditions using this approach, including Dirichlet- and Neumann boundary conditions (however, Robin boundary conditions do not belong to this class). We refer the reader to [90] or [91, Section 2] for details.

Let us finish this section with a quick discussion about what happens if one takes other metrics than the discrete H^1 metric (85) on $H_{x;\tau}(M)$. Initially, it may seem that the continuous version (84) is the more natural choice; however, there are results that indicate otherwise. Namely, Lim shows (under further curvature assumptions) that for a continuous function f on $C([0,t],M)$, one has

$$\lim_{|\tau|\to 0} \fint_{H_{x;\tau}(M)} \exp\big(-S_0[t;\gamma]\big) f(\gamma) d^{H^1}\gamma = \int_{C([0,t],M)} f(\gamma)\det\big(\mathrm{id}+K_\gamma\big) d\mathbb{P}^x(\gamma),$$

where K_γ is a certain trace-class integral operator on $L^2([0,t],\gamma^*TM)$ which is given in terms of the curvature along γ (see Def. 1.13 and Thm. 1.14 in [87]). Hence path integral formulas become considerably more complicated using this metric.

One can also consider the L^2 scalar product

$$(X,Y)_{L^2} = \int_0^1 \big\langle X(s), Y(s)\big\rangle ds \tag{86}$$

or its discretized version

$$(X,Y)_{\Sigma\text{-}L^2} = \sum_{j=1}^{N} \big\langle X(\tau_j), Y(\tau_j)\big\rangle \delta_j\tau \tag{87}$$

on $H_{x;\tau}(M)$. In terms of the discrete L^2-metric, one now has

$$\begin{aligned}&\exp(-t\Delta^V)u_0(x) = \\ &\lim_{|\tau|\to 0} C_\tau \fint_{H_{x;\tau}(M)} \exp\Big(-S_V[t;\gamma] - \frac{1}{6}\int_0^t \mathrm{scal}\big(\gamma(s)\big)ds\Big)\, u_0\big(\gamma(t)\big) d^{\Sigma\text{-}L^2}\gamma,\end{aligned} \tag{88}$$

where $C_\tau := \prod_{j=1}^N (\delta_j\tau)^{-n/2}$. Hence taking a different Riemannian metric on the path space (hence another volume form) results in a different pre-factor (which in this case also depends on the partition τ and not only on the dimension of the path space) and also a scalar curvature term appears (this result can be found in [4, 9]). In the case of the continuous L^2 metric, the pre-factor in front of the scalar curvature becomes $(3+2\sqrt{3})/60$ instead of $1/6$ and one gets an again different τ-dependent normalization factor (see Thm. 2.1 in [78]).

Conceptually, the L^2 metrics are somewhat less natural compared to the H^1 metric, since the action functional S_0 is not defined on L^2.

3.4.2 Approximation of the heat kernel

Let (M, g) be a compact connected Riemannian manifold and let $V : M \to \mathbb{R}$ be a smooth scalar potential. Much as the Feynman-Kac formula can be disintegrated using the Brownian bridge (or pinned Wiener) measures, resulting in a formula for $\exp(-t\Delta^V)(x,y)$, we should be able to obtain a path integral formula within our formalism for $\exp(-t\Delta^V)(x,y)$, too. We recall our convention

$$p(t,x,y) = \exp(-t\Delta)(x,y)$$

for the heat kernel of the Laplace-Beltrami operator on M (which is essentially-selfadjoint by the geodesic completeness of M). This time, the integration is expected to be over the spaces $H_{xy;\tau}(M)$ of piecewise geodesics that travel from x to y:

Definition 56. For a partition $\tau = \{0 = \tau_0 < \tau_1 < \cdots < \tau_N = t\}$ of the interval $[0,t]$, the space $H_{xy;\tau}(M)$ is the space of paths $\gamma \in H_{x;\tau}(M)$ with $\gamma(t) = y$ such that $\gamma|_{[\tau_{j-1},\tau_j]}$ is the unique shortest geodesic between its end points for each $j = 1, \dots, N$.

Clearly, via the evaluation map

$$\mathrm{ev}_\tau : H_{xy;\tau}(M) \longrightarrow M^{N-1}, \qquad \gamma \longmapsto \big(\gamma(\tau_1), \dots, \gamma(\tau_{N-1})\big)$$

maps $H_{xy;\tau}(M)$ injectively onto an open subset of M^{N-1}. This determines a manifold structure on $H_{xy;\tau}(M)$ (compare also [94, Lemma 16.1]). Let $H^\circ_{x;\tau}(M) \subseteq H_{x;\tau}(M)$ be the subset of paths γ such that $\gamma|_{[\tau_{j-1},\tau_j]}$ is the unique shortest geodesic between its end points for each $j = 1, \dots, N$ (it turns out that Theorem 55 also holds when one replaces $H_{x;\tau}(M)$ by $H^\circ_{x;\tau}(M)$, compare [9, Theorem 5.1]). Let us equip the spaces $H_{xy;\tau}(M)$ with a measure $d^{\Sigma\text{-}H^1}\gamma$ that satisfies the co-area formula

$$\int_{H^\circ_{x;\tau}(M)} f(\gamma)\, d^{\Sigma\text{-}H^1}\gamma = t^{-n/2} \int_M \int_{H_{xy;\tau}(M)} f(\gamma)\, d^{\Sigma\text{-}H^1}\gamma\, dy \tag{89}$$

for continuous bounded functions f on $H^\circ_{x;\tau}(M)$. Such a measure is constructed in Section 2.2.1 of [91] and it is explicitly given by (setting $x_0 := x$, $x_N := y$)

$$\begin{aligned} &\int_{H_{xy;\tau}(M)} f\big(\gamma(\tau_1), \dots, \gamma(\tau_{N-1})\big) d^{\Sigma\text{-}H^1}\gamma \\ &:= t^{n/2} \int_{M^{N-1}} f(x_1, \dots, x_{N-1}) \left(\prod_{j=1}^N J(x_{j-1}, x_j)(\delta_j\tau)^{n/2} \right)^{-1} dx_1 \cdots dx_{N-1}, \end{aligned} \tag{90}$$

where $J(x,y)$ is the Jacobian of the exponential map, given, for x, y close enough, by $J(x,y) = |\det d\exp_x|_{\dot\gamma_{xy}(0)}|$ (and γ_{xy} is the unique shortest geodesic traveling from x to y in time 1). For this measure, one then obtains from Theorem 55 that

$$(4\pi t)^{-n/2} \fint_{H_{xy;\tau}(M)} \exp\big(-S_0[t;\gamma]\big) d^{\Sigma\text{-}H^1}\gamma \longrightarrow p(t,x,y) \tag{91}$$

as $|\tau| \to 0$, where the convergence holds in the weak-$*$-topology of the space of finite Borel measures on M (with respect to one of the variables). Namely, if we define the approximate heat operator P_τ on $C^0(M)$ by the formula

$$(P_\tau u_0)(x) := \fint_{H^\circ_{x;\tau}(M)} \exp\big(-S_0[t;\gamma]\big) u_0\big(\gamma(t)\big) d^{\Sigma\text{-}H^1}\gamma$$

for a partition $\tau = \{0 = \tau_0 < \tau_1 < \cdots < \tau_N = t\}$ and the approximate heat kernel by

$$p(\tau,x,y) := (4\pi t)^{-n/2} \fint_{H_{xy;\tau}(M)} \exp\big(-S_0[t;\gamma]\big) d^{\Sigma\text{-}H^1},$$

then for each $x \in M$ and each $u_0 \in C^0(M)$,

$$\big(p(\tau,x,-),u_0\big)_{L^2} \overset{(89)}{=} (\delta_x, P_\tau u_0)_{L^2} \overset{\text{Theorem 55}}{\longrightarrow} (\delta_x, \mathrm{e}^{-t\Delta}u_0)_{L^2} = \big(p(t,x,-),u_0\big)_{L^2}.$$

Of course, this result is not at all satisfactory, and one would like the convergence in eq. (91) to hold pointwise. A partial result can be found in [9], who prove at least that there exists some sequence $\tau^{(k)}$ of partitions with $|\tau^{(k)}| \to 0$ for which eq. (91) holds pointwise uniformly for $(x,y) \in M \times M$. This result follows from Theorem 55 with a little more effort. However, even the following stronger result is true [91, Theorem 2.2.7].

Theorem 57 (M Ludewig). *Let (M,g) be a compact connected Riemannian manifold, $\dim M = n$, and let $V : M \to \mathbb{R}$ be a smooth scalar potential. Then we have*

$$\exp(-t\Delta^V)(x,y) = \lim_{|\tau|\to 0} (4\pi t)^{-n/2} \fint_{H_{xy;\tau}(M)} \exp\big(-S_V[t;\gamma]\big) d^{\Sigma\text{-}H^1}\gamma,$$

where the slash over the integral sign denotes division by $(4\pi)^{\dim(H_{xy;\tau}(M))/2}$. The convergence is uniform over $M \times M$.

This result was proved by the author using heat kernel approximation and it actually implies Theorem 55: The latter follows from integrating over the y variable and using co-area formula (89). The method is inspired by the paper [8], where a similar result is proved for discretized L^2-metric (86) on $H_{xy;\tau}(M)$ (the restriction of which to $H_{xy;\tau}(M)$ trivially satisfies the corresponding co-area formula (89) with L^2 replaced by H^1). One obtains that $p(t,x,y)$ is also given by

$$\lim_{|\tau|\to 0} (4\pi t)^{-n/2} C_\tau \fint_{H_{xy;\tau}(M)} \exp\left(-S_0[t;\gamma] - \frac{1}{6}\int_0^t \mathrm{scal}\big(\gamma(s)\big)ds\right) d^{\Sigma\text{-}L^2}\gamma,$$

with C_τ as in eq. (88). The idea is to use the semigroup of the heat kernel,

$$e^{-t\Delta^V}(x,y) = \big(e^{-\delta_1\tau\Delta^V} * \cdots * e^{-\delta_N\tau\Delta^V}\big)(x,y) \tag{92}$$

for any partition $\tau = \{0 = \tau_0 < \tau_1 < \cdots < \tau_N = t\}$ of the interval $[0,t]$, and then to replace the kernels $e^{-\delta_j\tau\Delta^V}$ by suitable approximations to discover a path integral on the right-hand side (here the symbel $*$ denotes convolution of kernels). The mentioned approximation result is then the following (for reference, see [91, Theorem 2.1.12]; compare also Theorem 1.2 in [92]).

Theorem 58 (M Ludewig). *Let (M,g) be a compact connected Riemannian manifold and let $k_t, \ell_t \in L^\infty(M\times M)$ be two time-dependent kernels. For $T, R > 0$ and $m \in \mathbb{N}$, suppose that there exist $\gamma_1, \gamma_2 \in \mathbb{R}$ and $c, \nu, \alpha_1, \ldots, \alpha_m, \beta_1, \ldots, \beta_m \geq 0$ satisfying $\alpha_i + \beta_i/2 \geq 1+\nu$ for each $1 \leq i \leq m$ such that for all $0 < t \leq T$ and all $x, y \in M$, we have*

$$|\ell_t(x,y)|, |k_t(x,y)| \leq e^{\gamma_1 t + \gamma_2 d(x,y)^2} p(t,x,y), \tag{93}$$

and for all $0 < t \leq T$ and all $x, y \in M$ with $d(x,y) < R$, we have

$$|k_t(x,y) - \ell_t(x,y)| \leq c\sum_{j=1}^{m} t^{\alpha_j} d(x,y)^{\beta_j} K_t(x,y). \tag{94}$$

Then there exist constants $C, \delta > 0$ such that for each partition $\tau = \{0 = \tau_0 < \tau_1 < \cdots < \tau_N = t\}$ of intervals $[0,t]$ with $0 < t \leq T$ and $|\tau| \leq \delta t$, we have

$$\left|k_{\delta_1\tau} * \cdots * k_{\delta_N\tau} - \ell_{\delta_1\tau} * \cdots * \ell_{\delta_N\tau}\right| \leq Ct^{1-\beta/2}\,|\tau|^\nu K_t$$

uniformly on $M \times M$. Here, $\beta := \max_{1\leq i\leq m}\beta_i$.

This can be used as follows to prove Theorem 57. Set $k_t(x,y) := e^{-t\Delta^V}(x,y)$ and set $\ell_t(x,y)$ equal to

$$\ell_t(x,y) := \frac{\exp\Big(-S_V[t;\gamma] + \frac{1}{12}\int_0^t \mathrm{scal}\big(\gamma_{xy;t}(s)\big)ds - \frac{1}{12}\mathrm{Ric}\big(\dot\gamma_{xy;t}(0), \dot\gamma_{xy;t}(0)\big)\Big)}{(4\pi t)^{n/2} J(x,y)},$$

where $\gamma_{xy;t}$ is the unique shortest minimizing geodesic between x and y parametrized by $[0,t]$ (which is defined for almost all pairs $(x,y) \in M\times M$). Using the short-time asymptotic expansion of the heat kernel (see e.g. [18, 51, 92]), one then verifies that k_t and ℓ_t satisfy the assumptions of Theorem 58 with $m = 3$, $\alpha_1 = 2$, $\beta_1 = 0$, $\alpha_2 = 1$, $\beta_2 = 1$, $\alpha_3 = 0$, $\beta_3 = 0$, and $\nu = 1/2$. Hence there exists $C > 0$ such that

$$\left|e^{-t\Delta^V}(x,y) - \ell_{\delta_1\tau}(x,y) * \cdots * \ell_{\delta_N\tau}(x,y)\right| \leq C\left(\frac{|\tau|}{t}\right)^{1/2} p(t,x,y)$$

for all t small enough and all partitions of the interval $[0, t]$ fine enough, uniformly in $(x, y) \in M \times M$. Using definition (90) of the discrete volume on $H_{xy;\tau}(M)$, we then obtain

$$\big(\ell_{\delta_1\tau} * \cdots * \ell_{\delta_N\tau}\big)(x, y) = (4\pi t)^{-n/2} \fint_{H_{xy;\tau}(M)} \exp\big(-S_V[t; \gamma]\big)\, F_\tau(\gamma)\, d^{\Sigma\text{-}H^1}\gamma, \tag{95}$$

where F_τ is the function

$$F_\tau(\gamma) := \exp\left(\frac{1}{12}\int_0^t \mathrm{scal}\big(\gamma(s)\big)ds - \frac{1}{12}\sum_{j=1}^N \mathrm{Ric}(\delta_j\gamma, \delta_j\gamma)\right).$$

Here $\delta_j\gamma := \dot\gamma(\tau_{j-1}+)\delta_j\tau$, $j = 1, \ldots, N$, are the "increments" of γ. One now discovers that the scalar curvature integral is the quadratic variation of Brownian motion, while the right-hand side is the discrete approximation of this. We therefore have

$$\lim_{|\tau|\to 0} \sum_{j=1}^N \mathrm{Ric}(\delta_j\gamma, \delta_j\gamma) = \int_0^t \mathrm{scal}\big(\gamma(s)\big)ds,$$

with the convergence being in measure with respect to the Brownian motion measure (compare Prop. 3.23 in [45]). Using this, one can show that one can drop the term F_τ in the integrand of eq. (95) and still end up with the same result in the limit $|\tau| \to 0$.

3.4.3 Short-time asymptotics of path integrals

In our previous discussions, the action functional S_0 and the H^1 metric played a prominent role. Even though it did not appear explicitly in the above, we always had the space

$$H_{xy}(M) := \big\{\gamma \in C([0, 1], M) \mid \gamma(0) = x, \gamma(1) = y, S_0[1; \gamma] < \infty\big\}$$

of finite action paths lurking in the background. This is an infinite-dimensional manifold modelled on a Hilbert space, and the continuous H^1 metric (84) turns it into an infinite-dimensional Riemannian manifold. It is equal to the space of paths which have Sobolev regularity H^1 in local charts and by definition is the largest path space where the action functional S_0 is defined. In this section, we will discuss the relations between the heat kernel representations of the path integrals discussed above und this infinite-dimensional path space. By rescaling the paths, we can always arrange the domain of definition of our paths to be the interval $[0, 1]$, which we will be doing henceforth, where

$$S_0[\gamma] := S_0[1; \gamma],$$

for any appropriate path $\gamma : [0,1] \to M$. This rescaling has the advantage that the dependence on the t parameter becomes more explicit in the formulas. Looking at Theorem 57, we may get the idea to write the heat kernel $p(t,x,y)$ as an integral over the infinite-dimensional manifold $H_{xy}(M)$,

$$p(t,x,y) \overset{\text{formally}}{=} (4\pi t)^{-n/2} \fint_{H_{xy}(M)} \exp\big(-S_0[\gamma]/t\big)\, d^{H^1}\gamma, \tag{96}$$

where the $1/t$ in the exponent comes from the rescaling of the domain of definition of the paths. Of course, this is only formal, as can be seen by the fact that the normalization constant indicated by the slash over the integral sign should now be $(4\pi t)^{\dim(H_{xy}(M))/2} = (4\pi t)^{\infty}$. However, taking formula (96) seriously for the moment, we notice that the integral has the form of a Laplace integral, which can be evaluated using the method of stationary phase as $t \to 0$ (see e.g. [44, Section 1.2]). Suppose that x and y are close to each other, so that there is a unique minimizing geodesic $\gamma_{xy} \in H_{xy}(M)$ connecting the two. Then this is the unique nondegenerate minimum of the action functional on $H_{xy}(M)$, and we have $S_0[\gamma_{xy}] = d(x,y)^2/2$. The method of stationary phase therefore asserts that

$$(4\pi t)^{-n/2} \fint_{H_{xy}(M)} \exp\big(-S_0[\gamma]/t\big)\, d^{H^1}\gamma \sim \frac{e^{-d(x,y)^2/2t}}{(4\pi t)^{n/2}} \det\big(\nabla^2 S_0|_{\gamma_{xy}}\big)^{-1/2}, \tag{97}$$

where the asymptotic relation means that the quotient of the two sides converges to one as $t \to 0$. Here the Hessian of the action S_0 comes into play, which is well known to essentially be given by the Jacobi differential equation,

$$\begin{aligned}\nabla^2 S_0|_\gamma[X,Y] &= \int_0^1 \Big(\big\langle \nabla_s X(s), \nabla_s Y(s)\big\rangle + \big\langle R\big(\dot\gamma(s), X(s)\big)\dot\gamma(s), Y(s)\big\rangle\Big) ds \\ &= (X,Y)_{H^1} + (\mathscr{R}_\gamma X, Y)_{L^2},\end{aligned}$$

where we set $\mathscr{R}_\gamma(s) := R(\dot\gamma(s), -)\dot\gamma(s)$ for the *Jacobi endomorphism* (see e.g. [94, Theorem 13.1]). Dualizing with respect to the H^1 metric, the Hessian $\nabla^2 S_0|_\gamma$ is therefore given by the operator $\mathrm{id} + (-\nabla_s^2)^{-1}\mathscr{R}_\gamma$. The "perturbation" $(-\nabla_s^2)^{-1}\mathscr{R}_\gamma$ is easily seen to be trace class so that the H^1-determinant appearing in eq. (97) is well-defined as a Hilbert space determinant. It is therefore natural to ask if eq. (97) indeed gives the correct heat kernel asymptotics. This turns out to be the case, even in the degenerate case, when x, y are in each other's cut locus.

Theorem 59. *Let (M,g) be a compact connected Riemannian manifold and let $n = \dim(M)$. For $x, y \in M$, suppose that the set $\Gamma_{xy}^{\min}$ of minimizing geodesics between x and y is a k-dimensional submanifold of $H_{xy}(M)$ which is nondegenerate in the sense that* *for* $\gamma \in \Gamma_{xy}^{\min}$, *the Hessian* $\nabla^2 S_0|_\gamma$ *is nondegenerate when restricted to the normal space* $N_\gamma \Gamma_{xy}^{\min}$ *of* $\Gamma_{xy}^{\min}$ *in* $H_{xy}(M)$. *Then we have*

$$\lim_{|\tau|\to 0} (4\pi t)^{n/2+k/2} e^{d(x,y)^2/2t} p(t,x,y) = \int_{\Gamma_{xy}^{\min}} \det\left(\nabla^2 S_0|_{N_\gamma \Gamma_{xy}^{\min}}\right)^{-1/2} d^{H^1}\gamma,$$

where we integrate with respect to the Riemannian volume measure on $\Gamma_{xy}^{\min}$ *determined by the continuous* H^1 *metric.*

These are exactly the asymptotics that are to be expected when one formally takes the stationary phase expansion of path integral (96), also in the degenerate case when $\dim(\Gamma_{xy}^{\min}) > 0$ (compare the appendix in [92]). Of course, if x and y are close, then $\Gamma_{xy}^{\min} = \{\gamma_{xy}\}$ and the integral is just the evaluation at γ_{xy}, so that Theorem 59 reduces to the result formally obtained in (97). In this case, it is furthermore known that the heat kernel has the short-time asymptotic expansion

$$p(t,x,y) \sim (4\pi t)^{-n/2} e^{-d(x,y)^2/2t} J(x,y)^{-1/2}$$

involving the Jacobian of the Riemannian exponential map $J(x,y)$ that already appeared in eq. (90). Comparing with Theorem 59, one therefore obtains the following corollary.

Corollary 60. *Let* (M,g) *be a compact connected Riemannian manifold. Thn the Jacobian of the exponential map* $J(x,y) = \det(d\exp_x|_{\dot\gamma_{xy}(0)})$ *is equal to the Hilbert space determinant of the action functional on the tangent space of* $H_{xy}(M)$ *at* γ_{xy},

$$J(x,y) = \det\left(\nabla^2 S_0|_{\gamma_{xy}}\right).$$

where the determinant is taken with respect to the continuous H^1 *metric* (84).

At first glance, one could think that Theorem 59 follows by evaluating the finite-dimensional path integrals over $H_{xy;\tau}(M)$ in Theorem 57 with the method of stationary phase and then taking the limit $|\tau| \to 0$. A problem is however that Theorem 57 gives no control over the time-uniformity of the approximation, i.e. there is no reason one should be allowed to exchange taking the mesh-limit $|\tau| \to 0$ and the time-limit $t \to 0$. This can be fixed though by using Theorem 58, which provides careful error estimates of the approximation. More concretely, one proves a different theorem on approximating the heat kernel by integrals over $H_{xy;\tau}(M)$ which is time-uniform; the downside is that one picks up a more complicated integrand involving curvature terms which then disappears again in the short-time limit. For details, to [89] or [91].

In applications from physics, path integrals are often regularized using zeta determinants, see e.g. [69] or [124]. In our case, instead of taking the determinant of

$\nabla^2 S_0|_\gamma$ with respect to the H^1 metric, i.e. the Hilbert space determinant of the operator id + $(\nabla_s^2)^{-1}\mathscr{R}_\gamma$, one can also take the zeta-regularized determinant of the operator $\nabla_s^2 + \mathscr{R}_\gamma$ on $L^2([0,1], \gamma^* TM)$ (with Dirichlet boundary conditions, since our paths have fixed endpoints). This leads to the following result.

Theorem 61 (M Ludewig). *Under the assumptions of Theorem 59, we equivalently have*

$$\lim_{|\tau|\to 0} (4\pi t)^{n/2+k/2} e^{d(x,y)^2/2t} p(t,x,y) = \int_{\Gamma_{xy}^{\min}} \frac{\det_\zeta(-\nabla_s^2)^{1/2}}{\det'_\zeta\left(-\nabla_s^2 + \mathscr{R}_\gamma\right)^{1/2}} d^{H^1}\gamma,$$

where we integrate with respect to the Riemannian volume measure on $\Gamma_{xy}^{\min}$ with respect to the continuous H^1 metric.

For a proof and further discussion of the formula in Theorem 61, we again refer to [89] or [91, Section 3.2] .

Remark 62. Here, for an unbounded self-adjoint operator L on a Hilbert space with non-negative eigenvalues, the zeta-function is defined by

$$\zeta_L(z) = \sum_{\lambda>0} \lambda^{-z},$$

where the sum goes over the positive eigenvalues of L; it converges for Re(z) large if the eigenvalues grow sufficiently fast. For elliptic differential operators (with suitable boundary conditions), one can show that this is the case and that ζ_L of has a meromorphic extension to the complex plane which is regular at zero. The zeta-regularized determinant of L is then defined by $\det_\zeta(L) = \exp(-\zeta'(0))$ if L has only positive eigenvalues and $\det_\zeta(L) = 0$ otherwise. In the case that L has zero eigenvalues, we furthermore define $\det'_\zeta(L) = \exp(-\zeta'(0))$. Notice that $\det'_\zeta\left(-\nabla_s^2 + \mathscr{R}_\gamma\right)$ in Theorem 61 is the regularized analog of $\det(\nabla^2 S_0|_{N_\gamma \Gamma_{xy}^{\min}})$ in Theorem 59.

Remark 63. The zeta determinant of the operator $-\nabla_s^2$ on $L^2([0,1], \gamma^* TM)$ with Dirichlet boundary conditions can be easily calculated to be $\det_\zeta(-\nabla_s^2) = 2^n$.

3.5 Resolvents (J Brüning)

We have seen that the resolvent of a nonnegative self-adjoint operator, say H, is equivalent to the heat semigroup (cf. Remark 2 in Sec. 3.2) under suitable transformations. This extends also to asymptotic expansions of the respective traces if they happen to exist; in particular, the Cauchy integral produces the heat trace from the resolvent trace. To be specific, we assume that the resolvent of the abstract operator H satisfies the condition

$$R_H(z^2)^l \text{ is trace class, for some } l \in \mathbb{N}. \tag{98}$$

This implies that spec(H) is purely discrete or equivalently, that $H - \lambda$ is a Fredholm operator for all $\lambda \in \mathbb{Z}$. We want to explain how much the resolvent construction for geometric operators defined on smooth compact RTM spaces equipped with a conical metric is tied to its stratified and controlled structure; for definiteness, *we will now assume that H is the Laplace-Beltrami operator on the regular part of a smooth compact RTM space.* If $X = W_{\text{reg}}$, the resolvent can be locally constructed by well-known pseudodifferential techniques; hence we may assume that $X \subset W_{\text{sing}}$ is a smooth manifold of dimension h. We fix a point $x_0 \in X$ and a chart (U_{x_0}, ψ_{x_0}) such that $\sigma_{U_{x_0}} : T_{U_{x_0}} \to U_{x_0}$ is trivial with fiber L_{x_0}, such that $\dim L_{x_0} =: \nu = m - h - 1$. In local coordinates near $(0, 0) \in \mathbb{R}^h \times \mathbb{R}_>$ the metric takes the form

$$g = g_{U_{x_0}} + dt^2 + t^2 g_{L_{x_0}},$$

where $g_{L_{x_0}}$ may depend on x and t. Thus we can write the Laplace-Beltrami operator in these coordinates as

$$\Delta_{x_0} = P_h - \frac{\partial^2}{\partial t^2} + A(x, t),$$

where P_h is a second-order elliptic differential operator on $\mathbb{R}^h$ and $A(x, t)$ is a symmetric and semibounded second-order elliptic operator on $L_{x_0\,\text{reg}}$. We extend Δ_{x_0} smoothly to a second-order elliptic operator, $\Delta_{x_0,\infty}$, on all of $\mathbb{R}^h \times \mathbb{R}_+$ such that it equals its value at (0,0) outside a compact set and acts on the space $C_c^\infty\left(\mathbb{R}^h \times \mathbb{R}_+, C_c^\infty(L_{x_0,\text{reg}})\right)$. We extend $A(x, t)$ as its Friedrichs extension (preserving the notation), such that we obtain a family of semibounded self-adjoint operators in the Hilbert space $\mathscr{H} := L^2(L_{x_0})$. We impose the following conditions on $A(x, t)$ (which are clearly satisfied if L_{x_0} is smooth):

$$A(x,t) \geq -\frac{1}{4}, \text{ and } \left(A(x,t) + \frac{1}{4}\right)^{1/2} \text{ is smooth in both variables;} \tag{99}$$

$$A(x,t)(A(0,0)+1)^{-1} \text{ is a smooth family of bounded operators in } \mathscr{H}; \tag{100}$$

$$A(0,0) \text{ is in the Schatten class of order } p \text{ if } p > \nu/2. \tag{101}$$

Now it is reasonable to try and construct the resolvent via the pseudodifferential calculus with operator-valued symbols, since in this case, the Calderón-Vaillancourt Theorem allows to estimate operator norms in terms of norms of the symbol and its derivatives [34]. To construct a first approximation to the resolvent, we write $\hat{P_h}$ for the symbol of P_h, and $A_0 := A(0,0)$, and put

$$G_0(|\hat{P_h}(x,\xi)|^2 + \lambda) := \left(-\frac{\partial^2}{\partial t^2} + t^{-2}A(x,0) + |\hat{P_h}(x,\xi)|^2 + \lambda\right)^{-1},$$

such that the desired approximation becomes

$$\mathscr{G}_0(\lambda)s(x,t) := (2\pi)^{-h}\int_{\mathbb{R}^h} e^{i\langle x,\xi\rangle} G_0(|\hat{P_h}(x,\xi)|^2 + \lambda)\hat{s}(x,\xi)d\xi. \tag{102}$$

In the sequel, we will write $z^2 := |\hat{P_h}(x,\xi)|^2 + \lambda$.

The second-order ordinary differential operator

$$B := -\frac{\partial^2}{\partial t^2} + t^{-2}A_0$$

will play a decisive role in the subsequent analysis. If $A_0 = aI$ for $a \in \mathbb{R}$, then B is unbounded from below iff $a < -\frac{1}{4}$ and essentially self-adjoint iff $a \geq 3/4$ in the Hilbert space $L^2(\mathbb{R})$. Hence small eigenvalues will play a role and we are forced to split

$$A_0 = A_{<\Lambda} \oplus A_{>\Lambda}, \tag{103}$$

where $\Lambda \in \mathbb{R}$ must be chosen sufficiently large in the reolvent set of A_0. This splitting can be achieved orthogonally, even with x-dependence. Then for $A_{>\Lambda}$, the necessary a priori estimates can be achieved to construct the resolvent of $\Delta_>$. For $A_{<\Lambda}$, however, we have to invoke the explicit form of the resolvent that can be written in terms of the modified Bessel functions, as a matrix-valued operator with kernel

$$R_{A_<}(z^2)(t_1,t_2) = (t_1t_2)^{\frac{1}{2}} I_\nu(t_1z)K_\nu(t_2z),$$

where ν is a matrix.

The ensuing analysis proceeds by constructing a convergent Neumann series, in showing that the difference between the full resolvent, $\mathscr{G}(z^2)$ of $\underline{\Delta}_0$, and its leading approximation in eq. (102) is sufficiently small in norm. The intricate details for the case $\kappa(W) = 2$ can be found in [31]; the case of arbitrary complexity is still more complicated, but doable. The details will be published soon.

4 Spectrum and geometry

In this section, we discuss some aspects of the interplay between the spectral theory of geometric operators on Riemannian manifolds and its geometry. The focus of our research is twofold: complete Riemannian manifolds on the one hand and compact

TM spaces on the other. (These are a priori disjoint subjects since a complete Riemannian manifold cannot be the regular part of a TM space with $\kappa > 1$.) Most results will be based on the techniques described in Section 3.

4.1 Spectral theory

4.1.1 Covariant Schrödinger operators (B Güneysu)

Let us start by looking at the regularity of eigensections of covariant Schrödinger operators. The following result follows immediately from Theorem 54 and Theorem 51, noting that for every self-adjoint lower-bounded operator H, every $f \in \mathrm{Dom}(H)$, and every $\lambda \in \mathbb{R}$ one has implication

$$Hf = \lambda f \Rightarrow f = \exp(t\lambda)\exp(-tH)f.$$

For the moment, let

$$(E, h, \nabla, V) \longrightarrow (M, g)$$

again be a covariant Schrödinger semigroup with a potential $V \in L^2_{\mathrm{loc}}(M, \mathrm{End}(E))$, where M is an arbitrary possibly incomplete connected Riemannian manifold.

Theorem 64 (B Güneysu).

a) *Assume* $|V| \in \mathscr{K}(M, g)$. *Then all eigensections of* H^V_∇ *are continuous.*
b) *Assume* $|V_-| \in \mathscr{K}(M)$. *Then all eigensections of* H^V_∇ *are locally bounded.*
c) *Assume* $|V_-| \in \mathscr{K}(M)$ *and that* (M, g) *is weakly ultracontractive in the sense that*

$$\sup_{x,y\in M} p(t_0, x, y) < \infty \quad \text{for some } t_0 > 0,$$

then each L^{q_1}*-eigensection of* H^V_∇ *lies in* L^{q_2}, *whenever* $1 \le q_1 \le q_2 \le \infty$.

Concerning the discreteness of the spectrum of covariant Schrödinger operators on incomplete Riemannian manifolds, we can now conclude:

Theorem 65 (F Bei, B Güneysu). *Let* $|V_-| \in \mathscr{K}(M, g)$.

a) *Let* M *be the regular part of an irreducible complex projective variety* $X \subset \mathbb{CP}^n$ *which has a real dimension* $m > 2$, *and let* g *be the Kähler metric on* M *which is induced by the Fubini-Study metric on* X. *Then* H^V_∇ *has a purely discrete spectrum, and if we number its eigenvalues in increasing order counting multiplicities,*

$$\lambda_0 \le \lambda_1 \le \cdots \to \infty, \tag{104}$$

then there exists constants $C_1, C_2 \geq 0$, $k_0 \in \mathbb{N}$ *such that for all* $k \geq k_0$ *one has*

$$\lambda_k \geq C_1 k^{2/m} + C_2. \tag{105}$$

b) *Let* M *be the regular part of a smooth TM space and let* g *be a conical metric on* M *of type* $(1, \dots, 1)$. *Then* H^V_∇ *has purely discrete spectrum, and if we number its eigenvalues as in eq.* (104), *then we again have eq.* (105).

Indeed, this becomes an immediate consequence of Theorem 53 and a standard argument from Tauberian theory, once one notices that in both situations (a) and (b) of Theorem 65 the Riemannian manifolds have a finite volume and that by the results of [12] and the references therein, again in both situations (a) and (b) of Theorem 65 one has

$$\sup_{x,y \in M} p(t,x,y) \leq Ct^{-m/2} \text{ for all small } t > 0.$$

Another important application of covariant Schrödinger semigroups is provided by the following result on essential self-adjointness [67].

Theorem 66 (B Güneysu, O Post). *Assume that* (M,g) *is geodesically complete and that* $|V_-| \in \mathcal{K}(M)$. *Then* $\nabla^\dagger \nabla + V$ *is essentially self-adjoint.*

Let us briefly explain the connection between semigroups and the proof of Theorem 66: Instead of “directly” proving the essential self-adjointness of $\nabla^\dagger \nabla + V$ (which leads to delicate integration by parts arguments [21]), one proves instead that $C^\infty_c(M,E)$ is an operator core for H^V_∇. Here, one essential difficulty is to prove that the compactly supported elements of $\mathrm{Dom}(H^V_\nabla)$, to be denoted with $\mathrm{Dom}_c(H^V_\nabla)$, can be approximated in the graph norm by the locally bounded elements of $\mathrm{Dom}_c(H^V_\nabla)$, to be denoted with $\mathrm{Dom}_{c,L^\infty_{\mathrm{loc}}}(H^V_\nabla)$. But given $f \in \mathrm{Dom}_c(H^V_\nabla)$, Theorem 51 implies that $\exp\big(-tH^V_\nabla\big)f$ is *locally bounded* for all $t > 0$, and it is then easy to pick a compactly supported smooth function χ such that $\chi \exp\big(-tH^V_\nabla\big)f \to f$ as $t \to 0+$ in the graph norm.

We have seen from abstract facts (cf. Theorem 31) that the compactness of the operator $V(H^\nabla_V|_{V=0} + 1)^{-1}$ plays a very important role in spectral theory. Concerning this compactness, we have established the following result in [30]:

Theorem 67 (J Brüning, B Güneysu). *Let* $m = \dim(M) \leq 3$. *Assume that* $V = V_1 + V_2$, *where* $|V_2| \in L^\infty(M)$, *and that there exists a heat kernel control pair* (F, L) *for* (M,g) *such that*

$$\int_M \big(|V_1|^2 + 1_{\{|V_2|>c\}}\big) F d\mathrm{vol} < \infty \quad \text{for all } c > 0. \tag{106}$$

Then one has the following stability of the essential spectrum:

$$spec_{ess}(H_{\nabla}^{V}|_{V=0}) = spec_{ess}(H_{\nabla}^{V}).$$

The proof of this result again relies heavily on the Kato-Simon inequality, or more precisely, its version for resolvents (73). Namely, it then suffices to establish the scalar case, which is straightforward. We remark that there is also a result of this type for dimensions > 3, which however is rather technical and requires some control (ultracontractiveness) on the geometry of (M, g). The surprising aspect of the above result is its validity for *every* Riemannian manifold. We refer the interested reader to [30] and also to [68] for more details.

The abstract Theorem 34 has indicated the importance of scattering theory within spectral theory. In this spirit, our next result is a geometric scattering result for two operators, Δ_{j,g_k}, $k = 1, 2$. To formulate this result, we fix for a moment a Riemannian metric g on M. If $\tilde{g}$ is another metric on M which is quasi-isometric to g, then we denote with

$$I_{g,\tilde{g}} : \Omega_{L^2}(M, g) \longrightarrow \Omega_{L^2}(M, \tilde{g}), \ \omega \longmapsto \omega.$$

the canonical identification operator. Let $\psi : M \to \mathbb{R}$ be smooth, so that the conformally equivalent metric $g_\psi := e^{2\psi} g$ is quasi-isometric to g if and only if ψ is bounded. For any $K > 0$ and any function $h : M \to (0, \infty)$, we introduce the following notation: $\mathscr{M}_{K,h}(M)$ stands for the space of complete metrics $\tilde{g}$ on M with $\min\{1, r_{\tilde{g}}\} \geq h$, and with curvature endomorphism bounded from below by $-K$. Here, $x \mapsto r_{\tilde{g}}(x)$ denotes the $W^{1,p}$-harmonic radius of (M, g') (with p chosen appropriately depending on the dimension of M). Given a Borel function $h : M \to (0, \infty)$, the conformal factor ψ will be called an *h-scattering perturbation of g*, if

$$\int_M d(g, \psi)(x) h^{-(\dim(M)+2)}(x) \mathrm{vol}_g(dx) < \infty, \tag{107}$$

where now

$$d(g, \psi)(x) := \max\left\{\sinh(2|\psi(x)|), |d\psi(x)|_g\right\}, \quad x \in M. \tag{108}$$

With these preparatory marks, we can formulate the following result which is proved in [13], (partially) generalizing earlier results for functions in [70, 96]:

Theorem 68 (F Bei, B Güneysu, J Müller). *Let $j \in \{0, \dots, m\}$, let $\psi : M \to \mathbb{R}$ be smooth with ψ, $|d\psi|_g$ bounded, and assume that $g, g_\psi \in \mathscr{M}_{K,h}(M)$ for some pair (K, h), in a way that ψ is an h-scattering perturbation of g. Then the wave operators $W_{\pm}(\Delta_{j,g_\psi}, \Delta_{j,g}, I_{g,\tilde{g}})$*

exist and are complete. In particular, the operators $\Delta_{j,g_\psi,ac}$ *and* $\Delta_{j,g,ac}$ *are unitarily equivalent and*

$$spec_{ac}(\Delta_{j,g_\psi}) = spec_{ac}(\Delta_{j,g}).$$

This result has been used in [13] to explicitly calculate the absolutely continuous spectrum for a new class of Riemannian manifolds. Again, Kato-Simon inequalities play an essential role in the proof of Theorem 68.

4.1.2 Laplace-type operators (F Bei)

If (M, g) is a Riemannian manifold, we have seen that the de Rham differential d_* is the building block of the Hodge Laplacian $\underline{\Delta}$. Therefore, the theory of Hilbert complexes is a fundamental tool for the spectral theory of this operator on incomplete Riemannian manifolds. The following results show the existence of a Hilbert complex which sits between the minimal and the maximal Hilbert complex, interpolating in a certain sense between them, and satisfying Poincare duality, in contrast to the former two complexes.

Theorem 69 (F Bei). *Let* (M, g) *be an open, oriented, and incomplete Riemannian manifold of dimension m. Then, for each* $i = 0, \ldots, m$, *we have the following isomorphism:*

$$Ker(d_{max,i})/\overline{Ran(d_{min,i-1})} \cong Ker(d_{max,m-i})/\overline{Ran(d_{min,m-i-1})}.$$

Assume now that, for each $i = 0, \ldots, m$, $Ran(d_{min,i})$ *is closed in* $L^2\Omega^{i+1}(M, g)$. *Then there exists a Hilbert complex* $(L^2\Omega^i(M, g)), d_{\mathfrak{M},i})$ *which satisfies the following properties for each* $i = 0, \ldots, m$:

1. $Dom(d_{min,i}) \subseteq Dom(d_{\mathfrak{M},i}) \subseteq Dom(d_{max,i})$, *that is* $d_{max,i}$ *is an extension of* $d_{\mathfrak{M},i}$ *which in turn is an extension of* $d_{min,i}$.
2. $Ran(d_{\mathfrak{M},i})$ *is closed in* $L^2\Omega^{i+1}(M, g)$.
3. *If we denote by* $H^i_{2,\mathfrak{M}}(M, g)$ *the cohomology of the Hilbert complex* $(L^2\Omega^i(M, g), d_{\mathfrak{M},i})$ *then we have:*

$$H^i_{2,\mathfrak{M}}(M, g) = Ker(d_{max,i})/Ran(d_{min,i})$$

and

$$H^i_{2,\mathfrak{M}}(M, g) \cong H^{m-i}_{2,\mathfrak{M}}(M, g).$$

4. *There exists a well-defined and nondegenerate pairing:*

$$H^i_{2,\mathfrak{M}}(M, g) \times H^{m-i}_{2,\mathfrak{M}}(M, g) \longrightarrow \mathbb{R}, \ ([\omega], [\eta]) \longmapsto \int_M \omega \wedge \eta.$$

Moreover, we have the following result.

Theorem 70 (F Bei). *Let (M, g) be an open, oriented and incomplete Riemannian manifold of dimension m. Suppose that, for each $i = 0, \dots, m$, $Ran(d_{min,i})$ is closed in $L^2\Omega^{i+1}(M, g)$. Let $(L^2\Omega^i(M, g), d_{\mathfrak{M},i})$ be the Hilbert complex built in Theorem 69. Assume that $(L^2\Omega^i(M, g), d_{\mathfrak{M},i})$ is a Fredholm complex. Then:*
1. *Any Hilbert complex $(L^2\Omega^i(M, g), D_i)$ extending $(\Omega^i_c(M), d_i)$ is a Fredholm complex.*
2. *For every $i = 0, \dots, m$ the quotient of the domain of $d_{max,i}$ with the domain of $d_{min,i}$, that is*

$$Dom(d_{max,i})/Dom(d_{min,i})$$

is a finite-dimensional vector space.

As an example, consider a compact and oriented smooth TM space W whose regular part M is endowed with a conical metric (of arbitrary type). Then Theorem 69 applies to (M, g) (that is, each $\mathrm{Ran}(d_{\mathrm{min},i})$ is closed in $L^2\Omega^{i+1}(M, g)$), because from [16] we know that $H^k_{2,\mathrm{min}}(M, g)$ is finite dimensional, implying that $\mathrm{Ran}(d_{k,\mathrm{min}})$ is closed. Moreover, if for example W is compact and oriented with only isolated singularities and g is any metric on M quasi-isometric to a conic metric or to a metric horn, then the assumptions of Theorem 70 are satisfied by (M, g). This follows for example from results presented in [27, 82].

Let now $\mathscr{W} = \big(W, \mathscr{S}(W), \mathscr{T}(W), g\big)$ be a compact RTM space. The techniques developed in [33] show that $\underline{\Delta} = \underline{\Delta}_g$, the (Friedrichs realization of the) Hodge Laplacian on $(M = W_{\mathrm{reg}}, g)$, has discrete spectrum and that $R^l_{\underline{\Delta}}(z^2)$ is trace class for $l > m/2$ and $\mathrm{Re} z^2 > 0$. But more is true; there is an asymptotic expansion of the type

$$\mathrm{tr}\, R_{\underline{\Delta}}(z^2)^l \sim_{z \to +\infty} \sum_{j \in \mathbb{Z}_+, 0 \le k \le k(j)} a_{jk}(l) z^{\alpha_j - 2l} \log^k z. \tag{109}$$

Here $(\alpha_j)_{j \in \mathbb{Z}_+}$ is a sequence of complex numbers with $\mathrm{Re}\, \alpha_j \to -\infty$. This expansion arises from the Singular Asymptotics Lemma [31] that uses pseudodifferential techniques and the fact that RTM spaces have a very useful smooth scaling towards each point of a singular stratum. This scaling is a priori only smoothly stratified, but the resolvent turns out to be smoothing even in approaching a singular stratum. For our specific operator $\underline{\Delta}$ and in the case of $\kappa(W) = 2$, the expansion (109) takes the following form. We denote the resolvent of $\underline{\Delta}$ again by $R_{\underline{\Delta}}(z^2)$, $z > 0$, and denote the kernel by $R_{\underline{\Delta}}(t, t'; x, x'; z^2)$. We use again $m = \dim M$, $h = \dim X$, $v = \dim L$ and denote by f a smooth cut-off function which equals 1 near (0,0). Then we obtain

$$\mathrm{tr}\, f R^l_{\underline{\Delta}}(z^2) = \int_0^\infty \int_{\mathbb{R}^h} f(t, x)\, \mathrm{tr}_H R^l_{\underline{\Delta}}(t, t; x, x; z^2)\, dx dt.$$

The somewhat surprising smoothness at 0 is explained by the scaling relation

$$\mathrm{tr}_H R^l_{\underline{\Delta}}(t,t;x,x;z^2) = t^{2l-h-1}\mathrm{tr}_H R^l_{\underline{\Delta}}(1,1;x,x;(tz)^2),$$

together with the "integrability condition", eq. (1.2b) in [32]. The expansion then takes the form (cf. [Theorem 5.2] in [32])

$$\mathrm{tr}\, R_{\underline{\Delta}}(z^2)^l \sim_{z\to+\infty} z^{m-2l}\sum_{j\geq 0} a_j z^{-j} + \sum_{j\geq 2l-h} b_j z^{-j}\log z. \tag{110}$$

As mentioned before, the expansion of the heat trace and the residues of the spectral zeta function associated with $\underline{\Delta}$ are appropriate linear combinations in the (a_j) and (b_j). The leading-order asymptotics determine the dimension of W and the eigenvalue asymptotics which extends the first and oldest statement in spectral geometry: Weyl's theorem. The main task in this context is now the geometric description of all coefficients in the expansion. We will come back to this question below.

4.2 Index theory (J Brüning)

The Atiyah-Singer index theorem for elliptic differential operators acting on the smooth sections of a smooth vector bundle over a smooth closed manifold, expresses the index as the integral of a characteristic differential form which can be computed locally. It turned out that it suffices to prove the theorem for first-order operators, like the twisted signature operator, and it was then a natural next step to study the index problem for such operators on manifolds with boundary. This was done in a well-known paper by Atiyah, Patodi, and Singer (APS), cf. [5], with the somewhat surprising result that the corresponding operator has to be defined by *nonlocal* boundary conditions, and contributes a *non-local* term to the index formula. Both are defined in terms of *spectral data* of a first order self-adjoint elliptic operator on the boundary, depending on a *boundary metric*. The complexity of the statement explains why it took a decade after the appearance of the Atiyah-Singer theorem to find the following striking result.

Theorem 71. *Let $W = W_{reg} \cup W_{sing}$ be a compact oriented manifold with boundary. Let $E, F \to W$ be smooth vector bundles and let*

$$D : C^\infty(W_{reg}, E) \longrightarrow C^\infty(W_{reg}, F)$$

be a first-order elliptic differential operator. Let, in addition,

$$A : C^\infty(W_{sing}, E) \longrightarrow C^\infty(W_{sing}, F)$$

be a first-order elliptic differential operator that is self-adjoint *in the Hilbert space $L^2(W_{sing}, E|W_{sing}; \mu)$, formed with respect to a Hermitian metric on the vector bundles*

$E|W_{sing}$ and a measure μ on W_{sing}. Then we can form the spectral projection, P_A, of A onto $\mathbb{R}_{\geq}$. Assume further that, in a suitable collar with boundary defining function t, we have the identity

$$D = \hat{D}(dt)\left(\frac{\partial}{\partial t} + A\right), \tag{111}$$

where $\hat{D}$ denotes the symbol of D. Then we define the domain of D as

$$Dom(D) := \left\{s \in C_c^{\infty}(W_{reg}, E) \ : P_A\left(s|(\{0\} \times W_{sing})\right) = 0\right\}. \tag{112}$$

*Next we introduce the η-*function *of A,*

$$\eta_A(\zeta) := \sum_{\lambda \in specA\setminus\{0\}} \lambda\,|\lambda\,|^{-\zeta}, \ Re\,\zeta >> 0;$$

then η_A extends meromorphically to $\mathbb{C}$ and is regular at 0. Denote, finally, the index form, i.e. the Atiyah-Singer integrand of D, by ω_D. Then the index of D is given by the formula

$$index\,D = \int_{W_{reg}} \omega_D - \frac{dim\ ker\,A + \eta_A(0)}{2}. \tag{113}$$

This result suggests that the index theory of first-order operators, like the Dirac operators and, in particular, the signature operator, should admit an extension to compact oriented RTM spaces. However, the fairly complicated statement of the theorem and the non-local character of the difference

$$\text{index } D - \int_{W_{\text{reg}}} \omega_D,$$

which is referred to as the "index defect", indicate that the resulting theory will be nontrivial. Since we must expect that both the singular stratum and its link – which reduces to a point in the case just treated – will contribute to a generalized eta-term. In particular, we should define self-adjoint operators, A_X, on each singular stratum. Of course, the appropriate definition of D by boundary conditions on W_{sing} is intertwined with these operators. The role of Hilbert complexes for the Hodge Laplacian, as discussed above, will also be important.

This theory is still far from being complete. For the case $\kappa(W) = 2$ there exist results for Dirac-type operators and the signature operator in particular, cf. [2, 22], under certain (topological) conditions making the operator essentially self-adjoint. For the signature operator, D_s, on a even-dimensional Riemannian manifold with boundary, the operator A turns out to be the "odd signature operator" $D_{o,s}$ and we obtain the following formula for the signature of W, to be denoted by σ_W. (cf. Theorem 4.14 in [5]),

$$\sigma_W = \int_M L(TM) - \frac{1}{2}\eta_A(0). \tag{114}$$

The extension to $\kappa(W) = 2$ in this case reads as follows. We write $M := W_{\mathrm{reg}}$, $B := W_{\mathrm{sing}}$, $\dim B := h$, $D_B := D_{s,B}$ respectively, with $D_{o,s,B}$ depending on whether h is even or odd. We denote the generic fiber of the bundle $\pi_B : \rho_B^{-1}(1) \to B$ by L and by $\mathscr{H}_L \to B$ the vector bundle of harmonic forms on the fiber; then $D_B^{\mathscr{H}_L}$ denotes the operator D_B twisted with the fiber harmonic forms. Finally, we denote by $\tilde{\eta}$ the generalization of η_A to fiber bundles, established by Bismut and Cheeger in [19]. Then we can state

Theorem 72 (J Brüning). *Assume that W satisfies the Witt condition, that is $\mathscr{H}_L^{v/2} = 0$. Then*

$$\sigma_W = \int_M L(TM) - \int_B L(TB) \wedge \tilde{\eta} - \frac{1}{2}\eta(D_B^{\mathscr{H}_L}), \tag{115}$$

where L is the Hirzebruch genus.

In the special case of orbit spaces of compact Lie group actions a result for RTM spaces of arbitrary complexity is known [24], but the index formula is still somewhat implicit. Also, if we treat the case with an arbitrary operator D, then it is unclear what the topological or metric interpretation of the index could be.

There is an inverse question of some interest: given an invariant of a TM space, can one find an operator having this invariant as index? Lott [88] has asked this question for the orbit space, W, of a semifree $\mathbb{S}^1$-action on a compact orientable 4k+1-dimensional manifold, M, and a "signature" which he defined for it, cf. loc.cit. Def. 3. The singular strata in this case are the components of the fixed point set, $M^{\mathbb{S}^1}$, and Lott's signature, $\sigma_{\mathbb{S}^1\backslash M}$, is given as the signature of a natural quadratic form on the basic forms with compact support in W_{reg}. Lott then proves the formula

$$\sigma_{\mathbb{S}^1\backslash M} = \int_{\mathbb{S}^1\backslash M} L\big(T(\mathbb{S}^1\backslash M)\big) - \frac{1}{2}\,\eta(D_{o,s,M^{\mathbb{S}^1}});$$

note that in this case $\dim M^{\mathbb{S}^1}$ is odd dimensional. This formula looks similar to the APS formula but is not readily recognized as such. Juan Orduz, a PhD student in project C7, finished his thesis in December 2016, with the following theorem as his main result.

Theorem 73 (J Orduz). *Denote by ω the signature involution and by D the Hodge-de Rham operator on $\mathbb{S}^1\backslash M$. Then there is a potential, Q, such that*

$$D + Q \text{ anticommutes with } \omega, \tag{116}$$

and

$$\operatorname{index}(D + Q)^+ = \sigma_{\mathbb{S}^1\backslash M}.$$

The proof starts with an old result by Brüning and Heintze [23], which says that the de Hodge-de Rham operator acting on invariant forms (say) can be pushed down to the orbit space as an essentially self-adjoint operator that does not, however, anticommute with the signature involution; the main work is then to find the potential Q which is in fact essentially unique.

Even though this theorem is suspected to hold in full generality, the proof applies, strictly speaking, only if the Witt condition is satisfied, where already the operator D is essentially self-adjoint.

Finally, we mention another application of the ideas of Atiyah, Patodi, and Singer, to the index problem on complete manifolds. Consider a complete and connected Riemannian manifold, (M, g), with Levi-Civita connection ∇^{TM} and curvature tensor R, and also a Dirac bundle,

$$(E, h, c, \nabla) \longrightarrow (M, g)$$

with D^c_∇ its Dirac operator, having domain of definition $H^1(M, E) = H^1_0(M, E)$. We will say that D^c_∇ is *nonparabolic at infinity* iff there is a Hilbert space, $H^1(M, E) \subset K \subset H^1_{\text{loc}}(M, E)$, such that D^c_∇ extends to a Fredholm operator $D_{\text{ext}} : K \to L^2(M, E)$ (this notion should not be confused with the *nonparabolicity of a Riemannian manifold*). We will also assume that there is a self-adjoint involution α on E that anticommutes with $D := D^c_\nabla$ such that $D = D^+ \oplus D^-$. Then we obtain the following result (Theorem 1.17 in [6].

Theorem 74 (W Ballmann, J Brüning, G Carron). *If M as above has at most finitely many ends, D is nonparabolic at infinity, and ω^+_D is integrable, then*

$$\operatorname{index} D^+_{ext} = \int_M \omega_{D^+} + \sum_{\mathscr{C}} \operatorname{Corr}(\mathscr{C}),$$

where the correction term $\operatorname{Corr}(\mathscr{C})$ depends only on the end.

If we further assume the existence of a distance function, F, on each end, we can split the end at the hypersurfaces $F^{-1}(T)$ and introduce APS-type complementary boundary conditions along the inner and outer boundaries respectively. Then in many situations, we obtain satisfying index formulae by letting $T \to \infty$. As an example we present the signature formula for a quotient, X, of complex hyperbolic space of even dimension n by a "neat" lattice. In the formula of the theorem, $|\Gamma_{\mathscr{C}}|$ is a basic invariant of the fundamental group, $\Gamma_{\mathscr{C}}$, of the cusp $\mathscr{C}$, for each $\mathscr{C}$ (this is Theorem 1.23 in [6].

Theorem 75. *On X as above, the signature operator is a Fredholm operator with index*

$$\sigma_X = \frac{vol\, X}{vol\, \mathbb{C}P^2} + 2^n \zeta(1-n) \sum_{\mathscr{C}} |\Gamma_{\mathscr{C}}| + \nu(-1)^{n/2}\left(\binom{n-2}{n/2} - \binom{n-2}{n/2-1}\right), \tag{117}$$

where ν *is the number of cusps.*

Finally, the same method is applicable to index calculations in the Melrose "calculus of fibered cusps". Bodo Graumann, a diploma student in project C7 has reproved the results in [80] in his thesis of 2017, using the above "calculus of Dirac systems", as the authors call it.

4.3 Spectral geometry (F Bei, J Brüning, B Güneysu)

4.3.1 Heat semigroups

For the moment, let (M, g) be a connected possibly incomplete Riemannian manifold. An important spectral geometric application of our results about heat semigroups is the following dynamical characterization of the Riemannian total variation, which generalizes de Giorgi's original Euclidean result [41]:

Theorem 76 (B Güneysu, D Pallara). *Let (M, g) be geodesically complete and assume that Ric, considered as a potential on $T^*M \to M$, satisfies $|Ric_-| \in \mathscr{K}(M)$ (for example, Ric could be assumed to be bounded from below by a constant). Then for every $f \in L^1(M)$ one has*

$$Var(f) = \lim_{t \to 0+} \left\| d e^{-t\Delta} f \right\|_{L^\infty \Omega^1(M,g)}, \tag{118}$$

where the possibly infinite quantity

$$Var(f) := \sup\left\{ \left| \int_M \overline{f(x)} d^\dagger \alpha(x) dvol(x) \right| : \alpha \in \Omega^1_{C_c}(M), \|\alpha\|_{L^\infty\Omega^1(M)} \leq 1 \right\}$$

denotes the Riemannian total variation of f.

The above result stems from [66], generalizing the earlier Riemannian results [35, 95] considerably. Again Theorem 51 plays an essential role. Namely, under the assumptions of Theorem 76, the semigroup $\exp(-t\Delta_1)$ maps bounded 1-forms to bounded 1-forms, with a controllable operator norm. Together with the commutation rule $de^{-t\Delta} = e^{-t\Delta_1}d$ and a duality argument, this turns out to be enough to estimate $de^{-t\Delta}f$.

Let us now turn to some probabilistic considerations. As we have noted, it may happen that under the Brownian motion measure, paths can explode in a finite time. As we have noted, this phenomenon does *not* happen if and only if one has

$$\int_M p(t,x,y)d\mathrm{vol}(y) = 1 \quad \text{for all } (t,x) \in (0,\infty) \times M. \tag{119}$$

This motives the following definition.

Definition 77. (M,g) is called *stochastically complete*, if one has (119).

If (M,g) is geodesically complete, then there exists the following very general volume criterion for stochastic completeness by Grigor'yan [54]:

Theorem 78. *If (M,g) is geodesically complete with*

$$\int_1^\infty \frac{s}{\log vol(B(x_0,s))} ds = \infty \quad \text{for some } x_0 \in M, \tag{120}$$

then (M,g) is stochastically complete.

For example, in combination with the Cheeger-Gromov volume estimates the last result immediately implies that geodesically complete Riemannian manifolds with Ricci curvature bounded from below by a constant are stochastically complete. More generally, one can allow quadratic (with respect to the distance function) lower Ricci bounds.

Another important question is whether or not one has

$$\int_0^\infty p(t,x,y)dt < \infty. \tag{121}$$

Again this question is closely linked to probability theory, for one has [57]:

Theorem 79. *The following properties are equivalent:*

(i) *One has eq. (121).*
(ii) *All Brownian motions on (M,g) are transient, in the sense that for all $x \in M$ and all precompact sets $U \subset M$ one has*

$$\mathbb{P}^x\{\text{there exists } s > 0 \text{ such that for all } t > s \text{ one has } X_t \notin U\} = 1,$$

that is, all Brownian motions on (M, g) eventually leave each precompact set, almost surely.

(iii) *There exists a nonnegative Green's function for the Laplace equation on (M, g). Furthermore, if some of the above properties is satisfied, then*

$$G : M \times M \longrightarrow (0, \infty], \quad G(x, y) := \int_0^\infty p(t, x, y) dt$$

is the pointwise minimal nonnegative Green's function for the Laplace equation with respect to (M, g).

This result justifies:

Definition 80. (M, g) is called *nonparabolic* if the equivalent conditions of Theorem 79 are satisfied, and otherwise *parabolic*.

It is also a classical fact that parabolicity implies stochastic completeness. Theorem 79 is of fundamental importance in quantum physics. Namely, if one wants to generalize the Euclidean Coulomb potential $(x, y) \mapsto 1/|x - y|$ to Riemannian manifolds, this potential should not only be a fundamental solution of the Laplace equation, but it also should have no sign changes, so that one can actually have a clear notion of the "repulsion" of equally charged particles, and of the "attraction" of differently charges particles. This was the motivation for the following definition which stems from [61]:

Definition 81. If (M, g) is nonparabolic, then the function G from Theorem 79 is called the *Coulomb potential on* (M, g).

Many geometric stability results for hydrogen-type atoms have been established in [46, 61], generalizing their well-known Euclidean analogs [86].

As for stochastic completeness, if (M, g) is geodesically complete, then there exists the following very general volume test for parabolicity [54] by Grigor'yan.

Theorem 82. *If (M, g) is geodesically complete with*

$$\int_1^\infty \frac{s}{vol(B(x_0, s))} ds = \infty \quad \textit{for some } x_0 \in M,$$

then (M, g) is parabolic.

On the other hand, we have seen that many natural Riemannian metrics are geodesically incomplete. In this context, we have recently established a new result for the regular parts of smooth TM spaces and their natural metrics [15]:

Theorem 83 (F. Bei, B. Güneysu). *Assume M is the regular part of a smooth TM space and that g is a conical metric on M of type $(1,\dots,1)$. Then (M,g) is parabolic, in particular, stochastically complete.*

This result has been proved in [15]. We refer the reader to [83], where P. Li and G. Tian have proved an analogous result for the Bergmann metric on an an algebraic variety.

4.3.2 Resolvents (Jochen Brüning)

We return to the resolvent expansion for the Hodge Laplacian, $\underline{\Delta}$, on a compact orientable RTM space, W, i.e. a smooth orientable TM space equipped with a conical metric on its regular part. If the resolvent trace can be expanded as in eq. (110), then its coefficients will contain vital geometric information. In fact, in the smooth compact case it is well known that the coefficients are $O(m)$-invariant polynomials in the curvature tensor and its covariant derivatives which, however, get more and more complicated. Many interesting properties are known, nevertheless; in particular, that spec $\underline{\Delta}$ does *not determine* the isometry class of (M,g).

In the realm of compact TM spaces with conical singularities, we can ask a more modest question, however:

Can we detect from the spectrum of $\underline{\Delta}$ whether or not W is actually singular?

Consider first the case of isolated conical singularities, i.e. a special case of $\kappa = 2$. A pointed tubular neighborhood of a singular stratum, $\{p\}$, is isomorphic to $(0,\varepsilon)\times N$ with (N,g_N) a closed, connected, and orientable Riemannian manifold of dimension $n = m-1$, with metric

$$g_p := dt^2 + t^2 g_N.$$

It is easily seen that the metric g_p is smooth at p if and only if (N,g_N) is the round unit sphere. If $n = 1$, then singularities are possible, depending on the area of $N = \mathbb{S}^1_r$, $r > 0$. A good counter-example is given by a surface of rotation generated by an arc in the positive quadrant which starts at 0 but is not tangent to the x-axis. For the more difficult case of algebraic curves, Brüning and Lesch have shown that one can detect their singular nature from the resovent expansion, cf. [27].

The expansion of eq. (110) contains a logarithmic term which is treated in some detail in Sec. 7 of [32]. This term turns out to be local; if it does not vanish, then the TM space W will certainly be singular. The constant term, however, contains the standard local plus a nonlocal term, cf. p. 422 in [27]. So, even if the logarithmic term vanishes, another non-local term, b, remains which also implies a singularity if it does not vanish. It would be detectable since the local contribution is the usual one; conversely, we would like to know whether its vanishing already implies regularity, i.e. that (N,g_N) is the round sphere.

The case $\dim N = 1$ is settled by what we said, the next case with a comparable expansion, $n = 3$, looks already much more complicated. Asilya Suleymanova, a PhD student in the team of C7, has finished her PhD thesis on this topic exactly in the last month of the CRC. She proved the following interesting results.

Theorem 84 (A. Suleymanova). *If W as above has dimension 4, and if the logarithmic term and the nonlocal constant term both vanish, then (N, g_N) must be a spherical space form. If it is cyclic, then it must be the round unit sphere.*

This theorem rests on the fact that the vanishing of the logarithmic term is equivalent to a linear identity involving the heat trace coefficients of N, and on a clever analysis of this identity. In higher (even) dimensions, the computations seem to become hopelessly complex, and the procedure just described will not work any more. A. Suleymanova has illustrated this phenomenon by analyzing links (N, g_N) with constant sectional curvature and showed the following.

Theorem 85 (A. Suleymanova). *Let W be an even dimensional smooth TM space with an isolated conical singularity, with link (N, g_N). Assume that (N, g_N) has constant sectional curvature S_N. Then the logarithmic term is given by an explicit polynomial in S_N of degree $\frac{n+1}{2}$. 1 is always a root, but the number of roots is larger than one if $n > 3$.*

Thus, the answer to our question is by no means easy, and it might well be negative. So the problem remains very interesting.

Acknowledgments: The Collaborative Research Center 647 *Space-Time-Matter* created a very fruitful research environment in Berlin, attracting many short-time visitors and a large number of PhD students and Postdocs who stayed here for a longer time, as the main “engine of progress”. In addition, the locals had ample opportunities to visit colleagues and conferences and thus to share the many interesting developments which had an impact on their research. Finally, the interaction between mathematicians and physicists produced a particular impetus to exchange and compare ideas with mutually unfamiliar origin, which shall be extended and deepened in the future.

Bibliography

[1] Aizenman M, Simon B. Brownian motion and Harnack inequality for Schrödinger operators. Comm Pure Appl Math 1982;35(2):209–73.

[2] Albin P, Gell-Redman J. The index of Dirac operators on incomplete edge spaces. Symmetry Integrability Geom Methods Appl 2016;12:45 pp.

[3] Ambrosio L, Fusco N, Pallara D. Functions of bounded variation and free discontinuity problems. Oxford Mathematical Monographs. New York: The Clarendon Press, Oxford University Press, 2000.

[4] Anderson L, Driver B. Finite dimensional approximations to Wiener measure and path integral formulas on manifolds. J Funct Anal 1999;165:430–98.
[5] Atiyah MF, Patodi VK, Singer IM. Spectral asymmetry and Riemannian geometry. I. Math Proc Cambridge Philos Soc 1975;77:43–69.
[6] Ballmann W, Bruning J, Carron G. Index theorems on manifolds with straight ends. Compos Math 2012;148(6):1897–968.
[7] Banagl M. Toplogical invariants of stratified spaces. Berlin, Heidelberg, New York: Springer, 2007.
[8] Bär C. Renormalized integrals and a path integral formula for the heat kernel on a manifold. Contemp Math 2012;584:179–97.
[9] Bär C, Pfäffle F. Path integrals on manifolds by finite dimensional approximation. J Reine Angew Math 2008;625:29–57.
[10] Bär C, Pfäffle F. Wiener measures on Riemannian manifolds and the Feynman-Kac formula. Matematica Contemporanea 2011;40:37–90.
[11] Bei F. On the L2-Poincaré duality for incomplete Riemannian manifolds: a general construction with applications. J Topol Anal 2016;8(1):151–86.
[12] Bei F. Sobolev spaces and Bochner Laplacian on complex projective varieties and stratified pseudomanifolds. J Geom Anal 2016;27(1):746–96.
[13] Bei F, Güneysu B, Müller J. Scattering theory of the Hodge Laplacian under a conformal perturbation. J Spec Theor 2017;7(1):235–67.
[14] Bei F. L2-de Rham-Hodge and L2-Atiyah-Bott-Lefschetz theorems on stratified pseudomanifolds. PhD thesis, Sapienza, Universita di Roma, 2012.
[15] Bei F, Güneysu B. q-parabolicity of stratified pseudomanifolds and other singular spaces. Ann Global Anal Geom 2017;51(3):267–86.
[16] Bei F. General perversities and L2 de Rham and Hodge theorems for stratified pseudomanifolds. Bull Sci Math 2014;138(1):2–40.
[17] Berline N, Getzler E, Vergne M. Heat kernels and Dirac operators. 2nd ed. Grundlehren Text Editions. Berlin: Springer–Verlag, 1992.
[18] Berline N, Getzler E, Vergne M. Heat kernels and Dirac operators. Berlin, Heidelberg, New York: Springer, 2004.
[19] Bismut J.-M, Cheeger J. η-invariants and their adiabatic limits. J Amer Math Soc 1989;2;33–70.
[20] Brasselet JP, Hector G, Saralegi M. L2-cohomologie des espaces stratifié. Manuscripta Math 1992;76 :21–32.
[21] Braverman M, Milatovich O, Shubin M. Essential self-adjointness of Schrödinger-type operators on manifolds. Russian Math Surv 2002;57(4):641–92.
[22] Brüning J. The signature operator on manifolds with a conical singular stratum. Asterisque 2009;328:1–44.
[23] Brüning J, Heintze E. Representations of compact Lie groups and elliptic operators. Invent Math 1978;719(2):169–203.
[24] Brüning J, Richardson K, Kamber F. The equivariant index theorem for transversally elliptic operators and the basic index theorem for Riemannian foliations. Electron Res Announc Math Sci 2010;17:138–54.
[25] Broderix K, Hundertmark D, Leschke H. Continuity properties of Schrödinger semigroups with magnetic fields. Rev Math Phys 2000;12(2):181–225.
[26] Brüning J, Geyler V, Pankrashkin K. Continuity properties of integral kernels associated with Schrödinger operators on manifolds. Ann Henri Poincare 2007;8 (4):781–816.
[27] Brüning J, Lesch M. Kähler-Hodge theory for conformal complex cones. Geom Funct Anal 1993;3(5):439–73.
[28] Brüning J, Lesch M. Hilbert complexes. J Funct Anal 1992;108(1):88–132.

[29] Brüning J, Lesch M. On the spectral geometry of algebraic curves. J Reine Angew Math 1996;474:25–66.

[30] Brüning J, Güneysu B. Heat kernel estimates and the relative compactness of perturbations by potentials. Preprint.

[31] Brüning J, Seeley R.T. Regular singular asymptotics. J Funct Anal 1985;58:133–48.

[32] Brüning J, Seeley R.T. The resolvent expansion for second order regular singular operators. J Funct Anal 1987;73:369–429.

[33] Brüning J, Seeley R.T. The expansion of the resolvent near a singular stratum of conical type. J Funct Anal 1991;95:255–290.

[34] Calderón A, Vaillancourt R. On the boundedness of pseudodifferential operators. J Math Soc Japan 1971;73:374–8.

[35] Carbonaro A, Mauceri G. A note on bounded variation and heat semigroup on Riemannian manifolds. Bull Aus Math Soc 2007;76:155–60.

[36] Cheeger J. On the Hodge theory of Riemannian pseudomanifolds. Proc Symp Pure Math 1980;36:91–146.

[37] Cheeger J. Spectral geometry of singular Riemannian spaces. J Diff Geom 1983;18:575–657.

[38] Chernoff P.R. Essential self-adjointness of powers of generators of hyperbolic equations. J Funct Anal 1973;12:401–14.

[39] Chernoff P.R. Note on product formulas for operator semigroups. J Funct Anal 1986;2:238–42.

[40] Coulhon T, Sikora A. Gaussian heat kernel upper bounds via the Phragmen-Lindelöf theorem. Proc Lond Math Soc (3) 2008;96(2):507–44.

[41] De Giorgi E. Su una teoria generale della misura (r - 1)-dimensionale in uno spazio ad r dimensioni. Ann Mat Pura Appl 1954;36:191–213.

[42] Driver B.K, Thalmaier A. Heat equation derivative formulas for vector bundles. J Funct Anal 2001;183(1):42–108.

[43] Dugundji J. Topology. Dubuque, IA: Wm. C. Brown Publishers, 1989.

[44] Duistermaat JJ. Fourier integral operators. Berlin, Heidelberg, New York: Springer, 1996.

[45] Emery E. Stochastic analysis in manifolds. Universitext. Berlin, Heidelberg, New York: Springer, 1989.

[46] Enciso A. Coulomb systems on Riemannian manifolds and stability of matter. Ann Henri Poincare 2011;12(4):723–41.

[47] Feynman R, Hibbs A.R. Quantum mechanics and path integrals. New York, St. Louis, San Francisco: McGraw-Hill Companies, 1965.

[48] Fröhlich J, Lieb E. H, Loss M. Stability of Coulomb systems with magnetic fields. I. The one-electron atom. Commun Math Phys 1986;104(2):251–70.

[49] Gaffney M.P. The conservation property of the heat equation on Riemannian manifolds. Commun Pure Appl Math 1959;12:1–11.

[50] Gibson C, Wirthmuller K, du Plessis A, Looijenga E. Topological stability of smooth mappings. Lecture Notes in Mathematics 552. Springer, 1976.

[51] Gilkey P. Invariance theory the heat equation and the Atiyah-Singer index theorem. Boca Raton: CRC Press, 1995.

[52] Goresky M, MacPherson R. Introduction homology theory. Topology 1980;19 :135–62.

[53] Goresky M. Introduction to the papers of R. Thom and J. Mather. Bull AMS 2012;49:469–74.

[54] Grigor'yan A. Heat kernel and analysis on manifolds. AMS/IP Studies in Advanced Mathematics, 47. Providence, RI: American Mathematical Society; Boston, MA: International Press, 2009.

[55] Grigor'yan A. Heat kernels on weighted manifolds and applications. The ubiquitous heat kernel. Contemporary Mathematics 398, 93–191, Providence, RI: American Mathematical Society, 2006.

[56] Grigor'yan A, Noguchi M. The heat kernel on hyperbolic space. Bull Lond Math Soc 1998;30(6):643–50.
[57] Grigor'yan A. Analytic and geometric background of recurrence and non-explosion of the Brownian motion on Riemannian manifolds. Bull Am Math Soc 1999;36 :135–249.
[58] Grummt R, Kolb M. Essential selfadjointness of singular magnetic Schrödinger operators on Riemannian manifolds. J Math Anal Appl 2012;388(1):480–9.
[59] Güneysu B, Pigola S. The Calderon-Zygmund inequality and Sobolev spaces on noncompact Riemannian manifolds. Adv Math 2015;281:353–93.
[60] Güneysu B. Kato's inequality and form boundedness of Kato potentials on arbitrary Riemannian manifolds. Proc Am Math Soc 2014;142(4):1289–1300.
[61] Güneysu B. Nonrelativistic hydrogen type stability problems on nonparabolic 3-manifolds. Ann Henri Poincare 2012;13(7):1557–73.
[62] Güneysu B. Sequences of Laplacian cut-off functions. J Geom Anal 2016;26 (1):171–84.
[63] Güneysu B, Guidetti D, Pallara D. L1-elliptic regularity and H=W on the whole Lp-scale on arbitrary manifolds. Preprint, 2015.
[64] Güneysu B. On generalized Schrödinger semigroups. J Funct Anal 2012;262 (11):4639–674.
[65] Güneysu B. Heat kernels in the context of Kato potentials on arbitrary manifolds. To appear in Potential Analysis, 2016.
[66] Güneysu B, Pallara D. Functions with bounded variation on a class of Riemannian manifolds with Ricci curvature unbounded from below. Math Ann 2015;363(3–4):1307–31.
[67] Güneysu B, Post O. Path integrals and the essential self-adjointness of differential operators on noncompact manifolds. Math Z 2013;275(1–2):331–48.
[68] Güneysu B. Covariant Schrödinger semigroups on noncompact Riemannian manifolds. To appear in: Operator theory: advances and applications.
[69] Hawking SW. Zeta function regularization of path integrals in curved spacetime. Commun Math Phys 1977;55:133–48.
[70] Hempel R, Post O, Weder R. On open scattering channels for manifolds with ends. J Funct Anal 2014;266(9):5526–583.
[71] Hess H, Schrader R, Uhlenbrock D. A. Kato's inequality and the spectral distribution of Laplacians on compact Riemannian manifolds. J Diff Geom 1980;15(1):27–37 (1981).
[72] Hess H, Schrader R, Uhlenbrock D. A. Domination of semigroups and generalization of Kato's inequality. Duke Math J 1977;44(4):893–904.
[73] Hinz A. M, Stolz G. Polynomial boundedness of eigensolutions and the spectrum of Schrödinger operators. Math Ann 1992;294(2):195–211.
[74] Hsu E. P. Stochastic analysis on manifolds. Graduate Studies in Mathematics 38. Providence RI: American Mathematical Society, 2002.
[75] Kato T. Perturbation theory for linear operators. Reprint of the 1980 edition. Classics in Mathematics. Berlin: Springer, 1995.
[76] Kato T. Schrödinger operators with singular potentials. Israel J Math 1972;13;135–48.
[77] Kuwae K, Takahashi M. Kato class measures of symmetric Markov processes under heat kernel estimates. J Funct Anal 2007;250(1):86–113.
[78] Laetsch T. An approximation to Wiener measure and quantization of the hamiltonian on manifolds with non-positive sectional curvature. J Funct Anal 2013;265(8):1667–727.
[79] Lawson H.B. Jr, Michelsohn M.-L. Spin geometry. Princeton Mathematical Series, 38. Princeton: Princeton University Press, 1989.
[80] Leichtnam E, Mazzeo R, Piazza P. The index of Dirac operators on manifolds with fibered boundaries. Bull Belg Math Soc Simon Stevin 2006;13(5):845–55.
[81] Leinfelder H, Simader C.G. Schrödinger operators with singular magnetic vector potentials. Math Z 1981;176(1):1–19.

[82] Lesch M, Peyerimhoff N. On index formulas for manifolds with metric horns. Comm Part Diff Eq 1998;23(3–4):649–84.
[83] Li P, Tian G. On the heat kernel of the Bergmann metric on algebraic varieties. J Am Math Soc 1995;8(4):857–77.
[84] Li P, Yau S.-T. On the parabolic kernel of the Schrödinger operator. Acta Math 1986;156(3–4):153–201.
[85] Lieb E. H, Loss M. Stability of Coulomb systems with magnetic fields. II. The many-electron atom and the one-electron molecule. Comm Math Phys 1986;104(2):271–82.
[86] Lieb E. H, Seiringer R. The stability of matter in quantum mechanics. Cambridge: Cambridge University Press, 2010.
[87] Lim A. Path integrals on a compact manifold with non-negative curvature. Rev Math Phys 2007;19(9):967–1044.
[88] Lott J. Signatures and higher signatures of S^1-quotients. Math Ann 2000;316:617–57.
[89] Ludewig M. Heat kernel asymptotics path integrals and infinite-dimensional determinants, [arXiv:1607.05891], 2016.
[90] Ludewig M. Path integrals on manifolds with boundary, [arXiv:1607.05151], 2016.
[91] Ludewig M. Path integrals on manifolds with boundary and their asymptotic expansions. PhD thesis, Universität Potsdam, Potsdam, 2016.
[92] Ludewig M. Strong short time asymptotics and convolution approximation of the heat kernel, [arXiv:1607.05152], 2016.
[93] J. MatherNotes on topological stability. Bull AMS 2012;49:475–506.
[94] Milnor J. Morse theory. Princeton: Princeton University Press, 1963.
[95] Miranda Jr M, Pallara D, Paronetto F, Preunkert M. Heat semigroup and functions of bounded variation on Riemannian manifolds. J Reine Angew Math 2007;613:99–119.
[96] Müller W, Salomonsen G. Scattering theory for the Laplacian on manifolds with bounded curvature. J Funct Anal 2007;253(1):158–206.
[97] Øksendal B. Stochastic differential equations. Berlin, Heidelberg, New York: Springer, 2007.
[98] Ouhabaz E.-M, Stollmann P, Sturm K.-T, Voigt J. The Feller property for absorption semigroups. J Funct Anal 1996;138(2):351–78.
[99] Nicolaescu L.I. Lectures on the geometry of manifolds. 2nd ed. Hackensack, NJ: World Scientific Publishing Co. Pte. Ltd., 2007.
[100] Reed M, Simon B. Methods of modern mathematical physics. I. Functional analysis. New York–London: Academic Press, 1972.
[101] Reed M, Simon B. Methods of modern mathematical physics. IV. Analysis of operators. New York–London: Academic Press Inc. 1978.
[102] Salamon D. Spin geometry and Seiberg-Witten equations. Unpublished manuscript (ETH Zürich 1999). Available at: https//people.math.ethz.ch/ salamon/ PREPRINTS/witsei.pdf.
[103] Saloff-Coste L. Aspects of Sobolev-type inequalities. London Mathematical Society Lecture Note Series, 289. Cambridge: Cambridge University Press, 2002.
[104] Schoen R, Yau S.-T. Lectures on differential geometry. Conference Proceedings and Lecture Notes in Geometry and Topology I. Cambridge MA: International Press, 1994.
[105] Simon B. Schrödinger semigroups. Bull Am Math Soc (N.S.) 1982;7(3):447–526.
[106] Simon B. Lower semicontinuity of positive quadratic forms. Proc Roy Soc Edinburgh Sect 1977/78;A 79(3–4):267–73.
[107] Simon B. An abstract Kato's inequality for generators of positivity preserving semigroups. Indiana Univ Math J 1977;26(6):1067–73.
[108] Simon B. A canonical decomposition for quadratic forms with applications to monotone convergence theorems. J Funct Anal 1978;28(3):377–85.
[109] Simon B. Kato's inequality and the comparison of semigroups. J Funct Anal 1979;32(1):97–101.

[110] Simon B. Functional integration and quantum physics. 2nd ed. Providence, RI: AMS Chelsea Publishing, 2005.
[111] Smolyanov O, Weizsäcker Hv, Wittich O. Chernoff's theorem and discrete time approximations of Brownian motion on manifolds. Potential Anal 2007;26(1):1–29.
[112] Stollmann P, Voigt J. Perturbation of Dirichlet forms by measures. Potent Anal. 1996;5(2):109–38.
[113] Sturm K.-T. Schrödinger semigroups on manifolds. J Funct Anal 1993;118 (2):309–50.
[114] Sturm K.-T. Heat kernel bounds on manifolds. Math Ann 1992;292(1):149–62.
[115] Sturm K-T. Analysis on local Dirichlet spaces. I. Recurrence conservativeness and Lp-Liouville properties. J Reine Angew Math 1994;456:173–96.
[116] Strichartz R. S. Analysis of the Laplacian on the complete Riemannian manifold. J Funct Anal 1983;52(1):48–79.
[117] Thom R. Ensembles et morphismes stratifiés. Bull AMS 1969;75:240–84.
[118] Verona A. Stratified mappings – structure and triangulability. Lecture Notes in Mathematics 1102. Springer, 1984.
[119] Warner, F. Foundations of differentiable manifolds and Lie groups. Scott Foresman and Company, 1971.
[120] Weidmann J. Lineare Operatoren in Hilberträumen. Mathematische Leitfäden. Stuttgart: B. G. Teubner, 1976.
[121] Weidmann J. Lineare Operatoren in Hilberträumen. Teil 1. Grundlagen. Mathematische Leitfäden. Stuttgart: B. G. Teubner, 2000.
[122] Weidmann J. Lineare Operatoren in Hilberträumen. Teil II. Anwendungen. Mathematische Leitfäden. Stuttgart: B. G. Teubner, 2003.
[123] Whitney H. Complexes of manifolds. Proc Nat Acad Sci USA 1947;33:10–11.
[124] Witten E. Index of Dirac operators. In Deligne P, et al. editor Quantum fields and strings: A course for mathematicians. Providence: American Mathematical Society, 1999.
[125] Yau S.-T. Some function-theoretic properties of complete Riemannian manifold and their applications to geometry. Indiana Univ Math J 1976;25(7):659–70.
[126] Yau S. T. On the heat kernel of a complete Riemannian manifold. J Math Pures Appl (9) 1978;57(2):191–201.

Klaus Ecker, Bernold Fiedler et al.

Singularities and long-time behavior in nonlinear evolution equations and general relativity

Abstract: Many central problems in geometry, topology and mathematical physics reduce to questions regarding the behavior of solutions of nonlinear evolution equations. Examples are Thurston's classification of compact 3-manifolds based on Hamilton's Ricci flow and the work of Christodoulou and Klainerman on the nonlinear stability of Minkowski spaces. Examples of hyperbolic equations include the Einstein field equations of general relativity as well as semilinear wave equations. Ricci flow and semilinear reaction-advection-diffusion equations are of parabolic type. In all these equations, the global dynamical behavior of bounded solutions for large times is of significant interest. Specific questions concern the convergence to equilibria; the existence of periodic, homoclinic, and heteroclinic solutions; and the existence and geometric structure of global attractors. On the other hand, many solutions develop singularities in finite time. The singularities have to be analysed in detail before attempting to extend solutions beyond their singularities or to understand their geometry in conjunction with globally bounded solutions. In this context we have been particularly interested in global qualitative descriptions of blow-up and grow-up profiles.

Keywords: geometric flows, minimal and prescribed mean curvature surfaces, singularity analysis, entropy, monotonicity formula, global attractors and regular cell complexes, meanders and zero numbers, Einstein constraints and symmetry breaking bifurcations, Belinskii-Khalatnikov-Lifshitz conjecture, blow-up in parabolic partial differential equations

Mathematics Subject Classification 2010: 28A75, 53C44, 53A10, 53Z05, 65N30, 35B41, 37G40, 57N50, 83C57, 83F05

DOI 10.1515/9783110452150-008

1 Nonlinear partial differential equations with applications in geometry, topology and mathematical physics (Ecker[1] et al.)

1.1 Introduction

Over the last five decades, nonlinear partial differential equations have played an important role in the solution of problems which arise in differential geometry, differential and geometric topology, the calculus of variations and in mathematical physics, especially in the theory of mathematical relativity. Let us only mention the central importance of minimal surface theory for the proof of the positive mass theorem due to Schoen and Yau [183] and the role of Hamilton's Ricci flow [105] in settling Thurston's geometrisation conjecture [200] (which includes the Poincaré conjecture) for orientable, closed 3-manifolds. Based on [105] as well as subsequent work (see for instance [51, 106, 110, 112, 133, 188]), this conjecture was settled in 2002 and 2003 by Perelman [164–166] (see also [140]).

Here we shall concentrate on elliptic and parabolic equations. Among the elliptic equations we will consider more specifically those arising as Euler–Lagrange equations of certain energy functionals as well as some fully nonlinear equations. The parabolic equations we study are usually steepest descent flows for those energy functionals in a suitable sense. We shall also present several results where additional constraints or boundary conditions are imposed.

Specific examples of such elliptic equations include semilinear equations in which the Laplacian of a function is given by a power of this function, minimal (mean curvature zero) surfaces and surfaces of more generally prescribed mean curvature, Einstein metrics, harmonic maps, Yang–Mills fields and Willmore surfaces. Except for the last which is of fourth order in the derivatives of the solution, these are all second-order problems. Their parabolic counterparts are scalar reaction diffusion equations, the mean curvature flow, the Ricci flow (and a certain normalised version of it), the harmonic map heat flow, the Yang–Mills flow and the Willmore flow. Let us state some of these flow equations here. The corresponding elliptic versions are obtained by ignoring the time derivative.

1.1.1 Reaction-diffusion equation

The simplest reaction-diffusion equation is the scalar equation

$$\left(\frac{\partial}{\partial t} - \Delta\right) u = |u|^{p-1} u \tag{1}$$

1 Klaus Ecker, Free University of Berlin, kecker@zedat.fu-berlin.de

for $p > 1$. It arises as the gradient flow for a functional involving the energy density

$$e(u) := \frac{1}{2}|Du|^2 - \frac{1}{p+1}u^{p+1}.$$

Reaction-diffusion type equations appear naturally, albeit often in tensorial form, in most other parabolic examples considered here. Moreover, they are also studied extensively by the research group within project C8 led by Bernold Fiedler (see Section 2).

1.1.2 Mean curvature flow

In mean curvature flow (curve shortening flow if $n = 1$) n-dimensional hypersurfaces $(M_t)_{t\in I}$ in $\mathbb{R}^{n+1}$ are moved at every point in the direction of their mean curvature vector at that point. If we parametrise M_t over a fixed n-dimensional manifold M^n with immersion $F_t := F(\cdot, t)$ and write $x = F(p, t)$ for points $x \in M_t$, then the equation is usually written as

$$\frac{\partial x}{\partial t} = \vec{H}(x), \tag{2}$$

where $\vec{H}(x)$ denotes the mean curvature vector of M_t at x and t ranges over a certain time interval. This equation can also be expressed as

$$\frac{\partial x}{\partial t} = \Delta_{M_t} x,$$

which formally resembles the standard heat equation. It is nonlinear since the Laplace–Beltrami operator on M_t depends nonlinearly on the first spatial derivatives of the immersion maps F_t.

Mean curvature flow is the gradient (steepest descent) flow for the energy given by the area functional. It was introduced by Ken Brakke in 1978 [32]. He studied the motion of generalised surfaces in the context of geometric measure theory. In 1984, Huisken [123] considered smooth, closed, convex hypersurfaces evolving by mean curvature flow (for the case $n \geq 2$), and Gage and Hamilton [93] considered curves evolving by the curve-shortening flow. Huisken and co-authors (see for instance [64, 65, 124, 127, 128]) used mean curvature flow to prove a number of important results about hypersurfaces which are in some sense analogous to results about Ricci flow proved by Hamilton and other authors. There are also mean curvature flow techniques available now which are analogous to Perelman's surgery methods for Ricci flow. However, in contrast to the Ricci flow case, these so far require additional assumptions on the Weingarten map (second fundamental form) of the initial hypersurface [37, 129, 130]. Mean curvature flow was also used to find maximal and prescribed spacelike hypersurfaces in Lorentzian manifolds as well as in the construction of optimal foliations in cosmological spacetimes [65] (see also [96]).

1.1.3 Ricci flow

The Ricci flow introduced by Hamilton in [105] evolves a time-dependent family of metrics on a fixed manifold by its Ricci tensor; more precisely, the family of metrics $g(t)$ satisfies the equation

$$\frac{\partial}{\partial t}g(t) = -2\,\mathrm{Ric}(g(t))\,, \tag{3}$$

where $\mathrm{Ric}(g(t))$ denotes the Ricci tensor corresponding to $g(t)$. Perelman [164] has shown that Ricci flow is a gradient flow for a certain energy functional involving an integral of the scalar curvature and a so-called dilation field, the latter giving rise to a one-parameter family of diffeomorphisms of the manifold. There is a strong relation between the Ricci flow of three-dimensional homogeneous metrics [132] as well as Thurston's eight-model geometries featuring in his geometrisation conjecture and the homogeneous cosmological models studied by the research group within project C8 led by Bernold Fiedler (see Section 2).

1.1.4 Harmonic map heat flow

The harmonic map heat flow evolves a family of maps $(u(\cdot,t))_{t\in I}$ between two fixed Riemannian manifolds M and N. If N is compact, it can be isometrically embedded into some Euclidean space, in which case the flow is governed by the equation

$$\left(\frac{\partial}{\partial t} - \Delta_M\right)u = A_N(u)(du, du)\,, \tag{4}$$

where $A_N(u(x,t))$ for $x \in M$ denotes the second fundamental form of N at the point $u(x,t) \in N$ and where du denotes the differential of u. For instance, if $N = \mathbb{S}^n$, then the right-hand side takes the form $|du|^2u$. Harmonic map heat flow is the gradient flow of the Dirichlet energy given by 1/2 times the L^2-integral of du. It was first studied in 1964 by Eells and Sampson [67] and used to establish the existence of harmonic maps from compact manifolds into manifolds with non-positive sectional curvatures (see also [135]). In [192, 193], the first major contributions to the regularity theory without the above assumption on the target manifold were achieved.

1.1.5 Yang–Mills flow

The Yang–Mills flow for a one-parameter family of connections is

$$\frac{\partial A}{\partial t} = \mathrm{div}\,F_A\,, \tag{5}$$

where F_A denotes the curvature of the connection form A. It is the steepest descent flow for the Yang–Mills energy, which is proportional to the L^2-integral of F_A.

Important results are due to Struwe [194] in four dimensions and due to Chen and Shen [46] in dimensions greater than four, to name just a few.

1.1.6 Willmore flow

The Willmore flow for surfaces immersed in Euclidean space with immersion denoted by x as for the mean curvature flow is given by the equation

$$\frac{\partial x}{\partial t} = -\Delta_{M_t} H + Q(A)\,, \tag{6}$$

where H denotes the mean curvature of M_t and $Q(A)$ is a cubic expression in the second fundamental form A of M_t. This flow is the steepest descent flow for the Willmore functional $\frac{1}{4}\int_M |\vec{H}|^2$, where $\vec{H}$ is the mean curvature vector of M. This flow has been extensively studied by Kuwert and Schätzle [142, 143].

1.2 Basic techniques

We now describe some important techniques related to specific research of the group, which will be described in Section 1.3.

1.2.1 Maximum principles and Bochner-type formulas

Probably the most basic tool for elliptic and parabolic equations is the weak maximum principle. Let us state the scalar weak maximum principle in its simplest form.

Proposition 1 (Weak maximum principle). *Let M be a compact, n-dimensional manifold with a smooth one-parameter family of metrics $(g(t))_{t\in[t_1,t_0]}$ and denote the Laplacian and gradient with respect to $g(t)$ by $\Delta = \Delta_{g(t)}$ and $\nabla = \nabla_{g(t)}$ respectively. Suppose $f : M\times[t_1, t_0] \to \mathbb{R}$ is sufficiently smooth on (t_1, t_0) and continuous on $[t_1, t_0]$ and satisfies the inequality*

$$\left(\frac{\partial}{\partial t} - \Delta\right) f(t) \le 0$$

for $t \in (t_1, t_0]$. Then $\max_M f(t) \le \max_M f(t_1)$ for all $t \in [t_1, t_0]$.

Already in this simple form the maximum principle is applicable to all the evolution equations listed above except for the Willmore flow. Similar versions hold for more general scalar parabolic equations on Riemannian manifolds, and there is also an important tensorial version of the maximum principle [51, 106].

Other important tools are Bochner-type formulas, which are the basis for the derivation of the evolution equations of higher-order quantities. Let us explain this for the simplest example, namely for solutions of the standard heat equation on compact

Riemannian manifolds. The simplest Bochner formula for a sufficiently smooth real valued function u on a compact Riemannian manifold (M, g) states that

$$\Delta|\nabla u|^2 = 2|\mathrm{Hess}\, u|^2 + 2\,\mathrm{Ric}(\nabla u, \nabla u) + 2\langle\nabla(\Delta u), \nabla u\rangle, \tag{7}$$

where ∇, Δ, Ric and $\langle\cdot,\cdot\rangle$ denote the gradient, the Laplacian, the Ricci curvature and the metric pairing on M. If u satisfies the standard heat equation, this immediately implies

$$\left(\frac{\partial}{\partial t} - \Delta\right)|\nabla u|^2 = -2\,|\mathrm{Hess}\, u|^2 - 2\,\mathrm{Ric}(\nabla u, \nabla u)\,. \tag{8}$$

Let us now explain how the weak maximum principle is used (see also [107]). If u satisfies the heat equation on M, then the identity

$$\left(\frac{\partial}{\partial t} - \Delta\right)u^2 = -2|\nabla u|^2 \tag{9}$$

holds. Because the manifold (M, g) is assumed to be compact, there is a constant $k \geq 0$ such that $\mathrm{Ric} \geq -k$. Therefore, eq. (8) implies the inequality

$$\left(\frac{\partial}{\partial t} - \Delta\right)|\nabla u|^2 \leq 2k\,|\nabla u|^2\,. \tag{10}$$

Combining eqs. (9) and (10), we infer that the function $f := |\nabla u|^2 + ku^2$ satisfies the assumption of the weak maximum principle. We conclude that $\max_M |\nabla u(t)|^2$ is bounded in terms of k and $\max_M u^2(0)$ but independent of t.

We shall later present several more sophisticated maximum principle arguments as we study evolution equations and inequalities for so-called higher-order quantities. At this stage, the only higher-order quantity we have encountered is the expression $|\nabla u|^2$ for solutions u of the heat equation on (M, g).

It is a common feature of solutions of the equations discussed in the previous section that no singularities can form (globally or locally) if certain quantities associated to the solutions stay (globally or locally) bounded. These associated quantities usually satisfy differential inequalities of the form

$$\left(\frac{\partial}{\partial t} - \Delta\right)f \leq \mathrm{RHS}(f)$$

(here f is a scalar function), where Δ is the appropriate Laplacian appearing also in the original equation, and the right-hand side satisfies $|\mathrm{RHS}(f)| \leq C\,(1 + |f|^p)$ for some $p \geq 0$ and uniform constant C. Note that $C \leq 0$ is admissible. There are also tensor identities of the form

$$\left(\frac{\partial}{\partial t} - \Delta\right)T = \mathrm{RHS}(T)\,, \tag{11}$$

with $|\mathrm{RHS}(T)| \leq C(1+|T|^p)$ for some $p \geq 0$ and uniform constant C, where $|T|$ denotes an appropriate tensor norm. Hence, the equation for T can be in some sense considered as a tensorial reaction-diffusion equation.

For the scalar reaction-diffusion equation (1), the crucial quantity is the solution u itself, whereas for the mean curvature flow (2) it is the Weingarten map A of the evolving hypersurfaces. By a calculation in [123] using Simons' identity [190] – which can be thought of as a Bochner-type identity, since $A = d\nu$ for a choice of unit normal field to M_t – the latter quantity satisfies the tensorial equation

$$\left(\frac{\partial}{\partial t} - \Delta\right) A = |A|^2 A\,, \tag{12}$$

so $p = 3$ here. If one instead considers $f := |A|^2$, then

$$\left(\frac{\partial}{\partial t} - \Delta\right) |A|^2 = -2|\nabla A|^2 + 2|A|^4\,, \tag{13}$$

where ∇A denotes the covariant derivative tensor of A. We therefore obtain $\mathrm{RHS}(f) \leq 2f^2$ by simply ignoring the negative term on the right-hand side. A more involved calculation [123] yields the inequality

$$\left(\frac{\partial}{\partial t} - \Delta\right) |\nabla A|^2 \leq -2|\nabla^2 A|^2 + c(n)|A|^2|\nabla A|^2\,, \tag{14}$$

where ∇A and $\nabla^2 A$ denote the covariant derivatives and covariant Hessian of A respectively. Analogous inequalities hold for higher-order covariant derivatives of A. Note that inequality (14) implies that if we already know that $|A|^2 \leq c_0$ on M_t independent of t, then the right-hand side of eq. (14) for $f := |\nabla A|^2$ satisfies $\mathrm{RHS}(f) \leq c(n, c_0)f$.

For the Ricci flow (3) one considers the curvature operator $\mathscr{R}$ of the evolving metrics. For a 3-manifold, this is given by a (3×3)-matrix depending on space and time. The quantity $\mathscr{R}$ satisfies

$$\left(\frac{\partial}{\partial t} - \Delta\right) \mathscr{R} = \mathrm{RHS}(\mathscr{R}) \tag{15}$$

with $|\mathrm{RHS}(\mathscr{R})| \leq c(n)|\mathscr{R}|^2$.

For the harmonic map heat flow (4), one chooses $f := |du|^2$. A Bochner-type calculation analogous to, but more involved than, the one presented for the standard heat equation (7) implies the important inequality

$$\left(\frac{\partial}{\partial t} - \Delta\right) |du|^2 \leq c(n, M)|du|^2 + c(n, N)|du|^4\,.$$

If N has non-positive sectional curvatures, then one can achieve this with a constant $c(n, N) \leq 0$. This was central for the work of Eells and Sampson [67] as, in the case where M is compact, the weak maximum principle applied to $f := e^{-c(n,M)t}|du|^2$ implies that $|du|^2$ cannot become unbounded in finite time. It then follows from standard parabolic partial differential equation (PDE) methods that the solution has to exist for all times. For the Yang–Mills heat flow (5), the curvature form satisfies a similar inequality to eq. (15) with $p = 2$.

The algebraic structure of the right-hand side of eq. (11) also plays a crucial role, not merely its growth behavior. For instance, Hamilton's tensor maximum principle [106, 123] applied to eq. (12) yields that convexity of compact hypersurfaces is preserved under mean curvature flow. There are stronger versions of this statement which will be discussed in Section 1.3.1. Similarly, identity (15) implies positive Ricci curvature on compact 3-manifolds is preserved [105]. More importantly, after diagonalising the curvature operator with eigenvalues λ_1, λ_2 and λ_3 (these correspond for 3-manifolds exactly to the sectional curvatures) the right-hand side of eq. (15) has the precise form

$$\begin{pmatrix} \lambda_1^2 + \lambda_2\lambda_3 \\ \lambda_2^2 + \lambda_1\lambda_3 \\ \lambda_3^2 + \lambda_1\lambda_2 \end{pmatrix}.$$

This was used in the proof of the famous Hamilton–Ivey result, which states that near points where the norm of the curvature operator tends to infinity as some finite time approaches, the sectional curvatures become asymptotically non-negative in a rescaled sense [51, 110, 112, 133].

Let us explain some more involved applications of the maximum principle for mean curvature flow. Assume that $|A|^2 \leq c_0$ on M_t uniformly in t. Then by combining eqs. (13) and (14) similarly as we combined eqs. (9) and (10) for the heat equation, one finds that the function $f := |\nabla A|^2 + \Lambda |A|^2$ (for suitable Λ depending on n and c_0) satisfies the assumption of the weak maximum principle. There are analogous functions involving higher covariant derivatives of A; thus, by an inductive argument [123], we are led to the following result.

Theorem 2 (Smoothness estimates). *Let $(M_t)_{t\in(t_1,t_0)}$ be a compact, smooth, properly immersed solution of mean curvature flow satisfying*

$$\max_{t\in(t_1,t_0)} \max_{M_t} |A|^2 \leq c_0 .$$

Then for every $m \in \mathbb{N}$ there exists a constant c_m depending on n, m as well as $\max_{1\leq k\leq m} \sup_{M_0} |\nabla^k A|^2$ such that

$$\max_{t\in(t_1,t_0)} \max_{M_t} |\nabla^m A|^2 \leq c_m .$$

Here $\nabla^k A$ denotes the $(k+2)$-tensor of the kth covariant derivatives of the second fundamental form A.

There is also a local version of the smoothness estimate [60, 65]. This is an analogue of an earlier result due to Shi [188] for Ricci flow. For both flows, these estimates are fundamental since they provide the basis for local subsequential convergence in most smooth rescaling arguments.

Theorem 3 (Local smoothness estimates). *Let (M_t) be a smooth, properly immersed solution of mean curvature flow in $B_\rho(x_0) \times (t_0 - \rho^2, t_0)$ which satisfies the estimate $|A(x)|^2 \leq c_0/\rho^2$ for all $x \in M_t \cap B_\rho(x_0)$ and $t \in (t_0 - \rho^2, t_0)$. Then for every $m \geq 1$ there is a constant c_m depending on n, m and c_0 such that for all $x \in M_t \cap B_{\rho/2}(x_0)$ and $t \in (t_0 - \rho^2/4, t_0)$ there holds $|\nabla^m A(x)|^2 \leq c_m/\rho^{2(m+1)}$.*

In the case $m = 1$ the proof again uses eqs. (13) and (14) to show that under the assumption on $|A|^2 \leq c_0$ in the theorem there is a $\Lambda > 0$ depending on c_0 such that the function $f := |\nabla A|^2 \left(\Lambda + |A|^2\right)$ satisfies an inequality of the form

$$\left(\frac{\partial}{\partial t} - \Delta\right) f \leq -\delta f^2 + K\,, \tag{16}$$

where $\delta > 0$ and K depend only on n and c_0. The bounds on the higher-order derivatives follow by an induction argument. Note that if a non-negative function f satisfies eq. (16), then the product of f with a suitable test function satisfies the assumption of the weak maximum principle. This theorem implies (using standard methods from analysis) that as long as the second fundamental form is bounded up to some time $t_0 > 0$ near a point $x_0 \in \mathbb{R}^{n+1}$ no singularity can form near this point at time t_0.

1.2.2 Monotonicity formulas

The following important monotonicity formula for mean curvature flow was proved by Huisken [124]. Similar formulas for other geometric heat flows were established in [46, 97, 108, 109, 193]. An analogue of this formula for Ricci flow was established by Perelman [164] using his so-called ℓ-distance. Define

$$\Phi_{(x_0,t_0)}(x,t) := \frac{1}{(4\pi(t_0 - t))^{n/2}} \exp\left(-\frac{|x - x_0|^2}{4(t_0 - t)}\right)$$

for $(x_0, t_0) \in \mathbb{R}^{n+1} \times I$, $I \subset \mathbb{R}$ an open interval, $x \in \mathbb{R}^{n+1}$ and $t < t_0$.

Theorem 4 (Huisken's monotonicity formula). *Let $(x_0, t_0) \in \mathbb{R}^{n+1} \times I$ and $(M_t)_{t\in I}$ be a smooth solution of mean curvature flow for which*

$$\int_{M_t} \Phi_{(x_0,t_0)} < \infty$$

for all $t \in I$ with $t < t_0$. Then for these times

$$\frac{d}{dt}\int_{M_t} \Phi_{(x_0,t_0)} = -\int_{M_t} \left|\vec{H}(x) - \frac{(x-x_0)^{\perp}}{2(t-t_0)}\right|^2 \Phi_{(x_0,t_0)} \leq 0\,. \tag{17}$$

In [64], a more general version of this monotonicity formula is proved; namely, if a nonnegative function u satisfies

$$\left(\frac{d}{dt} - \Delta_{M_t}\right) u \leq 0,$$

then under suitable growth conditions on M_0 in extrinsic balls and u at infinity,

$$\frac{d}{dt}\int_{M_t} u\,\Phi_{(x_0,t_0)} \leq 0\,.$$

This can be used to prove a non-compact version of the weak maximum principle for mean curvature flow.

The monotonicity formula also holds for the generalised mean curvature evolution defined by Brakke and is one of the central tools in the regularity theory of this flow. By the monotonicity for the above integral expression, the so-called *Gaussian density* is defined by

$$\Theta(\mathcal{M}, x_0, t_0) := \lim_{t \nearrow t_0} \int_{M_t} \Phi_{(x_0,t_0)}\,, \tag{18}$$

where

$$\mathcal{M} := \bigcup_{t\in I} M_t \times \{t\}$$

is the spacetime track of the flow that exists at all points $(x_0, t_0) \in \mathbb{R}^{n+1} \times I$. One easily shows that $\Theta(\mathcal{M}, x_0, t_0) = 1$ if the flow $\mathcal{M}$ is smooth at that point. By the local version of Huisken's formula, Θ can also be defined locally.

White [209] has shown that there exists a constant $\varepsilon > 0$ depending only on n such that $\Theta(\mathcal{M}, x_0, t_0) \geq 1+\varepsilon$ at all singular points of the flow. This can be used to give estimates for the size of the singular set (see for instance [60, 208]). Similar so-called epsilon regularity theorems hold for all the above geometric flows (even the Willmore flow, which is of fourth order) and also for their corresponding elliptic counterparts. More precisely, there are suitable local scaled energy quantities which near singularities deviate by a fixed amount from the value near smooth points. If there is some finite total energy – in the mean curvature flow for closed hypersurfaces this quantity is the spacetime integral of H^2 – then there cannot be too many singular points as the fixed local energy contributions at these would add up to exceed the finite total energy.

We note that a straightforward geometric calculation reveals that the right-hand side of the monotonicity formula (17) vanishes exactly when (M_t) is a solution of mean curvature flow which is homothetically shrinking about (x_0, t_0), that is, $M_t - x_0 = \sqrt{t_0 - t}\left(M_{t_0 - 1} - x_0\right)$ for $x \in M_t$ and $t < t_0$. Examples for $(x_0, t_0) = (0, 0)$ are given by the cylinders $S^{n-k}_{\sqrt{-2(n-k)t}} \times \mathbb{R}^k$ for $t < 0$ and for $0 \le k \le n$. For $k = 0$ these are shrinking spheres. Huisken [125] has shown that all embedded, homothetically shrinking solutions with non-negative mean curvature are of this form.

1.2.3 Rescaling arguments

Let us explain how singularities are studied by considering for simplicity only those at the first singular time $t_0 > 0$, say. Let $x_0 \in \mathbb{R}^{n+1}$ be such a singular point for $(M_t)_{t\in(0,t_0)}$. By the smoothness estimates, the norm of the second fundamental form will locally tend to infinity near x_0 as $t \nearrow t_0$. In order to study the flow near x_0, we consider the rescaled flow

$$M^j_s := \frac{1}{\lambda_j}\left(M_{\lambda_j^2 s + t_j} - x_j\right)$$

for a sequence $\lambda_j \searrow 0$, where $s \in (-\lambda_j^{-2} t_j, \lambda_j^{-2}(t_0 - t_j))$. If one chooses the sequences (x_j), (t_j) and (λ_j) appropriately, one can arrange that the second fundamental forms of the rescaled solutions (M^j_s) be bounded independently of j on their respective time intervals. Since $t_j \nearrow t_0$, we have $-\lambda_j^{-2} t_j \to -\infty$. The (local) smoothness estimates imply (local) uniform bounds on the derivatives of the second fundamental form of all orders independently of j. One can therefore find a subsequence of rescaled solutions which converges smoothly on compact subsets of spacetime to a smooth solution $(M'_s)_{s\in(-\infty,s_0)}$ for some $s_0 \ge 0$. Such solutions are called *ancient* solutions as they have existed forever. The classification of ancient solutions is thus a very important topic that will be considered in Section 1.3.1.

If the second fundamental form satisfies the so-called *type-I* condition, namely, that $\sup_{M_t} |A|^2 \le C(t_0 - t)^{-1}$ for C independently of t, then one can choose $x_j = x_0$ in the above rescaling process. One can then apply Huisken's monotonicity formula to the rescaled solutions to conclude that the ancient limiting solution is a homothetically shrinking solution of mean curvature flow. This procedure (i.e., choosing $x_j = x_0$) also works without the type I assumption; however, in this case, the rescaled solutions will in general not converge to a smooth limiting solution but merely to a Brakke-type solution (which we know again must be homothetically shrinking).

If the original solution is of *type II* (that is, not of type I) one can obtain a smooth rescaling limit (M'_s) which is defined for all $s \in (-\infty, \infty)$. These are called *eternal* solutions. The best-known eternal solution is the family of curves (Γ_t) defined by the graphs of the functions given by $u(x, t) = -\log\cos x + t$. This collection of curves is called the *grim reaper*. Higher-dimensional analogues are the solutions given by the family $G_t = \Gamma_t \times \mathbb{R}^{n-1}$. Note that there are ancient, not homothetically shrinking, solutions of mean curvature flow which at some finite time shrink to a point and become asymptotically round in the process. Rescaling limits (M_s') which equal the above family (G_s) cannot occur if the initial hypersurface is embedded and has positive mean curvature. This was proved by Sheng and Wang [187] and Andrews [11]. In the case of immersed closed planar curves with curvature $\kappa > 0$, Angenent [18] has shown that cusps can form in finite time which, after a suitable type II rescaling process, produce the grim reaper curve. This phenomenon is an example where a solution collapses near a singularity, as on the rescaling limit there is no lower bound on the ratio of minimum and maximum curvature. There are also other ways of describing this phenomenon (see Sections 1.2.5 and 1.3.1). It is a very important open problem to determine whether this kind of collapse can occur in case the initial hypersurface is embedded but does not necessarily have positive mean curvature.

Let us now consider solutions u of eq. (1). For $r > 0$ and $\beta = 1/(p-1)$, $p > 1$, the rescaled function defined by

$$u_r(x,t) := r^{2\beta} u(rx, r^2 t)$$

is again a solution of eq. (1). One easily checks that the energy density scales like

$$e(u_r)(x,t) = r^{\gamma} e(u)(rx, r^2 t),$$

where $\gamma = 4\beta + 2 = 2(p+1)/(p-1)$. For solutions of the heat equation we have $\beta = 0$, so $\gamma = 2$ in this case.

We can now state the scaled monotonicity formula of Giga and Kohn [97] for eq. (1). There is an analogue of this formula, albeit an unscaled version, for the harmonic map heat flow due to Struwe [193] (see also [61]). In the latter case the natural choice is $\gamma = 2$.

Theorem 5 (Monotonicity formula of Giga and Kohn). *Let u be a solution of eq. (1) on $\mathbb{R}^n \times \mathbb{R}_-$. Define $w(z,\tau) := (-t)^{\beta} u(z,t)$ with $x = \sqrt{-t}\, z$ and $\tau = \log(-t)$. (z and τ are often called similarity variables.) Then there is a Lyapunov functional $\mathscr{L}$ acting on $w(\cdot,\tau)$ satisfying*

$$\frac{d}{d\tau}\mathscr{L}(w(\cdot,\tau)) = -\int_{\mathbb{R}^n} \left(\frac{\partial w}{\partial z}(z,\tau)\right)^2 \exp\left(-\frac{|z|^2}{4}\right) dz \le 0.$$

Note that as $t \nearrow 0$ we have $\tau \nearrow \infty$. In the presence of a uniform bound for w, which is analogous to the type I condition for mean curvature flow, standard PDE theory guarantees smoothness of w for all $\tau > 0$, and one therefore expects as $\tau \to \infty$ to obtain at least subsequential convergence of $w(\cdot,\tau)$ to an equilibrium of eq. (1) written in similarity variables. Such convergence would imply that our solution u of eq. (1) has the asymptotic form

$$u(x,t) \approx \frac{1}{(-t)^{\beta}} v\left(\frac{x}{\sqrt{-t}}\right),$$

where $v(z) = w(z,-1)$ satisfies an elliptic equation. This is related to research carried out in Section 2.

1.2.4 Local monotonicity and mean value formulas

For the elliptic problems (harmonic maps, minimal surfaces and Yang–Mills fields) there are also monotonicity formulas. Here, one considers integrals of the energy density over balls divided by an appropriate power of the radius of the ball. The monotonicity formulas then state that these ratios are monotonically increasing in the radius.

For an n-dimensional minimal submanifold $M \subset \mathbb{R}^{n+k}$ one has

$$\frac{d}{dr} \frac{\operatorname{area}(M \cap B_r(x_0))}{r^n} \geq 0$$

for any ball $B_r(x_0) \subset \mathbb{R}^{n+k}$ [189]. More generally, if $u \geq 0$ satisfies $\Delta u \geq 0$ on a minimal submanifold M then one has

$$\frac{d}{dr}\left(\frac{1}{r^n} \int\limits_{M \cap B_r(x_0)} u\right) \geq 0$$

in those balls. If M is C^1 at $x_0 \in M$ this implies the mean value formula

$$u(x_0) \leq \frac{1}{\omega_n r^n} \int\limits_{M \cap B_r(x_0)} u,$$

where ω_n denotes the volume of the n-dimensional unit ball. This generalises the mean value formula for subharmonic functions on $\mathbb{R}^n$.

For harmonic maps $u : M \to N$ one considers $r^{-(n-2)} \int_{B_r(x_0)} |du|^2$, where $B_r(x_0)$ is a geodesic ball inside M [182]. There is an analogous formula for Yang–Mills fields [171].

Observe that standard balls in Euclidean space are super-level sets of the fundamental solution of the Laplacian. Analogously, Watson [207] defined super-level sets of the fundamental solution of the heat operator termed *heat balls* see also [71, 72]. Note that the function defined by $\Phi_{(x_0,t_0)}$ for fixed $(x_0, t_0) \in \mathbb{R}^{n+1} \times \mathbb{R}$ which is featured in Huisken's monotonicity formula (17) is the standard backward heat kernel when restricted to $\mathbb{R}^n$. Watson defined for all $r > 0$ and $(x_0, t_0) \in \mathbb{R}^n \times \mathbb{R}$

$$E_r(x_0, t_0) := \left\{(x,t) \in \mathbb{R}^n \times \mathbb{R} \; : \; t < t_0, \; \Phi_{(x_0,t_0)}(x,t) > \frac{1}{r^n}\right\}.$$

He then proved that if a function u satisfies

$$\left(\frac{\partial}{\partial t} - \Delta\right) u \leq 0$$

on $\mathbb{R}^n$ and times less than t_0, then u satisfies the monotonicity formula

$$\frac{d}{dr}\left(\frac{1}{r^n} \iint\limits_{E_r(x_0,t_0)} u(x,t) \frac{|x-x_0|^2}{4(t-t_0)^2}\, dx\, dt\right) \geq 0.$$

He thereby obtained an analogue of the mean value formula for sub-harmonic functions, namely

$$u(x_0,t_0) \leq \frac{1}{r^n} \iint\limits_{E_r(x_0,t_0)} u(x,t) \frac{|x-x_0|^2}{4(t-t_0)^2}\, dx\, dt$$

for all $r > 0$ and $(x_0, t_0) \in \mathbb{R}^n \times \mathbb{R}$.

Versions of such mean value formulas are already found in [90]. There are now numerous generalisations of Watson's formula even for nonlinear diffusion equations. For more general linear elliptic operators we refer to [94]. We mention in particular the local monotonicity formula for reaction diffusion equations and for the harmonic map heat flow [61] as well as for the Ricci flow [66]. In the following we present the relevant formula for the mean curvature flow [61] which is a local analogue of Huisken's monotonicity formula and of its generalisation featuring subsolutions of the heat operator on mean curvature flow [64].

For mean curvature flow we consider the analogous heat ball set but now for $x \in \mathbb{R}^{n+1}$ as in Huisken's formula. For simplicity let us centre the heat kernel at $(x_0, t_0) = (0,0)$, that is consider $\Phi = \Phi_{(0,0)}$. We furthermore define $\psi_r := \log(r^n\Phi) = \psi + n\log r$ (that is, $\psi = \log\Phi$) and abbreviate $E_r = E_r(0,0)$. Suppose $u \geq 0$ satisfies

$$\left(\frac{d}{dt} - \Delta\right) u \leq 0.$$

Then for the spacetime track $\mathcal{M}$ of the flow the local monotonicity formula

$$\frac{d}{dr}\left(\frac{1}{r^n} \iint\limits_{\mathcal{M}\cap E_r} u\left(|\nabla^{M_t}\psi|^2 + H^2\psi_r\right) d\mu_t\, dt\right) \geq 0$$

holds, where $d\mu_t$ denotes the surface element of M_t. If the solution of mean curvature flow is smooth at $(0,0) \in \mathbb{R}^{n+1} \times \mathbb{R}$, it leads to the mean value inequality

$$u(0,0) \leq \frac{1}{r^n} \iint\limits_{\mathcal{M}\cap E_r} u\left(|\nabla^{M_t}\psi|^2 + H^2\psi_r\right) d\mu_t\, dt.$$

If $M_t = \mathbb{R}^n$ for all $t < 0$, this reduces to Watson's formula. One can also show that

$$\lim_{r\to 0} \frac{1}{r^n} \iint\limits_{\mathscr{M}\cap E_r} \left(|\nabla^{M_t}\psi|^2 + H^2\psi_r\right) d\mu_t\, dt = \Theta(\mathscr{M}, 0, 0),$$

where Θ is the Gaussian density we encountered in eq. (18).

1.2.5 Non-collapsing

For the Ricci flow of closed 3-manifolds, Perelman [164] has proved that collapse cannot occur on a finite time interval, without imposing any further conditions on the solution. He gave two proofs of this fact. Let us mention one, as we shall in Section 1.3.2 describe an approach for mean curvature flow which is modelled on his. Perelman considered the functional

$$\mathscr{W}(g, f, \tau) := \int\limits_M \left(\tau(|\nabla f|^2 + R) + f - (n+1)\right) u\, dV$$

and associated entropy

$$\mu(g, \tau) := \inf \left\{ \mathscr{W}(g, f, \tau) \;:\; \int\limits_M u\, dV = 1 \right\}$$

for $\tau > 0$ and a smooth function f on a closed $(n+1)$-dimensional Riemannian manifold (M, g), where R is the scalar curvature of (M, g), and

$$u := \frac{e^{-f}}{(4\pi\tau)^{(n+1)/2}}.$$

His ingenious realisation was that when $\tau(t) > 0$ satisfies $\frac{\partial\tau}{\partial t} = -1$ and $(M, g(t))$ evolves by the Ricci flow then

$$\frac{d}{dt}\mu(g(t), \tau(t)) \geq 0,$$

where the time derivative has to be appropriately interpreted. An important consequence of this entropy formula is a lower volume ratio bound for solutions of Ricci flow on a closed manifold for a finite time interval $[0, T)$ asserting the existence of a constant $\kappa > 0$, only depending on n, T and $g(0)$, such that the inequality

$$\frac{V_t(B_r^t(x_0))}{r^{n+1}} \geq \kappa$$

holds for all $t \in [0, T)$ and $r \in [0, \sqrt{T})$ for balls $B_r^t(x_0)$ (with respect to $g(t)$) in which the inequality $r^2|Rm| \leq 1$ for the Riemann tensor of $g(t)$ holds. This lower volume ratio bound rules out certain collapsed metrics as rescaling limits near singularities of Ricci flow, such as products of Euclidean spaces with the so-called *cigar soliton* solution of Ricci flow given by $M = \mathbb{R}^2$ with the metric

$$ds^2 = \frac{dx^2 + dy^2}{1 + x^2 + y^2}\,.$$

The cigar solution is the Ricci flow analogue of the grim reaper for mean curvature flow.

On the other hand, when the mean curvature is positive, Sheng and Wang [187] (see also [11]) have proved an extrinsic non-collapsing estimate which rules out collapsing solutions such as the grim planes described above. Namely, if we define the *inscribed radius* $\bar{r}(x)$ at each point $x \in M_t = \partial\Omega_t$ as the radius of the largest ball contained in Ω_t which is tangent to M_t at x, then $r(x)H(x)$ is bounded from below uniformly in time. These ideas are further developed in Section 1.3.1.

1.2.6 Sobolev inequalities

The starting point of smooth mean curvature analysis is the following theorem of Huisken [123]. This is an analogue of an earlier result of Hamilton [105] for the Ricci flow on 3-manifolds with positive Ricci curvature and of a result by Gage and Hamilton for the curve shortening flow [93].

Theorem 6. *Let M_0 be an n-dimensional smooth, closed, convex hypersurface in $\mathbb{R}^{n+1}$ and $n \geq 2$. Then there exists a smooth solution of mean curvature flow on an interval $(0, T)$ for $T < \infty$ which contracts smoothly to a point as $t \nearrow T$. After rescaling this solution appropriately (for instance by keeping the area of the hypersurface fixed) it converges smoothly to a round sphere.*

The main ingredient in the proof of this theorem – the so-called *umbilical estimate* – is due to the fact that the quantity which is estimated below vanishes exactly when all principal curvatures of the hypersurface are equal to each other at all points. An application of the maximum principle as in [105] appears to be not possible. For details, refer to Section 1.3.1. The main technique used in the proof of this and related theorems (see also Section 1.3.1) include Sobolev inequalities and de Giorgi–Nash–Moser/Stampacchia iteration. Such techniques also replace the maximum principle arguments which are not applicable in the case of Willmore flow (see [142, 143]). The relevant Sobolev inequality is the one due to Michael and Simon [159], which states that

$$\left(\int_M |f|^{n/(n-1)}\right)^{(n-1)/n} \leq c(n) \int_M \left(|\nabla f| + |\vec{H}||f|\right)$$

for n-dimensional submanifolds in Euclidean space $\mathbb{R}^{n+k}$ with mean curvature vector $\vec{H}$ and all $f \in C_0^1(M)$. Here it is crucial that, unlike the Sobolev inequality for

Riemannian manifolds, the Sobolev constant does not depend on the hypersurface. In mean curvature flow, one applies this inequality to the evolving surfaces $(M_t)_{t\in[0,T]}$. If, for instance, these are compact, then one has $\frac{d}{dt}\,\mathrm{area}(M_t) = -\int_{M_t}|\vec{H}|^2$ and thus $\int_0^T\int_{M_t}|\vec{H}|^2 \le \mathrm{area}(M_0) < \infty$. Therefore, one can easily control the L^1-integral term on the right-hand side of the Sobolev inequality during the evolution.

1.3 Specific techniques and recent results

In this section we present the results obtained by members of the geometric analysis group within project C8 in recent years.

1.3.1 Singularity analysis for some geometric flows (Huisken,[2] Langford[3])

Let $(M_t)_{t\in I}$ be a family of smooth, compact and immersed hypersurfaces moving by mean curvature (2). We will call a hypersurface *mean convex* if it is two-sided and, with respect to one of the two choices of unit normal, ν, the mean curvature $H := -\vec{H}\cdot\nu$ is non-negative. Since the mean curvature satisfies the equation

$$\left(\frac{\partial}{\partial t} - \Delta\right)H = |A|^2 H,$$

where Δ is the Laplace–Beltrami operator of the induced metric, the weak maximum principle implies that mean convexity is preserved by the flow. This is the most basic case of a very general principle: Since the second fundamental tensor A (defined with respect to ν) satisfies the reaction-diffusion system

$$\left(\frac{\partial}{\partial t} - \Delta\right)A = |A|^2 A,$$

a version of the *tensor* maximum principle implies that any convex, symmetric cone of principal curvatures is preserved. An important corollary is the preservation of (uniform) *k-convexity*, where we say that a hypersurface is called *k-convex* for some $k \in \{1,\dots,n\}$ if the sum $\kappa_1 + \cdots + \kappa_k$ of its smallest k principal curvatures is everywhere non-negative, and *uniformly k-convex* if it is mean convex and there is a constant $\alpha > 0$ such that $\kappa_1 + \cdots + \kappa_k \ge \alpha H$. The k-convexity condition interpolates between mean convexity ($k = n$) and convexity[4] ($k = 1$).

---

2 Gerhard Huisken, University of Tübingen, gerhard.huisken@uni-tuebingen.de

3 Mathew Langford, University of Tennessee, Knoxville, mlangfo5@tennessee.edu

4 What we refer to as *convexity* is usually referred to as *local convexity*, the term *convex* being reserved for boundaries of convex bodies. Because M is compact, the two notions are equivalent if F is an embedding or if $n \ge 2$.

One-sided curvature pinching.

Mean curvature flow of convex hypersurfaces in dimensions $n \geq 2$ was studied in a groundbreaking paper of Huisken [123] where several new techniques were introduced, and it was proved that such hypersurfaces shrink to "round" points. A crucial part of the analysis was the umbilic estimate, which states that given any $\varepsilon > 0$, there is a constant C_ε (which depends only on ε and the initial data) such that

$$|\mathring{A}|^2 \leq \varepsilon H^2 + C_\varepsilon, \tag{19}$$

where $\mathring{A}$ denotes the trace free part of A. In particular, this implies that the *scale-invariant* ratio $|\mathring{A}|^2/H^2$ vanishes at a singularity of the flow. To prove such an estimate, one attempts to bound the function

$$f_\sigma := \frac{|\mathring{A}|^2}{H^2} H^\sigma$$

for some small $\sigma > 0$. Huisken achieves this by bounding L^p-norms of f_σ for large p and $\sigma \approx p^{-\frac{1}{2}}$ and applying a Stampacchia-type iteration with the help of the Michael–Simon Sobolev inequality. This method turns out to be quite robust, and variants of the argument were later applied to obtain curvature estimates in the non-convex setting. The next breakthrough was the *convexity estimate* [127, 128], which states that if the initial immersion is mean convex, then given any $\varepsilon > 0$ there is a constant C_ε depending only on ε and the initial data such that

$$\kappa_1 \geq -\varepsilon H - C_\varepsilon. \tag{20}$$

This estimate implies that the scale invariant tensor A/H is non-negative definite at a singularity, which is a strong restriction on the geometry of the hypersurface near a singularity. Interpolation between eqs. (19) and (20) are the *m-cylindrical estimates* [12, 129, 130]: If the initial immersion is $(m+1)$-convex, then given any $\varepsilon > 0$ there is a constant C_ε depending only on ε and the initial data such that

$$|A|^2 - \frac{H^2}{n-m} \leq \varepsilon H^2 + C_\varepsilon. \tag{21}$$

On an $(m+1)$-convex hypersurface, the left-hand side of eq. (21) is non-positive only at points which are either strictly m-convex, $\kappa_1 + \cdots + \kappa_m > 0$, or *m-cylindrical*, $0 = \kappa_1 = \cdots = \kappa_m$ and $\kappa_{m+1} = \cdots = \kappa_n$. This is particularly restrictive in the case $m = 1$ and $n \geq 3$ (see [129]). Note that the case $m = 0$ is just the umbilic estimate (19) and the case $m = n - 1$ follows from the convexity estimate (20).

Recently, by similar methods, Langford [147] obtained a slightly sharper version of the cylindrical estimates, which has some interesting applications. Namely, given an $(m+1)$-convex solution of mean curvature flow and any $\varepsilon > 0$, there is a constant C_ε depending only on ε and the initial data such that

$$\kappa_n - \frac{H}{n-m} \leq \varepsilon H + C_\varepsilon \,. \tag{22}$$

The case $m = 0$ is equivalent to the umbilic estimate (19) and the case $m = n-1$ follows from the convexity estimate (20); however, the remaining cases are new. Moreover, (22) implies the cylindrical estimate (21).

Note that the estimates (21) and (22) are asymptotically sharp in the sense that there is a sequence of compact, $(m+1)$-convex solutions of mean curvature flow converging to the shrinking cylinder solution $\mathbb{R}^m \times S^{n-m}_{\sqrt{-2(n-m)t}}$ which, in each case, achieves equality in the limit with $\varepsilon = C_\varepsilon = 0$.

Our main motivation for proving the "m-convexity" estimate (22) is that it constitutes the "boundary case" of an analogous "inscribed curvature" estimate, which is described subsequently.

The inscribed curvature

Next, we turn our attention to embedded solutions of eq. (2). Let $M = \partial\Omega \subset \mathbb{R}^{n+1}$ be a properly embedded hypersurface bounding a precompact open set $\Omega \subset \mathbb{R}^{n+1}$. Equip M with the outward pointing unit normal. Then the *inscribed curvature* $\overline{k}(x)$ of a point $x \in M$ is defined as the curvature of the boundary of the largest ball which is contained in Ω and whose boundary touches M at x [14]. A straightforward calculation [11] reveals that

$$\overline{k}(x) = \sup_{y \in M \setminus \{x\}} k(x,y)\,, \tag{23}$$

where

$$k(x,y) := \frac{2\,\langle x-y, \nu(x)\rangle_{\mathbb{R}^{n+1}}}{\|x-y\|^2_{\mathbb{R}^{n+1}}}\,.$$

Similarly, one can define the exscribed curvature $\underline{k}(x)$ at x as the (signed) boundary curvature of the largest ball, halfspace or ball complement having exterior contact at x. In that case, one obtains $\underline{k}(x) = \inf_{y \in M \setminus \{x\}} k(x,y)$. Note that switching the normal switches $\overline{k}$ and $\underline{k}$. Observing that either the supremum in eq. (23) is attained, or else

$$\overline{k}(x) = \limsup_{y \to x} k(x,y) = \sup_{y \in T_xM} A_x(y,y)/g_x(y,y)\,,$$

allows one to obtain derivative identities (in, say, the viscosity sense) for the Lipschitz function $\overline{k}$ (and similarly for $\underline{k}$) by analysing the smooth "two-point functions" $k(x,y)$ and $A_x(y,y)/g_x(y,y)$. In particular, along a solution of mean curvature flow,

$$\left(\frac{\partial}{\partial t} - \Delta\right)\overline{k} \le |A|^2\overline{k}$$

in the viscosity sense [14]. It follows that mean convex solutions of mean curvature flow are *interior non-collapsing*: $\overline{k}$ can be compared from above with H uniformly in time along a mean convex mean curvature flow [11, 187]. A Stampacchia iteration argument similar to those described above reveals that this bound improves at a singularity [36]. Precisely, for any $\varepsilon > 0$ there is a constant C_ε depending only on ε and the initial data such that

$$\overline{k} - H \le \varepsilon H + C_\varepsilon .$$

This estimate is sharp due to the fact that equality ($\overline{k} \equiv H$) holds on a cylinder $\mathbb{R}^{n-1} \times \mathbb{S}^1$. Similar arguments yield an "exterior" non-collapsing estimate, which can be improved to a sharp lower bound for $\underline{k}/H$.

Recently, Langford was able to refine this estimate in the class of $(m+1)$-convex flows [147]. In that case, for any $\varepsilon > 0$ there is a constant C_ε depending only on ε and the initial data such that

$$\overline{k} - \frac{H}{n-m} \le \varepsilon H + C_\varepsilon .$$

As mentioned above, the "boundary case" $\overline{k} = \kappa_n$ is the estimate (22), which reduces the proof to the "interior case" $\overline{k}(x) = k(x, y)$ for some $y \in M$.

As mentioned, the Stampacchia iteration method used to prove these estimates is very robust. Indeed, one can apply it, on the one hand, to obtain analogous estimates for flows by fully nonlinear curvature flows; this was first achieved in the thesis [146] and the related articles [12, 15, 16].

Flows by nonlinear functions of curvature

More generally, one can consider solutions $F : M^n \times [0, T) \to \mathbb{R}^{n+1}$ of the evolution equation

$$\frac{\partial F}{\partial t}(p, t) = -\eta(p, t)\nu(p, t) ,$$

where η is given by a smooth, symmetric function $f : \mathbb{R}^n \to \mathbb{R}$ of the principal curvatures of the solution. The equation becomes parabolic (in an appropriate sense) if f is monotone increasing in each argument. One considers the case that f is 1-homogeneous so that F scales like the curvature. Such flows have similar properties to the mean curvature flow; however, the nonlinearity of f introduces several technical difficulties and, in general, solutions can behave quite badly. On the other hand, if $n = 2$, or under certain additional concavity assumptions for f, we are able to obtain several interesting results. On the one hand, flows with concave speed functions tend to preserve scale-invariant upper bounds for A and $\overline{k}$. This allows one to recover analogous m-convexity and inscribed curvature estimates [14, 146, 148] (see also [37]). On the other hand, flows with convex speed functions tend to preserve scale-invariant lower bounds for A and $\underline{k}$. In this case, one can obtain convexity and exscribed curvature estimates [12, 14, 15, 148]. In case $n = 2$, no additional requirements on the speed (other than positivity) are necessary to obtain sharp estimates for κ_1 and κ_2 [16, 146]; however, it remains an open problem (with some interesting applications) to obtain inscribed and exscribed curvature estimates for general (1-homogeneous) surface flows. The large class of flows with concave and "inverse-concave" speeds, which includes, for example, all of the flows by 1-homogeneous roots of ratios of elementary symmetric polynomials, admits two-sided curvature pinching and non-collapsing estimates when the initial hypersurface is convex [13, 146, 148].

Ancient solutions

Finally, let us briefly mention the application of the above techniques to ancient solutions. To keep the discussion simple, we only consider solutions of the mean curvature flow, although similar considerations apply to the class of flows mentioned above. Recall that *ancient solutions* are those solutions defined for time intervals I of the form $(-\infty, T)$ with $T < \infty$ (since M is taken to be compact). Without loss of generality, we assume that $T = 1$. In principle, this property should be extremely rigid, since diffusion has had an arbitrarily long time to take effect. Indeed, when $n = 1$, there are only two compact, convex, embedded ancient solutions (up to rigid motions): the shrinking circle and the Angenent oval [56]. For $n \geq 2$, a complete classification remains open; however, some recent breakthroughs have been made. For instance, the shrinking sphere is the only *non-collapsing* ancient solution which is either uniformly convex or of type I curvature decay[5] [114]. In fact, the same statement was proved by Huisken and Sinestrari [130] for convex ancient solutions in dimensions $n \geq 2$ without the non-collapsing assumption. An analogous rigidity result for ancient solutions of the Ricci flow satisfying a suitable curvature pinching condition was proved by Brendle, Huisken and Sinestrari [38]. In fact, they show that such solutions must have constant sectional curvature.

The proof of the result in [130] is based on a clever modification of the proof of the umbilic estimate. The same idea applies to the cylindrical estimates [130] so that convex, uniformly $(m+1)$-convex ancient solutions must satisfy

$$\sup_{M\times(-\infty,1)} \left(|A|^2 - \frac{H^2}{n-m} \right) \leq 0 .$$

5 We say that an ancient solution of eq. (2) has *type I curvature decay* if $\limsup_{t\to-\infty} \sqrt{-t}\, \max_{M\times\{t\}} H < \infty$.

If $m = 0$, the solution must be a shrinking sphere. By the strong maximum principle, the inequality is strict in the remaining cases. It was left open whether or not the techniques apply to the convexity estimate. Making use of the breakthrough in [130] and the main idea from [15], Langford shows that this is indeed the case [147]: Any mean convex ancient solution for which the ratio κ_1/H is bounded from below satisfies $\kappa_1 \geq 0$. In fact, the technique also applies to the other pinching estimates discussed above, and we obtain optimal estimates for the largest principal curvature and the inscribed and exscribed radii for ancient k-convex solutions of mean curvature flow as long as their ratio to the mean curvature does not degenerate as $t \to -\infty$.

1.3.2 Entropy methods for mean curvature flow (Ecker)

Given a bounded open subset $\Omega \subset \mathbb{R}^{n+1}$ with smooth boundary, a smooth function $f : \overline{\Omega} \to \mathbb{R}$ and $\tau > 0$, the quantity

$$\mathscr{W}_{H_{\partial\Omega}}(\Omega, f, \tau) := \int\limits_{\Omega} \left(\tau|\nabla f|^2 + f - (n+1)\right) u\, dx + 2\tau \int\limits_{\partial\Omega} H_{\partial\Omega} u\, dS$$

and the associated entropy

$$\mu_{H_{\partial\Omega}}(\Omega, \tau) := \inf\left\{ \mathscr{W}_{H_{\partial\Omega}}(\Omega, f, \tau) \;:\; \int\limits_{\Omega} u\, dx = 1 \right\},$$

where $H_{\partial\Omega}$ denotes the mean curvature of $\partial\Omega$ and

$$u := \frac{e^{-f}}{(4\pi\tau)^{(n+1)/2}},$$

were studied in [62]. These generalise expressions introduced by Perelman in [164] (discussed in Section 1.2.5).

In [62], the following situation, which is analogous to Perelman's, was investigated: Consider a family of bounded open regions $(\Omega_t)_{t\in(0,T)}$ in $\mathbb{R}^{n+1}$ with smooth boundary hypersurfaces $M_t = \partial\Omega_t$ evolving by mean curvature flow (note that as boundaries of sets the hypersurfaces M_t are embedded). The unit normal to M_t points out of Ω_t in this setting. In [62], it was proved that if $\tau = \tau(t) > 0$ satisfies $\frac{\partial\tau}{\partial t} = -1$, then the expression

$$\frac{d}{dt}\mu_{H_{M_t}}(\Omega_t, \tau(t))$$

interpreted in an appropriate sense is greater than or equal to the sum of a a non-negative volume integral and an integral over M_t involving an expression that occurs naturally in Hamilton's Harnack inequality for the mean curvature of convex hypersurfaces evolving by mean curvature [111]. In Perelman's case, the corresponding derivative of his entropy is non-negative.

Ecker's programme aims at showing that the above time-derivative is indeed non-negative or at least bounded from below uniformly in time. It was proved in [62] that if this were true, the evolving enclosed domains were bounded and $T < \infty$, then there is a constant $\kappa > 0$ depending only on n, Ω_0, T, $\sup_{M_0} |H|$ and c_1 such that

$$\frac{|\Omega_t \cap B_r(x_0)|}{r^{n+1}} \geq \kappa$$

holds for all $t \in [0, T)$ and $r \in (0, \sqrt{T}]$ in balls $B_r(x_0) \subset \mathbb{R}^{n+1}$ satisfying the conditions $|\Omega_t \cap B_{r/2}(x_0)| > 0$ and

$$\frac{|\Omega_t \cap B_r(x_0)| + r^2 \int_{M_t \cap B_r(x_0)} |H_{M_t}|\, dS}{|\Omega_t \cap B_{r/2}(x_0)|} \leq c_1.$$

This lower bound on the volume ratio is termed *local non-collapsing of volume.*

Because the lower volume ratio bound is scale invariant, it is also valid for any smooth limit of suitably rescaled solutions (as considered in Section 1.2.3) of the flow consisting of smooth, compact embedded hypersurfaces, but now for all radii $r > 0$ as long as the other conditions still hold for the balls $B_r(x_0)$ we consider. It is furthermore shown in [62] how the latter information can be used to rule out certain degenerate rescaling limits which in turn provides invaluable information on the singularity structure of solutions of our evolution equation. For instance, this would rule out products of the grim reaper curve with Euclidean spaces as possible rescaling limits and would therefore provide a very different approach to proving non-collapsing behavior to the methods adopted in [11, 187].

In 2014, Ecker [63] succeeded in relating the derivative of the above entropy to quite different but also very natural boundary integrals. This time he used the Euler–Lagrange equations satisfied by minimisers of the entropy for fixed τ. Remarkably, the Hessian matrix of the solutions of the Euler–Lagrange equation for the entropy functional is a solution of a similar PDE to the one satisfied by the logarithm of positive solutions of the standard heat kernel (this implies the so-called log-concavity results in certain situations). Relevant Harnack inequalities resulting from these PDEs are proved in [108, 152]. In [63], the following result is proved among many others.

Theorem 7. *Let $(M_t)_{t\in(0,T)} = (\partial\Omega_t)_{t\in(0,T)}$ be a solution of mean curvature flow consisting of compact hypersurfaces and suppose $\tau(t)$ satisfies $\frac{\partial \tau}{\partial t} = -1$. Then the inequality*

$$\frac{d}{dt}\mu_{H_{M_t}}(\Omega_t, \tau(t)) \geq \int_{\Omega_t} \left(\frac{n+1}{2\tau(t)} - \Delta f\right) u\, dx$$

$$+ 2\tau(t) \int_{M_t} H_{M_t} \left(|A_{M_t}|^2 - \frac{1}{2\tau(t)} + \nabla^2 f(\nu, \nu) - \frac{1}{2\tau(t)}\right) u\, dS$$

holds for all $t \in (0, T)$, where f is a minimiser for $\mu_{H_{M_t}}(\Omega_t, \tau(t))$ and A_{M_t} is the second fundamental form of M_t.

It seems surprising that the volume integrand is exactly the expression featuring in the Li–Yau Harnack inequality [152], and the last two terms in the boundary integral appear naturally in Hamilton's matrix Harnack inequality for the linear heat equation.

A new approach in this continuing programme attempts to adapt methods in Chapter 9 of [164] involving calculations for the ℓ-distance, where, rather than showing non-negativity of time derivatives, these are related to second spatial derivatives of ℓ. This is motivated by the observation that some of the terms in the above boundary integral appear to arise by taking a second variation of a suitable functional. The ultimate goal of this programme is to remove the mean-convexity condition which so far has been essential in non-collapsing arguments. Note that in Ricci flow no additional condition is required to prove non-collapsing in finite time.

1.3.3 Monotonicity formulas for some geometric flows (Afuni[6])

In Sections 1.2.2 and 1.2.4, we discussed various monotonicity formulas for geometric variational problems and geometric evolution equations and stressed their importance for the regularity theory of solutions of these equations. Furthermore, we presented and partially listed localised versions of these formulas for geometric evolution equations. They hold on super-level sets of appropriately scaled backward heat kernels. Afuni [1] adapted the results in [61] to the Yang–Mills flow (5). Moreover, he generalised all the above-mentioned (local and global) monotonicity formulas to more abstract geometric settings, in particular allowing the underlying geometry to evolve with the flow in question [2–4]. Afuni's formulas also generalise those in [156].

1.3.4 Regularity of the Brakke flow (Lahiri[7])

In his thesis [144], Lahiri gives a corrected and simplified proof of Brakke's regularity theorems [32]. He uses the original approach built upon an explicit construction of families of graphs via convolution with the heat kernel. Under certain flatness assumptions, these families move almost by mean curvature flow. He fixes the numerous, often non-trivial, gaps in Brakke's arguments and improves some of his calculations and estimates. Among other things, he makes use of techniques and results developed in more recent years, most notably Huisken's monotonicity formula [124]. This way he maintained Brakke's central idea of approximating mean curvature flow by linear heat diffusion. Moreover, he elaborates on this approach to make it more adaptive for potential further applications.

Suppose $(M_t)_{t\in[t_1,t_2]}$ is a family of n-manifolds in $\mathbb{R}^{n+k}$ moving by smooth mean curvature flow and let μ_t be the induced measure. Using the evolution of the area element [123], one can show that

6 Ahmad Afuni, Free University of Berlin, afuni@math.fu-berlin.de
7 Ananda Lahiri, Max Planck Institute for Gravitational Physics (Albert Einstein Institute) Potsdam-Golm, ananda.lahiri@aei.mpg.de

$$\int \phi(\cdot, s_2)\, d\mu_{s_2} - \int \phi(\cdot, s_1)\, d\mu_{s_1}$$

$$= \int_{s_1}^{s_2} \left(- \int |\vec{H}|^2 \phi \, d\mu_t + \int \vec{H} \cdot D\phi \, d\mu_t + \int \frac{\partial \phi}{\partial t}(\cdot, t)\, d\mu_t \right) dt \tag{24}$$

for all $\phi \in C^1\left([t_1, t_2] \times \mathbb{R}^{n+k}, \mathbb{R}^+\right)$ with $\bigcup_{t\in(t_1,t_2)}\{\phi(t,\cdot) > 0\}$ compact. Following Brakke [32], equation (24) can be generalised to families of integral varifolds. A solution of this more general flow is called a Brakke flow. In [144], criteria are given which imply that a Brakke flow is actually induced by a smooth mean curvature flow. Lahiri's main results are:

1. A comparison theorem relating mean curvature flow and heat flow. This is a generalised version of Brakke [32].
2. If a Brakke flow in some region is contained in a narrow enough slab and also the area ratio in suitable balls is controlled by certain bounds, then in a smaller region it is a smooth graphical mean curvature flow. This was done by Brakke in [32]. However, his proof contains a major gap in the usage of the clearing out lemma, which he used to obtain height bounds. Lahiri corrects Brakke's argument and also gives an alternative proof with a height estimate derived from Huisken's monotonicity formula.
3. Brakke's general regularity theorem [32] says that at a time where no mass drop occurs, the singular set of a Brakke flow has measure zero. Lahiri provides a streamlined version of this proof incorporating ideas from [60], which make it shorter and more transparent.

For results 2 and 3 there are alternative versions due to Kasai and Tonegawa [138, 202]. They consider a more general flow, using techniques different to both Brakke's and Lahiri's. In particular, the explicit constructions via the heat kernel are replaced by an indirect blow-up argument.

1.3.5 Constrained curve flows (Dittberner[8])

For her thesis [59], Dittberner is working on two planar constrained curve flows. This is curve shortening flow in the plane with a constraint which forces the curve to maintain a certain property. In the first flow, the area-preserving curve shortening flow (APCSF), the length of a planar evolving curve decreases while keeping the enclosed area fixed. In the second flow, the length-preserving curve flow (LPCF), the evolving planar curve has fixed length while increasing the enclosed area. More precisely, let $\Sigma_0 = \gamma_0(\mathbb{S}^1)$ be an embedded $C^{3,\alpha}$-curve in $\mathbb{R}^2$. One then seeks a one-parameter family of maps $\gamma(\cdot, t) = \gamma_t : \mathbb{S}^1 \to \mathbb{R}^2$ satisfying the evolution equation

$$\frac{\partial \gamma}{\partial t}(p, t) = \left(\mathrm{h}(t) - \kappa(p, t)\right)\nu(p, t)$$

8 Friederike Dittberner, Free University of Berlin, dittberner@math.fu-berlin.de

for $p \in \mathbb{S}^1$ and $t \in [0, T)$, and the initial condition $\gamma(\mathbb{S}^1, 0) = \gamma_0(\mathbb{S}^1)$, where ν is the outward unit normal to $\Sigma_t = \gamma_t(\mathbb{S}^1)$, $\kappa(p, t)$ is the curvature at p and t, and h is given by either

$$\mathrm{h}(t) = \frac{2\pi}{|\Sigma_t|} \text{ (APCSF), or } \mathrm{h}(t) = \frac{1}{2\pi} \int_{\mathbb{S}^1} \kappa^2 \, ds_t \text{ (LPCF).}$$

The APCSF was first studied by Gage [92]. He proved that the solution exists globally and converges to a round circle provided that the initial surface Σ_0 is convex, smooth and embedded. The same was done for the LPCF by Pihan [168]. Using the distance comparison principle by Huisken [126], Dittberner [59] succeeded in showing that the curve Σ_t stays embedded for all $t \in [0, T)$ if the initial embedding Σ_0 satisfies $\int_p^q \kappa(s, 0) ds_0 \geq -\pi$ for all $p, q \in \mathbb{S}^1$. From this follows preservation of embeddedness for a short time, and the grim reaper as a rescaled singularity can be excluded.

1.3.6 Non-smooth initial data for curve shortening flow (Lauer[9])

In the study of partial differential equations a fundamental question is when, and in what sense, a solution exists for low regularity initial data. This is the main theme of Lauer's work. For example, what is the evolution (when taking the perspective of the level-set flow) of the Koch snowflake or the topologist's sine curve, both simple enough initial data to describe, but ones for which previous methods, which at the very least require writing the initial data locally as a graph, have failed to explain. In a recent paper [151], Lauer studies the question of existence and uniqueness of curve shortening flow when the initial data is a Jordan curve on $\mathbb{S}^2$. The main result is that the curve shortening flow is well-defined and smooth when the initial data is a measure zero Jordan curve.

The positive measure case is also well understood: If the initial data is a so-called Osgood curve, then for sufficiently small t the level-set flow is an annulus with smooth boundary (the boundaries then evolve as they would under the classical curve shortening flow). It follows that there are examples of Jordan curves on $\mathbb{S}^2$ whose evolution converges to either a hemisphere of $\mathbb{S}^2$ or becomes $\mathbb{S}^2$ itself in finite time.

This result is a generalisation of Lauer's work in [150] where the planar case was studied. There it was shown that the level-set flow of any locally connected compact set vanishes instantly, evolves as a smooth closed curve or is a smooth annulus for short time. The results of [150] thus include arbitrary Jordan curves and provided the first examples of a set $\gamma \subset \mathbb{R}^2$ with $\dim_{\mathscr{H}}(\gamma) > 1$ that evolves into a smooth curve. On the other hand [150], the question of whether or not locally connected sets in $\mathbb{R}^2$ satisfy the same trichotomy is left open. In another recent paper, Lam and Lauer [145] have proved that the topologist's sine curve instantly evolves into a smooth closed curve,

9 Joseph Lauer, lauer@zedat.fu-berlin.de

thus giving the first example of a domain whose boundary is not locally-connected but still evolves into a smooth closed curve.

Generalising from $\mathbb{R}^2$

The main obstacle in generalising the work of [150] to curve shortening flow in a general surface (Σ, g) is the extension of a length estimate for smooth curves whose proof takes up a majority of [150]. The estimate depends on a crude geometric quantity, which Lauer calls the r-multiplicity, for which straightforward generalisations readily exist. A key step involves straightening a foliation of a thin strip with complete leaves that are linear at infinity. Translating solutions to curve shortening flow, the grim reapers, are used in a fundamental way in the argument. Indeed, much of the argument in [150] extends easily to $\mathbb{S}^2$. There are two main differences.

1. In [150], the initial leaves of the foliations are linear at infinity. This is of course impossible on $\mathbb{S}^2$. Lauer introduces the concept of a (C, θ)-spacing, which provides a scale so that the foliations can be constructed using curves that are nearly linear on a set of some definite size.
2. Grim reapers are used in the planar case, but there are no translating solutions on $\mathbb{S}^2$. Thus, Lauer replaces them with a suitable solution of a Dirichlet problem. The analysis is more complicated since the solutions are not explicit.

Since the former difference can be generalised easily, the extension to arbitrary surfaces now relies on being able to do the latter in general. The existence of an appropriate evolving arc is proved by Allen, Layne and Tsukhara in [9], where only the constant curvature case is considered (thus avoiding some serious Riemannian geometry). The long-term existence portion of their result follows closely the argument given by Huisken [126] to prove the analogous result for the evolution of an arc between two parallel lines in the plane. Carrying out this programme in a general surface is likely the last step in extending the results of [150, 151] to their most general setting. Work is ongoing.

Higher dimensions

It is worth pointing out that the behavior described here is specifically one-dimensional. Indeed, an example has been given [161] of a finite area topological surface in $\mathbb{R}^3$, whose level-set flow does not evolve to be smooth. On the other hand, Hershkovits [121] has proved that certain Reifenberg sets in $\mathbb{R}^{n+1}$, which can be fractal and hence have Hausdorff dimension greater than n, do evolve to be smooth.

1.3.7 Ricci flow on a class of non-compact warped product manifolds and Gaussian estimates (Marxen[10])

In Marxen's thesis [157], he examines the evolution of the manifold $M = \mathbb{R} \times N$ with warped product metric $h = f_0^2 dx^2 + g_0^2 g_N$ under Ricci flow (3), where (N, g_N) is a flat,

10 Tobias Marxen, University of Hannover, tobias.marxen@math.uni-hannover.de

complete, connected Riemannian manifold of dimension $n \geq 2$, and $f_0, g_0 : \mathbb{R} \to \mathbb{R}$ are C^∞, positive (dx^2 denotes the standard metric on $\mathbb{R}$) and are chosen such that (M, h) is complete and has bounded curvature. He shows preservation of the warped product structure, long-time existence and that the solution is of type III, that is the norm of the curvature tensor decays like $C/(1+t)$ for $t \to \infty$. If g_0 is bounded and the metrics are rescaled in such a way that a fixed fibre $\{y\} \times N$ is isometric to (N, g_N) for all times, he succeeded in proving convergence for each r of the set of all points whose distance to $\{y\} \times N$ is $\leq r$ to a flat cylinder $[-r, r] \times N$. Furthermore, he shows that, if additionally $\lim_{x \to \pm\infty} g_0(x) = 0$, $(M, h(0))$ has finite volume and (N, g_N) is homogeneous, the solution collapses, that is the injectivity radius goes to 0 uniformly while the curvatures stay bounded. This result also holds for the normalised (volume-preserving) Ricci flow.

Marxen also considers Ricci flow on general complete, non-compact manifolds. Under appropriate curvature assumptions, he proved upper Gaussian estimates for the geometric quantities $|Rm|^2$, $|\nabla Rm|^2$ and $|T|^2$ (Rm denotes the Riemann tensor and T the traceless Ricci tensor) if these have compact support at initial time $t = 0$. Applying the Gaussian estimates to the warped product manifold $\mathbb{R} \times N$ yields, if the ends are hyperbolic (all sectional curvatures $k < 0$) at $t = 0$ and if N is compact, that the ends have asymptotically constant curvature $-(2nt - 1/k)^{-1}$ at each fixed positive time $t > 0$, including a quantitative estimate that measures how much the curvatures deviate from this constant. Moreover, $-(2nt - 1/k)^{-1}$ is exactly the rate at which the curvatures go to 0 in hyperbolic space H^{n+1} under Ricci flow. Finally, it is shown in [157], using the behavior of zeros of solutions of linear parabolic PDEs of second order on $\mathbb{R}$ and under appropriate additional assumptions, that the ends of the warped product manifold $\mathbb{R} \times N$ have negative curvature at each positive time $t > 0$ if they are hyperbolic at $t = 0$.

1.3.8 Minimal and prescribed mean curvature surfaces (Bourni,[11] Langford, Volkmann[12])

Regularity estimates for almost minimal surfaces

In joint work with Tinaglia [28–30], Bourni studies surfaces with “small" (rather than zero) mean curvature and proves various regularity results which were known to hold for minimal surfaces, such as curvature bounds and embeddedness. The “smallness” of the mean curvature is defined via a Sobolev ($W^{k,p}$)- or an L^p-norm. Their work provides optimal conditions on the mean curvature for the existing results to generalise to the non-minimal case and sheds more light on the estimates for minimal surfaces. In [28, 30], they study surfaces in $\mathbb{R}^3$ with bounded *total curvature* and

11 Theodora Bourni, University of Tennessee, Knoxville, tbourni@tennessee.edu
12 Alexander Volkmann, alexv@gmx.de

appropriately bounded mean curvature. In particular, they show that if the surface has bounded total curvature and the L^p-norm of the mean curvature, $p > 2$, is sufficiently small, then the surface is graphical away from its boundary. In addition, if the derivative of the mean curvature is small (in L^p, for $p > 2$), then $|A|$ is actually pointwise bounded at interior points. Their results generalise some previous curvature estimates in the case of minimal surfaces, in particular, the celebrated results of Colding and Minicozzi [54] and Choi and Schoen [48].

In [29], they obtain density estimates for compact surfaces immersed in $\mathbb{R}^n$ with total boundary curvature less than 4π and with sufficiently small L^p-norm of the mean curvature, $p \geq 2$. In fact, they show that these estimates hold for compact branched immersions. In particular, these density estimates imply that such surfaces are embedded up to and including the boundary. Their results generalise the main results in [70] for minimal surfaces. They furthermore show how one can apply their estimates to describe the geometry and topology of such surfaces in the 2-dimensional case. More specifically, they prove that for $n = 3$ and $p > 2$ the norm of the second fundamental form is bounded and use this curvature estimate to give a uniform upper bound for the genus of such surfaces. For the case of minimal surfaces, such results were derived in [201].

Regularity theory for varifolds

One of the most famous and remarkable regularity results, mainly due to its great generality, is Allard's regularity theorem [7]. He showed that a k-dimensional rectifiable varifold, under appropriate assumptions on its first variation and measure, is a $C^{1,\alpha}$-manifold for some $\alpha \in (0, 1)$. Even though Allard's theorem is very general, it requires the existence of a (weak) mean curvature. Therefore, the estimates of his theorem cannot be applied to manifolds where the normal is merely Hölder continuous. Similarly, when a boundary exists, its normal needs to be differentiable almost everywhere for Allard's boundary regularity theorem [8] to apply. In [24], Bourni was able to show that Allard's boundary regularity theorem still holds in the case of $C^{1,\alpha}$-boundaries for any $\alpha \in (0, 1]$. It is worth noting here that in [8], even though all the estimates depend only on the $C^{1,1}$-norm of the boundary, it is always assumed that the boundary is smooth, whereas in [24] the smoothness hypothesis is dropped; in particular, the proof requires no higher than $C^{1,\alpha}$-regularity of the boundary.

Bourni and Volkmann [31] managed to also resolve the issue in the interior. In particular, they proved that Allard's regularity theorem holds for rectifiable varifolds without assuming the existence of a weak mean curvature, but instead assuming a weaker condition on the first variation of the area. This condition is satisfied by $C^{1,\alpha}$-manifolds and is implied by the hypotheses of Allard's regularity theorem. They furthermore combined this result with the boundary regularity of [24] to include boundaries and thus provide a complete Allard-type $C^{1,\alpha}$-regularity theory for varifolds with generalised normal of class $C^{0,\alpha}$- and $C^{1,\alpha}$-boundaries.

Aspects of the Plateau problem

The Plateau problem concerns the existence of an area minimising surface with boundary in a given curve in an ambient manifold M. There are two main versions of the Plateau problem: minimising area amongst surfaces with fixed genus (area minimising in a fixed topological class), and minimising area without any restriction on the genus (absolutely area minimising case). Bourni and Coskunuzer [25] study the genus of absolutely area minimising surfaces Σ in a compact, orientable, strictly mean convex 3-manifold M with boundary $\partial\Sigma$ a simple closed curve lying in ∂M. The initial goal was to examine how "large" the class of boundary curves is such that any embedded area minimising surface bounded by them has genus at least g for an arbitrary given value g. In [25], they answer this question by showing that these curves are generic in the space of nullhomologous (i.e. that bound at least one surface) simple closed curves.

This theorem led to a series of very interesting results concerning not only area minimising but also minimal surfaces in M with boundary in ∂M. For instance, the authors show that curves which bound more than one minimal surface are generic for a strictly mean convex 3-manifold M. Furthermore, they establish the existence of non-embedded stable minimal surfaces (and in particular stable minimal disks) with a simple closed curve in ∂M as boundary for any strictly mean convex 3-manifold M. This result generalises Hall's [104] result and resolves a conjecture of Meeks.

Null mean curvature flow and marginally outer trapped surfaces (MOTS)

Bourni and Moore [27] study the evolution of hypersurfaces in spacetime initial data sets by their *null mean curvature*. They develop a theory of weak solutions using the level-set approach, with the aim of providing a new way of finding MOTS. Their main results in [27] can be described as follows. Let (M^{n+1}, g, K) be an initial data set in a Lorentzian spacetime and consider a two-sided closed and bounded hypersurface $\Sigma^n \subset M^{n+1}$ with globally defined outer unit normal vector field ν in M. Given a smooth hypersurface immersion $F_0 : \Sigma \to M$, the evolution of $\Sigma_0 := F_0(\Sigma)$ by null mean curvature is the one-parameter family of smooth immersions $F : \Sigma \times [0, T) \to M$ satisfying

$$\frac{\partial F}{\partial t}(p, t) = -(H + P)(p, t)\nu(p, t) \tag{25}$$

for $p \in \Sigma$ and $t \geq 0$ and the initial condition $F(\Sigma, 0) = F_0(\Sigma)$, where $H := \operatorname{div}_{\Sigma_t}(\nu)$ denotes the mean curvature of $\Sigma_t := F(\Sigma, t)$ in M and $P := \operatorname{tr}_{\Sigma_t} K$ is the trace of K over the tangent space of Σ_t.

Analogous to the behavior of solutions to mean curvature flow (2), singularities develop in general, as the null mean curvature of solutions of eq. (25) will tend to infinity at some points. This motivates their development of a theory of weak solutions to the classical flow (25), which they implement to investigate the limit of a hypersurface moving under null mean curvature flow. To develop the weak formulation for the classical evolution (25), they use the level-set method and employ the method of *elliptic regularisation*. This leads to the study of solutions u_ε of the strictly elliptic equation

$$\operatorname{div}\left(\frac{\nabla u_\varepsilon}{\sqrt{|\nabla u_\varepsilon|^2+1}}\right)-\left(g^{ij}-\frac{\nabla^i u_\varepsilon \nabla^j u_\varepsilon}{|u_\varepsilon|^2+1}\right)K_{ij}=-\frac{1}{\varepsilon\sqrt{|\nabla u_\varepsilon|^2+1}}. \tag{26}$$

Their main theorem in [27] roughly says that a smooth solution of eq. (26) exists and blows up on the *outermost* MOTS (for any $\varepsilon > 0$ small enough). They additionally show that one can extract a subsequence of $\{\varepsilon u_\varepsilon\}$ converging (as $\varepsilon \to 0$) locally uniformly to a limit u which blows up on a *generalised* MOTS; that is, it satisfies $H + P = 0$ at points where the normal and the mean curvature are defined (which actually holds almost everywhere). Then, the hypersurfaces $\Sigma_t := \{x \mid u(x) = t\}$ (given by the level sets of u) constitute the *weak* or *level-set* solution of (25) and, as $t \to \infty$, they converge to a generalised MOTS.

Translating solutions of mean curvature flow

Bourni and Langford [26] have recently shown that any strictly mean convex translating solution of mean curvature flow of dimension $n \geq 3$ which admits a cylindrical estimate and a corresponding gradient estimate is rotationally symmetric. As a consequence, any translating solution of the mean curvature flow which arises as a blow-up limit of a two-convex mean curvature flow of compact immersed hypersurfaces of dimension $n \geq 3$ is rotationally symmetric. The proof also works for a more general class of translator equations. As a particular application, it yields an analogous result for a class of flows of embedded hypersurfaces it includes the flow of two-convex hypersurfaces by the two-harmonic mean curvature.

1.3.9 Free boundary problems governed by mean curvature (Volkmann)

In [204], Volkmann considers the following three free boundary value problems for hypersurfaces that are governed by the mean curvature of the hypersurface:

A monotonicity formula for free boundary surfaces with respect to the unit ball

He proves a monotonicity identity for compact surfaces with free boundaries inside the boundary of the unit ball in $\mathbb{R}^n$ that have square integrable mean curvature. As a consequence, this yields a Li–Yau-type inequality in this setting, thereby generalising results of Oliveira and Soret [176] and Fraser and Schoen [89]. Then he derives some sharp geometric inequalities for compact surfaces with free boundaries inside arbitrary orientable support surfaces of class C^2. Furthermore, he obtains a sharp lower bound for the L^1-tangent-point energy of closed curves in $\mathbb{R}^3$, thereby answering a question raised by Strzelecki, Szumańska and von der Mosel [195].

Relative isoperimetric properties of asymptotically flat support surfaces

Volkmann defines a notion of mass for asymptotically flat hypersurfaces M of Euclidean space and proves a positive mass theorem in all dimensions. Then he establishes a free boundary version of an obstruction discovered by Schoen and Yau in their proof of the positive mass theorem [183], which was refined by Eichmair and Metzger [68], and very recently by Carlotto [40]: Positive mean curvature of $M \subset \mathbb{R}^3$ is not compatible with the existence of (certain) stable free boundary minimal surfaces. He then uses this to prove that given a compact subset K of $\mathbb{R}^3$, all volume-preserving stable free boundary constant mean curvature surfaces with respect to M of sufficiently large boundary length will avoid K, thereby obtaining a free boundary version of the main result in [68]. Finally, inspired by ideas of Eichmair and Metzger [69] Volkmann proves the existence of arbitrarily large isoperimetric regions relative to M.

Weak solutions of nonlinear mean curvature flow with Neumann boundary condition

Volkmann proposes a new flow approach to obtain relative isoperimetric inequalities. As a first step in this programme, he develops a weak level-set formulation for mean curvature flow (2) and positive powers of mean curvature flow with Neumann boundary condition. He proves the existence of weak solutions under natural conditions on the supporting surface and derives some properties of the evolving surfaces. The case of surfaces without boundary has been treated by Schulze [186].

1.3.10 Second-order elliptic systems with variational structure (Damialis[13])

Alikakos [5] introduced a stress tensor with respect to which such systems admit divergence-free conditions. A similar but less general approach (based on Pohožaev-type identities) was previously developed by Gui [100] who, as an application, rigorously derived Young's law for the contact angles at triple junctions of interfaces on the plane for the elliptic vector-valued Allen–Cahn equation. We note here the relation between the Allen–Cahn equation and minimal surfaces (see, e.g., Ilmanen [131]). The vector-valued analogue describes the coexistence of more than two phases and is related to systems of minimal surfaces (or systems of constant mean curvature surfaces in the anisotropic case).

The result of Gui was recently generalised in [6] where Young's law is derived at triodes of interfaces in three-dimensional space, again for the elliptic vector-valued Allen–Cahn equation

$$\Delta u - \nabla_u W(u) = 0\,,$$

13 Apostolos Damialis, apostolos.damialis@fu-berlin.de

but now for maps $u : \mathbb{R}^3 \to \mathbb{R}^3$ and a triple-well potential $W : \mathbb{R}^3 \to \mathbb{R}$ along with appropriate hypotheses describing the asymptotic behavior of solutions away from the triple junction and the interfaces. The proof involves a careful application of the divergence theorem utilising the divergence-free form of the equation via the aforementioned stress tensor. The generalisation from two to three (or higher) dimensions is not trivial since the singular set is now connected and a surgery has to be performed in order to isolate the singularities appearing in the surface integration. We note that once Young's law is known for the contact angles at triodes, Plateau's law for the angles at the quadruple junction that four triodes form, where six interfaces meet at a point, follows easily by elementary geometric considerations. The main advantage of their approach lies in the fact that it can be adapted to also attack the problem in higher dimensions. An obvious (but nontrivial) further generalisation would be to apply an inductive procedure reducing the problem to lower-dimensional ones by utilising the hierarchical structure of the singular set.

1.3.11 The angular momentum-mass inequality in general relativity (Cha[14])

The standard picture of gravitational collapse and the (weak) cosmic censorship conjecture [49, 163] suggest that there exists a certain relation between mass, angular momentum, charge, and the area of the black hole. Cha focusses on the relation between the angular momentum and mass, which is called the angular momentum-mass inequality. It has been established that this inequality holds for a large class of asymptotically flat, axially symmetric and maximal (mean curvature zero) initial data for the Einstein equations [53, 55, 185]. In [43], a reduction argument is introduced by which the general case is reduced to the maximal case, assuming that a canonical system of elliptic PDE possesses a solution, motivated by previous reduction arguments [33, 34, 184]. This result can be extended to the angular momentum-mass-charge inequality, and finally to a lower bound for black hole area in terms of mass, charge, and angular momentum under the same assumptions [42].

In [45], it is shown how to reduce several inequalities including the Penrose inequality and the angular momentum-mass inequality for asymptotically hyperboloidal data to the known maximal (or time symmetric) case in the asymptotically flat setting whenever a certain geometrically natural system of elliptic equations admits a solution. Furthermore, motivated by Chrusciel and Tod [52], it is shown in [44] how to extend this procedure to include the AdS-hyperbolic initial data. The solvability of the system of elliptic equations which is proposed here has been a very important issue to resolve, as it will lead to a proof of the angular momentum-mass inequality for a large class of (non-maximal) initial data. In fact, in [41] it is shown that there exists a solution to the system in the near maximal case.

14 Ye Sle Cha, Free University of Berlin, ycha@zedat.fu-berlin.de

1.3.12 Area-preserving Willmore flow in asymptotically Schwarzschild space (Jachan[15])

Jachan [134] studies an evolution equation related to the Willmore energy W, which from a PDE point of view is a parabolic quasilinear fourth-order equation with a non-local term. For a closed 2-surface Σ immersed in an ambient 3-manifold M, the energy W is defined by

$$W(\Sigma) := \frac{1}{4} \int_{\Sigma} H^2 \, d\mu ,$$

where H denotes the mean curvature of Σ. In the Euclidean case $M = \mathbb{R}^3$, the energy W is invariant under rescalings of Σ, and minimisers of W are round coordinate spheres of arbitrary radius [210]. In contrast, minimisers of W subject to *prescribed surface area* are round spheres of appropriate radius. In non-Euclidean M, these minimisers can be regarded as non-Euclidean analogues of spheres of fixed radius, which makes them interesting for a variety of applications. More specifically, Jachan investigates solutions $(F_t(\Sigma))_{t\geq 0} = (\Sigma_t)_{t\geq 0}$ of the area-preserving Willmore flow

$$\frac{\partial F}{\partial t}(p,t) = \left[\Delta_{\Sigma_t} H(p,t) + Q(p,t) + \lambda H(p,t)\right] \nu(p,t) \qquad (27)$$

for $p \in \Sigma$ and $t \geq 0$ and initial condition $F(\Sigma, 0) = \Sigma_0$. This flow preserves the surface area of Σ_t while $W(\Sigma_t)$ is decreasing and thus serves as a tool in the search for minimisers in the sense described above. Here, Δ and ν denote the Laplace–Beltrami operator and a fixed unit normal vector field on Σ_t, and Q is a polynomial in lower order quantities. The non-local term $\lambda = \lambda(\Sigma_t)$ is defined in terms of various L^p-norms of curvature quantities on Σ_t.

The flow (27) is an extension of the classical L^2-gradient flow of W, which has been investigated for $\Sigma_t \subset \mathbb{R}^n$ by Kuwert and Schätzle [142, 143]. The necessary adaptation and generalisation of the techniques established in those papers there constitutes the core part of [134]. Most notably, technical difficulties arising from the ambient curvature of M as well as from the term λ need to be tackled. Also, adequate blow-up machinery in non-flat M had to be set up in order to carry out singularity analysis.

The focus in [134] lies on ambient manifolds M that are asymptotically Schwarzschild, modelling the neighbourhood of a static black hole up to small deviations. A short time existence result for eq. (27) is established, as well as a quantification of "short time" in terms of curvature concentration on Σ_0. Furthermore, long-time results are proved stating that the solution of (27) exists for all positive times and converges in a subsequential sense to a minimiser as described above provided $W(\Sigma_0)$ is suitably small. For these results to hold, the solution is required to

15 Felix Jachan, Dresden University of Technology, felix.jachan@tu-dresden.de

stay outside the high curvature portion near the origin of M. Complementary to the above results, for $M = \mathbb{R}^3$, smooth convergence of the solution to a round coordinate sphere is established, resulting in a full adaptation of the results in [142, 143] to eq. (27) in $\mathbb{R}^3$.

1.3.13 Multilinear formulation of differential geometry (Huisken)

Arnlind, Hoppe, and Huisken [20] prove that many aspects of the differential geometry of embedded Riemannian manifolds can be formulated in terms of multi-linear algebraic structures on the space of smooth functions. In particular, they find algebraic expressions for Weingarten's formula, the Ricci curvature, and the Codazzi–Mainardi equations. For matrix analogues of embedded surfaces, they define discrete curvatures and Euler characteristics, and a non-commutative Gauss–Bonnet theorem is shown to follow. They also derive simple expressions for the discrete Gauss curvature in terms of matrices representing the embedding coordinates, and explicit examples are provided. Furthermore, they illustrate the fact that techniques from differential geometry can carry over to matrix analogues by proving that a bound on the discrete Gauss curvature implies a bound on the eigenvalues of the discrete Laplace operator.

In [19], Arnlind and Huisken show that the pseudo-Riemannian geometry of submanifolds can be formulated in terms of higher order multi-linear maps. In particular, a Poisson bracket formulation of almost (para-)Kähler geometry is obtained.

1.3.14 Intrinsic discretisation error estimates for geodesic finite elements (Hardering[16])

The thesis of Hardering [113] provides a systematic investigation of the geodesic finite element discretisation of elliptic and parabolic variational problems for maps from Euclidean space to a Riemannian manifold. Recently, geodesic finite elements have been developed and experimentally studied providing a conforming and objective, i.e., isometry-invariant, method of arbitrary order [177–180]. In [113], Hardering generalises some results from the basic theory of finite element methods for the minimisation of $W^{1,2}$-elliptic energies to functions with smooth Riemannian manifold codomains independently of embeddings into Euclidean space or choice of coordinate systems. The problem is discretised using geodesic finite elements; the results however carry over to other methods fulfilling the appropriate approximation qualities.

In order to obtain meaningful error bounds independent of coordinate systems or embeddings, different Sobolev-norms and notions of distance are introduced, and

16 Hanne Hardering, Dresden University of Technology, hanne.hardering@tu-dresden.de

their equivalence in restricted sets are shown. This makes use of properties of the tangent bundle as well as estimates for Jacobi fields. After recovering well-posedness of the definition of geodesic finite elements for sufficiently regular manifolds, classical approximation qualities such as interpolation errors and inverse estimates are rephrased in terms of the introduced intrinsic concepts and proved. Using a nonlinear version of Céa's Lemma combined with the interpolation error estimates, Hardering obtains a $W^{1,2}$-discretisation error estimate in a restricted set. It is then shown that this restricted solution indeed correlates to a local solution. Using a generalisation of the Aubin–Nitsche Lemma, L^2-error estimates for (morally) semi-linear energies are provided. Besides static problems, L^2-gradient flows for elliptic energies are also discussed. They are discretised by a method of time layers consisting of an implicit Euler scheme for the time discretisation and geodesic finite elements for the space discretisation. Using error estimates from [10] for the time discretisation, $W^{1,2}$- and L^2-error estimates for the fully discrete scheme are shown under the condition that smooth solutions to the continuous problems exist. Those ideas that lead to the $W^{1,2}$-discretisation error estimate for elliptic energies have been published in [99].

2 Global dynamics, blow-up and Bianchi cosmology (Fiedler[17] et al.)

2.1 Introduction

Many central problems in geometry, topology and mathematical physics reduce to questions concerning the behavior of solutions of nonlinear evolution equations. Parabolic partial differential equations (PDEs) play a key role and exhibit a rich structure. Three basic scenarios are possible.

First, under suitable dissipativity conditions, long-time existence is guaranteed and a low-dimensional global attractor of uniformly bounded eternal solutions exists. The structure of such global attractors details long-time behavior. Some of those structures are encoded in meanders – which open doors to a broad variety of algebraic structures far beyond the original PDE scope of the attractor question.

Second, long-time existence without dissipativity features trajectories which escape to infinity in infinite time (grow-up). In some cases it is possible to construct compactifications of an unbounded attractor containing, for example, equilibria at infinity. As an example we show how equivariant bifurcations give rise to black hole initial data that is not spherically symmetric.

Third, singularities may develop after finite time. In other words, some solutions escape to infinity within finite time (blow-up). Some of the ramifications are

17 Bernold Fiedler, Free University of Berlin, fiedler@math.fu-berlin.de

explored in complex time, here. Cosmologically, the most meaningful singularity is the big bang. We address the Bianchi models: a system of ordinary differential equations which represent the homogenous, but anisotropic, Einstein equations. The Belinskii–Khalatnikov–Lifshitz (BKL) conjecture favours them as models for a space-like singularity. We obtain stability results for a primordial universe tumbling between different Kasner states, each of which represents a self-similarly expanding universe. Furthermore we address the BKL question of particle horizons. To conclude, we embed general relativity in the broader context of Hořava–Lifshitz gravity to study the influence on the tumbling behavior between the Kasner states and the resulting chaoticity.

2.2 Global attractors of parabolic equations

In this section we survey recent progress on the global dynamics, mostly of the scalar parabolic semilinear equation

$$u_t = u_{xx} + f\,. \tag{28}$$

The emerging global theory gives rise to surprising topological structures, as finite regular cell complexes, in Section 2.2.1, and to combinatorial algebraic structures like Temperley–Lieb algebras, in Section 2.2.2. An application to axisymmetric, but not spherically symmetric, initial data for black holes which satisfy the Einstein constraint is given in Section 2.2.4. The question of stability of stationary solutions is addressed in Section 2.2.3.

For simplicity we consider the PDE (28) under Neumann $\mathcal{N}$ or periodic $\mathcal{P}$ boundary conditions, on the unit interval $0 < x < 1$. Other separated linear boundary conditions can be treated analogously to the Neumann case. The quasilinear uniformly parabolic case may be easily believed to yield identical results, but the long necessary groundwork is not complete at this stage.

2.2.1 Sturm global attractors (Fiedler)

We consider the PDE (28) for several classes of nonlinearities f. In particular, we survey our results [80–86, 175]. We denote these classes by $\mathcal{N}(x, u, u_x)$, $\mathcal{N}(u, u_x)$, ... for $f = f(x, u, u_x)$, $f = f(u, u_x)$, ... under Neumann boundary conditions, and by $\mathcal{P}(x, u, u_x)$, $\mathcal{P}(u, u_x)$, ... in the x-periodic case. Dirichlet or other separated boundary conditions of mixed type can be treated analogously to the Neumann case. We always assume $f \in C^1$ to be dissipative, i.e., the set of bounded *eternal solutions* $u(t, x)$, $t \in \mathbb{R}$ is itself bounded in a suitable Sobolev-type interpolation space $u(t, \cdot) \in X \hookrightarrow C^1$. Explicit assumptions on f which guarantee dissipativity involve, for example, sign conditions $f(x, u, u_x) \cdot u < 0$, for large $|u|$, and subquadratic growth in u_x. The bounded set $\mathcal{A}_f \subseteq X$ of spatial profiles $u(t, \cdot)$ on bounded eternal solutions is called the *global attractor* of (28). The concept originated with Ladyzhenskaya's work on the two-dimensional Navier–Stokes equation.

For general surveys on global attractors $\mathcal{A}$, see for example [47, 103]. The general theory establishes properties like existence, global attractivity (hence the name), compactness, finite dimensionality in the fractal box counting sense, and sometimes connectivity. From such a general, and hence necessarily superficial, point of view, the global attractors $\mathcal{A}_f$ of our model class (28) are just an easy exercise.

However, our objective is a much deeper global understanding of the detailed global structure of the global attractors $\mathcal{A}_f$. Let us address the general Neumann case $f \in \mathcal{N}(x, u, u_x)$ first. By results of Zelenyak and Matano, the PDE (28) then possesses a gradient-like Morse structure due to a decreasing Lyapunov, or energy, or Morse, functional V. Throughout we assume all Neumann equilibrium solutions

$$0 = v_{xx} + f \tag{29}$$

to be hyperbolic, i.e., to be quadratically nondegenerate critical points of V. Then

$$\mathcal{A}_f = \mathcal{E}_f \cup \mathcal{H}_f \tag{30}$$

consists of the finitely many equilibria $v \in \mathcal{E}_f$ and *heteroclinic orbits* $u(t, \cdot) \in \mathcal{H}_f$ *connecting* them:

$$u(t, \cdot) \to v_{\pm} \in \mathcal{E}_f \quad \text{for} \quad t \to \pm\infty\,, \tag{31}$$

with equilibria $v_+ \neq v_-$. We call $\mathcal{A}_f$ a *Sturm global attractor*, to emphasise the particular origin from the one-dimensional parabolic PDE class (28). Information on the vertex pairs (v_-, v_+) which do possess a heteroclinic edge (31), and on the pairs which don't, can be encoded in the (directed, acyclic and loop-free) *connection graph* $\mathcal{C}_f$; see [74] and Fig. 1(a).

As was discovered by Fusco and Rocha [91], a permutation σ given by the boundary values of the equilibria $\mathcal{E}_f = \{v_1, \ldots, v_N\}$ is the key to a global analysis of the Sturm global attractor $\mathcal{A}_f$ and its heteroclinic connection graph $\mathcal{C}_f$. The *Sturm permutation* $\sigma = \sigma_f$ encodes the orderings of the equilibria,

$$\begin{aligned} v_1 &< v_2 < \cdots < v_N && \text{at} \quad x = 0\,, \\ v_{\sigma(1)} &< v_{\sigma(2)} < \cdots < v_{\sigma(N)} && \text{at} \quad x = 1\,, \end{aligned} \tag{32}$$

at the respective Neumann boundaries. More precisely, the Sturm permutations σ_f determine all *unstable dimensions*, alias *Morse indices*, $i = i(v_k)$ of the equilibria $v_k \in \mathcal{E}$; see [91]. This is a linear ODE question of Sturm–Liouville type; hence, the name Sturm permutation for σ_f. Moreover, σ_f determines the (PDE) heteroclinic connection graph $\mathcal{C}_f$, explicitly and constructively, in a combinatorial and algorithmic style [75]. Global

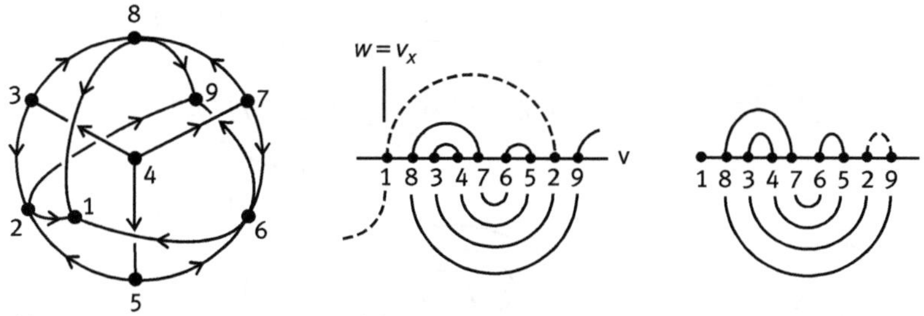

(a) Sturm global attractor (b) open meander (c) closed meander

Figure 1: *(a) Sturm global attractor $\mathcal{A}_f$ and connection graph $\mathcal{C}_f$ of a PDE* (28) *with nine hyperbolic equilibria v, labeled by* $1, \dots, 9 = N$. *One-dimensional heteroclinic orbits between equilibria of adjacent Morse index are indicated by arrows. (b) Open meander, at $x = 1$, which leads to the nine equilibrium solutions v by shooting from the Neumann boundary condition at $x = 0$. (c) Closed meander which arises by replacing the dashed upper arch* $(1, 2)$, *in (b), with the new dashed upper arch* $(9, 2)$; *see [74].*

Sturm attractors $\mathcal{A}_f$ and $\mathcal{A}_g$ with the same Sturm permutation $\sigma_f = \sigma_g$ are in fact C^0 orbit equivalent, i.e. homeomorphic in X under a homeomorphism which maps eternal PDE orbits $\{u(t, \cdot), t \in \mathbb{R}\} \subseteq \mathcal{A}_f$ to eternal orbits in $\mathcal{A}_g$; see [79].

Not all permutations σ are Sturm permutations σ_f arising from the Neumann class $f \in \mathcal{N}(x, u, u_x)$ of (28). In fact, $\sigma = \sigma_f$ for some f, if and only if the permutation $\sigma \in S_N$ is a meander permutation which, in addition, is dissipative and Morse [78]. Following Arnold, here, we call σ a *meander permutation* if σ describes the transverse crossings of a planar C^1 Jordan curve, or "meandering river", oriented from south-west to north-east, with the horizontal axis, or "road". Labelling the crossings along the "river", sequentially, the meander permutation σ is defined by the labels $\sigma(1), \sigma(2), \dots, \sigma(N)$ read along the road, left to right. We call a meander σ *dissipative* if $\sigma(1) = 1$ and $\sigma(N) = N$. We call σ *Morse* if the integers

$$i_k := \sum_{j=1}^{k-1} (-1)^{j+1} \operatorname{sign}\left(\sigma^{-1}(j+1) - \sigma^{-1}(j)\right) \tag{33}$$

are non-negative, for all $k = 1, \dots, N$. Geometrically, this forbids a net left winding of the river, as counted from the first crossing; see Figs. 1(b) and 2 for many examples. The PDE connection graphs $\mathcal{C}_f$ then follow, in each case, without further PDE analysis.

It is not difficult to see why Sturm permutations $\sigma = \sigma_f$ satisfy these three properties. Dissipativeness of σ follows because $\mathcal{E}_f$ must contain extreme equilibria $v_1(x) < v_2(x), \dots, v_{N-1}(x) < v_N(x)$, for all $0 \le x \le 1$, by dissipativeness of (28) and parabolic comparison. The Morse property $i_k \ge 0$ follows because $i_k = i(v_k) \ge 0$ are the non-negative Morse indices (see [91]). The meander property follows by ODE shooting for the equilibrium boundary value problem (29) in the (v, v_x) plane. Indeed,

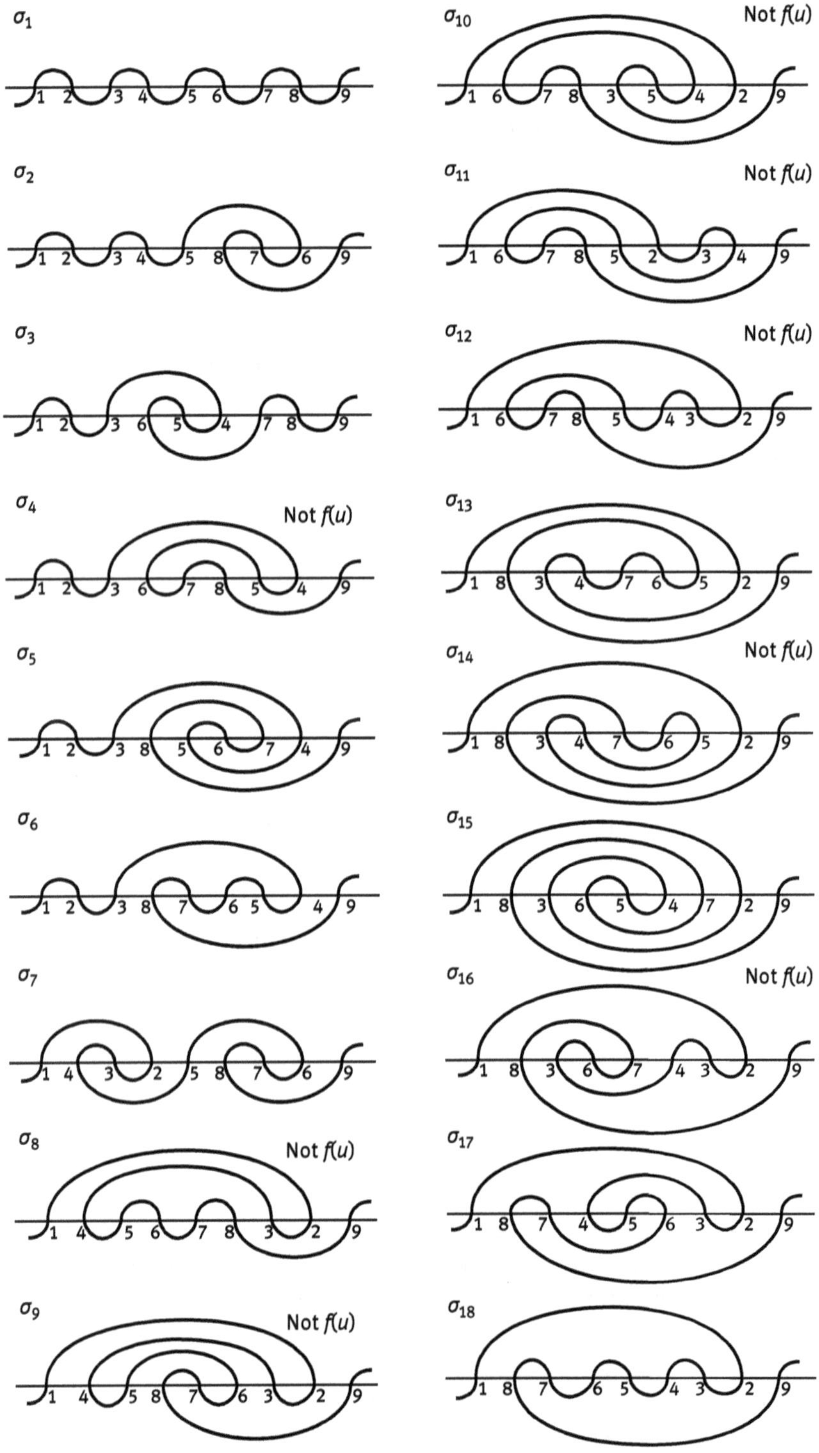

Figure 2: *All Sturm permutations in S(9), up to trivial symmetries, and their meanders. We indicate all cases which cannot be realised by nonlinearities* $f = f(u)$.

the "river" is the image of the horizontal Neumann v-axis $\{v_x = 0\}$, at $x = 0$, propagated to $x = 1$ under the ODE-flow (29). The "road" is the horizontal Neumann v-axis $\{v_x = 0\}$, at $x = 1$. See Section 2.2.2 for a deeper discussion of meanders. The converse claim, how all dissipative Morse meanders σ are actually Sturm permutations $\sigma = \sigma_f$ for some f, is beyond the limitations of this introductory survey (see [78]). The deceptively elementary description of Sturm permutations σ_f should not conceal the formidable complications arising from the meander property. Already Gauss got intrigued by the related question of self-intersecting orders in planar curves; see [95]. Rescaling $x = 1$ to become large, the evident complexities of chaotic stroboscope maps for the periodically forced pendulum or the forced van der Pol oscillator are also lurking here as special cases.

A key technical ingredient to the results indicated above are nodal properties, i.e. a refinement of the parabolic comparison principle in one space dimension, or in radially symmetric settings. Indeed, let $u_1(t, x)$ and $u_2(t, x)$ be any two non-identical solutions of eq. (28), in $X \hookrightarrow C^1$, and let the *zero number* $z(\varphi) \leq \infty$ count the strict sign changes of any continuous x-profile $\varphi : [0, 1] \to \mathbb{R}$. Then the zero number

$$t \mapsto z\big(u_1(t, \cdot) - u_2(t, \cdot)\big) \tag{34}$$

is finite, for any $t > 0$. Moreover z is non-increasing with t, and drops strictly at any multiple zeros of $x \mapsto u_1(t, x) - u_2(t, x)$; see [17]. Early versions of this, in the linear case, in fact date back to Sturm and convinced us to name the nonlinear ramifications after him; see [197].

It is the zero number dropping property (34) which generates surprising rigidity in the global Sturm attractors $\mathcal{A}_f$ and enables a detailed global analysis. For example, stable and unstable manifolds of hyperbolic equilibria v_+ and v_- intersect transversely, automatically, along any heteroclinic PDE orbit $u(t, \cdot) \to v_\pm$, for $t \to \pm\infty$, in $\mathcal{H}_f$; see [17, 120]. In particular, this implies local structural stability of Sturm attractors and, moreover, a strict dropping of Morse indices,

$$i(v_-) > z\big(u(t, \cdot) - v_-\big) \geq z(v_+ - v_-) \geq z\big(v_+ - u(t, \cdot)\big) \geq i(v_+)\,. \tag{35}$$

By a (standard) transitivity property of transverse heteroclinicity and a (Sturm specific) cascading principle, it turns out, in fact, that the full heteroclinic connection graph $\mathcal{C}_f$ is generated by the edges (v_-, v_+) with adjacent Morse index. Saddle-saddle heteroclinic orbits $i(v_-) = i(v_+)$ are excluded by Morse dropping (35). This precludes global topological ambiguities which, for example, homotopy-invariant Conley–Morse theory is not allowed to discard. Indeed the positive integer polynomial terms $(1 + t)\,q(t)$ in the Morse (in)equalities, alias connecting homomorphisms or Conley–Franzosa connection matrices, always need to accommodate global reshufflings of the connection graph due to non-generic saddle-saddle connections. This is where nodal arguments involving the zero number $z(u_1 - u_2)$ step in.

We summarise more recent results on Sturm global attractors $\mathcal{A}_f$ and their characterising Sturm permutations σ_f for the parabolic PDE (28) under Neumann ($\mathcal{N}$) and periodic ($\mathcal{P}$) boundary conditions, next. We conclude with recent progress in our global geometric understanding of $\mathcal{A}_f$ as regular topological cell complexes.

Little information is provided, in our approach, on the global Sturm attractor $\mathcal{A}_f$ of particular nonlinearities $f \in \mathcal{N}(x, u, u_x)$. The reason is purely on the ODE pendulum side (29). It is not an easy task to determine the equilibrium boundary order of $v \in \mathcal{E}_f$, as encoded in the Sturm permutation σ_f, for particular nonlinearities f. Indeed we address the whole class of dissipative nonlinearities $f = f(x, u, u_x)$, rather than any particular instance.

As a partial remedy, we also consider the subclass $\mathcal{N}(u)$ of nonlinearities $f = f(u)$, for which the equilibrium ODE (29) becomes x-autonomous Hamiltonian. See [85] and the survey [175] for an explicit combinatorial characterisation of this subclass of Sturm permutations σ_f. For a complete listing with $N = 9$ equilibria, see Fig. 2. The case $f \in \mathcal{N}(u, u_x^2)$ of nonlinearities $f(u,p) = f(u,-p)$ which are even in $p = u_x$ is only superficially more general. The equilibrium ODE (29) is then only reversible in x, but not necessarily Hamiltonian. Still, reversibility implies integrability here. The ODE (29), but not the PDE (28), can be smoothly transformed to become Hamiltonian. The original Sturm permutation σ_f becomes an involution, $\sigma_f = \sigma_f^{-1}$. Moreover σ_f can be represented by a Hamiltonian nonlinearity $g \in \mathcal{N}(u)$ such that $\sigma_f = \sigma_g$. In particular, the Sturm global attractors $\mathcal{A}_f$ in the larger x-reversible class $\mathcal{N}(u, u_x^2)$ coincide with those of the x-reversible Hamiltonian class $\mathcal{N}(u)$.

Next, we turn to the case $f \in \mathcal{P}(x, u, u_x)$ of periodic boundary conditions $x \in S^1$. The nodal dropping property (34) of zero numbers prevails. It has long been known, however, that any planar autonomous ODE vector field embeds into this larger class of PDEs [181]. In particular, saddle-saddle heteroclinic and homoclinic trajectories may arise, and automatic transversality of stable and unstable manifolds fails. In the S^1-equivariant case of x-independent non-linearities $f \in \mathcal{P}(u, u_x)$, on the other hand, automatic transversality does follow from mere (normal) hyperbolicity of equilibria and non-constant rotating waves $u(t, x) = U(x - ct)$. In fact, the global attractor then decomposes into equilibria $\mathcal{E}$, rotating waves $\mathcal{R}$, and their heteroclinic orbits $\mathcal{H}$ as

$$\mathcal{A}_f = \mathcal{E}_f \cup \mathcal{R}_f \cup \mathcal{H}_f\,. \tag{36}$$

In this periodic class, the PDE (28) therefore defines a Morse–Smale system. In the x-reversible, i.e. $O(2)$-equivariant, class $f \in \mathcal{P}(u, u_x^2)$ an explicit Lyapunov function has been constructed; see [85]. Then all "rotating" waves are frozen, i.e. of wave speed $c = 0$, and time-periodic solutions other than equilibria cannot arise.

In [86], we have developed an explicit algorithm to determine the heteroclinic connection graph in the general S^1-equivariant case $f_0 \in \mathcal{P}(u, u_x)$. The basic procedure consisted in a homotopy f_τ from f_0 to $f_1 \in \mathcal{P}(u, u_x^2)$ which preserved (normal) hyperbolicity of all equilibria and rotating waves, freezing the latter to wave speed

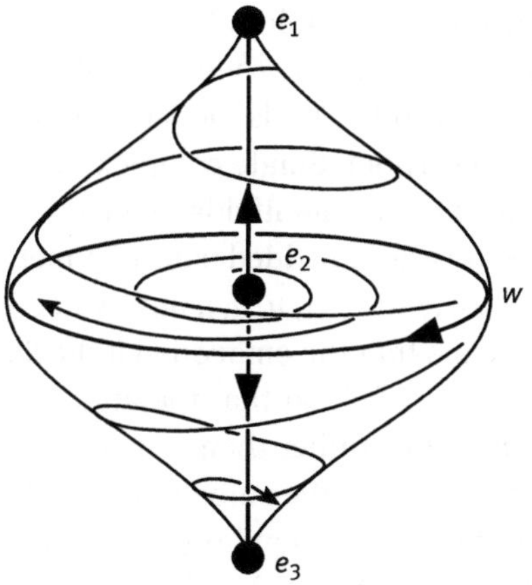

Figure 3: *A geometric representation of the Sturm spindle attractor with three equilibria and one rotating wave; see also [141].*

$c = 0$ at $\tau = 1$. We were then able to derive the heteroclinic orbits in $\mathscr{P}$ from those in the reversible Neumann class $f_1 \in \mathscr{N}(u, u_x^2)$ or, equivalently, the Hamiltonian Neumann class $f_1 \in \mathscr{N}(u)$.

As an explicit example, we mention the *Sturm spindle* attractor $\mathscr{A}$ of Fig. 3, which consists of three equilibria e_k of Morse indices $i(e_1) = i(e_3) = 0$, $i(e_2) = 3$ and one rotating wave w with (strong) unstable dimension one. Surprisingly, the same global attractor $\mathscr{A}$ has also been encountered by Krisztin and Walther in a quite different setting of delay differential equations

$$\dot{u}(t) = g\big(u(t), u(t-1)\big)\,, \tag{37}$$

for positive monotone feedback g in the delayed argument; see [141]. There is a nodal property similar to the zero number z of (35) at work, here. The delay class is not exhaustively classified; it might turn out to be a subclass of the parabolic class, together with the closely related class of negative monotone delayed feedback. This striking analogy remains unexplored at present.

At the end of this section, we return to the general Neumann class $f \in \mathscr{N}(x, u, u_x)$. We have a combinatorial description of this PDE class in terms of ODE shooting meanders, the Sturm permutations σ_f, and the explicit derivation of the heteroclinic connection graph $\mathscr{C}_f$. We did not yet succeed, however, to a priori characterise the class of all connection graphs $\mathscr{C}_f$ or, more ambitiously, the topological properties of the global Sturm attractors $\mathscr{A}_f$ themselves – other than by mindless successive enumeration of all σ_f. However, some promising steps have been achieved.

We recall the gradient-like structure which implies a general decomposition $\mathscr{A} = \mathscr{E} \cup \mathscr{H}$ of the global attractor into (hyperbolic) equilibria $\mathscr{E}$ and their heteroclinic orbits $\mathscr{H}$. In other words,

$$\mathscr{A} = \bigcup_{v \in \mathscr{E}} W^u(v) \tag{38}$$

decomposes into the disjoint union of the unstable manifolds $W^u(v)$ of all equilibria v. Here $W^u(v)$ consists of all ancient trajectories $u(t,\cdot)$ in $\mathcal{A}$ which limit onto v, for $t \to -\infty$. In particular, $v \in W^u(v)$. Note that dim $W^u(v) = i(v)$ is the Morse index of v, and $W^u(v) = \{v\}$ for $i(v) = 0$. We call eq. (38) the Thom–Smale complex of $\mathcal{A}$. In general, the boundaries $\partial W^u = (\text{clos } W^u) \setminus W^u$ need not be manifolds. In fact, (38) need not even be a cell complex, even though the $W^u(v)$ are embedded balls of dimensions $i(v)$. In the class of Sturm global attractors $\mathcal{A} = \mathcal{A}_f, f \in \mathcal{N}(x, u, u_x)$, however, the decomposition (38) is a *regular cell complex*. This means that the embedded $i(v)$-cells $W^u(v)$ possess embedded sphere boundaries $\partial W^u(v) \cong S^{i(v)-1}$, so that the attaching maps onto the boundary complexes are homeomorphisms, rather than just continuous maps. The proof requires a theorem of Schoenflies type for the embeddings of the closures clos $W^u(v)$; see [84]. For the resulting Thom–Smale cell complex (38), see [83].

For a more concrete example, the case of a single closed Sturm 3-ball

$$\mathcal{A}_f = \text{clos } W^u(0) = W^u(0) \cup S^2 . \tag{39}$$

with trivial equilibrium $i(0) = 3$ has been addressed in [83]. It has been shown how any finite regular cell decomposition of the Schoenflies boundary $\partial W^u(0) = S^2$ can be realised, topologically, as the Thom–Smale complex

$$S^2 = \bigcup_{v \in \mathcal{E} \setminus \{0\}} W^u(v) \tag{40}$$

of the remaining equilibria v with Morse index $i(v) \leq 2$, in the Neumann class $f \in \mathcal{N}(x, u, u_x)$.

A more careful analysis of this example, however, also keeps track of the two-dimensional *fast unstable manifold* $W^{uu}(0) \subset W^u(0)$ of those solutions $u(t,\cdot)$ in W^u which do not decay to v at the slowest possible exponential rate in $W^u(0)$, for $t \to -\infty$. By our Schoenflies results $\partial W^{uu}(0) = S^1$ decomposes S^2, as an “equator”, into two (closed) hemispheres $S^2_\pm$. Each closed hemisphere, separately, is realisable as a planar Sturm attractor $\mathcal{A}_\pm = \mathcal{A}_{f_\pm}$ for some non-linearities $f_\pm \in \mathcal{N}(x, u, u_x)$. To prove our Sturm realisation (40) in [83] we took one of the hemispheres to be a single closed two-cell. The general question of admissible, i.e. Sturm realisable, hemisphere decompositions is unexpectedly delicate. As a caveat, we just mention that the stable extreme equilibria v_1, v_{27} in any dissipative Sturm realisation of the 3d solid octahedron complex $\mathbb{O}$ of 27 equilibria must be chosen as adjacent corners of $\mathbb{O}$, rather than antipodally. For adjacent corners v_1, v_{27}, the two hemispheres $S^2_\pm$ are allowed to decompose S^2 into 1+7 or 2+6 octahedral faces, but not into 3+5 or 4+4 faces, by the fast unstable manifold $W^{uu}(0)$.

To complete our ambitious goal of a complete description of the Sturmian Thom–Smale complexes (38), an inductive process comes to mind. Suppose all Sturm global attractors $\mathcal{A}_f, f \in \mathcal{N}(x, u, u_x)$, have been understood for dimensions up to $n-1$. Two steps remain. In a first step, we have to understand the sphere boundary of single Sturm balls $\mathcal{A} = \operatorname{clos} W^u(0)$, for $i(0) = n$, as gluings of closed hemisphere complexes $\mathcal{A}_\pm$ of dimension $n-1$, which are assumed to be known by induction hypothesis. In a second step, we have to understand how these new n-cells attach to each other, and to lower-dimensional cells, to form a general Sturm global attractor of dimension n. Again, the respective hemisphere decompositions should play a crucial role here. At present, we have understood step 2 only up to the planar case $n = 2$; see [80–82]. Step 1 for $n = 3$ is imminent.

2.2.2 Meanders (Fiedler, Karnauhova,[18] Liebscher[19])

In the previous section, we introduced meander curves in the plane as shooting curves of parabolic PDEs (28) in one space dimension; see in particular eqs. (29) and (32) and Figs. 1 and 2. Quite independently from this ODE and PDE context, meanders also arise as trajectories of Cartesian billiards [74], and as representations of monomials in Temperley–Lieb algebras [58]. For simplicity, we now represent meander curves, or equivalently meander permutations, by $[N/2]$ standardised upper and lower half-circular arches joining at vertices $1, \dots, N$ along the horizontal axis. Also we replace the open meanders from south-west to north-east, of Section 2.2.1, by their closed Jordan curve counterparts. See Fig. 1(c) for one simple closing procedure.

More generally, we now regard closed meanders $\mathcal{M}$ as the pattern created by one or several disjoint closed smooth Jordan curves in the plane as they intersect a given horizontal line transversely. Let $c(\mathcal{M}) \geq 1$ count the connected components of $\mathcal{M}$. In view of Sturm global attractors, of course, the case $c(\mathcal{M}) = 1$ of connected meanders generated by a single closed Jordan curve is of primary interest.

Several new results have been obtained based on rainbows. Here a *proper upper rainbow* consists of a number α of concentric and adjacently nested upper arches in any meander. It is understood that α is chosen maximal in all cases. An *upper rainbow* consists of one or several neighbouring proper upper rainbows. Lower rainbows are defined analogously. Note that upper and lower rainbows may themselves be nested, giving rise to a forest of their sizes α. A closed meander without rainbow nesting is called a *seaweed meander*. Symbolically, a seaweed meander can be described as

$$\mathcal{M}(\alpha_1, \dots, \alpha_{m_+} \,|\, \beta_1, \dots, \beta_{m_-}), \tag{41}$$

where $\alpha_1, \dots, \alpha_{m_+}$ are the sizes of the neighbouring proper upper rainbows, and $\beta_1, \dots, \beta_{m_-}$ enumerate the proper lower rainbows, left to right. See Fig. 4(a) for an example. Trivially

$$\alpha_1 + \dots + \alpha_{m_+} = \beta_1 + \dots + \beta_{m_-} = N/2 \tag{42}$$

---

18 Anna Karnauhova, Free University of Berlin, anna_kaesfe@web.de

19 Stefan Liebscher, TNG Technology Consulting GmbH, contact@stefan-liebscher.de

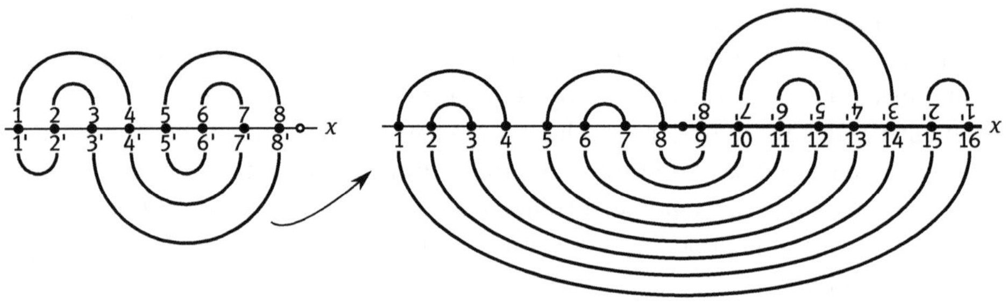

(a) Seaweed meander (b) Cleaved bi-rainbow meander

Figure 4: *(a) A seaweed meander $\mathcal{M}(2,2\,|\,1,3)$ with $N = 8$ intersections. (b) The associated cleaved bi-rainbow meander $\mathcal{M}_*(2,2,3,1\,|\,8)$ with $2N = 16$ intersections. Note how upper arches (i,j) of $\mathcal{M}$ have been preserved. Lower arches (i,j) of $\mathcal{M}$ have been converted to upper arches $(2N+1-i,\ 2N+1-j)$ of $\mathcal{M}_*$; see [74].*

for any closed meander of N vertices. The special seaweed case $m_- = 1$ of a single lower rainbow of size $\beta = N/2$ is called a *bi-rainbow meander*. The case where the upper rainbows may be arbitrarily nested, but there is only one single lower rainbow $\beta = N/2$, is called *lower rainbow meander*; see Fig. 4(b). For example any closed meander with N vertices can be "opened" to a closed lower rainbow meander with $2N$ vertices and the same number of connected components. Indeed we just duplicate the original vertices $j \mapsto j, j'$ and open up the horizontal axis to accommodate the vertices in the order $1, \dots, N, N', \dots, 1'$; see again Fig. 4. Recall how only connected meanders relate to an interpretation as ODE shooting curves for PDE Sturm global attractors. Lower rainbow meanders of the above general type were related to Cartesian billiards by Fiedler and Castañeda; see [74] and Fig. 5.

In fact, there exists a plane billiard with only horizontal and vertical billiard boundaries, with vertices all on the Cartesian integer grid and path slopes of $\pm 45°$ between half-integer boundary points, such that the connectivities of paths in the path and in the meander coincide. As an almost immediate consequence one obtains the following list for the number $c(\alpha_1, \dots, \alpha_m)$ of connected components of bi-rainbows $\mathcal{M}(\alpha_1, \dots, \alpha_n\,|\,\beta)$ in terms of greatest common divisors (gcd):

$$\begin{aligned} c(\alpha_1) &= \alpha_1\,, \\ c(\alpha_1, \alpha_2) &= \gcd(\alpha_1, \alpha_2)\,, \\ c(\alpha_1, \alpha_2, \alpha_3) &= \gcd(\alpha_1 + \alpha_2, \alpha_2 + \alpha_3)\,. \end{aligned} \tag{43}$$

The dissertation of Karnauhova [136] establishes how such formulas cannot be extended to $m \geq 4$. In fact c cannot be written as the gcd of any two polynomials f_1, f_2 in $\alpha_1, \dots, \alpha_m$ with integer coefficients. In [137] this obstacle is overcome, replacing the algorithm "gcd" by an explicit recursion of related logarithmic complexity in terms of the data $\alpha_1, \dots, \alpha_m$.

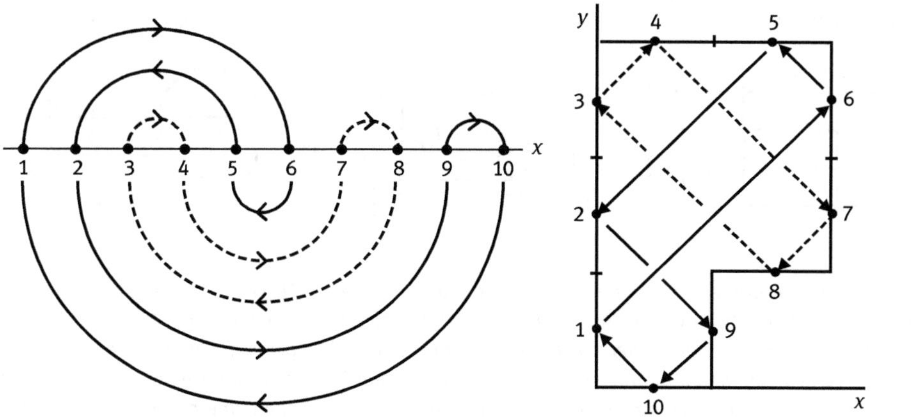

(a) Closed (multi-)meander (b) Non-transitive Cartesian billiard

Figure 5: *(a) A bi-rainbow meander with two connected components, one distinguished by dashed arches. (b) A corresponding non-transitive plane Cartesian billiard. One closed flight path is distinguished by dashed lines; see [74].*

Encouraged by suggestions of Matthias Staudacher (projects C4 “Structure of Quantum Field Theory: Hopf Algebras versus Integrability” and C5 “AdS/CFT Correspondence: Integrable Structures and Observables”) and earlier hints by Eberhard Zeidler, we also began to explore relations between meanders and Temperley-Lieb Hopf algebras $TL_N(\tau)$; see [74]. Algebraically, these are given by N multiplicative generators $e_0 = 1, e_2, \dots, e_{N-1}$ with the three relations

$$\begin{aligned} e_i^2 &= \tau e_i\,, \\ e_i e_j &= e_j e_i\,, \\ e_i e_{i\pm 1} e_i &= e_i\,, \end{aligned} \tag{44}$$

for all appropriate $i, j > 0$ with $|i - j| > 1$. See Fig. 6 for a representation of the Temperley–Lieb generators $e_i \in TL_N(\tau)$ by N-strand diagrams. Multiplication of generators is realised by horizontal concatenation of the generator diagrams, up to strand isotopy. Isolas effect multiplication by τ, each. As illustrated in Fig. 7 for

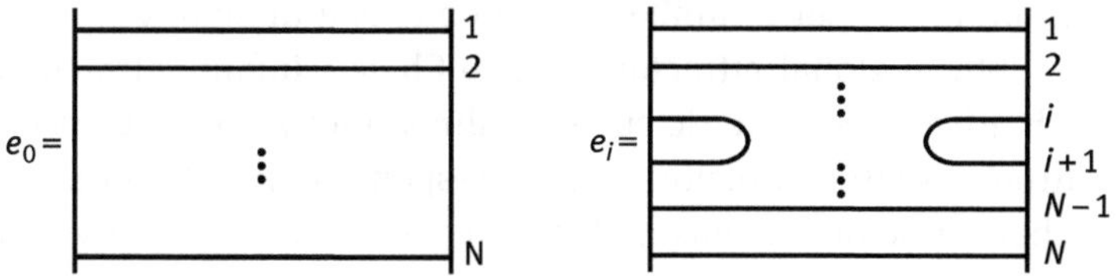

Figure 6: *Diagram representation of Temperley–Lieb generators $e_0 = 1$ and e_i, for $i = 1, \dots, N-1$; see [74].*

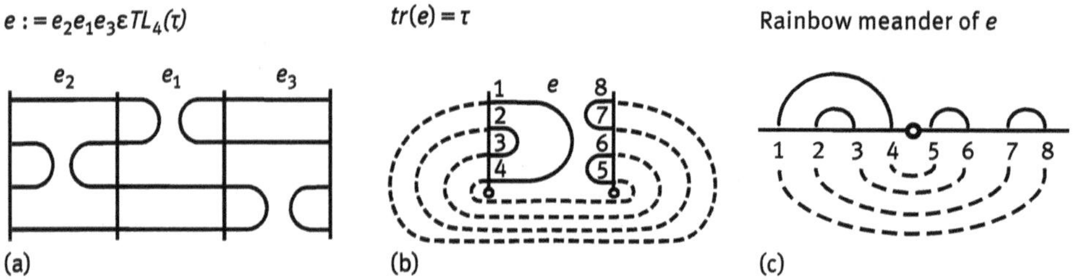

Figure 7: *(a) Diagram representation of the monomial $e = e_2e_1e_3 \in TL_4(\tau)$. (b) Homotopy equivalent diagram of e with dashed exterior strands matching right and left ends of the same strand index. Note the number $k = 1$ of exterior connected components of the closed diagram. (c) Joining the lower ends "o" of the vertical strand boundaries of e and opening the boundaries to become horizontal, we obtain an equivalent lower rainbow meander with eight intersections and the same number of connected components as in (b); see [74].*

$e := e_2e_1e_3 \in TL_4(\tau)$, Temperley–Lieb monomials e correspond to lower rainbow meanders $\mathcal{M}[e]$. The number c of connected components of $\mathcal{M}[e]$ is then related to the *Markov trace* tr_F of e:

$$\mathrm{tr}_F(e) = \tau^{c(\mathcal{M}[e])}, \tag{45}$$

see [58]. For further comments and a relation to the classical Young-Baxter equations, see also the dissertation by Karnauhova [136] and the references there.

Cases (41) and (42) of seaweed meanders $\mathcal{M}(\alpha_1, \dots, \alpha_{m_+} \mid \beta_1, \dots, \beta_{m_-})$ is related to the index of seaweed algebras, as studied by Dergachev and Kirillov [57]. The seaweed algebra of type $(\alpha_1, \dots, \alpha_{m_+} \mid \beta_1, \dots, \beta_{m_-})$ is the $\mathrm{gl}(N/2)$ subalgebra of matrices which preserve the vector spaces V_j spanned by unit vectors $e_1, \dots, e_{\alpha_1+\dots+\alpha_j}$ and W_j spanned by $e_{\beta_1+\dots+\beta_j+1}, \dots, e_{N/2}$. In fact, the seaweed meander corresponds to the Cartesian billiard on the domain of admissible nonzero entries of the seaweed matrices. The index of the algebra is the minimal kernel dimension of the map $A \mapsto f[A, \cdot]$, for any linear functional f in the dual algebra. By [57] this index coincides with the number $c(\mathcal{M})$ of connected components of the meander $\mathcal{M}$.

In her dissertation, Karnauhova considers the maximally disconnected case $\alpha_j = \beta_j$ of seaweed algebras with maximal index $N/2$, generated by trivial 2-cycles of σ, for the closed meander $\mathcal{M}$; see [136]. She then shifts the lower β-rainbows one step to the right. Ratcheting into the upper α-rainbows, she obtains a dissipative Morse meander σ_f, and hence a Sturm global attractor $\mathcal{A}_f$. The Chafee–Infante attractors [120], for example, which arise for symmetric cubic nonlinearities f and have been the first Sturm global attractors studied historically, correspond to $m_\pm = 1$ and $\alpha = \beta$. They are the Sturm global attractors of maximal dimension $\alpha = N/2$ for any given (odd) number $N + 1$ of equilibria. The same construction in fact works for general, but identical, upper and lower arch configurations. Although the resulting global attractors are all pitchforkable, and thus have been known "in principle" by Fusco and Rocha [91], the new approach is elegant and geometric: The Sturm global attractors can be obtained explicitly by successive double cone suspensions and elementary gluing concatenations.

Fiedler [73] embarked on a different approach to explore the relation between the chaotic dynamics of (x-periodically forced) equilibrium equations (29) with nonlinearity f and the Sturm permutations σ_f of the resulting shooting meanders. Homoclinic tangles with their associated shift dynamics are taken as a paradigm. In this direction, meander permutations have been defined for standard hyperbolic linear $SL_2(\mathbb{Z})$ torus automorphisms. The meander permutations compare the orderings of homoclinic points along the one-dimensional stable and unstable manifolds of their limits, respectively. As such, they are new topological invariants – with intriguing relations to continued fraction expansions and quadratic number fields.

2.2.3 Stability, instability and obstacles (Väth[20])

Consider the higher-dimensional analogue of eq. (28), that is, for semilinear equations

$$u_t = \Delta u + f(u)$$

on a domain $\Omega \subseteq \mathbb{R}^N$ with $N \geq 1$. More generally, consider systems $u: \Omega \to \mathbb{R}^n$

$$u_t = Lu + f(u) \tag{46}$$

with an elliptic differential operator L and associated Dirichlet, Neumann or mixed boundary values. Then the naturally associated function spaces $W^{1,2}_\Gamma(\Omega)$ and $L^2(\Omega)$ are not continuously embedded into $C(\overline{\Omega})$. Therefore, differentiability of the corresponding superposition operator $F(u)(x) = f(u(x))$ fails in $L^2(\Omega)$ (except in the linear case). This imposes an unrealistically restrictive growth condition for f in case $F: W^{1,2}_\Gamma(\Omega) \to L^2(\Omega)$. Moreover, no analogue of the zero number (34) with a property like eq. (35) is available. Nevertheless, we are interested in the long-term behavior of the equation in these two function spaces.

A particularly natural question is whether a "formally" linearised stability of some stationary solution u_0 implies asymptotic stability in these spaces. In other words, suppose the spectrum of the "formal" linearisation $L + f'(u_0)$ lies in the complex left half-plane. Do all solutions starting close to u_0 in the topology of the space $W^{1,2}_\Gamma(\Omega)$ or $L^2(\Omega)$ indeed remain close to u_0 and eventually tend to u_0 in the corresponding topology? Let A denote the operator naturally associated to $-L$ and the boundary conditions. Suppose A solves Kato's square root problem, that is, if $A^{1/2}: W^{1,2}_\Gamma(\Omega) \to L^2(\Omega)$ is an isomorphism. Then $W^{1,2}_\Gamma(\Omega) = D(A^{1/2})$. The asymptotic stability in $W^{1,2}_\Gamma(\Omega)$ is known for $F: W^{1,2}_\Gamma(\Omega) \to L^2(\Omega)$; see e.g. [119]. Gurevich and Väth [102] have relaxed this

20 Martin Väth, Free University of Berlin, martin@mvath.de

hypothesis to a subcritical growth condition on f. Moreover, they obtained an asymptotic stability result in $L^2(\Omega)$. The method of proof was to consider extrapolated spaces and extrapolated semi-groups which were obtained by two different procedures. As a side result, they have shown that these extrapolation procedures coincide if and only if A solves Kato's square root problem.

One particular motivation to consider the topology of the spaces $W^{1/2}_\Gamma(\Omega)$ and $L^2(\Omega)$ is that the following question so far can be addressed only in these topologies. Suppose that the problem (46) is endowed with some obstacle, i.e., the values of u on some parts of Ω or of its boundary are forced to stay within a certain cone by some constraint. One of our surprising results is that the principle of linear asymptotic stability may then fail in $W^{1,2}_\Gamma(\Omega)$. The proof is based on an instability criterion for variational inequalities and a rather general formula relating the fixed point index of the flow of the obstacle problem with the topological degree of a certain associated operator [139]. For specific examples, see [139, 203].

2.2.4 Symmetric and asymmetric black hole data (Fiedler, Hell,[21] Smith[22])

The Cauchy initial value problem for the relativistic Einstein equations requires initial data that satisfy the Einstein constraint. For black holes these constraints take the form of an equation for the metric g with prescribed scalar curvature R on an asymptotically flat Riemannian 3-manifold M; see Fig. 8 for an illustration and [174] for a general background. For simplicity, we consider a collar region $M = [r_0, r_1] \times S^2$ with

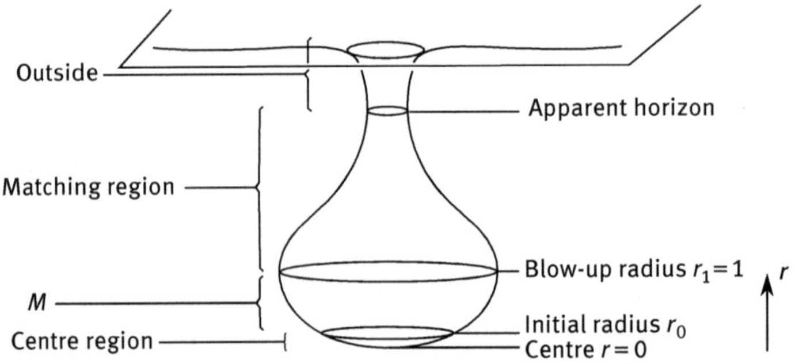

Figure 8: *A 2d-caricature of black hole initial data for the Einstein equations. The vertical variable is the radius r. Each horizontal section $\{r = \text{constant}\}$ in this figure is a circle, which represents a two-sphere S^2. The region outside the apparent horizon (and above it in this picture) was treated in the literature. In [76] we construct the region M between an initial radius $r_0 > 0$ and a blow-up radius $r_1 = 1$ whose section $\{r_1\} \times S^2$ is a critical point of the area functional. The matching and center regions remain to be constructed; see [74].*

21 Juliette Hell, Free University of Berlin, jhell@zedat.fu-berlin.de
22 Brian Smith, briansmith2865@gmail.com

standard 2-sphere cross-section. In a suitable coordinate foliation, the metric can be expressed as

$$g = u^2 dr^2 + r^2 \omega \tag{47}$$

with the standard metric ω on S^2. For example, consider prescribed scalar curvatures R such that $\lambda := r^2 R - 2$ is a constant bifurcation parameter. Then $u = u(r, x)$ can be rescaled to

$$u(r,x) = \left(\tfrac{\lambda}{2}\left(\tfrac{1}{r} - 1\right)\right)^{-1/2}\left(v\left(-\tfrac{1}{2}\log(1-r), x\right) + 1\right), \tag{48}$$

and the Einstein constraints on the radial metric coefficient u take the quasilinear parabolic form

$$v_t = (1+v)^2\left(\Delta v + \lambda f(v)\right). \tag{49}$$

Here $v = v(t,x)$, $x \in S^2$, Δ is the Laplace–Beltrami operator on S^2, and $f(v) = v - \frac{1}{2}v^2/(1+v)$ defines a cubic nonlinearity.

In [76], we obtain spherically anisotropic solutions v for isotropically prescribed scalar curvature R. Equilibria $v(b,x) = v_*(x) \not\equiv$ const provide self-similar anisotropic metrics g. Heteroclinic trajectories $u(t,x) \to v_\pm(x)$, for $t \to \pm\infty$, are asymptotically self-similar for $r \to 0$ and $r \to 1$. The metric g remains smooth, in fact, in the latter limit. We obtain these solutions by an $O(3)$ symmetry-breaking bifurcation analysis of the scaled quasilinear PDE in a center manifold at the trivial isotropic equilibrium $v_* \equiv 0$.

Symmetry-breaking bifurcations occur at $\lambda_\ell = \ell(\ell+1)$, for $l = 0, 1, 2, \ldots$, in the spherical harmonics kernels V_ℓ of dimension $2\ell + 1$. See Fig. 9 for the bifurcating branches of non-constant equilibria v, and Fig. 10 for their remaining reduced, nonspherical spatial isotropies. Heteroclinic trajectories follow, between v_* and 0, by standard bifurcation theory in the one-dimensional isotropy subspaces of each kernel V_ℓ. For heteroclinic trajectories between equilibria of different isotropies see also [50].

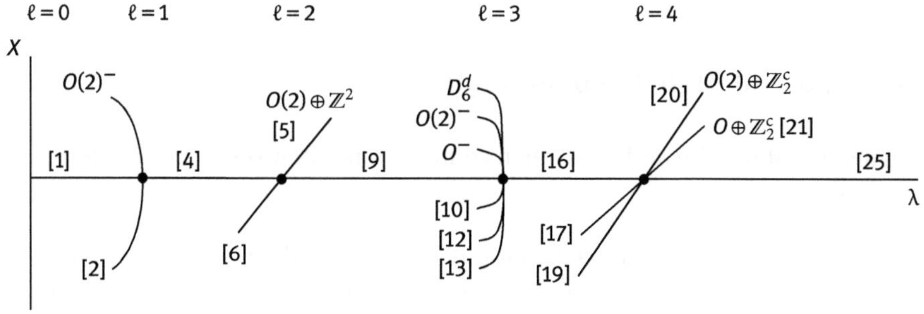

Figure 9: *Local transcritical and pitchfork bifurcation branches $(\lambda_K(s), v_K(s))$ from $\lambda_K(0) = \lambda_\ell = \ell(\ell+1)$, $v_K(0) = 0$ with maximal isotropies K as indicated. Numbers in brackets indicate co-dimensions of the strong unstable manifolds of their normally hyperbolic group orbits; see [76].*

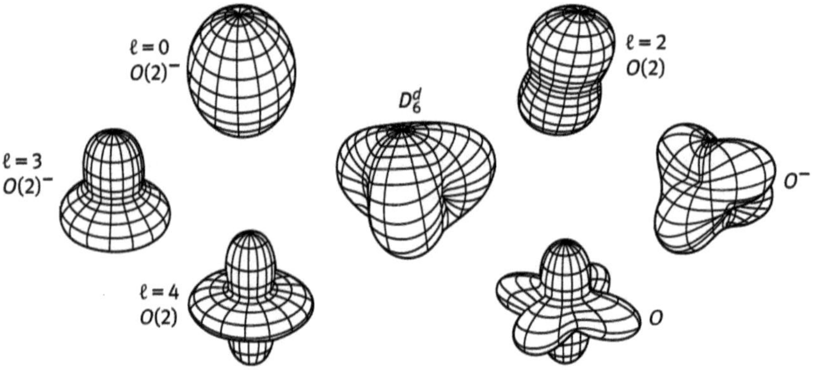

Figure 10: *The graphs of the bifurcating symmetry-breaking equilibria v_K over the sphere S^2, with isotropies K as indicated; see [86].*

The results above are of a local nature, near the spherically isotropic trivial equilibrium $v_* \equiv 0$. The fully isotropic case $\ell = 0$ was treated in [191] previously, by more direct methods. In practice specific computations, particularly of the unstable dimensions and the bifurcation directions, are limited to low ℓ and have been carried out for $\ell \leq 4$, only. Axisymmetric solutions $v = v(t, \vartheta)$, however, which depend only on the azimuthal angle ϑ of $x \in S^2$, are known to bifurcate for all $\ell = 0, 1, 2, \ldots$. Clearly, (49) becomes "spatially" one-dimensional,

$$v_t = (1 + v)^2 \big(v_{\vartheta\vartheta} + \cot \vartheta \, v_\vartheta + \lambda f(v)\big) , \tag{50}$$

with Neumann boundary on $0 < \vartheta < \pi$. The dissertation of Lappicy explores these versions of quasilinear Sturm parabolic equations with a singular spatial coefficients, by the global methods of Section 2.2.1; see [149]. In particular, this yields a plethora of axisymmetric, but not fully spherically isotropic, equilibrium solutions and their heteroclinic trajectories as parts of black hole initial data that satisfy the Einstein constraint.

2.3 Blow-up and Bianchi cosmology

2.3.1 Dynamics at infinity and blow-up in complex time (Ben-Gal,[23] Fiedler, Hell, Stuke[24])

In Section 2.2, a class of dissipative equations (28) was studied, where long-time existence and boundedness of the solutions was guaranteed. In the present section,

23 Nitsan Ben-Gal, bengal.nitsan@gmail.com
24 Hannes Stuke, Free University of Berlin, hannes.stuke@gmail.com

we focus on equations which develop singularities in finite or infinite time. In other words, some trajectories escape to infinity. In order to study the dynamics at infinity, one possible approach uses *Poincaré "compactification"*: The solution space is projected centrally onto a tangent upper hemisphere. More precisely the hemisphere at its north pole is tangent to the solution space at its origin. Arbitrarily far points in the solution space are thereby mapped to the equator of the hemisphere, also called the *sphere at infinity*. For details, see [117, 118, 167]. We keep the name Poincaré "compactification" which originates from ODE theory, where the solution space is $\mathbb{R}^n$ and the compactified space is a compact n-hemisphere. In the PDE context, the solution space is infinite dimensional and its "compactification" is a bounded hemisphere, which is not compact.

As an elementary but instructive example, consider the linear equation:

$$u_t = u_{xx} + bu\,, \tag{51}$$

for $0 < x < \pi$, with Neumann boundary conditions; see [117]. Equation (51) is not dissipative for positive b: some trajectories escape to infinity, in infinite time. The class of equations at the boundary between dissipativity and finite time blow-up is called *slowly nondissipative* in [23]: They admit at least one immortal trajectory escaping to infinity in forward time. The linear example (51) admits only one bounded equilibrium, located at the origin, and generates infinitely many equilibria at infinity. Those are located at the intersections of the one-dimensional eigenspaces, associated with the eigenfunctions of the Laplacian, with the sphere at infinity. Let us denote the equilibria at infinity by $\pm\Phi_m$, $m \in \mathbb{N}$. For all $b \neq n^2$ which are not square integers, all equilibria, both bounded and at infinity, are hyperbolic. Furthermore the dynamics within the sphere at infinity has a Chafee–Infante structure; see Section 2.2.1. More precisely, the two equilibria $\pm\Phi_0$ at infinity are stable. The equilibria $\pm\Phi_m$ at infinity each possess an m-dimensional unstable manifold that consists of heteroclinic orbits to the equilibria $\pm\Phi_j$, $j \in \{0, \dots, m-1\}$. The unstable manifold of the origin is of finite dimension $[\sqrt{b}]+1$, where $[\sqrt{b}]$ denotes the integer part of $\sqrt{b}$. This unstable manifold consists of heteroclinic orbits to the equilibria $\pm\Phi_i$, $i \in \{0, \dots, [\sqrt{b}]\}$. The stable manifold of the equilibrium at the origin is infinite dimensional and consists of heteroclinic orbits from the equilibria $\pm\Phi_m$ at infinity, $m > [\sqrt{b}]$, to the origin. This describes the global unbounded attractor for eq. (51).

When eq. (51) is augmented by a smooth bounded nonlinearity $g = g(u)$,

$$u_t = u_{xx} + bu + g\,, \tag{52}$$

the dynamics at infinity remains essentially unchanged. In her dissertation [23], Ben-Gal adapts the techniques of Section 2.2.1 to study the global heteroclinic structure of eq. (52). In particular, she proves the existence of a finite-dimensional unbounded inertial manifold containing the so called *unbounded* (or *noncompact*) *attractor*. This object consists of all bounded equilibria (generically hyperbolic), their finite-dimensional unstable manifolds, the finitely many equilibria at infinity (also generically hyperbolic) to which those unstable manifolds connect, and the finite-dimensional unstable manifolds of those equilibria at infinity. Theorems similar to the dissipative case reveal the heteroclinic structure of the unbounded global attractor of slowly nondissipative equations. Pimentel and Rocha generalised some of these results to bounded nonlinearities $g(x, u, u_x)$; see [169, 170].

We consider blow-up in finite time for eq. (28), $f = f(u)$, next. For smooth right hand side with polynomial growth, time can be rescaled such that trajectories do not run into equilibria at infinity in finite time. The resulting semiflow on the closed Poincaré hemisphere is non-singular. Conley index theory then provides heteroclinic orbits between equilibria, or general isolated invariant sets. Isolated invariant sets are contained in isolating neighbourhoods. An isolating neighbourhood is a closed neighbourhood whose boundary points eventually leave the neighbourhood in forward or backward time direction. For details, see [117] and the references there.

It turns out that even for some polynomial ordinary differential equations, equilibria at infinity might not be isolated invariant sets. The left-hand side of Fig. 11 illustrates such an equilibrium at infinity. This equilibrium is isolated, invariant and stable within the sphere at infinity. Towards the interior of the Poincaré hemisphere, however, it shows two petals of nested homoclinic loops. Therefore, it does not possess small isolating neighbourhoods. The complement of any small neighbourhood of the degenerate equilibrium, however, is an isolating neighbourhood. An invariant set with this property is called *of isolated invariant complement*. Although the complementary invariant set isolated by such a neighbourhood is possibly very complex, its Conley index is well-defined. Poincaré duality then defines a meaningful Conley index for the degenerate equilibrium at infinity. An invariant set at infinity attracts some finite initial conditions, if and only if it is not a repeller. This can be characterised in terms of Conley index. Therefore, this tool allows us to determine which part of the dynamics in the sphere at infinity is reachable by grow-up or blow-up solutions.

Through a standard construction illustrated in the right-hand side of Fig. 11, and described in detail in [117, 118], it is possible to replace the invariant set of isolated invariant complement by a benign *ersatz* with a well-defined Conley index. This ersatz can be used for the analysis of the global heteroclinic structure via classical Conley index theory.

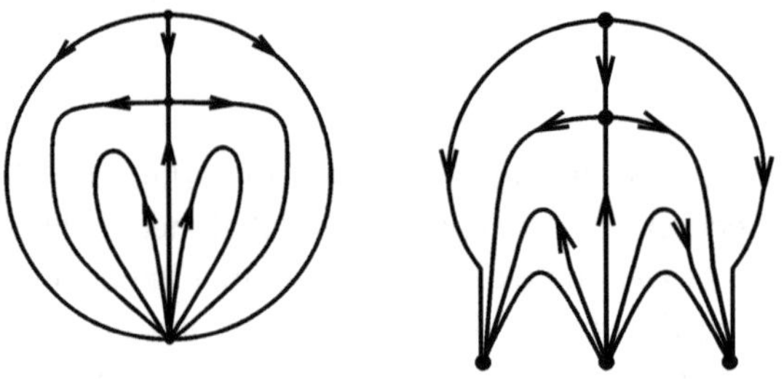

Figure 11: *Left: the lower equilibrium at infinity is not an isolated invariant set but is of isolated invariant complement. Right: the ersatz infinity for that equilibrium.*

Based on the idea of self-similar scaling and the Poincaré sphere at infinity, Fiedler has contributed to an unexpected application in a certain renormalisation approach to the experimentally observed formation of spin liquids in carbon nanotubes; see [122]. Renormalisation has been suggested to overcome the limitations of mean-field descriptions. In its most simple form, homogeneously quadratic ODE systems

$$\frac{dx}{dt} = -\nabla_x V \tag{53}$$

were addressed. The variational structure features a cubic functional

$$V(x) = -\frac{1}{3}\sum_{i=1}^{N} x_i(x^T A_i x) \tag{54}$$

and real symmetric $N \times N$-matrices A_i. As a result, we obtain self-similar and asymptotically self-similar blow-up solutions as well as their blow-up rates. These results provide a rational foundation for the renormalisation group approach to large, strongly correlated electron systems.

For fast nonlinear diffusion in several space dimensions the phenomenon of extinction of u, alias blow-up of $1/u$, has been addressed by Fila and Stuke [87, 88]. As a very simple paradigm for blow-up itself, consider the one-dimensional semilinear heat equation

$$u_t = u_{xx} + u^2 \tag{55}$$

on a bounded or unbounded interval. For real u, x and *real time* t, this equation has already been well-studied in the literature [172]. The solution is analytic in time and admits a local analytic continuation into the *complex time plane*. For Dirichlet boundary conditions, there exists a unique positive equilibrium u^* which is one-dimensionally unstable. Solutions in the one-dimensional analytic unstable manifold of u^* possess a real solution with complete blow-up in finite real time. Motivated by analyticity, Stuke studied extensions to complex, rather than real, time t. Of course, this entails complex u. Following an idea of Masuda [158] and Guo et al. [101], Stuke attempts to bypass real-time blow-up at $t = T$ by an excursion into the complex plane; see Fig. 12. Stuke proves that solutions on the complex unstable manifold of u^* stay bounded on time paths $\Gamma_p^\pm$ parallel to the real axis. The solutions converge to zero: They are heteroclinic from u^* to zero, along $\Gamma_p^\pm$, and locally foliate the nonsingular part of the complex one-dimensional unstable manifold W_*^u of u^*. After some real time $t = T + \delta > T$, at the latest, the solutions $\Gamma_c^\pm$ extend back to the real axis from, both the upper and the lower half-planes. Complete real blow-up, however, prevents these conjugate complex analytic branches to coincide, at real time overlap $t > T + \delta$.

Travelling pulse waves $u = U(x - ct)$ with wave speed c, and self-similar solutions, open a second approach to blow-up in complex time. Specifically, they correspond

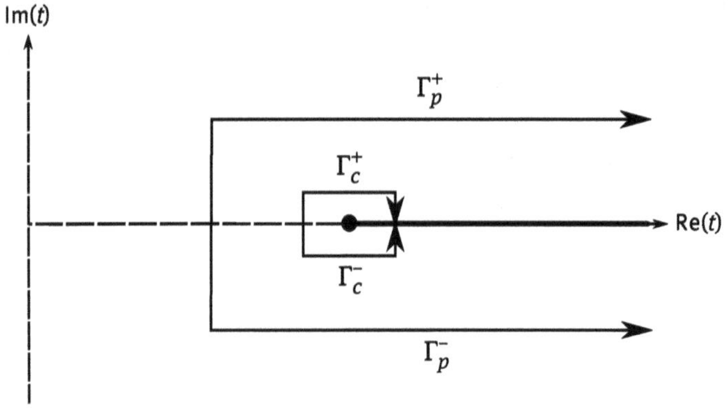

Figure 12: *The complex time domain in which we consider the continuation $\Gamma_p^\pm$, $\Gamma_c^\pm$ of blow-up solutions. The dot indicates the blow-up time T. After t = T we introduce a cut along the positive real axis, drawn as a bold line.*

to ODE solutions which are homoclinic to zero. The standing wave $u(t,x) = U(x) = -6/x^2$ with wave speed $c = 0$, for example, is an explicit solution of (55). In general, homoclinic travelling waves $u = U(\psi)$, $\psi = x - ct$, possess local expansions in ψ and $\log \psi$.

For the branched real blow-up on the complex unstable manifold W_*^u of u^*, matters are not quite that simple. Indeed a "center manifold", rather than a fully hyperbolic structure, appears in rescaled variables. This heralds essential singularities beyond $\log t$, and sectors with exponentially flat correction terms. For further details see the dissertation [196].

2.3.2 The tumbling universe (Liebscher, Härterich,[25] Georgi[26])

Arguably, the big bang is the largest imaginable blow-up. For us it is the initial singularity of a purely gravitational cosmological model. We briefly describe the general setting, and the exact reduction to ODE models of Bianchi class. In the next section, we present our own results.

Cosmological models of general relativity are solutions of the full Einstein equations

$$\mathrm{Ric}(g) - \tfrac{1}{2}\mathrm{Scal}(g)g = T, \tag{56}$$

which relate the geometry of spacetime – the curvature of a four-dimensional Lorentzian metric g – to the matter content encoded in the stress-energy tensor T, with or without a cosmological constant. The Einstein PDEs are of hyperbolic type and usually

25 Jörg Härterich, Ruhr University Bochum
26 Marc Gerorgi, Deutsche Bank, Berlin

coupled to kinetic equations of matter. A complete understanding of the full PDE system (56) is currently out of reach. Therefore, additional symmetries are often assumed to discuss special solutions and their perturbations.

The simplest model – and core of the standard cosmological model still in use – is the Friedmann model of spatially homogeneous and isotropic spacetime. This assumption of a six-dimensional symmetry group allows a reduction of (56) to a single scalar ODE that determines the expansion rate of the universe. This expansion rate can be compared with measurements of the Hubble constant and with the consequences of large-scale thermodynamics of the matter part.

The second simplest class are *Bianchi models* of spatially homogeneous but anisotropic spacetime. In other words, the spacetime is assumed to be foliated into spatial hypersurfaces given by the orbits of a three-dimensional symmetry group. Belinskii, Khalatnikov and Lifshitz (BKL) described the dynamics of cosmological models near the big-bang singularity as vacuum dominated, local and oscillatory [22]. This suggests to approximate the early universe by the dynamics of homogeneous, but not necessarily isotropic, vacuum spacetime.

The Einstein system (56) with a non-tilted, perfect-fluid matter model can be reduced to a five-dimensional ODE system in expansion-reduced variables,

$$\begin{aligned}
N_1' &= (q - 4\Sigma_+)N_1\,,\\
N_2' &= (q + 2\Sigma_+ + 2\sqrt{3}\Sigma_-)N_2\,,\\
N_3' &= (q + 2\Sigma_+ - 2\sqrt{3}\Sigma_-)N_3\,,\\
\Sigma_\pm' &= (q - 2)\Sigma_\pm - 3S_\pm\,,
\end{aligned} \tag{57}$$

with the abbreviations

$$\begin{aligned}
S_+ &= \tfrac{1}{2}\left((N_2 - N_3)^2 - N_1\,(2N_1 - N_2 - N_3)\right),\\
S_- &= \tfrac{1}{2}\sqrt{3}\,(N_3 - N_2)\,(N_1 - N_2 - N_3)\,,\\
q &= 2\left(\Sigma_+^2 + \Sigma_-^2\right) + \tfrac{1}{2}(3\gamma - 2)\Omega\,,\\
\Omega &= 1 - \Sigma_+^2 - \Sigma_-^2 - K\,,\\
K &= \tfrac{3}{4}\left(N_1^2 + N_2^2 + N_3^2 - 2\,(N_1N_2 + N_2N_3 + N_3N_1)\right).
\end{aligned} \tag{58}$$

This system is due to Wainwright and Hsu [206]. The initial big-bang singularity is approached in the rescaled time limit $t \to -\infty$. The variables N_k describe the curvature of the spatial hypersurfaces. Their signs determine the Lie-algebra type of the associated spatial symmetry imposed by the homogeneity assumption. Due to Bianchi's classification of three-dimensional Lie algebras – the tangent spaces to the assumed symmetry group – these models are called Bianchi models, although they have been introduced by Gödel [98] and Taub [199]. The shear variables $\Sigma_\pm$ relate to the rescaled eigenvalues of the second fundamental form of the spatial hypersurfaces. The matter density Ω is non-negative, and the vacuum boundary $\Omega = 0$ is invariant. The coefficient $\gamma < 2$ determines the equation of state of the perfect fluid, e.g. $\gamma = 4/3$ for radiation and $\gamma = 1$ for dust. See also [205] for further details on this dynamics approach to cosmology, and [116] for a survey on Bianchi models and open questions.

The most prominent features of system (57) are the *Kasner circle* $\mathcal{K}$ of equilibria,

$$\mathcal{K} = \left\{ \Sigma_+^2 + \Sigma_-^2 = 1,\ N_1 = N_2 = N_3 = 0 \right\}, \tag{59}$$

and the invariant *Kasner caps*

$$\mathcal{H}_k^{\pm} = \left\{ \Sigma_+^2 + \Sigma_-^2 = 1 - N_k^2,\ \pm N_k > 0,\ N_{k+1} = N_{k-1} = 0,\ k \bmod 3 \right\}. \tag{60}$$

The Kasner caps are filled with heteroclinic orbits between equilibria on $\mathcal{K}$; see Fig. 13. Equilibria of eq. (57) indicate self-similar blow-up in the big-bang limit which occurs at rescaled time $t \to -\infty$. The projections of the heteroclinic orbits to the Σ-plane lie on straight lines through the corners of a circumscribed equilateral triangle; see Fig. 14. In reversed physical time, these sequences define the *Kasner map* as an expansive map from the Kasner circle onto its double cover; see Fig. 14 and Section 2.3.3.

Concatenation of the heteroclinic orbits in the Kasner caps, i.e. iteration of the Kasner map, defines a subshift of finite type. Misner [160] introduced the name *Mixmaster* for the heuristic picture of generic trajectories following these formal heteroclinic sequences close to the big-bang singularity: The universe tumbles between various close to self-similar solutions given by the various Kasner equilibria.

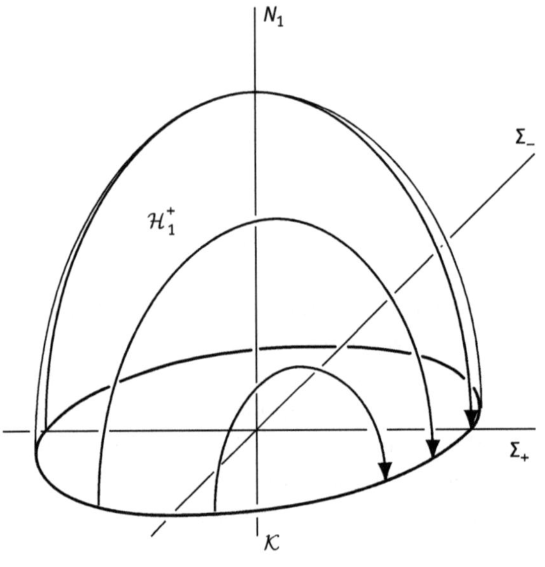

Figure 13: *Heteroclinic cap of vacuum Bianchi II solutions to the Kasner circle $\mathcal{K}$. Remaining caps are given by rotations over $2\pi/3$ in $\Sigma_\pm$ and cyclic permutations of $\{N_1, N_2, N_3\}$.*

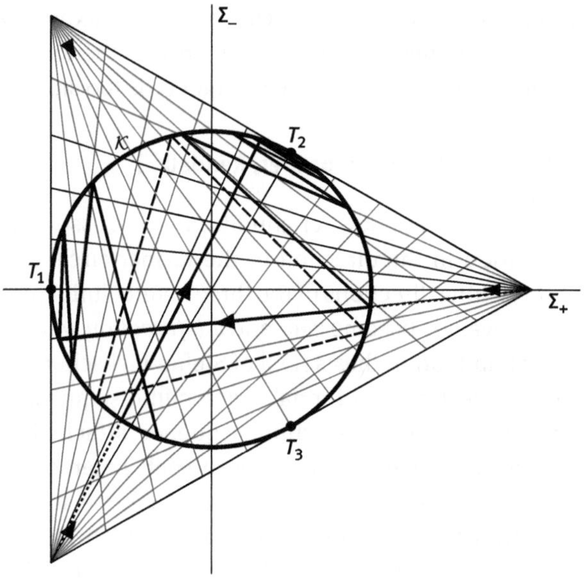

Figure 14: *The Kasner map and the three Taub points* T_1, T_2, T_3 *given by* $N_1 = N_2 = N_3 = 0$, $(\Sigma_+ + i\Sigma_-)^3 = -1$.

A crucial question is the *locality* of the BKL/Mixmaster picture. In the (non-vacuum) Friedmann model, the particle horizon (backward light cone) remains finite in expansion reduced variables, close to the big-bang singularity. This bounded particle horizon also bounds the domain of dependence by particle interaction, and therefore represents locality. In the Bianchi model (57), the particle horizon corresponds to the *Mixmaster integral*

$$I = \int_{-\infty}^{0} \left(\sqrt{N_1 N_2} + \sqrt{N_2 N_3} + \sqrt{N_1 N_3} \right) \mathrm{d}t\,, \tag{61}$$

see [115]. Locality holds true, if and only if this integral remains finite. Obvious counterexamples are the solutions on the line of Taub equilibria

$$N_1 = 0 \neq N_2 = N_3\,, \qquad \Sigma_- = 0\,, \quad \Sigma_+ = -1\,, \qquad \Omega = 0\,, \tag{62}$$

where the Mixmaster integral is $+\infty$. It is currently not known whether non-trivial counterexamples with $I = \infty$ exist. The first non-trivial positive examples with $I < \infty$ are quite recent [154]. More precisely, the longstanding BKL/Mixmaster conjecture therefore claims that $I < \infty$ for generic initial conditions $N_j(0)$, $\Sigma_\pm(0)$ in the respective Bianchi class [22, 115]. Here *genericity* can be understood either as *residual*, in the topological sense of Baire category, or as *almost everywhere* with respect to Lebesgue measure. In the Bianchi type IX of class A, where all $N_j > 0$, Ringström [173] has shown

Baire and Lebesgue generic convergence to the Kasner caps. Of course, the Mixmaster integral I vanishes on the Kasner caps. The mere generic convergence to the Kasner caps, however, does not imply generic finiteness of I if this convergence is too slow.

2.3.3 Kasner dynamics (Brehm,[27] Buchner,[28] Hell, Liebscher, Rendall[29])

In the vacuum model (57), $\Omega = 0$, each non-Taub Kasner equilibrium possesses one trivial eigenvalue zero of the linearisation in the direction of the Kasner circle and three non-trivial eigenvalues in the remaining directions of the three heteroclinic caps. At the Taub points, two non-trivial eigenvalues become zero and a *bifurcation without parameters* occurs, in the sense of the habilitation thesis [153] of Stefan Liebscher.

To clarify the relation of generic trajectories to formal heteroclinic sequences, we ask for the set of points converging to the heteroclinic sequence attached to a given Kasner equilibrium. This is a difficult question because each heteroclinic step, by itself, already consumes infinite rescaled time t. We might hope for an invariant codimension-one foliation transverse to the Kasner circle, which in particular contains the local strong unstable and strong stable invariant manifolds of each Kasner equilibrium. Each fibre would then become the stable set of the Kasner equilibrium, in backwards time $t \to -\infty$, as the big-bang singularity is approached.

However, there are topological obstructions: The sets of initial conditions converging to the Taub points, backwards, were classified in [173] and form manifolds of codimension two. Thus, if the heteroclinic sequence of a Kasner equilibrium terminates at a Taub point, this equilibrium cannot possess an attached stable fibre of codimension one. In particular, any transverse backwards stable foliation has to miss at least a dense subset of the Kasner circle.

Stable foliations along Kasner sequences away from Taub points have first been addressed by Liebscher et al. [154], as follows. Consider a Kasner equilibrium such that its attached (unique, backwards) heteroclinic sequence does not accumulate at a Taub point. Then this equilibrium possesses a unique, locally invariant, codimension-one, stable Lipschitz leaf. The leaves of equilibria in the same heteroclinic sequence form a backwards invariant continuous foliation. The distance in which the trajectories pass the equilibria in this foliation tends to zero exponentially.

The proof basically consists of three parts. First, a suitable eigenvalue property guarantees an arbitrarily strong contraction, transversely to the Kasner circle, and an arbitrarily small drift along the Kasner circle, locally. Second, the global passage along a heteroclinic orbit in the Kasner caps is modelled by a diffeomorphism close to the

---

27 Bernhard Brehm, Free University of Berlin, bbrehm@math.fu-berlin.de

28 Johannes Buchner, Institute for International Political Economy (IPE) Berlin, jbuchner@zedat.fu-berlin.de

29 Alan D. Rendall, University of Mainz, rendall@uni-mainz.de

Kasner map. This provides a bounded expansion along the Kasner circle, away from Taub points. Third, the combination of both maps yields a hyperbolic map between suitably chosen cross sections near the Kasner circle: Contraction transverse to the Kasner circle is inherited from the local passage, expansion along the Kasner circle is inherited from the global passage. The stable leaf is then obtained as the fixed point of a graph transform for the above construction.

Liebscher et al. [155], generalise this approach to weaker eigenvalue conditions. This applies to non-vacuum Bianchi systems with arbitrary ideal fluids, and to Einstein–Maxwell equations of Bianchi type VI_0. In the Einstein–Maxwell system, one of the heteroclinic caps is no longer induced by the geometry but rather by the dynamics of the magnetic field.

An alternative approach in [21], and the dissertation [39] by Buchner, utilise Takens linearisation [198] to study the local passage. Indeed, after linearisation, contraction and absence of drift follow trivially. However, the linearisation requires strong spectral non-resonance conditions. Non-resonance imposes severe limitations on this approach and on possible generalisations. In [21], for example, all formal sequences which are periodic or accumulate to periodic sequences (or the Taub points) are excluded to avoid resonances. In [39], Buchner has studied the non-resonance conditions more carefully, and certain periodic sequences are now admissible. His results on local passages extend to Bianchi class-B systems.

The techniques described above require to avoid the Taub points. This is their main restriction. In fact, the Kasner equilibria which, under iteration of the Kasner map, do not accumulate at Taub points form a (albeit uncountable) meagre set of measure zero. They are not generic, in any sense, but possess a finite Mixmaster integral (61) and thus a finite particle horizon.

Brehm, in his dissertation [35], is the first to answer the BKL/Mixmaster question in the Lebesgue sense. This problem had remained open for more than four decades. Brehm obtains two new results in the study of the Wainwright–Hsu system (57), (58) for the vacuum case $\Omega = 0$. The first result extends Ringström's attractor result to Bianchi type VIII vacuum solutions. The second result asserts finiteness $I < \infty$ of the Mixmaster integral (61) for Lebesgue almost all initial conditions.

Consider the Bianchi type IX first, where all $N_j > 0$ initially (and then for all times). The *Taub space* $\{N_2 = N_3,\ \Sigma_- = 0\}$ is time invariant, together with its two rotated "cousins" given by cyclic permutations of the indices $\{1, 2, 3\}$. Ringström [173] has shown convergence to the Kasner caps, for initial conditions outside the Taub spaces, in the time-rescaled big-bang limit $t \to -\infty$. The proof proceeded by averaging over rotations around the invariant Taub spaces, which are known explicitly.

The Bianchi type VIII is defined by $N_2, N_3 > 0 > N_1$ (up to index permutations). The above Taub space remains invariant, but its two cousins with $N_1 = N_2$ and $N_1 = N_3$ are now incompatible with $N_2, N_3 > 0 > N_1$ and have disappeared. Even worse: all solutions with $N_3 = 0$ and $N_2 > 0 > N_1$ converge to the Kasner-cicle $\mathcal{K}$ in *both* time-directions $t \to \pm\infty$ and fail to lie on the Kasner-caps. This poses a serious obstacle to the attractivity of the caps. For these reasons, attractor results have remained elusive in the Bianchi VIII setting.

For both Bianchi VIII ($N_2, N_3 > 0 > N_1$) and Bianchi IX ($N_1, N_2, N_3 > 0$), the attractor problem for Bianchi class A vacuum ($\Omega = 0$) has been settled by Brehm in his dissertation [35] as follows. The dynamics in the big-bang limit $t \to -\infty$ falls into one of the following three mutually exclusive classes:

(a) The solution converges to the three Kasner caps and has at least one α-limit point on the Kasner circle, which is distinct from the three Taub points T_j of Fig. 14 and their antipodal points $Q_j = -T_j$.

(b) The solution is contained in the invariant Taub subspace $\{N_2 = N_3, \Sigma_- = 0\}$, or, in the case of Bianchi IX, one of its two cousins.

(c_2) The solution has exactly two α-limit points on the Kasner circle, which are T_2 and $Q_2 = -T_2$. It has $\limsup |N_1N_3| > 0 = \liminf |N_1N_3|$, for $t \to -\infty$, and $\lim |N_1N_2| = \lim |N_2N_3| = 0$.

(c_3) The analogon of (c_2) with the indices 2 and 3 interchanged.

Case (a) applies for an open set of initial conditions which are a neighbourhood of the three caps minus the three Taub spaces. Furthermore, for any solution in case (a), the following non-Mixmaster integral J remains finite:

$$J = \int_{-\infty}^{0} (|N_1N_2| + |N_2N_3| + |N_3N_1|)\,\mathrm{d}t < \infty . \tag{63}$$

The cases (c_2) and (c_3) are impossible in the case of Bianchi IX; whether they can occur at all is currently unknown. Taken together with suitable measure-theoretic estimates, one can see that case (a) applies in a set of initial conditions, both of fat Baire category and of full Lebesgue measure. In other words (b), (c_2), (c_3) occur for a meagre set of Lebesgue-measure zero, only. Estimate (63) for the non-Mixmaster integral J is novel also in the Bianchi IX case. This is insufficient to bound the Mixmaster integral I of eq. (61), but nevertheless promising: from $\lim_{t\to-\infty} |N_iN_j| = 0$, to $\int_{-\infty}^{0} |N_iN_j|\mathrm{d}t < \infty$, and finally towards the goal $\int_{-\infty}^{0} \sqrt{|N_iN_j|}\mathrm{d}t < \infty$. The proof in Bianchi IX, where only the $J < \infty$ estimate (63) is novel compared to [173], proceeds along the lines of Ringström, via a refined quantitative averaging over rotations around the three invariant Taub subspaces combined with a more careful gluing of the different averaging regimes. In the Bianchi VIII case, the two missing invariant Taub subspaces are replaced by $\{|N_1| = |N_3|, \Sigma_- = \sqrt{3}\Sigma_+\}$ and its index-exchanged cousin. These two replacement Taub subspaces are invariant only up to leading order; this suffices for the averaging arguments and yields the trichotomy (a), (b), (c). The second main result in [35] establishes boundedness $I < \infty$ of the Mixmaster integral (61), for Lebesgue almost every initial condition in Bianchi VIII and IX vacuum. In other words, particle horizons form towards the singularity and the blow-up is local in the sense of BKL.

The above considerations reside within the general relativity (GR) framework of a purely gravitational Einstein universe. A larger framework is provided by the Hořava-Lifshitz (HL) models. HL gravity has been proposed as a candidate for the theory of quantum gravity; see [162] and the references there. An outlook, based on observations by Uggla, is the following. HL gravity breaks relativistic first principles and introduces anisotropic scalings between space and time. How do such modifications of the first principles affect the initial singularity and the structure of the BKL conjecture? As an attempt to answer this question, we address the question of the dynamics on its first level of complexity, i.e. the discrete dynamics contained in the heteroclinic chains arising in HL. Similarly to the Bianchi models for general relativity, it is possible to formulate HL models as a system of ODEs with constraints that admit a circle of equilibria, which we call the Kasner circle by analogy. It corresponds to the Bianchi type I model for HL. On the next level of the symmetry hierarchy, Bianchi type II, we find three invariant hemispheres of heteroclinic orbits that intersect only at their boundary, the Kasner circle, for HL and GR alike. On each hemisphere, the projections of the heteroclinic orbits to the plane of the Kasner circle emanate from a single point, as a straightforward calculation shows. Those three points of emanation form an equilateral triangle whose center coincide with the center of the circle of equilibria: for the HL parameter $v = 1/2$ corresponding to GR, the Kasner circle of equilibria is inscribed in the triangle, while this is not the case any more when the parameter changes to $v \neq 1/2$; see Fig. 15. Concatenation of heteroclinic orbits on different hemispheres form heteroclinic chains, just as they did in the GR case. However, the discrete dynamics on the Kasner circle induced by those heteroclinic chains change.

The BKL conjecture for GR describes the asymptotic behavior towards the big-bang singularity. This conjecture can be seen as an extrapolation from the dynamics of the heteroclinic chains to the continuous dynamics near the invariant hemispheres. Studying the discrete dynamics for the HL Bianchi models reveals how the BKL conjecture should be modified in this context.

Let us focus on the discrete dynamics induced by the heteroclinic chains, which map each equilibrium of the Kasner circle to the end of the heteroclinic leaving from it. Each point of emanation expands an arc of the Kasner circle enclosed by the two tangents to the circle through this point, by this HL Kasner map. Varying v amounts to changing the distance between the points of emanation of the heteroclinics and the circle of equilibria. This changes the size of the three expanding arcs and the factor of expansion; see Fig. 15. Two very different behaviors appear, depending on whether v is below or above 1/2.

Let us first discuss the case $1/2 < v < 1$. As can be seen in Fig. 15, right, the three expanding arcs are separated by three arcs of stable fix points at which heteroclinic chains end, due to the lack of heteroclinic orbits leaving from there. Generic initial conditions (both in the sense of Baire and Lebesgue) on the Kasner circle will lead to finite heteroclinic chains that end in a stable arc. Hence, an analogue to

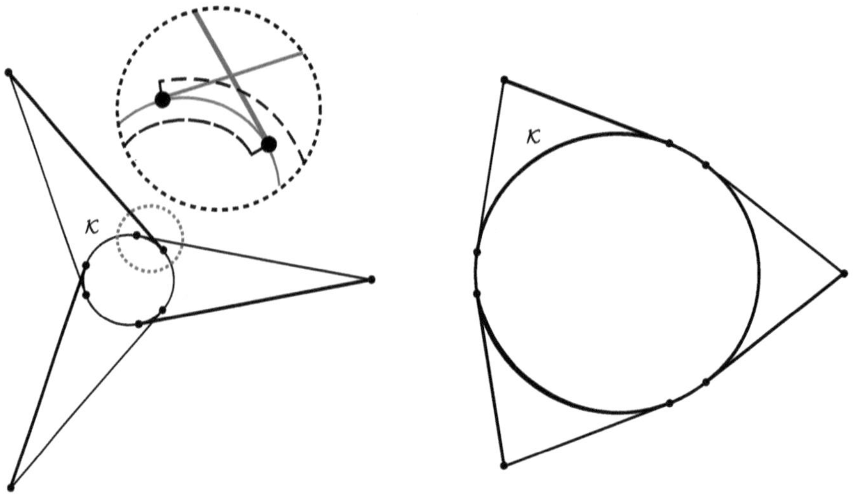

Figure 15: *The Kasner circle of equilibria, the three expanding arcs and the three points of emanation for the HL models. Left* $0 < v < 1/2$*; the dashed circle zooms into one of the three overlapping regions, where the overlapping sectors are indicated by dashed lines. Right* $1/2 < v < 1$.

the BKL conjecture should state that the asymptotic behavior of the system is, in general, convergence to a single equilibrium in a stable arc. This behavior is very different from the GR case where oscillatory chaos is present in the asymptotic behavior towards the big-bang singularity. However, not all initial conditions on the Kasner circle lead to finite heteroclinic chains. An invariant Cantor set of measure zero leads to heteroclinic chains that never enter the stable arcs, and the discrete dynamics on this Cantor set is chaotic. The proofs of these facts use symbolic dynamics and equivalence to a subshift of finite type, known to be chaotic. Hence, chaos has not completely vanished from the system for $1/2 < v < 1$, but it is not generic.

Let us now discuss the opposite case $0 < v < 1/2$. We see from Fig. 15, left, how the three expanding arcs now overlap, pairwise, while no stable arc is present. Therefore, the continuation of a heteroclinic chain that hits the overlap region is not unique: there are two heteroclinics leaving from there. In other words, the map describing the discrete dynamics is not well-defined and the concept of iterated function systems (IFS) comes into play. The IFS we consider is a collection of maps whose domains of definition cover the Kasner circle. The notion of chaotic discrete dynamical systems is generalised to IFS using the Hausdorff distance between sets. Iterations of a well-defined map are said to be chaotic if three conditions are fulfilled: sensitivity to initial conditions, topological mixing, and density of periodic orbits. In the case of an IFS, a single initial condition generates several trajectories. Therefore, each iteration of an IFS is a set instead of a single state. The three above conditions for chaos translate to IFS when distance between iterations is translated to Hausdorff distance between sets of possible iterations. The IFS arising for the HL Bianchi models with $0 < \nu < 1/2$ are chaotic in this sense, mostly due to the expansion property. Roughly speaking, this means that there are chaotic realisations of the IFS. This property only concerns the heteroclinic chains. Even if the HL models with $0 < \nu < 1/2$ are chaotic, like the GR case $\nu = 1/2$, their chaotic structure is quite different. The concept of a BKL era (i.e. phases of oscillations between two neighbouring expanding arcs) and the growth of their length along heteroclinic chains shape the chaos in the GR case, via the continued fraction expansion. In the HL for $0 < \nu < 1/2$, BKL erae do not play such a role: Their lengths are bounded above and the fine structure associated to their rate of growth disappears. The value $\nu = 1/2$ associated to GR can be seen as a critical transitional value where the dynamics bifurcates from non-generic chaos to generic chaos. The importance of erae is a feature that does not survive below $\nu = 1/2$. Hence, the perturbation of GR along the one-parameter family of HL will lead to a breakdown of the BKL conjecture.

This concludes our extended survey of some global dynamics aspects, from geometric and algebraic aspects of parabolic global attractors and their blow-up to black hole data and exact chaotic solutions of the Einstein equations in the setting of Bianchi cosmology. Not only many of the particular problems but, much broader, our overall perspective on these problems we owe, deeply, to the unique framework provided by the Sonderforschungsbereich 647 "Space – Time – Matter" of the Deutsche Forschungsgemeinschaft.

Bibliography

[1] Afuni A. Local monotonicity for the Yang–Mills–Higgs flow. Calculus Var Partial Differ Equ 2016;55(1):1–14.

[2] Afuni A. Heat ball formulæ for k-forms on evolving manifolds. Adv Calculus Var (to appear). DOI:10.1515/acv-2017-0026, http://ax0.github.io/preprints/hbfevm.pdf.

[3] Afuni A. Local energy inequalities for mean curvature flow into evolving ambient spaces. Preprint, http://ax0.github.io/preprints/mcfpaper.pdf.

[4] Afuni A. Energy identities and monotonicity for evolving k-forms on moving Riemannian spaces. J Evol Equ (to appear). DOI:10.1007/s00028-017-0410-6, http://ax0.github.io/preprints/evform.pdf.

[5] Alikakos ND. Some basic facts on the system $\Delta u - W_u(u) = 0$. Proc. Amer. Math. Soc., 2011;139(1):153–162.

[6] Alikakos ND, Antonopoulos P, Damialis A. Plateau angle conditions for the vector-valued Allen–Cahn equation. SIAM J Math Analysis 2013;45(6):3823–3837.

[7] Allard WK. On the first variation of a varifold. Ann Math 1972;95:417–491.

[8] Allard WK. On the first variation of a varifold: Boundary behavior. Ann Math 1975;101: 418–446.

[9] Allen PT, Layne A, Tsukahara K. The Dirichlet problem for curve shortening flow. Preprint: arXiv:1208.3510v2, 2012.

[10] Ambrosio L, Gigli N, Savaré G. Gradient flows in metric spaces and in the space of probability measures, 2nd ed. Basel: Lect Math ETH Zürich Birkhäuser, 2008.

[11] Andrews B. Noncollapsing in mean-convex mean curvature flow. Geom Topol 2012;16(3): 1413–1418.
[12] Andrews B, Langford M. Cylindrical estimates for hypersurfaces moving by convex curvature functions. Analysis PDE 2014;7(5):1091–1107.
[13] Andrews B, Langford M. Two-sided non-collapsing curvature flows. Ann Sc Norm Super Pisa Cl Sci 2016;15(5):543–60. Preprint: arXiv:1310.0717.
[14] Andrews B, Langford M, McCoy J. Non-collapsing in fully non-linear curvature flows. Ann Inst H Poincaré Analyse Non Linéaire 2013;30(1):23–32.
[15] Andrews B, Langford M, McCoy J. Convexity estimates for hypersurfaces moving by convex curvature functions. Analysis PDE 2014;7(2):407–433.
[16] Andrews B, Langford M, McCoy J. Convexity estimates for surfaces moving by curvature functions. J Diff Geom 2015;99(1):47–75.
[17] Angenent SB. The Morse–Smale property for a semilinear parabolic equation. J Diff Eq 1986;62:427–442.
[18] Angenent SB. On the formation of singularities in the curve shortening flow. J Diff Geom 1991;33(3):601–633.
[19] Arnlind J, Huisken G. Pseudo-Riemannian geometry in terms of multi-linear brackets. Lett Math Phys 2014;104(12):1507–1521.
[20] Arnlind J, Hoppe J, Huisken G. Multi-linear formulation of differential geometry and matrix regularizations. J Diff Geom 2012;91(1):1–39.
[21] Béguin F. Aperiodic oscillatory asymptotic behavior for some Bianchi spacetimes. Class Quantum Grav 2010;27:185005.
[22] Belinskii VA, Khalatnikov IM, Lifshitz EM. Oscillatory approach to a singular point in the relativistic cosmology. Adv Phys 1970; 19:525–573.
[23] Ben-Gal N. Grow-up solutions and heteroclinics to infinity for scalar parabolic PDEs. PhD thesis, Brown University, 2009.
[24] Bourni T. Allard-type boundary regularity for $C^{1,\alpha}$ boundaries. Adv Calc Var 2016;9(2): 143–161.
[25] Bourni T, Coskunuzer B. Area minimizing surfaces in mean convex 3-manifolds. J Reine Angew Math 2015;704:135–167.
[26] Bourni T, Langford M. Type-II singularities of two-convex immersed mean curvature flow. In preparation, 2016.
[27] Bourni T, Moore K. Null mean curvature flow and outermost MOTS. Preprint: arXiv:1503.04023, 2015.
[28] Bourni T, Tinaglia G. Curvature estimates for surfaces with bounded mean curvature. Trans Amer Math Soc 2012;364(11):5813–5828.
[29] Bourni T, Tinaglia G. Density estimates for compact surfaces with total boundary curvature less than 4_. Comm PDEs 2012;37(10):1870–1886.
[30] Bourni T, Tinaglia G. $C^{1,\alpha}$-regularity for surfaces with $H \in L^p$. Ann Global Analysis Geom 2014;46(2):159–186.
[31] Bourni T, Volkmann A. An Allard type regularity theorem for varifolds with a Hölder condition on the first variation. Calc Var 2016. Preprint: arXiv:1311.3963.
[32] Brakke KA. The motion of a surface by its mean curvature. Math Notes Princeton University Press, 1978.
[33] Bray HL, Khuri MA. A jang equation approach to the Penrose inequality. Discrete Contin. Dyn Syst 2010;27(2):741–766.
[34] Bray HL, Khuri MA. P.d.e.'s which imply the Penrose conjecture. Asian J Math 2011;15(4):557–610.

[35] Brehm B. Bianchi VIII and IX vacuum cosmologies: Almost every solution forms particle horizons and converges to the Mixmaster attractor. Doctoral thesis, Freie Universität Berlin 2017. Preprint: arXiv:1606.08058.
[36] Brendle S. A sharp bound for the inscribed radius under mean curvature flow. Invent Math 2015;202(1):217–237.
[37] Brendle S, Huisken G. Mean curvature flow with surgery of mean convex surfaces in R3. Invent Math 2015;203(2):615–654.
[38] Brendle S, Huisken G, Sinestrari C. Ancient solutions to the Ricci flow with pinched curvature. Duke Math J 2011;158(3):537–551.
[39] Buchner J. Ancient dynamics of the Einstein equations and the Tumbling Universe. Doctoral thesis, Freie Universität Berlin, 2013.
[40] Carlotto A. Rigidity of stable minimal hypersurfaces in asymptotically flat spaces. Preprint: arXiv:1403.6459v2, 2014.
[41] Cha YS. On geometric inequalities for near maximal axially symmetric initial data. In preparation, 2016.
[42] Cha YS, Khuri MA. Deformations of charged axially symmetric initial data and the mass-angular momentum-charge inequality. Ann Henri Poincaré 2014;16(12):2881–2918.
[43] Cha YS, Khuri MA. Deformations of axially symmetric initial data and the mass-angular momentum inequality. Ann Henri Poincaré 2015;16(3):841–896.
[44] Cha YS, Khuri MA. Geometric inequalities for asymptotically AdS initial data. In preparation, 2016;24.
[45] Cha YS, Khuri MA, Sakovich A. Reduction arguments for geometric inequalities associated with asymptotically hyperboloidal slices. Class Quantum Grav 2016;33(3):035009.
[46] Chen Y, Shen CL. Monotonicity formula and small action regularity for Yang-Mills flows in higher dimensions. Calc Var 1994;2(4):389–403.
[47] Chepyzhov VV, Vishik MI. Attractors of Equations of Mathematical Physics. Amer Math Soc Colloq Publ 2002;49.
[48] Choi HI, Schoen R. The space of minimal embeddings of a surface into a three-dimensional manifold of positive Ricci curvature. Invent Math 1985;81(3):387–394.
[49] Choquet-Bruhat Y. General relativity and the Einstein equations. Oxford Mathematical Monographs. Oxford University Press, 2009.
[50] Chossat P, Lauterbach R. Methods in Equivariant Bifurcations and Dynamical Systems. Singapore: World Scientific, 2000.
[51] Chow B, Lu P, Ni L. Hamilton's Ricci Flow, volume 77 of Mathematical Surveys and Monographs. Amer Math Soc Providence, RI, 2006.
[52] Chrusciel PT, Tod P. An angular momentum bound at null infinity. Adv Theor Math Phys 2009;13:1317–1334.
[53] Chrusciel PT, Li Y, Weinstein G. Mass and angular-momentum inequalities for axi-symmetric initial data sets. II. Angular momentum. Ann Phys 2008;323:2591–2613.
[54] Colding TH, Minicozzi, II WP. The space of embedded minimal surfaces of fixed genus in a 3-manifold. II. Multi-valued graphs in disks. Ann Math 2004;160(1):69–92.
[55] Dain S. Proof of the angular momentum-mass inequality for axisymmetric black hole. J Diff Geom 2008;79:33–67.
[56] Daskalopoulos P, Hamilton RS, Sesum N. Classification of compact ancient solutions to the curve shortening flow. J Diff Geom 2010;84(3):455–464.
[57] Dergachev V, Kirillov A. Index of Lie algebras of seaweed type. J Lie Theory 2000;10(2): 331–343.
[58] Di Francesco P, Golinelli O, Guitter E. Meanders and the Temperley–Lieb algebra. Comm Math Phys 1997;186(1):1–59.

[59] Dittberner F. Constrained curvature flows. Doctoral thesis, Freie Universität Berlin, 2016. In preparation.
[60] Ecker K. Regularity theory for mean curvature flow, volume 57 of progress in nonlinear differential equations and their applications. Basel: Birkhäuser, 2004.
[61] Ecker K. Local monotonicity formulas for some nonlinear diffusion equations. Calc Var 2005;23(1):67–81.
[62] Ecker K. A formula relating entropy monotonicity to Harnack inequalities. Comm Analysis Geom 2008;15(5):1025–1061.
[63] Ecker K. Entropy and differential Harnack type formulas for evolving domains. Calc Var 2014;50(1):455–480.
[64] Ecker K, Huisken G. Mean curvature evolution of entire graphs. Ann Math 1989;130(3): 453–471.
[65] Ecker K, Huisken G. Interior estimates for hypersurfaces moving by mean curvature. Invent Math 1991;105(1):547–569.
[66] Ecker K, Knopf D, Ni L, Topping P. Local monotonicity and mean value formulas for evolving Riemannian manifolds. J Reine Angew Math 2008;616:89–130.
[67] Eells J, Sampson JH. Harmonic mappings of Riemannian manifolds. Amer J Math 1964;86(1):109–160.
[68] Eichmair M, Metzger J. On large volume preserving stable CMC surfaces in initial data sets. J Diff Geom 2012;91(1):81–102.
[69] Eichmair M, Metzger J. Unique isoperimetric foliations of asymptotically flat manifolds in all dimensions. Invent Math 2013;194(3):591–630.
[70] Ekholm T, White B, Wienholtz D. Embeddedness of minimal surfaces with total boundary curvature at most 4π. Ann Math 2002;155(1):209–234.
[71] Evans LC. Partial differential equations, volume 19 of Graduate Studies in Mathematics, 2nd ed. RI: Amer Math Soc Providence, 2010.
[72] Evans LC, Gariepy RF. Wiener's criterion for the heat equation. Arch Rational Mech Analysis 1982;78(4):293–314.
[73] Fiedler B. Reversibility, continued fractions, and infinite meander permutations of planar homoclinic orbits in linear hyperbolic Anosov maps. Int J Bifurcation Chaos 2014;24(8): 1440008, 12.
[74] Fiedler B, Castañeda P. Rainbow meanders and Cartesian billiards. São Paulo J Math Sci 2012:6(2):247–275.
[75] Fiedler B, Grotta-Ragazzo C, Rocha C. An explicit Lyapunov function for reflection symmetric parabolic partial differential equations on the circle. Russ Math Surv 2014;69(3): 419–433.
[76] Fiedler B, Hell J, Smith B. Anisotropic Einstein data with isotropic non negative prescribed scalar curvature. Ann Inst H. Poincaré Analyse Non Linéaire 2015;32(2):401–428.
[77] Fiedler B, Rocha C. Heteroclinic orbits of semilinear parabolic equations. J Diff Eq 1996;125(1):239–281.
[78] Fiedler B, Rocha C. Realization of meander permutations by boundary value problems. J Diff Eq 1999;156(2):282–308.
[79] Fiedler B, Rocha C. Orbit equivalence of global attractors of semilinear parabolic differential equations. Trans Amer Math Soc 2000;352(1):257–284.
[80] Fiedler B, Rocha C. Connectivity and design of planar global attractors of Sturm type. II: Connection graphs. J Diff Eq 2008;244(5):1255–1286.
[81] Fiedler B, Rocha C. Connectivity and design of planar global attractors of Sturm type. I: Bipolar orientations and Hamiltonian paths. J Reine Angew Math 2009;635:71–96.
[82] Fiedler B, Rocha C. Connectivity and design of planar global attractors of Sturm type. III: Small and platonic examples. J Dyn Diff Eq 2010;22(2):121–162.

[83] Fiedler B, Rocha C. Nonlinear Sturm global attractors: Unstable manifold decompositions as regular CW-complexes. Discrete Contin Dyn Syst 2014;34(12):5099–5122.
[84] Fiedler B, Rocha C. Schoenflies spheres as boundaries of bounded unstable manifolds in gradient Sturm systems. J Dyn Diff Eq 2015;27(3):597–626.
[85] Fiedler B, Rocha C, Wolfrum M. A permutation characterization of Sturm global attractors of Hamiltonian type. J Diff Eq 2012;252(1):588–623.
[86] Fiedler B, Rocha C, Wolfrum M. Sturm global attractors for S^1-equivariant parabolic equations. Netw Heterog Media 2012;7(4):617–659.
[87] Fila M, Stuke H. Special asymptotics for a critical fast diffusion equation. Discrete Contin Dyn Syst 2014;7:725–735.
[88] Fila M, Stuke H. Very slow stabilization for a nonlinear Fokker–Planck equation. Nonlin Analysis 2016;138:93–110.
[89] Fraser A, Schoen R. The first Steklov eigenvalue, conformal geometry, and minimal surfaces. Adv Math 2011;226(5):4011–4030.
[90] Fulks W. A mean value theorem for the heat equation. Proc Amer Math Soc 1966;17(1): 6–11.
[91] Fusco G, Rocha C. A permutation related to the dynamics of a scalar parabolic PDE. J Diff Eq 1991;91(1):111–137.
[92] Gage M. On an area preserving evolution equation for plane curves. In: Dennis M. De- Turck, editor. Nonlinear problems in geometry. Proceedings of the Special Session on Nonlinear Problems in Geometry held at the 820th Meeting of the American Mathematical Society in Mobile, Alabama, Providence, Rhode Island: American Mathematical Society (AMS), May 3–4, 1985.
[93] Gage M, Hamilton RS. The heat equation shrinking convex plane curves. J Diff Geom 1986;23:69–96.
[94] Garofalo N. Calculus of variations and partial differential equations: Proceedings of a conference held in Trento, Italy June 16–21, 1986, volume 1340 of Lect. Notes Math., chapter New estimates of the fundamental solution and Wiener's criterion for parabolic equations with variable coefficients. Berlin, Heidelberg: Springer, 1988:70–83.
[95] Gauss CF. Carl Friedrich Gauss Werke VIII: Zur Geometria Situs. 1840, 271–286, Berlin, Heidelberg, ca.: Springer.
[96] Gerhardt C. H-surfaces in Lorentzian manifolds. Comm Math Phys 1983;89:523–553.
[97] Giga Y, Kohn RV. Asymptotically self-similar blow-up of semilinear heat equations. Comm. Pure Appl Math 1985;38:297–319.
[98] Gödel K. An example of a new type of cosmological solution of Einstein's field equations of gravitation. Rev Mod Phys 1949;21(3):447–450. Reprint: Gen. Relativ. Gravitation 32:1409–1417, 2000.
[99] Grohs P, Hardering H, Sander O. Optimal a priori discretization error bounds for geodesic finite elements. F. C. M. 2015;15(6):1357–1411.
[100] Gui C. Hamiltonian identities for elliptic partial differential equations. J Funct Analysis 2008;254(4):904–933.
[101] Guo JS, Ninomiya H, Shimojo M, Yanagida E. Convergence and blow-up of solutions for a complex-valued heat equation with a quadratic nonlinearity. Trans Amer Math Soc 2013;365(5):2447–2467.
[102] Gurevich P, Väth M. Stability for semilinear parabolic problems in L2 and W1;2. Z Analysis Anwend 2016;35(3):333–357.
[103] Hale JK. Asymptotic Behavior of Dissipative Systems. Math. Surveys and Monographs, 25. RI: Amer Math Soc Providence 1988;22(1):175–183.
[104] Hall P. Two topological examples in minimal surface theory. J Diff Geom 1984;19:475–481.
[105] Hamilton RS. Three-manifolds with positive Ricci curvature. J Diff Geom 1982;17(2): 255–306.

[106] Hamilton RS. Four-manifolds with positive curvature operator. J Diff Geom 1986;24(4): 153–179.
[107] Hamilton RS. Lecture notes. In CBMS conference, Hawaii, 1989.
[108] Hamilton RS. A matrix Harnack estimate for the heat equation. Comm Analysis Geom 1993;1:113–126.
[109] Hamilton RS. Monotonicity formulas for parabolic flows on manifolds. Comm Analysis Geom 1993;1:127–137.
[110] Hamilton RS. The formation of singularities in the Ricci flow. Surveys in Differential Geometry 1995; 2:7–136.
[111] Hamilton RS. Harnack estimate for the mean curvature flow. J Diff Geom 1995;41(1): 215–226.
[112] Hamilton RS. Non-singular solutions of the Ricci flow on three-manifolds. Comm Analysis Geom 1999;7(4):695–729.
[113] Hardering H. Intrinsic discretization error bounds for geodesic finite elements. Doctoral thesis, Freie Universität Berlin, 2015.
[114] Haslhofer R, Hershkovitz O. Ancient solutions of the mean curvature flow. To appear in Comm. Analysis Geom., 2016. Preprint: arXiv:1308.4095.
[115] Heinzle JM, Ringström H. Future asymptotics of vacuum Bianchi type VI0 solutions. Class Quantum Grav 2009;26:145001, 19.
[116] Heinzle JM, Uggla C. Mixmaster: Fact and belief. Class Quantum Grav 2009;26: 075016, 34.
[117] Hell J. Conley index at Infinity. Doctoral thesis, Freie Universität Berlin, 2009.
[118] Hell J. Conley index at infinity. Topol Meth Nonlin Analysis 2013;42:137–167.
[119] Henry DB. Geometric Theory of Semilinear Parabolic Equations. Lect. Notes Math. Number 840. Berlin, New York: Springer, 1981.
[120] Henry DB. Some infinite-dimensional Morse–Smale systems defined by parabolic partial differential equations. J Diff Eq 1985;59:165–205.
[121] Hershkovits O. Mean curvature flow of Reifenberg sets. Preprint: arXiv:1412.4799, 2014.
[122] Hsu SB, Fiedler B, Lin HH. Classification of potential flows under renormalization group transformation. Discrete Contin. Dyn Syst Ser B 2016;21(2):437–446.
[123] Huisken G. Flow by mean curvature of convex surfaces into spheres. J Diff Geom 1984;20 (1):237–266.
[124] Huisken G. Asymptotic behavior for singularities of the mean curvature flow. J Diff Geom 1990;31(1):285–299.
[125] Huisken G. Local and global behaviour of hypersurfaces moving by mean curvature. In Differential geometry: Partial differential equations on manifolds (Los Angeles, CA, 1990), volume 54 of Proc. Sympos. Pure Math., pages 175–191. Amer Math Soc. Providence, RI, 1993.
[126] Huisken G. A distance comparison principle for evolving curves. Asian J Math 1995;2(1): 127–134.
[127] Huisken G, Sinestrari C. Convexity estimates for mean curvature flow and singularities of mean convex surfaces. Acta Math 1999;183(1):45–70.
[128] Huisken G, Sinestrari C. Mean curvature flow singularities for mean convex surfaces. Calc Var 1999;8(1):1–14.
[129] Huisken G, Sinestrari C. Mean curvature flow with surgeries of two-convex hypersurfaces. Invent Math 2008;175(1):137–221.
[130] Huisken G, Sinestrari C. Convex ancient solutions of the mean curvature flow. J Diff Geom 2015;101(2):267–287.
[131] Ilmanen T. Convergence of the Allen–Cahn equation to Brakke's motion by mean curvature. J Diff Geom 1993;38(2):417–461.
[132] Isenberg J, Jackson M. Ricci flow of locally homogeneous geometries on closed manifolds. J Diff Geom 1992;35(3):723–741.
[133] Ivey T. Ricci solitons on compact three-manifolds. Diff Geom Appl 1993;3(4):301–307.

[134] Jachan F. Flächeninhaltserhaltender Willmore-Fluss im asymptotisch Schwarzschildschen. Doctoral thesis, Freie Universität Berlin, 2014.
[135] Jost J. Harmonic mappings between Riemannian manifolds, volume 4 of Proceedings of the CMA. Australian National University, 1983.
[136] Karnauhova A. Meanders. De Gruyter, Berlin, 2017.
[137] Karnauhova A, Liebscher S. Connected components of meanders: I. Bi-rainbow meanders. To appear in Discr Cont Dyn Syst A 2017;37(9):4835–4856.
[138] Kasai K, Tonegawa Y. A general regularity theory for weak mean curvature flow. Calc Var 2014;50(1):1–68.
[139] Kim IS, Väth M. The Krasnosel'skiı–Quittner formula and instability of a reaction-diffusion system with unilateral obstacles. Dynamics of PDEs 2014;11(3):229–250.
[140] Kleiner B, Lott J. Notes on Perelman's papers. Topology and Geometry 2008;12:2587–2858.
[141] Krisztin T, Walther HO. Unique periodic orbits for delayed positive feedback and the global attractor. J Dyn Diff Eq 2001;13(1):1–57.
[142] Kuwert E, Schätzle R. The Willmore flow with small initial energy. J Diff Geom 2001;57(3):409–441.
[143] Kuwert E, Schätzle R. Gradient flow for the Willmore functional. Comm Analysis Geom 2002;10(2):307–339.
[144] Lahiri A. Regularity of the brakke flow. Doctoral thesis, Freie Universität Berlin, 2014.
[145] Lam C, Lauer J. The level-set flow of the topologist's sine curve is smooth. Preprint: arXiv:1601.02442, 2016.
[146] Langford M. Motion of hypersurfaces by curvature. PhD thesis, Australian National University, 2014.
[147] Langford M. Inscribed curvature pinching for mean curvature flows. In preparation, 2016.
[148] Langford M, Lynch S. Inscribed and exscribed curvature pinching for fully non-linear curvature flows. In preparation, 2016.
[149] Lappicy P. Einstein's Constraints: A Dynamical Approach. Doctoral thesis, Freie Universität Berlin, 2017.
[150] Lauer J. A new length estimate and low regularity initial data. Geom Funct Analysis 2013;23:1934–1961.
[151] Lauer J. The evolution of Jordan curves on S2 by curve shortening flow. Preprint: arXiv:1601.05704, 2016.
[152] Li P, Yau ST. On the parabolic kernel of the Schrödinger operator. Acta Math 1986;156:153–201.
[153] Liebscher S. Bifurcation without Parameters, volume 2117 of Lect. Notes Math. Berlin, New York: Springer, 2015.
[154] Liebscher S, Härterich J, Webster K, Georgi M. Ancient dynamics in Bianchi models: Approach to periodic cycles. Comm. Math Phys 2011;305:59–83.
[155] Liebscher S, Rendall AD, Tchapnda SB. Oscillatory singularities in Bianchi models with magnetic fields. Ann Henri Poincaré 2013;14(5):1043–1075.
[156] Magni A, Mantegazza C, Tsatis E. Flow by mean curvature inside a moving ambient space. J Evol Equ 2013;13(3):561–576.
[157] Marxen T. Ricci Fluss auf einer Klasse nichtkompakter warped product Mannigfaltigkeiten und Gaußsche Abschätzungen. Doctoral thesis, Freie Universität Berlin, 2015.
[158] Masuda K. Analytic solutions of some nonlinear diffusion equations. Math Zeitschrift 1984;187(1):61–73.
[159] Michael JH, Simon LM. Sobolev and mean-value inequalities on generalized submanifolds of Rn. Comm Pure Appl Math 1973;26(3):361–379.
[160] Misner CW. Mixmaster universe. Phys Rev Lett 1969;22:1071–1074.

[161] Miura T. An example of a mean-convex mean curvature flow developing infinitely many singular epochs. J. G. A. 2016;26(4):3019–3026.
[162] Mukohyama S. Horava–Lifshitz cosmology: A review. Class Quantum Grav 2010;27:223101, 25.
[163] Penrose R. Naked singularities. Ann New York Acad Sci 1973;224:125–134.
[164] Perelman GJ. The entropy formula for the Ricci flow and its geometric applications. Preprint: arXiv:math/0211159v1, 2002.
[165] Perelman GJ. Finite extinction time for the solutions to the Ricci flow on certain threemanifolds. Preprint: arXiv:math/0307245v1, 2003.
[166] Perelman GJ. Ricci flow with surgery on three-manifolds. Preprint: arXiv:math/0303109v1, 2003.
[167] Perko L. Differential equations and dynamical systems. Texts Appl Math Springer 2001.
[168] Pihan DM. A length preserving geometric heat flow for curves. PhD thesis, University of Melbourne, Australia, 1998.
[169] Pimentel JFS, Rocha C. Noncompact global attractors for scalar reaction-diffusion equations. São Paulo J Math Sc 2015;9:299–310.
[170] Pimentel JFS, Rocha C. A permutation related to non-compact global attractors for slowly non-dissipative systems. J Dyn Diff Eq 2016;28:1–28.
[171] Price P. A monotonicity formula for Yang–Mills fields. Manuscripta Math 1983;43(2):131–166.
[172] Quittner P, Souplet P. Superlinear parabolic problems: Blow-up, global existence and steady states. Basel: Birkhäuser, 2007.
[173] Ringström H. The Bianchi IX attractor. Ann Henri Poincaré 2001;2(3):405–500.
[174] Ringström H. The Cauchy problem in general relativity. ESI Lectures in Mathematics and Physics. Eur Math Soc Zürich 2009.
[175] Rocha C, Fiedler B. Sturm global attractors of Hamiltonian type for semilinear parabolic equations. RIMS Kôkyûroku Bessatsu 2014;1881:139–157.
[176] Ros A, Vergasta E. Stability for hypersurfaces of constant mean curvature with free boundary. Geom Dedicata 1995;56(1):19–33.
[177] Sander O. Geodesic finite elements for Cosserat rods. Int J Num Meth Eng 2010;82(13): 1645–1670.
[178] Sander O. Geodesic finite elements on simplicial grids. Int J Num Meth Eng 2012;92(12): 999–1025.
[179] Sander O. Geodesic finite elements of higher order. IMA J Numer Analysis 2016;36(1): 238–266.
[180] Sander O, Neff P, Bîrsan M. Numerical treatment of a geometrically nonlinear planar Cosserat shell model. Comput Mech 2016;57(5):817–841.
[181] Sandstede B, Fiedler B. Dynamics of periodically forced parabolic equations on the circle. Ergodic Theory Dyn Syst 1992;12(3):559–571.
[182] Schoen R, Uhlenbeck K. A regularity theory for harmonic maps. J Diff Geom 1982;17(2):307–335.
[183] Schoen R, Yau ST. On the proof of the positive mass conjecture in general relativity. Comm Math Phys 1979;65(1):45–76.
[184] Schoen R, Yau ST. Proof of the positive mass theorem II. Comm Math Phys 1981;79: 231–260.
[185] Schoen R, Zhou X. Convexity of reduced energy and mass angular momentum inequalities. Ann Henri Poincaré 2013;14:1747–1773.
[186] Schulze F. Nonlinear evolution by mean curvature and isoperimetric inequalities. J Diff Geom 2008;79(2):197–241.
[187] Sheng W, Wang XJ. Singularity profile in the mean curvature flow. Meth Appl Analysis 2009;16(2):139–155.

[188] Shi WX. Deforming the metric on complete Riemannian manifolds. J Diff Geom 1989;30(1):223–301.

[189] Simon LM. Lectures on geometric measure theory, volume 3 of Proceedings of the CMA. Australian National University, 1983.

[190] Simons JH. Minimal varieties in Riemannian manifolds. Ann Math 1968;88:62–105.

[191] Smith B. Blow-up in the parabolic scalar curvature equation. Preprint: arXiv:0705.3774v2, 2012.

[192] Struwe M. On the evolution of harmonic mappings of Riemannian surfaces. Comm Math Helv 1985;80:558–581.

[193] Struwe M. On the evolution of harmonic maps in higher dimensions. J Diff Geom 1988;28: 485–502.

[194] Struwe M. The Yang-Mills flow in four dimensions. Calc Var 1994;2(2):123–150.

[195] Strzelecki P, Szumanska M, von der Mosel H. On some knot energies involving Menger curvature. Topol Appl 2013;160(13):1507–1529.

[196] Stuke H. Blow-up in complex time. Doctoral thesis, Freie Universität Berlin, 2017.

[197] Sturm C. Mémoire sur une classe d'équations à différences partielles. J Math Pure Appl 1836;1:373–444.

[198] Takens F. Partially hyperbolic fixed points. Topology 1971;10:133–147.

[199] Taub AH. Empty space-times admitting a three parameter group of motions. Ann Math 1951;53:472–490. Reprint: Gen. Relativ. Gravitation 36:2699–2719, 2004.

[200] Thurston WP. Three dimensional manifolds, Kleinian groups and hyperbolic geometry. Bull Amer Math Soc (N.S.) 1982;6(3):357–381.

[201] Tinaglia G. Curvature estimates for minimal surfaces with total boundary curvature less than 4_. Proc Amer Math Soc 2009;137(7):2445–2450.

[202] Tonegawa Y. A second derivative Hölder estimate for weak mean curvature flow. Adv Calc Var 2014;7(1):91–138.

[203] Väth M. Instability of Turing type for a reaction-diffusion system with unilateral obstacles modeled by variational inequalities. Math Bohem 2014;139(2):195–211.

[204] Volkmann A. Free boundary problems governed by mean curvature. Doctoral thesis, Freie Universität Berlin, 2014.

[205] Wainwright J, Ellis GFR, editors. Dynamical Systems in Cosmology. Cambridge, UK: Cambridge University Press, 1997.

[206] Wainwright J, Hsu L. A dynamical systems approach to Bianchi cosmologies: Orthogonal models of class A. Class Quantum Grav 1989;6(10):1409–1431.

[207] Watson NA. A theory of subtemperatures in several variables. Proc London Math Soc 1973;26:385–417.

[208] White B. The size of the singular set in mean curvature flow of mean-convex sets. J Amer Math Soc 2000;13(3):665–695.

[209] White B. A local regularity theorem for classical mean curvature flow. Ann Math 2005;161: 1487–1519.

[210] Willmore TJ. Riemannian geometry. Oxford science publications. Oxford: Oxford University Press, 1996.

Index

DOI 10.1515/9783110452150-009

Bei Fragen zur Produktsicherheit wenden Sie sich bitte an:
If you have any questions regarding product safety,
please contact:

Walter de Gruyter GmbH
Genthiner Straße 13
10785 Berlin
productsafety@degruyterbrill.com